Christian Reichardt

Solvents and Solvent Effects in Organic Chemistry

Second, revised and enlarged edition

Prof. Dr. Christian Reichardt
Fachbereich Chemie
der Philipps-Universität Marburg
Hans-Meerwein-Straße
D-3550 Marburg
Federal Republic of Germany

Editorial Director: Dr. H. F. Ebel
Copy Editor: Dr. Edeline Wentrup-Byrne, Brisbane/Australia
Production Manager: Dipl.-Ing. (FH) Hans Jörg Maier

Library of Congress Card No. applied for

A CIP catalogue record for this book is available
from the British Library.

Deutsche Bibliothek Cataloguing in Publication Data

Reichardt, Christian:
Solvents and solvent effects in organic chemistry / Christian
Reichardt. – 2., rev. and enl. ed. – Weinheim ; Basel (Schweiz)
; Cambridge ; New York, NY : VCH, 1988
1. Aufl. im Verl. Chemie, Weinheim, New York
1. Aufl. u.d.T.: Reichardt, Christian: Solvent effects in organic
chemistry
ISBN 3-527-26805-7 (Weinheim ...) Pb.
ISBN 0-89573-684-5 (Cambridge ...) Pb.

Composition: SDT – Satz- und Daten-Technik GmbH, D-6232 Bad Soden a. Ts.
Printer: betz-druck gmbh, D-6100 Darmstadt 12
Bookbinder: Josef Spinner, Großbuchbinderei GmbH, D-7583 Ottersweier
Printed in the Federal Republic of Germany

Christian Reichardt

Solvents and Solvent Effects in Organic Chemistry

Distribution:
VCH Verlagsgesellschaft, P.O. Box 1260/1280, D-6940 Weinheim (Federal Republic of Germany)
Switzerland: VCH Verlags-AG, P.O. Box, CH-4020 Basel (Switzerland)
Great Britain and Ireland: VCH Publishers (UK) Ltd., 8 Wellington Court, Wellington Street, Cambridge CB1 1HW (Great Britain)
USA and Canada: VCH Publishers, Suite 909, 220 East 23rd Street, New York, NY 10010-4606 (USA)

ISBN 3-527-26805-7 (VCH Verlagsgesellschaft) ISBN 0-89573-684-5 (VCH Publishers)

To Maria
and in memory of my parents

Preface to the Second Edition

The response to the first English edition of this book, published in 1979, has been both gratifying and encouraging. Its mixed character, lying between that of a monograph and a textbook, has obviously made it attractive to both the industrial and academic chemist as well as the advanced student of chemistry.

During the last eight years the study of solvent effects on both chemical reactions and absorption spectra has made much progress, and numerous interesting and fascinating examples have been described in the literature. In particular, the study of ionic reactions in the gas phase – now possible due to new experimental techniques – has allowed direct comparisons between gas-phase and solution reactions. This has led to a greater understanding of solution reactions. Consequently, Chapters 4 and 5 have been enlarged to include a description of ionic gas-phase reactions compared to their solution counterparts.

The number of well-studied solvent-dependent processes, *i.e.* reactions and absorptions in solution, has increased greatly since 1979. Only a representative selection of the more instructive, recently studied examples could be included in this second edition.

The search for empirical parameters of solvent polarity and their applications in multiparameter equations has recently been intensified, thus making it necessary to rewrite large parts of Chapter 7.

Special attention has been given to the chemical and physical properties of organic solvents commonly used in daily laboratory work. Therefore, all Appendix Tables have been improved; some have been completely replaced by new ones. A new well-referenced table on solvent-drying has been added (Table A-3). Chapter 3 has been enlarged, in particular by the inclusion of solvent classifications using multivariate statistical methods (Section 3.5). All these amendments justify the change in the title of the book to *Solvents and Solvent Effects in Organic Chemistry*.

The references have been up-dated to cover literature appearing up to the first part of 1987. New references were added to the end of the respective reference list of each chapter from the first edition.

Consistent use of the nomenclature, symbols, terms, and SI units recommended by the IUPAC commissions has also been made in the second edition.*)

I am very indebted to many colleagues for corrections, comments, and valuable suggestions. Especially helpful suggestions came from Professors H.-D. Försterling, Marburg, J. Shorter, Hull/England, and R. I. Zalewski, Poznań/Poland, to whom I am very grateful. For critical reading of the whole manuscript and the improvement of my English I again thank Dr. Edeline Wentrup-Byrne, now living in Brisbane/Australia. Dr. P.-V. Rinze, Marburg, and his son Lars helped me with the author index. Finally, I would like to thank my wife Maria for her sympathetic assistance during the preparation of this edition and for her help with the indices.

Marburg (Lahn), Spring 1988 — Christian Reichardt

---

* *Cf.* Pure Appl. Chem. *51*, 1 (1979); ibid. *53*, 753 (1981); ibid. *55*, 1281 (1983); ibid. *57*, 105 (1985).

Preface to the First Edition

The organic chemist usually works with compounds which possess labile covalent bonds and are relatively involatile, thereby often rendering the gas-phase unsuitable as a reaction medium. Of the thousands of reactions known to occur in solution only few have been studied in the gas-phase, even though a description of reaction mechanisms is much simpler for the gas-phase. The frequent necessity of carrying out reactions in the presence of a more or less inert solvent results in two main obstacles: The reaction depends on a larger number of parameters than in the gas-phase. Consequently, the experimental results can often be only qualitatively interpreted because the state of aggregation in the liquid phase has so far been insufficiently studied. On the other hand, the fact that the interaction forces in solution are much stronger and more varied than in the gas-phase, permits to affect the properties and reactivities of the solute in manifold modes.

Thus, whenever a chemist wishes to carry out a chemical reaction he not only has to take into consideration the right reaction partners, the proper reaction vessels, and the appropriate reaction temperature. One of the most important features for the success of the planned reaction is the selection of a suitable solvent. Since solvent effects on chemical reactivity have been known for more than a century, most chemists are now familiar with the fact that solvents may have a strong influence on reaction rates and equilibria. Today, there are about three hundred common solvents available, nothing to say of the infinite number of solvent mixtures. Hence the chemist needs, in addition to his intuition, some general rules and guiding-principles for this often difficult choice.

The present book is based on an earlier paperback "Lösungsmitteleffekte in der organischen Chemie" [1], which, though following the same lay-out, has been completely rewritten, greatly expanded, and brought up to date. The book is directed both toward the industrial and academic chemist and particularly the advanced student of chemistry, who on the one hand needs objective criteria for the proper choice of solvent but on the other hand wishes to draw conclusions about reaction mechanisms from the observed solvent effects.

A knowledge of the physico-chemical principles of solvent effects is required for proper bench-work. Therefore, a description of the intermolecular interactions between dissolved molecules and solvent is presented first, followed by a classification of solvents derived therefrom. Then follows a detailed description of the influence of solvents on chemical equilibria, reaction rates, and spectral properties of solutes. Finally, empirical parameters of solvent polarity are given, and in an appendix guidelines to the everyday choice of solvents are given in a series of Tables and Figures.

The number of solvent systems and their associated solvent effects examined is so enormous that a complete description of all aspects would fill several volumes. For example, in Chemical Abstracts, volume 85 (1976), approximately eleven articles per week were quoted in which the words "Solvent effects on . . ." appeared in the title. In the present book only a few important and relatively well-defined areas of general importance have been selected. The book has been written from the point of view of practical use for the organic chemist rather than from a completely theoretical one.

In the selection of the literature more recent reviews were taken into account mainly. Original papers were cited in particular from the didactic point of view rather than priority, importance or completeness. This book, therefore, does not only have the character of a monograph but also to some extent that of a textbook. In order to help the reader in his use of the literature cited, complete titles of the review articles quoted are given. The literature up until December 1977 has been considered together with a few papers from 1978. The use of symbols follows the recommendations of the Symbols Committee of the Royal Society, London, 1971 [2].

I am very grateful to Professor Karl Dimroth, Marburg, who first stimulated my interest in solvent effects in organic chemistry. I am indebted to Professors W. H. Pirkle, Urbana/Illinois, D. Seebach, Zürich/Switzerland, J. Shorter, Hull/England, and numerous other colleagues for helpful advice and information. Thanks are also due to the authors and publishers of copyrighted materials reproduced with their permission (*cf.* Figure and Table credits on page 495). For the careful translation and improvement of the English manuscript I thank Dr. Edeline Wentrup-Byrne, Marburg. Without the assistance and patience of my wife Maria, this book would not have been written.

Marburg (Lahn), Summer 1978 — Christian Reichardt

References

[1] C. Reichardt: *Lösungsmitteleffekte in der organischen Chemie*. 2nd edition. Verlag Chemie, Weinheim 1973;
Effets de solvant en chimie organique (translation of the first-mentioned title into French, by I. Tkatchenko), Flammarion, Paris 1971;
Rastvoriteli v organicheskoi khimii (translation of the first-mentioned title into Russian, by E. R. Zakhsa), Izdatel'stvo Khimiya, Leningrad 1973.

[2] *Quantities, Units, and Symbols*, issued by The Symbols Committee of the Royal Society, London, in 1971.

Contents

List of Abbreviations

Abbreviations and Recommended Values of Some Fundamental Constants and Numbers[a)]

N_A	Avogadro constant	$6.0220 \cdot 10^{23}\ mol^{-1}$
c	speed of light in vacuum	$2.9979 \cdot 10^{8}\ m \cdot s^{-1}$
e	elementary charge	$1.6022 \cdot 10^{-19}$ C
h	Planck constant	$6.6262 \cdot 10^{-34}$ J · s
R	gas constant	$8.3144\ J \cdot K^{-1} \cdot mol^{-1}$ (or $0.08206\ dm^3 \cdot atm \cdot K^{-1} \cdot mol^{-1}$)
k_B	Boltzmann constant (R/N_A)	$1.3807 \cdot 10^{-23}\ J \cdot K^{-1}$
V_0	standard molar volume of an ideal gas (at 273.15 K and 1 standard atmosphere pressure)	$22.414\ dm^3 \cdot mol^{-1}$
T_0	zero of the Celsius scale	273.15 K
π	ratio of the circumference to the diameter of a circle	3.1416
e	exponential number and base of natural logarithms (ln)	2.7183
ln 10	natural logarithm of ten (ln x = ln 10 · lg x; lg = decadic logarithm)	2.303

Abbreviations and Symbols for Units[a)]

bar	bar (10^5 Pa)	pressure
cg/g	centigram/gram	weight percent
cl/l	centilitre/litre	volume percent
cmol/mol	centimol/mol	mole percent
cm	centimetre (10^{-2} m)	length
cm^3	cubic centimetre (millilitre; $10^{-6}\ m^3$)	volume
C	coulomb	electric charge
°C	degrees centigrade (Celsius)	temperature
dm^3	cubic decimetre (litre; $10^{-3}\ m^3$)	volume
J	joule	energy
kJ	kilojoule (10^3 J)	energy
K	kelvin	temperature

l	litre (1 dm^3; 10^{-3} m^3)	volume
m	metre	length
min	minute	time
mol	mole	amount of substance
MPa	megapascal (10^6 Pa)	pressure
mT	millitesla (10^{-3} T)	magnetic flux density
nm	nanometre (10^{-9} m)	length
Pa	pascal	pressure
ppm	parts per million	
s	second	time

Abbreviations and Symbols for Properties[b)]

a_i	activity of solute i	
$A(^1H)$	ESR hyperfine splitting constant (splitting caused by ^1H)	mT
A_j	the solvent's anion-solvating tendency or 'acity' (Swain)	
AN	acceptor number (Gutmann)	
α	electric polarizability of a molecule	m^3
α	empirical parameter of solvent hydrogen-bond donor acidity (Taft and Kamlet)	
B	empirical parameter of solvent Lewis basicity (Koppel and Palm)	
B_{MeOD}	empirical parameter of solvent Lewis basicity, based on IR measurements (Koppel and Palm)	
B_{PhOH}	empirical parameter of solvent Lewis basicity, based on IR measurements (Koppel and Paju; Makitra)	
B_j	the solvent's cation-solvating tendency or 'basity' (Swain)	
β	empirical parameter of solvent hydrogen-bond acceptor basicity (Taft and Kamlet)	
c	cohesive pressure (cohesive energy density) of a solvent	MPa
c_i	molar concentration of solute i	$mol \cdot l^{-1}$

C_A, C_B	Lewis acidity and Lewis basicity parameter (Drago)	
cmc	critical micelle concentration	$mol \cdot l^{-1}$
D_{HA}	molar bond dissociation energy for the bond between H and A	$kJ \cdot mol^{-1}$
D_π	empirical parameter of solvent Lewis basicity, based on a 1,3-dipolar cycloaddition reaction (Nagai *et al.*)	
DN	donor number (Gutmann)	$kcal \cdot mol^{-1}$
DN^N	normalized donor number (Marcus)	
δ, δ_H	solubility parameter of Hildebrand	$MPa^{1/2}$
δ	chemical shift of NMR signals	ppm
δ	solvent polarizability correction term (Taft and Kamlet)	
E	energy, molar energy	$kJ \cdot mol^{-1}$
E	electric field strength	$V \cdot m^{-1}$
E	enol constant (K. H. Meyer)	
E	empirical parameter of solvent Lewis acidity (Koppel and Palm)	
E_A	activation energy in the Arrhenius equation	$kJ \cdot mol^{-1}$
E_A, E_B	Lewis acidity and Lewis basicity parameter (Drago)	
EA	electron affinity	$kJ \cdot mol^{-1}$
E_B^N	empirical parameter of solvent Lewis basicity, based on the $n \rightarrow \pi^*$ absorption of an aminyloxide radical (Mukerjee; Wrona)	
E_K	empirical parameter of solvent polarity, based on the $d \rightarrow \pi^*$ absorption of a molybdenum complex (Walther)	$kcal \cdot mol^{-1}$
E_{MLCT}^*	empirical parameter of solvent polarity, based on the $d \rightarrow \pi^*$ absorption of a tungsten complex (Lees)	
E_T	transition energy, excitation energy	$kcal \cdot mol^{-1}$ or $kJ \cdot mol^{-1}$
$E_T(30)$	empirical parameter of solvent polarity, based on the intramolecular CT absorption of a pyridinium-*N*-phenoxide betaine dye (Dimroth and Reichardt)	$kcal \cdot mol^{-1}$
E_T^N	normalized $E_T(30)$-parameters (Reichardt)	

E_T^{SO}	empirical parameter of solvent polarity, based on the $n \rightarrow \pi^*$ absorption of an *S*-oxide (Walter)	$kcal \cdot mol^{-1}$
EPA	electron pair acceptor	
EPD	electron pair donor	
ε_0	absolute permittivity of the vacuum	$8.8542 \cdot 10^{-12} C^2 \cdot J^{-2} \cdot m^{-1}$
ε_r	relative permittivity, dielectric constant	
Φ	empirical parameter of solvent polarity, based on the $n \rightarrow \pi^*$ absorption of ketones (Dubois)	
G	empirical parameter of solvent polarity, based on IR measurements (Schleyer and Allerhand)	
ΔG°	standard molar Gibbs energy change	$kJ \cdot mol^{-1}$
$\Delta G^{\neq}$	standard molar Gibbs energy of activation	$kJ \cdot mol^{-1}$
ΔG°_{solv}	standard molar Gibbs energy of solvation	$kJ \cdot mol^{-1}$
ΔG°_{hydr}	standard molar Gibbs energy of hydration	$kJ \cdot mol^{-1}$
$\Delta G_t^\circ(X, O \rightarrow S)$, $\Delta G_t^\circ(X, W \rightarrow S)$	standard molar Gibbs energy of transfer of a solute X from a reference solvent (O) or water (W) to another solvent (S)	$kJ \cdot mol^{-1}$
γ_i	activity coefficient of solute i	
ΔH°	standard molar enthalpy change	$kJ \cdot mol^{-1}$
$\Delta H^{\neq}$	standard molar enthalpy of activation	$kJ \cdot mol^{-1}$
ΔH_v	molar enthalpy (heat) of vapourization	$kJ \cdot mol^{-1}$
H_0	acidity function (Hammett)	
HBA	hydrogen-bond acceptor	
HBD	hydrogen-bond donor	
HOMO	highest occupied molecular orbital	
I, *IP*	ionization potential	$kJ \cdot mol^{-1}$
I	gas-chromatographic retention index (Kováts)	
J	NMR coupling constant	Hz
k	rate constant	s^{-1} (monomolecular reactions); $dm^3 \cdot mol^{-1} \cdot s^{-1}$ (bimolecular reactions)

k_0	rate constant in a reference solvent or in the gas phase	s^{-1} (monomolecular reactions); $dm^3 \cdot mol^{-1} \cdot s^{-1}$ (bimolecular reactions)
k_0	in Hammett equations the rate constant of the unsubstituted substrate	s^{-1} (monomolecular reactions); $dm^3 \cdot mol^{-1} \cdot s^{-1}$ (bimolecular reactions)
K	equilibrium constant	
K_a, K_b	acid and base ionization constants	$dm^3 \cdot mol^{-1}$ or $mol \cdot dm^{-3}$
K_{auto}	autoionization ion product, autoprotolysis constant	$mol^2 \cdot dm^{-6}$
K_{Assoc}	equilibrium constant of an association reaction	
K_{Dissoc}	equilibrium constant of a dissociation reaction	
K_{ion}	equilibrium constant of an ionization reaction	
K_T	equilibrium constant of a tautomeric equilibrium	
$K_{o/w}$	1-octanol/water partition coefficient (Hansch and Leo)	
KB	kauri-butanol number	
L	desmotropic constant (K. H. Meyer)	
LUMO	lowest unoccupied molecular orbital	
λ	wavelength	nm (10^{-9} m)
m	mass of a particle	g
M_r	relative molecular mass of a substance ('molecular weight')	
M	miscibility number (Godfrey)	
MH	microscopic hydrophobicity parameter of substituents (Menger)	
μ	empirical parameter of solvent softness (Marcus)	
μ	permanent dipole moment of a molecule	C m (or D)
μ_{ind}	induced dipole moment of a molecule	C m (or D)
μ_i°	standard chemical potential of solute i	$kJ \cdot mol^{-1}$

μ_i^{∞}	standard chemical potential of solute i at infinite dilution	$kJ \cdot mol^{-1}$
n, n_D	refractive index (at sodium D line)	
N	empirical parameter of solvent nucleophilicity (Winstein and Grunwald)	
N_+	nucleophilic parameter for (nucleophile + solvent)-systems (Ritchie)	
ν	frequency	Hz, s^{-1}
ν°	frequency in the gas phase or in a reference solvent	Hz, s^{-1}
$\tilde{\nu}$	wavenumber ($1/\lambda$)	cm^{-1}
Ω	empirical parameter of solvent polarity, based on a Diels-Alder cycloaddition reaction (Berson)	
p	pressure	Pa, bar
P	measure of solvent polarizability (Koppel and Palm)	
P	empirical parameter of solvent polarity, based on ^{19}F-NMR measurements (Taft)	
PA	proton affinity	$kJ \cdot mol^{-1}$
Py	empirical parameter of solvent polarity, based on the $\pi^* \rightarrow \pi$ emission of pyrene (Winnik)	
$P_{o/w}$	1-octanol/water partition coefficient (Hansch and Leo)	
pH	$-\lg [H_3O^{\oplus}]$, $-\lg c_{H_3O^{\oplus}}$	
pK	$-\lg K$	
π	internal pressure of a solvent	MPa
π^*	empirical parameter of solvent dipolarity/polarizability (Taft and Kamlet)	
π_x	hydrophobicity parameter of substituent X in C_6H_5X (Hansch)	
r	radius of sphere representing an ion or a cavity	cm
r	distance between centres of ions or molecules	cm
ϱ	density (mass divided by volume)	$g \cdot cm^{-3}$
ϱ	reaction constant (Hammett)	
ϱ_A	absorption constant (Hammett)	
S	generalized for solvent	

S	empirical parameter of solvent polarity, based on the Z-values (Brownstein)	
$\mathcal{S}$	$\lg k_2$ for the Menschutkin reaction of tri-*n*-propylamine with iodomethane (Drougard and Decroocq)	
ΔS°	standard molar entropy change	$J \cdot K^{-1} \cdot mol^{-1}$
$\Delta S^{\neq}$	standard molar entropy of activation	$J \cdot K^{-1} \cdot mol^{-1}$
Sp	solvophobic power of a solvent (Abraham)	
σ	substituent constant (Hammett)	
σ	NMR screening constant	
t	centigrade temperature	°C
T	absolute temperature	K
t_{mp}	melting point	°C
t_{bp}	boiling point	°C
U	potential energy of interactions in a system	kJ
ΔU_v	molar energy of vapourization	$kJ \cdot mol^{-1}$
V_i	molar volume of i	$cm^3 \cdot mol^{-1}$
V_m	molar volume	$cm^3 \cdot mol^{-1}$
$\Delta V^{\neq}$	molar volume of activation	$cm^3 \cdot mol^{-1}$
x_i	mole fraction of i	$cmol \cdot mol^{-1}$
X	empirical parameter of solvent polarity, based on an S_E2 reaction (Gielen and Nasielski)	
χ_R, χ_B	empirical parameters of solvent polarity, based on the $\pi \rightarrow \pi^*$ absorption of merocyanine dyes (Brooker)	$kcal \cdot mol^{-1}$
${}^{O}y_X^S$, ${}^{W}y_X^S$	solvent-transfer activity coefficient of a solute X from a reference solvent (O) or water (W) to another solvent (S)	
Y	empirical parameter of solvent ionizing power, based on *t*-butyl chloride solvolysis (Winstein and Grunwald)	
Y_{OTs}	empirical parameter of solvent ionizing power, based on 2-adamantyl tosylate solvolysis (Schleyer and Bentley)	
Y	measure of solvent polarization (Koppel and Palm)	
z_i	charge number of an ion i	positive for cations, negative for anions

Z	empirical parameter of solvent polarity, based on the intermolecular CT absorption of a pyridinium iodide (Kosower)	$kcal \cdot mol^{-1}$

---

[a] A manual of symbols and terminology for physicochemical quantities and units together with recommended values of fundamental constants is given in: Pure and Applied Chemistry *51*, 3...41 (1979).

[b] A glossary of terms used in physical organic chemistry is given in: Pure and Applied Chemistry *55*, 1281...1371 (1983).

"*Agite, Auditores ornatissimi, transeamus alacres ad aliud negotii! quum enim sic satis excusserimus ea quatuor Instrumenta artis, et naturae, quae modo relinquimus, videamus quintum genus horum, quod ipsi Chemiae fere proprium censetur, cui certe Chemistae principem locum prae omnibus assignant, in quo se jactant, serioque triumphant, cui artis suae, prae aliis omnibus effectus mirificos adscribunt. Atque illud quidem Menstruum vocaverunt.*" *)

Hermannus Boerhaave (1668–1738)
De menstruis dictis in chemia, in:
Elementa Chemiae (1733) [1,2].

1 Introduction

The development of our knowledge of solutions reflects to some extent the development of chemistry itself [3]. Of all known substances, water was the first to be considered as a solvent. As far back as the time of the Greek philosophers there was speculation about the nature of solution and dissolution. The Greek alchemists considered all chemically active liquids under the name "Divine water". In this context the word "water" was used to designate everything liquid or dissolved.

The alchemist's search for a universal solvent, the so-called "Alkahest" or "Menstruum universale", as it was called by Paracelsus (1493–1541), indicates the importance given to solvents and the process of dissolution. Although the eager search of the chemists of the 15th to 18th centuries did not in fact lead to the discovery of any "Alkahest", the numerous experiments performed led to the uncovering of new solvents, new reactions, and new compounds**). From these experiences arose the earliest chemical rule that "like dissolves like" *(similia similibus solvuntur)*. However, at that time, the words solution and dissolution comprised all operations leading to a liquid product and it was still a long way to the conceptual distinction between the physical dissolution of a salt or of sugar in water, and the chemical change of a substrate by dissolution, for example, of a metal in an acid. Thus, in the so-called chemiatry period (iatrochemistry period), it was believed that the nature of a substance was fundamentally lost upon dissolution. Van Helmont (1577–1644) was the first to strongly oppose this contention. He claimed that the dissolved substance had not disappeared, but was present in the solution, although in aqueous form, and could be recovered [4]. This line of thought reached a climax in the

* "Well then, my dear listeners, let us proceed with fervor to another problem! Having sufficiently analyzed in this manner the four resources of science and nature, which we are about to leave (*i.e.* fire, water, air, and earth) we must consider a fifth element which can almost be considered the most essential part of chemistry itself, which chemists boastfully, no doubt with reason, prefer above all others, and because of which they triumphantly celebrate, and to which they attribute above all others the marvellous effects of their science. And this they call the solvent (menstruum)."

** Even if the once famous scholar J.B. Van Helmont (1577–1644) claimed to have prepared this "Alkahest" in a phial, together with the adherents of the alkahest theory he was ridiculed by his contemporaries who asked in which vessel he has stored this universal solvent.

theory of osmotic pressure by van't Hoff (1852–1911) [5] and the theory of electrolytic dissociation by Arrhenius (1859–1927) [6].

The influence of solvents on the rates of chemical reactions [7, 8] was first noted by Berthelot and Péan de Saint-Gilles in 1862 in connection with their studies on esterification of acetic acid with ethanol: "...l'éthérification est entravée et ralentie par l'emploi des dissolvants neutres étrangers à la réaction" [9]*. After thorough studies on the reaction of trialkylamines with haloalkanes, Menschutkin in 1890 concluded that a reaction cannot be separated from the medium in which it is performed [10]. In a letter to Prof. Louis Henry he wrote in 1890: "Or, l'expérience montre que ces dissolvants exercent sur la vitesse de combinaison une influence considérable. Si nous représentons par 1 la constante de vitesse de la réaction précitée dans l'hexane C_6H_{14}, cette constante pour la même combinaison dans $CH_3—CO—C_6H_5$, toutes choses égales d'ailleurs sera 847.7. La différence est énorme, mais, dans ce cas, elle n'atteint pas encore le maximum.... Vous voyez que les dissolvants, soi-disant indifférents ne sont pas inertes; ils modifient profondément l'acte de la combinaison chimique. Cet énoncé est riche en conséquences pour la théorie chimique des dissolutions" [26]**. Menschutkin also discovered that, in reactions between liquids, one of the reaction partners may constitute an unfavourable solvent. Thus, in the preparation of acetanilide, it is not without importance whether aniline is added to an excess of acetic acid, or *vice versa*, since aniline in this case is an unfavorable reaction medium. Menschutkin related the influence of solvents primarily to their chemical, not their physical properties.

The influence of solvents on chemical equilibria was discovered in 1896, simultaneously with the discovery of keto-enol tautomerism*** in 1,3-dicarbonyl compounds (Claisen [14]: acetyldibenzoylmethane and tribenzoylmethane; Wislicenus [15]: methyl and ethyl formylphenylacetate; Knorr [16]: ethyl dibenzoylsuccinate and ethyl diacetylsuccinate) and the nitro-isonitro tautomerism of primary and secondary nitro compounds (Hantzsch [17]: phenylnitromethane). Thus, Claisen wrote: "Es gibt Verbindungen, welche sowohl in der Form —C(OH)=C(—)—CO— wie in der Form —CO—CH(—)—CO— zu bestehen vermögen; von der Natur der angelagerten Reste, von der Temperatur, bei den gelösten Substanzen auch von der Art des Lösungsmittels hängt es ab, welche

* "...the esterification is disturbed and decelerated on addition of neutral solvents not belonging to the reaction" [9].

** "Now, experience shows that solvents exert considerable influence on reaction rates. If we represent the rate constant of the reaction to be studied in hexane C_6H_{14} by 1, then, all else being equal, this constant for the same reaction in $CH_3—CO—C_6H_5$ will be 847.7. The increase is enormous, but in this case it has not even reached its maximum. ... So you see that solvents, in spite of appearing at first to be indifferent, are by no means inert; they can greatly influence the course of chemical reactions. This statement is full of consequences for the chemical theory of dissolutions" [26].

*** The first observation of a tautomeric equilibrium was made in 1884 by Zincke at Marburg [11]. He observed that, surprisingly, the reaction of 1,4-naphthoquinone with phenylhydrazine gives the same products as that obtained from the coupling reaction of 1-napththol with benzene diazonium salts. This phenomenon, that the substrate can react either as phenylhydrazone or as a hydroxyazo compound, depending on the reaction circumstances, was called *Ortsisomerie* by Zincke [11]. Later on, the name *tautomerism*, with a different meaning however from that accepted today, was introduced by Laar [12]. For a description of the development of the concept of tautomerism see Ingold [13].

von den beiden Formen die beständigere ist" [14]*). The study of the keto-enol equilibrium of ethyl formylphenylacetate in eight solvents, led Wislicenus to the conclusion that the keto-form predominates in alcoholic solution, the enol-form in chloroform or benzene. He stated that the final ratio in which the two tautomeric forms coexist, must depend on the nature of the solvent and on its dissociating power, whereby he suggested that the dielectric constants were a possible measure of this "power". Stobbe was the first to review these results [18]. He divided the solvents into two groups according to their ability to isomerize tautomeric compounds. His classification reflects, to some extent, the modern division into protic and aprotic solvents. The effect of solvent on constitutional and tautomeric isomerization equilibria was later studied in detail by Dimroth [19] (using triazole derivatives, *e.g.* 5-amino-4-methoxycarbonyl-1-phenyl-1,2,3-triazole) and Meyer [20] (using ethyl acetoacetate).

It has long been known that UV/Vis absorption spectra may be influenced by the phase (gas or liquid) and that the solvent can bring about a change in the position, intensity, and shape of the absorption band**). Hantzsch later termed this phenomenon *solvatochromism****) [22]. The search for a relationship between solvent effect and solvent property led Kundt in 1878 to propose the rule, later called after him, that increasing dispersion (*i.e.* increasing index of refraction) is related to a shift of the absorption maximum towards longer wavelength [23]. This he established on the basis of UV/Vis absorption spectra of six dyestuffs, namely, chlorophyll, fuchsin, aniline green, cyanine, quinizarin, and egg yolk in twelve different solvents. The – albeit limited – validity of Kundt's rule, *e.g.* found in the cases of 4-hydroxyazobenzene [24] and acetone [25], led to the realization that the effect of solvent on dissolved molecules is a result of electrical fields. These fields in turn originate in the dipolar properties of the molecules in question [25]. The similarities in the relationships between solvent effects on reaction rate, equilibrium position, and absorption spectra has been related to the general solvating ability of the solvent in a fundamental paper by Scheibe *et al.* [25].

---

* "There are compounds capable of existence in the form $-\mathrm{C(OH)}=\overset{|}{\mathrm{C}}-\mathrm{CO}-$ as well as in the form $-\mathrm{CO}-\overset{|}{\mathrm{C}}\mathrm{H}-\mathrm{CO}-$; it depends on the nature of the substituents, the temperature, and for dissolved compounds, also on the nature of the solvent, which of the two forms will be the more stable" [14].

** A survey of older works of solvent effects on UV/Vis absorption spectra has been given by Sheppard [21].

*** It should be noted that the now generally accepted meaning of the term *solvatochromism* differs from that introduced by Hantzsch (*cf.* Section 6.2).

2 Solute–Solvent Interactions

2.1 Solutions

In a limited sense *solutions* are homogeneous liquid phases consisting of more than one substance in variable ratios, when for convenience one of the substances, which is called the *solvent* and may itself be a mixture, is treated differently from the other substances, which are called *solutes* [1]. Normally, the component which is in excess is called the solvent and the minor component(s) is the solute. When the sum of the mole fractions of the solutes is small compared to unity, the solution is called a *dilute solution**. A solution of solute substances in a solvent is treated as an *ideal dilute solution* when the solute activity coefficients γ are close to unity ($\gamma = 1$) [1, 171].

A solvent should not be considered a macroscopic continuum characterized only by physical constants such as density, dielectric constant, index of refraction *etc.*, but as a discontinuum which consists of individual, mutually interacting solvent molecules. According to the extent of these interactions, there are solvents with a pronounced internal structure (*e.g.* water) and others in which the interaction between the solvent molecules is small (*e.g.* hydrocarbons). The interaction between species in solvents (and in solutions) is on the one hand too big for it to be treated by the laws of the kinetic theory of gases; on the other hand it is too small for it to be treated by the laws of solid state physics. Thus, the solvent is neither an indifferent medium in which the dissolved material diffuses in order to distribute itself evenly and disorderly, nor does it possess an ordered structure resembling a crystal lattice. Nevertheless, the long-distance ordering in a crystal corresponds somewhat to the local ordering in a liquid. Thus, neither of the two possible models – gas and crystal model – can be applied to solutions without limitation. Between the two models there is such a wide spectrum of thinkable and experimentally established variants, that it is difficult at all to develop a generally valid model for liquids. Due to the complexity of the interactions, the structure of liquids – in contrast to that of gases and solids – is the least-known of all aggregation states. Therefore, the experimental and theoretical examination of the structure of liquids belongs to the most difficult tasks of physical chemistry [2–7, 172–174].

Any theory of the liquid state has to explain – among others – the following facts: Except for water, the molar volume of a liquid is roughly 10% greater than that of the corresponding solid. According to X-ray diffraction studies, a short-range order of solvent molecules persists in the liquid state and the nearest neighbour distances are almost the same as in the solid. The solvent molecules are not moving freely, as in the gaseous state, but instead move in the potential field of their neighbours. The potential energy of a liquid is higher than that of its solid by about 10%. Therefore, the heat of fusion is roughly 10% of the heat of sublimation. Each solvent molecule has an environment very much like that of a solid, but some of the nearest neighbours are replaced by holes. Roughly one neighbour molecule in ten is missing.

---

* The superscript ∞ attached to the symbol for a property of a solution denotes the property of an *infinitely dilute solution.*

Even for the most important solvent – water – the investigation of its inner fine structure is still the subject of current research [8–15, 15a]*). Numerous different models, *e.g.* the "flickering cluster model" of Franck and Wen [16], were developed to describe the structure of water. However, all these models prove themselves unapproachable for a complete description of the physico-chemical properties of water and an interpretation of its anomalies [11]. Fig 2-1 should make clear the complexity of the inner structure of a solvent in the example of water.

Liquid water consists both of bound ordered regions of a regular lattice, and regions where the water molecules are hydrogen-bonded in a random array; it is permeated by monomeric water and interspersed with random holes, lattice vacancies, and cages. There are chains and small polymers as well as bound, free, and trapped water molecules [9, 176].

In principle, organic solvents such as the protic ones should possess a similar complicated structure. However, whereas water has been thoroughly studied [17], the examination of the inner structure of organic solvents is still in the beginning [172, 177–179].

From the idea that the solvent only provides an indifferent reaction medium, comes the ***Ruggli-Ziegler dilution principle***, long known to the organic chemist. According to this principle, in the case of cyclization reactions, the desired intramolecular reaction will be favored over the undesired intermolecular reaction by high dilution with an inert solvent [18].

The assumption of forces of interaction between solvent and solute led, on the other hand, to the century old principle that "like dissolves like" *(similia similibus solvuntur)*, where the word "like" should not be too narrowly interpreted. In many cases the presence of similar functional groups in the molecules suffices. When a chemical similarity is present, the solution of the two components will usually have a structure similar to the one of the pure materials (*e.g.* alcohol-water mixtures [19]). This rule of thumb has only limited validity since there are many examples of solutions of chemically dissimilar compounds. For example, methanol and benzene, water and *N,N*-dimethylformamide, aniline and diethyl ether, or polystyrene and chloroform, are completely miscible at room temperature. On the other hand, insolubility can occur in spite of similarity of the two partners. Thus, polyvinylalcohol does not dissolve in ethanol, acetyl cellulose is insoluble in ethyl acetate, and polyacrylonitrile in acrylonitrile [20]. Between these two extremes there is a whole range of possibilities where the two materials dissolve each other to a limited extent. The system water/diethyl ether is such an example. Pure diethyl ether dissolves water to the extent of 15 mg/g at 25 °C, whereas water dissolves diethyl ether to the extent of 60 mg/g. When one of the two solvents is in large excess a homogeneous solution is obtained. Two phases occur when the ratio is beyond the limits of solubility.

Rather than the "like dissolves like" rule, it is the intermolecular interaction, between solvent and solute molecules, which determines the mutual solubility. A compound *A* dissolves in a solvent *B* only when the intermolecular forces of attraction K_{AA} and K_{BB} for the pure compounds can be overcome by the forces K_{AB} in solution [21].

* The amusing story of polywater, which excited the scientific community for a few years during the late 1960's and early 1970's, has been reviewed by Franks [175]. It turned out that polywater was not a new and more stable form of pure water, but merely dirty water. The strange properties of polywater are due to high concentrations of siliceous material dissolved from quartz capillaries in which it was produced.

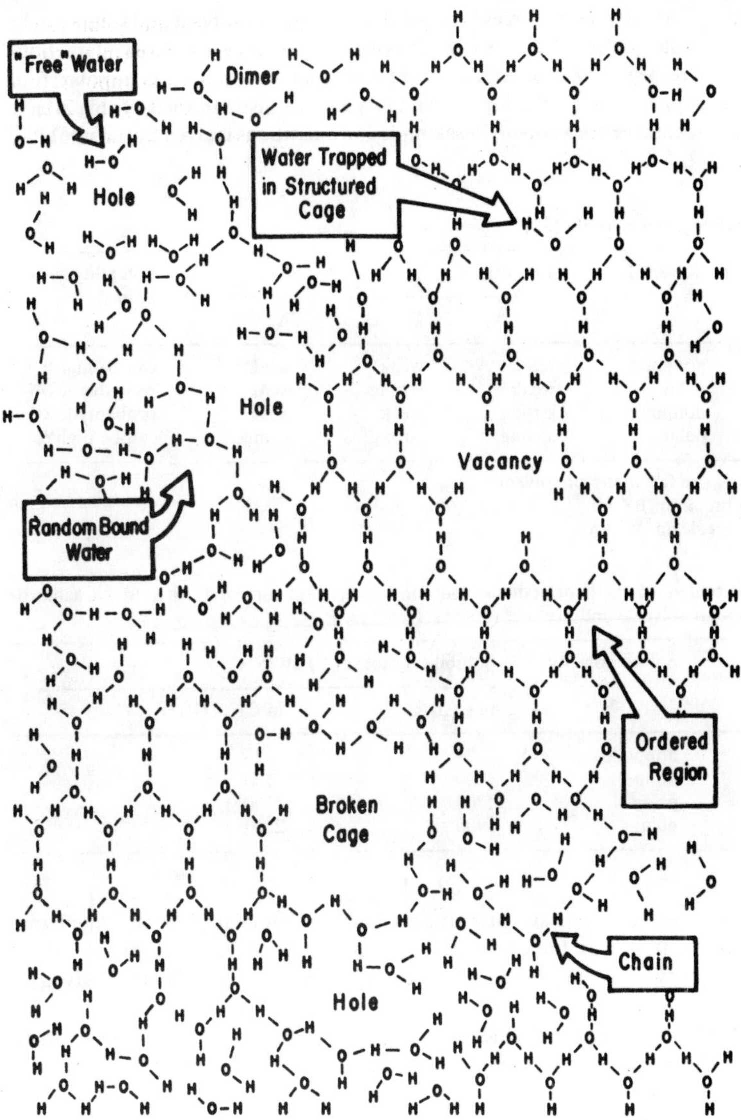

Fig. 2-1. Two-dimensional schematic diagram of the three-dimensional structure of liquid water [9].

The sum of the interaction forces between the molecules of solvent and solute can be related to the so-called *polarity**) of *A* and *B*. Denoting compounds with large interactions $A \cdots A$ or $B \cdots B$ respectively as polar, and those with small interactions as unpolar, four cases allowing a qualitative prediction of solubility can be distinguished (Table 2-1).

An experimental verification of these simple considerations is given by the solubility data in Table 2-2.

Table 2-1. Solubility and polarity [22].

Solute A	Solvent B	Interaction			Solubility of A in B
		A···A	B···B	A···B	
Nonpolar	nonpolar	weak	weak	weak	can be high[a)]
Nonpolar	polar	weak	strong	weak	probably low[b)]
Polar	nonpolar	strong	weak	weak	probably low[c)]
Polar	polar	strong	strong	strong	can be high[a)]

[a)] Not much change for solute or solvent.
[b)] Difficult to break up B···B.
[c)] Difficult to break up A···A.

Table 2-2. Solubilities of methane, ethane, chloromethane, and dimethyl ether in tetrachloromethane (nonpolar solvent) and acetone (polar solvent) [22].

Solute	Solute polarity	Solubility/(mol · m^{-3}) at 25 °C	
		in CCl_4	in CH_3COCH_3
CH_4	nonpolar	29	25
CH_3CH_3	nonpolar	220	130
CH_3Cl	polar	1700	2800
CH_3OCH_3	polar	1900	2200

The solubilities of ethane and methane are higher in unpolar tetrachloromethane, whereas the opposite is true for chloromethane and dimethyl ether. A survey of the reciprocal miscibility of some representative examples of organic solvents is given in Fig. 2-2.

The *solubility parameter* δ of Hildebrand [4, 24] as defined in Eq. (2-1) can often be used in estimating the solubility of non-electrolytes in organic solvents.

$$\delta = \left(\frac{\Delta U_v}{V_m}\right)^{\frac{1}{2}} = \left(\frac{\Delta H_v - R \cdot T}{V_m}\right)^{\frac{1}{2}} \tag{2-1}$$

In this equation V_m is the molar volume of the solvent, and ΔU_v and ΔH_v are the molar energy and the molar enthalpy (heat) of vaporization for a gas of zero pressure,

* For a more detailed definition of solvent polarity see Sections 3.2 and 7.1.

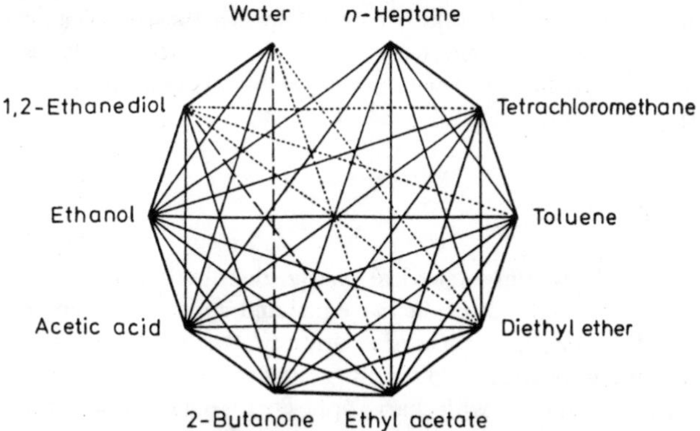

Fig. 2-2. Miscibility of organic solvents [23]. —— miscible in all proportions; ---- limited miscibility; ..... little miscibility; without line: immiscible.

respectively. δ is a solvent property which measures the work necessary to separate the solvent molecules (*i.e.* disruption and reorganization of solvent/solvent interactions) to create a suitably sized cavity, large enough to accommodate the solute. Accordingly, highly ordered self-associated solvents exhibit relatively large δ-values ($\delta = 0$ for the gas phase). As a rule, it has been found that a good solvent for a certain non-electrolyte has a δ-value close to that of the solute [20, 24, 25]; *cf.* Table 3-3 in Section 3.2 for a collection of δ-values. Often a mixture of two solvents, one having a δ-value higher and the other having a δ-value lower than that of the solute is a better solvent than each of the two solvents separately [24]; *cf.* also Section 3.2.

A nice example demonstrating mutual insolubility due to different δ-values has been described by Hildebrand [180], and was later improved [181]. A system of eight non-miscible liquid layers has been constructed. The eight layers in order of increasing densities are paraffin oil, silicon oil, water, aniline, perfluorodimethylcyclohexane, white phosphorus, gallium, and mercury. This system is stable indefinitely at 45 °C; this temperature is required to melt the gallium and phosphorus [181].

2.2 Intermolecular Forces [26, 27, 182–184]

Intermolecular forces are those which can occur between closed-shell molecules [26, 27]. These are also called van-der-Waals-forces, since van der Waals recognized them as the reason for the non-ideal behaviour of the real gases. Intermolecular forces are usually classified in two distinct categories. The first category comprises the so-called directional, induction, and dispersion forces which are non-specific and cannot be completely saturated (just as Coulomb forces between ions cannot). To the second group belong hydrogen bonding forces and the forces of charge transfer or electron-pair donor acceptor forces. The latter group are specific, directional forces which can be saturated and lead to stoichiometrical molecular compounds. For the sake of completeness, in the following the Coulomb forces between ions and electrically neutral molecules (with permanent dipole moment) will be considered first, even though they do not belong to the intermolecular forces in the narrower sense.

2.2.1 Ion-Dipole Forces [28, 185]

Electrically neutral molecules with an unsymmetrical charge distribution possess a permanent dipole moment μ. If the magnitude of the two equal and opposite charges of this molecular dipole is denoted by q, and the distance of separation l, the dipole moment is given by $\mu = q \cdot l$. When placed in the electrical field resulting from an ion, the dipole will orient itself so that the attractive end (the end with charge opposite to that of the ion) will be directed toward the ion, and the other repulsive end directed away. The potential energy of an ion-dipole interaction is given by

$$U_{\text{ion-dipole}} = -\frac{1}{4\pi \cdot \varepsilon_0} \cdot \frac{z \cdot e \cdot \mu \cdot \cos\theta}{r^2} \qquad (2\text{-}2)^{*)}$$

where ε_0 is the permittivity of the vacuum, $z \cdot e$ the charge on the ion, r the distance from the ion to the center of the dipole, and θ the dipole angle relative to the line r joining the ion and the center of the dipole. Cos $\theta = 1$ for $\theta = 0°$, *i.e.* in this case the dipole is positioned next to the ion in such a way that the ion and the separated charges of the dipole all lie on a line ((+ −) ⊕ or (− +) ⊖).

Only molecules possessing a permanent dipole moment should be called ***dipolar molecules***. Apart from a few hydrocarbons (*n*-hexane, cyclohexane, and benzene) and some symmetrical compounds (carbon disulfide, tetrachloromethane, and tetrachloroethene) all common organic solvents possess a permanent dipole moment between 0 and $18 \cdot 10^{-30}$ C m (*i.e.* Coulombmeter). Among the solvents listed in the Appendix, Table A-1, hexamethylphosphoric triamide is the one with the highest dipole moment ($\mu = 18.48 \cdot 10^{-30}$ C m), followed by propylene carbonate ($\mu = 16.7 \cdot 10^{-30}$ C m), and sulfolane ($\mu = 16.05 \cdot 10^{-30}$ C m).

Ion-dipole forces are important for solutions of ionic compounds in dipolar solvents where solvated species such as $Na(OH_2)_m^{\oplus}$ and $Cl(H_2O)_n^{\ominus}$ (for solutions of NaCl in H_2O) exist. In the case of some metal ions these solvated species can be sufficiently stable to be considered as discrete species such as $[Co(NH_3)_6]^{3\oplus}$ or $Ag(CH_3CN)_{2...4}^{\oplus}$.

2.2.2 Dipole-Dipole Forces [29]

Directional forces depend on the electrostatic interaction between molecules which possess a permanent dipole moment μ due to their unsymmetrical distribution of charge. When two dipolar molecules are optimally oriented to each other at a distance r as shown

* It should be noted that Eqs. (2-2) to (2-6) are valid only for gases; an exact application to solutions is not possible. Furthermore, Eqs. (2-2) to (2-6) are restricted to cases with $r \gg l$.

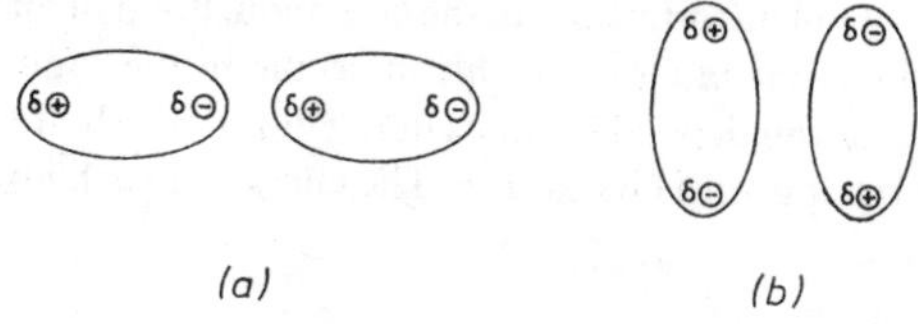

Fig. 2-3. (a) "Head-to-tail" arrangement of two dipole molecules; (b) Antiparallel arrangement of two dipole molecules.

in Fig. 2-3a, then the force of attraction is proportional to $1/r^3$. An alternative arrangement is the anti-parallel arrangement of the two dipoles as shown in Fig. 2-3b.

Unless the dipole molecules are very voluminous, the second arrangement is the more stable one. The two situations exist only when the attractive energy is larger than the thermal energies. Therefore, the thermal energy will normally prevent the dipoles from optimal orientation. If all possible orientations were equally probable, then attraction and repulsion would compensate each other. The fact that dipole orientations leading to attraction are statistically favored, leads to a net attraction which is strongly temperature dependent, according to Eq. (2-3) (k_B = Boltzmann constant; T = absolute temperature) [29].

$$U_{\text{dipole-dipole}} = -\frac{1}{4\pi \cdot \varepsilon_0} \cdot \frac{2\mu_1^2 \cdot \mu_2^2}{3k_B \cdot T \cdot r^6} \tag{2-3}$$

As the temperature increases, the interaction energy becomes less negative until at very high temperatures all dipole orientations are equally populated and the potential energy is zero.

Among other interaction forces, these dipole-dipole interactions are mainly responsible for the association of dipolar organic solvents such as dimethylsulfoxide [30] or *N,N*-dimethylformamide [31].

2.2.3 Dipole-Induced Dipole Forces [32]

The electric dipole of a molecule possessing a permanent dipole moment μ can induce a dipole moment in a neighbouring molecule. This induced moment always lies in the direction of the inducing dipole. Thus, attraction always exists between the two partners, which is independent of temperature. The induced dipole moment*) will be bigger the larger the polarizability α of the apolar molecule experiencing the induction of the permanent dipole is. The energy of interaction, averaged over all possible orientations of the permanent dipole, is given by Eq. (2-4) [32].

$$U_{\text{dipole-induced dipole}} = -\frac{1}{4\pi \cdot \varepsilon_0} \cdot \frac{\alpha_1 \cdot \mu_2^2 + \alpha_2 \cdot \mu_1^2}{r^6} \tag{2-4}$$

* The induced dipole moment is defined as $\mu_{\text{ind}} = 4\pi \cdot \varepsilon_0 \cdot \alpha \cdot E$ (ε_0 permittivity of vacuum; α electric polarizability of the molecule; E electric field strength).

Similarly, a charged particle such as an ion introduced into the neighbourhood of an uncharged, apolar molecule will distort the electron cloud of this molecule in the same way. The polarization of the neutral molecule will depend upon its inherent polarizability α, and on the polarizing field afforded by the ion with charge $z \cdot e$. The energy of such an interaction is given by Eq. (2-5).

$$U_{\text{ion-induced dipole}} = -\frac{1}{4\pi \cdot \varepsilon_0} \cdot \frac{z^2 \cdot e^2 \cdot \alpha}{2 \cdot r^4} \tag{2-5}$$

The importance of both of these interactions is limited to situations such as solutions of dipolar or ionic compounds in nonpolar solvents.

2.2.4 Instantaneous Dipole-Induced Dipole Forces [33, 34, 186]

Even in atoms and molecules possessing no permanent dipole moment, the continuous electronic movement results, at any instant, in a small dipole moment μ which can fluctuatingly polarize the electron system of the neighbouring atoms or molecules. This coupling causes the electronic movements to be synchronized in such a way that a mutual attraction results. The energy of such so-called dispersion interactions may be expressed as

$$U_{\text{dispersion}} = -\frac{3\alpha_1 \cdot \alpha_2}{2r^6} \cdot \left(\frac{I_1 \cdot I_2}{I_1 + I_2}\right) \tag{2-6a}$$

where α_1 and α_2 are the polarizabilities and I_1 and I_2 the ionization potentials of the two different interacting species [33]. When applied to two molecules of the same substance, Eq. (2-6a) reduces to Eq. (2-6b).

$$U_{\text{dispersion}} = -\frac{3\alpha^2 \cdot I}{4r^6} \tag{2-6b}$$

Dispersion forces are extremely short-range in action (depending on $1/r^6$!).

Dispersion forces are universal for all atoms and molecules, they alone are responsible for the aggregation of molecules which possess neither free charges nor electrical dipole moments. Due to the higher polarizability of π-electrons, especially strong dispersion forces exist between molecules with conjugated π-electron systems (*e.g.* aromatic hydrocarbons). But also for many other dipole molecules with high polarizability, the major part of the cohesion is due to dispersion forces. For example, the calculated cohesion energy of liquid 2-butanone at 40 °C consists of 8% orientational energy, 14% inductional energy, and 78% dispersion energy [35]. Two molecules with $\alpha = 3 \cdot 10^{-30}$ m^3, $I = 20 \cdot 10^{-19}$ J, and $r = 3 \cdot 10^{-10}$ m have an interaction potential of -11.3 kJ/mol (-2.7 kcal/mol) [35a]. These values of α, I, and the average intermolecular distance r correspond to those for liquid HCl. It is instructive to compare the magnitude of these dispersion forces to that of the dipole-dipole interactions. For two dipoles, both with dipole moments of $3.3 \cdot 10^{-30}$ C m (1.0 D), separated by a distance of $r = 3 \cdot 10^{-10}$ m and oriented as in Fig. 2-3a, the interaction energy is only -5.3 kJ/mol (-1.1 kcal/mol) [35a].

Thus, for HCl and most other compounds, the dispersion forces are considerably stronger than the dipole-dipole forces of nearest neighbour distance in the liquid state. However, at larger distances the dispersion energy falls off rapidly.

As a result of the α^2-term in Eq. (2-6b), dispersion forces increase rapidly with the molecular volume and the number of polarizable electrons. The polarizability α is connected with the molecular refraction and the index of refraction, according to the equation of Lorenz-Lorentz. Therefore, solvents with a large index of refraction, and hence large optical polarizability, should be capable of enjoying particularly strong dispersion forces. As indicated in Table A-1 (Appendix), all aromatic compounds possess relatively high indices of refraction, *e.g.* quinoline ($n = 1.6273$), iodobenzene ($n = 1.6200$), aniline ($n = 1.5863$), and diphenyl ether ($n = 1.5763$); of all organic solvents carbon disulfide has the highest index of refraction ($n = 1.6280$).

Solvents with high polarizability are often good solvators for anions which also possess high polarizability. This is due to the fact that the dispersional interactions between the solvents and the large, polarizable anions like $I_3^\ominus$, $I^\ominus$, $SCN^\ominus$ or the picrate anion are significantly larger than for the smaller anions like $F^\ominus$, $HO^\ominus$ or $R_2N^\ominus$ [36]. Fluorocarbons have unusually low boiling points because tightly held electrons in fluorine have only a small polarizability.

2.2.5 Hydrogen Bonding [37–46, 187–190]

Liquids possessing hydroxy groups or other groups with a hydrogen atom bound to an electronegative atom X are strongly associated and have abnormal boiling points. This observation led to the contention that particular intermolecular forces apply here. These are designated as hydrogen bridges, or hydrogen bonds, characterized by a co-ordinative divalency of the hydrogen atom involved. A general definition of the hydrogen bond is: when a covalently bound hydrogen atom forms a second bond to another atom, the second bond is referred to as a *hydrogen bond* [44].

The concept of hydrogen bonding was introduced in 1919 by Huggins [41]. The first definitive paper on hydrogen bonding – applied to the association of water molecules – was published in 1920 by Latimer and Rodebush [191]. All three were working in the Laboratory of G. N. Lewis, University of California, Berkeley/USA.

A hydrogen bond is formed by the interaction between the partners R—X—H and :Y—R′ according to Eq. (2-7).

$$\text{R-X-H} + \text{:Y-R'} \rightleftharpoons \text{R-X-H}\cdots\text{Y-R'} \tag{2-7}$$

R—X—H is the proton donor and :Y—R′ makes available an electron pair for the bridging bond. Thus, hydrogen bonding can be regarded as a preliminary step in a Brønsted acid-base reaction which would lead to a dipolar reaction product $\text{R—X}^\ominus \cdots \text{H—Y}^\oplus\text{—R'}$. X and Y are atoms of higher electronegativity than hydrogen (*e.g.* C, N, P, O, S, F, Cl, Br, I). Both inter- and intramolecular hydrogen bonding are possible, the latter when X and Y belong to the same molecule.

The most important electron pair donors (*i.e.* hydrogen bond acceptors) are the oxygen atoms in alcohols, ethers, and carbonyl compounds, as well as nitrogen atoms in amines and *N*-heterocycles. Hydroxy-, amino-, carboxyl- and amide groups are the most important proton donor groups. Strong hydrogen bonds are formed by the pairs O—H···O, O—H···N, and N—H···O, weaker ones by N—H···N, and the weakest by Cl_2C—H···O and Cl_2C—H···N. The π-electron systems of aromatic compounds, alkenes, and alkines can also act as weak hydrogen bond acceptors.

When two or more equal molecules associate, so-called *homo-intermolecular* hydrogen bonds are formed (Fig. 2-4). The association of different molecules (*e.g.* R—O—H···NR_3) results in *hetero-intermolecular* hydrogen bonds. The designation *homo-* and *heteromolecular* [192] as well as *homo-* and *heteroconjugated* hydrogen bond are also in use.

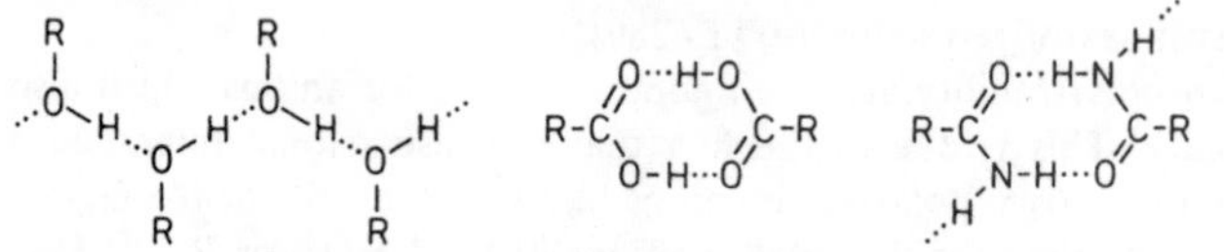

Fig. 2-4. Homo-intermolecular hydrogen bonds in alcohols, carboxylic acids, and amides (the hydrogen bonds are denoted by dotted lines).

Hydrogen bonds can be either *inter*molecular or *intra*molecular. Both types of hydrogen bonds are found in solutions of 2-nitrophenol, depending on the Lewis basicity of the solvent [298]. The intramolecularly hydrogen-bonded form exists in non-hydrogen-bonding solvents (*e.g.* cyclohexane, tetrachloromethane). 2-Nitrophenol breaks its intramolecular hydrogen bond to form an intermolecular one in electron pair donor (EPD) solvents (*e.g.* anisole, HMPT).

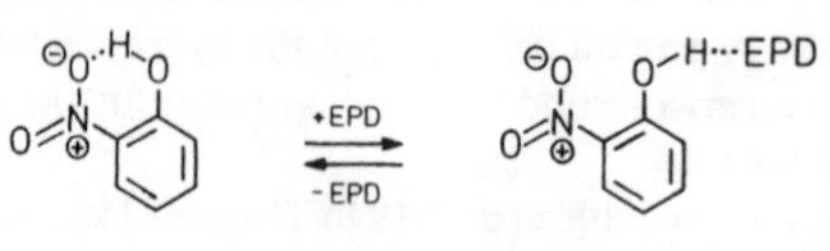

Circular hydrogen bonds have been found in the hexahydrate of α-cyclodextrin (cyclohexaamylose) [193]. Hydration water molecules and hydroxy groups of the macromolecule cooperate to form a network-like pattern with circular O—H···O hydrogen bonds. If the O—H···O hydrogen bonds run in the same direction, the circle is called

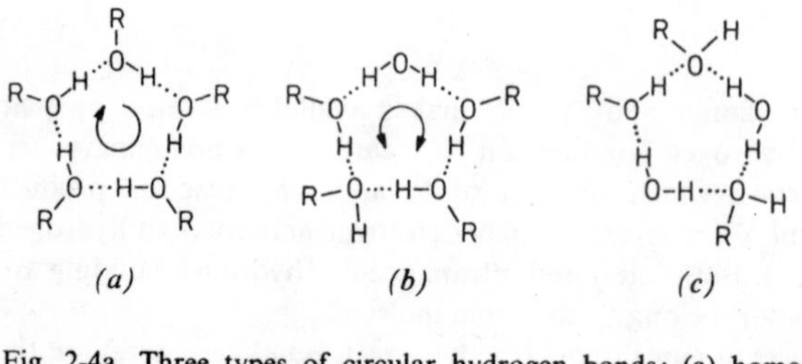

Fig. 2-4a. Three types of circular hydrogen bonds: (a) homodromic, (b) antidromic, and (c) heterodromic hydrogen bonds [193].

homodromic. Circles with the two counter-running chains are called *antidromic*, and circles with more randomly oriented chains are designated *heterodromic* [193]; *cf.* Fig. 2-4a. Such circular hydrogen bonds can be of importance with respect to the inner molecular structure of water and alcohols (*cf.* also Fig. 2-1).

The question of the exact geometry of hydrogen bonds (distances, angles, lone-pair directionality) has been reviewed [194].

The bond dissociation enthalpy for normal hydrogen bonds is *ca.* 13...42 kJ/mol (3...10 kcal/mol)*). For comparison, covalent single bonds have dissociation enthalpies of 210...420 kJ/mol (50...100 kcal/mol). Thus, hydrogen bonds are approx. ten times weaker than covalent single bonds, but also approx. ten times stronger than the nonspecific intermolecular interaction forces.

Hydrogen bonds are characterized by the following structural and spectroscopic features [189]: (a) The distances between the neighbouring atoms involved in the hydrogen bond [X and Y in Eq. (2-7)] are considerably smaller than the sum of their van-der-Waals radii; (b) The X—H bond length is increased and hydrogen bond formation causes its IR stretching mode to be shifted towards lower frequencies; (c) The dipolarity of the X—H bond increases on hydrogen bond formation, leading to a larger dipole moment of the complex than expected from vectorial addition of its dipolar components R—X—H and Y—R'; (d) Due to the reduced electron density at protons involved in hydrogen bonds, they are deshielded, resulting in substantial downfield shifts of their ^{1}H-NMR signals; (e) In hetero-molecular hydrogen bonds, a shift of the Brønsted acid/base equilibrium $R{-}X{-}H\cdots Y{-}R' \rightleftharpoons R{-}X^{\ominus}\cdots H{-}Y^{\oplus}{-}R'$ to the right-hand side with increasing solvent polarity is found (*cf.* Chapter 4.4.1 and reference [195] for a recent example).

Up until now there has been no general agreement as to the best description of the nature of the forces in the hydrogen bond [42–46]. The hydrogen bond can be described as dipole-dipole or as resonance interaction. Since hydrogen bonding occurs only when the hydrogen bond is bound to an electronegative atom, the first assumption concerning the nature of the hydrogen bond was that it consists of a dipole-dipole interaction such as $R{-}X^{\delta\ominus}{-}H^{\delta\oplus}\cdots Y^{\delta\ominus}{-}R'$. This viewpoint is supported by the fact that the strongest hydrogen bonds are formed in pairs in which the hydrogen is bonded to the most electronegative elements (*e.g.* $F{-}H\cdots F^{\ominus}$, $\Delta H = -155$ kJ/mol). The greater strength of the hydrogen bond compared with non-specific dipole-dipole interactions is due to the much smaller size of the hydrogen atom relative to any other atom, which allows it to approach another dipole more closely. This simple dipole model accounts for the usual linear geometry of the hydrogen bond, because a linear arrangement maximizes the attractive forces and minimizes the repulsion [39].

But there are reasons to believe that more is involved in hydrogen bonding than simply an exaggerated dipole-dipole interaction. The shortness of hydrogen bonds

* Bond dissociation enthalpies outside these limits are, however, known. Examples of weak, normal, and strong hydrogen bonds are found in the following pairs: phenol/benzene ($\Delta H = -5$ kJ/mol) [47], phenol/triethylamine ($\Delta H = -37$ kJ/mol) [47], and trichloroacetic acid/triphenylphosphane oxide ($\Delta H = -67$ kJ/mol) [48]. An extremely strong hydrogen bond is found in $Me_4N^+HF_2^-$ ($\Delta H = -155$ kJ/mol) [44]. The strength of a hydrogen bond correlates with the basicity of the proton-acceptor and the acidity of the proton-donor molecule. Compounds with very strong hydrogen bonds have been reviewed [190].

indicates considerable overlap of van-der-Waals-radii and this should lead to repulsive forces unless otherwise compensated. Also, the existence of symmetrical hydrogen bonds of the type $F^{\delta\ominus}\cdots H\cdots F^{\delta\ominus}$ cannot be explained in terms of the electrostatic model. When the X—Y distance is sufficiently short, an overlap of the orbitals of the X—H bond and the electron pair of :Y can lead to a covalent interaction. According to Eq. (2-8), this situation can be described by two contributing "protomeric" structures, which differ only in the position of the proton*).

$$R\text{-}X\text{-}H\cdots Y\text{-}R' \xleftrightarrow{p} R\text{-}X^{\ominus}\cdots H\text{-}Y^{\oplus}\text{-}R' \qquad (2\text{-}8)$$

The approximate quantum mechanical description of proton states by linear combination of these protomeric structures has been called *protomerism* (symbol *p*) [38]. It seems to be applicable to hydrogen bond systems, in which a proton transfer may occur between two potential minima of equal depth [38].

Solvents containing proton-donor groups are designated *protic* solvents [36] or *HBD* solvents [196]; solvents containing proton-acceptor groups are called *HBA* solvents [196]. The abbreviations *HBD* (hydrogen-bond donor) and *HBA* (hydrogen-bond acceptor) refer to donation and acceptance of the proton, and not to the electron pair involved in hydrogen bonding.

Solvents without proton-donor groups have been designated *aprotic* solvents [36]. However, this term is rather misleading, since, for example, solvents commonly referred to as *dipolar aprotic* (*e.g.* CH_3SOCH_3, CH_3CN, CH_3NO_2) are in fact *not* aprotic. In reactions where strong bases are employed, their protic character can be recognized. Therefore, the term *aprotic* solvents should be replaced by *nonhydroxylic* or better still by *non-HBD* solvents [197].

Typical protic or HBD solvents are water, ammonia, alcohols, carboxylic acids, and primary amides. Typical HBA solvents are amines, ethers, ketones, and sulfoxides. *Amphiprotic* solvents can act both as HBD and as HBA solvents simultaneously (*e.g.* water, alcohols, amides; *cf.* Fig. 2-4).

In *type-A* hydrogen bonding, the solute acts as HBA-base and the solvent as HBD-acid; in *type-B* hydrogen bonding, the roles are reversed [196].

Hydrogen bonding is responsible for the strong, temperature-dependent self- and hetero-association of amphiprotic solvents (*e.g.* water, alcohols, amides).

Hydrogen bonding plays a particularly important role in the interactions between anions and HBD solvents. Hence, HBD solvents are good anion solvators. Due to the small size of the hydrogen atom, small anions like $F^{\ominus}$, $Cl^{\ominus}$, or $HO^{\ominus}$ are more effectively solvated by such solvents than the larger ones, *e.g.* $I_3^{\ominus}$, $I^{\ominus}$, $SCN^{\ominus}$, or the picrate ion [36]. This is also one of the reasons why the Gibbs energy of hydration, ΔG_{solv}, of the halide ions decreases in the series $F^{\ominus} > Cl^{\ominus} > Br^{\ominus} > I^{\ominus}$ [49].

Hydrogen bonding is of paramount importance for the stabilization and the shape of large biological molecules in living organisms (*e.g.* cellulose, proteins, nucleic acids). For instance, the anaesthetic properties of some halogen-containing solvents such as

---

* The term "protomeric structure" was obviously introduced in analogy to the well-known "mesomeric structures", which are used to describe the electronic ground state of aromatic compounds such as benzene in terms of a resonance hybrid [38]:

chloroform, halothane (CF_3—CHClBr), and methoxyflurane (CH_3O—CF_2—$CHCl_2$) have been connected with their ability to hinder the formation of biologically important hydrogen bonds. This is shown in the following equilibrium [300]:

$$Cl_3C\text{-}H + \;>N\text{-}H\cdots O{=}C< \;\rightleftharpoons\; Cl_3C\text{-}H\cdots O{=}C< + \;>N\text{-}H$$

Anaesthetic — Peptide H-bond in Proteins

Halohydrocarbon solvents containing an acidic C—H bond, shift this equilibrium in favour of free or less associated species, thus perturbing the ion-channels which determine the permeability of neuron membranes to $K^{\oplus}/Na^{\oplus}$ ions in the nervous system. Hydrogen bonds play a decisive role on the structure and dimension of these ion-channels on which this permeability depends [300].

2.2.6 Electron Pair Donor-Electron Pair Acceptor Interactions (EPD/EPA Interactions) [50–59, 59a, 59b]

When tetrachloromethane solutions of yellow chloranil and colourless hexamethylbenzene are mixed, an intensely red solution is formed ($\lambda_{max} = 517$ nm [50]). This is due to the formation of a complex between the two components, and is only one example of a large number of so-called *electron pair donor/electron pair acceptor complexes (EPD/EPA complexes)* *). It is generally accepted that the characteristic long-wavelength absorptions of these EPD/EPA complexes are associated with an electron transfer from the donor to the acceptor molecule. Mulliken termed these absorptions "charge-transfer (CT) absorptions" [51].

A necessary condition for the formation of an additional bonding interaction between two valency-saturated molecules is the presence of an occupied molecular orbital of sufficiently high energy in the EPD-molecule, and the presence of a sufficiently low unoccupied orbital in the EPA-molecule **). Based on the type of orbitals involved in bonding interactions, all EPD-molecules can be divided into three groups [51, 53]: n-, σ-, and π-EPD. In the first group the energetically highest orbital is the lone pair of the n-electrons of heteroatoms (R_2O, R_3N, R_2SO), in the second the electron pair of a σ-bond

* Synonyms for *EPD/EPA complex* are *electron donor acceptor (EDA) complex* [50], *molecular complex* [57, 58], and *charge-transfer (CT) complex* [51]. Since normally the term *molecular complex* is only used for weak complexes between neutral molecules, and the appearance of a charge-transfer absorption band does not necessarily prove the existence of a stable complex, the more general expression *EPD/EPA complex*, proposed Gutmann [53], will be used here. This will comprise all complexes whose formation is due to an interaction between electron pair donors (Lewis bases) and electron pair acceptors (Lewis acids), irrespective of the stabilities of the complexes or the charges of the components.

** The fundamental difference between this EPD/EPA bonding interaction and a normal chemical bond is that in an ordinary chemical bond each atom supplies one electron to the bond, whereas in EPD/EPA bonding one molecule (the donor) supplies the pair of electrons, while the second molecule (the acceptor) provides the vacant molecular orbital.

(R—Hal, cyclopropane), and in the third the pair of π-electrons of unsaturated and aromatic compounds (alkenes, alkyl benzenes, polycyclic aromatics). Similarly, EPA-molecules can also be divided into three groups [51, 53]: v-, σ-, and π-EPA. The lowest orbital in the first group is a vacant valency-orbital of a metal atom ($Ag^{\oplus}$, certain organometallic compounds), in the second a nonbonding σ-orbital (I_2, Br_2, ICl), and in the third a system of π-bonds (aromatic and unsaturated compounds with electron-withdrawing substituents such as aromatic polynitro compounds, halobenzoquinones, tetracyanoethene). Because, in principle, any donor is able to form a complex with any acceptor, there exist nine different types of EPD/EPA complexes. The largest number of investigations have been concerned with complexes of type π-EPD/π-EPA (*cf.* the above-mentioned hexamethylbenzene/chloranil complex) and π-EPD/σ-EPA (*cf.* complexes of aromatic hydrocarbons and alkenes with halogens and interhalogens).

The reaction enthalpies, ΔH, for the formation of strong EPD/EPA complexes, often used as a measure of the bond energies, lie between -42 and -188 kJ/mol (-10 to -45 kcal/mol) [59]. n-EPD/v-EPA complexes are particular members of this group (*e.g.* $Et_2O—BF_3$, $\Delta H = -50$ kJ/mol or -11.9 kcal/mol [60]). For weak complexes, ΔH is usually larger than dispersion energies but smaller than approx. 42 kJ/mol (10 kcal/mol) [59]. π-EPD/π-EPA complexes between neutral molecules are examples ($-\Delta H = 0\ldots21$ kJ/mol or $0\ldots5$ kcal/mol), *e.g.* benzene/1,3,5-trinitrobenzene ($\Delta H = -8$ kJ/mol or -1.9 kcal/mol [57]).

No general agreement exists as to the relative importance of the different intermolecular forces in making up the EPD/EPA complexes. According to Mulliken's VB description of weak EPD/EPA complexes, the electronic ground state can be considered as a hybrid of two limiting structures (*a*) and (*b*) in Fig. 2-5.

$$D + A \rightleftharpoons (\underline{D\cdots A} \longleftrightarrow D^{\oplus}\cdots A^{\ominus}) \xrightarrow{h\nu_{CT}} (D\cdots A \longleftrightarrow \underline{D^{\oplus}\cdots A^{\ominus}})$$

(*a*) (*b*)

Fig. 2-5. Formation and optical excitation of an EPD/EPA-complex between donor D and acceptor A (the predominating mesomeric structure in ground and excited state is underlined).

The non-ionic structure (*a*) represents a state without any donor-acceptor interactions, in which only non-specific intermolecular forces hold D and A together. The mesomeric structure (*b*) characterizes a state in which an ionic bond has been formed by transfer of an electron from D to A. This electron transfer will be easier the lower the ionization potential of the donor [61, 63], and the higher the electron affinity of the acceptor [62, 63]. The ionic limiting structure (*b*) is relatively energy-rich and contributes only slightly to the ground state. Nevertheless, this small contribution is sufficient in establishing an extra bonding interaction in addition to the non-specific van-der-Waals-forces. However, newer investigations have shown that these charge-transfer forces are weaker than previously believed, and the classical van-der-Waals-forces (including electrostatic forces) suffice in explaining the stabilities of EPD/EPA complexes [59, 64, 198]. The relative importance of contributions from the electrostatic and charge-transfer forces in the ground state of EPD/EPA complexes has been studied by many authors. For a review see reference [183; Vol. 1, p. 6ff.]. It seems that both electrostatic and charge-transfer interactions are important in the ground state of EPD/EPA complexes. Their relative contribution, however, varies widely in different EPD/EPA complexes [183].

Another description of EPD/EPA interactions, particularly useful for strong complexes, is based on the coordinative interaction between Lewis bases or nucleophiles (as EPD) and Lewis acids or electrophiles (as EPA) [53]. The intermolecular bonding is seen not as a hybrid of electrostatic and charge-transfer forces, but as one of electrostatic and covalent ones. The interaction of the acceptor A with the electron pair of the donor D is a result of an overlap of the orbitals of the two molecules; consequently, a finite electron density is created between the two partners according to Eq. (2-9).

$$\text{D:} + \text{A} \rightleftharpoons \text{D}^{\oplus}\text{—A}^{\ominus} \tag{2-9}$$

Hence, the structure $D^{\oplus}$—$A^{\ominus}$ is a covalent one and the EPD/EPA interaction between D and A can be described as a Lewis acid/base interaction [65].

Of the solvents, aromatic and olefinic hydrocarbons are π-donors (π-EPD); alcohols, ethers, amines, carboxamides, nitriles, ketones, sulfoxides and *N*- and *P*-oxides are *n*-donors (*n*-EPD); and haloalkanes are σ-donors (σ-EPD). Boron and antimony trihalides are acceptor solvents (v-EPA), as are halogens and mixed halogens (σ-EPA), and liquid sulfur dioxide (π-EPA). In principle, all solvents are amphoteric in this respect, *i.e.* they may act as donor (nucleophile) and acceptor (electrophile) simultaneously. For example, water can act as a donor (by means of the oxygen atom) as well as an acceptor (by forming hydrogen bonds). This is one of the reasons for the exceptional importance of water as a solvent.

n-Donor solvents are particularly important for the solvation of cations. Examples are hexamethylphosphoric triamide, pyridine, dimethyl sulfoxide, *N,N*-dimethylform-amide, acetone, methanol, and water. Their specific EPD-properties make them excellent cation solvators, and they are, therefore, good solvents for salts. They are also known as *coordinating solvents* [66]. The majority of inorganic reactions are carried out in coordinating solvents.

An empirical semiquantitative measure of the nucleophilic properties of EPD-solvents is provided by the so-called *donor number DN* (or *donicity*) of Gutmann [53, 67] (*cf.* also Section 7.2). This donor number has been defined as the negative ΔH-values for 1 : 1-adduct formation between antimony pentachloride and electron pair donor solvents (D) in dilute solution in the non-coordinating solvent 1,2-dichloroethane, according to Eq. (2-10)*.

$$\text{D:} + SbCl_5 \xrightarrow[\text{in Cl-CH}_2\text{CH}_2\text{-Cl}]{\text{room temp.}} \overset{\oplus}{\text{D}}\text{—}\overset{\ominus}{\text{Sb}}Cl_5 \tag{2-10}$$

Solvent Donor Number $DN = -\Delta H_{D-SbCl_5}/(\text{kcal} \cdot \text{mol}^{-1})$

* An analogous approach was first used by Lindqvist and Zackrisson [67a]. The authors established a series of EPD solvents calorimetrically, based on their increasing donor capacities relative to a standard acceptor ($SbCl_5$ or $SnCl_4$) with which the given donor is combined in 1,2-dichloroethane.

The linear relationship between $-\Delta H_{D-SbCl_5}$ and the logarithm of the corresponding equilibrium constant ($\lg K_{D-SbCl_5}$) shows that the entropy contributions are equal for all the studied acceptor/donor solvent reactions. Therefore, one is justified in considering the donor numbers as semiquantitative expressions for the degree of coordination interaction between EPD solvents and antimony pentachloride. Antimony pentachloride is regarded as an acceptor on the borderline between hard and soft Lewis acids. A list of organic solvents ordered according to increasing donicity is given in Table 2-3. From this it is seen that, for example, nitromethane and acetonitrile are weak donor solvents, whereas dimethylsulfoxide and triethylamine are very strong donors. The higher the donor number, the stronger the interaction between solvent and acceptor.

Unfortunately, donor numbers have been defined in the non-SI unit $kcal \cdot mol^{-1}$. Marcus has recently presented a scale of dimensionless, normalized donor numbers DN^N, which are defined according to $DN^N = DN/(38.8\ kcal \cdot mol^{-1})$[200]. The non-donor solvent 1,2-dichloroethane ($DN = DN^N = 0.0$) and the strong donor solvent hexamethyl-

Table 2-3. Donor numbers (donicities) *DN* [199, 200, 212, 241; and references cited therein] and normalized DN^N-values [200] of thirty-six organic EPD solvents[a)], determined calorimetrically in dilute 1,2-dichloroethane solutions at room temperature and valid for isolated EPD solvent molecules[b)].

Solvents	$DN/(kcal \cdot mol^{-1})$[c)]	DN^N[d)]
1,2-Dichloroethane *(reference solvent)*	*0.0*[e)]	*0.00*[e)]
Nitromethane	2.7	0.07
Nitrobenzene	8.1	0.21
Acetic anhydride	10.5	0.27
Cyanobenzene, Benzonitrile	11.9	0.31
Ethanenitrile, Acetonitrile	14.1	0.36
Tetrahydrothiophene-1,1-dioxide, Sulfolane	14.8	0.38
1,4-Dioxane	14.8	0.38
4-Methyl-1,3-dioxol-2-one, Propylene carbonate	15.1	0.39
(Cyanomethyl)benzene, Benzylcyanide	15.1	0.39
2-Methylpropanenitrile, *i*-Butanenitrile	15.4	0.40
Diethyl carbonate	16.0	0.41
Propanenitrile	16.1	0.41
1,3-Dioxol-2-one, Ethylene carbonate	16.4	0.42
Methyl acetate	16.5	0.43
Butanenitrile	16.6	0.43
3,3-Dimethyl-2-butanone, *t*-Butyl methyl ketone	17.0	0.44
Acetone	17.0	0.44
Ethyl acetate	17.1	0.44
3-Methyl-2-butanone, Methyl *i*-propyl ketone	17.1	0.44
2-Butanone	17.4	0.45
Diethyl ether	19.2	0.49
Tetrahydrofuran	20.0	0.52
Trimethyl phosphate	23.0	0.59
Tri-*n*-butyl phosphate	23.7	0.61
N,N-Dimethylformamide	26.6	0.69
1-Methylpyrrolidin-2-one	27.3	0.70
N,N-Dimethylacetamide	27.8	0.72
Tetramethylurea	29.6	0.76
Dimethyl sulfoxide	29.8	0.77

Table 2-3. (Continued)

Solvents	$DN/(\text{kcal}\cdot\text{mol}^{-1})$ [c]	DN^{N} [d]
N,N-Diethylformamide	31.0	0.80
Triethylamine	31.7 [f]	0.82
N,N-Diethylacetamide	32.1	0.83
Pyridine	33.1	0.85
Hexamethylphosphoric triamide	38.8	*1.00* [g]
Tris(pyrrolidino)phosphane oxide	47.2 [h]	1.22

[a] A compilation of about 170 *DN*-values taken from different sources can be found in reference [200]. Further 14 *DN*-values, determined indirectly via ^{1}H-NMR shift of chloroform, are given in reference [293].

[b] As the basic donor numbers were measured in an inert diluent, they reflect the donicity of the isolated EPD solvent molecules. In neat, associated EPD solvents an increase in the donicity should occur [199]. For such highly-structured solvents (*e.g.* water, alcohols, amines) the term *bulk donicity* has been introduced [201] in order to rationalize the deviations of these solvents in plots of $^{23}Na^{\oplus}$-NMR shifts [202] and ESR parameters [203] *vs.* the donor numbers. Because of the great discrepancies which exist between the DN_{bulk}-values given in the literature, they are not included in this table. For a collection of bulk donicities, DN_{bulk}, *cf.* reference [200], Table II.

[c] For the definition of *DN cf.* Eq. (2-10). For conversion into SI-units: $1\ \text{kcal}\cdot\text{mol}^{-1} = 4.184\ \text{kJ}\cdot\text{mol}^{-1}$.

[d] $DN^{N} = DN/(38.8\ \text{kcal}\cdot\text{mol}^{-1})$ [200]; $DN = 38.8\ \text{kcal}\cdot\text{mol}^{-1}$ for hexamethylphosphoric triamide as reference solvent.

[e] Zero by definition.

[f] R. W. Taft, N. J. Pienta, M. J. Kamlet, and E. M. Arnett, J. Org. Chem. *46*, 661 (1981); after adjustment for difference in temperature and solvent [200].

[g] Unity by definition [200].

[h] Y. Ozari and J. Jagur-Grodzinsky, J. Chem. Soc., Chem. Commun. *1974*, 295 (determined in dichloromethane as inert solvent).

phosphoric triamide (HMPT: $DN = 38.8\ \text{kcal}\cdot\text{mol}^{-1}$; $DN^{N} = 1.0$) have been used to fix the scale. Although solvents with higher donicity than HMPT are known (*cf.* Table 2-3), it is expedient to choose the solvent with the highest directly (*i.e.* calorimetrically) determined *DN*-value so far as the second reference solvent [200]*. The DN^{N}-values are included in Table 2-3.

A visual estimate of the different donicities of EPD solvents can easily be made using the colour reaction with copper(II), nickel(II), or vanadyl(IV) complexes as acceptor solutes [204].

The donor number has proven very useful in coordination chemistry, since it can be correlated with other physical observables for such reactions, *e.g.* thermodynamical (ΔG or K), kinetic (rates), electrochemical (polarographic half-wave and redox potentials), and spectroscopical (chemical shifts of NMR-signals) [53, 67–69, 205–207].

The donor number approach has been criticized for conceptual [208] and experimental reasons [200, 209–212]. For this and other reasons other Lewis basicity parameters have been sought.

* The donor number of $38.8\ \text{kcal}\cdot\text{mol}^{-1}$ for HMPT was given by Gutmann [67]. It should be mentioned, however, that a much higher *DN*-value of $50.3\ \text{kcal}\cdot\text{mol}^{-1}$ has been recently measured for this solvent by Bollinger *et al.* [214]. This shows that serious problems arise in measuring the Lewis basicity of this EPD solvent towards $SbCl_5$.

Another remarkable Lewis basicity scale for 75 non-HBD solvents has been established recently by Gal and Maria [211, 212]. This involved very precise calorimetric measurements of the standard molar enthalpies of 1 : 1 adduct formation of EPD solvents with gaseous boron trifluoride, $\Delta H^{\circ}_{D-BF_3}$, in dilute dichloromethane solution at 25 °C, according to Eq. (2-10a).

$$D: + BF_3 \underset{\text{in } CH_2Cl_2}{\overset{25\,°C}{\rightleftharpoons}} \overset{\oplus}{D}-\overset{\ominus}{B}F_3 \qquad (2\text{-}10a)$$

A selection of $\Delta H^{\circ}_{D-BF_3}$-values is given in Table 2-4. This new Lewis basicity scale is more comprehensive and seems to be more reliable than the donor number scale. – A comparison of various Lewis basicity scales has recently been given by Persson [301].

Table 2-4. Molar enthalpies of complex formation between boron trifluoride and several non-HBD solvents, determined in dichloromethane at 25 °C, according to Eq. (2-10a) [211, 212].

Solvents	$-\Delta H^{\circ}_{D-BF_3}/(\text{kJ}\cdot\text{mol}^{-1})$ [a]
Dichloromethane	10.0
Nitrobenzene	35.79
Nitromethane	37.63
Tetrahydrothiophene-1,1-dioxide	51.32
Acetonitrile	60.39
Propylene carbonate	64.19
3-Pentanone	72.28
1,4-Dioxane	74.09
Ethyl acetate	75.55
Acetone	76.03
Di-*i*-propyl ether	76.61
Diethyl ether	78.77
Tetrahydrofuran	90.40
1,3-Dimethylimidazolidin-2-one, DMEU	98.93
Dimethyl sulfoxide	105.34
N,N,N',N'-Tetramethylurea	108.62
N,N-Dimethylformamide	110.49
3,4,5,6-Tetrahydro-1,3-dimethylpyrimidin-2(1*H*)-one, DMPU	112.13
1-Methylpyrrolidin-2-one	112.56
Hexamethylphosphoric triamide	117.53
Tris(pyrrolidino)phosphane oxide	122.52
Pyridine	128.08
Triethylamine	135.87
1-Methylpyrrolidine	139.51

[a] See reference [212] for a set of 75 $\Delta H^{\circ}_{D-BF_3}$-values. At present, $\Delta H^{\circ}_{D-BF_3}$-values for *ca.* 350 organic EPD compounds are known (J.-F. Gal and P.-C. Maria, private communication).

Persson, Sandström, and Goggin have recently proposed an empirical solvent scale, called D_S-scale, ranking the donor strength of 64 EPD solvents towards a soft acceptor such as mercury(II) bromide [303]. The D_S-values correspond to the wavenumber shift of the symmetric IR stretching vibration from gas-phase to solution of $HgBr_2$. An additional D_H-scale of donor strength towards hard acceptors (*e.g.* $Na^{\oplus}$) has been derived for 24 EPD solvents [303].

An analogous empirical quantity for characterizing the electrophilic properties of EPA solvents has been derived by Gutmann and coworkers from the ^{31}P-NMR chemical shifts produced by the electrophilic actions of acceptor solvents A in triethylphosphane oxide, according to Eq. (2-11) (*cf.* also Section 7.4) [70, 199, 207, 213].

$$(Et_3P{=}O \longleftrightarrow Et_3\overset{\oplus}{P}{-}\overset{\ominus}{O}) + A \rightleftharpoons Et_3\overset{\delta\oplus}{P}{\cdots}\overset{\delta\ominus}{O}{-}A \qquad (2\text{-}11)$$

$$AN = \frac{\delta_{corr}(A) - \delta_{corr}(n\text{-}C_6H_{14})}{\delta_{corr}(Et_3PO{-}SbCl_5) - \delta_{corr}(n\text{-}C_6H_{14})} \cdot 100 = \Delta\delta_{corr} \cdot 2.348/\text{ppm}$$

These quantities have been termed *acceptor number AN* (or *acceptivity*) and they were obtained from the relative ^{31}P-NMR chemical shift values δ_{corr} (*n*-hexane as reference solvent) related to those of the 1:1-adduct $Et_3PO{-}SbCl_5$ dissolved in 1,2-dichloroethane, which has been arbitrarily taken to have the value of 100. The acceptor numbers are dimensionless numbers expressing the acceptor property of a given solvent relative to those of $SbCl_5$, which is also the reference compound for assessing the donor numbers. A compilation of organic solvents in order of increasing acceptor numbers is given in Table 2-5.

Table 2-5. Acceptor numbers (acceptivities) *AN* [70, 213] of forty-eight organic EPA solvents, determined ^{31}P-NMR spectroscopically at 25 °C.

Solvents	*AN*[a]
n-Hexane *(reference solvent)*	*0.0*
Triethylamine	1.4
Diethyl ether	3.9
Tetrahydrofuran	8.0
Benzene	8.2
Tetrachloromethane	8.6
Ethyl acetate	9.3
Diethylamine	9.4
Tri-*n*-butyl phosphate	9.9
Diethylene glycol dimethyl ether	9.9
1,2-Dimethoxyethane	10.2
Hexamethylphosphoric acid triamide	10.6
Methyl acetate	10.7
1,4-Dioxane	10.8
Acetone	12.5
1-Methylpyrrolidin-2-one	13.3
N,N-Dimethylacetamide	13.6
Pyridine	14.2
Nitrobenzene	14.8
Cyanobenzene	15.5
N,N-Dimethylformamide	16.0
Trimethyl phosphate	16.3
1,2-Dichloroethane	16.7
4-Butyrolactone	17.3
Morpholine	17.5
4-Methyl-1,3-dioxol-2-one, Propylene carbonate	18.3

Table 2-5. (Continued)

Solvents	*AN*[a]
N,*N*-Dimethylthioformamide	18.8
Ethanenitrile, Acetonitrile	18.9
Tetrahydrothiophen-1,1-dioxide, Sulfolane	19.2
Dimethyl sulfoxide	19.3
Dichloromethane	20.4
Nitromethane	20.5
1,2-Diaminoethane	20.9
Chloroform	23.1
2-Methyl-2-propanol, *t*-Butanol	27.1
N-Methylformamide	32.1
2-Propanol	33.6
2-Aminoethanol	33.7
1-Butanol	36.8
1-Propanol	37.3
Ethanol	37.9
Formamide	39.8
Methanol	41.5
Acetic acid	52.9
2,2,2-Trifluoroethanol	53.3
Water	54.8
Formic acid	83.6
$Et_3PO \cdot SbCl_5$ *in 1,2-dichloroethane as reference compound*	*100.0*
Trifluoroacetic acid	105.3

[a] For the definition of *AN cf.* Eq. (2-11). All δ-values have been extrapolated to zero concentration and corrected for differences in volume susceptibilities.

Acceptor numbers are less than 10 for nonpolar non-HBD solvents, they vary between about 10...20 for dipolar non-HBD solvents, and they cover a wide range of about 25...105 for protic solvents (*cf.* Table 2-5). Surprisingly, benzene and tetrachloromethane have stronger electrophilic properties than diethyl ether and tetrahydrofuran. Acceptor numbers are also known for binary solvent mixtures [70, 213].

Using the neutral Fe(II) complex $[Fe(phen)_2(CN)_2]$, the different Lewis acidity of EPA solvents can easily be visualized by its colour change: solutions of this Fe(II) complex are blue in HMPT, violet in dichloromethane, red in ethanol, and yellow in trifluoroacetic acid [204].

Another approach to the estimation of EPD/EPA interactions between a Lewis-acid A and a Lewis-base B was given by Drago [71]. Drago proposed the four-parameter Eq. (2-12) to correlate the standard enthalpy of the reaction of an acceptor A with a donor B to give a neutral 1 : 1 adduct in an inert solvent (tetrachloromethane or *n*-hexane).

$$-\Delta H^{\circ}_{AB}/(\mathrm{kJ \cdot mol^{-1}}) = E_A \cdot E_B + C_A \cdot C_B \tag{2-12}$$

E_A and C_A are empirical acceptor parameters and E_B and C_B are empirical donor parameters. The *E* parameters are measures of the tendency of an acid or a base to participate in *e*lectrostatic interactions, while the *C* parameters are measures of their tendency to form *c*ovalent bonds.

The original set of E and C parameters has been determined mainly with the help of enthalpies of adduct formation of iodine and phenol as acceptors with alkylamines as donors. Subsequently, the best set of E and C parameters has been obtained by computer optimization of a large data base of enthalpies and four arbitrarily fixed reference values [71, 215]: $E_A = C_A = 1$ for iodine, $E_B = 1.32$ for *N,N*-dimethylacetamide, and $C_B = 7.40$ for diethyl sulfide. Table 2-6 gives a selection of E and C parameters for Lewis acids and bases commonly used as solvents.

Table 2-6. Some E and C parameters expressing Lewis acid/base strength according to Drago [217][a]; *cf.* Eq. (2-12).

Lewis acids	E_A	C_A	Lewis bases	E_B	C_B
$SbCl_5$	14.4 [b]	1.17[b]	$[(CH_3)_2N]_3PO$	1.52	3.55
BF_3 (g)	9.88	1.62	CH_3SOCH_3	1.34	2.85
$(CF_3)_2CHOH$	5.93	0.62	$CH_3CON(CH_3)_2$	*1.32* [c]	2.58
C_6H_5OH	4.33	0.44	C_5H_5N	1.17	6.40
CF_3CH_2OH	3.88	0.45	$CH_3CO_2C_2H_5$	0.975	1.74
$CHCl_3$	3.02	0.16	CH_3COCH_3	0.94	2.33
$(CH_3)_3C—OH$	2.04	0.30	$(C_2H_5)_2O$	0.94	3.25
H_2O	1.64	0.57	CH_3CN	0.89	1.34
I_2	*1.00*[c]	*1.00*[c]	$(C_2H_5)_2S$	0.34	*7.40*[c]
SO_2	0.92	0.81	C_6H_6	0.28	0.59

[a] For a more complete list see references [71, 215, 217].
[b] Corrected values; see reference [217].
[c] Used to fix the E/C-scale.

On the basis of these parameters, it is possible to predict the enthalpies of Lewis acid/base reactions, even those reactions which might be inaccessible experimentally, with remarkable accuracy (within $\pm 0.8\ kJ \cdot mol^{-1}$) [216].

Drago's E/C analysis and Gutmann's donor number approach [53, 67] have been compared [200, 217, 218]. Eq. (2-12) has been extended for specific and nonspecific interactions between solutes and polar solvents [219].

Finally, an attempt was made to establish a measure of the electron-donating and electron-accepting power of organic solvents by means of infrared [72, 73] and ^{1}H-NMR measurements [73]. Further empirical Lewis acid and base parameters will be discussed in Chapters 7.2...7.5.

2.2.7 Solvophobic Interactions [74–77, 220–224]

Hydrocarbons are extremely low soluble in water. Accordingly, the dissolution of a hydrocarbon in water is usually associated with an increase in the Gibbs energy G of the system ($\Delta G > 0$). Since it is known experimentally that the dissolution of a hydrocarbon in water is exothermic ($\Delta H < 0$) it follows from $\Delta G = \Delta H - T \cdot \Delta S$ that the entropy of the system must decrease. This can be interpreted as a consequence of the highly ordered structure of the water molecules around the dissolved hydrocarbon molecules. In other words, the water molecules are more tightly packed around the dissolved hydrocarbon molecules than in pure water. This is called a structure increase. If aqueous solutions of two hydrocarbons are mixed, the two hydrocarbons may form an aggregate with simultaneous partial reconstruction of the original undisturbed water structure. This is shown schematically in Fig. 2-6.

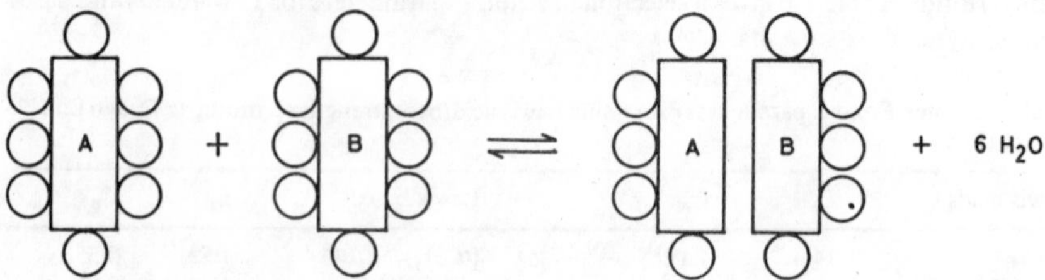

Fig. 2-6. The formation of a hydrophobic interaction between two hydrocarbon molecules A and B (the circles represent water molecules) [78].

Due to the contact between A and B, fewer water molecules are now in direct contact with the hydrocarbon molecules. Thus, the ordering influence of the hydrophobic molecules will be diminished and the entropy increases ($\Delta S > 0$). Although thermal energy is required for the destructuring of the hydration shells around A and B ($\Delta H > 0$), the free energy diminishes upon aggregation ($\Delta G < 0$). Therefore, it is energetically advantageous for apolar molecules or apolar groups in large molecules in water to aggregate with expulsion of water molecules from the hydration shells. This phenomenon is known as *hydrophobic interaction*. This expression reflects the thermodynamic disadvantage of a direct contact between hydrophobic and hydrophilic groups (*e.g.* alkyl side chains in proteins with water molecules). The system escapes this condition by clustering the hydrophobic groups*).

This hydrophobic interaction can be illustrated by considering the thermodynamic parameters for the dissolution of the archetypal apolar hydrocarbon methane in cyclohexane (an apolar, non-associated solvent) and in water (a polar, strongly self-associated solvent); Table 2-7 [225].

Table 2-7. Thermodynamic parameters for dissolution of gaseous methane in cyclohexane and water at 25 °C [225].

Solvents	$\Delta G_s^\circ/(\mathrm{kJ \cdot mol^{-1}})$	$\Delta H_s^\circ/(\mathrm{kJ \cdot mol^{-1}})$	$\Delta S_s^\circ/(\mathrm{J \cdot mol^{-1} \cdot K^{-1}})$
Water	26.4	−13.8	−134
Cyclohexane	14.2	− 2.5	− 54

* Glass beads can be used as an illustration of hydrophobic interactions. The glass beads, covered with dichlorodimethylsilane, can be regarded as solid hydrocarbon particles. Thus, only hydrophobic interactions are possible. In a structured solvent such as water or formamide, the beads cluster together. When the polarity of the solvent is decreased by addition of alcohols the clusters desintegrate [79].

The unfavorable Gibbs energy ($\Delta G_s^\circ \gg 0$) for the dissolution of methane in water is the result of a strongly negative entropy of solution ($\Delta S_s^\circ \ll 0$) which prevails over the favorable enthalpic contribution ($\Delta H_s^\circ < 0$). The negative enthalpy and entropy of transfer of methane from cyclohexane to water can be interpreted in terms of an increased degree of water-water hydrogen bonding in the solvation shell surrounding the apolar solute molecule. This is also called the *hydrophobic effect* [220]:

This hydrophobic effect was first postulated by Frank and Evans in 1945. They wrote: "The nature of deviation found for non-polar solutes in water leads to the idea that the water forms frozen patches or microscopic icebergs around such solute molecules. The word 'iceberg' represents a microscopic region, surrounding the solute molecule, in which water molecules are tied together in some sort of quasi-solid structure" [226].

In principle, such interactions should also apply to other solvents resembling water, and therefore the more general term *solvophobic interactions* has been proposed [80]. In fact, analogous water-like behaviour has been observed with self-associated solvents other than water, *e.g.* ethanol [81], glycerol [82], ethylammonium nitrate [227], and some dipolar non-HBD solvents [228].

Although it has been widely accepted that the hydrophobic interaction is 'entropy-driven', this classical view has to be modified in the light of some newer findings [79a, 227, 229–231]. Contrary to previous suggestions, it appears that the major contribution to the hydrophobic interaction between the methylene groups of *n*-alkanes is an enthalpic and not an entropic effect [230]. Consequently, the poor solubility of nonpolar solutes in water should be due to unfavorable enthalpy and not to unfavorable entropy [227, 231].

Solvophobic interactions are important in the aggregation of polymethine dyes [81] and the stabilization of particular conformations of polypeptides and proteins in aqueous solution [222, 232]. They also play an important role in the biochemical complexation between enzyme and substrate [77, 78, 83, 84].

Hydrophobicity parameters for organic substituents have been developed by Hansch *et al.*, using partitioning phenomena [296], and by Menger *et al.*, using kinetic measurements (hydrolysis of long-chain esters) [297]; *cf.* Section 7.2. Further results connected with the presence of solvophobic interactions in solutions are discussed in Chapters 2.5 and 5.4.8.

2.3 Solvation [49, 85–98, 98a, 233–241]

The term *solvation* refers to the surrounding of each dissolved molecule or ion by a shell of more or less tightly bound solvent molecules. This solvent shell is the result of intermolecular forces between solute and solvent. For aqueous solutions the term used is *hydration*. Intermolecular interactions between solvent molecules and ions are particularly important in solutions of electrolytes, since ions exert specially strong forces on solvent molecules. Crude electrostatic calculations show that the field experienced by nearest neighbours of dissolved ions is $10^6 \ldots 10^7$ V/cm. Fig. 2-7 shows a highly simplified picture of such an interaction between ions and dipolar solvent molecules.

The *solvation energy* is considered as the change in Gibbs energy when an ion or molecule is transferred from a vacuum (or the gas-phase) into a solvent. The Gibbs energy

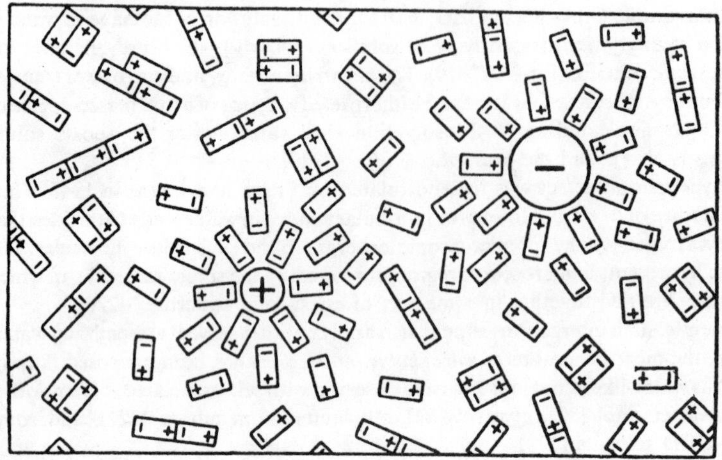

Fig. 2-7. Solvation of ions in a solvent consisting of dipolar molecules [99]. The charges of the dipolar molecules are in fact partial charges $\delta\oplus$ and $\delta\ominus$.

of solvation, $\Delta G^{\circ}_{\mathrm{solv}}$, a measure of the solvation ability of a particular solvent, is the result of a superimposition of four principal components of a different nature [100]:

(a) the cavitation energy linked with the hole which the dissolved molecule or ion produces in the solvent;
(b) the orientation energy corresponding to the phenomenon of partial orientation of the dipolar solvent molecules caused by the presence of the solvated molecule or ion (*cf.* Fig. 2-7);
(c) the isotropic interaction energy corresponding to the unspecific intermolecular forces with a long radius of activity (*i.e.* electrostatic, polarisation, and dispersion energy);
(d) the anisotropic interaction energy resulting from the specific formation of hydrogen bonds or electron pair donor/electron pair acceptor bonds at well localized points in the dissolved molecules.

The dissolution of a substance requires that not only the interaction energy of the solute molecules (for crystals the lattice energy*) be overcome but also the interaction energy between the solvent molecules themselves. This is compensated by the gain in Gibbs energy of solvation, $\Delta G^{\circ}_{\mathrm{solv}}$. The standard molar Gibbs energy of solvation, $\Delta G^{\circ}_{\mathrm{solv}}$, can be formulated as the difference between the Gibbs energy of solution, $\Delta G^{\circ}_{\mathrm{soln}}$, and the crystal lattice energy, $\Delta G^{\circ}_{\mathrm{latt}}$, as shown by means of the customary Born-Haber cycle in Fig. 2-8.

* The lattice energy is the work required to separate to infinity the elements of the lattice from their equilibrium position at 0 K. For ionic lattices of the alkali halides it is of the order 628...837 kJ/mol (150...200 kcal/mol) [49]. For molecular lattices of organic compounds such as benzene, naphthalene, and anthracene it is of the order 42...105 kJ/mol (10...25 kcal/mol) [101]. The experimental heat of sublimation of benzene is 44.6 kJ/mol (10.7 kcal/mol) [102].

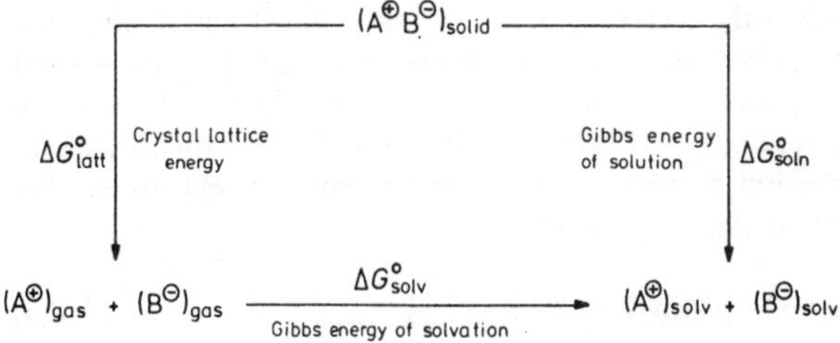

Fig. 2-8. The relationship between standard molar Gibbs energies of solvation and solution and the crystal lattice energy of an ionophore $A^{\oplus}B^{\ominus}$: $\Delta G^{\circ}_{solv} = \Delta G^{\circ}_{soln} - \Delta G^{\circ}_{latt}$.

If the liberated solvation energy is higher than the lattice energy, then the overall process of dissolution is exothermic. In the opposite case the system uses energy and the dissolution is endothermic. The values for NaCl are typical: lattice energy +766 kJ/mol, hydration energy −761 kJ/mol, and energy of solution +3.8 kJ/mol. The energies of solution are generally small because interaction within the crystal lattice is energetically similar to interaction with the solvent.

The Gibbs energies of solvation of individual ions cannot be directly measured but they can be calculated [49]. The Gibbs energies of hydration of some representative ions are collected in Table 2-8. From these it is seen that these values can be as high as bond energies or even higher (209...628 kJ/mol; 50...150 kcal/mol). Consequently, the solvent is often considered a direct reaction partner and should really be included in the reaction equation. The isolation of numerous solvates such as hydrates, alcoholates, etherates, and ammoniates, especially of inorganic or organometallic compounds, are examples. Between the two extremes, *viz.* the simple solvation resulting from weak intermolecular interactions, and the *bona fide* chemical modification of the substrate by the solvent, all other possibilities exist.

Table 2-8. Standard molar Gibbs energies of hydration, ΔG°_{hydr}, of some representative single ions at 25 °C [241, 242][a].

Cations	$\Delta G^{\circ}_{hydr}/(kJ \cdot mol^{-1})$	Anions	$\Delta G^{\circ}_{hydr}/(kJ \cdot mol^{-1})$
$H^{\oplus}$	−1056	$F^{\ominus}$	− 472
$Li^{\oplus}$	− 481	$Cl^{\ominus}$	− 347
$Na^{\oplus}$	− 375	$Br^{\ominus}$	− 321
$K^{\oplus}$	− 304	$I^{\ominus}$	− 283
$Mg^{\oplus\oplus}$	−1838	$HO^{\ominus}$	− 439
$Al^{\oplus\oplus\oplus}$	−4531	$SO_4^{\ominus\ominus}$	−1090

[a] For a comprehensive compilation of Gibbs energies of solvation see C.M. Criss and M. Salomon: *Thermodynamic Measurements – Interpretation of Thermodynamic Data.* In A.K. Covington and T. Dickinson (eds.): *Physical Chemistry of Organic Solvent Systems.* Plenum Press, London New York 1973, p. 253 ff. – *Cf.* also D.W. Smith: *Ionic Hydration Enthalpies.* J. Chem. Educ. *54*, 540 (1977). – A critical selection of standard molar heat capacities of hydration, $\Delta_{hyd}C^{\circ}_{p}/(J \cdot K^{-1} \cdot mol^{-1})$, of single ions has been given by M.H. Abraham and Y. Marcus, J. Chem. Soc., Faraday Trans. I *82*, 3255 (1986).

The most direct measure of the energetics of ion solvation is, without doubt, their *standard molar Gibbs energy of solvation*, *i.e.* transfer from the gas-phase to the solvent (*cf.* Fig. 2-8). However, this quantity is generally unknown, particularly for ions in nonaqueous solvents. Therefore, ΔG°_{solv} is advantageously replaced by the *standard molar Gibbs energy of transfer* of the ion X from water, W, as reference solvent, to another solvent, S, $\Delta G^\circ_t(X, W \rightarrow S)$, as defined by Eq. (2-12a):

$$\Delta G^\circ_t(X, W \rightarrow S) = \mu^\infty_X(\text{in S}) - \mu^\infty_X(\text{in W}) = R \cdot T \cdot \ln {}^W y^S_X \tag{2-12a}$$

μ^∞_X is the standard (*i.e.* infinite dilution) chemical potential of X and ${}^W y^S_X$ the so-called *solvent-transfer activity coefficient* of X.

In order to obtain the $\Delta G^\circ_t(X, W \rightarrow S)$ of individual ions from experimental data on complete electrolytes, the extrathermodynamic assumption that $\Delta G^\circ_t(Ph_4As^\oplus, W \rightarrow S) = \Delta G^\circ_t(Ph_4B^\ominus, W \rightarrow S)$ for all solvents has been made, using $Ph_4As^\oplus Ph_4B^\ominus$ as reference electrolyte ($Ph = C_6H_5$). This seems reasonable because the large symmetrical ions of tetraphenylarsonium tetraphenylborate are of comparable size, structure, and charge, and are, therefore, similarly solvated on transfer from one solvent to another. Arguments in favor for and against this extrathermodynamic assumption have been reviewed [235, 241, 243, 244]*.

Experimentally, the molar Gibbs energy of transfer of an anion $X^\ominus$ is obtained from the combined results of four solubility measurements, namely of the salts $Ph_4As^\oplus Ph_4B^\ominus$ and $Ph_4As^\oplus X^\ominus$ in water, W, and of the same salts in the solvent S. Then the Gibbs energy of transfer is ($Ph = C_6H_5$):

$$\begin{aligned}\Delta G^\circ_t(X^\ominus, W \rightarrow S) = R \cdot T[&2 \cdot \ln s(Ph_4AsX,W) - 2 \cdot \ln s(Ph_4AsX,S) \\ &+ \ln s(Ph_4AsPh_4B,S) - \ln s(Ph_4AsPh_4B,W)]\end{aligned} \tag{2-12b}$$

where s is the solubility, expressed in the molar scale ($mol \cdot l^{-1}$).

Table 2-9 collects selected values of $\Delta G^\circ_t(X, W \rightarrow S)$ obtained on this basis, taken from the extensive and critically evaluated compilation of Marcus [244]. A nice graphical representation of the changes in ΔG°_t, ΔH°_t, and ΔS°_t for the transfer of univalent single ions from water to other solvents has been given by Persson [301]. *Cf.* Chapter 5.5.3 for further discussions.

The following three aspects are also of importance in solvation: the stoichiometry of the solvate complexes (normally described by the coordination or solvation number), the

---

*Analogously, the following extrathermodynamic "reference electrolyte" assumptions are widely used:

$$\Delta H^\circ_t(Ph_4As^\oplus, W \rightarrow S) = \Delta H^\circ_t(Ph_4B^\ominus, W \rightarrow S),$$

and similarly

$$\Delta S^\circ_t(Ph_4As^\oplus, W \rightarrow S) = \Delta S^\circ_t(Ph_4B^\ominus, W \rightarrow S),$$

for the transfer from water to all solvents at any temperature [244]. This is equivalent to assuming that the molar Gibbs energy of transfer, $\Delta G^\circ_t(X, W \rightarrow S)$, at a given reference temperature (usually 298.15 K) is valid for all temperatures [244].

Table 2-9. Selected standard molar Gibbs energies of transfer of single ions X from water (W) to seven nonaqueous solvents (S), ΔG_t°(X, W → S)/(kJ · mol^{-1})[a)], at 25 °C (molar scale), taken from the compilation of Marcus [244].

X	S						
	CH_3OH	C_2H_5OH	CH_3COCH_3	$HCON(CH_3)_2$	CH_3CN	CH_3SOCH_3	$[(CH_3)_2N]_3PO$
$H^\oplus$	10.4	11.1		−18	46.4	−19.4	
$Li^\oplus$	4.4	11		−10	25	−15	
$Na^\oplus$	8.2	14		− 9.6	15.1	−13.4	
$K^\oplus$	9.6	16.4	4	−10.3	8.1	−13.0	−16
$Ag^\oplus$	6.6	4.9	9	−20.8	−23.2	−34.8	−44
$(CH_3)_4N^\oplus$	6	10.9	3	− 5.3	3	− 2	
$(C_6H_5)_4As^\oplus$ [b)]	−24.1	−21.2	−32	−38.5	−32.8	−37.4	−39
$F^\ominus$	16			51	71		
$Cl^\ominus$	13.2	20.2	57	48.3	42.1	40.3	58
$Br^\ominus$	11.1	18.2	42	36.2	31.3	27.4	46
$I^\ominus$	7.3	12.9	25	20.4	16.8	10.4	30
$CN^\ominus$	8.6	7	48	40	35	35	
$ClO_4^\ominus$	6.1	10		4	2		− 7
$(C_6H_5)_4B^\ominus$ [b)]	−24.1	−21.2	−32	−38.5	−32.8	−37.4	−39

[a)] A positive value of ΔG_t°(X, W → S) means that the ion is better solvated by water than by solvent S; a negative value means that the ion is more strongly solvated after transfer from water to solvent S.
[b)] See text for the so-called tetraphenylarsonium tetraphenylborate assumption.

lability of the solvate complexes (usually described by the rate of exchange of the molecules of the solvent shell with those of the bulk solvent), as well as the fine structure of the solvation shell (for water often described by the simple model of ion solvation of Frank and Wen [16]).

Coordination and solvation numbers reflect the simple idea that the solvation of ions or molecules consists of a coordination of solute and solvent molecules. The *coordination number* is defined as the number of solvent molecules in the first coordination sphere of an ion in solution [103]. This first coordination sphere is composed only of solvent molecules in contact with or in bonding distance of the ion such that no other solvent molecules are interposed between them and the ion. This kind of solvation is sometimes termed *primary* or *chemical solvation*. Coordination numbers, determined by different experimental techniques [103], range in water from approx. 4 for $Be^{2\oplus}$ to approx. 9 for $Th^{4\oplus}$, although the majority of the values are close to 6 (*e.g.* for $Al^{3\oplus}$).

The *solvation number* is defined as the number of solvent molecules per ion which remain attached to a given ion long enough to experience its translational movements [94, 97, 104]. The solvation number depends upon the reference ion and its assumed solvation number as well as upon the method of measurement. Depending on the method of measurement, solvent molecules loosely bound in the second or in a higher sphere are included. The partial ordering of more distant solvent molecules beyond the primary solvation shell is termed *secondary* or *physical solvation*. For example, mobility measurements indicate the number of solvent molecules moving with the ion, while dielectric measurements indicate only the number of solvent molecules in the first sphere. The solvation number of $Li^{\oplus}$ in water, determined using different electrolytic transference methods, varies therefore between 5 and 23. An inspection of the solvation numbers measured by electrolytic transport methods shows that the order of hydration numbers of the alkali metal cations is: $Li^{\oplus} > Na^{\oplus} > K^{\oplus} > Rb^{\oplus} > Cs^{\oplus}$. The alkaline earth metal cations are more highly solvated than the alkali cations ($Mg^{\oplus\oplus} > Ca^{\oplus\oplus} > Sr^{\oplus\oplus} > Ba^{\oplus\oplus}$). The more dilute the solution the greater the solvation of a given ion. The halogen anions are hydrated in the order $F^{\ominus} > Cl^{\ominus} > Br^{\ominus} > I^{\ominus}$. Therefore, as a rule it can be stated that the smaller the ion and the greater its charge, the more highly it is solvated [94, 97, 104]. Conductance data show, that the solvation number for a given ion varies strongly with the solvent. Thus, the solvation number of $Li^{\oplus}$ varies from 1.4 in sulfolane, 7 in methanol, 9 in acetonitrile to 21 in water. The conductance data indicate also that in all organic solvents used, the solvation of the alkali cations is in the order: $Li^{\oplus} > Na^{\oplus} > K^{\oplus} > Rb^{\oplus} > Cs^{\oplus}$. The order of solvation of the halogen anions in the organic solvents studied is, in general, $Cl^{\ominus} > Br^{\ominus} > I^{\ominus}$ [94, 97, 104].

Even in the case of strong interactions between solvent and solute, the life-time of each solvate is brief since there is continuous rotation or exchange of the solvent shell molecules. The time required for reorientation of hydrates in water is of the order $10^{-10} \ldots 10^{-11}$ s at 25 °C [91]. If the exchange between bulk solvent molecules and those in the inner solvation shell of an ion is slower than the NMR-time scale, then it is possible to observe two different resonance signals for the free and bound solvent. In this way it has been shown, using ^{17}O-NMR spectroscopy, that the hexaaquo hydration spheres of $Al^{3\oplus}$ and $Cr^{3\oplus}$, and the four water molecules bound by $Be^{2\oplus}$ exchange slower than 10^4/s, and the alkali metal cations faster than 10^4/s [96, 105].

In general, since solvent molecules directly bound to an ion have different chemical shifts from those of the bulk solvent, NMR-spectroscopy is a very useful method for studying solvation shells [106–111]. But if the exchange rates are too high, the NMR-signals coalesce to a single time-averaged resonance signal. It is usually assumed that solvent molecules in environments other than the first coordination sphere are exchanging at diffusion-controlled rates and therefore appear in the environmentally averaged bulk solvent resonance. A variety of different solvent nuclei have been used for this purpose: ^{1}H, ^{13}C, ^{17}O, and ^{31}P. As an example, the ^{1}H-NMR spectrum of 2.1 *M* aqueous solution of $Al(NO_3)_3$ at -40 °C shows two signals [112]. The low-field signal arises from the coordinated solvent, and the high-field resonance from the bulk solvent. Two ^{13}C–NMR signals are also observed for aqueous dimethylsulfoxide containing $AlCl_3$ at 30 °C, one for the bulk and one for the bound solvent (1.94 ppm upfield) [113].

The ^{1}H-NMR spectrum of an aqueous $Al(ClO_4)_3$ solution in [D_6] acetone shows nicely the two different signals of bulk water and hydration water in the $Al^{3\oplus}$ inner shell even at room temperature [245]. The addition of acetone slows down the proton exchange rate. A primary hydration number of six for $Al^{3\oplus}$ has been obtained in this way [245].

Another approach to the study of ion-solvent interactions involves the determination of the solvent effect on the resonance frequency of the solute ion, using nuclei of spin $I \neq 0$ such as ^{7}Li, ^{23}Na, ^{27}Al, ^{35}Cl, ^{59}Co, ^{69}Ga, ^{133}Cs, ^{195}Pt, and ^{205}Tl [106–111, 111a, 246, 247, 294]. $^{205}Tl^{\oplus}$ is an exceptionally sensitive ion [294]. In going from water to pyridine the change in resonance frequency is approximately 782 ppm (!) [114]. In comparison, the change in chemical shift for $^{23}Na^{\oplus}$ in these two solvents is only about 1.3 ppm [115]. Therefore, $^{205}Tl^{\oplus}$ and other ions are very useful probes for the study of solvation and solvent structure. The greater the Lewis basicity of the solvent, the higher the resonance frequency of the $^{205}Tl^{\oplus}$ ion. The increase in resonance frequency with increasing solvent Lewis basicity can be considered as a measure of the strength of interaction between the solute ion and solvent molecules [294].

A number of models have been developed to describe the fine structure of the solvent shells of ions and molecules. While the agreement with the experimental findings is more or less satisfactory, it is for the most part only qualitative (for reviews see references [85, 91, 94, 95, 98]). According to the influence of the solute on the solvent structure, two different types of solvent can be distinguished (Fig. 2-9) [98]. In the former case the pure solvent does not show a high degree of order. The directional properties of the dissolved ion dominate in a rather large region around the center and decrease gradually proceeding into the unperturbed bulk solvent. The solution consists of an ordered sphere – the primary solvation shell A – and the disordered bulk solvent B (Fig. 2-9a) [98].

In the latter case, the solvent possesses a highly ordered structure such as found in water. Frank and Wen [16] distinguish between three different regions in the solvent surrounding a solute. In the first coordination sphere A, the solvent molecules are strongly bound to the ion and therefore appear less mobile than the molecules in the bulk solvent. At some distance from the ion there exists the normal structure of the pure ordered solvent C. Between A and C, according to Frank and Wen [16], lies an intermediate region of disorder B, with highly mobile solvent molecules. This has been introduced in order to explain the "structure making" and "structure breaking" properties of ions of different charge and size in aqueous solutions. The concept of different regions around the dissolved ion was enlarged by Gurney [116], who introduced the term *cosphere* for the zone surrounding a spherical ion in which significant differences in structure and properties of solvent molecules are to be expected*). In contrast to the ordinary strong *positive hydration* of small spherical ions possessing a structure-making effect on the solvent molecules (*cf.* Fig. 2-9a), water molecules around a dissolved ion are in some cases more mobile than in pure water. In other words, the exchange frequency of water molecules around the ions is greater than in regions of pure water (*cf.* region B in Fig. 2-9b). This explains the experimental observation that aqueous solutions of certain

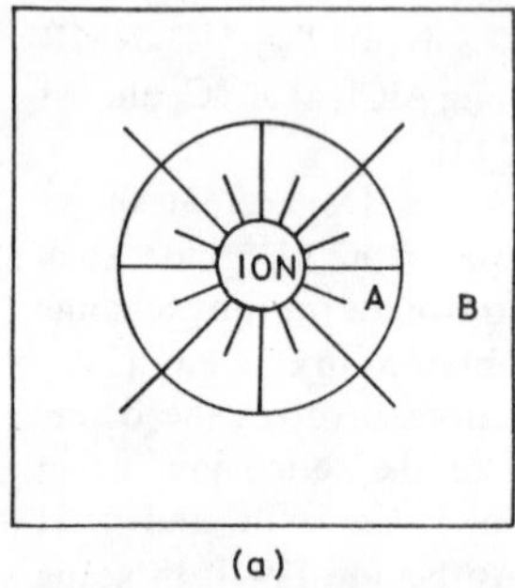

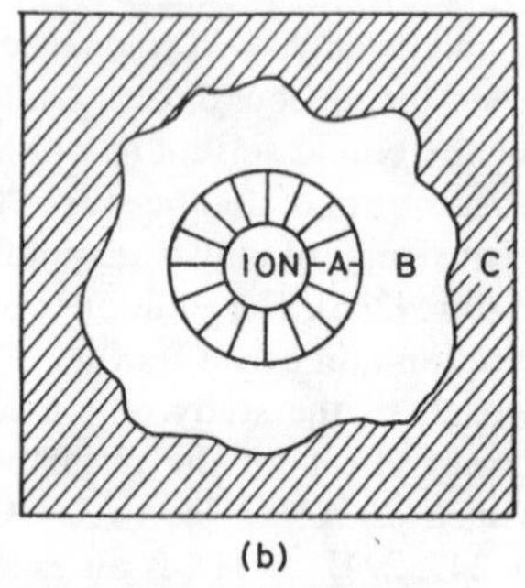

Fig. 2-9. Schematic multizone models for ion solvation in solvents: (a) with low degree of order such as hydrocarbons, consisting of solvation shell A and disordered bulk solvent B [98]; (b) in highly ordered solvents such as water, consisting of solvation shell A with immobilized solvent molecules, followed by a structure-broken region B, and the ordered bulk solvent C (Frank and Wen [16]).

salts such as potassium iodide show a greater fluidity than pure water at the same temperature. This effect has been called *negative hydration* [85] and it is connected with the structure-breaking effect of large single-charged spherical ions on solvent molecules [91, 117]. The structure-breaking effect of large ions is not restricted to water as a solvent. Ethylene glycol and glycerol are liquids which also show this effect for a number of salts which cause structure-breaking in water [117]. But, up until now, the correctness of the multizone models for ion solvation proposed by Frank and Wen [16] and others suffers from direct experimental proof [117]. Consequently, owing to the lack of detailed knowledge of the solvents' structure and of satisfactory molecular theories for associated liquids, all attempts for a detailed description of solvation shells are still imperfect.

The solubility of a dissolved non-electrolyte solute can be reduced by the addition of a salt. This phenomenon, known as the *salting-out effect*, is of practical importance for the isolation of organic compounds from their solutions. In the presence of a dissolved dissociated salt, a fraction of the solvent molecules suffers solvational interaction with the

---

* The formation in a liquid, of temporary solvent molecule groups which have some crystalline character has been called *cybotaxis* (Greek, κυβεύω, dice-play, τάξις, an arrangement) by Stewart [116a]; see also [116b]. A *cybotactic region* may then be defined as the volume around a solute molecule in which the ordering of the solvent molecules has been influenced by the solute, including both the first solvation shell and the transition region; *cf.* [129].

ions of the electrolyte, thus suffering diminished activity, and leading to salting-out of the dissolved non-electrolyte solute. In other words, the salting-out can be considered as the difference in solubility in two kinds of solvents, the ion-free and the ion-containing one [248].

2.4 Selective Solvation [89, 94, 96, 118–120, 241, 249, 250]

The description of solvation of ions and molecules in solvent mixtures is even more complicated. Besides the interaction between solvent and solute, the interaction between unlike solvent molecules plays an important supplementary role. This leads to large deviations from the ideal behaviour expected from Raoult's law of vapor pressure depression of binary mixtures. Even the description of the very complex structure of binary alcohol/water mixtures is a difficult undertaking [19].

From investigations of the solvation of ions and dipolar molecules in binary solvent mixtures it has been found that the ratio of the solvent components in the solvent shell may be different from that in the bulk solution. As expected, the solute surrounds itself preferably by the component of the mixture which leads to the more negative Gibbs energy of solvation, ΔG°_{solv}. The observation that the solvent shell has a composition other than the macroscopic ratio is termed *selective* or *preferential solvation* (*cf.* Fig. 2-10). These

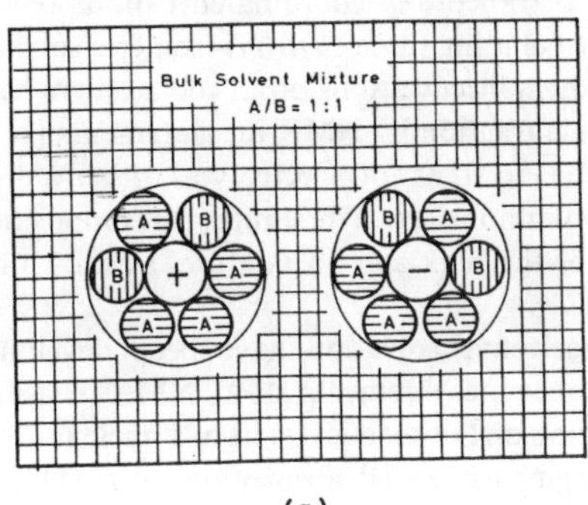

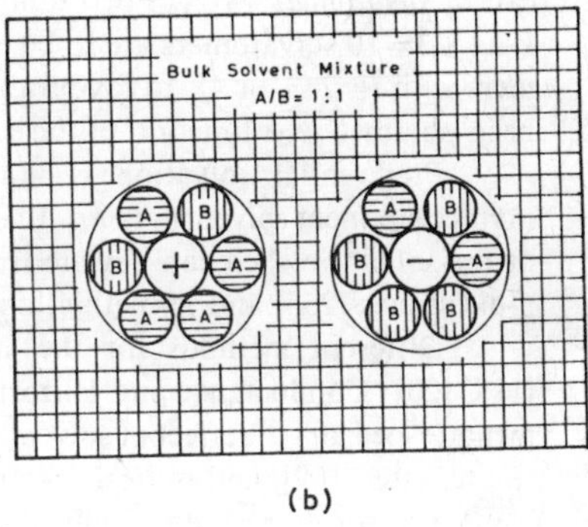

Fig. 2-10. Schematic model for the selective solvation of ions by one component of a binary 1:1-mixture of the solvents A and B [119].
(a) Homoselective solvation: both ions are preferentially solvated by the same solvent A.
(b) Heteroselective solvation: the cation is preferentially solvated by A and the anion by B.

terms are generally used to describe the molecular-microscopic local solute-induced inhomogeneity in a multicomponent solvent mixture. They include both (i) nonspecific solute/solvent association caused by dielectric enrichment in the solvent shell of solute ions or dipolar solute molecules, and (ii) specific solute/solvent association such as hydrogen-bonding or EPD/EPA interactions.

When in a mixture of two solvents, both ions of a binary salt are solvated preferably by the same solvent, the term applied is *homoselective solvation* (Fig. 2-10a). Similarly, the preferred solvation of the cation by one, and the anion by the other solvent, is termed *heteroselective solvation* (Fig. 2-10b) [119]. Thus, in a solution of silver nitrate in the binary solvent mixture acetonitrile/water, a preferential solvation of $Ag^{\oplus}$ by acetonitrile and of $NO_3^{\ominus}$ by water was observed (heteroselective solvation) [121]*). In contrast, in solutions of calcium chloride in water/methanol mixtures, both $Ca^{2\oplus}$ and $Cl^{\ominus}$ are solvated largely by water (homoselective solvation) [122]. $Zn^{2\oplus}$ (from $ZnCl_2$) in the solvent mixture water/hydrazine is preferentially solvated by hydrazine; in an acetonitrile/water mixture solvation is largely by water [123]. Ag_2SO_4 is heteroselectively solvated in methanol/dimethylsulfoxide mixtures: the silver ion is preferentially solvated by dimethylsulfoxide, whereas the sulfate ion is preferably solvated by methanol. The Ag_2SO_4 salt is only sparingly soluble in methanol and in dimethylsulfoxide. Its solubility is higher in mixtures of the two solvents than in the neat liquids, since both the cation and the anion can be solvated with the solvent component for which it has a greater affinity [123a]**). The $Cu^{\oplus}$ ion (from $CuClO_4$) shows strong preferential solvation by acetonitrile in acetonitrile/acetone mixtures which may be of interest in the hydrometallurgical purification of copper [252]. Even protons exhibit preferential solvation by amines in mixed water/amine ion clusters studied in the gas phase [253].

In a binary mixture of solvents S_1 and S_2, a cation $M^{z\oplus}$ with a coordination number k and charge $z^{\oplus}$ forms $(k+1)$ cations of the type $[M(S_1)_i(S_2)_{k-i}]^{z\oplus}$ with $i=0\ldots k$, differently solvated in the first solvation shell. These differently solvated species have been called *solvatomers* [254]. For example, with octrahedrally coordinated cations ($k=6$) $k+1+3=10$ solvatomers are to be expected (including three *cis/trans*-isomeric solvatomers with $i=2$, 3, or 4). In favourable cases the concentrations of all solvatomers have been obtained as a function of the solvent mole fraction by NMR measurements [254].

Preferential solvation is not restricted to ions of electrolytes dissolved in multicomponent solvent systems. Even for dipolar nonelectrolyte solutes the composition of the solvation shell can deviate from that of the bulk solvent mixture, as shown for β-disulfones [255] and *N*-methylthiourea [256].

Different methods for the study of selective solvation have been developed [118, 120]: Conductance and Hittorf transference measurements [119], NMR measurements (especially the effect of solvent composition on the chemical shift of a nucleus in the solute) [106–109], and optical spectra measurements like IR absorption shifts [111] or UV/Vis absorption shifts of solvatochromic dyes in binary solvent mixtures [124, 249].

A convenient measure of the degree of selective solvation is the bulk solvent composition at which both solvents of a binary mixture participate equally in the contact solvation shell. This is the solvent composition at which the NMR chemical shifts lie midway between the values for the two pure solvents. This composition has been called the *equisolvation* or *iso-solvation point* (usually expressed in mole fractions of one solvent) [125]. According to Fig. 2-10, this point describes the bulk solvent composition at which both solvents A and B participate equally in the solvation shell of the cation or the anion, respectively.

---

* The reasons for preferential solvation of $Ag^{\oplus}$ ions by acetronitrile in acetonitrile/water mixtures and the solvation shell structure of silver ions have been discussed recently [251].

** A comprehensive tabulation on selective solvation of ions in a number of binary solvent systems is given by Gordon [96] (p. 256).

A useful probe of the immediate chemical environment of solute ions is the NMR chemical shift of alkali metal ions obtained in binary solvent mixtures [111, 126, 295]. These measurements are based on the assumption that the chemical shift of the solute cation is determined in an additive fashion by the solvent molecules comprising the first solvation shell. For example (*cf.* Fig. 2-11), the iso-solvation point of $^{23}Na^{\oplus}$ in dimethyl sulfoxide/acetone mixtures occurs at $x \approx 0.21$ cmol/mol dimethyl sulfoxide, indicating the higher solvating ability of this solvent relative to acetone. As shown schematically in Fig. 2-11, the preferential solvation of $^{23}Na^{\oplus}$ by dimethyl sulfoxide displaces its chemical shift towards δ_{DMSO} and a deviation from the straight line is observed.

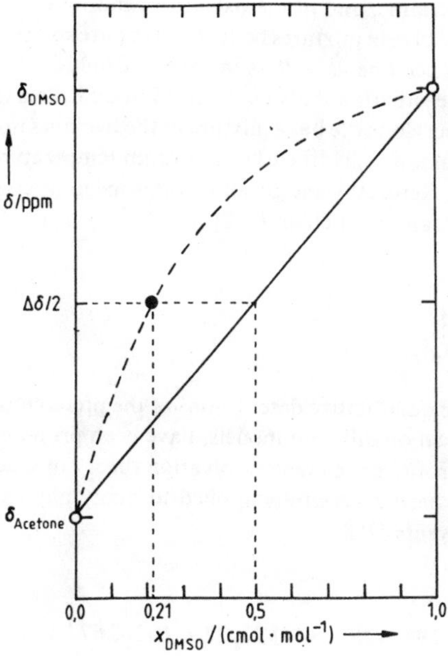

Fig. 2-11. NMR chemical shift of $^{23}Na^{\oplus}$ as a function of the mole fraction of dimethyl sulfoxide (DMSO) in a binary mixture of DMSO and acetone (according to [295]). *Straight line:* ideal case without preferential solvation, primary solvation shell of the same composition as the bulk solvent mixture. *Curved line:* real case with preferential solvation of $^{23}Na^{\oplus}$ by DMSO and iso-solvation point at x_{DMSO}/(cmol · mol^{-1}) ≈ 0.21, that is, the mole fraction of the bulk solvent for which the solvated ion chemical shift is the average of the shifts obtained in the pure solvents ($\Delta\delta = \delta_{DMSO} - \delta_{Acetone}$).

The iso-solvation points obtained from $^{23}Na^{\oplus}$ chemical shifts of sodium tetraphenylborate in different binary solvent mixtures indicate that the solvating ability of a series of organic solvents is in the following order: $CH_3SOCH_3 \gg CH_3NO_2$; pyridine $> CH_3NO_2$; $CH_3SOCH_3 > CH_3CN$; pyridine $> CH_3CN$; $C_6H_5CN > CH_3NO_2$; $CH_3SOCH_3 >$ pyridine [126].

The term selective solvation also applies when one and the same dipolar molecule is preferentially solvated at two different loci by two different solvents. An example is the

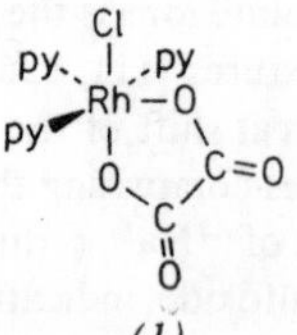

(1)

chloro-oxalato-tripyridine-rhodium(III) complex *(1)*, which dissolves in a 1 : 1 mixture of pyridine and water, but not in either pure water or pyridine [127]. Presumably, Gibbs energy of solvation large enough to overcome the lattice forces is attained only by selective solvation of the three pyridine ligands by pyridine, and of the oxalato ligand by water.

Many macromolecular compounds dissolve in mixtures better than in pure solvents [20]. Thus, polyvinylchloride is insoluble in acetone as well as in carbon disulfide, but soluble in a mixture of the two. The opposite situation is also known. Malodinitrile and *N,N*-dimethylformamide both dissolve polyacrylonitrile but a mixture of the two does not [20]. Soaps dissolve neither in ethylene glycol nor in hydrocarbons at room temperature but are quite soluble in a mixture of the two. Here, ethylene glycol solvates the ionic end, and the hydrocarbon the apolar end of the fatty acid chain [128].

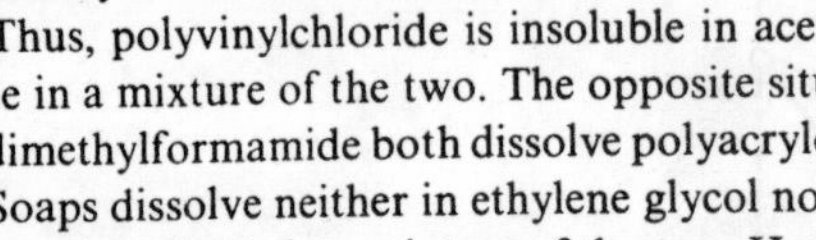

Hydrocarbon ···· CH_3-$(CH_2)_n$-$C^{\ominus}$ (O···H-O-CH_2 / O···H-O-CH_2)

Several attempts at producing suitable quantitative descriptions of the preferential solvation of ions and neutral molecules, based on different models, have been proposed [120, 257–261]. One of them, called the *co*mpetitive *p*referential *s*olvation theory of weak molecular interactions (COPS theory), has been successfully applied to many physico-chemical properties measured in mixed solvents [258].

2.5 Micellar Solvation (Solubilization) [96, 128, 130–132, 220, 262–267]

Special conditions are found in solutions of large cations and anions possessing a long unbranched hydrocarbon chain, *e.g.* CH_3—$(CH_2)_n$—$CO_2^{\ominus}M^{\oplus}$, CH_3—$(CH_2)_n$—$SO_3^{\ominus}M^{\oplus}$, or CH_3—$(CH_2)_n$—$N(CH_3)_3^{\oplus}X^{\ominus}$ (with $n > 7$). Such compounds are known as *amphiphiles*, reflecting the presence of distinct polar and nonpolar regions in the molecule. Salts of such large organic ions are often highly aggregated in dilute aqueous solution. The resulting structured aggregates, together with counterions localized near their periphery by coulomb forces, are termed *micelles**). Fig. 2-12 gives a schematic representation of the formation of a spherical micelle by an anionic amphiphile.

* The term *micelle* has been introduced in 1877 by Nägeli (from the Latin *mica*, a crumb) for a molecular organic aggregate of limited size without exact stoichiometry [270]. The existence of surfactant aggregates in aqueous soap solutions was established in 1896 by Krafft [271], and the first description of a surfactant micelle was given in 1913 by Reychler [272].

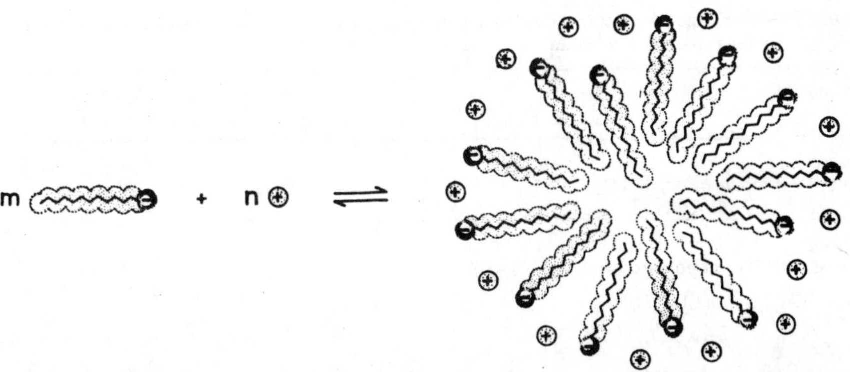

Fig. 2-12. Schematic two-dimensional representation of spherical micelle formation by an anionic amphiphile such as $CH_3—(CH_2)_{11}—CO_2^{\ominus}M^{\oplus}$ in water. The head group (⊖), the counterions (⊕), and the hydrocarbon chains are only schematically indicated to denote their relative position [96, 132]. The highly charged interface (ionic head groups plus bound counterions) between the micelle's hydrophobic core and the bulk solution is called the *Stern layer*.

The hydrophobic part of the aggregate molecules forms the core of the micelle while the polar head groups are located at the micelle-water interface in contact with the water molecules. Such micelles usually have average radii of 2...4 nm and contain 50...100 monomers in water [132]. Their geometric structure is usually roughly spherical or ellipsoidal. In non-aqueous nonpolar solvents, the micellar structures are generally the inverse of those formed in water. In these solvents the polar head groups form the interior of the micelle while the hydrocarbon chains of the ions are in contact with the nonpolar solvent [132].

At very low concentrations, ionic amphiphiles behave as normal strong electrolytes, but if the concentration is raised above the so-called *c*ritical *m*icelle *c*oncentration (*cmc*; usually $10^{-4}...10^{-2}$ mol·l^{-1}), spherical aggregates are formed. The driving force for micelle formation is believed to be the result of three intermolecular interactions: hydrophobic repulsion between the hydrocarbon chains and the aqueous environment (*cf.* Section 2.2.7), charge repulsion of the ionic head groups, and the van der Waals attraction between the alkyl chains (usually unbranched $C_8...C_{18}$ hydrocarbon chains).

Typical surfactants are listed in Table 2-10 along with their respective *cmc*-values and aggregation numbers [268].

In reality, micellar systems are more complex as shown by the simple static picture given in Fig. 2-12 (also known as the Hartley model). A more realistic picture of micellar structures has been given by Menger [269]. According to his "porous cluster" or "reef" model, micelles possess rugged, dynamic surfaces, water-filled pockets, nonradial distribution of chains, and random distribution of terminal methyl groups. A micelle is a dynamic molecular assembly which exists in equilibrium with its monomer, where monomer units are both leaving and entering the micelle. A monomer remains in a micelle only $10^{-8}...10^{-3}$ s depending on the chain length of the surfactant molecule. Another, so-called surfactant-block model of micelles has been given by Fromherz [273].

Table 2-10. Some typical surfactants (*surf*ace *act*ive *agents*) with their critical micell concentrations (*cmc*) and aggregation numbers in aqueous solutions at 25°C [268].

Surfactants	*cmc*/(mol·l⁻¹)	Aggregation number
Anionic		
Sodium *n*-dodecylsulfate (SDS) $CH_3—(CH_2)_{11}—OSO_3^{\ominus}Na^{\oplus}$	0.0081	62
Cationic		
n-Cetyl-trimethylammonium bromide (CTAB) $CH_3—(CH_2)_{15}—\overset{\oplus}{N}(CH_3)_3Br^{\ominus}$	0.0013	78
Nonionic		
Polyoxyethylene(6)dodecanol $CH_3—(CH_2)_{11}—(OCH_2CH_2)_6—OH$	0.00009	400
Zwitterionic		
N-*n*-Dodecyl-*N*,*N*-dimethyl-3-ammonio-1-propane sulfonate (SB-12) $CH_3—(CH_2)_{11}—\overset{\oplus}{N}(CH_3)_2—(CH_2)_3—SO_3^{\ominus}$	0.003	55

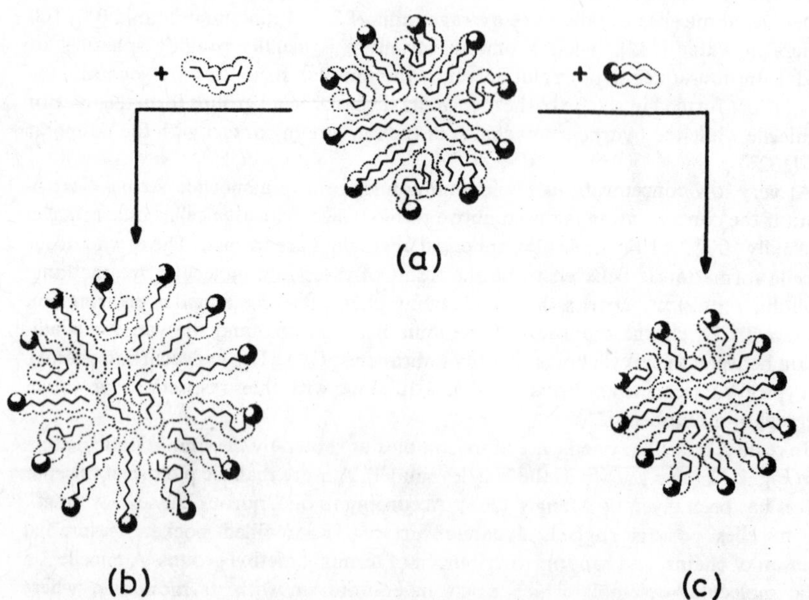

Fig. 2-13. Schematic two-dimensional representation of the solubilization of (b) *n*-nonane as a nonpolar substrate, and (c) 1-pentanol as another amphiphile, by a spherical ionic micelle (a) of an *n*-decanoic acid salt in water [130, 132].

The existence of micelles in solutions of large ions with hydrocarbon chains is responsible for the observation, that certain substances, normally unsoluble or only slightly soluble in a given solvent, dissolve very well on addition of a surfactant (detergent or tenside). This phenomenon is called *solubilization* and implies the formation of a thermodynamically stable isotropic solution of a normally slightly soluble substrate (the *solubilizate*) on the addition of a surfactant (the *solubilizer*) [128, 130–132]. Non-ionic, nonpolar solubilizates such as hydrocarbons can be trapped in the hydrocarbon core of the micelle. Other amphiphilic solutes are incorporated alongside the principal amphiphile and oriented radially, and small ionic species can be adsorbed on the surface of the micelle. Two modes of solubilizate incorporation are illustrated in Fig. 2-13.

Because the micellar interior is far from being rigid, a solubilized substrate is relative mobile. Like micelle formation, solubilization is a dynamic equilibrium process. Only one example of solubilization shall be mentioned. Addition of dodecylammonium propionate solubilizes as large molecule as vitamin B_{12a} (aquocobalamin) in benzene, in which normally it is completely insoluble [133]. Not only solubilities, but also rates and products of organic reactions can be affected by the addition of surfactants to the reaction medium. The alteration of chemical reactions by incorporation of reactant molecules into organized assemblies such as micelles has recently received considerable investigation [132, 263, 266, 274–277] (*cf.* also Section 5.4.8).

2.6 Ionization and Dissociation [49, 96, 134–139, 278, 279]

Solutions of non-electrolytes contain neutral molecules or atoms*) and are non-conductors. Solutions of electrolytes are good conductors due to the presence of anions and cations. The study of electrolytic solutions has shown, that electrolytes may be divided into two classes: *ionophores* and *ionogens* [134]. Ionophores (like alkali halides) are ionic in the crystalline state and they exist only as ions in the fused state as well as in dilute solutions. Ionogens (like hydrogen halides) are substances with molecular crystal lattices which form ions in solution only if a suitable reaction occurs with the solvent. Therefore, according to Eq. (2-13), a clear distinction must be made between the *ionization* step, which produces ion pairs by heterolysis of a covalent bond in ionogens, and the *dissociation* process, which produces free ions from associated ions [137].

$$(\mathrm{A{-}B})_{solv} \underset{}{\overset{K_{Ion}}{\rightleftharpoons}} (\mathrm{A^{\oplus}B^{\ominus}})_{solv} \underset{K_{Assoc}}{\overset{K_{Dissoc}}{\rightleftharpoons}} (\mathrm{A^{\oplus}})_{solv} + (\mathrm{B^{\ominus}})_{solv} \qquad (2\text{-}13)$$

Ionogen — Ion Pair — Free Ions

$$K_{Ion} = [\mathrm{A^{\oplus}B^{\ominus}}]/[\mathrm{A-B}] \qquad (2\text{-}14)$$

$$K_{Dissoc} = [\mathrm{A^{\oplus}}] \cdot [\mathrm{B^{\ominus}}]/[\mathrm{A^{\oplus}B^{\ominus}}] \qquad (2\text{-}15)$$

* An example of a monoatomic un-ionized substrate solution is that of mercury in air-free water, which contains zero-valent mercury atoms [140].

The index "solv" indicates that the species in parentheses are within one solvent cage.

Ionophores may exist in solution as an equilibrium mixture containing ion pairs and free ions. *Ion pairs* are defined as pairs of oppositely charged ions with a common solvation shell whose life-times are sufficiently long for the pairs to be recognizable kinetic entities in solution and those for which only electrostatic binding forces are assumed [135]. Experimentally, ion pairs behave as one unit in determining electric conductivity, kinetic behaviour, and some thermodynamic properties (*e.g.* activity coefficient; osmotic pressure) of electrolyte solutions. In an external electric field such paired ions do not move individually but reorient themselves as an electric dipole. The ion-pair concept was introduced in 1926 by Bjerrum [280] to account for the behaviour of ionophores in solvents of low dielectric constant.

It is possible to distinguish between free ions from associated and covalently bonded species by conductivity measurements, because only free ions are responsible for electrical conductivity in solution [136]. Spectrophotometric measurements distinguish between free ions and ion pairs on the one hand, and covalent molecules on the other, because in a first approximation the spectroscopic properties of ions are independent of the degree of association with the counterion [141]. The experimental equilibrium constant K_{exp}, obtained from conductance data, may then be related to the ionization and dissociation constants by Eq. (2-16).

$$K_{exp} = \frac{[A^{\oplus}] \cdot [B^{\ominus}]}{[A-B] + [A^{\oplus}B^{\ominus}]} = \frac{K_{Ion} \cdot K_{Dissoc}}{1 + K_{Ion}} \tag{2-16}$$

When the extent of ionization is small, then $K_{exp} = K_{Ion} \cdot K_{Dissoc} (K_{Ion} \ll 1$ or $[A^{\oplus}B^{\ominus}] \approx 0)$. For strong electrolytes, where $K_{Ion} \gg 1$, Eq. (2-16) reduces to $K_{exp} = K_{Dissoc}$.

The two steps of Eq. (2-13), ionization and dissociation, are influenced in different ways by solvents. The coulombic force of attraction between two oppositely charged ions is inversely proportional to the dielectric constant of the solvent, according to Eq. (2-17). Therefore, only solvents with sufficiently high dielectric constants will be capable

$$U_{ion-ion} = -\frac{1}{4\pi \cdot \varepsilon_0} \cdot \frac{z^{\oplus} \cdot z^{\ominus} \cdot e^2}{\varepsilon_r \cdot r} \tag{2-17}$$

(U = potential energy of an ion-ion interaction; $z \cdot e$ = charge on the ion; r = distance between the ions; ε_0, ε_r = permittivity of the vacuum and of the medium, resp.)

of reducing the strong electrostatic attraction between oppositely charged ions to such an extent that ion pairs can dissociate into free solvated ions. These solvents are usually called *dissociating solvents**). It is only by liberation of the standard molar Gibbs energy of

* Nernst [141a] and Thomson [141b] first showed independently, that solvents of high dielectric constant promote the dissociation of ionic solutes. The term "dissociating solvent" was first used by Beckmann [141c] in connection with his ebullioscopic determination of the molecular mass of dissolved substances. Later on the term "smenogenic solvent" was proposed by Fuoss for solvents of low dielectric constant which favor the formation of ion pairs. Conversely, "smenolytic solvents" are those whose dielectric constants are high enough to prevent ion association [134]. The latter two terms have, however, found little application.

solvation, ΔG°_{solv}, that the electrostatic interaction between oppositely charged ions can be overcome.

Ion association is only noticeable in aqueous solutions at very high concentrations because of the exceptionally high dielectric constant of water ($\varepsilon_r = 78.3$), but are found at much lower concentrations in alcohols, ketones, carboxylic acids, and ethers. In solvents of dielectric constants less than 10...15 practically no free ions are found (*e.g.* in hydrocarbons, chloroform, 1,4-dioxan, acetic acid); on the other hand, when the dielectric constant exceeds 40, ion associates barely exist (*e.g.* water, formic acid, formamide). In solvents of intermediate dielectric constant ($\varepsilon_r = 15...20$, *e.g.* ethanol, nitrobenzene, acetonitrile, acetone, *N,N*-dimethylformamide) the ratio between free and associated ions depends on the structure of the solvent as well as the electrolyte (*e.g.* ion size, charge distribution, hydrogen-bonded ion pairs, specific ion solvation, *etc.*) [96]. Thus, lithium halides in acetone ($\varepsilon_r = 20.7$) are very weak electrolytes, whereas tetraalkylammonium halides are strongly dissociated in the same solvent [142–144]. In solvents of very small dielectric constant like benzene ($\varepsilon_r = 2.3$) very large association constants are usually found. This indicates that most ion pairs in such solutions exist in the form of higher aggregates [96].

The ability of a solvent to transform the covalent bond of an ionogen into an ionic bond, *i.e.* its *ionizing power*, is not determined in the first instance by its dielectric constant. Rather the ionizing power of a solvent depends on its ability to function as an electron pair acceptor or donor [53, 137]. A dissociating solvent is not necessarily an ionizing one – and *vice versa*. In most cases ionization of bonds of the type $H^{\delta\oplus}-X^{\delta\ominus}$ (e.g. ionization of hydrogen halides), $R^{\delta\oplus}-X^{\delta\ominus}$ (*e.g.* ionization of haloalkanes in S_N1 reactions), or $M^{\delta\oplus}-R^{\delta\ominus}$ (*e.g.* ionization of organometallic compounds) is strongly assisted by electron pair donor (EPD) and electron pair acceptor (EPA) solvents (*cf.* Section 2.2.6), according to (R=H, alkyl):

$$\mathrm{EPD} \curvearrowright \overset{\delta\oplus}{R}-\overset{\delta\ominus}{X} \curvearrowright \mathrm{EPA}$$

$$\mathrm{EPD} \curvearrowright \overset{\delta\oplus}{M}-\overset{\delta\ominus}{R} \curvearrowright \mathrm{EPA}$$

The ionization of an ionogen can therefore be regarded as coordinative interaction between substrate and solvent [281]. The polarization of the covalent bond to be ionized can occur *via* a nucleophilic attack of the EPD solvent on the electropositive end of the bond, or by an electrophilic attack of an EPA solvent on the electronegative end. Both attacks can, of course, also occur simultaneously. The following examples are illustrative.

$$H_2O + \overset{\delta\oplus}{H}-\overset{\delta\ominus}{Cl} + H{-}O{-}H \rightleftharpoons H_2\overset{\oplus}{O}-H + \overset{\ominus}{Cl}\cdots H{-}O{-}H$$

$$(CH_3)_3\overset{\delta\oplus}{C}-\overset{\delta\ominus}{Cl} + H{-}O{-}R \rightleftharpoons (CH_3)_3C^{\oplus} + \overset{\ominus}{Cl}\cdots H{-}O{-}R$$

$$(Me_2N)_3P{=}O + Cl\overset{\delta\oplus}{Mg}-\overset{\delta\ominus}{CH_2}{-}C_6H_5 \rightleftharpoons (Me_2N)_3\overset{\oplus}{P}{-}O{-}MgCl + \overset{\ominus}{|CH_2}{-}C_6H_5$$

In EPD solvents ionization depends on the stabilization of the cation through coordination and, in some solvents, by solvation of the anions as well. In EPA solvents, the anion is stabilized through coordination and, to a lesser extent, additional solvation of the cation may occur.

An evaluation of the ionizing power of a solvent requires knowledge, not only of its coordinating abilities, but also of its dielectric constant. According to Eq. (2-13), solvents of high dielectric constant promote the dissociation of ion pairs. The consequential decrease in ion pair concentration displaces the ionization equilibrium in such a way that new ion pairs are formed from the substrate. Thus, a good ionizing solvent must not only be a good EPD or EPA solvent but also possess a high dielectric constant. The donor and acceptor properties of ionizing solvents can be described empirically in a quantitative way by donor numbers [67] or acceptor numbers [70] (*cf.* Section 2.2.6).

The extraordinary ionizing ability of water is above all due to the fact that it may act as an EPD as well as an EPA solvent. Thus water is both an ionizing and dissociating medium whereas nitromethane, nitrobenzene, acetonitrile, and sulfolane are mainly dissociating. *N,N*-Dimethylformamide, dimethyl sulfoxide, and pyridine are mildly dissociating but good ionizing solvents. Hexamethylphosphoric triamide is an excellent ionizing medium due to its exceptional donor properties, particularly in the case of metalcarbon bonds [145–146]. Alcohols and carboxylic acids as hydrogen-bond donors are good EPA solvents and therefore, good ionizing solvents for suitable substrates.

Chlorotriphenylmethane constitutes a classical example for distinguishing the ionizing and dissociating ability of a solvent. In 1902 Walden used it in liquid sulfur dioxide in the first demonstration of the existence of carbenium ions [147]. The colourless chlorotriphenylmethane dissolves in liquid sulfur dioxide ($\varepsilon_r = 15.6$ at 0 °C) with giving an intensive yellow color ($\lambda_{max} = 430$ nm). This is caused by a partial formation of ion pairs which do not conduct electricity. At low concentrations the ion pairs partially dissociate into free ions, which do conduct electricity [148, 149].

$$[(C_6H_5)_3C\text{-}Cl]_{SO_2} \underset{\text{colorless}}{} \xrightleftharpoons{K_{Ion}} [(C_6H_5)_3\overset{\oplus}{C}\ \overset{\ominus}{Cl}]_{SO_2} \xrightleftharpoons{K_{Dissoc}} [(C_6H_5)_3C^{\oplus}]_{SO_2} + [Cl^{\ominus}]_{SO_2} \tag{2-18}$$

colorless yellow yellow

$K_{Ion} = 1.46 \cdot 10^{-2}$ (0 °C); $K_{Dissoc} = 2.88 \cdot 10^{-3}$ mol/l (0 °C);
$K_{exp} = 4.1 \cdot 10^{-5}$ mol/l (0 °C) [148].

Sulfur dioxide is a π-electron pair acceptor. Until now the available explanation for the strong ionizing power of SO_2 is the formation of an EPD—EPA complex between the halide anion and the sulfur dioxide molecules [148]. Table 2-11 summarizes some of the available data for the comparative efficiency of various solvents in promoting the ionization of chlorotriphenylmethane [150].

The K_{Ion} of chlorotriphenylmethane varies in different solvents by at least a factor of 10^5. In the protic solvents *m*-cresol and formic acid whose dielectric constants are 11.8 and 58.5, respectively, chlorotriphenylmethane is strongly ionized but in the former only slightly dissociated. The remarkable ionizing power of phenols and carboxylic acids has been attributed to their EPA-properties, *i.e.* their ability to form a hydrogen bond between

Table 2-11. Ionization equilibrium constants K_{Ion} of chlorotriphenylmethane in various solvents at 0...25 °C [150]. *Cf.* also [282].

Solvents	ε_r [a] (at 0...30 °C)	$K_{Ion} \cdot 10^4$	References
Nitrobenzene	34.8 (25 °C)	Too low to measure (25 °C)[c]	[151]
Acetonitrile	35.9 (25 °C)	Too low to measure (25 °C)	[152]
Dichloromethane	8.9 (25 °C)	0.07	[153]
1,1,2,2-Tretrachloroethane	8.2 (20 °C)	0.48 (18.5 °C)	[154]
1,2-Dichloroethane	10.4 (25 °C)	0.56 (20 °C)	[154]
Nitromethane	35.9 (25 °C)	2.7 (25 °C)	[155]
Sulfur dioxide	15.6 (0 °C)[b]	146 (0 °C)	[148]
Formic acid	58.5 (16 °C)	3100 (20.5 °C)	[156]
m-Cresol	11.8 (25 °C)	5600[d] (18 °C)	[157]

[a] J.A. Riddick, W.B. Bunger, and T.K. Sakano: *Organic Solvents.* 4th edition, in A. Weissberger (ed.), *Techniques of Chemistry*, Vol. II. Wiley-Interscience, New York 1986.
[b] A.A. Maryott and E.R. Smith: *Table of Dielectric Constants of Pure Liquids.* NBS Circular 514, Washington 1951.
[c] Because nitrobenzene absorbs strongly at the wavelength of the carbenium ion maximum from chlorotriphenylmethane, this result was obtained with chloro-diphenyl-4-tolylmethane.
[d] This K_{Ion}-value corresponds to 36±4% ionization of chlorotriphenylmethane in *m*-cresol [157].

the hydroxyl group and the halide ion. Solvents with high dielectric constants but lacking pronounced EPA-properties, such as acetonitrile and nitrobenzene are barely capable of ionizing chlorotriphenylmethane. In the case of tri-4-anisylchloromethane the K_{Ion}-value in the EPA solvent sulfur dioxide at 0 °C is about $5 \cdot 10^{10}$ times greater than in nitrobenzene at 25 °C [151].

On the other hand, the ionization of chlorotriphenylmethane is also favored by EPD solvents. Since the developing carbenium ion is an electrophilic species, it readily interacts with nucleophilic solvents. Thus, the extent of ionization of chlorotriphenylmethane in nitrobenzene increases on the addition of aprotic EPD solvents in direct relation to the donor number [158]. *Cf.* reference [299] for a study of ionization and dissociation equilibria of other halotriphenylmethanes in solution (Ph_3C—X with X=F, Cl, Br).

Another remarkable example of the solvent effect on the ionization of ionogens is the Friedel-Crafts intermediate antimony pentachloride/4-toluoyl chloride. It can exist as two distinct well-defined adducts depending on the solvent from which it is recrystallized, the donor-acceptor complex *(2)* or the ionic salt *(3)* [159].

$$H_3C\text{-}C_6H_4\text{-}C(Cl){=}\overset{\oplus}{O}\text{-}\overset{\ominus}{Sb}Cl_5 \rightleftharpoons \left[H_3C\text{-}C_6H_4\text{-}\overset{\oplus}{C}{=}O\right] SbCl_6^{\ominus}$$

(2) *(3)*

The donor-acceptor complex *(2)* is isolated from tetrachloromethane solution ($\varepsilon_r = 2.2$), the ionic salt *(3)* from chloroform solution ($\varepsilon_r = 4.8$). When dissolved in chloroform, the donor-acceptor complex recrystallizes as the ionic salt. Similarly, the ionic salt is converted to the donor-acceptor complex when dissolved in tetrachloromethane. This result shows that in solution an equilibrium exists between the two forms. The isolation depends on the solvent used for recrystallization. Similar results were obtained in the case of the adduct between acetylchloride and aluminium trichloride, which is un-ionized in chloroform, but completely ionized in nitrobenzene [160].

Other nice examples of well-studied solvent-dependent ionization equilibria of ionogens are azidocycloheptatriene ⇌ tropylium azide [282, 283] and (triphenylcyclopropen-1-yl) (4-nitrophenyl)malononitrile *(2a)* ⇌ triphenylcyclopropenium dicyano-(4-nitrophenyl)methide *(3a)*, the latter being one of the first examples of direct heterolysis of a weak carbon-carbon bond to a carbocation and carbanion in solution [284].

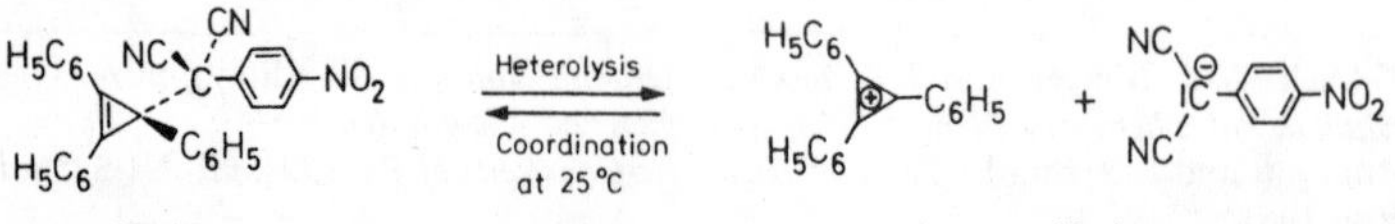

(2a) *(3a)*

When dissolved in nonpolar solvents such as benzene or diethyl ether, the colourless *(2a)* forms an equally colourless solution. However, in more polar solvents (*e.g.* acetone, acetonitrile), the deep-red colour of the resonance-stabilized carbanion of *(3a)* appeared ($\lambda = 475 \ldots 490$ nm), and its intensity increases with increasing solvent polarity. The carbon-carbon bond in *(2a)* can be broken merely by changing from a less polar to a more polar solvent. Cation and anion solvation provides the driving force for this heterolysis reaction, whereas solvent displacement is required for the reverse coordination reaction. The Gibbs energy for the heterolysis of *(2a)* correlates well with the reciprocal solvent dielectric constant in accordance with the Born electrostatic equation [285], except for EPD solvents such as dimethyl sulfoxide, which gave larger ΔG°_{het}-values than those expected from pure electrostatic solvation [284].

A similar solvent-induced carbon-carbon heterolysis has been observed with oligomers of 1-cycloheptatrienylidene-4-(dicyanomethylidene)-2,5-cyclohexadiene [291].

The first pure organic salt, $C_{48}H_{51}^{\oplus}C_{67}H_{39}^{\ominus}$, consisting solely of carbon and hydrogen atoms and fully ionised in the crystalline state and in solution, has been prepared recently by mixing tris-1-(5-isopropyl-3,8-dimethylazulenyl)cyclopropenylium perchlorate with potassium tris(7*H*-dibenzo[*c*, *g*]fluorenylidenemethyl)methide in tetrahydrofuran solution [292].

It should be mentioned that the ionization step in Eq. (2-13) is analogous to that involved in S_N1- and S_N2-reactions of aliphatic substrates. For example, in solvolytic reactions of haloalkanes, the process of going from a covalently bonded initial state to a dipolar or ionic activated complex (transition state), is similar to the ionization step in Eq. (2-13). Therefore, those solvent properties which promote ionization are also important in the estimation of solvent effects on nucleophilic displacement reactions [161] (*cf.* Section 5.4.1).

The ionization of an ionogen and its subsequent dissociation according to Eq. (2-13) can be further elaborated. Between the ion pair immediately formed on heterolysis of the covalent bond and the independently solvated free ions there are several steps of progressive loosening of the ion pair by penetration of solvent molecules between the ions. At least four varieties of ion interactions representing different stages of dissociation have been postulated [96, 134, 138, 141]; *cf.* Eq. (2-19) and Fig. 2-14.

Ionizing Solvents → Ionization; Dissociating Solvents → Dissociation

$$\underset{\text{Ionogen}}{(\mathrm{A{-}B})_{solv}} \underset{}{\overset{\text{Ionization}}{\rightleftharpoons}} \underset{\text{Contact Ion Pair}}{(\mathrm{A}^{\oplus}\mathrm{B}^{\ominus})_{solv}} \overset{\text{Dissociation}}{\rightleftharpoons} \underset{\text{Solvent-separated Ion Pair}}{(\mathrm{A}^{\oplus}/\!/\,\mathrm{B}^{\ominus})_{solv}} \tag{2-19}$$

$$\overset{\text{Dissociation}}{\rightleftharpoons} \underset{\text{Free Ions}}{(\mathrm{A}^{\oplus})_{solv} + (\mathrm{B}^{\ominus})_{solv}}$$

Based on the mutual geometric arrangement of the two ions and the solvent molecules, the following definitions of ion pairs have been given (*cf.* Fig. 2-14).

First, immediately after ionization, *contact ion pairs**) are formed where no solvent molecules intervene between the two ions that are in close contact. The contact ion pair constitutes an electric dipole having only one common primary solvation shell. The ion pair separated by the thickness of only one solvent molecule is called a *solvent-shared ion pair**). In solvent-shared ion pairs, the two ions already have their own primary solvation shells. These, however, interpenetrate each other. Contact and solvent-shared ion pairs are separated by an energy barrier which corresponds to the necessity of creating a void between the ions that grow to molecular size before a solvent molecule can occupy it. Further dissociation leads to *solvent-separated ion pairs**). Here, the primary solvation shells of both ions are in contact, so that some overlap of secondary and further solvation shells takes place. Increase in ion-solvating power and dielectric constant of the solvent favours solvent-shared and solvent-separated ion pairs. However, a clear experimental distinction between solvent-shared and solvent-separated ion pairs is not easily obtainable. Therefore, the designations solvent-shared and solvent-separated ion pairs are sometimes interchangable. Eventually, further dissociation of the two ions leads to *free*, *i.e.* unpaired solvated *ions* with independent primary and secondary solvation shells. The circumstances under which contact, solvent-shared, and solvent-separated ion pairs can exist as thermodynamically distinct species in solution are reviewed respectively by Swarcz [138] and Marcus [241].

Interestingly, recent theoretical calculations of Gibbs energy profiles for the separation of *tert*-butyl cation and chloride ion during the hydrolysis of 2-chloro-2-methylpropane have given support for the existence of a contact ion pair, while solvent-

* Some authors use the designations *intimate ion pair*, *internal ion pair* (Winstein [162]), *cage ion pair* (Kosower [129]), or *inner-sphere ion pair* (Marcus [241]) instead of *contact ion pair*, and *external ion pair* (Winstein [162]) or *outer-sphere ion pair* (Marcus [241]) for *solvent-shared* and *solvent-separated ion pairs*. The more general designation *tight* and *loose ion pair* (Swarcz [138]) implies, that in principle more than two different kinds of ion pairs may exist in solution. A recent IUPAC glossary recommends the designations *tight ion pair* (or *intimate* or *contact ion pair*) and *loose ion pair* [286].

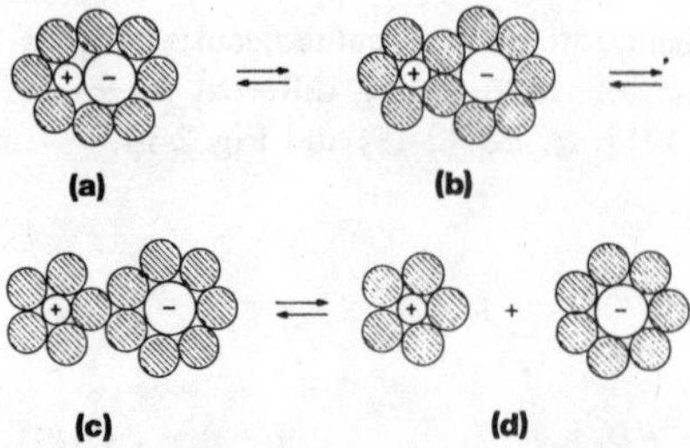

Fig. 2-14. Schematic representation of the equilibrium between (a) a solvated contact ion pair, (b) a solvent-shared ion pair, (c) a solvent-separated ion pair, and (d) unpaired solvated ions of a 1 : 1 ionophore in solution, according to reference [241]. Hatched circles represent solvent molecules of the primary solvation shell.

separated ion pairs and free, unpaired ions do not appear as energetically distinct species [302]. Preliminary Monte Carlo simulations predict the occurrence of a contact ion pair at a C—Cl distance of 29 pm and the onset of the solvent-separated ion-pair regime near 55 pm. A significant barrier of ca. 8 kJ/mol (2 kcal/mol) between the contact and solvent-separated ion pairs has been calculated [302].

The suggestion, that ion pairs may exist in more than one distinct form was made by Winstein [162] and Fuoss [163] in 1954, but direct evidence for the existence of contact and solvent-separated ion pairs came from UV/Vis-spectroscopic investigations of sodium fluorenide in tetrahydrofuran solution [141, 164]. Further evidence for the existence of a dynamic equilibrium between contact and solvent-separated ion pairs (*e.g.* hyperfine splitting of radical-anion ESR lines by cationic nuclei; electronic spectra of mesomeric anions; *etc.*) is summarized by Gordon [96], Szwarc [138], and Marcus [241]. Increasing association of ions in solution greatly affects their chemical behaviour. A large variety of possible ion-pair effects on rate constants, mechanism and stereochemistry is known, especially in reactions of ion pairs containing carbenium ions [161, 165] or carbanions [166, 168, 168a].

The observation that the rate of loss of optical activity during the solvolysis of certain chiral substrates $R^{\delta\oplus}-X^{\delta\ominus}$ exceeded the rate of acid production and the occurrence of a special salt effect, led to the postulation of two distinct ion-pair intermediates [162]. The basic Winstein solvolysis scheme is given by Eq. (2-20). According to this scheme,

$$(\overset{\delta\oplus}{R}-\overset{\delta\ominus}{X})_{solv} \rightleftharpoons (R^{\oplus}X^{\ominus})_{solv} \rightleftharpoons (R^{\oplus}//X^{\ominus})_{solv} \rightleftharpoons (R^{\oplus})_{solv} + (X^{\ominus})_{solv} \tag{2-20}$$

$$\begin{array}{ccc} (R^{\oplus}X^{\ominus})_{solv} & (R^{\oplus}//X^{\ominus})_{solv} & (R^{\oplus})_{solv} + (X^{\ominus})_{solv} \\ \downarrow & \downarrow & \downarrow \\ \text{products} & \text{products} & \text{products} \end{array}$$

the solvolysis products are not only obtained from free unpaired ions, but also from the two different ion pairs, depending on the solvent-dependent degree of dissociation.

An analogous scheme holds for the reactions of certain dipolar organometallics $R^{\delta\ominus}-M^{\delta\oplus}$, according to Eq. (2-21) [138, 167, 168, 168a].

$$\underset{\text{Higher Aggregates}}{(\overset{\delta\ominus}{R}-\overset{\delta\oplus}{M})_{n,solv}} \rightleftharpoons \underset{\text{Ionogen}}{(\overset{\delta\ominus}{R}-\overset{\delta\oplus}{M})_{solv}} \rightleftharpoons \underset{\text{Contact Ion Pair}}{(R^{\ominus}M^{\oplus})_{solv}} \tag{2-21}$$

$$\rightleftharpoons \underset{\text{Solvent-separated Ion Pair}}{(R^{\ominus}//M^{\oplus})_{solv}} \rightleftharpoons \underset{\text{Free Ions}}{(R^{\ominus})_{solv} + (M^{\oplus})_{solv}}$$

Whereas the spectral behavior of solvent-separated ion pairs and free ions is very similar, the UV/Vis-spectra of contact and solvent-separated ion pairs are usually different from each other, as has been shown with sodium fluorenide [141, 164]. Due to the penetration of solvent molecules between the ion-pair couples, the direct influence of the metal cation on the π-electron system of the carbanion is lost. With increasing dissociation, the absorption maximum of sodium fluorenide in tetrahydrofuran solution is shifted bathochromically in direction of the absorption maximum of the free fluorenide ion: $\lambda_{max} = 356$ nm $\rightarrow$ 373 nm $\rightarrow$ 374 nm, for the contact ion pairs, solvent-separated ion pairs, and free fluorenide ions, respectively [164]. The equilibrium between contact and solvent-separated ion pairs is shifted in the direction of increased dissociation by the addition of cation solvators such as EPD solvents. Thus, the proportion of solvent-separated ion pairs for sodium fluorenide at 25 °C in tetrahydrofuran is 5 cmol/mol, whereas in 1,2-dimethoxyethane, a better cation solvator, it is 95 cmol/mol. In strong EPD solvents such as dimethyl sulfoxide, hexamethylphosphoric triamide, or poly-ethyleneglycol dimethyl ethers, most of the fluorenide salts exist as solvent-separated ion pairs only. Small quantities of dimethyl sulfoxide, added to the sodium fluorenide solution in 1,4-dioxane, convert the contact ion pairs to dimethyl sulfoxide-separated ion pairs [141, 164].

Sodium naphthalenide behaves similarly when the solvent is changed from tetrahydrofuran to 1,2-dimethoxyethane. The formation of solvent-separated from contact ion pairs is shown by a dramatic simplification of the ESR-spectrum: the 100-line spectrum of the contact ion pair, due to the spin-spin coupling of the unpaired electron with the four equal hydrogen nuclei in the α- and β-positions, together with the sodium nucleus ($I = 3/2$), collapses to a 25-line spectrum as the interaction with the sodium ion is disrupted [169, 170].

Other illustrative examples of carbanionic ion-pair dissociation/aggregation are: lithium triphenylmethide, which exists as a tight ion pair in diethyl ether and as a solvent-separated ion pair in tetrahydrofuran as shown by UV/Vis-spectrophotometric measurements [287], and lithium 10-phenylnonafulvene-10-oxide, which exists as a tight ion pair *(2b)* in tetrahydrofuran solution and as a solvent-separated ion pair *(3b)* when hexamethylphosphoric triamide or dimethyl sulfoxide are added (^{1}H- and ^{13}C-NMR measurements) [288].

This second case is particularly interesting since the addition of an EPD solvent is connected with a shift from the olefinic nonafulvenoxide anion in *(2b)* to the aromatic benzoyl [9] annulene anion in *(3b)*. Without association of the lithium cation with the enolate oxygen atom, the negative charge is preferably delocalized in the [9] annulene ring. Therefore, the aromatic character of this ionophore depends on its ion-pair character [288].

(2b) ⇌ room temperature (3b)

The degree of aggregation of organolithium compounds (alkyl-, aryl-, and alkinyl-lithium compounds as well as lithium enolates) in dilute tetrahydrofuran solution at -108 °C has been recently determined by means of cryoscopic measurements [289] and high-field ^{1}H-NMR spectroscopy [290].

3 Classification of Solvents

Due to the physical and chemical differences between the numerous organic and inorganic solvents it is difficult to organise them in a useful scheme. Here we shall present five attempts at classification of solvents, which should prove useful to the chemist. Due to broad definitions some overlapping of these is unavoidable. As has been customary in the preceding chapters, non-aqueous organic solvents will receive particular attention [1–15, 103–108].

3.1 Classification of Solvents according to Chemical Constitution

Solvents can be classified according to their chemical bonds: (a) molecular liquids (molecule melts; covalent bonds only), (b) ionic liquids (molten salts; only ionic bonds), and (c) atomic liquids (low-melting metals like liquid mercury or liquid sodium; metallic bonds) [16]. Numerous transitions are possible by mixing solvents of these three classes (Fig. 3-1). However, research into this area is still far from exhausted.

The customary non-aqueous organic solvents belong to the group of molecular melts and, according to their chemical constitution, to the following classes of compounds (*cf.* Table A-1, Appendix): aliphatic and aromatic hydrocarbons and their halogen and nitro derivatives, alcohols, carboxylic acids, carboxylic esters, ethers, ketones, aldehydes, amines, nitriles, unsubstituted and substituted amides, sulfoxides, and sulfones. The classification of solvents according to chemical constitution allows certain qualitative predictions, summarized in the old rule "similia similibus solvuntur". In general, a compound dissolves far easier in a solvent possessing related functional groups than in one of a completely different nature. A proper choice of solvent, based on the knowledge of its chemical reactivity, helps to avoid undesired reactions between solute and solvent. For example, condensations should not be carried out in solvents possessing carbonyl groups (*e.g.* ketones) or hydrolyses in carboxylic esters, amides, or nitriles.

Liquid crystals or mesomorphic compounds occupy a special position [17–22, 22a, 109, 110]. Compounds capable of forming liquid crystals are long, flat, and fairly rigid

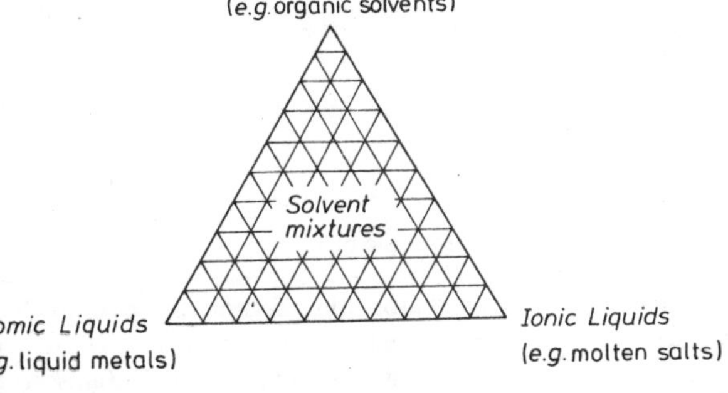

Fig. 3-1. Classification of solvents according to their characteristic chemical bonds [16].

along the axis of the molecule. Most known mesomorphic solvents are systems of the following general structure with polarizable aromatic nuclei held in a planar skeleton:

R[1]–⟨benzene⟩–[bridge]–⟨benzene⟩–R[2]

Common central bridges:
$—CH_2—CH_2$, —CH=CH—, —C≡C—,
—CH=N—, —CH=N(O)—, —CO—O—,
—N=N—, —N=N(O)—,

Common terminal substituents:
R—, RO—, HO—, R—CO—O—,
R—O—CO—O—,
R—O—$(CH_2)_n$—O—, H_2N—,
O_2N—,
N≡C—, Cl—, Br—, I—

Unlike normal isotropic liquids, which possess a completely random rearrangement of molecules, liquid crystals are considerably ordered. The degree of order in the latter lies somewhere between that of isotropic liquids and crystals. Liquid crystals are classified into lyotropic and thermotropic crystals depending on the way in which the mesomorphic phase is generated. *Lyotropic* liquid-crystalline solvents are formed by addition of controlled amounts of polar solvents to certain amphiphilic compounds. *Thermotropic* liquid-crystalline solvents, simply obtained by temperature variations, can be further classified into nematic, smectic, and cholesteric solvents depending on the type of molecular order present. In *nematic* mesophases the long molecular axes of the component molecules are arranged, on the average, parallel to one another. There is no further ordering present and these are the most fluid of liquid-crystalline solvents. *Cholesteric* mesophases are optically active nematic phases, additionally characterized by a gradual twist in orientational alignment as one proceeds through the bulk solvent, and forming a twisted helical macrostructure. In *smectic* mesophases the component molecules are further arranged in layers, with their long molecular axes parallel to one another and perpendicular to the plane of the layers. As result of this additional molecular ordering, smectic solvents are considered to be the most ordered and the less fluid liquid-crystalline solvents. A typical example of a liquid crystalline nematic solvent, at room temperature, is *N*-(4-Methoxybenzylidene)-4-*n*-butylaniline ("MBBA") [23]. At 21 °C the crystalline structure is lost and an ordered fluid, thermally stable up to about 48 °C, is formed (nematic range 21 to 48 °C).

H_3C–O–⟨benzene⟩–CH=N–⟨benzene⟩–$CH_2CH_2CH_2CH_3$

$$\text{Solid phase} \xrightleftharpoons{21\ °C} \text{Nematic phase} \xrightleftharpoons{48\ °C} \text{Isotropic phase}$$

Liquid crystals are usually excellent solvents for other organic compounds. Non-mesomorphic solute molecules may be incorporated into liquid crystalline solvents without destruction of the order prevailing in the liquid crystalline matrix. The anisotropic solute-solvent interaction leads to an appreciable orientation of the guest molecules with respect to the axis of preferred solvent alignment. The consequences may be useful as shown by the use of liquid crystals as anisotropic solvents for spectroscopic investigations of anisotropic molecular properties [166]. Ordered solvent phases such as liquid crystals have also been used as reaction media, particularly for photochemical reactions; *cf.* for example [111, 155, 163] and Section 5.5.9.

The second corner of the triangle in Fig. 3-1 is occupied by ionic liquids. Of these, molten salts are becoming increasingly important as solvents for inorganic as well as organic reactions [3, 24–30, 112–114]. High thermal stability, good electrical conductivity*), low viscosity, wide liquid range, low vapor pressure and the resulting possibility of high working temperatures, together with their excellent ability to dissolve salts and metals, make them extremely useful reaction media. Therefore, such systems are becoming increasingly important technologically. A further advantage of molten salts is their high heat conductivity which permits a very rapid dispersal of the heat of reaction.

Often these are the only solvents capable of dissolving salt-like hydrides such as NaH and CaH_2, carbides, nitrides, various oxides, sulfides, and cyanamides. Many metals form atomic dispersions in the melts of their halogenides, yielding extremely strong reducing solutions. The working temperature for oxide and silicate melts lies above 1500 °C, for normal salts between 100 and 1000 °C, and for eutectic mixtures it is often at even lower temperatures. Organic salts such as tetra-*n*-hexyl-ammoniumbenzoate are liquid even at room temperature. Table 3-1 gives a somewhat arbitrary selection of inorganic and organic salts and salt mixtures, together with their physical constants [27, 28].

Although the dielectric constants of molten salts are generally quite small ($\varepsilon_r = 2 \ldots 3$) they behave as strongly dissociating solvents! This is due to the ability of the solvent ions to exchange places with solute ions of the same charge.

Only four examples of the many reactions which can be advantageously carried out in molten salts, are given. When 1,1-dichloroethane is passed through a $ZnCl_2$/KCl-melt at 330 °C, vinyl chloride is formed in 97 cmol/mol yield by dehydrohalogenation. Likewise, the addition of hydrogen chloride to acetylene proceeds with 89 cmol/mol yield in the same melt. A combination of these two steps allows a ready synthesis of the technically important vinyl chloride from acetylene and 1,1-dichloroethane according to Eq. (3-1) [25, 31].

$$HC{\equiv}CH + H_3C{-}CHCl_2 \xrightarrow[\text{yield 68.5 cmol/mol}]{\text{in } ZnCl_2/KCl/HgCl_2} 2\ H_2C{=}CH{-}Cl \qquad (3\text{-}1)$$

Ionic liquids such as the eutectic $AgNO_3/KNO_3/AgCl$ (mp 113 °C) are the best solvents for effecting silver ion-catalyzed isomerizations in caged organic molecules, *e.g.* basketane → snoutane [31a], *cf.* Eq. (3-2).

$$\text{basketane-}CO_2CH_3 \xrightarrow[\text{130 °C, 12 h; yield 75 cmol/mol}]{\text{in } AgNO_3/KNO_3/AgCl} \text{snoutane-}CO_2CH_3 \qquad (3\text{-}2)$$

* Ionic melts possess electrical conductivities which are roughly a factor ten larger than those of concentrated aqueous solutions of strong electrolytes (ionophores).

Table 3-1. Melting points (t_{mp}), boiling points (t_{bp}), and liquid range $\Delta t = t_{bp} - t_{mp}$ of some fused inorganic and organic salt systems in °C [27, 28][a]

Fused salts[a]	t_{mp}/°C	t_{bp}/°C	Δt/°C
NaCl	808	1465	657
KCl	772	1407	635
LiCl	610	1382	772
KOH	360	—	—
NaOH	318	1390	1072
$NaNO_3$	310	380	70
$CH_3CO_2^{\ominus}K^{\oplus}$	295	—	—
$ZnCl_2$	283	732	449
$NaNH_2$	208	—	—
$ZnCl_2$(60)—NaCl(20)—KCl(20)	203	—	—
KCl(33)—$AlCl_3$(67)	128	—	—
$(n\text{-}C_6H_{13})_4N^{\oplus}I^{\ominus}$	105	—	—
$AlBr_3$	97.5	257	159.5
$LiNO_3$(25.8)—NH_4NO_3(66.7)—NH_4Cl(7.5)	86	—	—
$AgNO_3$(52)—$TlNO_3$(48)	82.5	—	—
$SbCl_3$	73.2	221	147.8
$(n\text{-}C_6H_{13})_4N^{\oplus}NO_3^{\ominus}$	69	—	—
$(n\text{-}C_4H_9)_3NH^{\oplus}NO_3^{\ominus}$[c]	21.5	119[d]	—
$n\text{-}C_4H_9$—$NH_3^{\oplus}SCN^{\ominus}$[c]	20.5	130[d]	—
CH_3—CH_2—$NH_3^{\oplus}NO_3^{\ominus}$[e]	12.5[c]	170[d]	—
$(n\text{-}C_6H_{13})_4N^{\oplus}C_6H_5CO_2^{\ominus}$[f]	−50[c]	—	—
$(n\text{-}C_6H_{13})(C_2H_5)_3N^{\oplus}(n\text{-}C_6H_{13})(C_2H_5)_3B^{\ominus}$[g]	> −75[c]	87[d]	—

[a] *Cf.* also the compilation of salt properties in A.J. Gordon and R.A. Ford: *The Chemist's Companion*. Wiley-Interscience, New York, London, Sydney, Toronto 1972, p. 40ff.
[b] By binary and ternary eutectics the figures in parentheses give the portions in cmol/mol.
[c] C.F. Poole, B.R. Kersten, S.S.J. Ho, M.E. Coddens, and K.G. Furton, J. Chromatogr. *352*, 407 (1986).
[d] Decomposition temperature; see reference[c].
[e] D. Mirejowsky and E.M. Arnett, J. Am. Chem. Soc. *105*, 1112 (1983).
[f] C.G. Swain, A. Ohno, D.K. Roe, R. Brown, and T. Maugh, J. Am. Chem. Soc. *89*, 2648 (1967); T.G. Coker, J. Ambrose, and G.J. Janz, ibid. *92*, 5293 (1970).
[g] For other liquid tetraalkylammonium tetraalkylborides see W.T. Ford, R.J. Hauri, and D.J. Hart, J. Org. Chem. *38*, 3916 (1973).

Molten sodium tetrachloroaluminate (a 1 : 1 mixture of NaCl and $AlCl_3$) is a good reaction medium for the Friedel-Crafts acylation reaction given in Eq. (3-3) [115].

(1) + $SOCl_2$, 30 min
(2) + $AlCl_3$ in C_6H_6, 5 h
(3) + Ice/H_2O, 30 min
yield 79 - 88 cmol/mol

in NaCl/$AlCl_3$ (1:1)
170 °C, 5 min
yield 96 cmol/mol

(3-3)

Whereas the classical procedure for the synthesis of 1-indanone from 3-phenylpropanoic acid consists of three reaction steps with a total reaction time of *ca.* six hours [116], the molten salt reaction is finished in five minutes and gives an even better yield [115].

A 1 : 2 mixture of 1-methyl-3-ethylimidazolium chloride and aluminum trichloride, an ionic liquid that melts below room temperature, has been recommended recently as solvent and catalyst for Friedel-Crafts alkylation and acylation reactions of aromatics [162], and as solvent for UV/Vis- and IR-spectroscopic investigations of transition metal halide complexes [167]. The corresponding 1-methyl-3-ethylimidazolium tetrachloroborate (as well as *n*-butylpyridinium tetrachloroborate) represent new molten salt solvent systems, stable and liquid at room temperature [169].

Organic molten salts such as tris-*n*-butyl-dodecylphosphonium halides (melting point below 40 °C) have been used as reaction media for nucleophilic aromatic substitution of aryl tosylates by halide ions [156].

The third corner of the triangle in Fig. 3-1 represents liquid metals, *e.g.* mercury or liquid sodium, which until now have received little attention as reaction media. Chemical reactions in liquid alkali metals have been reviewed recently [164].

3.2 Classification of Solvents using Physical Constants

The following physical constants can be used to characterize the properties of a solvent: melting and boiling point, vapor pressure, heat of vaporization, index of refraction, density, viscosity, surface tension, dipole moment, dielectric constant, polarizability, specific conductivity, *etc.* A compilation of data of usual organic solvents is given in Table A-1 (Appendix).

Solvents can be broadly classified as low, middle, or high boiling, *viz.* $t_{bp} < 100$ °C, 100...150 °C, or > 150 °C at 1 bar. Similarly, liquids can be classified according to their *evaporation number* using diethyl ether as reference (evaporation number = 1 at 20 °C and 65 cl/l relative air humidity). Thus, low volatility signifies evaporation numbers < 10, medium volatility 10...35, and high volatility > 35 [32]. Using viscosity as a criterion, solvents are of low viscosity when the dynamic viscosity < 2 mPa · s at 20 °C, medium viscosity (2...10 mPa · s), and high viscosity (> 10 mPa · s) [32].

The degree of association of molecules in a liquid can be estimated by means of its *Trouton constant* [117]. At the normal boiling temperature, T_{bp}, vaporization proceeds with standard molar changes of enthalpy, ΔH°_{bp}, and entropy, ΔS°_{bp}, from which Trouton's rule is derived as given in Eq. (3-4).

$$\Delta S^{\circ}_{bp} = \frac{\Delta H^{\circ}_{bp}}{T_{bp}} \approx 21 \text{ cal} \cdot \text{mol}^{-1} \cdot \text{K}^{-1} \text{ or } 88 \text{ J} \cdot \text{mol}^{-1} \cdot \text{K}^{-1} \qquad (3\text{-}4)^{*)}$$

This rule works best for apolar, quasi-spherical molecules. Large deviations occur when chemical association is involved (*e.g.* carboxylic acids), from molecular dipolarity (*e.g.* dimethyl sulfoxide), and from molecular asphericity (*e.g.* neopentane/*n*-pentane).

* Trouton's rule can also be written independent of units: $\Delta S^{\circ}_{bp}/R = \Delta H^{\circ}_{bp}/(R \cdot T_{bp}) \approx 11$.

Strongly associating solvents (*e.g.* HF, H_2O, NH_3, alcohols, carboxylic acids) have Trouton constants which are higher than the average value of 88 $J \cdot mol^{-1} \cdot K^{-1}$ found for non-associating solvents such as diethyl ether and benzene.

In this connection two other physical solvent properties are important: the *cohesive pressure c* (also called *cohesive energy density*) and the *internal pressure* π of a solvent [98–100].

The *cohesive pressure c* is a measure of the total molecular cohesion per unit volume, given by Eq. (3-5),

$$c = \frac{\Delta U_v}{V_m} = \frac{\Delta H_v - R \cdot T}{V_m} \tag{3-5}$$

where ΔU_v and ΔH_v are respectively the energy and enthalpy (heat) of vaporization of the solvent to a gas of zero pressure, and V_m is the molar volume of the solvent. On vaporization of a solvent to a non-interacting vapour, *all* intermolecular solvent-solvent interactions will be broken. Therefore, *c* represents the total strength of the intermolecular solvent structure. Cohesive pressure has very high values for solvents of high polarity and low values for nonpolar solvents such as perfluorohydrocarbons with weak interaction forces. Intermolecular hydrogen bonding in a solvent increases the cohesive pressure (*cf.* Table 3-2). Cohesive pressure is related to the energy required to create cavities in a liquid in order to accommodate solute molecules during the process of dissolution.

On the other hand, the *internal pressure* π is defined as the change in internal energy of a solvent as it undergoes a very small isothermal expansion, as seen in Eq. (3-6).

$$\pi = \left(\frac{\partial U}{\partial V_m}\right)_T \tag{3-6}$$

(U = molar internal energy; V_m = molar volume; T = absolute temperature). This small expansion does not necessarily disrupt all the intermolecular solvent-solvent interactions. The internal pressure results from the forces of attraction between solvent molecules exceeding the forces of repulsion, *i.e.* mainly dispersion and dipole-dipole interactions (*cf.* Table 3-2).

Although there is obviously a close connection between cohesive pressure and internal pressure, they are not equivalent as shown by the compilation of *c* and π-values for some selected organic solvents shown in Table 3-2 [99, 100, 154]. It has been assumed, that π is mainly a reflection of dispersion and dipole-dipole interactions within the solvent, whereas *c* additionally includes specific solvent-solvent interactions such as hydrogen bonding. Hydrogen bonding in a solvent increases the cohesive pressure, while the internal pressure is comparable to that of solvents without hydrogen bonding. Therefore, the hydrogen-bonding pressure or energy density contribution can be measured by the difference $(c - \pi)$ [99, 100]. Values of π approach those of *c* only for weakly polar solvents with dipole moments less than *ca.* $7 \cdot 10^{-30}$ C m (*ca.* 2 D) and without specific solvent/solvent interactions. As shown in Table 3-2, the ratios $n = \pi/c$ approach a value of unity for nonpolar solvents (*e.g.* hydrocarbons), but it can be less than or greater than unity for other solvents. High values of *n* are obtained for the noninteracting fluorohydrocarbons while at the other end of the solvent spectrum HBD solvents with very low values of *n* are found.

Table 3-2. Cohesive pressures (c), internal pressures (π), and their ratios $n = \pi/c$ for twenty-six organic solvents at 20 °C [99, 154].

Solvents	c/MPa[a)]	π/MPa[a)]	$n = \pi/c$
Perfluoro-*n*-heptane	136	220	1.62
Perfluoro(methylcyclohexane)	147	231	1.58
1,4-Dioxane	402	499	1.24
2,2,4-Trimethylpentane	200	236	1.18
Methylcyclohexane	260	297	1.14
Cyclohexane	285	326	1.14
Cyclohexanone	364	413	1.13
Tetrachloromethane	312	345	1.10
n-Hexane	225	239	1.06
Benzene	357	379	1.06
Toluene	337	355	1.05
Diethyl ether	251	264	1.05
1,2-Dichloroethane	416	427	1.03
Ethyl acetate	347	354	1.02
Chloroform	362	370	1.02
Acetophenone	456	457	1.00
Dichloromethane	414	408	0.98
Carbon disulfide	410	377	0.92
Acetone	398	337	0.85
t-Butanol	473	339	0.72
Acetonitrile	590	379	0.64
1-Butanol	485	300	0.62
1,2-Ethanediol	887	502	0.57
Ethanol	703	291	0.41
Methanol	887	285	0.32
Water[b)]	2302	151	0.07

[a)] 1 MPa = 1 J · cm^{-3} = 0.2390 cal · cm^{-3}.
[b)] At 25 °C; taken from reference [100].

The square root of the cohesive pressure c as defined in Eq. (3-5) has been termed the *solubility parameter* δ by Hildebrand and Scott [98] because of its value in correlating and predicting the solvency of solvents for non-electrolyte solutes [*cf.* Eqs. (2-1) and (5-77) in Sections 2.1 and 5.4.2, respectively]. Solvency is defined as the ability of solvents to dissolve a compound [118]. A selection of δ-values is given in Table 3-3.

The solvency increases as the δ-value of the solvent approaches that of the solute. Two liquids are miscible if their solubility parameters differ by no more than *ca.* 3 units.

Because the term *solubility parameter* is too restrictive for this quantity which can be used to correlate a wide range of physical and chemical properties (for instance *cf.* Section 5.4.2), the term *cohesion parameter* has been proposed by Barton [99]. The term solubility parameter suggests a close relationship between the phenomenon "solubility" or "miscibility" and that of "cohesion" or "vaporization". This seems to be reasonable, considering what happens in a mixing process: the like molecules of each component in a mixture are separated from one another to an infinite distance, comparable to what happens in the vaporization process. A comprehensive review on determination and application of solubility parameters has recently been given by Barton [99].

Table 3-3. Hildebrand solubility parameters, δ, of thirty organic solvents at 25 °C, taken from reference [99] (Table 2 in Chapter 8).

Solvents	$\delta/\mathrm{MPa}^{1/2}$ [a]	Solvents	$\delta/\mathrm{MPa}^{1/2}$ [a]
Water	47.9	Acetic acid	20.7
Formamide	39.3	1,4-Dioxane	20.5
N-Methylformamide	32.9	Carbon disulfide	20.4
1,2-Ethanediol	29.9	Cyclohexanone	20.3
Methanol	29.6	Acetone	20.2
Tetrahydrothiophene-1,1-dioxide	27.4	1,2-Dichloroethane	20.0
Ethanol	26.0	Chlorobenzene	19.4
N,N-Dimethylformamide	24.8	Chloroform	19.0
Dimethyl sulfoxide	24.5	Benzene	18.8
Acetonitrile	24.3	Ethyl acetate	18.6
1-Butanol	23.3	Tetrahydrofuran	18.6
Cyclohexanol	23.3	Tetrachloromethane	17.6
Pyridine	21.9	Cyclohexane	16.8
t-Butanol	21.7	*n*-Hexane	14.9
Aniline	21.1	Perfluoro-*n*-heptane	11.9

[a] $1\ \mathrm{MPa}^{1/2} = 1\ \mathrm{J}^{1/2} \cdot \mathrm{cm}^{-3/2} = 0.4889\ \mathrm{cal}^{1/2} \cdot \mathrm{cm}^{-3/2}$.

An alternate approach for the prediction of mutual miscibility of solvents has been given by Godfrey [119] (*cf.* also Appendix, Chapter A-1). As a measure of lipophilicity (*i.e.* affinity for oil-like substances) the so-called *miscibility numbers* (*M*-numbers, with values between 1 and 31) have been developed. These are serial numbers of 31 classes of organic solvents, ordered empirically by means of simple test tube miscibility experiments and critical solution temperature measurements. There is a close correlation between *M*-numbers and Hildebrand's δ-values [99].

The solvency of hydrocarbon solvents used in paint and lacquer formulations is empirically described by their *kauri-butanol-numbers*, *i.e.* the volume in milliliters at 25 °C of the solvent required to produce a defined degree of turbidity when added to 20 g of a standard solution of kauri resin in 1-butanol [120]. Standard values are $KB = 105$ for toluene and $KB = 40$ for *n*-heptane/toluene (75 : 25 cl/l). A high *KB*-number corresponds to high solvent power. An approximately linear relationship does exist between Hildebrand's δ-values and *KB*-numbers for hydrocarbons with $\mathrm{KB} > 35$ [99].

Solvents whose molecules possess a permanent *dipole moment* are designated *dipolar* as opposed to *apolar* or *nonpolar* for those lacking a dipole moment. Unfortunately, in the literature the terms "polar" and "apolar" or "nonpolar" are used indiscriminately to characterize a solvent by its dielectric constant as well as its permanent dipole moment, even though dipole moment and dielectric constant are not directly related. Molecules, which possess a centre of symmetry in all possible conformations, more than one *n*-fold axis of symmetry, or a plane of symmetry perpendicular to an *n*-fold axis of symmetry, cannot exhibit a permanent dipole moment for symmetry reasons. Therefore, only those molecules which belong to the point groups C_1, C_s, C_n, or C_{nv} can have a permanent dipole moment. The permanent dipole moments of organic solvents vary from 0 to $18.5 \cdot 10^{-30}$ C m (0 to 5.5 D); *cf.* Appendix, Table A-1. Values of dipole moments increase steadily on going from hydrocarbon solvents to solvents containing dipolar groups such as $C^{\delta+}\cdots O^{\delta-}$, $C^{\delta+}\cdots N^{\delta-}$, $N^{\delta+}\cdots O^{\delta-}$, $S^{\delta+}\cdots O^{\delta-}$, or $P^{\delta+}\cdots O^{\delta-}$. The orientation of dipolar solvent molecules around the solute molecule in the absence of specific solute/solvent interactions is largely determined by the dipole moment.

It should not be forgotten that the solution value of a solute dipole moment (μ_s) differs from its gas-phase value (μ_g) and depends on the nature of the solvent. For theoretical approaches in relating the difference ($\mu_s - \mu_g$) to various physical solvent parameters (*e.g.* ε_r, n_D, *etc.*) see reference [122].

The importance of electric moments of orders higher than two (dipoles) such as quadrupoles and even octupoles in solute/solvent interactions between multipolar molecules has been stressed [121]. Depending on the charge distribution, there exist multipoles (2^n-poles) such as monopoles (n=0; *e.g.* $Na^{\oplus}$, $Cl^{\ominus}$), dipoles (n=1: *e.g.* HF, H_2O), quadrupoles (n=2; *e.g.* CO_2, C_6H_6), octupoles (n=3; *e.g.* CH_4, CCl_4), and hexadecapoles (n=4; *e.g.* SF_6). According to Reisse [121], only neutral species with a spherical charge distribution (*e.g.* rare gases) should be designated as *apolar*. All others, with non-spherical charge distribution, should be called *polar*, *i.e.* dipolar, quadrupolar, octupolar, *etc.*, depending on the first non-zero electric moment. In this respect, methane and tetrachloromethane are polar molecules, as well as *cis*-(dipolar) and *trans*-1,2-dichloroethene (quadrupolar), and 1,3-dioxane (dipolar) and 1,4-dioxane (quadrupolar). It has been shown that the 2^n-polar contributions (n > 1) to solute/solvent interactions are in many cases non-negligible [121].

The *dielectric constants* play a particular role in the characterization of solvents. Their importance over other criteria is due to the simplicity of electrostatic models of solvation and they have become a useful measure of solvent polarity. In this connection it is important to realize, what exactly is represented by the macroscopic dielectric constant of a solvent (also called relative permittivity $\varepsilon_r = \varepsilon/\varepsilon_0$ where ε_0 is a constant, the permittivity of vacuum). Dielectric constants are determined by inserting the solvent between the two charged plates of a condensor. The strength of the electric field E between the plates is lower than the value E_0 measured when the plates are in a vacuum, and the ratio E_0/E gives the numerical value of the dielectric constant. If the solvent molecules do not have permanent dipole moments of their own, then the external field will separate the charge within the molecules thereby inducing dipoles. Molecules with induced or permanent dipoles are forced into an ordered arrangement by the charged plates, causing what is known as polarization. The larger the polarization, the larger the drop in the electric field strength. Therefore, the dielectric constant represents the ability of a solvent to separate charge and to orient its dipoles. The dielectric constants of organic solvents vary from about 2 (*e.g.* hydrocarbons) to about 180 (*e.g.* secondary amides); *cf.* Appendix, Table A-1. Solvents with large dielectric constants may act as dissociating solvents (*cf.* Section 2.6) and are therefore called *polar* solvents in contrast to the *apolar* or *nonpolar* solvents with low dielectric constants. Dielectric constant values often run parallel to the dissolving power of the solvent, because in the case of ionic solutes (*i.e.* ionophores) solvents of high dielectric constant facilitate dissolution by separating the ions.

Since both the dielectric constant ε_r and the dipole moment μ are important complementary solvent properties, it has been recommended that organic solvents should be classified according to their *electrostatic factor EF* (defined as the product of ε_r and μ) which takes into account the influence of both properties [101]. Considering the *EF* values and the structure of solvents a four-part classification of organic solvents has been established: hydrocarbon solvents (*EF* $0 \ldots 7 \cdot 10^{-30}$ C m), electron-donor solvents (*EF* $7 \ldots 70 \cdot 10^{-30}$ C m), hydroxylic solvents (*EF* $50 \ldots 170 \cdot 10^{-30}$ C m), and dipolar non-HBD solvents (*EF* $\geq 170 \cdot 10^{-30}$ C m) [99, 101].

As mentioned before, dielectric constants as well as dipole moments are often used in the quantitative characterization of *solvent polarity*. However, the characterization of a solvent by means of its "polarity" is an unsolved problem since the term "polarity" itself has, until now, not been precisely defined. As polarity one can understand a) the permanent dipole moment of a compound, b) its dielectric constant, or c) the sum of all those molecular properties responsible for all the interaction forces between solvent and solute molecules (*e.g.* Coulombic, directional, inductive, dispersion, hydrogen-bonding, and EPD/EPA interaction forces) [33]. The important thing concerning the so-called polarity of a solvent is its *overall solvation ability*. This in turn depends on the sum of all specific as well as non-specific interactions between solvent and solute. Therefore, in the following the term "solvent polarity" will be applied according to the definition c) above. It should be noted, however, that all interactions which lead to a chemical change of the solute are excluded under this definition (*e.g.* protonation, oxidation, reduction, and complexation).

Evidently, "solvent polarity", as so-defined, is badly described in a quantitative manner by means of individual physical constants such as dielectric constant, dipole moment, *etc.* It is no surprise therefore, that the macroscopic dielectric constants are an unsuitable measure of molecular-microscopic interactions. This has often been demonstrated experimentally. One reason is, that the molecular-microscopic dielectric constant of the solvent in the vicinity of the solute is lower than that for the bulk solvent, because solvent dipoles in a solvation shell are less free to orientate themselves in a direction imposed by charged condensor plates. In the extreme case, complete dielectric saturation can occur for solvent molecules around an ionic solute.

The failure of the solvent dielectric constant to represent solute/solvent interactions has led to the definition of polarity in terms of empirical parameters. Such attempts at obtaining better parameters of solvent polarity by choosing a solvent-dependent standard system and looking after the changes in parameters of that system when the solvent is changed (*e.g.* rate constants of solvent-dependent reactions or spectral shifts of solvatochromic dyes) are treated in Chapter 7.

Based solely on the chemical structure of solvent molecules, a non-empirical solvent polarity index, called the *first-order valence molecular connectivity index*, ${}^1\chi^v/f$, has been proposed by Kier [137]. ${}^1\chi^v$ is calculated from molecular connectivity indices assigned to each atom in a solvent molecule (the latter depending on the number of σ-, π- and n-electrons as well as bonded H-atoms), and f is the number of isolated functional groups present in the solvent molecule. ${}^1\chi^v/f$ is equal to 0.0 and 3.0 for water and cyclohexane, respectively. These purely calculated, non-empirical solvent polarity parameters correlate fairly well with some physical solvent properties [137].

Optically active organic solvents, consisting of chiral molecules*) (chiral solvents), which rotate the plane of linearly polarized light, have become increasingly important lately [35]. In principle, diastereomeric solvates are formed when a mixture of enantiomers

* A molecule is chiral and hence optically active, when it does not possess planes, centres, or alternating axes of symmetry (order $n > 2$) [34].

is dissolved in an optically active solvent. Consequently, these solvates should possess slightly different physical and chemical properties. In fact, optically active solvents have already found use in stereoselective syntheses [36–38, 123], as NMR shift reagents [39, 124], in the evaluation of the optical purity of enantiomers [40], and in the gas-chromatographic separation of enantiomers on chiral phases [41, 124, 125]. A selection of such optically active solvents is given in Table A-2 (Appendix).

Some recent examples of enantioselective syntheses in chiral solvents are: the high-pressure (10^4 bar) Wagner-Meerwein rearrangement of a racemic oxirane derivative in (−)-(2*S*, 3*S*)-diethyltartrate (optical yield 6.7%) [42]; the chlorination of substituted aziridines with *t*-butyl hypochlorite to give *N*-chloroaziridines in the presence of chiral trifluoromethylcarbinols (highest enantiomeric excess, *e.e.* = 28.7%) [126]; the Grignard reaction of 2,2-dimethylpropanal with phenylmagnesium bromide in (−)-menthyl methyl ether (*e.e.* = 19.4% [127]; and the addition of *n*-butyllithium to benzaldehyde in the presence of (+)-(*S*,*S*)-1,4-bis(dimethylamino)-2,3-dimethoxybutane (DDB) (optical yield 30%) [128], according to the following equation:

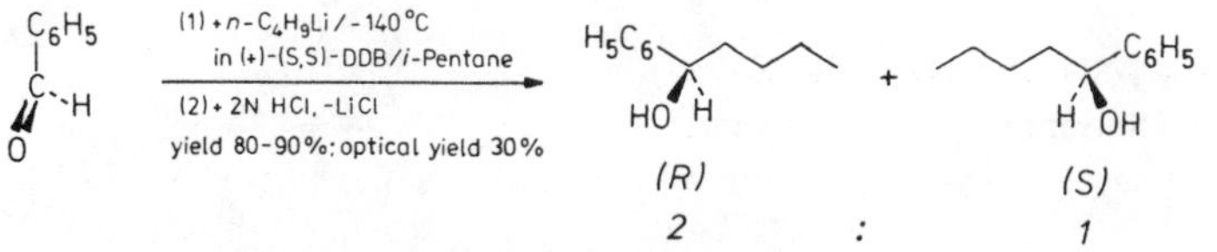

C_6H_5–CH=O (1) + n-C_4H_9Li / −140 °C in (+)-(S,S)-DDB/*i*-Pentane; (2) + 2N HCl, −LiCl; yield 80–90%; optical yield 30% → H_5C_6–CH(OH)–C_4H_9 (*R*) **2** : C_4H_9–CH(OH)–C_6H_5 (*S*) **1**

Aldol additions between *achiral* reactants in *chiral* solvents have also been examined [157–159]. Only low asymmetrical inductions (*e.e.* = 2...22%) have been found [158]. A twofold stereodifferentiation is observed in aldol additions between *chiral* reactants carried out in *chiral* solvents [159]. However, asymmetric inductions caused by chiral solvents or cosolvents are usually rather small [157].

3.3 Classification of Solvents in terms of Acid-Base Behaviour

3.3.1 Brønsted-Lowry Theory of Acids and Bases [43–50, 107, 168]

According to the Brønsted-Lowry definition, acids and bases are proton donators and -acceptors, respectively, as expressed in the following equilibrium

$$\underset{\text{Acid}}{HA^{z+1}} \rightleftharpoons \underset{\text{Conjugate Base}}{A^{z}} + H^{+} \tag{3-7}$$

where $z = 0, \pm 1, \ldots$ [51, 52]. Since, in solution, the isolated proton cannot exist [129], an acid-base reaction will take place only in the presence of a base possessing a higher proton affinity than the conjugate base A^z. As most solvents possess acid or base properties, the strengths of acids and bases depend on the medium in which they are dissolved.

The equilibrium shown in Eq. (3-8) will be established when an acid HA is dissolved in a basic solvent SH.

$$\underset{\text{Acid}}{HA^{z+1}} + \underset{\text{Solvent}}{SH} \rightleftharpoons \underset{\text{Lyonium Ion}}{SH_2^+} + A^z \qquad (3\text{-}8)$$

The strength of the acid HA in the solvent SH is given by the acidity contant K_a according to Eq. (3-9)*) [53].

$$K_a = \frac{[SH_2^+]\cdot[A^z]}{[HA^{z+1}]} \qquad (3\text{-}9)$$

In an acidic solvent SH, the acid-base equilibrium shown in Eq. (3-10) will be established.

$$\underset{\text{Base}}{A^z} + \underset{\text{Solvent}}{SH} \rightleftharpoons HA^{z+1} + \underset{\text{Lyate Ion}}{S^-} \qquad (3\text{-}10)$$

Increasing basicity or acidity of the solvent displaces the equilibria (3-8) and (3-10) to the right. The addition of these two equations gives a new equilibrium describing the self-ionization (autoprotolysis) of the solvent.

$$\underset{\text{Solvent}}{2\,SH} \rightleftharpoons \underset{\text{Lyonium Ion}}{SH_2^+} + \underset{\text{Lyate Ion}}{S^-} \qquad (3\text{-}11)$$

Eq. (3-11) reflects both the acidic and basic properties of a solvent, which are described quantitatively by the *autoprotolysis constant* K_{auto}.

$$K_{auto} = [SH_2^+]\cdot[S^-] \qquad (3\text{-}12)^{**)}$$

Autoprotolysis constants for some representative solvents are collected in Table A-12 (Appendix). The useful pH-range of a solvent increases as the autoprotolysis constant decreases. The smaller the autoprotolysis constant, the greater the range of acid or base strengths which can exist in a solvent***).

---

* Since in general the solvent SH is in large excess its concentration [SH] remains almost constant and is therefore included in the constant K_a.

** Since the solvent concentration [SH] remains practically constant due to the large excess of unionized solvent, this term is included in K_{auto}.

*** Thus, the magnitude of K_{auto} is an important criterion for the proper choice of solvent for titrations in non-aqueous solvents; *cf.* Section A-8 (Appendix).

Self-ionizing solvents possessing both acid and base characteristics (*e.g.* water) are designeted *amphiprotic solvents* in contrast to *aprotic solvents* which do not self-ionize to a measurable extent (*e.g.* aliphatic hydrocarbons, tetrachloromethane) [47, 53–55].

It is not possible to draw a sharp line between amphiprotic and aprotic solvents since, in practice, amphiprotic solvents with extremely small K_{auto}-values behave like aprotic solvents. It has been suggested that solvents with K_{auto}-values greater than 20 should be called aprotic rather than amphiprotic [44].

This classification of solvents was first proposed by Brønsted, who distinguished between four types of solvents on the basis of their acid and base properties [54]. Davies extended the Brønsted classification and distinguished between solvents with dielectric constants greater or smaller than 20, thus arriving at eight classes of solvents [47]. Kolthoff's classification in a slightly simplified fashion is given in Table 3-4 [56].

Table 3-4. Classification of organic solvents according to their Brønsted acid-base behaviour [56].

Solvent designation		Relative acidity[a]	Relative basicity[a]	Examples
Amphiprotic	Neutral	+	+	H_2O, CH_3OH, $(CH_3)_3COH$, $HOCH_2CH_2OH$, C_6H_5OH
	Protogenic	+	–	H_2SO_4, $HCOOH$, CH_3COOH
	Protophilic	–	+	NH_3, $HCONH_2$, $CH_3CONHCH_3$, $H_2N—CH_2CH_2—NH_2$
Aprotic	Dipolar Protophilic	–	+	$HCON(CH_3)_2$, CH_3SOCH_3, pyridine, 1,4-dioxane $(C_2H_5)_2O$, tetrahydrofuran
	Dipolar Protophobic	–	–	CH_3CN, CH_3COCH_3, CH_3NO_2, $C_6H_5NO_2$, sulfolane
	Inert	–	–	Aliphatic hydrocarbons, C_6H_6, $Cl—CH_2CH_2—Cl$, CCl_4

[a] – indicates weaker and + indicates stronger acid or base than water.

Water is the prototype of an amphiprotic solvent and all other solvents with similar acid-base properties are called *neutral solvents*. Solvents which are much stronger acids and much weaker bases than water are called *protogenic solvents*, while those which are much stronger bases and much weaker acids than water are designated *protophilic solvents*. This division is somewhat arbitrary since by agreement water is the reference which is defined as neutral.

From Eq. (3-8) it is seen that the ionization of an acid depends on the basicity of the solvent. In other words, the effective strength of an acid is greater, the higher the proton affinity of the medium. However, the ionization of the acid depends not only on the basicity of the solvent, but also on its dielectric constant and its ion-solvating ability. The dependence of the acidity and basicity constants of a compound on the basicity and acidity, respectively, of the solvents, leads to a distinction between *levelling* and *differentiating solvents* [49, 57, 58].

All mineral acids ionize to the same extent in aqueous solution: they are essentially completely ionized due to almost quantitative reaction with the base water. It would be a strange coincidence, however, if all these acids had exactly the same acid strength in spite of their different constitution. The explanation is that water exerts a levelling effect on the acid strengths. Fig. 3-2 shows that in water the strengths of all acids stronger than H_3O^+ are adjusted to that of H_3O^+ itself. Solvents exhibiting such behaviour are called *levelling solvents**). In order to establish the relative strengths of the mineral acids, it is necessary to perform the measurements in solvents of very low basicity and ionizing ability. Logically enough, such solvents are designated *differentiating solvents*. Fig. 3-2 makes it clear, that the strongest acid that can exist in a solvent is the lyonium ion $SH_2^{\oplus}$ (in case of (a) $H_3O^{\oplus}$), while the lyate ion, $S^{\ominus}$, is the strongest base that can exist in a solvent (in case of (b) $HO^{\ominus}$).

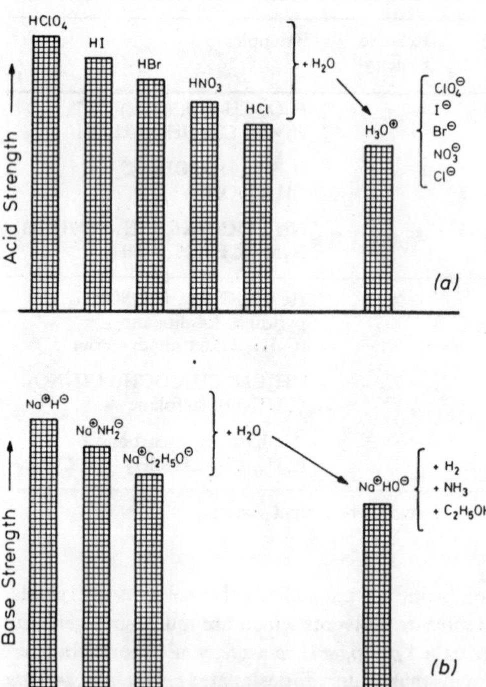

Fig. 3.2. Schematic description of the levelling effect of water on (a) acids and (b) bases in aqueous solution [2]. Relative orders of acidity and basicity are not invariable to changes of solvent and of the conjugated acid or base, respectively.

---

* The expression "levelling solvents" ("chemisch nivellierende Lösungsmittel") was introduced by A. Hantzsch, Z. Elektrochem. *29*, 221 (1923); see p. 234.

Thus, in methanol, hydrochloric acid is completely ionized, whereas nitric acid is only partly so. In the less basic formic acid, hydrochloric acid is also only partially ionized, whereas the first proton of sulfuric acid is still completely ionized. Acetonitrile is a very weak base and an exceptionally weak acid, thus, only slight levelling of acids and bases occurs in this solvent, making it a good differentiating solvent. Perchloric acid appears to be strong in acetonitrile, whereas other acids are differentiated [59]: $HBr > H_2SO_4 > HNO_3 > HCl$ and picric acid. When the solvent is a stronger base than water, its levelling effect will apply also to weaker acids. Thus, in liquid ammonia even the carboxylic acids are practically fully ionized.

Similarly, strong bases will also have equal basicities in sufficiently acidic solvents. As shown in Fig. (3-2b), all bases stronger than the $HO^{\ominus}$ ion are adjusted to the basicity of this ion in water. Consequently, the effective basicities of guanidines and carbanions cannot be measured in water, but only in less acidic differentiating solvents such as liquid ammonia or diethyl ether.

Evidently, in a given solvent the more highly ionized acid or base is also the strongest one. However, if the same acid is examined in different solvents, one finds, surprisingly enough, that the most acidic solution is the one in which the acid is the least ionized. For example, a solution of hydrochloric acid in benzene is a stronger acid than in aqueous solution; in the former, ionization is slight, in the latter, complete. In the aqueous solution, an indicator would have to compete with the base H_2O for the protons; however, in the benzene solution, competition involves only the much weaker base $Cl^{\ominus}$. Therefore, a larger proportion of the indicator would be transformed into its conjugate acid in benzene than in water, thus making the benzene solution of HCl the better proton donator.

Numerous acidity and basicity scales have been elaborated for water and other solvents. However, there is no one single scale of acidity and basicity, equally valid and useful for all types of solvents and applicable to both equilibrium and kinetic situations. Excellent reviews on different acidity functions are given by Boyd [60] and Bates [50].

In dilute aqueous solution the acidity is measured using pH-values. For concentrated acid solutions and non-aqueous acid solutions pH-values are no longer available. Hence, the *Hammett acidity function* H_0 is used as a measure of the acidity of such media [130]. The proton donor ability of an acid in such media is measured by studying the equilibria (*via* UV/Vis absorption spectra) of coloured indicators (a series of nitroanilines) which differ by one proton.

Brønsted acids stronger than pure (100%) sulfuric acid ($H_0 = -11.9$) are classified as *super acids* [131, 132]. Thus perchloric acid ($HClO_4$), fluorosulfonic acid ($F—SO_3H$), and trifluoromethanesulfonic acid ($CF_3—SO_3H$) are considered super acids. Even exceedingly weak basic solvents (*e.g.* carbonyl compounds; aromatic, olefinic, and saturated hydrocarbons) are protonated by these super acids to give the corresponding carbocations [131].

Streitwieser *et al.* [160] and Bordwell *et al.* [161] used the lyate ions of organic solvents such as cyclohexylamine and dimethyl sulfoxide in the determination of the C—H acidity of weak organic carbon acids. Using *super base* systems such as alkali-metal salts of cyclohexylamine (*i.e.* lithium and cesium cyclohexylamide) [160] and dimethyl sulfoxide (sodium dimsyl) [161] in an excess of these non-HBD solvents, relative acidity scales for weak carbon acids have been established. In this way, pK_a^{DMSO}-values for the ionization of over a thousend Brønsted acids in dimethyl sulfoxide have become available, covering a range of *ca.* 35 pK_a units [161]. The upper limits of measurements in cyclohexylamine and dimethyl sulfoxide are imposed by the acidities of the two solvents, corresponding to maximum determinable pK_a^{CHA} and pK_a^{DMSO}-values of *ca.* 39 and 32, respectively.

The acidity and basicity of solvents can be measured in different ways [49]. Besides the usual experimental methods of measuring acid-base equilibrium constants, another possible approach is the determination of solvent basicities and acidities by measuring the change in some physical property (like an IR or UV/Vis absorption or NMR chemical shift) of the molecules of a standard substrate when transferred from a reference solvent to another solvent. For example, the shift in wavenumber of the ≡C—H valence vibration band of phenylacetylene when transferred from tetrachloromethane to nineteen other solvents were measured, giving a relative order of basicity ranging from tetrachloromethane (low) to hexamethylphosphoric triamide (high) [61]. The basicity of solvents was also measured using the ^{1}H-NMR chemical shift of the chloroform proton $\Delta\delta_\infty(CHCl_3)$ obtained by extrapolation to infinite $CHCl_3$ dilution in the solvent in question and in an inert reference solvent (cyclohexane), respectively. The results, summarized in Table 3-5, establish an order of solvent basicity using chloroform as the standard substrate [62].

Table 3-5. Solvent order of increasing basicity relative to chloroform, measured using the relative ^{1}H-NMR chemical shift and extrapolated to infinite dilution with cyclohexane as reference solvent [62].

Solvents	$\Delta\delta_\infty(CHCl_3)$ [a] ppm
Cyclohexane	0.00
Tetrachloromethane	0.18
Chloroform	0.20
Nitromethane	0.47
Acetonitrile	0.56
1,4-Dioxane	0.63
Diethyl ether	0.74
Tetrahydrofuran	0.79
Acetone	0.92
Cyclohexanone	0.97
Triethylamine	1.22
N,N-Dimethylformamide	1.30
Dimethyl sulfoxide	1.32
Pyridine	1.56
Hexamethylphosphoric acid triamide	2.06

[a] $\Delta\delta_\infty(CHCl_3) = \delta_\infty(CHCl_3, \text{solvent}) - \delta_\infty(CHCl_3, \text{cyclohexane})$

Recent gas-phase studies of proton-transfer reactions with step-wise solvation of the reactants (*i.e.* incrementally addition of solvent molecules to form supermolecular clusters) have demonstrated that the acid/base behaviour of isolated solvent molecules can be dramatically different from their performance as bulk liquids. Water, the classical amphiprotic solvent, shall serve as an example.

In contrast to the acid/base behaviour of "polymeric" bulk water, "monomeric" water is a relatively weak acid and base in the gas-phase compared to its substituted derivatives (R—OH, R—O—R, *etc.*) whose conjugated base or acid ions are stabilized by polarization of the alkyl groups. The gas-phase basicity of water is 138 kJ/mol (33 kcal/mol) below that of ammonia. Its gas-phase acidity is comparable to that of propene and it is less acidic than phenol by about 167 kJ/mol (40 kcal/mol). With respect to the well-known acid/base properties of water, ammonia, and phenol in aqueous solution, one has to conclude that enormous solvation energies must contribute to the difference from the behaviour of isolated water molecules. *Cf.* Chapter 4.2.2 for further discussions and references.

3.3.2 Lewis Theory of Acids and Bases [43-65, 65a]

According to Lewis, acids are electron pair acceptors (EPA) and bases electron pair donors (EPD) connected through the following equilibrium [63, 65a]:

A	+	:D	⇌	A—D	(3-13)
Acid		Base		Acid-Base	
EPA		EPD		Complex	
Electrophile		Nucleophile			
Hard		Hard			
Soft		Soft			

[—— favorable combination (Hard–Hard, Soft–Soft); ------ unfavorable combination (Hard–Soft, Soft–Hard)]

The Lewis acid/base complex is formed *via* an overlap between a doubly occupied orbital of the donor D and a vacant orbital of the acceptor A (*cf.* also Section 2.2.6). This acid/base approach was extended by Pearson who divided Lewis acids and bases into two groups, hard and soft, according to their electronegativity and polarizability (principle of hard and soft acids and bases; HSAB concept) [66, 67]. Hard acids (*e.g.* $H^{\oplus}$, $Li^{\oplus}$, $Na^{\oplus}$, BF_3, $AlCl_3$, hydrogen-bond donors HX) and hard bases (*e.g.* $F^{\ominus}$, $Cl^{\ominus}$, $HO^{\ominus}$, $RO^{\ominus}$, H_2O, ROH, R_2O, NH_3) are those derived from small atoms with high electronegativity and generally of low polarizability. Soft acids (*e.g.* $Ag^{\oplus}$, $Hg^{\oplus}$, I_2, 1,3,5-trinitrobenzene, tetracyanoethene) and soft bases (*e.g.* $H^{\ominus}$, $I^{\ominus}$, $R^{\ominus}$, $RS^{\ominus}$, RSH, R_2S, alkenes, C_6H_6) are usually derived from large atoms with low electronegativity and are usually polarizable. The usefulness of this division arises from a simple rule concerning the stability of Lewis acid/base complexes: *hard acids prefer to coordinate to hard bases and soft acids to soft bases* [66, 67]. This HSAB concept describes a wide range of chemical phenomena in a qualitative way and has found many applications in organic chemistry [66–70] (for a criticism of the HSAB concept see [71, 72]). Quite recently, a review of advances in the HSAB concept and its theoretical justification has been given by Pearson [170]. Based on ionization potentials and electron affinities, even numerical values of absolute hardnesss, *i.e.* resistance to deformation or change in the electronic charge cloud, can now be assigned to various Lewis acids and bases.

Solvents can be classified as EPD or EPA according to their chemical constitution and reaction partners [65]. However, not all solvents come under this classification since *e.g.* aliphatic hydrocarbons possess neither EPD nor EPA properties. An EPD solvent solvates preferably electron-pair acceptor molecules or ions. The reverse is true for EPA solvents. In this respect, most solute/solvent interactions can be classified as generalized Lewis-acid/base reactions. A dipolar solvent molecule will always have an electron rich or basic site, and an electron poor or acidic site. Gutmann introduced so-called donor numbers, *DN*, and acceptor numbers, *AN*, as quantitative measures of the donor and acceptor strengths [65]; *cf.* Section 2.2.6 and Tables 2-3 and 2-4. Due to their coordinating ability, electron-pair donor and acceptor solvents are, in general, good ionizers; *cf.* Section 2.6.

The application of the HSAB concept to solutions leads to the rule, that hard solutes dissolve in hard solvents and soft solutes dissolve in soft solvents [66]. This rule can be considered as a modern version of "similia similibus solvuntur". For example, benzene is considered a very soft solvent since it contains only a basic function. Contrary to benzene, water is a very hard solvent, with respect to both its basic and acidic properties. It is the ideal solvent for hard bases and hard acids. The hardness of water is reduced by the introduction of alkyl substituents in proportion to the size of the alkyl group. In alcohols, therefore, softer solutes become soluble. Whereas in methanol oxalate salts are quite insoluble, the corresponding softer bisthiooxalate salts are quite soluble.

Since hydrogen-bonding is a hard acid-hard base interaction, small basic anions prefer specific solvation by protic solvents. Hence the reactivity of $F^{\ominus}$, $HO^{\ominus}$, or $CH_3O^{\ominus}$ is reduced most in going from a dipolar non-HBD solvent such as dimethylsulfoxide to a protic solvent like methanol. Dipolar non-HBD solvents are considered as fairly soft compared to water and alcohols [66].

Similar considerations can be made in the case of cation solvation. Small hard cations of high oxidation state will be preferably solvated by hard EPD solvents like H_2O or ROH. In principle, relative to the gas state, all ions become softer in the solute state as the result of solvation [66].

A quantitative scale describing the softness of solvents has been proposed quite recently by Marcus [171]. His μ-scale of solvent softness (from *malakos* = soft in Greek) is defined as the difference between the mean of the standard molar Gibbs energies of transfer of sodium and potassium ions from water (W) to a given solvent (S), $\Delta G_t^\circ(\mathrm{Me}^{\oplus}, \mathrm{W} \rightarrow \mathrm{S})/(\mathrm{kJ \cdot mol^{-1}})$, and the corresponding transfer energy for silver ions, divided by 100. Since water is a hard solvent, the Gibbs energy of transfer of ions from water as a reference solvent to other solvents should depend on the softness of these solvents to a different extent for hard and soft ions. Provided that the charge and size of the ions are equal, hard ions should prefer water and soft ions the softer solvents. This definition of μ was selected because the size of the soft $Ag^{\oplus}$ ion is intermediate between those of hard $Na^{\oplus}$ and $K^{\oplus}$. The degree of softness among solvents with oxygen, nitrogen, and sulfur donor atoms increases in the series *O*-donor (alcohols, ketones, amides) < *N*-donor (nitriles, pyridines, amines) < *S*-donor solvents (thioethers, thioamides) [171].

The concept of the superacidity of Brønsted acids has been extended to Lewis acids [131]. It is suggested that those Lewis acids stronger than anhydrous aluminium trichloride (the most commonly used Friedel-Crafts catalyst) should be designated as *super acids*. These superacidic Lewis acids include such higher-valence halides as antimony, arsenic, tantalum, and niobium pentafluorides. Frequently used conjugated Brønsted-Lewis super acids are $F—SO_3H/SbF_5$ (magic acid) and HF/SbF_5 (fluoroantimonic acid) [131]. These superacidic systems are considered to be *ca.* 10^{16} times stronger than 100 percent sulfuric acid [131].

3.4 Classification of Solvents in terms of Specific Solute/Solvent Interactions

Parker divided solvents into two groups according to their specific interactions with anions and cations, namely *dipolar aprotic solvents* and *protic solvents* [73]. The distinction lies principally in the dipolarity of the solvent molecules and their ability to form hydrogen bonds. The origin of this solvent classification was the experimental finding that certain S_N2 reactions at saturated carbon atoms involving anions as nucleophiles, are much faster in the so-called dipolar aprotic solvents than in protic solvents. This is because in dipolar aprotic solvents most anions are much less solvated and hence more reactive than in protic solvents [74]. It appears appropriate to add to these two groups a third one, namely, the *apolar aprotic solvents*, according to Fig. 3-3.

An apolar aprotic solvent is characterized by a low dielectric constant ($\varepsilon_r < 15$), a low dipole moment ($\mu < 8.3 \cdot 10^{-30}$ C m = 2.5 D), a low E_T^N-value (E_T^N *ca.* 0.0...0.3); *cf.* Table A-1, Appendix), and the inability to act as hydrogen-bond donor. Such solvents interact only slightly with the solute since only the non-specific directional, induction, and dispersion forces can operate. To this group belong aliphatic and aromatic hydrocarbons, their halogen derivatives, tertiary amines, and carbon disulfide.

In contrast, dipolar aprotic solvents*) possess large dielectric constants ($\varepsilon_r > 15$), sizeable dipole moments ($\mu > 8.3 \cdot 10^{-30}$ C m = 2.5 D), and average E_T^N-values of 0.3 to 0.5. These solvents do not act as hydrogen-bond donors since their C—H bonds are not strongly enough polarized. However, they are usually good EPD solvents and hence cation solvators due to the presence of lone electron pairs. Among the most important dipolar aprotic solvents are acetone, acetonitrile [75], benzonitrile, *N,N*-dimethylacetamide [76, 77], *N,N*-dimethylformamide [76–78], dimethylsulfone [79], dimethylsulfoxide [80–84], hexamethylphosphoric triamide [85], 1-methyl-2-pyrrolidinone [86], nitrobenzene, nitromethane [87], cyclic carbonates such as propylene carbonate (4-methyl-1,3-dioxol-2-one) [88], sulfolane (tetrahydrothiophen-1,1-dioxide) [89, 90, 90a], 1,1,3,3-tetramethylurea [91, 91a]· and tetrasubstituted cyclic ureas such as 3,4,5,6-tetrahydro-1,3-dimethyl-pyrimidin-2-(1*H*)-one (dimethyl propylene urea, DMPU) [133]. The latter is a suitable substitute for the carcinogenic hexamethylphosphoric triamide (*cf.* Table A-13) [134].

* Although widely used, the term *dipolar aprotic* solvent is in fact rather misleading. Solvents referred to as dipolar *aprotic* are in fact *not* aprotic. In reactions where strong bases are employed their protic character can be recognized. In dimethyl sulfoxide solution, the pK_a-values are, for CH_3NO_2 17.2, CH_3CN 31.3, CH_3SOCH_3 35, 1-methyl-2-pyrrolidone *ca.* 35, HMPT *ca.* 45 or above [135]. It has therefore been recommended by Bordwell *et al.* [135] that the designation *dipolar aprotic* for these solvents be replaced by *dipolar nonhydroxylic* or better still by *dipolar non-HBD* solvents. The abbreviations *HBD* (hydrogen-bond donor) and *HBA* (hydrogen-bond acceptor) refer to donation and acceptance of the proton, and not to the electron pair involved in hydrogen bonding (*cf.* Section 2.2.5).

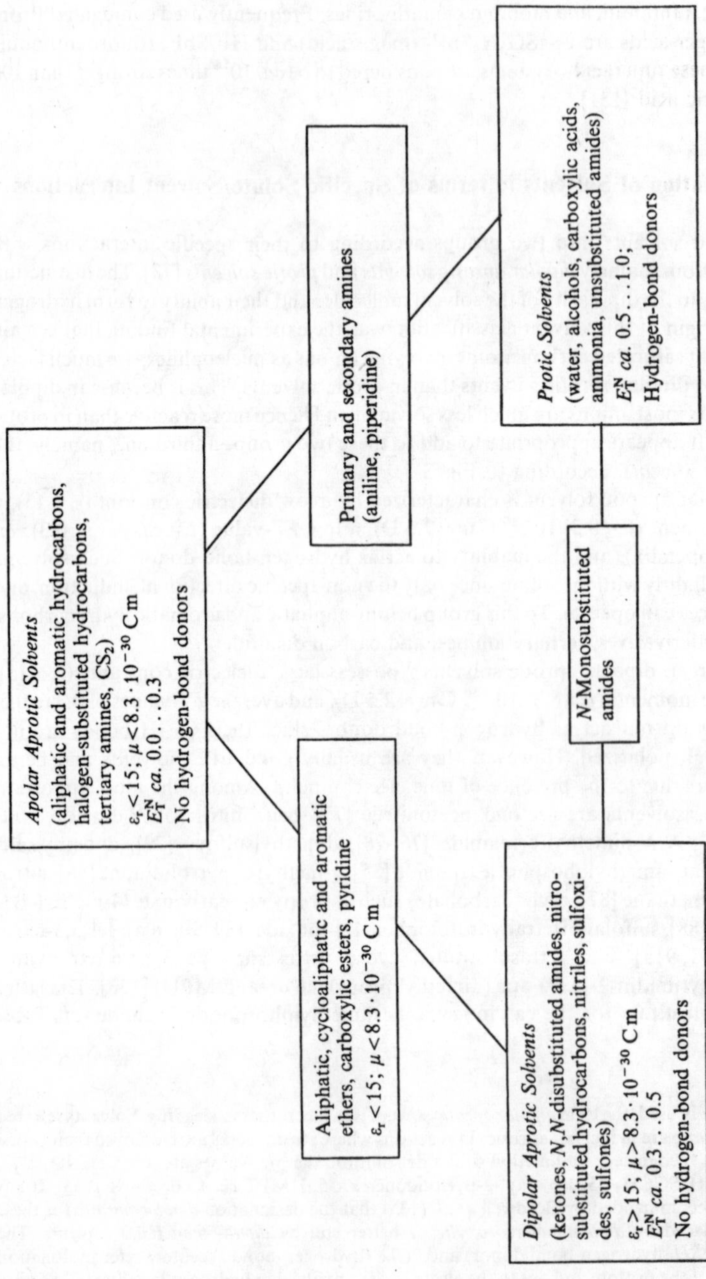

Fig. 3-3. An extension of Parker's classification of organic solvents [73] (concerning the E_T^N-values *cf.* Section 7.4).

Protic solvents contain hydrogen atoms bound to electronegative elements (F—H, —O—H, —N—H, *etc.*) and are, therefore, hydrogen-bond donors *i.e.* HBD solvents (*cf.* Section 2.2.5). With exception of acetic acid (and its homologues), the dielectric constants are usually larger than 15, and the E_T^N-values lie between 0.5 and 1.0, indicating these solvents as strongly polar. To this class of solvents belong water, ammonia, alcohols, carboxylic acids, and primary amides.

It is emphasized that this classification is not rigid. There are several solvents which cannot be unequivocally assigned to any one of these three groups, *e.g.* ethers, carboxylic esters, primary and secondary amines, and *N*-monosubstituted amides such as *N*-methylacetamide [91 b]. The choice of $\varepsilon_r = 15$ as the borderline is arbitrary but practical, since, in solvents with smaller ε_r-values, ion association occurs, so that the free solvated ions can no longer be observed (*cf.* Section 2.6). This division of solvents into three classes has mainly heuristic value. Its usefulness is due to the fact that special prominence is given to the dipolar aprotic solvents with their extraordinary specific ion solvation [73, 92–96].

Protic solvents are particularly good anion solvators due to their hydrogen-bonding ability [136]. This tendency is the more pronounced, the higher the charge density (*i.e.* ratio of charge to volume) of the anion to be solvated, and hence its hardness according to the HSAB-principle. It should be noted that, the stronger the solvation, the more the nucleophilic reactivity of the anion will be decreased. Therefore, in protic solvents, the strongest nucleophiles will be the ones with lower or more diffused charge density, *i.e.*, "soft anions" (*cf.* Section 5.4.1).

In contrast, in dipolar aprotic solvents, anion solvation is mainly *via* ion-dipole and ion-induced dipole forces. The latter are important for large polarizable soft anions, with low charge density, in soft dipolar aprotic solvents. Therefore, whereas these solvents are on the average bad anion solvators, they are usually the better, the larger and softer the anion. This has the consequence, that the reactivity of anions is exceptionally high in dipolar aprotic solvents, and the rate constants of S_N2 reactions can increase by several powers of ten when the solvent is changed from protic to dipolar aprotic (*cf.* Section 5.4.2).

The observation that protic solvents are far better anion solvators than dipolar aprotic solvents, and that the reverse is true for cation solvation, has led to extremely valuable rules for the selection of solvents for specific reactions [73, 92–97].

3.5 Classification of Solvents using Multivariate Statistical Methods

Chemical experience suggests that more than three solvent classes, as recognized by Parker [73] (*cf.* Section 3.4 and Fig. 3-3), may be necessary to classify solute/solvent interactions for a wide range of organic solvents. Multivariate statistical methods have therefore been used recently in the classification and selection of organic solvents [102, 138–143]. Compilations of their physicochemical constants (*e.g.* boiling points, molar volumes, heats of vaporization, dipole moments, dielectric constants, molar refractions, *etc.*) and sometimes additionally empirical parameters of solvent polarity (*cf.* Chapter 7) are used as basic data sets. The extraction of chemical information contained in such a data set, *i.e.* the detection of the relative importance of individual variables in determining the data structure, can be done by two statistical methods: multiple linear regression analysis

(MRA) [144] and factor (FA) or principal component analysis (PCA) [145]. This kind of analysis is part of the relatively new research field of *chemometrics**) [146, 147].

In MRA a dependent variable Y is described in terms of a series of explanatory variables $X_1 \ldots X_n$ as given in Eq. (3-14).

$$Y = a_1 \cdot X_1 + a_2 \cdot X_2 + \ldots + a_n \cdot X_n + b \tag{3-14}$$

It is assumed that all the explanatory variables are independent of each other and truly additive as well as relevant to the problem under study [144]. MRA has been widely used to establish linear Gibbs energy (LGE) relationships [144, 149, 150]. The Hammett equation is an example of the simplest form of MRA, namely bivariate statistical analysis. For applications of MRA to solvent effects on chemical reactions *cf.* Chapter 7.7.

The other statistical method used for seeking regularities in physicochemical data, FA, was first developed and used in psychometrics [145]. FA may be described as a mathematical method for seeking the simplest existing linear structure within a given set of multidimensional data. Starting with a matrix of such experimental data (descriptors), it is possible to extract, using sophisticated statistical methods, the minimum number of underlying, non-measurable variables (factors or principal components) which are necessary to describe this whole data set in multiple regression equations. After the number of factors (or components) have been found and their magnitude have been calculated for particular solvents, often a physical or chemical meaning emerges. Although these factors are pure mathematical constructs and do not necessarily embody a direct physical significance, the advantage of FA is that an otherwise hidden physical or chemical interpretation will emerge. The two methods, FA and PCA, are coincident when PCA is used after normalizing the data set. The mathematics of fitting FA and PCA to a matrix of chemical data is well described in the literature [145, 151], and the capabilities of PCA in different fields of pattern recognition have been reviewed [152].

A clear geometrical description of solvent classification using FA or PCA can be given as follows: Common descriptors used for the classification of solvents are their physicochemical constants and empirical parameters of solvent polarity (*cf.* Chapter 7). Each descriptor defines a coordinate axis in a coordinate system. If m descriptors are used as the basic data set, they will therefore define an m-dimensional space in which each solvent can be described by a point (with coordinates equal to the m descriptors). The whole set of different solvents will then define a swarm of points in the m-dimensional descriptor space. If only three descriptors are involved, say boiling point, dipole moment, and dielectric constant, a simple right-angled three-dimensional coordinate system, as given in Fig. 3-4, results with t_{bp}, μ, and ε_r plotted along the x, y, and z axes, respectively.

FA/PCA now constitutes a projection of this swarm of points down to a space of lower dimensions in such a way that the first component vector (factor F_1) describes the

---

* *Chemometrics* has been defined as the application of mathematical and statistical methods to chemical measurements, in particular in providing maximum chemical information through the analysis of chemical data. Because of the enormous increase in generating analytical data, analytical chemists were among the first to use chemometrical methods extensively [148]. The first paper mentioning the name *chemometrics* was from Wold and was published in 1972: S. Wold, Kem. Tidskr. *84*, 34 (1972).

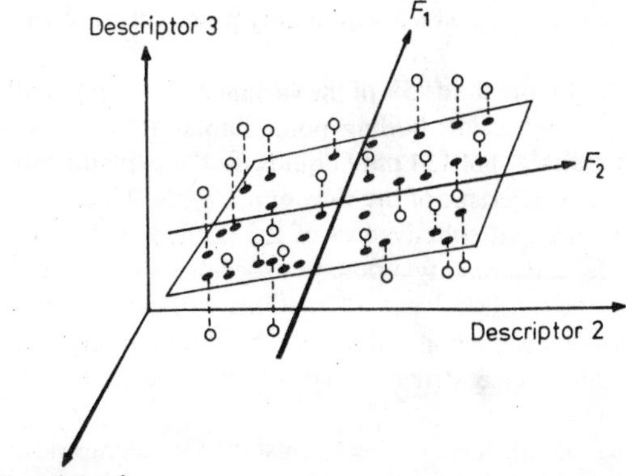

Fig. 3-4. Geometrical representation of FA. The three-dimensional property space is defined by three solvent descriptors (*e.g.* t_{bp}, μ, and ε_r) and filled with 30 solvent points, some of them already lying in the plane defined by the two factors F_1 and F_2 (according to [139] and [142]).

direction through the swarm showing the largest variation in the data. The second component (factor F_2) shows the next largest variation, *etc.* The supposition that the components (factors) should be independent of each other means, that their vectors must be at right angle to each other, *i.e.* mutually orthogonal. To the extent that the solvent points in Fig. 3-4 fall into the plane defined by F_1 and F_2 the position of an individual solvent now needs only two coordinates, instead of the original three, for its localization. Thus, the intrinsic dimensionality of the three sets of solvent property data is reduced to two.

The coordinates of each solvent point are (i) the factor (or principal component) scores F, and (ii) the factor (or principal component) loadings L. They give the information necessary to reconstitute the original physical properties D of any solvent according to Eq. (3-15).

$$D = F_1 \cdot L_1 + F_2 \cdot L_2 + \ldots + F_n \cdot L_n \tag{3-15}$$

Eigenvectors and eigenvalues are the products of calculation at the beginning. They characterise the property of the square matrix (correlation or covariance) derived from the initial data matrix, and they allow to calculate the factor scores F and factor loadings L, respectively.

The advantage of this empirical model is that the systematic variation in the solvent data can now be described using fewer variables than in the original data set. Eventually, an attempt is made to explain the factors F_1 and F_2, which themselves define a new coordinate system, by considering an underlying physical or chemical meaning (*e.g.* polarity, polarizability, or Lewis acidity/basicity of the solvent molecule), thus leading finally to a new solvent classification.

Martin *et al.* [102] were the first to apply FA to solvent classification. A factor analysis for 18 organic solvents with 18 physicochemical parameters led to a solvent classification similar to Parker's classification [73], which was mainly based on chemical intuition (*cf.* Fig. 3-3).

Using PCA, Cramer [139] found that more than 95% of the variances in six physical properties (activity coefficient, partition coefficient, boiling point, molar refractivity, molar volume, and molar vaporization enthalpy) of 114 pure liquids can be explained in terms of only two parameters which are characteristic of the solvent molecule. These two factors are correlated to the molecular bulk and cohesiveness of the individual solvent molecules, the interaction of which depends mainly upon nonspecific, weak intermolecular forces. This is closely related to solute/solvent interactions without specific, strong interactions. With these factors, experimental values of 18 common physical properties for 139 additional liquids of diverse structure have been predicted with surprising accuracy [139].

FA of data matrices containing 35 physicochemical constants and empirical parameters of solvent polarity (*cf.* Chapter 7) for 85 solvents have been carried out by Svoboda *et al.* [140]. An orthogonal set of four parameters was extracted from these data, which can be correlated to solvent polarity as expressed by the Kirkwood function $(\varepsilon_r - 1)/(2\varepsilon_r + 1)$, to solvent polarizability as expressed by the refractive index function $(n^2 - 1)/(n^2 + 1)$, as well as to the solvent Lewis acidity and basicity. That means, four solvent parameters are generally needed for the quantitative empirical description of solvent effects on chemical reactions and light absorptions: two are needed to describe the nonspecific solvation of polar and dispersion character, and two to describe specific solvation of electrophilic and nucleophilic character. For correlations of solvent effects using only one empirical solvent parameter, the best parameters are the $E_T(30)$-values, which are derived from the UV/Vis absorption of a solvatochromic dye (*cf.* Sections 6.2.1 and 7.4).

Elguero *et al.* [141], have reduced Palm's analogous tetraparametric model for the multiple correlation of solvent effects [*cf.* Eq. (7-48) in Section 7.7], to a triparametric one with two factors explaining 94% of the data variance given in an original set of four descriptors [Y, P, E, and B of Eq. (7-48) in Section 7.7] for 51 solvents.

A data matrix of eight common descriptors of solvent properties for 82 solvents were analysed with PCA by Carlson *et al.* [142]. The eight descriptors are melting point, boiling point, density, dielectric constant, dipole moment, refractive index, $E_T(30)$ [*cf.* Eq. (7-27) in Section 7.4], and lg P (*i.e.* the logarithm of the equilibrium partition coefficient of a solvent between 1-octanol and water at 25 °C [153]). Using a two components model of the whole data set [one component explained principally by ε_r, μ, and $E_T(30)$, the other strongly correlated to the refractive index], different strategies for a systematic selection of solvents for chemical reactions are proposed. This has been applied to the Willgerodt-Kindler reaction between acetophenone and sulfur in the presence of morpholine, in order to reveal the influence of different solvents on the optimum reaction conditions (other variables: amount sulfur/ketone, amount morpholine/ketone, and reaction temperature) [165].

Application of PCA to a set of five thermodynamic and spectroscopic basicity-dependent properties of 22 organic non-HBD solvents (related to hydrogen-bonding, proton-transfer, and interaction with hard and soft Lewis acids) by Maria *et al.* [143] has led to the interesting result that already two factors are sufficient to account for about 95% of the total variance of the solvent data. Physical significance has been given to these two principal factors by correlating them with intrinsic gas-phase affinities of the solvent molecules toward the proton and the potassium cation. A blend of electrostatic and charge-transfer (or electron-delocalization) character can be attributed to the first factor F_1. The second factor F_2 corresponds to an essentially electrostatic character. A third factor F_3, of marginal importance only, arises in part from steric hindrance to acid/base complexation. Thus, the inherent dimensionality of the condensed-phase basicity of organic non-HBD molecules commonly used as solvents is essentially reduced to two [143].

The most ambitious approach to a general classification of solvents by PCA is that of Chastrette *et al.* [138]. His classification is based on the representation of 83 solvents as points in an eight-dimensional property space, using the Kirkwood function $(\varepsilon_r - 1)/(2\varepsilon_r + 1)$, molar refraction $V_m \cdot (n^2 - 1)/(n^2 - 2)$, Hildebrand's δ-parameter [*cf.* Eq. (2-1) in Section 2.1], refraction index, boiling point, dipole moment, and the energies of HOMO and LUMO as solvent descriptors. Five descriptors are the properties of bulk solvents, whereas the last three (μ, HOMO, and LUMO) are molecular properties. The calculated HOMO and LUMO energies of the solvents are included in the set of basic variables in order to take into account Lewis acid/base interactions between solute and solvent. Because some of the eight descriptors are linearly correlated to each other, the dimensionality needed to describe the space solvent classification should be lower than eight.

Indeed, it was possible to reduce the original eight-dimensional space by suppressing five principal components, providing an easily visualised three-dimensional solvent property space, with only an 18% loss of information. This subspace is defined by the principal components F_1 (strongly correlated with the molar refraction, refractive index, and HOMO energy), F_2 (strongly correlated with the Kirkwood function, dipole moment, and boiling point), and F_3 (strongly correlated with the LUMO energy). Therefore, F_1 can be interpreted as an index of the polarizability of the solvent; F_2 represents the polarity of the solvent; and F_3 can be explained by the electron affinity and Lewis acidity of the solvent. The Lewis basicity of the solvent seems to be included in F_1.

The 83 organic solvents have been grouped into *nine classes* by their clustering of principal component values, using a nonhierarchical multivariate taxonomy to progressively classify solvents by means of the discriminating power of the eight descriptors (*cf.* Fig. 3-5).

Classes (1)...(3) comprise dipolar aprotic*) solvents. The first class (AD) contains the usual protic solvents having a relative low dipolarity ($\mu \leq ca.\ 12 \cdot 10^{-30}$ C m). More dipolar aprotic solvents ($\mu \geq ca.\ 12 \cdot 10^{-30}$ C m) are found in the second class (AHD). A third class (AHDP) contains only two members, differing from the second by their high polarizability.

Classes (4)...(6) include apolar aprotic*) solvents. In classes (4) and (5), ARA and ARP, are found aromatic apolar ($\mu \approx 0 \ldots 4 \cdot 10^{-30}$ C m) and aromatic relatively dipolar

* Bordwell *et al.* [135] have pointed out that solvents referred to as *dipolar aprotic* are in fact not aprotic. In reactions employing strong bases their protic character can be recognized. Therefore, instead of *dipolar aprotic* the designation *dipolar nonhydroxylic* or better *dipolar non-HBD* solvents is strongly recommended. *Cf.* Section 2.2.5 and 3.4 (footnote). In order to avoid confusion, the nomenclature proposed by Chastrette *et al.* [138] is retained in Fig. 3-5.

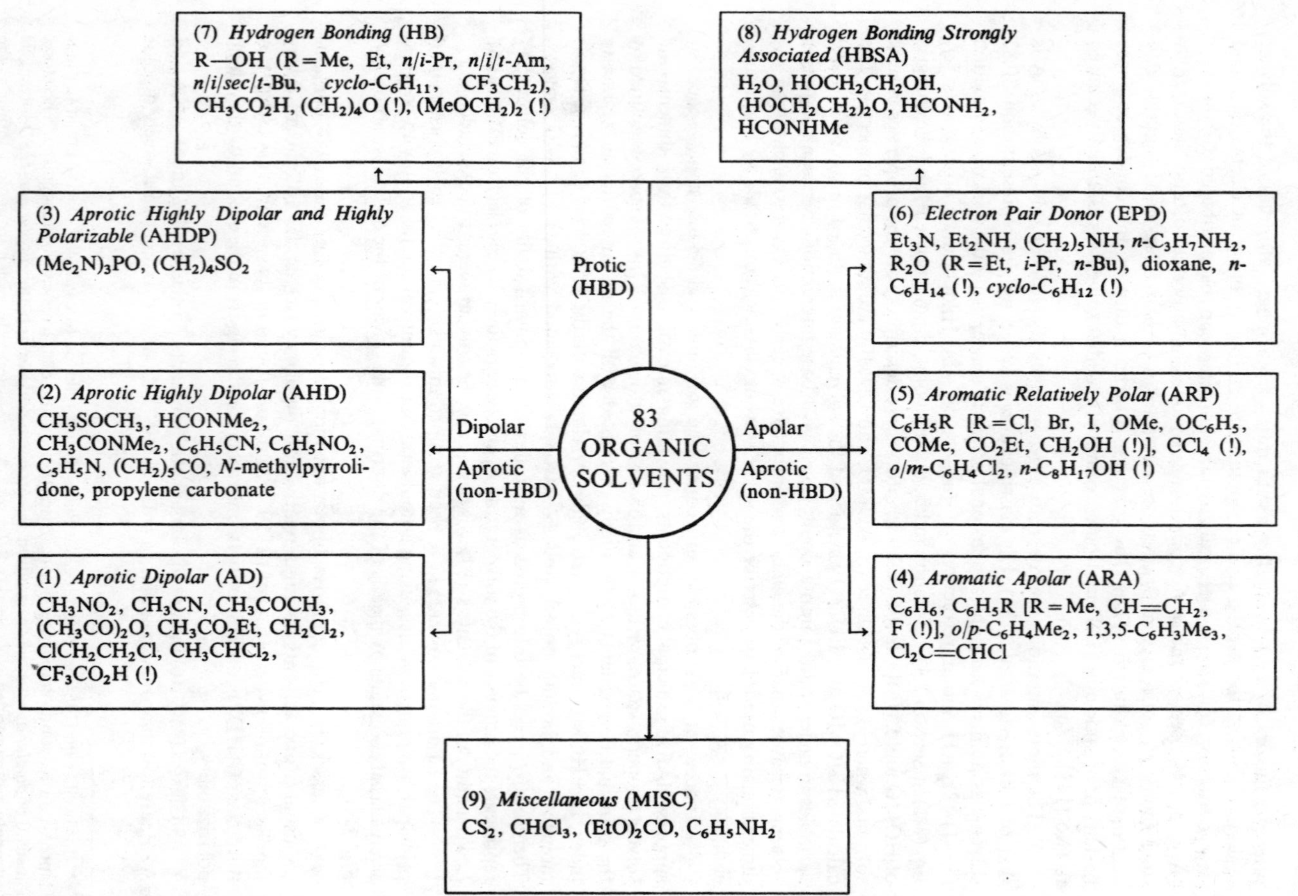

Fig. 3-5. Chastrette's classification of organic solvents [138].

solvents ($\mu \approx 4 \ldots 10 \cdot 10^{-30}$ C m). Class (6), called EPD, consists of solvents which are typical electron pair donors ($\mu \approx 4 \ldots 10 \cdot 10^{-30}$ C m).

Protic or HBD solvents are found in classes (7) and (8). These two classes of hydrogen bonding solvents (HB and HBSA) are clearly separated by the degree of their intermolecular association. If a total of ten solvent classes are established, water would then constitute a subclass of class HBSA.

Eventually, class (9), labelled MISC (from miscellaneous), consists of four solvents which have their high polarizability and nothing else in common.

It is remarkable, that this overall solvent classification, obtained entirely using statistical methods (PCA), correlates strongly with the chemist's intuition! Some of the solvent classes of Parker's scheme (*cf.* Fig. 3-3) are recovered in Fig. 3-5.

However, certain limitations are obvious. Some solvents, marked by (!) in Fig. 3-5, are not expected to be classed as found: Trifluoroacetic acid is classified with AD; benzyl alcohol, 1-octanol, and tetrachloromethane appear with the class ARP; *n*-hexane and cyclohexane with class EPD; tetrahydrofuran and 1,2-dimethoxyethane with class HB.

The reasons for some of these irregularities may be associated with the right choice of HOMO and LUMO of multifunctional solvents. For heteroatom-containing solvents such as benzyl alcohol, the HOMO of the π-orbital or the high-lying heteroatom lone-pair HOMO are available for solute/solvent interactions. The possible regiospecific interaction of the solvent frontier orbitals with solute molecules could mean, that a solvent may have two points in the descriptor space and hence two places in the resulting classification scheme, depending on the particular solute/solvent system under consideration. According to the classification in Fig. 3-5, benzyl alcohol is treated as an aromatic solvent, rather than as an alcohol.

This duality in the descriptor choice demonstrates the fallibility of searching for an unique, general valid classification of all organic solvents. Further work has to be done in order to clarify this point.

4 Solvent Effects on the Position of Homogeneous Chemical Equilibria

4.1 General Remarks

An equilibrium is homogeneous when all components are either exclusively in the gas phase or exclusively in solution. For gas-phase equilibria, the ratio of the product concentrations for end and starting materials is constant at a given temperature (law of mass action of Guldberg and Waage, 1867). When the reaction partners are dissolved, the standard molar Gibbs energy of solvation, ΔG°_{solv}, is liberated due to the intermolecular interactions between solvent and solute. In general, this quantity is different for starting and end products. Thus a displacement of the equilibrium can take place when going from the gas phase to solution [1–4]. An unchanged equilibrium constant can only be expected when ΔG°_{solv} is accidently the same for starting and end products.

The effect of the medium on the position of equilibrium can be considered from two points of view: (a) comparison of the gas-phase and solution equilibrium constants, and (b) comparison of the equilibrium constants for different solvents. Unfortunately, few equilibrium reactions have been studied both in the gas and liquid phases [5, 6]. These are primarily non-ionic reactions where the interaction between reacting molecules and solvent is relatively small (*e.g.* the Diels-Alder dimerization of cyclopentadiene). In the following chapter, therefore, equilibria which have been examined in solvents of different polarity will be the main topic considered (except for acid-base reactions described in Section 4.2.2).

Let us consider a simple isomerization reaction A $\rightleftharpoons$ B in the solvents I and II, whose abilities to solvate A and B are different. This corresponds to the Gibbs energy diagram shown in Fig. 4-1.

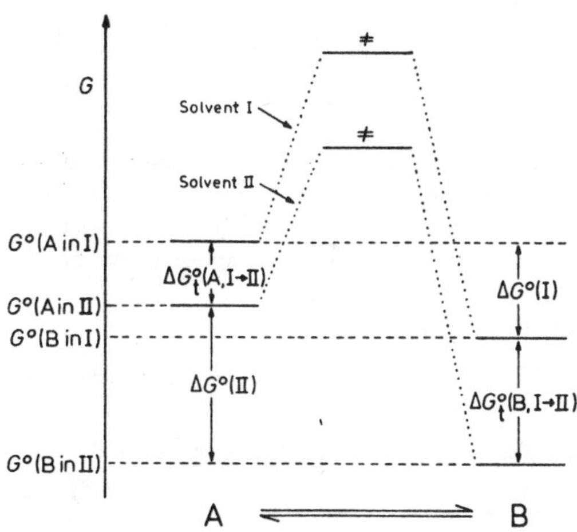

Fig. 4-1. One-dimensional Gibbs energy diagram for an equilibrium reaction A $\rightleftharpoons$ B in the solvents I and II. Ordinate: standard molar Gibbs energies of the reactants A and B in solvents I and II; Abscissa: not defined. ΔG°(I) and ΔG°(II): standard molar Gibbs energies of reaction in solvents I and II, respectively; ΔG°_{t}(A, I→II) and ΔG°_{t}(B, I→II): standard molar Gibbs energies of transfer of the solutes A and B from solvent I to solvent II, respectively [ΔG°_{t}(A, I→II) $= G^{\circ}$(A in I) $- G^{\circ}$(A in II), and ΔG°_{t}(B, I→II) $= G^{\circ}$(B in I) $- G^{\circ}$(B in II)], *cf.* Eq. (2-12a) in Section 2.3; $\neq$ = transition state.

From Fig. 4-1, Eq. (4-1) can be immediately derived,

$$\Delta G^\circ(II) + \Delta G_t^\circ(\text{A, I}\rightarrow\text{II}) = \Delta G_t^\circ(\text{B, I}\rightarrow\text{II}) + \Delta G^\circ(\text{I}) \tag{4-1}$$

which on rearrangement leads to Eq. (4-2) [102]:

$$\begin{aligned}\Delta G^\circ(\text{II}) - \Delta G^\circ(\text{I}) &= \Delta\Delta G^\circ(\text{I}\rightarrow\text{II}) \\ &= \Delta G_t^\circ(\text{B, I}\rightarrow\text{II}) - \Delta G_t^\circ(\text{A, I}\rightarrow\text{II})\end{aligned} \tag{4-2}$$

Since, for equilibria, the logarithm of the equilibrium constant is proportional to the standard molar Gibbs energy change, ΔG°, according to Eq. (4-3),

$$\Delta G^\circ = -R \cdot T \cdot \ln K \tag{4-3}$$

it follows from Eqs. (4-2) and (4-3) that the difference in the molar transfer Gibbs energies of educt A and product B, $\Delta\Delta G_t^\circ(\text{I}\rightarrow\text{II})$, determines the solvent effect on the position of this equilibrium. In the particular case of Fig. 4-1, $\Delta G_t^\circ(\text{B, I}\rightarrow\text{II}) > \Delta G_t^\circ(\text{A, I}\rightarrow\text{II})$, so that the equilibrium is displaced towards B when the solvent is changed from I to II.

The required standard molar Gibbs energies of transfer can be obtained from activity coefficient measurements, using Eq. (4-3a),

$$\Delta G_t^\circ(\text{X, I}\rightarrow\text{II}) = -R \cdot T \cdot \ln(\gamma_\text{I}/\gamma_\text{II}) \tag{4-3a}$$

in which γ refers to activity coefficients of solute X in solvents I and II. Methods used to obtain these activity coefficients have included vapour pressure, solubility, and distribution coefficient measurements [103]; *cf.* also Eq. (2-12b) and Table 2-9 in Section 2.3.

In studying solvent effects on equilibria, it is, in principle, not sufficient to investigate the ΔG° changes alone, because this term is determined by both an enthalpy and an entropy term according to Eq. (4-4).

$$\Delta G^\circ = \Delta H^\circ - T \cdot \Delta S^\circ \tag{4-4}$$

Transfer functions can also be defined for the thermodynamic state functions ΔH° and ΔS° [102]. The ease of calorimetric measurements has made the standard molar transfer enthalpy, $\Delta H_t^\circ(\text{X, I}\rightarrow\text{II})$, readily available. If both transfer Gibbs energies and transfer enthalpies are available it should be possible to achieve a complete dissection of the effect of solvents on the various thermodynamic parameters.

Four types of reaction control can be recognized:

(a) Cooperative effects, with $\Delta\Delta H^\circ$ and $T \cdot \Delta\Delta S^\circ$ having opposite signs. Then these two terms will be additive;

(b) Enthalpy-controlled effects, in which the two terms are opposed, but the enthalpy term is larger;

(c) Entropy-controlled effects, in which the two terms are opposite but the $T \cdot \Delta\Delta S^\circ$ is larger; and
(d) Compensating effects, in which the two terms are opposed but nearly equal.

A change in temperature may interconvert categories (b), (c), and (d).

A quantitative description of the solvent effects on equilibria is only possible in the most favorable and simple cases (*cf.* reference [7] for an example) due to the complexity of the intermolecular interactions between solute and solvent. Thus, in the following sections, these effects are studied only on a more qualitative basis, using acid/base, tautomeric, and other equilibria as examples.

4.2 Solvent Effects on Acid/Base Equilibria [8–13]

4.2.1 Brønsted Acids and Bases in Solution

As already emphasized in Section 3.3.1, the ionization equilibrium of an acid, Eq. (4-5), or of a base, Eq. (4-6), is affected by a solvent change, not only because of the

$$HA^{z+1} + SH \rightleftharpoons SH_2^{+} + A^{z} \tag{4-5}$$

$$A^{z} + SH \rightleftharpoons HA^{z+1} + S^{-} \tag{4-6}$$

acidity or basicity of the solvent, but also because of its dielectric constant, and the ability of the solvent to solvate the various species of Eqs. (4-5) and (4-6). A change in dielectric constant or in solvating ability can thus influence the acidity of an acid HA or the basicity of a base A. Thus, for example, the acidity constant K_a for carboxylic acids is up to 10^6 times larger in water ($\varepsilon_r = 78.3$) than in absolute ethanol ($\varepsilon_r = 24.6$) although water is only 15...20 times stronger a base than ethanol.

Eq. (4-5) shows the reaction of an acid with an amphiprotic solvent to form a solvated proton and the conjugate base of the acid, at an infinite distance from each other. Part of the energy required for this reaction comes from the electrostatic interactions between these ions and can be estimated from simple electrostatic theory. The electrostatic work necessary to charge isolated species HA, which are assumed to be spherical with radius r_{HA} and of charge $z_{HA} \cdot e$, in a structureless medium of dielectric constant ε_r, is given by the Born equation (4-7) (ε_0 = permittivity of vacuum; z = number of elementary charges e; N_A = Avogadro constant) [14].

$$\Delta G^\circ_{\text{electrostatic}} = \frac{1}{4\pi \cdot \varepsilon_0} \cdot \frac{N_A \cdot z_{HA}^2 \cdot e^2}{2 \cdot \varepsilon_r \cdot r_{HA}} \tag{4-7}$$

Application of this equation to acid/base reaction (4-5) leads to a net change in Gibbs energy per mol as shown in Eq. (4-8), if only pure electrostatic interactions are considered [8].

$$\Delta G^\circ_{\text{electrostatic}} = \frac{1}{4\pi \cdot \varepsilon_0} \cdot \frac{N_A \cdot e^2}{2 \cdot \varepsilon_r} \cdot \left(\frac{1}{r_{SH_2^{\oplus}}} + \frac{z_A^2}{r_A} - \frac{z_{HA}^2}{r_{HA}} \right) \tag{4-8}$$

Thus, the Gibbs energy difference $\Delta\Delta G$ for the ionization of a mol of HA in solvent 1 and solvent 2 with dielectric constants $\varepsilon_{r(1)}$ and $\varepsilon_{r(2)}$, respectively, provided that the radii of the reactants are the same in both solvents, is given by Eq. (4-9).

$$\Delta\Delta G^\circ = (\Delta G^\circ_{\text{electrostatic}})_2 - (\Delta G^\circ_{\text{electrostatic}})_1$$

$$= \frac{1}{4\pi\cdot\varepsilon_0}\cdot\frac{N_A\cdot e^2}{2}\cdot\left(\frac{1}{r_{SH_2^\oplus}}+\frac{z_A^2}{r_A}-\frac{z_{HA}^2}{r_{HA}}\right)\cdot\left(\frac{1}{\varepsilon_{r(2)}}-\frac{1}{\varepsilon_{r(1)}}\right) \tag{4-9}$$

Because the Gibbs energy of reaction, ΔG, is related to the equilibrium constant K_a, according to Eq. (4-3), Eq. (4-9) can be converted into Eq. (4-10).

$$\ln\frac{(K_a)_2}{(K_a)_1} = -\frac{1}{4\pi\cdot\varepsilon_0}\cdot\frac{N_A\cdot e^2}{2\cdot RT}\cdot\left(\frac{1}{r_{SH_2^\oplus}}+\frac{z_A^2}{r_A}-\frac{z_{HA}^2}{r_{HA}}\right)\cdot\left(\frac{1}{\varepsilon_{r(2)}}-\frac{1}{\varepsilon_{r(1)}}\right) \tag{4-10}$$

Eq. (4-10) can be used only for solvents of equal acid and base strength, because only the effect of the solvent dielectric constant on the degree of ionization is considered. Under these conditions, Eq. (4-10) predicts that the logarithm of the ionization constant K_a of HA should be inversely proportional to the dielectric constant of the solvent in which HA is dissolved. Because of these restrictions, Eq. (4-10) can be expected to yield only semiquantitative results. But, it allows us to predict qualitatively how the charge type of an acid affects the ionization constant in solvents of different dielectric constants.

If the acid HA in Eq. (4-5) has a charge of +1 ($z=0$; *e.g.* NH_4^+) the right-hand side of Eq. (4-10) involves only the difference between two reciprocal radii ($z_A=0$), and is often very nearly zero. Therefore, a change in the dielectric constant should in this case have only a small, if any, influence on the ionization equilibrium of an acid such as NH_4^+. This is plausible, because in this acid/base reaction (4-5) no charges are created or destroyed at all. Also, there is no electrostatic attraction between the positive charged species HA and the neutral conjugate base A. Indeed, the cationic acid NH_4^+ is only about 1/10 as strong in ethanol as it is in water according to the lower basicity of the solvent ethanol compared with water (*cf.* Table 4-1).

On the other hand, if HA is an uncharged acid ($z=-1$; *e.g.* CH_3—CO_2H) the right-hand side of Eq. (4-10) involves the sum of two reciprocal radii ($z_{HA}=0$) and a strong influence of the dielectric constant on the ionization equilibrium is expected. Because in acid/base reactions of this charge type, neutral molecules are converted into anions and cations, which attract each other, reaction (4-5) will shift to the right with an increase in dielectric constant of the solvent in which HA is dissolved. Ionization increases when ε_r increases. This rule is qualitatively verifiable for water and alcohols as solvents. In fact, acetic acid is only 10^{-6} times, approximately, as strong an acid in ethanol (low ε_r) as in water (high ε_r) (*cf.* Table 4-1).

In the case of an anionic acid ($z=-2$; *e.g.* HSO_4^-) the acid/base equilibrium (4-5) will be shifted to the right with an increase in dielectric constant of the solvent, in which HA is dissolved. Here, the ionization will increase much more quickly with an increase in dielectric constant than in the preceding case.

Table 4-1. Solvent influence on the acid/base equilibrium $HA^{z+1} + SH \rightleftharpoons SH_2^+ + A^z$ for various charge types[a].

z	Acid-base charge type	pK_a change with increasing basicity of SH	pK_a change with increasing dielectric constant of SH	Example	$\Delta pK_a = (pK_a)_{H_2O} - (pK_a)_{C_2H_5OH}$ [a, b]
0	$HA^{\oplus}/A^0$	decrease	no or small effect	$NH_4^{\oplus}/NH_3$	−1.2
				$H_5C_6NH_3^{\oplus}/H_5C_6NH_2$	−0.4
−1	$HA^0/A^{\ominus}$	decrease	decrease	$H_3C—CO_2H/H_3C—CO_2^{\ominus}$	−5.7
				Picric acid/Picric anion	−3.3
−2	$HA^{\ominus}/A^{2\ominus}$	decrease	decrease	$HO_2C—(CH_2)_2—CO_2^{\ominus}/{}^{\ominus}O_2C—(CH_2)_2—CO_2^{\ominus}$	−5.8[d]
				$HO_2C—CH{=}CH—CO_2^{\ominus}/{}^{\ominus}O_2C—CH{=}CH—CO_2^{\ominus}$ [c]	−5.5[d]

[a] *Cf.* reference [11] for more examples.
[b] Dielectric constants: $\varepsilon_r = 78.3$ (H_2O), $\varepsilon_r = 32.7$ (CH_3OH), and $\varepsilon_r = 24.6$ (C_2H_5OH) at 25 °C.
[c] Fumaric acid.
[d] $\Delta pK_a = (pK_a)_{H_2O} - (pK_a)_{CH_3OH}$ in this case.

Table 4-1 shows the predicted solvent influence on the acidity constants K_a for various charge types of acid/base pairs.

Analogous rules apply to the basicity constants K_b according to Eq. (4-6). For example, for a pair A^0/HA^+ ($z=0$; *e.g.* $NH_3 + ROH$) K_b increases as the dielectric constant of the solvent increases.

Because it is practically impossible to find solvents which differ only in their dielectric constants, and not in their basicity or acidity, the predictions, based on Eq. (4-10), are usually not in good agreement with the experimental results. In addition, the solvation capability or polarity of a solvent is not described by its dielectric constant alone. Besides the pure electrostatic Coulomb interactions, there exist other specific and unspecific interaction forces such as ion-dipole, dipole-dipole, hydrogen bonding, ion-pair formation, *etc.* Also, the model used to describe the coulombic effects on acid/base ionization constants ignores the actual shape and size of the individual ions. For example, the acidity constant of picric acid increases in contrast to the carboxylic acids by a factor of 1500 only for a solvent change ethanol to water (*cf.* Table 4-1). Because the negative charge of the picric anion is delocalized over a large molecule, the solvation enthalpy of this anion will be lower than that of the carboxylate anions. This means, that its stability does not change with the solvation capability of the solvent to the same extent as for the carboxylate anions, where the negative charge is more or less localized on only two oxygen atoms.

Using activity coefficients, it is possible to examine in greater detail the effect of solvents on acid and base strengths. Equilibrium constants expressed in terms of concentration are solvent-dependent. Solvent-independent so-called thermodynamic equilibrium constants are obtained when the concentration terms are replaced by activity terms. The thermodynamic equilibrium constant K for the reaction $HA \rightleftharpoons H^{\oplus} + A^{\ominus}$ is given by the following equation:

$$K = \frac{a_{H^+} \cdot a_{A^-}}{a_{HA}} = \frac{[H^+]\gamma_{H^+}[A^-]\gamma_{A^-}}{[HA]\gamma_{HA}} = K_a \cdot \frac{\gamma_{H^+} \cdot \gamma_{A^-}}{\gamma_{HA}} \tag{4-11}$$

In highly dilute aqueous solution the activity coefficients approach the value 1. That is, in aqueous solution K and K_a become practically equal at infinite dilution. If the equilibrium constant, expressed in concentration terms, is denoted by K_{SH} for the solvent SH, and K_{H_2O} is the value of K_a in water at infinite dilution, then it follows that

$$K = K_{SH} \cdot \frac{\gamma_{H^+} \cdot \gamma_{A^-}}{\gamma_{HA}} = K_{H_2O} \tag{4-12}$$

where the values of γ are the appropriate activity coefficients in the solvent SH. Analogously, for another acid HB with an acid/base equilibrium $HB \rightleftharpoons H^{\oplus} + B^{\ominus}$, Eq. (4-13) can be written.

$$K' = K'_{SH} \cdot \frac{\gamma_{H^+} \cdot \gamma_{B^-}}{\gamma_{HB}} = K'_{H_2O} \tag{4-13}$$

Division of Eq. (4-12) by Eq. (4-13) gives Eq. (4-14)

$$\frac{K_{SH} \cdot \gamma_{HB} \cdot \gamma_{A^-}}{K'_{SH} \cdot \gamma_{HA} \cdot \gamma_{B^-}} = \frac{K_{H_2O}}{K'_{H_2O}} \tag{4-14}$$

which can be rearranged to Eq. (4-15)

$$K_{SH} = K'_{SH} \cdot \frac{K_{H_2O}}{K'_{H_2O}} \cdot \frac{\gamma_{HA} \cdot \gamma_{B^-}}{\gamma_{HB} \cdot \gamma_{A^-}} \tag{4-15}$$

The acidity constant of a given acid in any solvent SH can then be calculated from Eq. (4-15) when its acidity constant in water at infinite dilution, the acidity constant of another acid HB in water and in solvent SH, and the acitivity term are all known. The values γ_{HA} and γ_{HB} can be obtained from solubilities, partial pressures, distribution coefficients, *etc.* The ratio $\gamma_{B^-}/\gamma_{A^-}$ can be determined potentiometrically, or from the solubility of salts. The values of K_{SH} calculated in this way correspond satisfactorily with the ones measured.

At this point only reference can be given to the various acidity scales for different series of solvents, such as those of Hammett [15] and Grunwald [16]. The acidity functions are introduced in order to have expressions, which are not affected by dielectric constant and which allow a quantitative comparison of acidity in different solvents. But it should be stated, that there exists no single scale of acidity or basicity, equally valid in all types of solvents and applicable to both equilibrium and kinetic situations [17, 109].

Medium effects on acid/base equilibria in aqueous solutions of strong acids have been analyzed not only in terms of Hammett acidity functions but also in terms of linear Gibbs energy relationships, first developed by Bunnett *et al.* [225]; *cf.* [226] for a recent review.

The ionisation constants of ca. 4500 acidic organic compounds in aqueous solution have been compiled [110] as well as the methods of determination [111] and prediction [112] of pK_a-values. Particular attention has been payed to C—H acidic compounds [113]. Whereas the ionisation constants of Brønsted acids and bases for aqueous solutions are well known, the corresponding pK_a-values for nonaqueous solutions are comparatively scarce.

4.2.2 Gas-Phase Acidities and Basicities [18–21, 114–118]

Acidities and basicities of organic compounds are expected to be different in the gas phase and in solution. Whereas in the gas phase acidity and basicity are intrinsic properties of the individual molecules, in the liquid state these properties belong to the phase as a whole due to the interaction between solute and solvent molecules. Solution-phase acidities and basicities reflect solvent effects as well as the intrinsic proton-donating and -accepting power of the solute. Therefore, acidities and basicities measured under solvent-free conditions must be available before the interplay between solute properties and solvent effects can be studied. Efforts in this direction have not made much progress until recently due to the lack of methods for measurements of gas-phase acidities and basicities.

At present, gas-phase acidities and basicities for many organic compounds are available primarily due to the development within the past 20 years of three new experimental techniques: pulsed high pressure (*i.e.* 0.1...1300 Pa) mass spectrometry (HPMS) [22, 23, 118], flowing afterglow (FA) technique with a fast flowing gas like helium in the pressure range of *ca.* $10^{-1}...10^{-2}$ Pa [119], and pulsed electron beam, trapped ion cell, ion cyclotron resonance (ICR) spectrometry, carried out at *ca.* $10^{-6}...10^{-3}$ Pa [24–26, 115].

The development of these ion-molecule equilibrium measurements has completely changed the status of acid/base reactions (and of other reactions; *cf.* Section 5.2) in the gas phase. It is now possible to compare the complex and poorly understood situation in solution with the simple state in the gas phase. It is also possible to determine the acidity of all acids in the gas phase, from the weakest such as methane to the strongest. In solution, however, due to the levelling effect of the solvent or solubility problems, only a certain range of acids can be measured in a given solvent.

Using these methods, relative intrinsic acidities and basicities of molecules in the gas phase have been determined by measuring equilibrium constants, $K = [A^{\ominus}] \cdot [BH]/[HA] \cdot [B^{\ominus}]$, for proton-transfer reactions such as

$$(HA)_{gas} + (B^{\ominus})_{gas} \rightleftharpoons (A^{\ominus})_{gas} + (BH)_{gas} \tag{4-16}$$

Similar measurements for proton-transfer reactions such as (S = solvent molecules)

$$(HA)_{gas} + (B^{\ominus} \cdot S_n)_{gas} \rightleftharpoons (A^{\ominus} \cdot S_n)_{gas} + (BH)_{gas} \tag{4-17}$$

have allowed the scrutiny of the solvent influence as a function of stepwise solvation of the participating ions by sequential addition of solvent molecules (n = from 0 up to 4...9), thus, bridging the gap between the gas-phase and solution reaction.

The standard molar Gibbs energy change for reaction (4-16), $\Delta G^\circ = -RT \cdot \ln K$, is then a measure of the relative acidity of HA and BH (or of the relative basicity of $B^\ominus$ and $A^\ominus$). Series of acids and bases have been studied to establish a scale of relative acidities in the same manner as pK_a-values are determined in solution.

In addition, an absolute intrinsic acidity or basicity scale as the case may be, corresponding to the reaction (4-18),

$$(HA)_{gas} \rightleftharpoons (H^\oplus)_{gas} + (A^\ominus)_{gas} \tag{4-18}$$

can be established by incorporating certain standard reactions such as $H_2 \rightleftharpoons H^\oplus + H^\ominus$ and $HF \rightleftharpoons H^\oplus + F^\ominus$ for which ΔH° and ΔG° can be calculated from available data into the relative scale. These absolute gas-phase acidities are conveniently expressed in terms of the proton affinity, PA, of the anion $A^\ominus$, defined as the standard molar enthalpy change ΔH° for the gaseous deprotonation reaction (4-18): $PA(A^\ominus) = -\Delta H^\circ$. For most simple cases the change in ΔS° for this reaction is about the same. Thus, ΔH° (and ΔG°) varies in about the same way as the proton affinity does.

From consideration of the following thermodynamic cycle,

$H{-}A \longrightarrow H^\bullet + A^\bullet$	D_{HA} (homolytic dissociation energy of the H—A bond)
$A^\bullet + e^\ominus \longrightarrow A^\ominus$	$-EA_A$ (electron affinity of radical A)
$H^\bullet \longrightarrow H^\oplus + e^\ominus$	IP_H (ionization potential of atomic hydrogen, *i.e.* 1312 kJ/mol)
$H{-}A \longrightarrow H^\oplus + A^\ominus$	

the proton affinity of $A^\ominus$ is given by Eq. (4-19):

$$PA = D_{HA} - EA_A + IP_H \tag{4-19}$$

Since IP_H is a constant for every reaction, for the sake of simplicity this term is usually omitted; the acidities are then expressed as $D_{HA} - EA_A$. The acid strength increases as this difference decreases.

Continuous scales of gas-phase proton affinities for numerous organic compounds can be found in references [115, 120–122].

The gas-phase acidity orders differ in most cases dramatically from those observed in solution since the Gibbs energies of solvation (*ca.* 200–600 kJ/mol; *cf.* Table 2-8 in Section 2.3) are much larger than the intrinsic acidity differences for most pairs of compounds. Thus, the relative acidities in solution are often dictated by the differential Gibbs energies of solvation rather than by the intrinsic properties of the solute molecules.

The solution and gas-phase acidities of C—H acids are of particular interest because of the wide structural variations which are possible in this class of compounds [113, 123]. A qualitative ordering of a selection of C—H acids (and some O—H acids for comparison) gives the following sequence of increasing acidity in the gas phase [120]:

$CH_4 < H_2O < CH_3OH < C_6H_5CH_3 < HC{\equiv}CH < CH_3SOCH_3 < CH_3CN$

$< CH_3COCH_3 < CH_3CHO < C_6H_5COCH_3 < CH_3NO_2 <$ Cyclopentadiene

$<$ Fluorene $< CH_3CO_2H < CH_2(CN)_2$

Surprisingly enough, toluene is more acidic than water in the gas phase but *ca.* 20 orders of magnitude less acidic in solution. Thus, in the gas phase the reaction of $HO^{\ominus}$ with toluene proceeds rapidly, with the release of energy, to yield charge-delocalized $C_6H_5CH_2^{\ominus}$ and H_2O. In aqueous solution the reaction goes in the opposite direction, converting the benzyl anion into toluene by water, and the latter giving charge-localized $HO^{\ominus}$.

Contrary to the behaviour in solution, malononitrile is a stronger acid than acetic acid in the gas phase; and fluorene, which is almost 10^5 times less acidic than cyclopentadiene in solution, becomes the stronger of the two in the gas phase. The reason for the reversed cyclopentadiene/fluorene acidity order is, the better solvation of the smaller cyclopentadiene anion relative to the large fluorenyl anion with the more delocalized negative charge, in going from the gas phase to solution. Whereas the larger, more charge-delocalized anion is preferably produced in the gas phase, in solution, the smaller anion with the higher charge density, *i.e.* the better solvated anion, is favoured.

A study of some C—H acids in dimethyl sulfoxide solution led to an acidity order which is nearly parallel to the order in the gas phase, whereas in protic solvents the order is different [116, 124, 125]. This result exhibits the importance of specific solute/solvent interactions such as hydrogen bonding for the comparison of acid/base equilibria measured in the gas phase and in solution.

Some features of the solution pK_a-scale are retained in the gas phase. For instance, the acidity of hydrogen halides increases in the order $HF < HCl < HBr < HI$, both in aqueous solution and in the gas phase [120].

The discovery of Brauman and Blair in 1968 [34] that the acidity of aliphatic alcohols is completely reversed on going from bulk solution to the gas phase was a landmark in the interpretation of solvent and substituent effects on acid/base equilibria. The gas-phase acidity of alcohols increases in the following order [34, 125, 126]:

$H\text{-}OH < CH_3\text{-}OH < CH_3CH_2\text{-}OH < CH_3CH_2CH_2\text{-}OH < (CH_3)_2CH\text{-}OH$

$< CH_3CH_2CH_2CH_2\text{-}OH < (CH_3)_3C\text{-}OH < Me_3CCH_2\text{-}OH < C_6H_5CH_2\text{-}OH$

$< (Me_3C)_2CH\text{-}OH < CH_3\text{-}SH < CH_3CH_2\text{-}SH \lll C_6H_5\text{-}OH$

In the gas phase, tertiary alcohols are more acidic than secondary alcohols, and these in turn are stronger acids than primary alcohols. In other words, the anion $R—CH_2O^{\ominus}$ is a stronger base than both $R_2CH—O^{\ominus}$ and $R_3C—O^{\ominus}$. This is in striking contrast to the solution behaviour, where introduction of alkyl groups at the OH-bearing carbon atom causes a significant increase in basicity of alkoxide anions [35, 127].

Prior to the availability of gas-phase data, the solution order of acidities was taken as evidence of anion destabilization through electron donation by methyl groups. Now it is clear that the liquid-phase ordering is due entirely to differential solvation of the reactants of Eq. (4-18). The effects of alkyl groups on gas-phase acidities have been considered in terms of inductive and polarization effects. It is now established that the stabilization of anions by alkyl substitution is due mostly to the role played by polarization forces between the negatively charged centre and the alkyl groups. A charged atom in an isolated ion in the gas-phase has only its attached alkyl groups to interact with, whereas in solution the polarizability of the surrounding solvent molecules is an additional factor in stabilizing the ion. Since polarizability generally increases with molecular volume, all alkyl groups are more polarizable than the hydrogen atom, and can stabilize a nearby charge whether that charge is negative or positive. Hence, the gas-phase acidities (and basicities) of alcohols should increase the greater the number and size of the alkyl groups. The reversed acidity order of alcohols obtained in bulk solution ($RCH_2—OH > R_2CH—OH > R_3C—OH$) can be explained by the assumption that, stabilization of the alkoxide ions through hydrogen bonding should be much better with the sterically less hindered $RCH_2—O^\ominus$ than with $R_2CH—O^\ominus$ and $R_3C—O^\ominus$ [128]. Since the Gibbs energies of solvation of the alkoxide ions are large compared to the difference in ionization energies according to Eq. (4-18), a reversal of the acidity order easily occurs in going from the gas phase to solution. Thus, the solution acidity order of alcohols is an artefact, and does not represent any intrinsic property of the alcohol molecules.

By quantitatively measuring the equilibrium given in Eq. (4-20), it has been shown [129] that the reversal of the relative acidities of methanol and ethanol on going from the

$$CH_3O^\ominus\cdots HOCH_3 + C_2H_5OH \rightleftharpoons C_2H_5O^\ominus\cdots HOCH_3 + CH_3OH \tag{4-20}$$

gas phase to bulk solvent is almost half completed with the first molecule of solvation. The Gibbs energy of solvation of $CH_3O^\ominus$ using one molecule of CH_3OH is approximately 71 kJ/mol (17 kcal/mol) [130]. Thus, the first molecule of solvent causes the behaviour of the alkoxide ion to become already "solution-like". In other cases [*cf.* Eq. (4-17)] more solvent molecules are necessary for the reactivity to approach that of the bulk solvated form. Obviously, the first few solvent molecules can contribute most of the total solvation energy. These results also show that the solvating abilities of isolated solvent molecules can be very different from their performance as bulk liquids. For example, the extremely low gas-phase acidity and basicity of "monomeric" water is completely different from its acid/base behaviour as "polymeric" water, *i.e.* as the classical amphiprotic solvent.

Other well-studied cases for acidity order changes induced by differential solvation are substituted phenols [131] and halo-substituted carboxylic acides [34a, 132]. For a comparison of the acid/base behaviour of oxygen versus sulfur acids and bases (*e.g.* R—OH/R—SH) in the gas-phase and in solution *cf.* reference [214].

By comparing the gas-phase and solution acidities of substituted phenols it has been found that, the solvent not only plays a dominant role in controlling the phenol acidity but also in modifying the effects of substitutents on this acidity. Modifications due to the differential substituent solvation can significantly change the order of substituent effects on phenol acidity [131]. An analogous solvent influence on the substituent effects on gas-phase and solution basicity were found in the series of 4-substituted pyridines [33].

The gas-phase acidity order of haloacetic acids $X—CH_2—CO_2H$ is: $F < Cl < Br$, *i.e.* the order is reversed compared to that in aqueous solution [34a, 132]. Thus, the well-known aqueous acidity order is not caused by the increasing inductive substituent effect ($Br < Cl < F$), as is generally assumed, but rather by solvation effects.

α-Amino acids such as glycine which are known to exist as zwitterions in the crystalline state and in aqueous solution, are not zwitterionic in the gas phase [133]. By measuring the gas-phase acidity and basicity of glycine it has been found that, glycine exists in the gas phase as a non-ionic molecule $H_2N—CH_2—CO_2H$! Although the zwitterionic form of α-amino acids predominates in dipolar non-HBD solvents, the ratio of zwitterionic to uncharged form is with 2...40 in dimethyl sulfoxide much smaller than the corresponding values of $10^4 \ldots 10^5$ in aqueous solution [219]. This large difference can be explained by the greater solvation of the carboxylate group in water compared to dimethyl sulfoxide, with solvation of the ammonium group being similar in the two solvents [219].

A historically interesting example of solvent influence on basicity which has puzzled chemists for a long time will be used to conclude this Section.

It has long been known that the proton-acceptor abilities of alkylamines in aqueous solution, as expressed by their pK_b-values, are in the order $NH_3 < RNH_2 < R_2NH > R_3N$ (!) [27]. The unexpected reduced base strength for tertiary alkylamines is found for all the common alkyl groups. However, if the basicities of methylamines are determined in the gas phase, they increase monotonically from ammonia to trimethylamine as expected from theoretical considerations: $NH_3 < CH_3NH_2 < (CH_3)_2NH < (CH_3)_3N$ [28–30, 115]. Therefore, the "basicity anomaly" of tertiary alkylamines in aqueous solution must have to do with the differential solvation of the reacting species of the corresponding acid/base equilibrium.

The reaction between the ammonium ion, $NH_4^{\oplus}$, and trimethylamine, $(CH_3)_3N$, analogous to Eq. (4-16), has been studied by pulsed ICR mass spectrometry [115]. The Gibbs energy diagram in Fig. 4-2 describes what happens to the reactants in going from the

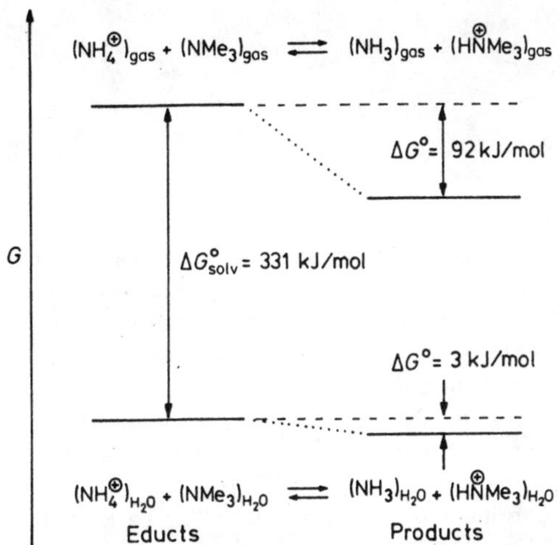

Fig. 4-2. One-dimensional Gibbs energy diagram for the acid/base equilibrium reaction between ammonia and trimethylamine in the gas phase (top) and in aqueous solution (bottom) [115].

gas phase to aqueous solution. In the gas phase the products are stabler than the educts by 92 kJ/mol. In aqueous solution, the Gibbs energy of reaction falls to only 3 kJ/mol because of preferential solvation of the $NH_4^{\oplus}$ ion by hydrogen bonding. This means that at equilibrium in aqueous solution the concentration of the ammonium ion is higher by a factor of 10^{15} than it is in the gas phase.

Why is trimethylamine a much stronger base than ammonia in the gas phase? The increase in base strength with the increasing number of alkyl substituents at the amine nitrogen atom can be considered in terms of inductive and polarization effects. The electric field of the positive charge permeates the space around the alkylammonium ion and distorts the electron clouds of the alkyl groups. Both the inductive and polarization effect make $Me_3NH^{\oplus}$ a stabler ion than $NH_4^{\oplus}$. In particular, the replacement of hydrogen atoms by the larger, more polarizable alkyl groups stabilizes charged centres. The observation that the gas-phase acidity of alcohols increases with alkyl substitution (see above) shows that alkyl groups can stabilize a charge of either sign, withdrawing or donating electrons as needed.

What happens to the same reaction in aqueous solution? Whereas the neutral reactants, ammonia and trimethylamine, are hydrated about equally well, the ammonium ion is hydrated much more strongly than is $Me_3NH^{\oplus}$. As shown by Eqs. (4-21) and (4-22), solvation through hydrogen bonding will tend to increase the base strength of all amines

$$R\text{-}\bar{N}H_2 + H_3O^{\oplus} \underset{\text{in } H_2O}{\rightleftharpoons} R\text{-}\overset{\oplus}{N}(\text{-H}\cdots OH_2)_3 \qquad (4\text{-}21)$$

$$R_3N| + H_3O^{\oplus} \underset{\text{in } H_2O}{\rightleftharpoons} R_3\overset{\oplus}{N}\text{-H}\cdots OH_2 \qquad (4\text{-}22)$$

in aqueous solution because the positively charged ammonium ions will be better solvated than the uncharged amines [31]. However, the solvation through hydrogen bonding will decrease with increasing alkyl substitution; *cf.* Eq. (4-22). The ammonium ion can be stabilized by four hydrogen bonds, whereas $Me_3NH^{\oplus}$ has only one acidic hydrogen atom.

The apparent basicity "anomaly" of alkylamines can now be understood in terms of two opposing influences, one base strengthening (due to increasing alkylation of the amine), and the other base weakening (because of reduced solvation of the ammonium ions with increasing alkylation).

Why are aniline ($pK_b = 9.4$) and pyridine ($pK_b = 8.8$) so much less basic than ammonia ($pK_b = 4.8$) in aqueous solution? For a long time, students of organic chemistry have been given reasons in terms of lone-pair delocalization and sp^2/sp^3 hybridization of the nitrogen atom. From gas-phase studies it is now clear that aniline and pyridine are inherently much stronger bases than ammonia [116]*). Therefore, the reversed basicity order obtained in aqueous solution is caused by differential hydration! However, if more suitable reference compounds are used, cyclohexylamine ($pK_b = 3.3$) for aniline and piperidine ($pK_b = 2.9$) for pyridine, then aromatic amines are indeed less basic than saturated ones in aqueous solution. Thus, according to Arnett [213], "the right idea has been promoted over the years but for the wrong reasons, a not infrequent occurrence in chemistry".

Some general rules for solvation effects on acid/base equilibria as transferred from the gas phase to solution have been collected recently by Arnett [213].

4.3 Solvent Effects on Tautomeric Equilibria

4.3.1 Solvent Effects on Keto/Enol Equilibria [36–43, 134]

In general, 1.3-dicarbonyl compounds, which include β-dialdehydes, β-ketoaldehydes, β-diketones, and β-ketocarboxylic esters, may exist in solution or as pure compound in three tautomeric forms**): the diketo form *(4a)*, the *cis*-enolic *(4b)*, and the *trans*-enolic form *(4c)*.

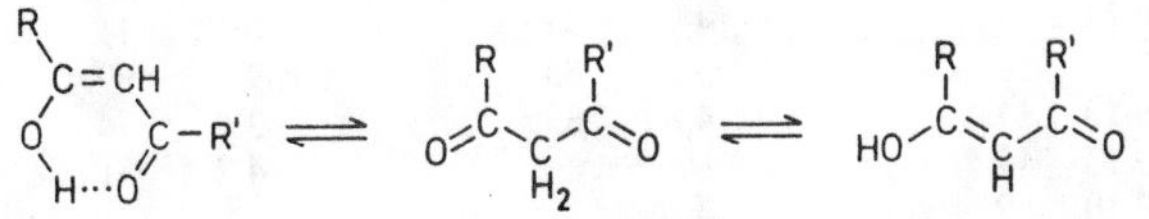

(4b) ⇌ *(4a)* ⇌ *(4c)*

Open-chain 1,3-dicarbonyl compounds are observed in the *trans*-enolic form only in rare cases [43] (for examples see references [44, 45]). When the *trans*-enolic form is excluded, the keto/enol equilibrium constant K_T is given by Eq. (4-23).

$$K_T = \frac{[\text{enol}]}{[\text{diketo}]} \tag{4-23}$$

In solution, open-chain 1,3-dicarbonyl compounds enolize practically exclusively to the *cis*-enolic form *(4b)*, which is stabilized by intramolecular hydrogen bonding. In contrast, cyclic 1,3-dicarbonyl compounds (*e.g.* cycloalkane-1,3-diones [46]), can give either *trans*-enols (for small rings) or *cis*-enols (for large rings). As the diketo form usually

* The much greater gas-phase base strength of aniline (and cyclohexylamine) compared with ammonia is due to the polarizability of the large carbocyclic residue. However, aniline is less basic than cyclohexylamine in the gas-phase as well as in aqueous solution. The nitrogen lone-pair in aniline, unlike that in cyclohexylamine, is conjugated with the aromatic π-system of the benzene ring and, to some extent, delocalized. Protonation of the aniline nitrogen atom localizes this electron pair and causes some loss of delocalization energy.

** If R≠R′, two *cis*-enolic and two *trans*-enolic forms can exist.

is more dipolar than the chelated *cis*-enolic form, the keto/enol ratio often depends on solvent polarity. This will be discussed in more detail for the cases of ethyl acetoacetate and acetylacetone [47–50, 134, 135].

The equilibrium constants, measured by ^{1}H-NMR spectroscopy, of ethyl acetoacetate and acetylacetone [47, 48, 134] (Table 4-2) indicate for these *cis*-enolizing 1,3-dicarbonyl compounds a higher enol content in apolar aprotic than in dipolar protic or dipolar aprotic solvents.

Table 4-2. Equilibrium constants and mole fractions of enol tautomers of ethyl acetoacetate *(4a)* (R $=CH_3$, R′=OEt; K_T), acetylacetone *(4a)* (R=R′=CH_3; K_T'), and 5,5-dimethyl-1,3-cyclohexanedione *(5a)* (K_T''), determined ^{1}H-NMR spectroscopically at *ca.* 20 °C at solute concentrations of *ca.* $10^{-3} \ldots 10^{-2}$ mol/l, *i.e.* under conditions unperturbed by self-association of the solute [134], values for pure solutes excepted [47].

Solvents (deuterated)	K_T	x(enol) (cmol · mol^{-1})	K_T'	x(enol) (cmol · mol^{-1})	K_T''	x(enol) (cmol · mol^{-1})
Gas phase[a)]	0.74	42.5	11.7	92	–	–
Cyclohexane	1.65	62	42	98	–	–
Tetrahydrofuran	0.40	29	7.2	88	–	–
Toluene	0.39	28	10	91	0.08	7
Tetrachloromethane	0.29	22.5	29	97	–	–
Benzene	0.26	21	14.7	94	0.12	11
Ethanol	0.14	12	5.8	85	169	99.4
1,4-Dioxane	0.13	11.5	4.8	83	2.8	74
Acetone	0.13	11.5	–	–	4.2	81
Pyridine	0.10	9	3.7	79	–	–
Chloroform	0.09	8	5.94	86	0.05	5
Dichloromethane	0.09	8	4.2	81	–	–
Pure solute[b)]	0.081	7.5	4.3	81	–	–
Methanol	0.07	6.5	2.9	74	148	99.3
Water	0.07	6.5	0.23	19	19	95
Dimethyl sulfoxide	0.05	5	2.0	67	94	99.0

[a] Values at 40 °C; see reference [39].
[b] Values at 33 °C; see reference [47].

The ratios obtained in the apolar aprotic solvents approach the gas-phase values [39]. In principle, on dissolution of β-dicarbonyl compounds in solvents of low polarity, the percentage of the *cis*-enolic form increases, whereas polar solvents displace the equilibrium towards the diketo form. Although at first sight it is surprising that increasing solvent polarity diminishes the enol content, this is understandable in terms of intramolecular chelatization of the enol. The enol form is the least polar of the two tautomers because, in the enol form intramolecular hydrogen bonding helps reduce the dipole-dipole repulsion of the carbonyl groups, which is unreduced in the diketo form. Furthermore, the enol stabilization due to the intramolecular hydrogen bonding will be more pronounced when intermolecular hydrogen bonding with the solvent does not compete. Thus, a change to a more polar solvent, with a tendency towards intermolecular hydrogen bonding (EPD solvents) is generally associated with a decline in enol content.

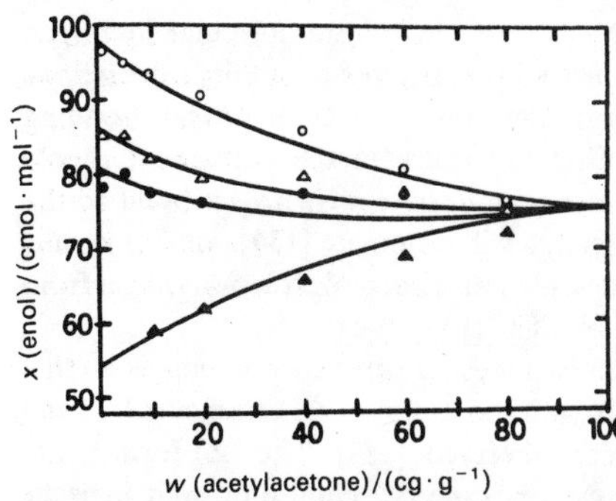

Fig. 4-3. Effect of solvent and concentration on the keto/enol equilibrium of acetylacetone in four solvents of different polarity at 37±2 °C: CCl_4 (○), $CHCl_3$ (△), CH_2Cl_2 (●), and $HCON(CH_3)_2$ (▲) [50].

In agreement with this, the enol content also depends strongly on the initial concentration of the 1,3-dicarbonyl compound; see Fig. 4-3 [50].

As the dipolar 1,3-dicarbonyl compound acetylacetone is progressively diluted with apolar solvents, the enol content increases. Conversely, progressive dilution with a dipolar aprotic EPD solvent such as *N,N*-dimethylformamide reduces the enol content of the acetylacetone solution.

In contrast to the *cis*-enolizing 1,3-dicarbonyl compounds, the *trans*-enolizing cycloalkane-1,3-diones with 4...6 membered rings show exactly the opposite dependence on solvent polarity [46]. In these compounds, intramolecular hydrogen bonding is excluded on steric grounds. For example, 5,5-dimethyl-1,3-cyclohexanedione *(5a,b)* is 95% enolized in aqueous solution [51]. However, in dilute solution in toluene, an apolar solvent, it is only 7% enolized [52, 134]; x(enol) = 7 cmol/mol; *cf.* Table 4-2.

(5a) ⇌ *(5b)*

(6a) ⇌ *(6b)*

Another example is the *β*-ketonitrile *(6a,b)*. Because of the linearity of the cyano group, a cyclic structure with an intramolecular hydrogen bond is impossible. As predicted, it is found that the enol content is greater in polar than in apolar solvents [53].

In general, for the protomer pairs in which the enol cannot form an intramolecular hydrogen bond such as *(4a)* ⇌ *(4c)*, the tautomeric equilibrium seems to be controlled almost completely by the hydrogen-bond acceptor property (Lewis basicity) of the solvent. EPD solvents enhance the enol content strongly; *cf.* *(5a)* in Table 4-2.

For the protomer pairs such as *(4a)* ⇌ *(4b)*, in which intramolecular hydrogen bonding is possible, the solute/solvent effect due to dipolarity/polarizability interactions dominates, although differential stabilization of the tautomers by hydrogen bonding remains significant. If there exists a substantial difference in the permanent dipole moments of both tautomers, and both tautomers can donate a hydrogen bond to the solvent, the dipole/dipole solute/solvent interactions will dominate [134]. Similar results have been obtained for acetoacetic acid itself, for which the enol mole fraction ranges from less than 2 cmol/mol in D_2O to 49 cmol/mol in CCl_4 [136].

Open-chain β-keto carboxylic esters with two mesityl substituents such as methyl 3-hydroxy-2,3-dimesityl-2-propenoate exist in solution only as (*Z*)-isomer *(4b)* and (*E*)-isomer *(4c)*; no keto form *(4a)* has been observed [223]. The (*Z*)-form exists predominantly in nonpolar solvents such as cyclohexane (90 cmol/mol) and benzene (87 cmol/mol). Increasing solvent polarity shifts this (*Z*)/(*E*)-equilibrium in favour of the more polar (*E*)-isomer, up to 76 cmol/mol (*E*)-form in ethanol [223]. The introduction of mesityl substituents does even stabilize enols of simple monocarbonyl compounds such as 2,2-dimesitylethenol, $Mes_2C{=}CH{-}OH$ [224].

The influence of solvents on tautomeric equilibria has been tried to relate to the solubility of both tautomers. Formally, the dissolution of a solute in a saturated solution may be regarded as an equilibrium:

crystals + solvent ⇌ dilute saturated solution

The energy of the solid phase will be independent of the solvent, and thus differences in solubility on going from one solvent to another will be a measure of the solvent effect on the Gibbs energy of the dissolved compound [1, 2]. Table 4-3 shows, that the variation in the equilibrium constants for the keto/enol tautomerization of 3-benzoyl camphor *(7a,b)*, in a series of solvents, is consistent with the solubilities of the two tautomers.

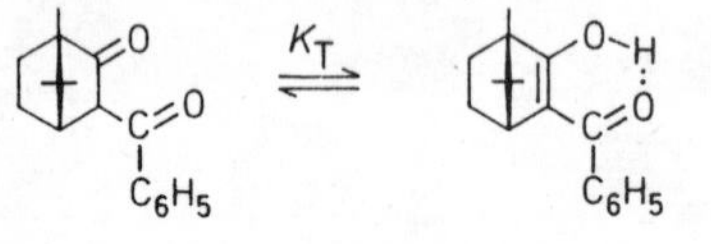

K_T

(7a) (7b)

The van't-Hoff-Dimroth relationship (4-24) [37, 54] states that two interconvertible isomers are in equilibrium when the ratio of their concentrations is proportional with that of their solubilities *S* in the appropriate solvent. In Eq. (4-24) *G* is a solvent-independent constant, characteristic for the 1,3-dicarbonyl compound. Therefore, the equilibrium constant, K_T

$$K_T = \frac{[\text{enol}]}{[\text{diketo}]} = G \cdot \frac{S_{\text{enol}}}{S_{\text{diketo}}} \tag{4-24}$$

equals *G* multiplied by the solubility ratio. From this it follows that the concentration of an enol will be at a maximum in the solvent where it has the highest relative solubility.

Table 4-3. Keto/enol equilibrium of 3-benzoyl camphor in five solvents at 0 °C [37].

Solvents	K_T [a]	$\frac{S_{enol}}{S_{diketo}}$ [b]	G
Diethyl ether	6.81	6.39	1.06
Ethyl acetate	1.98	1.81	1.09
Ethanol	1.67	1.57	1.06
Methanol	0.87	0.75	1.15
Acetone	0.85	0.80	1.06

[a] $K_T = [\text{enol}]/[\text{diketo}]$.
[b] Enol solubility/diketo solubility.

The applicability of Eq. (4-24) was demonstrated by O. Dimroth using 3-benzoyl camphor *(7a,b)* as an example [37]. This is particularly favorable, since both the diketo and the enol form are separately isolable, and the solubility of each is readily determined due to the slowness of their tautomerization. Although the ratio of the S-terms varies by a factor of 8, the value of G is constant within an error of $\pm 5\%$. This is understandable, since for each 1,3-dicarbonyl compound, the same characteristic difference between diketo and enol forms should in principle be observed.

The solubility is a reflection of how well the diketo and the enol form is solvated. Deviations from Eq. (4-24) are due mainly to the fact that the equation holds only for dilute solutions. At high concentrations the solubility is no longer determined by solvent-solute interactions alone, but self-solvation (*i.e.* interactions between the dissolved molecules themselves) now plays a role. Thus, as already stated, the enol content of a solution depends also on the concentration of the 1,3-dicarbonyl compound [55].

From a thermodynamic point of view, the keto/enol equilibrium is determined by the change in Gibbs energy, ΔG°, which in turn is comprised of the enthalpy ΔH° and entropy ΔS° of enolization; *cf.* Eq. (4-4). Therefore, the position of equilibrium in solution will be determined by the differences ΔH°_{solv} and ΔS°_{solv} between keto and enol forms, according to Eq. (4-25).

$$\Delta G^\circ_{keto-enol} = \Delta H^\circ_{keto-enol} \pm \Delta H^\circ_{solv} - T \cdot \Delta S^\circ_{keto-enol} - T \cdot \Delta S^\circ_{solv} \qquad (4\text{-}25)$$

The values of $\Delta H_{keto-enol}$ and $T \cdot \Delta S_{keto-enol}$ can be determined from the temperature dependence of the equilibria in the gas phase [39]. However, the magnitudes and signs of the solvation terms are not well known. ΔH_{solv} will be either negative or positive when changing from an unpolar to a polar solvent depending on which of the tautomers is the more polar. The resulting change in the equilibrium can be compensated by the corresponding entropy change. In an unpolar solvent, stronger solvation leads to a higher degree of order for the solvent molecules and, hence, to a decrease in the entropy. In contrast, polar solvent molecules show a high degree of order already in the absence of a dipolar solute. The thermodynamic parameters for the enolization reaction of acetylacetone in seven solvents have recently been determined. The molar Gibbs energy of reaction varies from -0.25 kJ/mol in methanol to -9.1 kJ/mol in cyclohexane at 25 °C [134].

Since these complications have prevented a quantitative estimation of the enthalpies and entropies of keto and enol forms in different solvents, a quantitative relationship between ΔG or K_T, and the solvating ability or polarity of the solvent (expressed in physical characteristics such as the dielectric constant ε_r, the dipole moment μ, the refraction index n, *etc.*) has not been possible either. Such equations would have the form of Eq. (4-26):

$$\Delta G^\circ = f(\varepsilon_r, \mu, n, \ldots) \quad \text{or} \quad K_T = f(\varepsilon_r, \mu, n, \ldots) \tag{4-26}$$

For pure electrostatic interactions between solvent and solute molecules, the Kirkwood Eq. (4-27) [56] is applicable (with dielectric constant ε_r; r and μ are the radius and the dipole moment of the solute molecule, respectively).

$$\Delta G^\circ = -\frac{1}{4\pi \cdot \varepsilon_0} \cdot \frac{\mu^2}{r^3} \cdot \frac{\varepsilon_r - 1}{2\varepsilon_r + 1} \tag{4-27}$$

Application of this equation to keto/enol equilibria gives Eq. (4-28), derived by Powling and Bernstein [57].

$$(\Delta H^\circ)_{gas} = (\Delta H^\circ)_{solution} + \frac{1}{4\pi \cdot \varepsilon_0} \cdot \left(\frac{\varepsilon_r - 1}{2\varepsilon_r + 1} \cdot \frac{\varrho}{M_r}\right) \cdot (\mu_1^2 - \mu_2^2) \tag{4-28}$$

M_r is the molar mass and ϱ is the density of the solvent, μ_1 and μ_2 are the dipole moments of the least and more stable isomers, respectively. The proposed linear function between the enthalpy of tautomerization and the solvent quantity $[(\varepsilon_r - 1)/(2\varepsilon_r + 1)]\varrho/M_r$ gives in the case of ethyl acetoacetate and acetylacetone approximate straight lines (omitting the alcohols), but with considerably scattering of the points [47]. This scattering is not surprising in view of the rather concentrated solutions used, the specific interactions expected in hydrogen-bonding solvents, and the possible entropy effects in these systems.

A wide variety of different theoretical (*e.g.* Kirkwood function) and empirical (*cf.* Chapter 7) parameters of solvent polarity have successfully been tested using multivariate statistical methods in order to model the solvent-induced changes in keto/enol equilibria [134].

Most simple monocarbonyl compounds are enolized to such a small extent, that it is difficult to determine reliably their enol content in solution [58, 137]. The enol content of acetone for example is about $6 \cdot 10^{-8}$ cmol/mol at equilibrium [137]. A remarkable example of a solvent-dependent keto/enol equilibrium of a monocarbonyl compound is 2-hydroxy-7-isopropyl-1,4-dimethylazulene (2-hydroxy-guajazulene) *(8a, b)* [59].

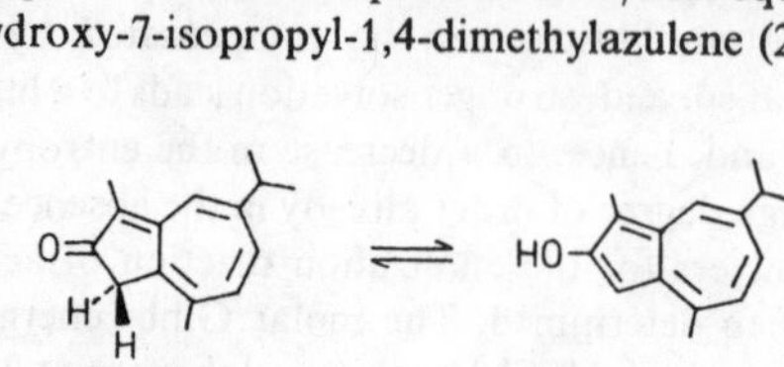

O= ⇌ HO–

(8a) *(8b)*

Whereas in chloroform and water, no enol form *(8b)* is detectable, the enol content is 5 cmol/mol in cyclohexane, 20 cmol/mol in methanol, 55 cmol/mol in 1,4-dioxane, and 95 cmol/mol in dimethyl sulfoxide. Apparently, the enol is stabilized in solvents which can act as hydrogen bond acceptors, while the keto form is favored in protic solvents acting as hydrogen bond donors. 6-Hydroxy-4,8-dimethylazulene behaves similarly [59a]. In polar solvents such as acetonitrile or dimethyl sulfoxide, the enolic azulenoid structure is exclusively observed, whereas in less polar solvents, dichloromethane or chloroform, a keto/enol equilibrium in a ratio of about 3 : 1 is detectable by ^{1}H-NMR measurements [59a].

In the case of the tautomerization between 9-anthrone, *(9a)*, and 9-anthranol, *(9b)*, the equilibrium lies practically completely on the side of the keto form *(9a)* in the gas phase as well as in inert solvents such as *iso*-octane and benzene; *e.g.* in benzene at 20 °C, the enol content is 0.25 cmol/mol [60, 134].

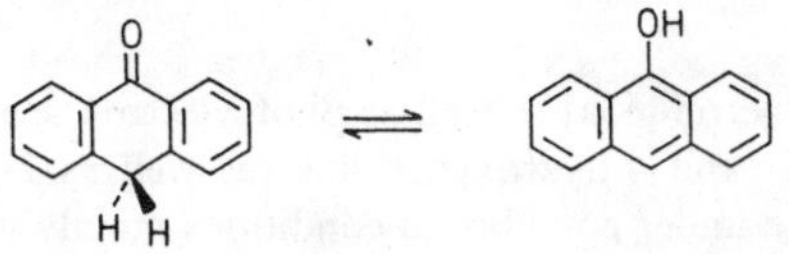

(9a) *(9b)*

Addition of increasing amounts of triethylamine to a benzene solution of *(9a)* leads to a gradual shift of the equilibrium towards the enol form *(9b)*. This can be interpreted in terms of hydrogen-bond formation between 9-anthranol and triethylamine. In hydrogen-bond accepting solvents such as *N,N*-dimethylformamide (enol content 56.5 cmol/mol at 20 °C), pyridine (58 cmol/mol), and dimethyl sulfoxide (61.5 cmol/mol) the anthranol content increases further [61, 134].

Particularly well-studied tautomeric keto/enol equilibria are those of 3-pyridinyl 2-picolyl ketone [138] and *t*-butyl 2-picolyl ketone [139].

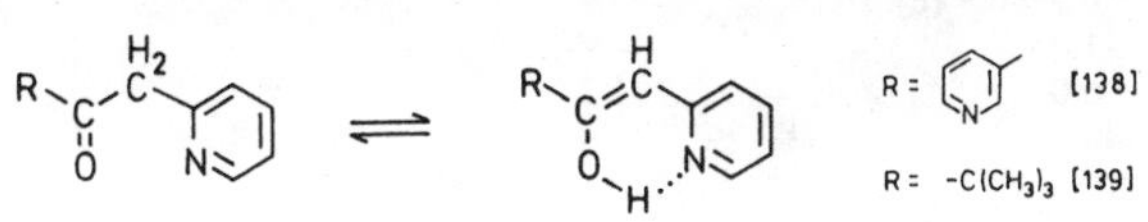

R = 3-pyridinyl [138]

R = $-C(CH_3)_3$ [139]

(10a) *(10b)*

In both cases, the tautomeric equilibria are shifted to the more dipolar keto form *(10a)* with increasing solvent polarity. For instance, *t*-butyl 2-picolyl ketone (R = *t*-butyl) is the only tautomer observed in 2,2,2-trifluoroethanol [139], and 3-pyridinyl 2-picolyl ketone (R = 3-pyridinyl) dominates in water with a mole fraction of 60 cmol/mol[138].

Further examples of solvent-dependent keto/enol equilibria are found in reference [43].

4.3.2 Solvent Effects on Other Tautomeric Equilibria [62–64, 140]

Solvent effects similar to those described for the keto/enol equilibria can also be found for other tautomerisms, *e.g.* lactam/lactim, azo/hydrazone, ring/chain equilibria, *etc.* [62–64]. The pecularities arising here can only be indicated by means of a few representative examples.

One of the classic studies of *lactam/lactim tautomerism* is the determination of the 2-hydroxypyridine *(11a)* ⇌ 2-pyridone *(11b)* equilibrium [63–65, 141–145].

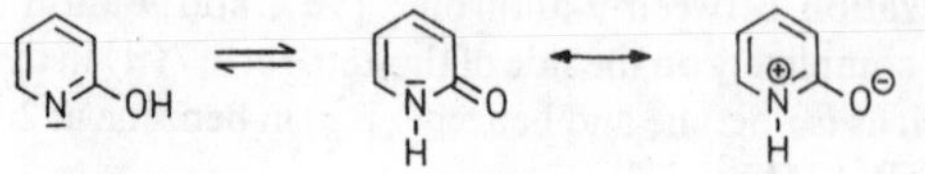

(11a) *(11b)* *(11b')*

IR and UV/Vis [65], mass spectrometric [65a] as well as photoelectron spectroscopic measurements [65b] reveal that 2- and 4-hydroxypyridine (as well as 2- and 4-mercaptopyridine) exist in the gas phase under equilibrium conditions mainly in the hydroxy- (resp. mercapto-) forms. This is contrary to the situation in solution. In solution, in most solvents the tautomeric pyridone-form predominates in the 2- and 4-series, the equilibrium constant depending on the polarity of the solvent [66, 67]. The gas-phase and solution equilibrium constants of 2- and 4-hydroxypyridine are given in Table 4-4.

Table 4-4. Gas-phase and solution equilibrium constants $K_T = [NH]/[OH]$ of 2- and 4-hydroxypyridine at 25...30 °C unless otherwise stated [65, 67].

Solvents	2-Hydroxypyridine K_T	4-Hydroxypyridine K_T
Vapour [a]	0.4 ± 0.25 (by UV) [c] 0.5 ± 0.3 (by IR) [d]	<0.1 [d]
Cyclohexane [b]	1.7	—
Chloroform [b]	6.0	1.3
Acetonitrile [b]	148	4.6
Water [a]	910	1900

[a] Reference [65]. [c] At 130 °C.
[b] Reference [67]. [d] At 250 °C.

The gas-phase equilibrium constants differ from those in aqueous solution by as much as 10^4! The large differences between the stabilities of the tautomeric forms in the gas phase and in solution once more reveal the dominant influence of solvation on relative molecular stabilities.

By considering the equilibrium *(11a)* ⇌ *(11b)* in solvents of varying polarity it has been found that increasing solvent polarity shifts the equilibrium towards the pyridone-form. This form is more dipolar than the hydroxy-form due to the contribution of the charge-separated mesomeric form *(11b')*. Furthermore, the hydrogen-bonding ability of the solvent plays an important role since hydrogen-bond donors tend to stabilize the oxo-form, whereas hydrogen-bond acceptors stabilize the hydrogen-form. For example, the oxo-form of 6-chloro-4-methyl-2-hydroxypyridine predominates in water, while in various other solvents the hydroxy-form mainly exists: 67 cmol/mol in methanol, 56 cmol/mol in chloroform, 96 cmol/mol in dimethyl sulfoxide, and 95 cmol/mol in cyclohexane [66].

The precise determination of the protomeric equilibrium constants K_T for 2(4)-hydroxypyridine ⇌ 2(4)-pyridone is rather difficult because of self-association even in highly diluted solution of the tautomers in nonpolar solvents such as cyclohexane. Self-associated tautomers may have K_T-values which are substantially different from those of the unassociated isomers [142].

A quantitative model for the differential solvation of the hydroxypyridine/pyridone tautomer pair in terms of reaction-field and hydrogen-bonding effects, using multivariate regression analysis, has been given by Beak *et al.* [141]. *Ab initio* calculations correctly predict the greater stability of 2-hydroxypyridine *(11a)* as compared to 2-pyridone *(11b)* in the gas-phase [144].

In contrast to mono-hydroxypyridines and mono-hydroxypyrimidines, the lactam-lactim equilibria of uracils are not found to be markedly influenced by solvent polarity [143].

In the vapour phase, both 2- and 4-hydroxyquinoline exist as the NH-forms, *i.e.* as 2- and 4-quinolones, in contrast to the results found for 2- and 4-hydroxypyridines [145].

The lactam/lactim tautomerism of hydroxamic acids and their O-alkyl and O-acyl derivatives have also been studied [146]. Hydroxamic acids exist in the solid state and in polar solvents as the lactam tautomer only, whereas in nonpolar solvents the hydroximic tautomer is also present.

Di-(2-quinolyl)methane forms a solvent-dependent tautomeric equilibrium between a colourless form *(12a)* and a coloured, hydrogen-bonded form *(12b)* in solution [68].

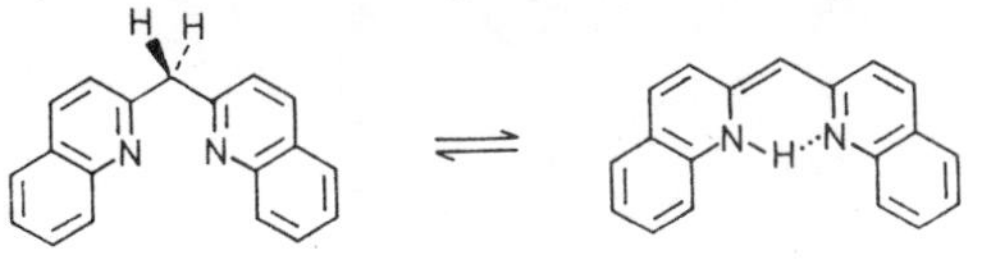

(12a) ⇌ *(12b)*

The thermodynamic data for the transformation of the two forms have been estimated and are given in Table 4-5.

In hydrogen-bond donor solvents such as alcohols and chloroform the tautomeric equilibrium is shifted in favor of the colourless form *(12a)* more than in other solvents. This is obviously due to the formation of hydrogen bonds between *(12a)* and these protic solvents. In aprotic solvents, ΔH° is negative and the reaction is exothermic. Since, however, all ΔG° values are positive, the negative value of ΔH° must be over-compensated by a positive entropy change; *cf.* Eq. (4-4).

This entropy decrease for the formation of the coloured form *(12b)* may be due to the planarization of the molecule (formation of the N—H···N bridge) and the fixation of

Table 4-5. Thermodynamic data for the tautomeric conversion *(12a)* → *(12b)* of di(2-quinolyl)methane at 20 °C [68].

Solvents	ΔG° [a)] (kJ · mol⁻¹)	ΔH° [a)] (kJ · mol⁻¹)	$T \cdot \Delta S^{\circ}$ [a)] (kJ · mol⁻¹)
Ethanol	+12.6	+ 9.6	− 2.9
Chloroform	+ 9.6	+ 8.4	− 1.3
tert-Butanol	+ 8.8	+ 6.7	− 2.1
Benzene	+ 6.3	− 0.4	− 6.7
Tetrachloromethane	+ 5.4	− 1.7	− 7.1
N,N-Dimethylformamide	+ 4.6	− 3.3	− 7.9
n-Heptane	+ 4.2	− 6.3	−10.5
Carbon disulfide	+ 2.9	−10.0	−13.0

[a] The energy which is added to the system is considered to be positive, and that given up by the system to be negative.

the di-*cis*-form *(12b)*, while in *(12a)* there is free rotation around the central C—C bond. In protic solvents, the colourless form *(12a)* is already stabilized by hydrogen bonding and $T \cdot \Delta S^{\circ}$ is small. Therefore, the ΔG° values are greater in these solvents and ΔH° is positive [68].

The ketimine *(13a)*, prepared from desoxybenzoine and aniline, is also subject to a solvent-dependent tautomerism called *imine/enamine tautomerism*. The enamine content of a solution of *(13a)* increases in the order tetrachloromethane (31 cmol/mol at 35 °C), [D_5]pyridine (47.5 cmol/mol at 55 °C), and [D_6]dimethyl sulfoxide (67 cmol/mol at 55 °C) [69]. Hydrogen-bond acceptor solvents favor the enamine form *(13b)* due to hydrogen-bonding, whereas in less polar and apolar solvents the equilibrium is shifted towards the imine form *(13a)* [69].

$$H_5C_6{-}CH_2{-}C(C_6H_5){=}N{-}C_6H_5 \rightleftharpoons H_5C_6{-}CH{=}C(C_6H_5){-}NH{-}C_6H_5$$

(13a) *(13b)*

Other remarkable cases of solvent-dependent imine/enamine tautomerism have been reported by Ahlbrecht *et al.* [147], Scheffold *et al.* [69a], and Pérez-Ossorio *et al.* [211].

Compounds capable of a solvent-dependent *amino/imino tautomerism* are 3-methylcytosine *(14a, b)* and 1-alkyladenines [69b]. It has been shown by IR and UV/Vis spectroscopy that, in all cases, the imino-forms such as *(14b)* predominate in nonpolar media (*e.g.* 1,4-dioxane). However, the content of the amino-form *(14a)* increases with increasing solvent polarity, and in aqueous solution the amino-form predominates [69b]. Further interesting examples of solvent-dependent tautomeric amino/imino equilibria are given in references [148, 149].

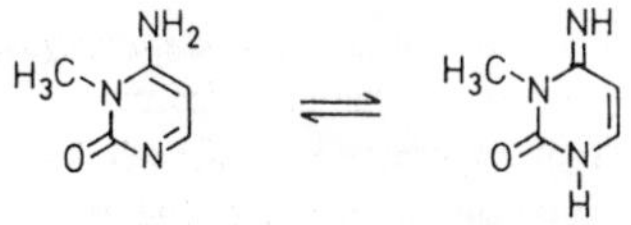

(14a) (14b)

An extreme case is the *nitrone/hydroxylamine tautomerism* between 2-methylindolenine-*N*-oxide *(15a)* and 2-methyl-*N*-hydroxyindole *(15b)* [70]. The position of this equilibrium depends strongly on the proton-accepting and donating abilities of the solvent: in pyridine and acetonitrile there exists 0 cmol/mol of form *(15a)*, 33 cmol/mol in tetrachloromethane, but 100 cmol/mol in phenol. Thus it is possible to observe either *(15a)* (in phenol) or *(15b)* (in pyridine) depending on the solvent used.

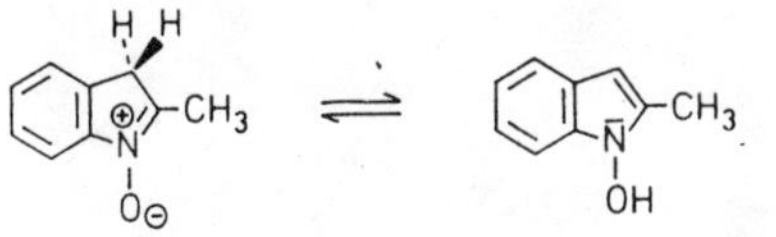

(15a) (15b)

The position of the *N*-oxide/*N*-hydroxy equilibrium of 1-hydroxybenzotriazole, which is similar in structure to *(15b)*, is also solvent-dependent: approximately 6, 11, 18, and 26 cmol/mol of the *N*-oxide form is present at equilibrium in dimethyl sulfoxide, acetone, formamide, and methanol, respectively [218].

Because azo dyes are of commercial importance as colouring materials, the *azo/hydrazone tautomerism* of hydroxy-substituted azo compounds has been intensively studied [71, 228]. In the case of 4-phenylazo-1-naphthol *(16a)* an increase in the solvent polarity displaces the tautomeric equilibrium towards the more dipolar quinone hydrazone-form *(16b)* [72–74, 74a, 74b, 150–154]. In addition the NH and OH groups of both tautomers are capable of forming hydrogen bonds with suitable solvents. Due to the

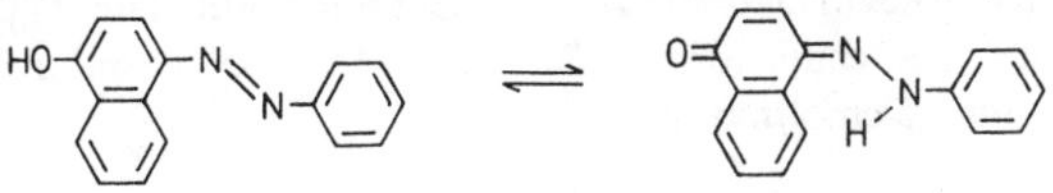

(16a) (16b)

stronger hydrogen-bond donor ability of the OH group compared with that of the NH group it will be to a different degree. Thus, the formation of hydrogen bonds with HBA solvents such as pyridine should mainly stabilize the azo form, whereas the basic imino group in the hydrazone form should be more stabilized in HBD solvents such as chloroform or acetic acid. Using UV/Vis spectroscopic measurements, the following order of increasing proportions of the hydrazone form *(16b)* has been found in solution (mole fractions in cmol/mol): pyridine (15) < acetone (30) < ethanol (31) < methanol (40) < benzene (56) < chloroform (79) < acetic acid (89) [151]. That is, the azo form *(16a)* is indeed stabilized in pyridine, acetone, ethanol, and methanol, whereas the hydrazone form dominates in chloroform and acetic acid. This is relative to the equilibrium in benzene which is used as reference solvent. In *N,N*-dimethylformamide and dimethyl sulfoxide solutions, the azo dye *(16)* is converted into its mesomeric anion, due to the high basicity and high dielectric constants of these solvents [151]. Similar results have been obtained with 4-*alkyl*azo-1-naphthols [152]. Quantum-chemical calculations have shown that the azo form *(16a)* should be the more stable isomer in the gas phase [153, 154].

Relative independent of solvent polarity, 4-nitrosophenol *(17a)* exists in solution mainly in the 1,4-benzoquinone monoxime-form *(17b)* [75, 76]: *ca.* 83 cmol/mol monoxime in 95 cl/l aqueous ethanol [75], *ca.* 86 cmol/mol in 1,4-dioxane, and *ca.* 75 cmol/mol in acetone (at 20 °C) [76].

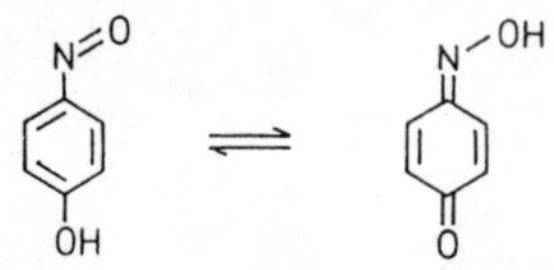

(17a) ⇌ *(17b)*

In case of 2-nitrosophenols such as 2,4-dialkyl-6-nitrosophenols, the tautomeric equilibrium is shifted towards the 1,2-benzoquinone monoxime-form with increasing polarity of the solvent [76a].

The reversible tautomeric equilibrium between phosphane oxides and ylides with a P—OH bond, *e.g.* $R_2P(=O)-CHR'_2 \rightleftharpoons R_2P(-OH)=CR'_2$ ($R = C_6H_5$; $R' = p\text{-}Cl-C_6H_4-SO_2$), have been found to be solvent-dependent [155]. The more dipolar phosphane oxide dominates in polar solvents (90 cmol/mol in dichloromethane at 25 °C), whereas in HBA solvents such as tetrahydrofuran the P—OH form is favoured (*ca.* 54 cmol/mol).

Finally, four different examples of solvent-dependent ***ring/chain tautomerism*** should be mentioned [77, 210]. The equilibrium between phthalaldehydic acid *(18a)* and phthalide *(18b)*, which in decalin lies in favor of *(18b)* (90 cmol/mol phthalide at 20 °C), is strongly shifted towards the open-chain form *(18a)* in the hydrogen-bond accepting solvent dimethyl sulfoxide (only 5 cmol/mol phthalide) [78].

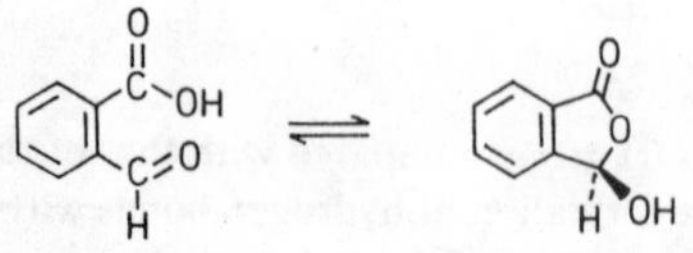

(18a) ⇌ *(18b)*

In the case of 5-hydroxy-2-pentanone, there is a slight preference for the open-chain form *(19b)* over the cyclic hemiketal *(19a)* (in [D_{12}]cyclohexane 55 cmol/mol and in [D_6]dimethyl sulfoxide 61 cmol/mol *(19b)*) in most organic solvents [79].

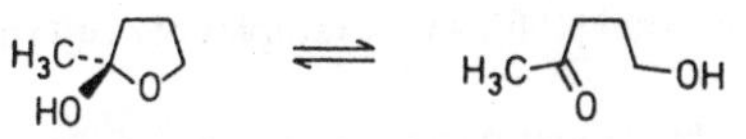

(19a) ⇌ *(19b)*

An increase in solvent polarity further favors the open-chain tautomer; in water there is no evidence for any cyclic form [79]. This is remarkable in view of the fact that the furanose/pyranose equilibria of sugars, which are interconverted through the open-chain form, are also solvent-dependent [80, 81, 159]. Arabinose, for example, in [D_5] pyridine consists of 66 cmol/mol pyranose form ($\alpha:\beta=33:33$) and 34 cmol/mol furanose form ($\alpha:\beta=21:13$), compared with 95.5 cmol/mol pyranose form ($\alpha:\beta=60:35.5$) and 4.5 cmol/mol furanose form ($\alpha:\beta=2.5:2.0$) in deuterium oxide as solvent [80, 159].

In solution, the hydroxylaminomethylation product of 2-naphthol and acetaldehyde exhibits an equilibrium between the hydroxynitrone-form *(20a)* and the cyclic hydroxylamine-form *(20b)*, the position of which depends on the solvent: 29 cmol/mol hydroxynitrone-form in [D_6]dimethyl sulfoxide, and 94 cmol/mol in [D_4]methanol [80a].

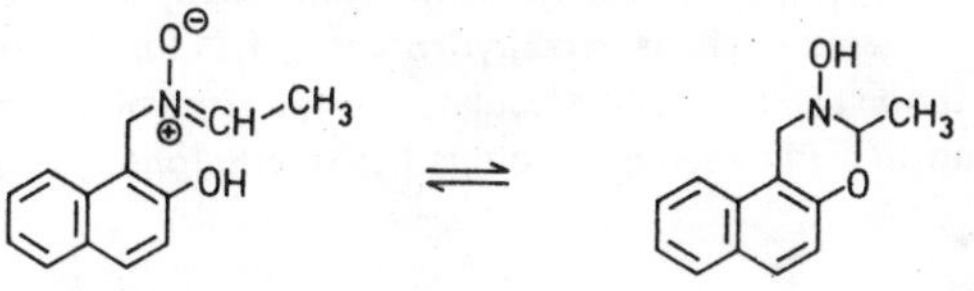

(20a) ⇌ *(20b)*

The ring/chain tautomeric equilibrium between (2-hydroxyphenylimino)phosphorane *(21a)* and 1,3,2-benzoxazaphospholine *(21b)* has been studied in thirteen solvents by NMR spectroscopy [80b]. This equilibrium is shifted towards the ring-form *(21b)* in hydrogen-bond acceptor solvents (*e.g.* tris-*n*-propylamine, dimethyl sulfoxide), compared to inert solvents such as benzene or acetone. Therefore, depending on substituents and solvents, it is possible to prepare either iminophosphoranes or benzoxazaphospholines [80b].

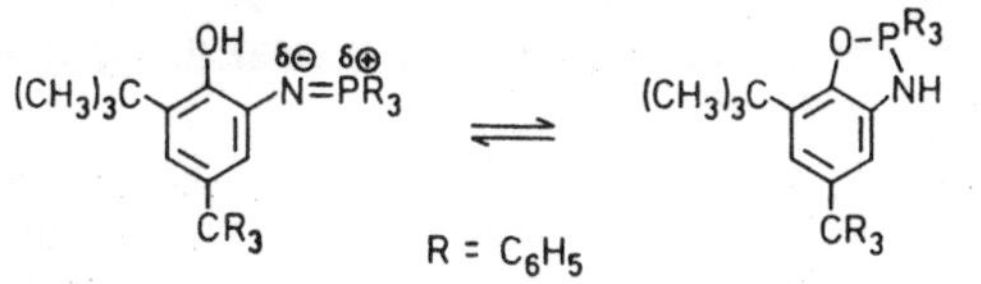

R = C_6H_5

(21a) ⇌ *(21b)*

Further remarkable examples of solvent-dependent ring/chain tautomeric equilibria can be found in references [156–158, 210].

The position of *metallotropic tautomeric equilibria* can also be strongly solvent-dependent [160–163]. Metallotropic transformations of the σ,σ-type are related to prototropic tautomeric equilibria whereby the mobile hydrogen atom is replaced by an organometallic group. Metallotropic reactions also include π,π- and σ,π-transitions depending on the nature of the bond formed by the metal [160]. Two examples of σ,σ-type metallotropic equilibria will conclude this Section.

The metallotropic C → O transition of the trimethylstannyl group of liquid trimethylphenacyltin *(22a)* to give the O-isomer *(22b)* has been found to be solvent-dependent [161].

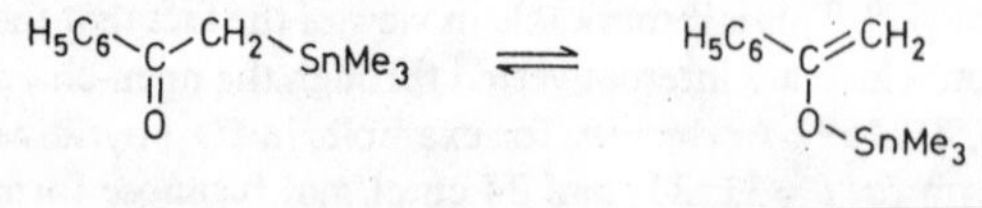

H_5C_6–C(=O)–CH_2–$SnMe_3$ ⇌ H_5C_6–C(=CH_2)–O–$SnMe_3$

(22a) *(22b)*

As expected, the relative concentration of the more dipolar C-isomer *(22a)* increases with increasing solvent polarity (concentration of *(22a)* in cmol/mol): C_6H_{12} (74) < pure liquid (78) < C_6H_6 (81) < $ClCH_2CH_2Cl$ (83) < $CHCl_3$ (95) < CH_3COCH_3 (>99) [161].

The dynamic exchange of the trimethylstannyl substituent in 1,3-cyclopentadienyl-trimethylstannan *(23a)* (and in 1,3,5,7-cyclononatetraenyl trimethylstannan) proceeds in less polar, weakly coordinating solvents such as tetrahydrofuran, 1,2-dimethoxy-ethane, chloroform, and dichloromethane *via* an intramolecular, orbital-symmetry controlled sigmatropic reaction. Addition of EPD solvents such as hexamethylphosphoric

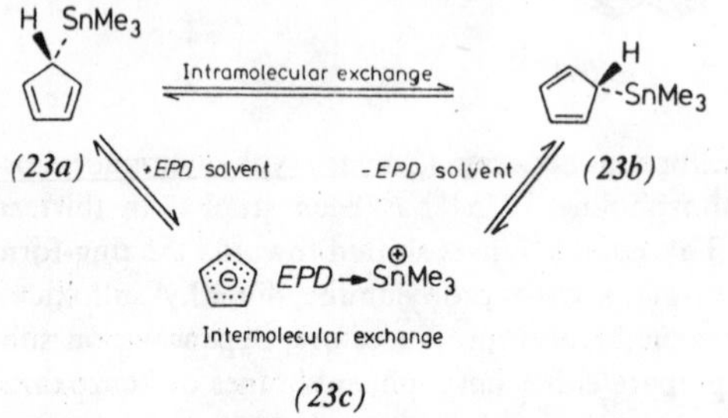

H SnMe3 — Intramolecular exchange — H SnMe3

(23a) +EPD solvent −EPD solvent *(23b)*

EPD → SnMe3 (⊕), cyclopentadienide (⊖)

Intermolecular exchange

(23c)

triamide or *N,N*-dimethylformamide to the tetrahydrofuran solution of *(23a)* shifts the equilibrium in favour of the ion pair *(23c)*, thus facilitating an intermolecular, dissociative mechanism for the substituent exchange [162].

In general, polar but weakly coordinating solvents will facilitate intramolecular metallotropic processes. Solvents exhibiting both high polarity and high coordinating capacity (EPD solvents), capable of inducing the heterolysis of the carbon-metal bond, should accelerate metallotropic processes *via* an intermolecular dissociative mechanism.

4.4 Solvent Effects on Other Equilibria

Not only tautomeric equilibria are subject to considerable solvent effects. Other equilibria such as rotational and conformational equilibria [81–83], *cis/trans* (or *E/Z*) isomerization, valence isomerization [84], ionization, dissociation, and association [85] (some of which are considered in Section 2.6), complex equilibria [86], acid/base equilibria *etc.*, are also strongly affected by the medium. Only a small number of representative examples will be considered in this Section in order to give an idea of how solvents can affect these different kinds of equilibria.

4.4.1 Solvent Effects on Brønsted Acid/Base Equilibria [8–13, 104–108, 163]

Sections 3.3.1 and 4.2.1 dealt with Brønsted acid/base equilibria in which the solvent itself is involved in the chemical reaction either as acid or as base. This Section now describes some examples of solvent effects on proton-transfer (PT) reactions in which the solvent does not intervene directly as a reaction partner. New interest in the investigation of such acid/base equilibria in non-aqueous solvents has been generated by the pioneering work of Barrow *et al.* [164]. He studied the acid/base reactions between carboxylic acids and amines in tetrachloromethane and chloroform.

According to Eq. (4-29), protons can be transferred from Brønsted acids A—H to bases |B, *via* the hydrogen-bonded covalent and ionic complexes *(a)* and *(b)*, depending on both the relative acidity resp. basicity strength of A—H and |B and the solvation capability of the surrounding medium. Eq. (4-29) is simplified because not only 1:1 complexes but 1:2 and higher complexes can be formed in solution.

$$\text{A-H} + \text{|B} \underset{}{\overset{\text{Association}}{\rightleftharpoons}} \underset{(a)}{\text{A-H}\cdots\text{B}} \underset{\text{Transfer}}{\overset{\text{Proton}}{\rightleftharpoons}} \underset{(b)}{\text{A}^{\ominus}\cdots\text{H-B}^{\oplus}} \overset{\text{Dissociation}}{\rightleftharpoons} \text{A}^{\ominus} + \text{H-B}^{\oplus} \qquad (4\text{-}29)$$

The solvent can influence all three steps of Eq. (4-29): the association, the proton-transfer, and the dissociation step. The main factor which determines the position of the acid/base equilibrium given in Eq. (4-29) is the differential solvation of the covalent and the ionic hydrogen-bonded complexes *(a)* and *(b)*. Both the hydrogen-bonded complex *(a)* and the proton-transfer ion pair *(b)* have been observed in the systems 4-nitrophenol/triethylamine [165], picric acid/triethylamine [166], chloro-substituted phenols/*N*-methylpiperidine or *n*-octylamine [167], and trifluoroacetic acid/pyridine [168]. With increasing solvent polarity the proton-transfer equilibrium *(a)* ⇌ *(b)* is shifted in favour of the ionic structure *(b)*. New thermodynamic parameters ($K, \Delta H^{\circ}$) for the formation of hydrogen-bonded complexes of phenol with various bases in different solvents can be found in reference [209]. The strengths of hydrogen bonds between solutes (protonated amines and phenolate ions) in aqueous solution have been studied recently [220]. Formation of such solute/solute hydrogen bonds in water as solvent requires that competition from hydrogen-bonding of the HBD and HBA molecules to 55 M water must be overcome.

A simple example of an *intra*molecular proton-transfer reaction (thus avoiding the association and dissociation step) is given by the Mannich base *(24)* [169, 170], which can be considered as an analogue of the corresponding intermolecular complexes between phenols and amines [163]. UV/Vis and IR spectroscopic measurements show that this proton-transfer equilibrium is shifted to the right-hand side with increasing solvent polarity (concentration of *(24b)* in cmol/mol): CCl_4 (0) < $CHCl_3$ (15) < CH_2Cl_2 (26)

$< ClCH_2CH_2Cl$ (30) $< CH_3CN$ (40) [170]. A linear relationship exists between ln K_{PT} and Onsager's reaction field parameter $(\varepsilon_r - 1)/(2\varepsilon_r + 1)$, demonstrating the presence of nonspecific solute/solvent interactions only in these non-HBD solvents.

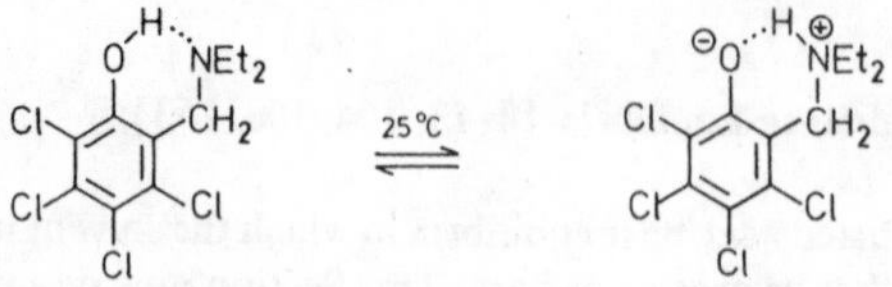

(24a) ⇌ (24b) [25 °C]

Another rather simple example is the acid/base reaction between tropolone and triethylamine, which has been studied using IR and ^{1}H-NMR spectroscopy in various solvents [171].

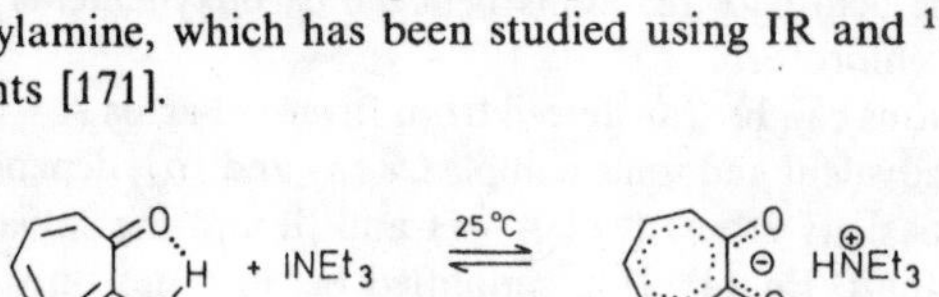

Tropolone + NEt_3 ⇌ [25 °C] tropolonate $HNEt_3^{\oplus}$

In non-HBD solvents such as *n*-heptane, tetrachloromethane, diethyl ether, deuterochloroform, and dimethyl sulfoxide, tropolone transfers its proton to triethylamine to give an ion pair which is in equilibrium with the non-associated reactants. There is no formation of a hydrogen-bonded complex between tropolone and triethylamine because of the fact that tropolone itself is intramolecularly hydrogen-bonded. The extent of the ion pair formation increases with solvent polarity. In polar HBD solvents such as ethanol, methanol, and water, this proton-transfer equilibrium is shifted completely towards the formation of triethylammonium tropolonate [171].

A peculiar example of a solvent-dependent regiospecific proton-transfer equilibrium is found for 4-amino-5-methylacridine *(25)* [172]. In aqueous hydrochloric acid, the ring nitrogen atom of *(25)* is protonated to give *(25a)*, whereas in ethanolic hydrochloric acid the primary amino group does preferentially accept the proton to give *(25b)*. Without a methyl group in the 5-position (*i.e.* with 4-aminoacridine) only the ring nitrogen atom is protonated in both solvents.

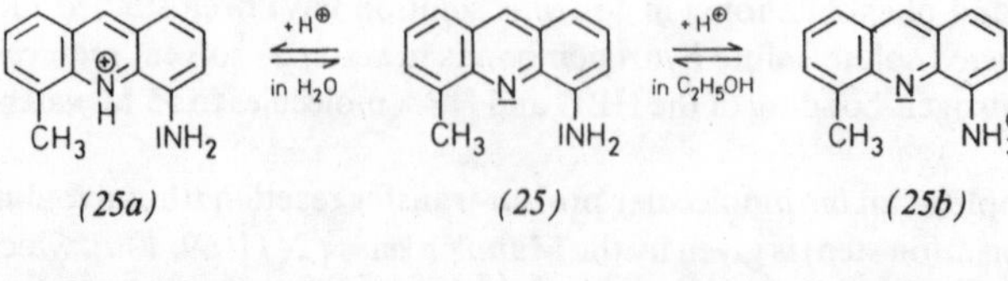

(25a) ⇌ [$+H^{\oplus}$, in H_2O] *(25)* ⇌ [$+H^{\oplus}$, in C_2H_5OH] *(25b)*

Obviously, the 5-methyl substituent sterically impedes the ethanol solvation of the $NH^{\oplus}$-form *(25a)*, thus favouring reaction at the more exposed 4-amino group. In water, protonation of the ring nitrogen atom and solvation of the resulting $NH^{\oplus}$-form *(25a)* by the smaller water molecules can take place despite the methyl group [172].

4.4.2 Solvent Effects on Lewis Acid/Base Equilibria [106–108, 173, 174]

Sections 2.2.6 and 2.6 dealt with Lewis acid/base equilibria in which principally the solvent itself is involved in the chemical reaction either as Lewis acid (EPA solvents) or as Lewis base (EPD solvents). This Section includes some recent examples of solvent-dependent Lewis acid/base equilibria in which the solvent is not directly involved as the reaction partner, but as the surrounding and interacting medium.

Formally analogous to Eq. (4-29), Eq. (4-30) describes in a simplified manner the reaction between a Lewis acid $A^{\oplus}$ and a Lewis base $|B^{\ominus}$, via tight ion pairs (which sometimes can be considered as EPD/EPA complexes), to give the covalent ionogen A—B [*cf.* also Eq. (2-13) in Section 2.6].

$$\underset{\text{Free Ions}}{A^{\oplus} + |B^{\ominus}} \underset{\text{Dissociation}}{\overset{\text{Association}}{\rightleftharpoons}} \underset{\text{Tight Ion Pair}}{\underset{(a)}{A^{\oplus}|B^{\ominus}}} \underset{\text{Ionization}}{\overset{\text{Electron transfer}}{\rightleftharpoons}} \underset{\text{Ionogen}}{\underset{(b)}{A-B}} \qquad (4\text{-}30)$$

The position of this equilibrium depends on the electrophilicity or nucleophilicity of $A^{\oplus}$ and $|B^{\ominus}$, respectively, as well as the solvation capability of the surrounding medium. The solvent can influence the association as well as the electron-transfer step (or in the reverse reaction the ionization and dissociation step). The position of the Lewis acid/base equilibrium given in Eq. (4-30) will depend mainly on the differential solvation of the ionic and covalent species *(a)* and *(b)*.

A simple example of an *intra*molecular Lewis acid/base reaction (thus avoiding the association step) is the xanthene dye rhodamine B, which exists in solution either in the red-coloured zwitterionic form *(26a)* or as the colourless lactonic form *(26b)* [175, 221, 222]. Solutions of rhodamine B in non-HBD solvents such as dimethyl

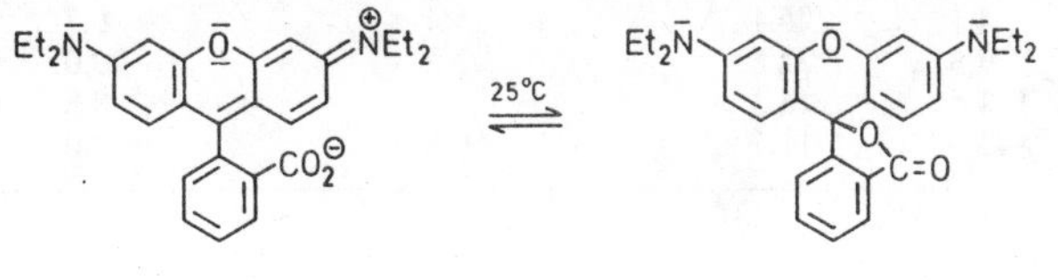

Et₂N̄ ... Ō ... $\overset{\oplus}{N}Et_2$; $CO_2^{\ominus}$ $\xrightleftharpoons{25\,°C}$ Et₂N̄ ... Ō ... N̄Et₂ ; O, C=O

λ_{max} = 543 nm (in EtOH)

(26a) *(26b)*

sulfoxide, *N,N*-dimethylformamide, 1,4-dioxane, pyridine, and hexamethylphosphoric triamide are entirely colourless, indicating complete conversion into the inner lactone *(26b)*. Protic solvents stabilize the zwitterion and shift the equilibrium toward the highly coloured zwitterionic form (concentration of *(26a)* in cmol/mol): $(CH_3)_3COH$ (1.6) < *n*-C_3H_7OH (65.2) < C_2H_5OH (70.6) < H_2O (81.5) < $HCONH_2$ (88.5) < CH_3OH (89.2) < CF_3CH_2OH (94.6) [221]. The conversion of *(26b)* to *(26a)* is sensitive to the presence of hydroxy groups to such an extent that a white piece of cellulose turns red on contact with a colourless solution of *(26b)* [175]. Addition of acids to solutions of rhodamine B in either non-HBD or protic solvents produces an intensely coloured cation (λ_{max} = 553 nm in ethanol [222]) by protonation of the carboxylate group of *(26a)* [221, 222].

Analogous solvent-dependent intramolecular Lewis acid/base equilibria between lactonic and zwitterionic forms have been found also for the xanthene dyes fluorescein and eosin [176].

The colourless spiropyran *(27b)* is another important example of an intramolecular Lewis acid/base equilibrium. In solution it is in equilibrium with the coloured zwitterionic species *(27a)* [99].

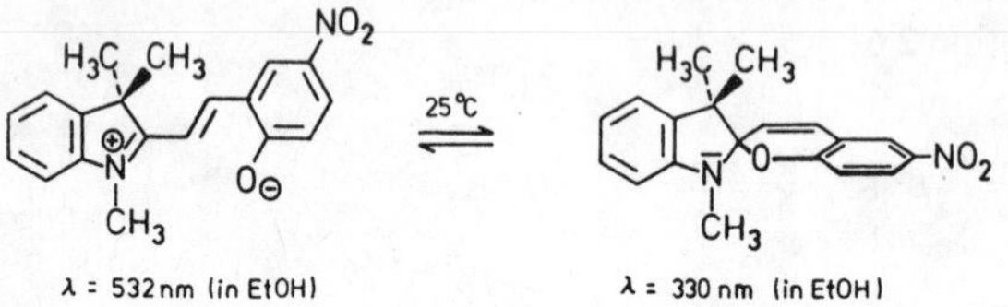

25 °C

λ = 532 nm (in EtOH) λ = 330 nm (in EtOH)

(27a) *(27b)*

The thermodynamic data presented in Table 4-6 reveal that the change in equilibrium constant is in the direction anticipated, with more ionization occurring as the solvent becomes more polar. Accordingly, the Gibbs energy change decreases monotonically with increasing solvent polarity.

Table 4-6. Thermodynamic data for the intramolecular Lewis acid/base reaction *(27a)* ⇌ *(27b)* at 25 °C [99].

Solvents	$K \cdot 10^5$ a)	$\Delta G°$ ($kJ \cdot mol^{-1}$)	$\Delta H°$ ($kJ \cdot mol^{-1}$)	$T \cdot \Delta S°$ ($kJ \cdot mol^{-1}$)
Benzene	4.1	25.1	18.4	− 6.7
Chloroform	9.8	22.6	2.2	−20.0
Acetone	134	16.3	7.5	− 8.8
Ethanol	843	11.7	8.8	− 3.0

a K = [*(27a)*]/[*(27b)*].

The negative entropy changes observed in all solvents are a result of an ordering of solvent molecules in the environment of the zwitterionic form. Since polar solvents are *per se* more structured than apolar solvents, proportionally less negative entropy changes are obtained in more polar solvents such as ethanol.

A further illustrative example is the *inter*molecular Lewis acid/base reaction between tropylium and isothiocyanate ions *via* tight ion pairs *(28a)* to give 7-isothiocyanatocycloheptatriene *(28b)* [177].

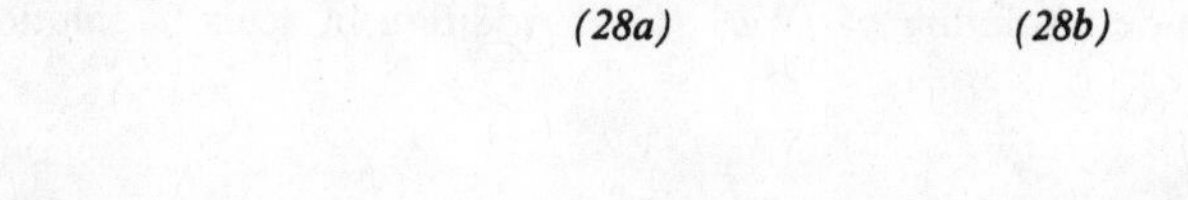

(28a) *(28b)*

In solution both the ionic and covalent form of tropylium isothiocyanate have been directly observed by low temperature ^{13}C- and ^{1}H-NMR spectroscopy. In deuteriochloroform and in diethyl ether, the covalent form *(28b)* is exclusively present below −10 °C. With increasing temperature and with increasing solvent polarity (addition of CD_3CN to the solution of *(28b)* in $CDCl_3$), the relative concentration of the ionic *(28a)* increases. In pure acetonitrile the ionic form *(28a)* dominates. Evidence that the ionic form is a tight ion pair in these solvents is given by NMR and UV/Vis spectra (*e.g.* the occurrence of a charge-transfer absorption between isothiocyanate and tropylium ion). Accordingly, the Gibbs energy of activation for the random migration of the isothiocyanato group around the cycloheptatriene ring decreases with increasing solvent polarity [177]. That is, the stabilization of the ionic *(28a)* by polar solvents corresponds to a similar stabilization of the preceding dipolar activated complex.

Analogous results have been obtained for the Lewis acid/base equilibrium between ionic tropylium azide and covalent 7-azidocycloheptatriene [178]. Again, in less polar solvents such as deuteriochloroform and even [D_6]acetone, no ionization to give the tropylium and azide ions could be detected. Dipolar liquid sulfur dioxide, however, induces complete ionization at low temperature (−70 °C).

The examples mentioned above are characterized by heterolysis of C—O or C—N bonds. In conclusion, a solvent-dependent Lewis acid/base reaction between carbocations and carbanions, produced by heterolysis of a weak C—C bond, is presented (*cf.* also Section 2.6).

A large variety of different combinations of charge-delocalized carbenium ions with carbanions has been investigated in order to find a well-balanced equilibrium mixture of free ions or ion-pairs and a neutral covalent product with clean C—C bond formation in solution [179]. The equilibrium between (4-nitrophenyl)malononitrile anions and trianisylmethyl cations [179] or triphenylcyclopropenium cations [180] was finally found to be the most useful. The Lewis acid/base equilibrium between the (4-nitrophenyl)malononitrile anion and the triphenylcyclopropenium cation has been already discussed in Section 2.6 as an example for solvent-dependent ionization reactions. The first-mentioned reaction can be described as:

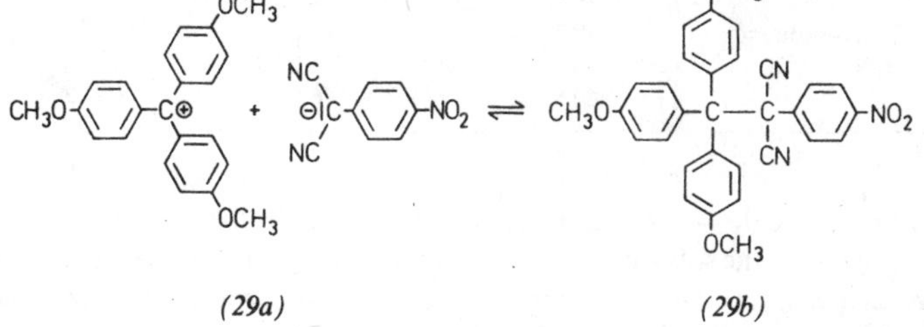

(29a) *(29b)*

A [D_2]dichloromethane solution at room temperature contains 70 cmol/mol ionic *(29a)* and 30 cmol/mol covalent *(29b)*. With decreasing solvent polarity the equilibrium is shifted toward the right-hand side (concentration of *(29b)* in cmol/mol): CD_2Cl_2 (30) $<CD_3CN$ (55) $<CD_3COCD_3$ (85) $<(CD_2)_4O$ (100). In [D_8]tetrahydrofuran solution only the covalent *(29b)* is present [179].

4.4.3 Solvent Effects on Conformational Equilibria [81–83, 181–184]

Changing the medium has a particular effect on various conformational and rotational equilibria [83, 181–184]. Because the Gibbs energy differences between conformational isomers are almost always very small (*ca.* 0...13 kJ/mol) and the solvation enthalpies of dipolar solutes are at least as large and often much larger than this, the medium can affect conformational equilibria very considerably. It is often found that one conformer or rotamer is predominant in one medium but not in another. This has led to the long-established rule that the conformer (rotamer) of higher dipole moment is more favored in media of high dielectric constant [83].

For example, the standard molar Gibbs energy for the *rotational* equilibrium *(30a)* $\rightleftharpoons$ *(30b)* of chloroacetaldehyde is strongly solvent-dependent, as shown in Table 4-7 [87].

36 °C

(30a) *(30b)*

Table 4-7.- Solvent dependence of the relative rotamer population (mole fraction of *(30b)*) and the standard molar Gibbs energy differences between rotamers of chloroacetaldehyde at 36 °C [87].

Solvents	*x (30b)* (cmol · mol^{-1})	$\Delta G°$ (kJ · mol^{-1}) for *(30a)* $\rightleftharpoons$ *(30b)*
trans-Decalin	44	−1.26
Cyclohexane	45	−1.30
Tetrachloromethane	47	−1.46
Chloroform	55	−2.34
Benzene	58	−2.68
Dichloromethane	61	−2.97
Acetone	72	−4.18
Acetonitrile	76	−4.60
N,N-Dimethylformamide	79	−5.23
Dimethyl sulfoxide	84	−6.07
Formamide	85	−6.28

Inspection of Table 4-7 reveals a substantial increase in the more dipolar rotamer *(30b)* as the polarity of the solvent increases. In saturated hydrocarbon solvents (the least dipolar solvents used), *(30b)* is only favored by *ca.* 1.3 kJ/mol, whereas in formamide (the most dipolar solvent used) it is favored by *ca.* 6 kJ/mol. In view of the higher dipole moment of *(30b)* over *(30a)* this appears reasonable [87].

Another remarkable example is the medium effect on the rotational equilibrium of ethoxycarbonylmethylene triphenylphosphorane *(31a)* $\rightleftharpoons$ *(31b)*. As the polarity of the solvent increases, the equilibrium shifts in the direction of the *s-trans*-isomer *(31b)*, as shown by the equilibrium constants presented in Table 4-8.

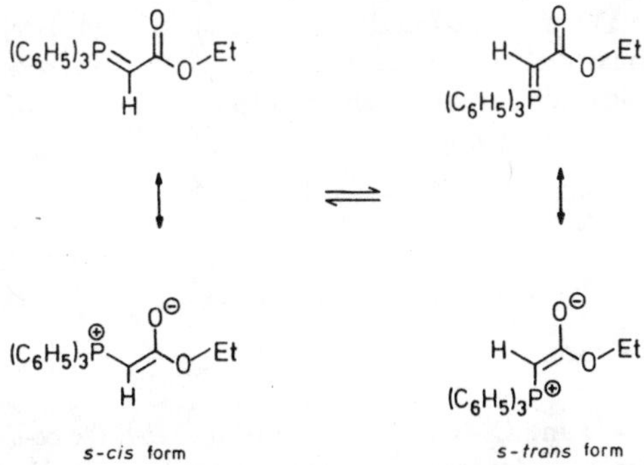

s-cis form (31a) ⇌ s-trans form (31b)

Table 4-8. Equilibrium constants for the *s-cis/s-trans*-isomerization reaction *(31a)* ⇌ *(31b)* of ethoxycarbonylmethylene triphenylphosphorane in various solvents at −10...0 °C [95].

Solvents (deuterated)	$K = [s\text{-}cis]/[s\text{-}trans]$
Tetrachloromethane	6.2
Benzene	5.9
Nitromethane	2.4
Acetonitrile	2.3
Chloroform	1.8
Chloroform/Methanol (5:1)	0.83
Chloroform/LiBr	0.57

In the *s-cis* rotamer *(31a)* the $P^{\oplus}/O^{\ominus}$ attraction is maximized. Increasing dielectric constant of the solvent reduces the attraction of the opposite charges, resulting in an increase in the *s-trans* rotamer *(31b)*. An exception is found with chloroform as solvent ($\varepsilon_r = 4.8$). Here the *s-trans* isomer is stabilized more effectively than in the more polar acetonitrile ($\varepsilon_r = 37.5$) and nitromethane ($\varepsilon_r = 35.9$). This is obviously due to the fact, that chloroform associates with the negative oxygen atom through hydrogen bonding. Addition of a HBD solvent such as methanol to the chloroform solution enhances the *s-cis* → *s-trans* conversion to an even greater extent. Addition of the extremely polar lithium bromide favours the formation of the *s-trans* isomer furthermore [95]. Analogous results were obtained with formylbenzyliden triphenylphosphorane [96].

Further well-studied recent examples of solvent-dependent rotational equilibria are furfural [185], the *N,N*-dimethylamides of furoic and thenoic acids [186], benzil monoimines [187], and methyl 2-, 3-, and 4-fluorobenzoates [188].

Alicyclic compounds exhibiting *conformational* isomerism are also subject to considerable medium effects [80–83, 182, 184].

One interesting example is (+)-*trans*-2-chloro-5-methylcyclohexanone *(32)*. The sign of the Cotton effect in its ORD spectrum is reversed by transferring it from water to *n*-heptane (molecular rotations at 330 nm +382° and −1486°, respectively) [189]. This can be ascribed to a diaxial/diequatorial conformational equilibrium *(32a)* ⇌ *(32b)* which lies more to the left-hand side in *n*-heptane than in water.

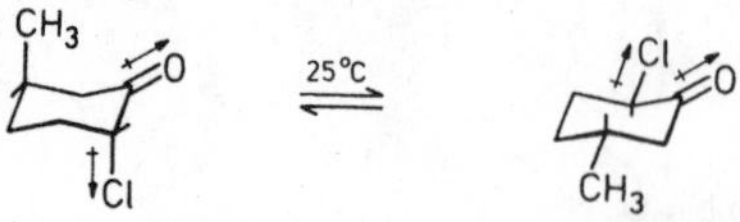

(32a) ⇌ (25 °C) *(32b)*

Due to the nearly parallel C=O and C—Cl dipole vectors in *(32b)*, the *ee*-isomer must have the larger net dipole moment and is therefore better solvated in polar solvents. In nonpolar solvents the electrostatic repulsion between the two equatorial C=O and C—Cl dipoles in *(32b)* is unfavourable, and the molecule escapes from this situation to give the *aa*-isomer *(32a)* even at the cost of nonbonded axial repulsions [189].

The majority of results obtained for other cyclohexanone derivatives such as 4-methoxycyclohexanone [94a], a bridged 4-oxacyclohexanone [94b], 2- and 4-halocyclohexanones [190], and 2-chlorocyclohexanone [191] have been similar. The conformational Gibbs energy differences $\Delta G°(e \rightarrow a)$ for 4-methoxycyclohexanone have been determined in thirty-four solvents. It shows a marked sensitivity to solvent change, the axial conformer being the more stable in all but the most dipolar HBD solvents [94a]. It is interesting to note that although the equatorial conformer is the least dipolar one, it is stabilized by polar solvents more effectively than the axial isomer. This is due to the larger quadrupole moment of the equatorial conformer [197].

In the case of diaxial/diequatorial equilibria of *trans*-1,2-dihalocyclohexanes (with and without a 4-*tert*-butyl group), $\Delta G°(aa \rightarrow ee)$ also shows a pronounced dependence on the medium, varying from about 4 kJ/mol in apolar solvents to about −2 kJ/mol in polar solvents [90–94, 192–194]. This is mainly due to the very different dipole moments of the two conformers. In general, the more dipolar diequatorial isomer is favoured in polar solvents.

Another remarkable example is provided by the observation of the strong influence of solvents on the conformation of phencyclidine *(33)*, a drug developed as an anesthetic, but later withdrawn because of its psychotomimetic effects [195].

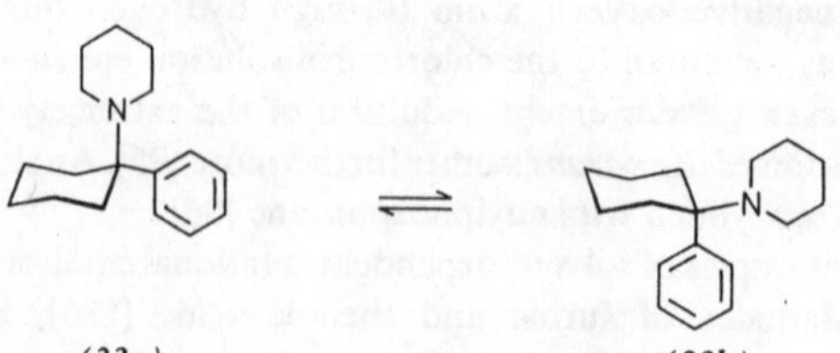

(33a) ⇌ *(33b)*

In dipolar non-HBD solvents such as [D$_6$]acetone and [D$_3$]acetonitrile the equilibrium is more to the left-hand side than in the less polar [D$_2$]dichloromethane. Upon transfer from CD_2Cl_2 to an HBD solvent (CD_2Cl_2/CD_3OD 1 : 2 cl/l), the equilibrium is shifted substantially to the right-hand side. The ratio *(33b)*/*(33a)* varies from 99:1 in CD_2Cl_2/CD_3OD at −80 °C to 1:1 in CD_3COCD_3/CD_3CN at room temperature. Obviously, conformer *(33b)* with the piperidine ring in the equatorial position is stabilized by hydrogen-bonding in HBD solvents; similar stabilization of the axial piperidine ring in *(33a)* is sterically inhibited. Consequently, it can be expected that the structure of this drug is subject to change on passing through a cell membrane [195].

A thoroughly examined case namely that of a heterocycloalkane which involves an axial/equatorial conformational change shall conclude this Section.

The position of the acid-catalyzed equilibrium between *cis-* *(34a)* and *trans*-2-isopropyl-5-methoxy-1,3-dioxane *(34b)* has been determined in seventeen different solvents; *cf.* Table 4-9 [89].

OCH_3 (H⊕) 25 °C OCH_3

(34a) axial *cis* isomer $\mu = 9.5 \cdot 10^{-30}$ Cm

(34b) equatorial *trans* isomer $\mu = 4.3 \cdot 10^{-30}$ Cm

Table 4-9. Solvent dependence of standard molar Gibbs energy differences between *cis*/*trans*-isomers of 2-isopropyl-5-methoxy-1,3-dioxane at 25 °C [89].

Solvents	$\Delta G^\circ_{OCH_3}/(kJ \cdot mol^{-1})$ for *(34a)* ⇌ *(34b)* [a]
n-Hexane	−4.44
Cyclohexane	−4.31
Tetrachloromethane	−3.77
1,3,5-Trimethylbenzene	−3.64
tert-Butylbenzene	−3.47
Diethyl ether	−3.47
Toluene	−2.97
Tetrahydrofuran	−2.72
Benzene	−2.47
1,1,1-Trichloroethane	−2,43
Acetone	−1.42
Nitrobenzene	−0.84
Deuteriochloroform	−0.79
Chloroform	−0.67
Dichloromethane	−0.42
Methanol	−0.13
Acetonitrile	+0.04

[a] $K = [trans]/[cis]$.

Inspection of Table 4-9 discloses that the axial *cis* isomer *(34a)*, which is the conformer with the higher dipole moment, becomes more favoured as the solvent polarity increases. In the most polar solvent studied, acetonitrile, ΔG° is nearly zero. Benzene, toluene, chloroform, dichloromethane, and methanol are seen to behave as more polar solvents than their dielectric constants would lead one to predict. The deviation for chloroform was particularly difficult to explain (for a full discussion see reference [89]). In general, good correlations between $\Delta G^{\circ}_{OCH_3}$-values and other solvent-dependent phenomena such as absorption maxima of solvatochromic dyes, rate constants of reactions involving dipolar activated complexes, *etc.*, were obtained [89]. Because of this, it was recommended that the solvent scale obtained should be used as an empirical scale of solvent polarity, useful for the prediction of medium effects on other solvent-dependent reaction rates or equilibria [89] (*cf.* Section 7.2).

The so-called *anomeric effect*, *i.e.* that polar substituents located on a carbon α to a hetero atom of six-membered heterocyclic rings prefer the axial position, has been shown to be solvent-dependent [83]. Recent remarkable examples for this phenomenon are *trans*-2,3-bis(trimethylsiloxy)- and *trans*-2,3-bis(*tert*-butoxy)-1,4-dioxanes as well as the corresponding 2,5-disubstituted 1,4-dioxanes. Surprisingly, their diaxial $\rightleftharpoons$ diequatorial conformational equilibria are shifted in the opposite direction with increasing solvent polarity [217]. That is,contrary to widely accepted views, polar solvents can reinforce the anomeric effect, at least in some of the above-mentioned cases [217].

Numerous attempts have been made to calculate relative conformer energies in solution, using physical properties of solutes and solvents, in order to get theoretical procedures or models with predictive ability [83, 88, 182, 188, 190, 192, 196–198]. The methods used include quantum-chemical calculations (*e.g.* [198]), statistical mechanics and molecular dynamics calculations (*e.g.* [182]), direct dipole-dipole methods (*e.g.* [83]), and reaction field methods based on Onsager's theory [199] of dipole molecules in the condensed phase (*e.g.* [83, 88, 188, 190, 194, 197]). In general, a quantitative description of solvent effects on conformational equilibria can be given on the basis of this methods, except in cases where specific solute-solvent interactions occur.

According to the reaction field method, the electrostatic stabilization of a solute molecule, located in the centre of a spherical cavity, with dipole moment μ and radius r in a solvent modeled as a uniform dielectric with relative permittivity ε_r can be expressed as in Eq. (4-31) (ε_0 = permittivity of vacuum).

$$\Delta G^{\circ}_{solv} = G^{\circ}_{vapour} - G^{\circ}_{solution}$$

$$= -\frac{1}{4\pi \cdot \varepsilon_0} \cdot \frac{\varepsilon_r - 1}{2\varepsilon_r + 1} \cdot \frac{\mu^2}{r^3} \tag{4-31}$$

The difference in the Gibbs energy of solvation, ΔG°_{solv}, for two species in equilibrium, A $\rightleftharpoons$ B, is then given by Eq. (4-32) by assuming they have the same size.

$$\Delta\Delta G^{\circ}_{solv} = -\frac{1}{4\pi \cdot \varepsilon_0} \cdot \frac{\varepsilon_r - 1}{2\varepsilon_r + 1} \cdot \left(\frac{\mu_A^2}{r^3} - \frac{\mu_B^2}{r^3}\right) \tag{4-32}$$

Quantitatively, Eq. (4-32) predicts that the more dipolar isomer will be preferentially stabilized in more polar media. Quantitatively, the expression significantly overestimates the solvent effects obtained experimentally for conformational equilibria [182]. Further modifications are necessary, *e.g.* adjustment for backpolarization of the solute by its own reaction field, inclusion of the effect of the solute's quadrupole moment on the reaction field [197], *etc*. Specific solute-solvent interactions such as with HBD solvents cannot be treated with this reaction field theory. For a more detailed discussion *cf.* references [83, 182].

A purely empirical correlation between the $\Delta\Delta G^\circ_{solv}$ of equilibria such as A $\rightleftharpoons$ B and solvent polarity has been given by the parabolic Eq. (4-33), where $X = (\varepsilon_r - 1)/(2\varepsilon_r + 1)$ and, in the majority of cases, $C = 0.5 = \lim_{\varepsilon_r \to \infty} X$ [196].

$$\Delta\Delta G^\circ_{solv} = A + B \cdot \sqrt{C - X} \tag{4-33}$$

If $X = 0$, the A parameter should be regarded as the extreme $\Delta\Delta G^\circ_{solv}$-value for solvents with infinitely large polarity, and the B parameter could be considered as a measure of the susceptibility of the equilibrium to changes of solvent polarity. Eq. (4-33) was successfully applied to various conformational and tautomeric equilibria [196].

4.4.4 Solvent Effects on *Cis/Trans* or *E/Z* Isomerization Equilibria [200]

One of the simplest examples for a *cis/trans* or *E/Z* isomerization equilibrium is represented by *trans*- *(35a)* and *cis*-1,2-dichloroethene *(35b)*. In the gas phase at 185 °C, the equilibrium mixture contains 63.5 cmol/mol of the thermodynamically more stable *cis*-isomer *(35b)* [201].

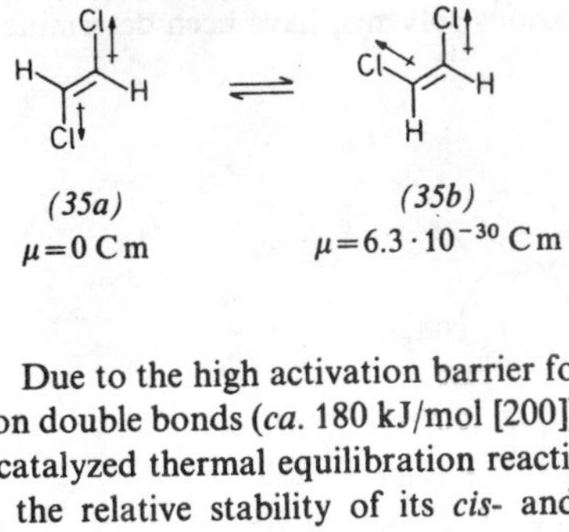

(35a) $\mu = 0$ Cm $\rightleftharpoons$ *(35b)* $\mu = 6.3 \cdot 10^{-30}$ Cm

Due to the high activation barrier for *cis/trans* isomerization reactions at carbon-carbon double bonds (*ca.* 180 kJ/mol [200]), it is often impossible to measure directly the non-catalyzed thermal equilibration reaction in solution. For 1,2-dichloroethene, however, the relative stability of its *cis*- and *trans*-isomer in various solvents has been determined by means of calorimetric measurements of heats of solution [202]. Surprisingly, these measurements show a quite similar solvent effect on both diastereomers even though the *cis*-isomer is a dipolar molecule and the *trans*-isomer is not. Therefore, the position of this *cis/trans* equilibrium should not be very solvent-dependent.

One reason for this at first sight unexpected result is given by the fact that probably 70...90% of the solute/solvent interaction term is caused by London dispersion forces which are more or less equal for the *cis*- and *trans*-isomer. Another important reason is that one has to take into account higher electric moments: the *trans*-isomer has a quadrupole moment, and the *cis*-isomer also has moments of a higher order than two. Calculations of solute/solvent interactions of both diastereomers using a reaction field model led to the conclusion that the quadrupolar contribution of the *trans*-isomer is comparable to the dipolar contribution of the *cis*-isomer. It has been pointed out that the neglect of solute/solvent interactions implying higher electric moments than the dipole moment can lead to completely false conclusions [202].

Unlike *cis/trans* isomers, the activation barrier separating *s-cis* and *s-trans* isomers are usually small (*ca.* 40...50 kJ/mol); the Gibbs energy differences for *s-cis* and *s-trans* isomers are also small (*ca.* 4...20 kJ/mol). For example, an easier measurable *cis/trans* isomerization reaction can be carried out with 3-*tert*-butylaminopropenal *(36)* [203].

(36a)
E-s-E form

(36b)
Z-s-Z form

According to its ^{13}C- and ^{1}H-NMR spectrum, this vinylogous amide exists as the *E-s-E* form in polar solvents such as [D_4]methanol, and as a mixture of *Z-s-Z* and *E-s-E* isomers in nonpolar solvents such as deuterochloroform (30 cmol/mol *(36a)* and 70 cmol/mol *(36b)*). As expected, the more dipolar *E-s-E* form is stabilized in polar solvents (dipole moment of the related *E-s-E* 3-dimethylaminopropenal $21 \cdot 10^{-30}$ C m).

Finally, a *cis/trans* isomerization reaction of a heterocycloalkane will be presented. The *cis/trans* isomer ratios of 2,3-dibenzoyl-1-benzylaziridine *(37)* ($R = CH_2C_6H_5$), when equilibrated by base catalysis at 33 °C in various solvents, have been determined using ^{1}H-NMR measurements [97].

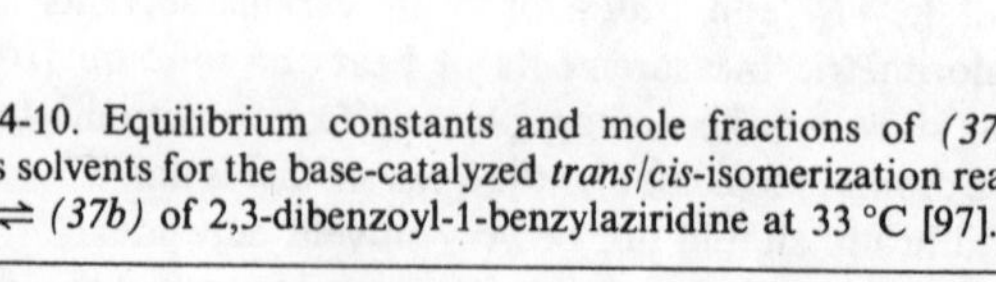

(37a)
$\mu = 8.7 \cdot 10^{-30}$ C m

(37b)
$\mu = 15.8 \cdot 10^{-30}$ C m

Table 4-10. Equilibrium constants and mole fractions of *(37b)* in various solvents for the base-catalyzed *trans/cis*-isomerization reaction *(37a)* ⇌ *(37b)* of 2,3-dibenzoyl-1-benzylaziridine at 33 °C [97].

Solvents	x*(37b)*/(cmol · mol^{-1})	K = [*cis*]/[*trans*]
tert-Butanol	24	0.32
Ethanol	45	0.82
Methanol	62	1.63
Dimethyl sulfoxide	84	5.25

This isomerization reaction passes through *cis/trans*-isomeric open-chain azomethine-ylides, which arise from conrotatory ring-opening at the C—C bond [98]. The equilibrium constants range from 5.25 in dimethyl sulfoxide to 0.32 in *tert*-butanol as shown in Table 4-10 and approximately parallel the polarity of the solvents used. The more dipolar *cis*-aziridine *(37b)* is the more stable isomer in polar solvents, the reverse is true for the less dipolar *trans*-aziridine *(37a)*. There is an isomer ratio of 1.63 in methanol (corresponding to 62 cmol/mol *cis*-isomer); this is in close agreement with the separately determined solubility ratio of 1.50 (corresponding to 60 cmol/mol *cis*-isomer). This close agreement of the ratios of solubilities of the two isomers with their equilibrium constants is a modern example of a long recognized phenomenon [37]; *cf.* Table 4-3 and Eq. (4-24).

4.4.5 Solvent Effects on Valence Isomerization Equilibria [84]

Valence isomerization reactions interconvert so-called valence isomers by simple reorganization of some of the bonding electrons, without any atom migration. Since both valence isomers may have different structures and thus different physical properties (*e.g.* different dipole moments) it is to be expected that solvents should influence the equilibrium between non-degenerated isomers.

For example, ^{1}H-NMR measurements showed that the azido/tetrazole equilibrium of thiazolo[2.3-e]tetrazole *(38b)* is considerably affected by the medium [100].

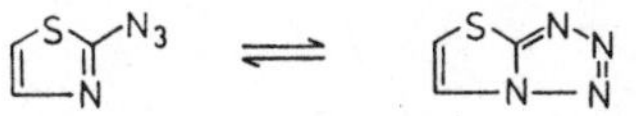

(38a) ⇌ *(38b)*

Whereas in the gas phase and in nonpolar solvents such as tetrachloromethane and benzene, the 2-azidothiazole *(38a)* is the more stable isomer, in polar solvents such as dimethyl sulfoxide and hexamethylphosphoric triamide the bicyclic valence isomer *(38b)* is the dominating species [100]. This result is in line with the fact that the dipole moment of phenyl azide ($\mu = 5.2 \cdot 10^{-30}$ C m) is smaller than that of cyclic 1,2,3-benzotriazole ($\mu = 13.7 \cdot 10^{-30}$ C m). Similar results have been obtained with the valence isomers of 3-azidopyrazine-1-oxide [204].

Another remarkable example is 2,4,6-tri-*tert*-butyl-*N*-thiosulfinylaniline *(39a)* which is in equilibrium with its valence isomer *(39b)*, while in the solid state only *(39b)* exists [205]. The equilibrium ratio *(39b)*/*(39a)* is subject to a considerable solvent effect:

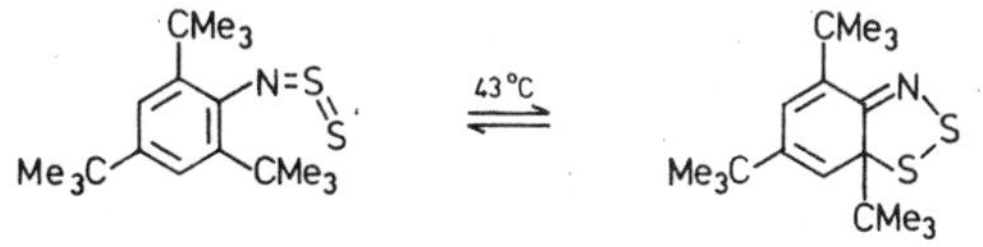

(39a) $\mu \approx 5 \cdot 10^{-30}$ C m ⇌ (43 °C) *(39b)* $\mu \approx 10 \cdot 10^{-30}$ C m

the mole fraction of *(39b)* in *n*-hexane solution is 86 cmol/mol and in acetonitrile solution 96 cmol/mol. Polar solvents enhance the preference of the cyclic isomer *(39b)*, mainly due to its larger dipole moment [205]. This is noteworthy because the aromaticity of the benzene ring is destroyed in *(39b)*.

In the case of the oxepin/benzene oxide valence isomerization *(40a)* ⇌ *(40b)* it has been found by UV/Vis measurements, that with *iso*-octane as solvent, only about 30 cmol/mol benzene oxide is present, whereas in water/methanol (85 : 15 cl/l) the benzene oxide portion rises to about 90 cmol/mol [101]. Increasing solvent polarity shifts this

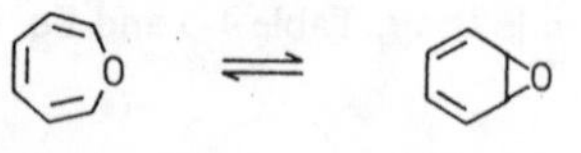

(40a) *(40b)*

equilibrium towards the more dipolar benzene oxide isomer, in agreement with quantum-chemical MNDO calculations [206]. These calculations have shown, that in the gas phase *(40a)* is more stable than *(40b)*, but *(40b)* is more stabilized in polar media.

A further recent example is the reversible valence isomerization equilibrium between the dipolar 8,8-diformylheptafulvene *(41a)* and the less dipolar 8a*H*-cyclohepta[b]furan-3-carbaldehyde *(41b)* [212]. In the solvent deuterium oxide using

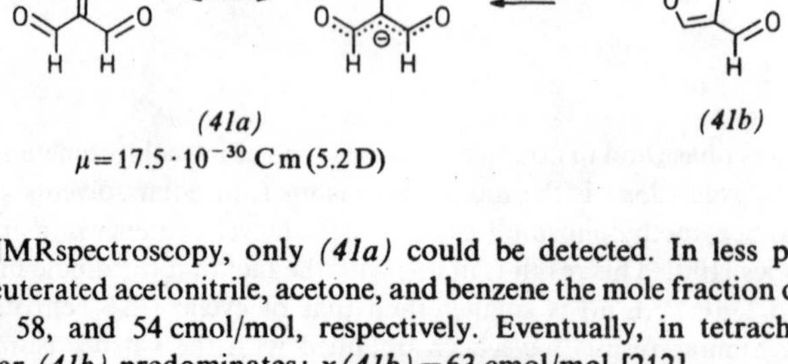

(41a) *(41b)*

$\mu = 17.5 \cdot 10^{-30}$ Cm (5.2 D)

^{1}H-NMRspectroscopy, only *(41a)* could be detected. In less polar solvents such as perdeuterated acetonitrile, acetone, and benzene the mole fraction of *(41a)* decreases to $x = 69$, 58, and 54 cmol/mol, respectively. Eventually, in tetrachloromethane the less dipolar *(41b)* predominates; x*(41b)* = 63 cmol/mol [212].

Finally, the allylcarbinyl/cyclopropylcarbinyl rearrangement *(42a)* ⇌ *(42b)* should be mentioned as a striking example of solvent-dependent valence isomerizations [207].

The deep-red diphenylcyclopropylcarbinyl lithium *(42b)*, which is stable in tetrahydrofuran solution, opened completely to the colourless γ,γ-diphenylallylcarbinyl

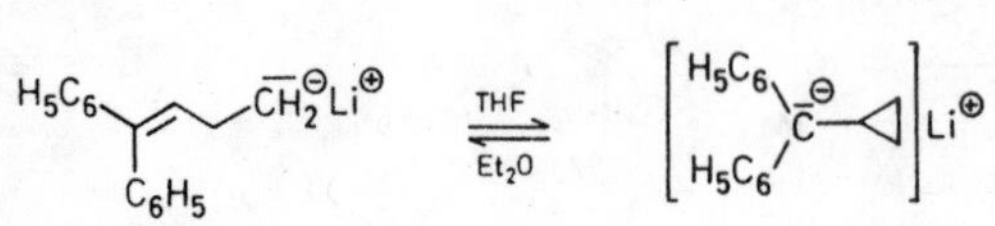

H_5C_6 / C_6H_5 $\overline{C}H_2^{\ominus}Li^{\oplus}$ THF ⇌ Et_2O $[(H_5C_6)_2\overline{C}^{\ominus}$–cyclopropyl$]$ $Li^{\oplus}$

(42a) *(42b)*

lithium in diethyl ether. The *retro* rearrangement to *(42b)* can be achieved simply by adding tetrahydrofuran to the solution of *(42a)* in diethyl ether [207]. Obviously, tetrahydrofuran which is the better cation-solvating EPD solvent makes the carbon-lithium bond in *(42a)* more ionic, thus favouring the ring closure to *(42b)* which has a more delocalized negative charge.

A closely related example, the equilibrium between tight and solvent-separated ion pairs of lithium 10-phenylnonafulvene-10-oxide, has already been given in Section 2.6 (formulas *(2b)* and *(3b)*). Depending on the solvent-influenced association with the lithium cation, the anion exists either as aromatic benzoyl [9]annulene anion or as olefinic nonafulveneoxide anion [208].

4.4.6 Solvent Effects on Electron-Transfer Equilibria

Examples of solvent effects on electron-transfer equilibria between organic species (*i.e.* redox and disproportionation reactions) are rather scarce.

A nice example is the disproportionation reaction of the 1-ethyl-4-methoxycarbonylpyridinyl radical *(43)* which leads to the ion pair *(43a)*/*(43b)* [215, 216].

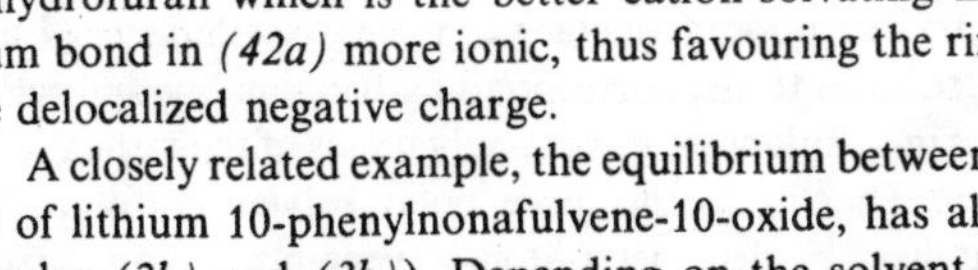

CO_2CH_3 CO_2CH_3 CO_2CH_3

2 N ⇌ ⊕N + ⊖ N

C_2H_5 C_2H_5 C_2H_5

(43) *(43a)* *(43b)*

Solvent	$HCON(CH_3)_2$	CH_3CN	C_2H_5OH	$HCONH_2$	H_2O
$K = \frac{[\text{ion pair}]}{[\text{racical}]}$	$4 \cdot 10^{-13}$	10^{-13}	10^{-8}	10^{-6}	$5 \cdot 10^{-5}$

As expected, an increase in solvent polarity shifts this equilibrium from the left to the right-hand side due to the better solvation of the ion pair as compared to the neutral radical [215].

From equilibrium constants for transfer from water to vapour, determined by dynamic vapour pressure measurements, the hydrophilic character of *p*-benzoquinone and *p*-hydroquinone has been estimated [227]. *p*-Benzoquinone is about 3.2 orders of magnitude less strongly solvated by water than is *p*-hydroquinone. Because *p*-hydroquinone is so much more strongly solvated than *p*-benzoquinone, its reducing power in water is less by *ca.* 18 kJ/mol (0.2 Volt) than it would be in a medium of unit dielectric constant. Therefore, the redox potential of biologically important hydroquinone/quinone systems should be strongly affected by the surrounding medium, particularly if the corresponding electron-transfer reaction is part of an electron transport chain embedded in mitochondrial inner membranes [227].

Quite recently, an interesting solvent-dependent, reversible electron-transfer equilibrium between a pair of resonance-stabilized carbocations and carbanions and the corresponding carbon radicals has been described [229]. Addition of bis(4-dimethylaminophenyl)-phenylcarbenium tetrafluoroborate (malachite green) to sodium tris(4-nitrophenyl)-methide leads in tetrahydrofuran to the corresponding triphenylmethyl radicals by single electron transfer, whereas in sulfolane as solvent only traces of the radicals could be detected by ESR spectroscopy. That is, in the more polar solvent sulfolane both carbocation and carbanion are stabilized by electrostatic ion/solvent interactions, whereas in the less polar tetrahydrofuran single electron transfer from the carbanion to the carbocation occurs readily to produce trityl radicals. Dilution of the radical solution in tetrahydrofuran with sulfolane leads to a sharp decrease of the ESR signal and to an increase of the carbanion absorption in the UV/Vis spectrum, in accordance with a reversible electron-transfer equilibrium [229].

5 Solvent Effects on the Rate of Homogeneous Chemical Reactions

5.1 General Remarks [1–29, 452–463]

A change of solvent can considerably change both the rate and order of homogeneous chemical reactions. Already in 1890, Menschutkin demonstrated in his classical study on the quaternization of triethylamine with iodoethane in 23 solvents, that the rate of reaction varied remarkably depending on the choice of solvent. In diethyl ether, the rate was four times faster than in *n*-hexane, 36 times faster in benzene, 280 times faster in methanol, and 742 times faster in benzyl alcohol [30]. Thus, by means of a proper choice of solvent, decisive acceleration or deceleration of a chemical reaction can be achieved. This can be of great practical importance either in the laboratory or in chemical industry. In some extreme cases, rate accelerations by a factor of up to *ca.* 10^9 (!) can be achieved solely by a solvent change [31]. Therefore, it is very important to establish rules and theories, enabling a rational selection of solvent and design for chemical synthesis.

The dependence of the reaction rate on the medium can in principle be approached from two points of view: (a) comparison of the rates of reaction in the gas phase and in solution, and (b) comparison of the rates of reaction in different solvents.

Until recently, only very few reactions which occur in solution have also been sufficiently examined in the gas phase [32]. Therefore, the comparison of solvent effects has been essentially limited to method (b). Reactions which have been studied according to method (a) are mainly non-ionic reactions, *i.e.* reactions without any charge separation or charge dispersion during the activation process, such as, for example, pericyclic reactions where neutral reactants produce neutral products. Reactions which follow ionic mechanisms with considerable charge separation or charge dispersion during activation, such as, for example, proton-transfer or S_N2 ion-molecule reactions with charged species as reactants, have until recently been investigated only using method (b).

Three new experimental techniques, developed within the past twenty years, now make possible to also study ionic reactions in the gas phase. These are pulsed ion-cyclotron-resonance (ICR) mass spectrometry, pulsed high-pressure mass spectrometry (HPMS), and the flowing afterglow (FA) technique [469–478; *cf.* also the references given in Section 4.2.2]. Although their approaches are quite independent, the results obtained for acid/base and other ionic reactions agree within an experimental error of 0.4...1.3 kJ/mol (0.1...0.3 kcal/mol) and are considered as reliable as those obtained in solution.

In solution, ions are produced by the heterolysis of covalent bonds in ionogens. This ionization reaction is favored by solvents due to their cooperative EPD and EPA properties (*cf.* Section 2.6). In the gas phase, however, ionization of neutral molecules to form free ions is rarely observed because this reaction is very endothermic. For example, in order to ionize gaseous H—Cl into $H^{\oplus}$ and $Cl^{\ominus}$, an energy of 1393 kJ/mol (333 kcal/mol) must be provided. This considerably exceeds the 431 kJ/mol (103 kcal/mol) needed to homolytically cleave H—Cl into hydrogen and chlorine atoms. Thus, for the creation of isolated ions in the gas phase, energy must be supplied by some means other than solvation with EPD/EPA solvents. The most widely used method is ionization by electron impact of gaseous molecules such as employed in mass spectrometry. However, whereas ions in solution are stabilized by their solvation shells, gaseous ions are immediately destroyed when they strike a solid surface. Therefore, special care has to be taken to restrict the motion of gaseous ions and to hold them long enough so that their chemical reactivity can be studied. Several methods have been developed for storing gaseous ions. In ICR mass spectrometry, a static magnetic trap is used [469].

The study of reactions of isolated ions and molecules in the gas phase without interference from solvents has led to very surprising results. Gas-phase studies of proton-transfer and nucleophilic substitution reactions permit the measurement of the intrinsic properties of the bare reactants and makes it possible to distinguish these basic properties from effects attributable to solvation. Furthermore, these studies provide a direct comparison of gas-phase and solution reactivities of ionic reactants. It has long been assumed that solvation retards the rates of ion-molecule reactions. Now, using these new techniques, the dramatic results obtained make it possible to show the extent of this retardation. For example, in a typical S_N2 ion-molecule reaction in the gas phase the substrates react by a factor of 10^{10} times faster than when they are dissolved in acetone, and by a factor of 10^{15} (!) times faster than in water (*cf.* Table 5-2 in Section 5.2).

Clearly, the effect of the solvent on a chemical reaction is much larger than previously assumed. In solution, the behaviour of ions and molecules is dictated mainly by the solvent and only to a lesser extent by their intrinsic properties. This will be elaborated on in subsequent Sections. A comparison of gas-phase reactivities and solution reactivities will be given in Section 5.2.

There are, in principle, two ways in which solvents can affect the reaction rates of homogeneous chemical reactions: *via* static, or equilibrium, solvent effects and *via* dynamic, or frictional, solvent effects [463, 465].

The static influence of solvents on rate constants can be understood in terms of the transition-state theory. According to this theory, solvents can modify the Gibbs energy of activation (as well as the corresponding activation enthalpies, activation entropies, and activation volumes) by differential solvation of the reactants and the activated complex. Reaction rates are very sensitive to barrier heights. For example, a change of only 8.4 kJ/mol (2 kcal/mol) in an activation barrier can alter the reaction rate at room temperature by a factor of 31. Here, it is implicitly assumed that, the required reorientational relaxation of solvent molecules during the activation process is sufficiently fast and the activated complex is in thermal equilibrium with the solvent, due to the frequent collisions of the reacting system with the surrounding solvent molecules. The Hughes-Ingold rules of solvent effects on reaction rates (*cf.* Section 5.3.1) are based on equilibrium solvation of the activated complex.

But this equilibrium hypothesis is not necessarily valid for rapid chemical reactions. This brings us to the second way in which solvents can influence reaction rates namely *via* dynamic or frictional effects. For broad-barrier reactions in strongly dipolar, slowly relaxing solvents, non-equilibrium solvation of the activated complex can occur and the solvent reorientation may also influence the reaction rate. In the case of slow solvent relaxation, significant dynamic contributions to the experimentally determined activation parameters, which are completely absent in the conventional transition-state theory, can exist. In the extreme case, solvent reorientation becomes rate limiting and the transition-state theory breaks down. In this situation, rate constants will depend on the solvent dynamics, and will vary with friction, *i.e.* with some measure of the coupling of the solvent such as density, internal pressure, or viscosity. In the opposite regime of reactions, *i.e.* those with sharp barriers and weakly dipolar, rapidly relaxing solvents, reaction rates are fairly well described by the transition-state theory. At present, it seems to be difficult to quantitatively separate the influence of the dynamic and static solvent effects on the reaction rate. The importance of these less familiar frictional solvent effects on reaction rates has only recently been stressed [463, 465, 466], although the first theoretical treatment of such dynamic solvent effects was given by Kramers already in 1940 [479].

Because the transition-state theory is essential for an even qualitative understanding of solvent effects on reaction rates, some important features of this theory will be outlined below.

Arrhenius' classical theory of reaction kinetics is based on the assumption, that the starting materials (reactants) have to overcome an energy barrier, the activation energy, in order to be transformed into the products. This picture has been developed and made more explicit in the Theory of Absolute Reaction Rates [2–5, 7, 8, 11, 24, 464–466]. The influence of solvent on reaction rates is best treated by means of this theory – also known as Transition-State Theory, developed almost simultaneously in 1935 by Eyring and Evans and Polanyi [464, 465].

Consider a reaction between the starting compounds A and B, and suppose that during the course of the reaction these two first form an activated complex, which then decomposes to the end products, C and D. The reaction can then be described as follows:

$$\underset{\text{Reactants}}{A+B} \rightleftharpoons \underset{\substack{\text{Activated}\\ \text{Complex}}}{(AB)^{\neq}} \rightarrow \underset{\text{Products}}{C+D} \qquad (5\text{-}1)$$

The theory of absolute reaction rates is based on the following assumptions:

(a) The reactants are visualized as being in quasi-equilibrium with the activated complex. The corresponding quasi-thermodynamic equilibrium constant is given by Eq. (5-2) (with a = activities, [ ] = molar concentrations, and γ = activity coefficients).

$$K^{\neq} = \frac{a_{(AB)^{\neq}}}{a_A \cdot a_B} = \frac{[(AB)^{\neq}]}{[A] \cdot [B]} \cdot \frac{\gamma^{\neq}}{\gamma_A \cdot \gamma_B} \qquad (5\text{-}2)$$

The use of activity coefficients takes into account deviations from ideal behaviour in solution. The activity coefficients usually refer to the standard state of infinite dilution for the solutes: $\lim_{c \to 0} \gamma = 1$.

(b) The formation of the products C and D does not significantly affect the equilibrium between reactants and activated complex.

(c) The activated complex has all the properties of a normal molecule (with N atoms) except that one vibrational degree, of the $3N$-6 vibrational degrees of freedom, is transformed into a translational degree of freedom, which leads to the decomposition of the activated complex.

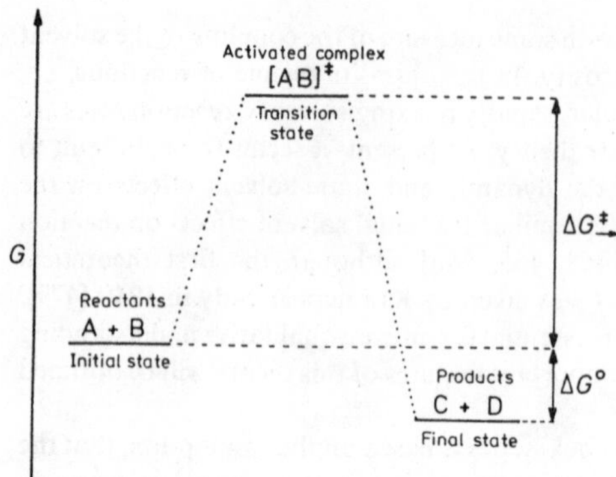

Fig. 5-1. One-dimensional Gibbs energy diagram for reaction (5-1) in solution. Ordinate: relative standard molar Gibbs energies of reactants, activated complex, and products; Abscissa*): not defined, expresses only the sequence of reactants, activated complex, and products as they occur in the chemical reaction. ΔG°: standard molar Gibbs energy of the reaction; $\Delta G^{\neq}$: standard molar Gibbs energy of activation for the reaction from the left to the right.

(d) The activated complex exists at the top of an energy barrier as shown in Fig. 5-1. The activated complex represents that point in the progress of the reaction at which the re-formation of reactants is as likely as the formation of products. The region in the neighborhood of this maximum is also called the "transition state"**).

(e) The net reaction rate is determined by the rate at which the activated complex passes over the energy barrier in the direction of product formation. The probability of a forward passage through the activated complex is given by a transmission coefficient, which is usually assumed to approach unity.

(f) The change in the Gibbs energy of activation in going from the gas phase to solution or from one solvent to another is evaluated as the relative modification in Gibbs energy by differential solvation of the reactants and the activated complex – whereby it is implicitly assumed that in solution reactants and activated complex are in thermal equilibrium with the solvent. This equilibrium hypothesis will not be valid for rapid

---

* The so-called *reaction coordinate* – often used as abscissa in such diagrams – is essentially a molecular-microscopic quantity: In the case of a unimolecular reaction it is the internuclear distance; in case of a bimolecular reaction it represents a translational degree of freedom and in the transition state a normal mode of vibration which leads to the decomposition of the activated complex. ΔG° and $\Delta G^{\neq}$ are, however, macroscopic thermodynamic state functions, which are experimentally available only at the maxima and minima and not at configurations between them. Therefore, in order to avoid the indiscriminate mixing of macroscopic and molecular-microscopic quantities, the one-dimensional presentation of Fig. 5-1 is preferable to the smooth curves often used and called Gibbs energy "profiles". In Fig. 5-1 the reaction coordinate has no meaning. The abscissa expresses only the sequence of reactants, activated complex, reaction intermediates, and products as they occur in chemical reactions. *Cf.* also Fig. 4-1 in Chapter 4.

** The terms *activated complex* and *transition state* are often wrongly regarded as synonymous. *Activated complex* refers to the real molecular entity at the point of maximum Gibbs energy on the reaction path, whereas *transition state* describes only the set of states or energy levels at this point. The *transition state* has no physical existence; it is a multidimensional mathematical relationship between potential energy and atomic configuration. *Cf.* J. Chem. Educ. *64*, 208 (1987).

chemical reactions with slow reorientational relaxation of the solvent molecules. For broad-barrier reactions in slowly relaxing solvents, solvent reorientation becomes rate limiting and the transition-state theory may break down [463, 465, 466].

Because the reaction rate is assumed to be proportional to the concentration of the activated complex, the specific rate constant k is proportional to $K^{\neq}$. Using statistical calculations the proportionality factor can be estimated as $k_B \cdot T/h$. Provided that each activated complex passing the transition state actually becomes product (transmission coefficient close to unity), and that the activity coefficients are close to unity, the specific rate constant of the elementary reaction is given by Eq. (5-3),

$$k = \frac{k_B \cdot T}{h} \cdot K^{\neq} \cdot (1\ \text{mol} \cdot \text{l}^{-1})^{1-n} = \frac{R \cdot T}{N_A \cdot h} \cdot K^{\neq} \cdot (1\ \text{mol} \cdot \text{l}^{-1})^{1-n} \tag{5-3}$$

where k is the rate constant including units with the dimensions $(\text{mol} \cdot \text{l}^{-1})^{1-n} \cdot (\text{time})^{-1}$; k_B the Boltzmann constant; h the Planck constant; T the absolute temperature; R the gas constant; N_A the Avogadro constant; $K^{\neq}$ the quasi-thermodynamic equilibrium constant related to a hypothetical unit molar concentration standard state for the postulated equilibrium between reactant(s) and activated complex; and n the molecularity and kinetic order of the elementary reaction (mainly, 1 or 2).

This simple equation predicts that the rate constant for any chemical reaction consists of an equilibrium constant multiplied by an universal frequency factor, $k_B \cdot T/h$, which varies only with the temperature*).

Since $K^{\neq}$ represents an equilibrium constant, it is possible to define quantities like $\Delta G^{\neq}$, $\Delta H^{\neq}$, and $\Delta S^{\neq}$ according to Eqs. (5-4) and (5-5), which are called Gibbs energy, enthalpy, and entropy of activation, respectively.

$$\Delta G^{\neq} = -R \cdot T \cdot \ln K^{\neq} \tag{5-4}$$

$$\Delta G^{\neq} = \Delta H^{\neq} - T \cdot \Delta S^{\neq} \tag{5-5}$$

From Eqs. (5-3), (5-4), and (5-5), it follows that the specific rate constant for the standard state ($\gamma = 1$) can also be given by Eq. (5-6).

$$k = \frac{R \cdot T}{N_A \cdot h} \cdot e^{-\frac{\Delta G^{\neq}}{R \cdot T}} \cdot (1\ \text{mol} \cdot \text{l}^{-1})^{1-n}$$

$$= \frac{R \cdot T}{N_A \cdot h} \cdot e^{-\frac{\Delta H^{\neq}}{R \cdot T}} \cdot e^{\frac{\Delta S^{\neq}}{R}} \cdot (1\ \text{mol} \cdot \text{l}^{-1})^{1-n} \tag{5-6}$$

* $k_B \cdot T/h \approx 6 \cdot 10^{12}\ \text{s}^{-1}$ at 25 °C.

Eq. (5-6) is assumed to be valid for reactions in solution although it is probably applicable only to an ideal gas system. According to this equation, the smaller $\Delta G^{\neq}$, *i.e.* the difference between the Gibbs energies of reactants and activated complex, the greater the reaction rate of a chemical reaction.

In principle, when studying solvent effects on reaction rates, it is not sufficient to investigate only the $\Delta G^{\neq}$ change, because according to Eq. (5-5) this term is determined by both an enthalpy and an entropy term. There are four types of reaction rate control [41]:

(a) Cooperative effects, with $\Delta\Delta H^{\neq}$ and $T \cdot \Delta\Delta S^{\neq}$ having opposite signs. Then these two terms will be additive;

(b) Enthalpy-controlled reactions, in which the two terms are opposed (*i.e.* $\Delta\Delta H^{\neq}$ and $T \cdot \Delta\Delta S^{\neq}$ have equal signs), but the activation enthalpy term is larger;

(c) Entropy-controlled reactions, in which the two terms oppose, but the $T \cdot \Delta\Delta S^{\neq}$ term is larger; and

(d) Compensating effects, in which the two terms are opposed, but nearly equal.

If reaction (5-1) takes place in solution, then the initial reactants, as well as the activated complex, will be solvated to a different extent, according to the solvating power of the solvent used. This differential solvation can retard or accelerate a reaction in the manner described in Fig. 5-2.

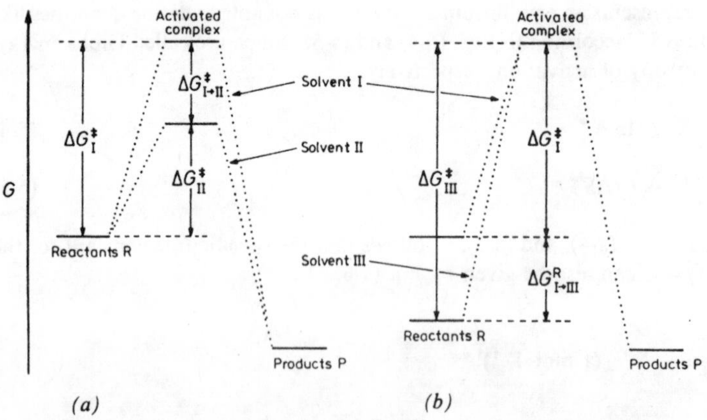

Fig. 5-2. One-dimensional Gibbs energy diagram for a chemical reaction in three different solvents I, II, and III (*cf.* Fig. 5-1). (a) Reaction with non-solvated (solvent I) and solvated (solvent II) activated complex (preferential solvation of the activated complex); (b) Reaction with non-solvated (solvent I) and solvated (solvent III) reactants (preferential solvation of the reactants). $\Delta G_{I}^{\neq}$, $\Delta G_{II}^{\neq}$, and $\Delta G_{III}^{\neq}$: standard molar Gibbs energies of activation in solvents I, II, and III, respectively; $\Delta G_{I \to II}^{\neq}$ and $\Delta G_{I \to III}^{R}$: standard molar Gibbs energies of the transfer of the activated complex and the reactants R from solvent I to solvents II and III, respectively.

$\Delta G_I^{\neq}$ in Fig. (5-2a) and (5-2b) represents the Gibbs energy of activation for a given chemical reaction in an ideal solvent I. In such a case neither the reactants or the activated complex are solvated. If in another solvent II only the activated complex is solvated, then $\Delta G_{II}^{\neq}$ in Fig. (5-2a) results. The Gibbs energy of activation is reduced by the Gibbs energy of transfer, $\Delta G_{I\to II}^{\neq}$, with a consequent rate acceleration. If on the other hand, only the initial reactants are solvated as happens in solvent III, then $\Delta G_{III}^{\neq}$ in Fig. (5-2b) results. The Gibbs energy of activation is increased by the Gibbs energy of transfer, $\Delta G_{I\to III}^{R}$, with a consequent rate deceleration. The solvation of the products does not have any influence on the reaction rate. Because in reality the initial reactants as well as the activated complex are solvated, but usually to a different extent, the difference of both Gibbs transfer energies determines the reaction rate in solution.

A dissection, therefore, of the reaction rate solvent effects into initial and transition state contributions would lead to more direct information concerning the nature of the solvent involvement [453, 467].

Fig. 5-3 shows the Gibbs energy diagram for a chemical reaction carried out in two different solvents I and II. The standard molar Gibbs energy of the reactants R in solvent I is designated G_I^R and in solvent II as G_{II}^R. The difference in Gibbs energies between the two solvents, $(G_I^R - G_{II}^R)$, is termed the Gibbs energy of transfer, $\Delta G_{I\to II}^{R}$. Similarly, for the activated complex one obtains $\Delta G_{I\to II}^{\neq} = G_I^{\neq} - G_{II}^{\neq}$.

The difference in Gibbs energies of activation for the reaction in the two solvents is designated $\Delta\Delta G^{\neq}$, and from Fig. 5-3 it follows that

$$\Delta\Delta G^{\neq} = \Delta G_I^{\neq} - \Delta G_{II}^{\neq} = (G_I^{\neq} - G_I^R) - (G_{II}^{\neq} - G_{II}^R) \tag{5-7}$$

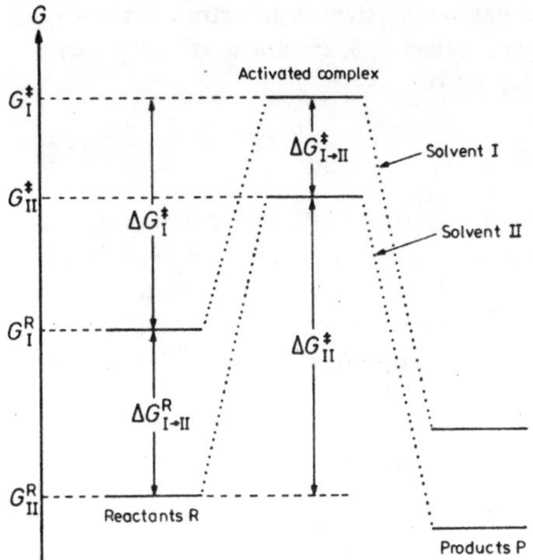

Fig. 5-3. One-dimensional Gibbs energy diagram for a chemical reaction in two different solvents I and II (*cf.* Figs. 5-1 and 5-2). $\Delta G_I^{\neq}$ and $\Delta G_{II}^{\neq}$: standard molar Gibbs energies of activation in solvents I and II; $\Delta G_{I\to II}^{R}$ and $\Delta G_{I\to II}^{\neq}$: standard molar Gibbs energies of transfer of the reactants R and the activated complex from solvent I to solvent II, respectively.

this can be simplified to give

$$\Delta\Delta G^{\neq} = \Delta G^{\neq}_{\text{I}\to\text{II}} - \Delta G^{\text{R}}_{\text{I}\to\text{II}} \tag{5-8}$$

From Eq. (5-8) it is apparent that $\Delta G^{\neq}_{\text{I}\to\text{II}}$ can be evaluated from the measurable Gibbs energies of transfer of the reactants, $\Delta G^{\text{R}}_{\text{I}\to\text{II}}$, in conjunction with the measured kinetic activation parameters, $\Delta\Delta G^{\neq}$.

The required Gibbs transfer energies, $\Delta G^{\text{R}}_{\text{I}\to\text{II}}$, can be obtained from activity coefficient measurements according to Eq. (5-8a),

$$\Delta G^{\text{i}}_{\text{I}\to\text{II}} = -R\cdot T\cdot\ln\left(\gamma_{\text{I}}/\gamma_{\text{II}}\right) \tag{5-8a}$$

in which γ refers to activity coefficients of solute i in solvents I and II. The γ-values can be obtained from vapor pressure, solubility, and distribution coefficient measurements [23, 467].

Consequently, by combining the thermodynamic and kinetic measurements values for $\Delta G^{\neq}_{\text{I}\to\text{II}}$ and $\Delta G^{\text{R}}_{\text{I}\to\text{II}}$ can be obtained. Both terms can be positive (destabilization), negative (stabilization), or zero (no solvent effect). When both terms have the same sign, a balancing situation occurs, with opposite sign a reinforcing situation. The effect of the solvent on the rate can be expected to be largest in the reinforcing situation and smallest in the balancing situation [467].

Because of the complicated interactions between solvents and solutes, the prediction of solvent effects on reaction rates, and the correlation of these effects with intrinsic solvent properties, is very difficult. Nevertheless, many authors have tried to establish – empirically or theoretically – correlations between rate constants or Gibbs energies of activation and characteristic solvent parameters such as dielectric constant, ε_r, dipole moment, μ, refractive index, n, solubility parameter, δ, empirical solvent polarity parameters, *etc*., as schematically shown by Eq. (5-9).

$$k \quad \text{or} \quad \Delta G^{\neq} = f(\varepsilon_r, \mu, n, \delta, \ldots) \tag{5-9}$$

An early attempt in this direction was O. Dimroth's effort in correlating rate constants k with the solubility S of the reactants in the solvents used [42]. While investigating the intramolecular rearrangement of 5-hydroxy-1-phenyl-1,2,3-triazole-4-carboxylic esters in various solvents, he found, in agreement with a rule formulated by van't Hoff [43], that the rate constants are inversely proportional to the solubility of the rearranging isomers; *cf*. Eq. (5-10).

$$k = const.\cdot\frac{1}{S} \tag{5-10}$$

Relationships such as this have only a very limited application [2]. More modern attempts to correlate the solvent influence on reaction rates with physical and empirical parameters of the solvents can be found in Chapters 5.4 and 7.6, respectively.

The following Chapters deal first with gas-phase reactivities and then with the most important qualitative and quantitative general relationships between reaction rates and solvent properties. The next Chapter (5.2) begins with a comparison of reactions in the gas phase and in solution, thus demonstrating the huge changes in reaction rates obtained in going from the gas phase to solution. The following Chapter (5.3) deals with the qualitative Hughes-Ingold rules (and their limitations) concerning solvent effects on substitution and elimination reactions, based on a sub-classification of these reactions according to the different charge-types of the initial reactants [16]. After that, further analogous examples of solvent effects on organic reactions, using a classification of reactions given by Kosower [15] and Reichardt [468], are added. Attempts at establishing quantitative relationships between reaction rates and physical solvent properties are then reviewed (Chapter 5.4), followed by a treatment of some specific solvation effects on reaction rates (Chapter 5.5).

5.2 Gas-Phase Reactivities

In the gas phase, bond fission is invariably homolytic and complications from solvents are absent. Reactants follow their reaction route in splendid isolation which depends only on the intrinsic (inherent) properties of the reactant molecules. On the other hand, bond fission in solution is generally heterolytic due to the EPD/EPA properties of the solvents. The ubiquitous solvent molecules perturb the reactants in their reaction course, sometimes to such an extent that the solvent is entirely responsible for the observed rate constants.

Accordingly, in the gas phase, reactions without charge separation or charge dispersion are preferred, *e.g.* radical-producing and pericyclic reactions. In solution, on the other hand, reactions involving charge separation and charge dispersion can also be carried out, *e.g.* ionization and S_N2 ion-molecule reactions. Reactions can only occur in the gas phase *and* in solution, when the intermolecular interactions between reactants and solvents are so weak that the non-ionic gas phase reaction mechanism is virtually unchanged on transferring the reactants to solution.

An example of such a reaction, which has the same mechanism in the gas phase and in solution, is the bimolecular Diels-Alder cycloaddition reaction of cyclopentadiene to *endo*-dicyclopentadiene – a reaction between neutral reactants to give a neutral product. As the Arrhenius activation energies and the rate constants of this reaction in Table 5-1 show, there is only a slight variation on going from the gas phase to solution. The rate constants vary by only a factor of *ca.* 3, accompanied by a corresponding small change in the activation energies.

The homolytic thermolysis of di-*tert*-butyl peroxide [36, 37] and diacetyl peroxide [38, 39] in the gas phase and in a variety of solvents, have roughly the same rate constants and activation energies both in the gas phase and in solution.

On the contrary, reactions involving charge separation or charge dispersion during the activation process depend strongly on the reaction medium, due to the strong intermolecular interactions between the ionic and dipolar reactants and the solvent. Examples include S_N1 and S_N2 reactions, elimination reactions, isomerizations involving polar and charged groups, as well as proton- and electron-transfer reactions.

A typical example of such reactions is the exothermic S_N2 nucleophilic displacement reaction $Cl^{\ominus} + CH_3{-}Br \rightarrow Cl{-}CH_3 + Br^{\ominus}$. Table 5-2 provides a comparison of Arrhenius activation energies and specific rate constants for this Finkelstein reaction in both the

Table 5-1. Comparison of the activation energies and specific rate constants for the bimolecular Diels-Alder cycloaddition reaction of cyclopentadiene giving *endo*-dicyclopentadiene in the gas phase and in solution at 20 °C [3, 33, 34].

Medium	E_a (kJ · mol^{-1})	k_2 (l · mol^{-1} · s^{-1})	k_2^{rel}
Gas phase	69.9	$6.9 \cdot 10^{-7}$	1.2
Ethanol	68.6	$19 \cdot 10^{-7}$	3.4
Nitrobenzene	63.2	$13 \cdot 10^{-7}$	2.3
Paraffin oil	72.8	$9.8 \cdot 10^{-7}$	1.8
Carbon disulfide	70.7	$9.3 \cdot 10^{-7}$	1.7
Tetrachloromethane	71.5	$7.9 \cdot 10^{-7}$	1.4
Benzene	68.6	$6.6 \cdot 10^{-7}$	1.2
Neat liquid	67.8	$5.6 \cdot 10^{-7}$	*1.0*

Table 5-2. Comparison of the activation energies and specific rate constants for the bimolecular S_N2 ion-molecule reaction $Cl^{\ominus} + CH_3{-}Br \rightarrow Cl{-}CH_3 + Br^{\ominus}$ in the gas phase and in solution at 25 °C [480].

Medium	E_a (kJ · mol^{-1})	k_2 (cm^3 · molecule^{-1} · s^{-1})	k_2^{rel}
Gas phase (at 24 °C)	11	$2.1 \cdot 10^{-11}$	$2.5 \cdot 10^{15}$
Acetone[a)]	66	$5.5 \cdot 10^{-21}$	$6.6 \cdot 10^{5}$
N,N-Dimethylformamide[a)]	75	$8.3 \cdot 10^{-22}$	$1.0 \cdot 10^{5}$
Methanol[b)]		$1.0 \cdot 10^{-26}$	1.2
Water[c)]	103	$8.3 \cdot 10^{-27}$	*1.0*

[a] D. Cook and A.J. Parker, J. Chem. Soc., Part B *1968*, 142.
[b] R. Alexander, E.C.F. Ko, A.J. Parker, and T.J. Broxton, J. Am. Chem. Soc. *90*, 5049 (1968).
[c] R.H. Bathgate and E.A. Moelwyn-Hughes, J. Chem. Soc. *1959*, 2642.

gas phase and solution. The new techniques described above (*cf.* Sections 4.2.2 and 5.1) have made it possible to determine the rate constant of this ion-molecule reaction in the absence of any solvent molecules in the gas phase. The result is surprising: On going from a protic solvent to a non-HBD solvent and then further to the gas phase, the ratio of the rate constants is approximately $1 : 10^5 : 10^{15}$! The activation energy of this S_N2 reaction in water is about ten times larger than in the gas phase. The suppression of the S_N2 rate constant in aqueous solution by up to 15 orders of magnitude demonstrates the vital role of the solvent.

A detailed molecular-level understanding of the role of solvation on the nature of S_N2 reaction pathways has been revealed only recently. Fig. 5-4 compares the gas-phase S_N2 enthalpy diagram with two minima, first proposed by Brauman *et al.* [474], with the more familiar single transition-state diagram obtained in solution.

The unexpected gas-phase double-minima diagram can be best explained as follows: As the reactants approach one another, long-range ion-dipole and ion-induced dipole interactions first produce loose ion-molecule association complexes or clusters. This is related to a decrease in enthalpy prior to any chemical barrier produced by orbital overlap between the reactants. For reasons of symmetry, an analogous drop in enthalpy must exist

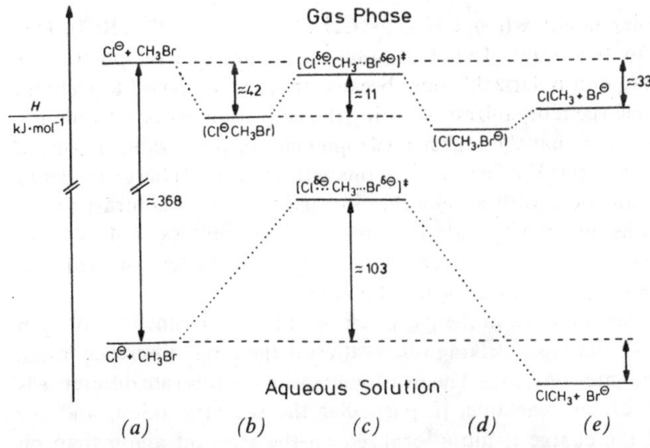

Fig. 5-4. Schematic one-dimensional enthalpy diagram for the exothermic bimolecular Finkelstein reaction $Cl^{\ominus} + CH_3{-}Br \rightarrow Cl{-}CH_3 + Br^{\ominus}$ in the gas phase and in aqueous solution [469, 474, 476]. Ordinate: standard molar enthalpies of (a) the reactants, (b, d) loose ion-molecule clusters held together by ion-dipole and ion-induced dipole forces, (c) the activated complex, and (e) the products. Abscissa: not defined, expresses only the sequence of (a)...(e) as they occur in the chemical reaction.

on the product side. Because the neutral reactant and product molecules will, in general, have different dipole moments and polarizabilities, the two minima will also be different. Only in the case of degenerate, identity S_N2 reactions ($X^{\ominus} + CH_3{-}X \rightarrow X{-}CH_3 + X^{\ominus}$), the enthalpy of the two minima will be equal.

Then, during the activation process, the enthalpy increases up to the central transition state level. The height of this central, intrinsic barrier is the major factor responsible for the inherent reactivities of the various reactant combinations. It is important to note that the enthalpy of the activated complex is less than that of the reactants. In other words, the enthalpy barrier for the formation of the activated complex is smaller than the lowering in enthalpy due to the attraction between the reactant ion and the dipolar reactant molecule.

This S_N2 reaction model, containing pre- and post-ion/dipole clusters that are lower in enthalpy than both the initial reactants and final products, and separated by an intrinsic barrier which localizes the transition state, has proven useful in application to other nucleophilic displacement reactions. Theoretical studies of S_N2 reaction pathways have also yielded double-minima enthalpy diagrams of the type shown in Fig. 5-4 [481–483].

Variations in the nucleophile ($X^{\ominus}$), leaving group ($Y^{\ominus}$), and alkyl substrate (R) involved in S_N2 reactions ($X^{\ominus} + R{-}Y \rightarrow X{-}R + Y^{\ominus}$) has led to a wide range of intrinsic reactivities in the absence of the complicating effects of solvents. These reactivities have been discussed in terms of nucleophilicity, leaving group ability, and steric hindrance [474–477]. For example, the following order of increasing anionic nucleophilicity has been found in the gas phase: $Br^{\ominus} < CN^{\ominus} < Cl^{\ominus} \ll CH_3S^{\ominus} < CH_3O^{\ominus} \approx F^{\ominus} < HO^{\ominus}$ [474, 480]. According to this sequence, the intrinsic nucleophilicities follow the reverse order of the polarizabilities (*e.g.* $CH_3O^{\ominus} < CH_3S^{\ominus}$ and $F^{\ominus} < Cl^{\ominus} < Br^{\ominus}$). This is exactly the opposite to the results found in solution studies. There, polarizable nucleophiles are better than non-polarizable ones because they are supposed to respond better to the demand for charge reorganization during the activation process. From the gas-phase results it can be seen that the higher nucleophilicity of polarizable anions in solution is purely a solvation effect! Furthermore, anions with a localized charge are better gas-phase nucleophiles than those with a delocalized charge. Thus, in contrast to the phenyl anion benzyl anions and the cyanide ion are poor nucleophiles [474, 484]. In solution, however, anions with a localized charge are generally better solvated and therefore less reactive than anions with delocalized charge.

By comparing the reaction rates in the gas phase with those obtained in solution (*cf.* Table 5-2 and Fig. 5-4), the most striking observation is the large difference in the absolute magnitudes of the rate constants. The principal reason for this rate difference is the differential solvation of the reactants, in particular the reactant anion, and the activated complex. Since the charge is more localized on the reactant anion than on the activated complex, the former will be better solvated than the charge-delocalized activated complex. This results in a greater decrease in reactant enthalpy compared to that of the activated complex. Whereas the gas-phase activation barrier lies lower than the enthalpy of the reactants, the differential solvation of the reactants and the activated complex causes an increase in this activation barrier to a value higher than the reactants (*cf.* Fig. 5-4).

This differential solvation of reactants and activated complex is greater for protic solvents because protic, *i.e.* HBD solvents, are more sensitive to charge delocalization than aprotic, *i.e.* non-HBD solvents, due to reduced hydrogen-bonding with increasing charge delocalization. This is the main reason for the large rate enhancements in dipolar non-HBD solvents relative to protic solvents (*cf.* Table 5-2) [6].

Before the reactant anion and the dipolar reactant molecule can come into contact in solution, the solvation shell around the two reactants must be at least partially disrupted. Therefore, the reaction rate in solution is determined primarily by the amount of energy needed to destroy the solvation shells and not by the intrinsic properties of the reactants. The resistance to breaking the hydration shell corresponds to an activation barrier of about 103 kJ/mol for the S_N2 reaction given in Table 5-2 and Fig. 5.4.

In addition to the dramatic difference in absolute rates between gas-phase and solution reactions, there are also differences and even reversals in relative rates. For instance, the gas-phase order of increasing nucleophilicity of halogen anions is $I^{\ominus} < Br^{\ominus} < Cl^{\ominus} < F^{\ominus}$, whereas in protic solvents the reverse order is found [6]. *Cf.* Section 5.5.2 for further discussion.

The arguments used here for S_N2 displacement reactions can be generally applied to other ionic reactions.

It has been demonstrated that only a small number of solvent molecules are needed to bridge most of the gap between the enthalpy diagram for nucleophilic displacement reactions in the gas phase and that in solution [475, 477, 485–488].

For instance, the S_N2 displacement reaction $HO^{\ominus} + CH_3{-}Br \rightarrow HO{-}CH_3 + Br^{\ominus}$ has been carried out in the gas phase as a function of stepwise solvation of the hydroxide ion [485–487]. As Table 5-3 shows, already on the addition of just one water molecule to the nucleophile a significant decrease in reactivity is observed. The overall decrease in

Table 5-3. Comparison of the activation energies and specific rate constants for the bimolecular S_N2 ion-molecule reaction $HO^{\ominus} + CH_3—Br \rightarrow HO—CH_3 + Br^{\ominus}$ in the gas phase and at various degrees of hydration of the hydroxide ion at *ca.* 23 °C [485].

Medium	Nucleophile	$\frac{E_a}{(kJ \cdot mol^{-1})}$	$\frac{k_2}{(cm^3 \cdot molecule^{-1} \cdot s^{-1})}$	k_2^{rel}
Gas phase	$HO^{\ominus}$	*ca.* 2[a]	$1.0 \cdot 10^{-9}$	$4.8 \cdot 10^{15}$
Gas phase	$HO^{\ominus} \cdot (H_2O)_1$		$6.3 \cdot 10^{-10}$	$3.0 \cdot 10^{15}$
Gas phase	$HO^{\ominus} \cdot (H_2O)_2$		$2.0 \cdot 10^{-12}$	$9.5 \cdot 10^{12}$
Gas phase	$HO^{\ominus} \cdot (H_2O)_3$		$\leq 2.0 \cdot 10^{-13}$	$\leq 9.5 \cdot 10^{11}$
Water[b]	$HO^{\ominus} \cdot (H_2O)_n$	96	$2.1 \cdot 10^{-25}$	*1.0*

[a] Estimated from the reaction probability per collision [480].
[b] At 25 °C; *cf.* E.A. Moelwyn-Hughes, Proc. Roy. Soc., Part A *196*, 540 (1949); R.H. Bathgate and E.A. Moelwyn-Hughes, J. Chem. Soc. *1953*, 2642.

reactivity amounts to at least three orders of magnitude with the addition of three molecules of water to the hydroxide ion (the solvation number of $n = 3$ represents the operational limit of the flowing-afterglow technique used for these measurements). The addition of the first solvent molecule leads to only a slight rate decrease but further solvent addition reduces the reactivity precipitously. Eventually, in aqueous solution, a further dramatic drop in reactivity of twelve orders of magnitude is observed, accompanied by a corresponding increase in activation energy; *cf.* Table 5-3. Obviously, the solvation capability, in particular the hydrogen-bond donor ability, of three-dimensional structured "polymeric" water and of "monomeric" water is quite different.

Such a decrease in reactivity with increasing solvation can be qualitatively accounted for in terms of the double-minimum enthalpy diagram proposed for S_N2 reactions by Brauman *et al.* [474]; *cf.* Fig. 5-4. Figure 5-5 gives a schematic representation of the changes in enthalpy for the gas-phase reaction of stepwise hydrated hydroxide ions with bromomethane [485]. According to this diagram, the S_N2 reaction proceeds *via* a three-step mechanism: (i) formation of a loosely bound cluster (b) from the reactants (a), (ii) isomerization of this cluster by methyl transfer with inversion of configuration *via* the activated complex (c) to cluster (d), and (iii) final dissociation of the loosely bound cluster (d) into products (e).

As the nucleophile becomes more and more solvated the differential solvation of the reactants, the cluster, and the activated complex leads to an activated complex which becomes relatively less stable and hence to an increasingly large central reaction barrier. While the height of this central barrier may remain below the enthalpy of the reactants at low hydration numbers, thus accounting for the high reaction efficiencies observed at this degree of hydration, further solvation eventually leads to a central barrier with an enthalpy larger than the reactants and thus to a dramatic decrease in reaction efficiency [485]*.

* Very recent calculations of the interaction energies of both reactants and activated complex of simple S_N2 reactions with clusters of water molecules have shown that about sixty or even more water molecules are needed to explain the huge rate difference of about twenty orders of magnitude between gas-phase and solution S_N2 reactions; *cf.* Y. S. Kong and M. S. Jhon, Theor. Chim. Acta *70*, 123 (1986).

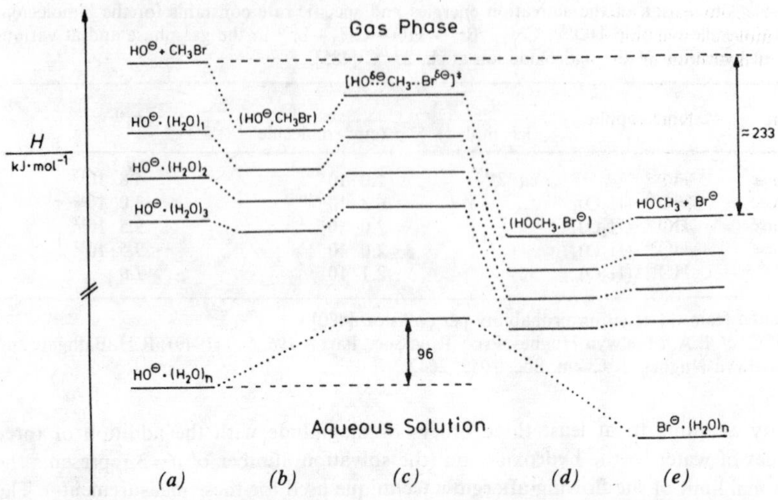

Fig. 5-5. Schematic one-dimensional relative enthalpy diagram for the exothermic bimolecular displacement reaction $HO^{\ominus} + CH_3—Br \rightarrow HO—CH_3 + Br^{\ominus}$ in the gas phase and at various degrees of hydration of the hydroxide ion [485]. Ordinate: standard molar enthalpies of (a) the reactants, (b, d) loose ion-molecule clusters held together by ion-dipole and ion-induced dipole forces, (c) the activated complex, and (e) the products. Abscissa: not defined, expresses only the sequence of (a)...(e) as they occur in the chemical reaction. The barrier heights ascribed to the activated complex at intermediate degrees of hydration were chosen to be qualitatively consistent with the experimental rate measurements; *cf.* Table 5-3 [485]. Possible hydration of the neutral reactant and product molecules, $CH_3—Br$ and $HO—CH_3$, are ignored. The barrier height ascribed to the activated complex in aqueous solution corresponds to the measured Arrhenius activation energy. A somewhat different picture of this S_N2 reaction in the gas phase which calls in question the simultaneous solvent-transfer from $HO^{\ominus}$ to $Br^{\ominus}$ is given in reference [487].

Considering the product side in Fig. 5-5, there are three possible reaction pathways open to the intermediate cluster (b), formed in the gas-phase reaction of hydrated hydroxide ions with bromomethane [475].

$$HO^{\ominus}\cdots H{-}OH \;\; CH_3{-}Br \quad (b) \rightarrow \begin{cases} (1) & Br^{\ominus} + HO(CH_3)\cdots H{-}OH \\ (2) & Br^{\ominus}\cdots H{-}OH + HO{-}CH_3 \\ (3) & Br^{\ominus}\cdots HO{-}CH_3 + H{-}OH \end{cases}$$

Option (1) gives the unsolvated and option (3) the product-solvated bromide ion. Formation of the hydrated bromide ion according to option (2) requires the concomitant transfer of a water molecule from the reactant nucleophile ($HO^{\ominus}$) to the leaving group ($Br^{\ominus}$) which can take place by simultaneous or sequential migration of H_2O and CH_3. Each of these three pathways should result in a different enthalpy diagram.

Recent measurements at room temperature have clearly shown that about 95% of the reaction between $HO^{\ominus} \cdot (H_2O)_1$ and CH_3—Br proceeds *via* path (1) to produce the unsolvated $Br^{\ominus}$ ion as the principal primary product and not the solvated species, $Br^{\ominus} \cdot H_2O$ and $Br^{\ominus} \cdot HOCH_3$ [487]. Thus, a sequential inversion (*i. e.* methyl transfer) and solvent transfer seems to be the preferred route for this particular reaction in the gas phase.

Recent *ab initio* SCF calculations of the degenerate S_N2 reaction $Cl^{\ominus} \cdot (H_2O)_n + CH_3—Cl \rightarrow Cl—CH_3 + Cl^{\ominus} \cdot (H_2O)_n$ provide further insight into the possible mechanisms of solvent transfer between reactant and product ions and the actual features of the intermediate enthalpy change at various degrees of solvation [489]. For the non-hydrated S_N2 reaction, the calculations produce a double-minima enthalpy diagram analogous to that given in Fig. 5-5. The singly-hydrated reaction may proceed via *simultaneous* migration of H_2O and CH_3 through a symmetric activated complex but also *sequentially* by CH_3 transfer with subsequent migration of the H_2O molecule or *vice versa*. The doubly-hydrated reaction involves two H_2O migrations and a CH_3 transfer which can take place one by one, two by one, or all three simultaneously. The most favorable path involves one molecule of H_2O being transferred first with little or no barrier, followed by CH_3 transfer and the transfer of the second H_2O molecule. For large solvation numbers, when the first solvation shell of $Cl^{\ominus}$ is completed, the initial interaction between the ion-molecule reactants first involves dehydration, thus introducing a new feature into the enthalpy diagram; *cf.* the lower part of Fig. 5-5. In other words, the rate constants do not decrease monotonically with increasing solvation number and the corresponding enthalpy diagrams do not transform systematically toward that for the reaction in solution [487].

In the experiments reported in this Section the reactions proceed in the absence of bulk solvent in an inert gaseous medium at low total pressures so that available solvent molecules are limited to those associated directly with the reactant ions. The main difference between these solvated-ion/molecule reactions in the gas phase and the corresponding reactions in solution is the kinetic role of the bulk solvent. Solvated-ion/molecule reactions carried out in the gas phase reveal the kinetic participation of the ion-solvate in the absence of bulk solvent [487].

Further examples of gas-phase measurements which reveal the influence of stepwise solvation of homo- and heteroconjugated anionic nucleophiles, $A^{\ominus} \cdot (AH)_n$ and $A^{\ominus} \cdot (BH)_n$, on the kinetics of S_N2 discplacement reactions with halomethanes can be found in references [475, 486]. In all cases, solvation of the reactant anion with up to three solvent molecules leads to a decrease in the rate constants by at least three orders of magnitude.

Not only reaction rates but also the kind of products obtained can be changed in going from the gas phase to solution. For instance, the reaction of the stepwise solvated hydroxide ion with acetonitrile follows three distinct pathways: bare and singly-solvated (in the gas phase) as well as bulk-solvated hydroxide ions react with acetonitrile to give three different sets of products as result of displacement, proton-transfer, and hydrolysis reactions [488]!

In conclusion, these gas-phase measurements provide new clues to the role of solvation in ion-molecule reactions. For the first time it is possible to study intrinsic reactivities and the extent to which the properties of gas-phase ion-molecule reactions transform to those of the corresponding reactions in solution. It is clear, however, that gas-phase solvated-ion/molecule reactions in which solvent molecules are transferred into the intermediate clusters by the nucleophile cannot be exact duplicates of solvated-ion/molecule reactions in solution in which solvated reactants exchange solvent molecules with the surrounding bulk solvent. Future gas-phase measurements and the study of solute/solvent molecule clusters [743] will surely shed further light on these rather complex solvation phenomena and help to bridge the gap between gaseous and condensed-phase chemistry.

5.3 Qualitative Theory of Solvent Effects on Reaction Rates

Organic reactions can be loosely grouped into three classes depending on the character of the activated complex through which these reactions can proceed: dipolar, isopolar, and free-radical transition-state reactions [15, 468].

Dipolar activated complexes differ considerably in charge separation or charge distribution from the initial reactants. *Dipolar transition-state reactions* with large solvent effects can be found amongst ionization, displacement, elimination, and fragmentation reactions such as:

$$\text{R-X} \rightleftharpoons [\overset{\delta\oplus}{\text{R}}\cdots\overset{\delta\ominus}{\text{X}}]^{\ddagger} \longrightarrow \text{products}$$

$$\text{R-M} \rightleftharpoons [\overset{\delta\ominus}{\text{R}}\cdots\overset{\delta\oplus}{\text{M}}]^{\ddagger} \longrightarrow \text{products}$$

$$\text{Y:} + \text{R-X} \rightleftharpoons [\overset{\delta\oplus}{\text{Y}}\cdots\text{R}\cdots\overset{\delta\ominus}{\text{X}}]^{\ddagger} \longrightarrow \text{products}$$

$$\text{Y-A-B-C-X} \rightleftharpoons [\overset{\delta\oplus}{\text{Y}}\cdots\text{A}\cdots\text{B}\cdots\text{C}\cdots\overset{\delta\ominus}{\text{X}}]^{\ddagger} \longrightarrow \text{products}$$

Isopolar activated complexes differ very little or not at all in charge separation or charge distribution from the corresponding initial reactants. *Isopolar transition-state reactions* with small or negligible solvent effects can be found amongst pericyclic reactions, two examples of which are Diels-Alder cycloaddition reactions and the Cope rearrangement of 1,5-hexadienes:

[Scheme: butadiene + R-substituted alkene ⇌ [cyclic transition state with R]‡ → product]

[Scheme: R-substituted 1,5-hexadiene ⇌ [cyclic transition state with R]‡ → product]

Free-radical activated complexes are formed by the creation of unpaired electrons during homolytic bond cleavage. *Free-radical transition-state reactions* with small or negligible solvent effects are found among radical-pair formation and atom-transfer reactions such as:

$$R{-}R \rightleftharpoons \left[\overset{\delta\odot}{R}\cdots\cdots\overset{\delta\odot}{R}\right]^{\ddagger} \longrightarrow R\odot + \odot R$$

$$R\odot + A{-}X \rightleftharpoons \left[\overset{\delta\odot}{R}\cdots A\cdots\overset{\delta\odot}{X}\right]^{\ddagger} \longrightarrow R{-}A + \odot X$$

In the following Sections, typical examples of solvent effects on dipolar, isopolar, and free-radical transition-state reactions are given.

5.3.1 The Hughes-Ingold Rules

The effect of solvent on aliphatic nucleophilic substitution and elimination reactions was investigated by Hughes and Ingold. They used a simple qualitative solvation model considering only pure electrostatic interactions between ions or dipolar molecules and solvent molecules in initial and transition states [16, 44]. Depending on whether the reaction species are neutral, positively or negatively charged, all nucleophilic substitution and elimination reactions may be divided into different charge types. Based on certain reasonable assumptions as to the extent of solvation to be expected in the presence of electric charges:

(a) increase in magnitude of charge will increase solvation;
(b) increase in dispersal of charge will decrease solvation; and
(c) destruction of charge will decrease solvation more than dispersal of charge,

the gross effect of the solvent on reactions of different charge types can be summarized as follows:

(a) An increase in solvent polarity results in an increase in the rates of those reactions in which the charge density is greater in the activated complex than in the initial reactant molecule(s).

(b) An increase in solvent polarity results in a decrease in the rates of those reactions in which the charge density is lower in the activated complex than in the initial reactant molecule(s).

(c) A change in solvent polarity will have a negligible effect on the rates of those reactions that involve little or no change in the charge density on going from reactant(s) to the activated complex.

In other words, a change to a more polar solvent will increase or decrease the reaction rate depending on whether the activated reaction complex is more or less dipolar than the initial reactants. In this respect the term "solvent polarity" was used synonymously with the power to solvate solute charges. It was assumed to increase with the dipole moment of the solvent molecules, and to decrease with increased thickness of shielding of the dipole charges.

For example, the reaction between two equally charged ions results in an increase in charge density during the activation process. Therefore, the reaction rate will increase with increasing solvent polarity. On the other hand, a reaction between oppositely charged ions will be slower in polar solvents, which are good ion solvators, because in this case a reduction of charge density occurs in going from the reactants to the activated complex. Furthermore, reactions where charge is either created or destroyed during the activation process, should be affected to a greater extent by a solvent polarity change than reactions in which there is only charge dispersal. Thus, the rates of substitution reactions involving charge creation or destruction are altered by a factor of $10^3 \ldots 10^6$ in going from water to ethanol, whereas S_N reactions involving charge dispersal are increased by only a factor of $3 \ldots 10$ when the solvent is changed from ethanol to water.

These Hughes-Ingold rules can be used for making qualitative predictions about the effect of solvent polarity on the rates of all heterolytic reactions of known mechanisms. For nucleophilic substitution reactions of type (5-11) and (5-12)

$$\text{R-X} \xrightleftharpoons{S_N1} [\overset{\delta\oplus}{\text{R}}\cdots\overset{\delta\ominus}{\text{X}}]^{\ddagger} \longrightarrow \text{R}^{\oplus} + :\text{X}^{\ominus} \xrightarrow{+:\text{Y}^{\ominus}} \text{R-Y} + :\text{X}^{\ominus} \tag{5-11}$$

$$\text{Y}:^{\ominus} + \text{R-X} \xrightleftharpoons{S_N2} [\overset{\delta\ominus}{\text{Y}}\cdots\text{R}\cdots\overset{\delta\ominus}{\text{X}}]^{\ddagger} \longrightarrow \text{Y-R} + :\text{X}^{\ominus} \tag{5-12)*}$$

the predictions are compiled in Table 5-4. The middle three columns show, for each charge type and mechanism, what happens to the charges on going from the initial to the transition state. The last column shows the predicted solvent effects. The conclusions

* S_N2 reactions of the type shown in Eq. (5-12) have been commonly represented as occurring by attack of a lone electron pair from $Y^{\ominus}$ at the backside of R with simultaneous displacement of $X^{\ominus}$. However, this representation is misleading because electrons shift singly, not in pairs! Therefore, Pross and Shaik recommend that synchronous S_N2 displacement reactions be represented in terms of $Y:^{\ominus} + R \cdot\cdot X \rightarrow Y \cdot\cdot R + :X^{\ominus}$, where a single electron from the donor anion $Y:^{\ominus}$ pairs with one electron from R, and an electron from R is transferred to X, accompanied by cleavage of the R—X bond. In certain cases, this process can merge with a single-electron transfer (SET) pathway wherein the product Y—R is formed by coupling of an intermediate geminate radical pair according to

$$Y:^{\ominus} + R \cdot\cdot X \xrightarrow{\text{SET}} Y^{\odot} + R \cdot\cdot X^{\odot\ominus} \longrightarrow Y \cdot\cdot R + :X^{\ominus}$$

That is, the act of shifting the single electron from Y to X may occur either with or without free-radical formation. Usually, the concerted non-radicaloid process is energetically favoured; *cf.* A. Pross, Acc. Chem. Res. *18*, 212 (1985); S. S. Shaik, Progr. Phys. Org. Chem. *15*, 197 (1985). For the recognition of SET pathways in organometallic reactions *cf.* the review of J. K. Kochi: *Organometallic Mechanisms and Catalysis*. Academic Press, New York 1978.

Table 5-4. Predicted solvent effects on rates of nucleophilic substitution reactions [16, 44–46].

Reaction type	Initial reactants	Activated complex	Charge alteration during activation	Effect of increased solvent polarity on rate[a)]
(a) S_N1	R—X	$R^{\delta+}\ldots X^{\delta-}$	Separation of unlike charges	Large increase
(b) S_N1	$R—X^+$	$R^{\delta+}\ldots X^{\delta+}$	Dispersal of charge	Small decrease
(c) S_N2	Y + R—X	$Y^{\delta+}\ldots R\ldots X^{\delta-}$	Separation of unlike charges	Large increase
(d) S_N2	Y^- + R—X	$Y^{\delta-}\ldots R\ldots X^{\delta-}$	Dispersal of charge	Small decrease
(e) S_N2	Y + $R—X^+$	$Y^{\delta+}\ldots R\ldots X^{\delta+}$	Dispersal of charge	Small decrease
(f) S_N2	Y^- + $R—X^+$	$Y^{\delta-}\ldots R\ldots X^{\delta+}$	Destruction of charge	Large decrease

[a] The terms "large" and "small" arise from the theory that the effect of the dispersal of charge should be notably smaller than the effect of its creation or destruction and have therefore only relative significance.

derived from Table 5-4 were experimentally confirmed in a large number of substitution reactions. Some typical examples shall demonstrate the application of these rules.

The solvolysis of 2-chloro-2-methylpropane is 335000 times faster in water than in the less polar solvent ethanol [40]; *cf.* reaction type (a) in Table 5-4.

$$(H_3C)_3C{-}Cl \underset{25\,°C}{\overset{k_1}{\rightleftharpoons}} \left[(H_3C)_3C^{\delta\oplus}\cdots Cl^{\delta\ominus}\right]^{\ddagger} \longrightarrow (H_3C)_3C^{\oplus} + Cl^{\ominus} \longrightarrow \text{products} \qquad (5\text{-}13)$$

$\mu = 2.9 \cdot 10^{-30}$ C m $\qquad \mu \approx 27 \cdot 10^{-30}$ C m

Solvent	C_2H_5OH	CH_3OH	$HCONH_2$	HCO_2H	H_2O
$k_1^{rel*)}$	1	9	430	12200	335000

Comparison of the solvolysis rate constants of 2-chloro-2-methylpropane obtained in water and in benzene solution reveals a rate acceleration of *ca.* 10^{11} with increasing solvent polarity [47]**. The solvolysis rate of 1-bromoadamantane in ethanol/water increases by a factor of 4900 when the volume fraction of water goes from 10 to 60 cl/l [48].

* Here and in the following examples only the relative rate constants, k^{rel}, with respect to the "slowest solvent" are given. The selected solvents are ordered from left to right with increasing polarity.

** In the case of 2-chloro-2-methylpropane which leads mainly to *t*-butanol and *t*-butyl ethers together with some *i*-butene, the term *solvolysis* is normally restricted to the reaction in water and other protic solvents. In non-HBD solvents, however, the only reaction product is *i*-butene. For convenience, the term *solvolysis* is often used in the literature to cover both types of reaction, solvolysis and dehydrohalogenation of 2-chloro-2-methylpropane.

Rearrangement reactions, where the first step corresponds to the ionization of a tertiary haloalkane, also proceed faster with increasing solvent polarity, as shown by the Wagner-Meerwein rearrangement of 3-chloro-2,2,3-trimethylnorbornane to 2-*exo*-chlorobornane [49].

$$\xrightarrow[20\,°C]{k_1} \left[\cdots \mathrm{Cl}^{\delta\ominus} \ \delta\oplus \right]^{\ddagger} \xrightarrow{-\mathrm{Cl}^{\ominus}} \oplus \xrightarrow{+\mathrm{Cl}^{\ominus}} \mathrm{Cl} \qquad (5\text{-}14)$$

Solvent	Petroleum ether	C_6H_5Cl	$C_6H_5NO_2$	CH_3CN	CH_3NO_2
k_1^{rel}	1	20	65	200	610

In agreement with the predictions for reaction type (b) in Table 5-4, the thermolysis of triethylsulfonium bromide takes place more slowly in polar solvents, *e.g.* alcohols, than in relative less polar solvents, *e.g.* acetone [50].

$$(H_5C_2)_2\overset{\oplus}{S}{-}C_2H_5 \underset{70\,°C}{\overset{k_1}{\rightleftharpoons}} \left[(H_5C_2)_2\overset{\delta\oplus}{S}\cdots\overset{\delta\oplus}{C_2H_5} \right]^{\ddagger} \longrightarrow (H_5C_2)_2S + \overset{\oplus}{C_2H_5} \xrightarrow{+\mathrm{Br}^{\ominus}} C_2H_5Br \qquad (5\text{-}15)$$

Solvent	CH_3COCH_3	$CHCl_3$	$C_6H_5NO_2$	$i\text{-}C_5H_{11}OH$	$C_6H_5CH_2OH$
k_1^{rel}	290	>230	180	4.5	1

Solvolysis rates of *tert*-butyldimethylsulfonium salts decrease with increasing solvent polarity [710]. Analogously, the rate of solvolysis of triethyloxonium salts in ethanol/water mixtures decreases with increasing water content [490]. Solvolysis rates of *N-tert*-alkylpyridinium salts such as 1-(1-methyl-1-phenylethyl)pyridinium perchlorate are almost independent of solvent polarity, whereas *N-sec*-alkylpyridinium salts exhibit small decreases in rate with increasing solvent polarity [710]. The non-creation of charge in the activation process of these type (b) reactions much reduces the influence of solvent polarity on rate; *cf.* Table 5-4.

An example of reaction type (c) in Table 5-4 is the well-known Menschutkin reaction [30] between tertiary amines and primary haloalkanes yielding quaternary ammonium salts. Its solvent dependence was studied very thoroughly by a number of investigators [51–65, 491–496]. For instance, the reaction of tris-*n*-propylamine with iodomethane at 20 °C is 120-times faster in diethyl ether, 13000-times faster in chloroform, and 110000-times faster in nitromethane than in *n*-hexane [60]. It has been estimated, that the activated complex of this Menschutkin reaction should have a dipole moment of *ca.* $29 \cdot 10^{-30}$ C m (8.7 Debye) [23, 64], which is much larger than the dipole moments of the reactant molecules (tris-*n*-propylamine $2.3 \cdot 10^{-30}$ C m = 0.70 Debye; iodomethane $5.5 \cdot 10^{-30}$ C m = 1.64 Debye) [64].

Activation parameter data as a function of solvent have been published for the reaction of triethylamine with iodoethane [59]. The reaction rates obtained in dipolar aprotic solvents together with the activation parameters are given in Table 5-5 [59].

$$(H_5C_2)_3N + H_2C(CH_3)\text{-}I \xrightarrow[50\,°C]{k_2} \left[(H_5C_2)_3\overset{\delta\oplus}{N}\cdots C(CH_3)H_2 \cdots \overset{\delta\ominus}{I} \right]^{\neq} \longrightarrow (H_5C_2)_3\overset{\oplus}{N}\text{-}CH_2CH_3 + I^{\ominus} \quad (5\text{-}16)$$

Table 5-5. Absolute and relative rate constants, Gibbs activation energies, activation enthalpies, and activation entropies of the Menschutkin reaction between triethylamine and iodoethane in twelve solvents at 50 °C [59].

Solvents	$k_2 \cdot 10^5$ / $(l \cdot mol^{-1} \cdot s^{-1})$	k_2^{rel}	$\Delta G^{\neq}$ / $(kJ \cdot mol^{-1})$	$\Delta H^{\neq}$ / $(kJ \cdot mol^{-1})$	$\Delta S^{\neq}$ / $(J \cdot mol^{-1} \cdot K^{-1})$
1,1,1-Trichloroethane	1.80	1	102.6	52.3	−156
Chlorocyclohexane	3.09	2	101.6	54.4	−146
Chlorobenzene	9.30	5	98.2	46.9	−159
1,1-Dichloroethane	11.9	7	97.1	48.1	−151
Chloroform	15.4	9	96.2	49.0	−146
1,2-Dichlorobenzene	18.1	10	96.7	49.4	−146
Acetone	31.7	18	94.1	47.3	−145
Cyclohexanone	33.7	19	94.8	51.5	−134
1,2-Dichloroethane	42.6	24	93.4	45.2	−149
Propionitrile	59.6	33	92.3	48.5	−135
Benzonitrile	76.5	43	92.6	49.0	−135
Nitrobenzene	93.4	52	92.0	49.4	−133

A solvent change from 1,1,1-trichloroethane to nitrobenzene causes a 52-fold rate acceleration for reaction (5-16) which corresponds to a decrease in $\Delta G^{\neq}$ of 10.6 kJ/mol. Bimolecular reactions such as (5-16), which produce charge during the activation process, usually show large negative entropies of activation. A negative entropy of activation indicates a greater degree of ordering in the transition state than in the initial state, due to an increase in solvation during the activation process. The $\Delta S^{\neq}$ values of Table 5-5 show, that the largest decrease of activation entropy is obtained in the less polar solvents. This observation can be rationalized because polar solvents will have some structure corresponding to the orientation of the dipolar solvent molecules due to intermolecular solvent-solvent interactions. In less polar solvents, however, which have only a small or no dipole moment, the solvent molecules will be relatively unoriented and consequently have a higher entropy. Thus, non-polar solvents will have a greater entropy loss as a result of increased solvation during the activation process. Consequently, reactions going through dipolar activated complexes should have a larger negative entropy of activation in less polar solvents than in polar solvents [226].

The Menschutkin-type S_N2 reaction of triphenylphosphane with iodomethane has been studied in thirteen solvents [500]. In propylene carbonate, this reaction is 245 times faster than in di-*i*-propyl ether. In agreement with the highly dipolar activated complex, large solvent-dependent negative activation volumes, $\Delta V^{\neq}$, have been obtained as result of charge-induced reorientation of the surrounding solvent molecules during the activation process; *cf.* also Section 5.5.11.

An example of reaction type (d) in Table 5-4 is the Finkelstein halide exchange reaction between iodomethane and radioactive labeled iodide ion. The rate constant for this reaction decreases by about 10^4 in the solvent change from less polar acetone to water [66].

$$\overset{*}{I}^{\ominus} + CH_3\text{-}I \underset{25\,°C}{\overset{k_2}{\rightleftharpoons}} \left[\overset{*}{I}^{\delta\ominus}\cdots CH_3\cdots I^{\delta\ominus}\right]^{\ddagger} \longrightarrow \overset{*}{I}\text{-}CH_3 + I^{\ominus} \qquad (5\text{-}17)$$

Solvent	CH_3COCH_3	C_2H_5OH	$(CH_2OH)_2$	CH_3OH	H_2O
k_1^{rel}	13000	44	17	16	1

Other examples of this type of reaction are: S_N2 reactions between azide ion and 1-bromobutane [67], bromide ion and methyl tosylate [68], and bromide ion and iodoethane [497]. In changing the medium from non-HBD solvents (HMPT, 1-methylpyrrolidin-2-one) to methanol, the second-order rate constants decrease by a factor of $2 \cdot 10^5$ [67], $9 \cdot 10^4$ [68], and $1 \cdot 10^5$ [497], respectively. The large decrease in these rate constants in going from the less to the more polar solvent is not only governed by the difference in solvent polarity, as measured by dipole moment or dielectric constant, but also by the fact that the less polar solvents are dipolar aprotic and the more polar solvents are protic (*cf.* Section 5.5.2).

The second-order reaction between trimethylamine and the trimethylsulfonium ion gives the predicted rate decrease with increasing solvent polarity [69]; *cf.* reaction type (e) in Table 5-4.

$$(CH_3)_3N + CH_3\text{-}\overset{\oplus}{S}(CH_3)_2 \underset{45\,°C}{\overset{k_2}{\rightleftharpoons}} \left[(CH_3)_3\overset{\delta\oplus}{N}\cdots CH_3\cdots\overset{\delta\oplus}{S}(CH_3)_2\right]^{\ddagger} \longrightarrow (CH_3)_3\overset{\oplus}{N}\text{-}CH_3 + S(CH_3)_2 \qquad (5\text{-}18)$$

Solvent	CH_3NO_2	C_2H_5OH	CH_3OH	H_2O
k_2^{rel}	119	10	6	1

Obviously the initial reactants are more strongly solvated than the activated complex.

Finally, as an example of reaction type (f) in Table 5-4, the alkaline hydrolysis of the trimethylsulfonium ion demonstrates the predicted large rate decrease by increasing the proportion of water in an aqueous ethanolic reaction medium [70].

$$HO^{\ominus} + CH_3\text{-}\overset{\oplus}{S}(CH_3)_2 \underset{100\,°C}{\overset{k_2}{\rightleftharpoons}} \left[\overset{\delta\ominus}{HO}\cdots CH_3\cdots\overset{\delta\oplus}{S}(CH_3)_2\right]^{\ddagger} \longrightarrow HO\text{-}CH_3 + S(CH_3)_2 \qquad (5\text{-}19)$$

Solvent	Ethanol/water			
H_2O in cl/l	0	20	40	100
k_2^{rel}	19600	480	40	1

Similar findings were obtained for the alkaline decomposition of triarylsulfonium halides with ethoxide ion in aqueous ethanol at 120 °C [71]. The results indicate, that the decomposition rate is decreased about 10^6-fold by increasing the water content from 2.3 to 98.2 cmol/mol [71].

It should be mentioned that a solvent change affects not only the reaction rate, but also the reaction mechanism (*cf.* Section 5.5.7). The reaction mechanism of some haloalkanes goes from an S_N1-mechanism to S_N2, when the solvent is changed from aqueous ethanol to acetone. On the other hand, reactions of halomethanes, which proceed in aqueous ethanol by a S_N2-mechanism, can become S_N1 in more strongly ionizing solvents like formic acid. For a comparison of solvent effects on nucleophilic substitution reactions at primary, secondary and tertiary carbon atoms see reference [72].

The mechanisms of nucleophilic substitution and β-elimination reactions are closely parallel with regard to the rate-determining step. The two monomolecular reactions (S_N1 and E_1) have a common rate-controlling step, whereas in the two bimolecular reactions (S_N2 and E_2), the electron transfers from reagent to leaving group are similar. However, they pass through a larger chain of carbon atoms in elimination than in substitution reactions. Therefore, similar rules for solvent effects on monomolecular, (5-20), and bimolecular, (5-21), β-elimination reactions of different charge-type have been obtained by Hughes and Ingold [16, 44] (*cf.* Table 5-6).

$$\text{H-CR}_2\text{-CR}_2\text{-X} \overset{E_1}{\rightleftharpoons} [\text{H-CR}_2\text{-}\overset{\delta\oplus}{\text{C}}\text{R}_2\cdots\overset{\delta\ominus}{\text{X}}]^{\ddagger} \longrightarrow \text{H-CR}_2\text{-}\overset{\oplus}{\text{C}}\text{R}_2 + :\text{X}^{\ominus} \xrightarrow{+:\text{Y}^{\ominus}} \text{Y-H} + \text{R}_2\text{C=CR}_2 + :\text{X}^{\ominus} \quad (5\text{-}20)$$

$$\text{Y:}^{\ominus} + \text{H-CR}_2\text{-CR}_2\text{-X} \overset{E_2}{\rightleftharpoons} [\overset{\delta\ominus}{\text{Y}}\cdots\text{H}\cdots\text{CR}_2\text{=}\!\!\text{=}\text{CR}_2\cdots\overset{\delta\ominus}{\text{X}}]^{\ddagger} \longrightarrow \text{Y-H} + \text{R}_2\text{C=CR}_2 + :\text{X}^{\ominus} \quad (5\text{-}21)$$

The changes undergone by the charges during the activation process are shown in the three middle columns of Table 5-6. The last column shows the predicted solvent effects.

Owing to the similarity of solvent effects for E_1 and S_N1 reactions due to the same rate-controlling step, only some examples of E_2 reactions are mentioned. Typical examples for the reaction types (c) to (f) in Table 5-6 are shown in Eqs. (5-22) to (5-25).

$$(\text{CH}_3)_3\text{N} + (\text{CH}_3)_3\text{C-Br} \longrightarrow (\text{CH}_3)_3\overset{\oplus}{\text{N}}\text{-H} + \text{H}_2\text{C=C(CH}_3)_2 + \text{Br}^{\ominus} \quad (5\text{-}22)$$

$$\text{HO}^{\ominus} + \text{CH}_3\text{-CH}_2\text{-Br} \longrightarrow \text{H}_2\text{O} + \text{H}_2\text{C=CH}_2 + \text{Br}^{\ominus} \quad (5\text{-}23)$$

$$\text{H}_2\text{O} + \text{O}_2\text{N-C}_6\text{H}_4\text{-CH}_2\text{-CH}_2\text{-}\overset{\oplus}{\text{N}}\text{(CH}_3)_3 \longrightarrow \text{H}_3\text{O}^{\oplus} + \text{O}_2\text{N-C}_6\text{H}_4\text{-CH=CH}_2 + \text{N(CH}_3)_3 \quad (5\text{-}24)$$

$$\text{C}_2\text{H}_5\text{O}^{\ominus} + \text{CH}_3\text{-CH}_2\text{-}\overset{\oplus}{\text{S}}\text{(CH}_3)_2 \longrightarrow \text{C}_2\text{H}_5\text{OH} + \text{H}_2\text{C=CH}_2 + \text{S(CH}_3)_2 \quad (5\text{-}25)$$

Table 5-6. Predicted solvent effects on rates of β-elimination reactions [16, 44, 73, 74].

Reaction type	Initial reactants	Activated complex	Charge alteration during activation	Effect of increased solvent polarity on rate[a]
(a) E_1	H–C–C–X	$\text{H–C–C}^{\delta\oplus}\cdots\text{X}^{\delta\ominus}$	Separation of unlike charges	Large increase
(b) E_1	$\text{H–C–C–X}^{\oplus}$	$\text{H–C–C}^{\delta\oplus}\cdots\text{X}^{\delta\oplus}$	Disperal of charge	Small decrease
(c) E_2	Y: + H–C–C–X	$\text{Y}^{\delta\oplus}\cdots\text{H}\cdots\text{C}\equiv\text{C}\cdots\text{X}^{\delta\ominus}$	Separation of unlike charges	Large increase
(d) E_2	$\text{Y:}^{\ominus}\text{ + H–C–C–X}$	$\text{Y}^{\delta\ominus}\cdots\text{H}\cdots\text{C}\equiv\text{C}\cdots\text{X}^{\delta\ominus}$	Dispersal of charge	Small decrease
(e) E_2	$\text{Y: + H–C–C–X}^{\oplus}$	$\text{Y}^{\delta\oplus}\cdots\text{H}\cdots\text{C}\equiv\text{C}\cdots\text{X}^{\delta\oplus}$	Dispersal of charge	Small decrease
(f) E_2	$\text{Y:}^{\ominus}\text{ + H–C–C–X}^{\oplus}$	$\text{Y}^{\delta\ominus}\cdots\text{H}\cdots\text{C}\equiv\text{C}\cdots\text{X}^{\delta\oplus}$	Destruction of charge	Large decrease

[a] The terms "large" and "small" arise from the theory that the effect of the dispersal of charge should be notably smaller than the effect of its creation or destruction and have therefore only relative significance.

In Eqs. (5-23) and (5-24) there is no net change of charge after reaction, but for Eq. (5-22) charge is created and for Eq. (5-25) it is destroyed. Further examples of observed effects for solvent changes on rates of mono- and bimolecular eliminations are given by Hughes and Ingold [16, 44]. In most cases studied, haloalkanes and 'onium salts in ethanol/water mixtures, the observed solvent effects are in the expected direction.

According to the Hughes-Ingold rules, in E_2 and S_N2 reactions the dispersal of charge is spread over more atoms than in E_1 and S_N1 reactions, so increase in solvent polarity will generally favor the E_1 and S_N1 mechanisms over the E_2 and S_N2 mechanisms. Thus, solvent change can alter not only the rate but also the reaction mechanism.

The competition between β-elimination and substitution reaction determines the proportion of alkene produced. The ratio of elimination to substitution is affected by the solvent as well as other factors (concentration and strength of the attacking base,

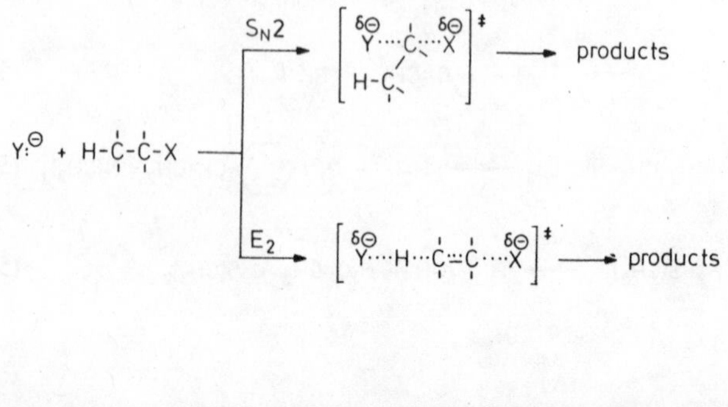

$$\text{Y:}^{\ominus} + \text{H–C–C–X} \xrightarrow{S_N2} \left[\begin{array}{c}\overset{\delta\ominus}{\text{Y}}\cdots\text{C}\cdots\overset{\delta\ominus}{\text{X}}\\ \text{H–C}\end{array}\right]^{\ddagger} \longrightarrow \text{products}$$

$$\text{Y:}^{\ominus} + \text{H–C–C–X} \xrightarrow{E_2} \left[\overset{\delta\ominus}{\text{Y}}\cdots\text{H}\cdots\text{C}\equiv\text{C}\cdots\overset{\delta\ominus}{\text{X}}\right]^{\ddagger} \longrightarrow \text{products}$$

temperature, structure of substrate and attacking base). Looking at the activated complexes for S_N2 and E_2 reactions, it can be seen that the charge is more widely dispersed in the elimination transition state, than in that of substitution.

From this it follows, that both reactions are decelerated with increasing solvent polarity, but due to the larger charge dispersal on activation in E_2 reactions, solvent stabilization of the activated complex of elimination is less than that for substitution. Therefore, E_2 eliminations are more decelerated than S_N2 substitutions with increasing solvent polarity, and the alkene yield should fall slightly on making this solvent change. Table 5-7 gives some examples about solvent effects on the proportion in which alkene is formed in β-elimination reactions.

Table 5-7. Observed solvent effects on the proportion of alkene formed in mono- and bimolecular β-eliminations in ethanol/water mixtures [16, 44].

		Water in ethanol in cl/l			
		0	20	40	100
(a) *Bimolecular reactions*	$(E_2 + S_N2)$				
$HO^{\ominus} + i\text{-}C_3H_7\text{—}Br$	(55 °C)	71	59	54	—
$HO^{\ominus} + (H_5C_2)_3S^{\oplus}$	(100 °C)	—	100	100	86
(b) *Monomolecular reactions*	$(E_1 + S_N1)$				
$t\text{-}C_4H_9Br$	(25 °C)	19.0	12.6	—	—
$t\text{-}C_5H_{11}\text{—}Br$	(25 °C)	36.3	26.2	—	—
$t\text{-}C_5H_{11}\text{—}S(CH_3)_2^{\oplus}$	(50 °C)	64.4	47.8	—	—

Although the solvent effects are small, the alkene formation diminishes as predicted with increasing water content (corresponding to increased solvent polarity). The S_N2/E_2 reaction of 2-phenylpropyl tosylate with sodium cyanide (in hexamethylphosphoric triamide and in *N,N*-dimethylformamide as solvents at 100 °C) gives α-methylstyrene (elimination product) and 1-cyano-2-phenylpropane (substitution product) [75]. It has been found in accordance with the predictions of the Hughes-Ingold rules, that the elimination/substitution ratio decreases as the polarity of the solvents (measured by the dielectric constant) increases [75].

Analogously, solvent effects on alkene formation in S_N1 and E_1 reactions can be predicted [16, 44]. Owing to the fact that the first step in both reactions, the heterolysis of the C—X bond, is exactly the same, we have to consider the activated complexes which lead to either alkene or substitution product.

$$\text{H-C-C-X} \xrightarrow{-:X^{\ominus}} \text{H-C-C}^{\oplus} \begin{cases} \xrightarrow[S_N1]{+:Y^{\ominus}} \left[\text{H-C-C}^{\delta\oplus}\cdots Y^{\delta\ominus}\right]^{\ddagger} \longrightarrow \text{products} \\ \xrightarrow[E_1]{+:Y^{\ominus}} \left[\overset{\delta\ominus}{Y}\cdots\text{H}\cdots\text{C}{=}\text{C}^{\delta\oplus}\right]^{\ddagger} \longrightarrow \text{products} \end{cases}$$

Both reactions are decelerated with increasing solvent polarity. But due to the greater dispersal of charge in the transition state of the E_1 reaction, the β-elimination is again more decelerated than the substitution reaction. This results in decreasing alkene formation with increasing solvent polarity. Similar results are obtained for reactions in which Y: represents a neutral base (*e.g.* H_2O). Some observed effects of solvent changes on alkene formation in E_1 and S_N1 reactions are given in Table 5-7. Experimental results are relatively scarce. Nevertheless, the predicted rate trends for all charge-types of 'onium salts and haloalkanes in protic media are obtained [16, 44]. Newer studies of the competition between substitution and elimination in solvolyses of various 2-halo-2-methylpropanes in acetic acid, ethanol, and water have shown, that decreasing solvent polarity favors alkene formation [76].

A further refinement of the Hughes-Ingold rules has been given by Westaway regarding the influence of solvents on the structure of S_N2 activated complexes [498]; *cf.* Eq. (5-12) and Table 5-4. His *solvation rule for S_N2 reactions* states that a change in solvent will not lead to a change in the structure of an S_N2 transition state if both the attacking nucleophile, Y, and the leaving group, X, have the same charge as in reaction types (d) and (e) in Table 5-4 (called *Type I* S_N2 reactions). A solvent change will, however, lead to a change in the structure of the S_N2 transition state when nucleophile and leaving group have opposite charges as in reaction types (c) and (f) in Table 5-4 (called *Type II* S_N2 reactions).

If the activated complex of *Type I* S_N2 reactions is transferred from the gas phase to solution, the solvent will interact nearly equally with the two partially charged groups Y and X, thus lowering the Gibbs energy of the activated complex. The Gibbs energy of the activated complex will vary from solvent to solvent and the rate constant changes with solvent, depending on the interaction between Y and X and the solvent. However, the relative charge density of Y and X will not be altered by a solvent change, and thus the structure of the activated complex will not be changed even though its Gibbs energy has been changed.

On the other hand, in activated complexes of *Type II* S_N2 reactions, the charges on Y and X are of opposite sign, and thus the nucleophile Y and the leaving group X will interact differently with the solvent. For instance, in protic solvents the negatively charged group will be preferentially solvated *via* hydrogen-bond interactions, whereas a much weaker ion-dipole type interaction will occur at the positively charged group. As a result, the difference in solute/solvent interactions at both ends of the activated complex will be large enough to cause a shift in charge density between Y and X along the Y...C...X bond axis of the S_N2 transition state. In this case, not only the Gibbs energy but also the structure of the activated complex will be altered from solvent to solvent.

In other words, whether or not an S_N2 reaction has a *tight* or *loose* activated complex will not only depend upon the nature of the reactants Y and R–X, in solution it will also be affected by the nature of the solvent. Better solvation of the activated complex of a type II S_N2 reaction by solvents with improved EPD/EPA properties will lead to a loosening of the activated complex. Transferring this activated complex from solution to the gas phase, with subsequent loss of the charge-separation stabilizing solvation, will therefore increase its tightness; *cf.* also [499].

This *solvation rule for S_N2 reactions* can be useful in predicting the influence of a change in solvent on the structure of activated complexes. It is in agreement with studies involving leaving group heavy atom and secondary α-deuterium kinetic isotope effects as well as theoretical calculations of solvent effects on transition-state structures. Possible limitations of this solvation rule have been discussed; *cf.* [498] and relevant references cited therein.

Although the Hughes-Ingold theory of solvent effects was first illustrated in the field of nucleophilic aliphatic substitutions and β-eliminations, it should be applicable to all other heterolytic reactions in solutions, where the activation processes are connected with creation, dispersal, or destruction of charge. Using Kosower's classification [15, 468], in the following Sections the solvent effects on other organic reactions will be discussed to give the reader a feeling for the expected solvent influence, and to help him in finding the right solvent for a particular reaction under study.

5.3.2 Solvent Effects for Dipolar Transition State Reactions

Reactions involving dipolar activated complexes are those reactions in which, compared to the initial state, there are considerable differences in charge distribution. In addition to S_N1/S_N2 and E_1/E_2 reactions as described in Section 5.3.1, solvent effects on the following reactions involving dipolar activated complexes have been studied: aromatic nucleophilic (S_NAr) and aromatic electrophilic (S_EAr) reactions; aliphatic electrophilic substitution (S_E1/S_E2) reactions; aliphatic electrophilic (A_E) and aliphatic nucleophilic addition (A_N) reactions; cycloaddition and cycloreversion reactions; aldol addition reactions; as well as rearrangement, fragmentation, and isomerization reactions. A selection of typical and particularly instructive examples, taken from a vast amount of examples given in the literature, shall demonstrate the usefulness of the simple qualitative Hughes-Ingold approach.

The simplest example of an aromatic nucleophilic substitution is the S_NAr reaction between 1-chloro-2,4-dinitrobenzene and piperidine, the two-step mechanism of which, given in Eq. (5-26), is now fully established [501–503]. It involves formation of a Meisenheimer-type zwitterionic intermediate *via* a dipolar activated complex, followed by spontaneous or base-catalysed elimination of HCl to give the reaction product. Regarding this last step, the solvent can modify the relative rates of the first and second step as well as the reaction order.

$$\mathrm{O_2N\text{-}C_6H_3(NO_2)\text{-}Cl} + \mathrm{HNC_5H_{10}} \underset{25\,^\circ C}{\overset{k_2}{\rightleftharpoons}} \left[\text{O}_2\text{N-}\ldots(\delta\ominus)\ldots\text{C(Cl)}\cdots\text{NH}(\delta\oplus)\ldots\right]^{\ddagger} \longrightarrow$$

$$\text{O}_2\text{N-}(\ominus)\text{C(Cl)-}\overset{\oplus}{\text{N}}\text{H}\ldots \xrightarrow{-\mathrm{HCl}} \mathrm{O_2N\text{-}C_6H_3(NO_2)\text{-}NC_5H_{10}} \quad (5\text{-}26)$$

Solvent	$c\text{-}C_6H_{12}$	C_6H_6	C_6H_5Cl	CH_3COCH_3	$HCON(CH_3)_2$	CH_3SOCH_3
k_2^{rel}	1	2	5	13	29	50

In agreement with the separation of unlike charges during the activation process, an increase in rate up to a factor of 50 with increasing solvent polarity has been found for reaction (5-26), carried out in thirteen non-HBD solvents [503]. The absence of base-catalysis suggests that specific solvent effects are negligible in non-HBD solvents. In protic solvents, however, specific solvation of the piperidine nucleophile leads to a diminution in rate with increasing HBD ability of the hydroxylic solvents [503].

As an example of a solvent-dependent electrophilic substitution reaction, the azo coupling (S_EAr) reaction of 4-nitrobenzenediazonium tetrafluoroborate with *N,N*-dimethylaniline is given in Eq. (5-27) [504]. According to the two-step arenium ion mechanism, the activation process of the rate-limiting first step is connected with the dispersion of the positive charge. This should lead to a decrease in rate with increasing solvent polarity.

$$(CH_3)_2N{-}C_6H_5 + {}^{\oplus}N{=}N{-}C_6H_4{-}NO_2 \xrightleftharpoons[20\,^{\circ}C]{k_2} \left[(CH_3)_2N{=}C_6H_5({\delta\oplus}){\cdots}\overset{\delta\oplus}{N}{=}N{-}C_6H_4{-}NO_2\right]^{\ddagger} \longrightarrow$$

$$(CH_3)_2N{=}C_6H_5(\oplus)(H){-}N{=}N{-}C_6H_4{-}NO_2 \xrightarrow{-H^{\oplus}} (CH_3)_2N{-}C_6H_4{-}N{=}N{-}C_6H_4{-}NO_2 \qquad (5\text{-}27)$$

Arenium ion

Solvent	CH_3NO_2	CH_3CN	CH_3COCH_3	$HCON(CH_3)_2$	CH_3SOCH_3	HMPT
k_2^{rel}	4748	509	110	7	4	1
DN	2.7	14.1	17.0	26.6	29.8	38.8

The solvent-induced change in rate is, however, much larger than expected from the relatively small difference in polarity between nitromethane and hexamethylphosphoric triamide. This together with the parallelism between rate decrease and increase in the solvent donor number *DN* (*cf.* Table 2-3 in Section 2.2.6) suggests that specific solvation and stabilization of the diazonium ion by EPD solvents play a dominant role in the reaction (5-27). Very likely, formation of an EPD/EPA complex between the reactants in a rapid preequilibrium step precedes the rate-controlling first step [504].

The addition of bases (*e.g.* tertiary amines) can induce proton abstraction from the arenium ion intermediate in the second reaction step. Investigation of base-catalysed azo coupling reactions with *N,N*-dialkylanilines in different organic solvents have shown that, the solvent itself can abstract the proton from the arenium ion intermediate. In solvents of relatively high basicity (CH_3OH, CH_3SOCH_3), neither base-catalysis nor a kinetic isotope effect was observed. In less basic solvents (CH_3NO_2), however, the reaction is strongly accelerated by base and exhibits a substantial deuterium isotope effect [505]. In other words, a change in solvent can also induce a change in the rate-limiting step of azo coupling reactions.

The influence of the reaction medium on azo coupling reactions has been reviewed by Zollinger *et al.* [506].

The S_E2 cleavage of tetramethyltin with iodine is an aliphatic electrophilic substitution reaction which is subject to strong solvent dependence [507–509]. The rate of iodinolysis of $(CH_3)_4Sn$ in acetonitrile at 25 °C is more than 10^5 times faster than that in tetrachloromethane (carried out at 50 °C owing to its slow rate in this solvent) [509].

$$\text{I-I} + \text{CH}_3\text{-Sn(CH}_3)_3 \underset{25\,°\text{C}}{\overset{k_2}{\rightleftharpoons}} \left[\overset{\delta\ominus}{\text{I}}\cdots\text{I}\cdots\text{CH}_3\cdots\overset{\delta\oplus}{\text{Sn}}(\text{CH}_3)_3\right]^{\ddagger} \longrightarrow$$

$$\text{I}^{\ominus} + \text{I-CH}_3 + {}^{\oplus}\text{Sn(CH}_3)_3 \longrightarrow \text{I-CH}_3 + \text{I-Sn(CH}_3)_3 \qquad (5\text{-}28)$$

Solvent	CCl_4	C_6H_5Cl	CH_2Cl_2	CH_3COCH_3	C_2H_5OH	CH_3CN
k_2^{rel}	1[a]	7.4	252	739	47800	274000

[a] at 50 °C.

The fact that the second-order rate constant is strongly solvent-dependent is in agreement with a highly dipolar activated complex as shown schematically in Eq. (5-28) by the S_E2(back) mechanism. Neither reactant has a dipole moment; the degree of charge separation in the activated complex is estimated to be between 0.8 and 1 [509]. A fast preequilibrium formation of an EPD/EPA complex between the reactants, followed by a rate-limiting iodinolysis of the alkyl metal by electron-transfer from the alkyl metal donor to the iodine moiety to give an intermediate radical ion pair, $(CH_3)_4Sn^{\oplus\cdot}$, $I_2^{\ominus\cdot}$, is also consistent with the observed solvent-dependence [509]. In S_E2 reactions, solvents not only affect the reaction rates, but also the dichotomy between S_E2(front), S_E2(back), and S_Ei mechanisms, as well as selectivity $S(R/CH_3)$ in the cleavage of unsymmetrical alkyl metals such as $(CH_3)_nSnR_{4-n}$ [508].

Solvent effects on the rates of organotin and organomercury alkyl exchange reactions have been studied and reviewed by Petrosyan [510]. It should be mentioned that the solvent influence on the reactivity of organometallic compounds was first studied by Ziegler *et al.* in 1929/30. He showed that organolithium compounds react faster in diethyl ether than in hydrocarbons such as benzene and cyclohexane [511].

Now, some addition reactions will be considered, the solvent dependence of which have been reviewed [77, 78]. Addition of uncharged electrophiles (*e.g.* Br_2, ArS—Cl, NO—Cl, R—CO_3H) to carbon-carbon multiple bonds leads to the development of a small, usually dispersed charge in the activated complex. In more polar solvents this is accompanied by a slight rate acceleration. In reactions with substantial charge development in the activated complex, larger rate accelerations with increasing solvent polarity are observed.

A strongly solvent-dependent electrophilic reaction is the addition of halogens to alkenes [79–81] and alkynes [81a]. In a rapid equilibrium a loose transitory EPD/EPA complex (1:1) between halogen and alkene is formed [512]. This is followed by the rate determining step, which involves an S_N1-like unimolecular ionization to form a halonium intermediate which can be either symmetrical or unsymmetrical. This then reacts with a nucleophile Nu: to give the products; *cf.* Eq. (5-29).

H, H, H, R + Br_2 $\overset{fast}{\rightleftharpoons}$ [Br_2 π-complex] $\underset{25\,°C}{\overset{k_2}{\rightleftharpoons}}$ [$Br^{\delta\ominus}$ ⋯ $Br^{\delta\oplus}$]‡

R = n-C_3H_7

$\xrightarrow{-Br^{\ominus}}$ [bromonium ion, $Br^{\oplus}$] $\xrightarrow{+:Nu}$ products (5-29)

Solvent	CCl_4	1,4-Dioxane	CH_3CO_2H	CH_3OH	H_2O
k_2^{rel}	1	51	4860	$1.6 \cdot 10^5$	$1.1 \cdot 10^{10}$

It has been shown, that in the case of bromine addition to 1-pentene in solvents of different polarity, the overall rate constant varies by a factor of 10^{10} (!) [81]. This dramatic solvent effect has been taken – together with other findings – as strong evidence for the so-called Ad_EC1-mechanism, which involves considerable charge separation in the activation step. It has also been demonstrated, that protic solvents enhance this addition by a specific electrophilic solvation of the anionic part of the activated complex [81]. It appears that in the rate-determining step of alkane brominations there is also a small specific nucleophilic solvent contribution [513]. In addition it should be mentioned, that not only the rate, but also the stereospecifity of this halogen addition is strongly solvent-dependent (*cf.* Section 5.5.7) [79, 81].

The rate of the analogous bromine addition to 1-hexyne is *ca.* 10^4 times faster in methanol/water (50:50 cl/l) than in acetic acid [81a].

In the same manner, the second-order rate constant for the reaction of cyclohexene with chloro-2,4-dinitrophenylsulfane increases with increasing solvent polarity [82].

Ar–S–Cl + [alkene] $\underset{35\,°C}{\overset{k_2}{\rightleftharpoons}}$ [$Ar\text{–}S^{\delta\oplus}$ ⋯ $Cl^{\delta\ominus}$]‡ ⟶ [$Ar\text{–}S^{\oplus}$ episulfonium ion] + $Cl^{\ominus}$ ⟶ [Ar–S, Cl adduct] (5-30)

Ar = 2,4-dinitrophenyl (NO_2, NO_2)

Solvent	CCl_4	$CHCl_3$	CH_3CO_2H	$ClCH_2CH_2Cl$	$C_6H_5NO_2$
k_2^{rel}	1	605	1370	1380	2800

Neutral reactants form a dipolar activated complex, which reacts through a thiiranium intermediate [83, 515] to give the products. Similar solvent effects have been observed in the addition of chloro-phenylsulfane to alkenes [514].

A reaction corresponding to Eq. (5-30) is the addition of nitrosyl chloride to alkenes such as cyclohexene or styrene [84, 85]. The reaction seems to be faster in polar solvents (*e.g.* nitrobenzene and chloroform) than in less polar solvents (*e.g.* toluene and tetrachloromethane). This is consistent with the view that the reaction involves an electrophilic attack of $NO^{\delta\oplus}—Cl^{\delta\ominus}$. The reaction was, however, found to be also very slow in diethyl ether. Presumably this is due to strong bonding of the NO^+ cation to the ether oxygen atom [84].

The epoxidation of alkenes with peroxycarboxylic acids gives the corresponding oxirane, by an electrophilic 1,1-addition mechanism as outlined in Eq. (5-31) [77, 86, 86a, 87, 516].

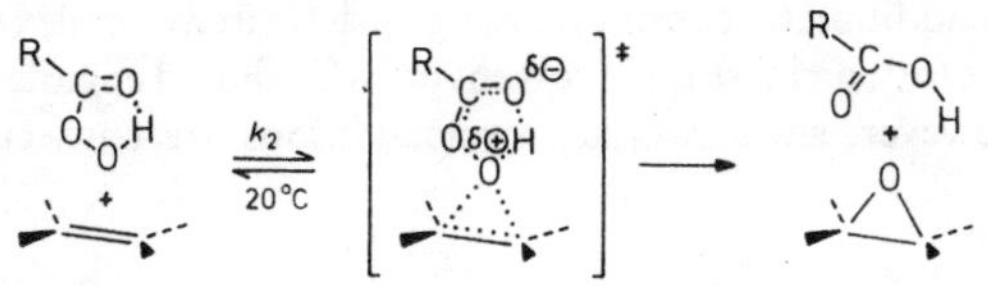

(5-31)

Solvent	$(C_2H_5)_2O$	1,4-Dioxane	n-C_6H_{14}	C_6H_6	CH_2Cl_2	$CHCl_3$
k_2^{rel} for $R=C_6H_5$ and cyclohexene [86a]	1	2.5	6.2	40	58	122

In the reaction of cyclohexene with peroxybenzoic acid ($R=C_6H_5$) [86a], and trimethylethene with peroxyacetic acid ($R=CH_3$) [87], a moderate increase in rates with increasing solvent polarity was observed. This is in agreement with the mechanism proposed in Eq. (5-31). The highest rates for these reactions are obtained in non-HBD halogen-containing solvents (CH_2Cl_2, $CHCl_3$). The slowest rates are found in solvents capable of intermolecular association with the peroxycarboxylic acid (ethers). The strength of intermolecular peracid-solvent interaction increases with increasing solvent basicity [78, 87].

The addition of peroxyacetic acid to the heteronuclear C=O double bond of cyclohexanone (*i.e.* the first step of the Baeyer-Villiger oxidation reaction) also exhibits a small solvent dependence [517].

Nucleophilic additions to carbon-carbon triple bonds are also subject to solvent influence as shown by the example given in Eq. (5-32) [88].

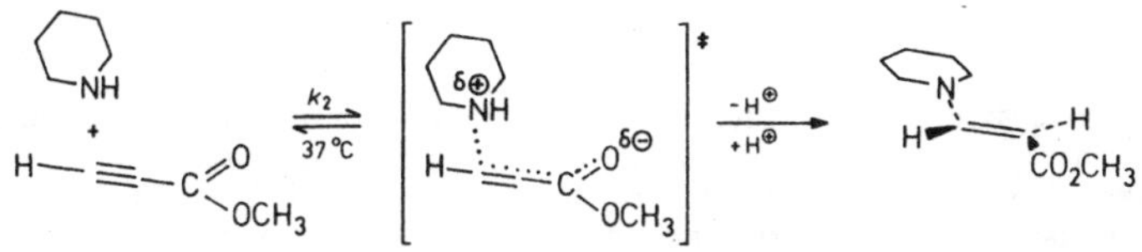

(5-32)

Solvent	c-C_6H_{12}	$(C_2H_5)_2O$	C_6H_6	CH_3OH	1,4-Dioxane	CH_3CN
k_2^{rel}	1	2	10	65	73	865

The second-order rate constant for the reaction between methoxycarbonylacetylene and piperidine increases with increasing solvent polarity, caused by the increased solvation of the strongly dipolar activated complex, which is formed from neutral molecules [88]. Analogous solvent effects have been observed for the nucleophilic addition of aziridine to 3-dimethylaminopropynal [89].

Concerted $[_\pi 2_s + _\pi 2_s]$ cycloadditions are in principle forbidden by orbital symmetry [90]. This restriction is bypassed when these reactions occur *via* zwitterions or biradicals, or by the symmetry-allowed $[_\pi 2_a + _\pi 2_s]$ process. Since cycloadditions proceeding through zwitterionic intermediates or dipolar activated complexes should be affected by solvent polarity, the investigation of the solvent effects on rates can be of considerable value when considering potential models for the activated complex and the reaction mechanism [91–93]. The possible solvent effects on one-step and two-step cycloaddition reactions are schematically shown in Fig. 5-6 [92]*⁾.

In the case of one-step cycloaddition reactions involving an activated complex with a different dipolarity than the reactants, an increase in solvent polarity should enhance the reaction rate (*cf.* Fig. 5-6a). However, since two-step cycloadditions are consecutive

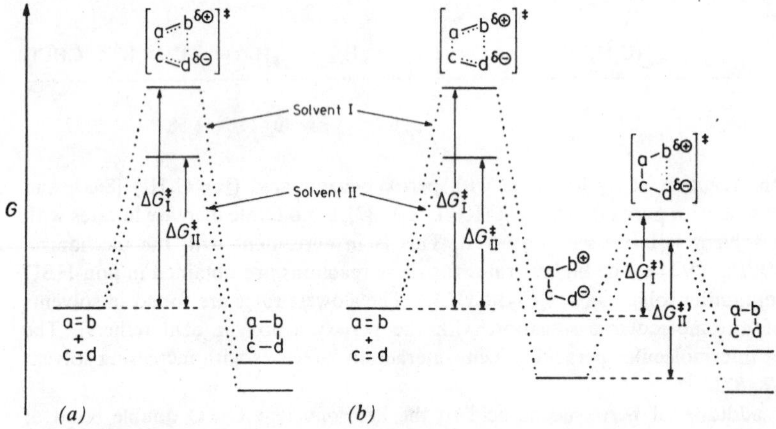

Fig. 5-6. Gibbs energy diagram for (a) one-step cycloaddition reactions with dipolar activated complex, and (b) two-step cycloaddition reactions with zwitterionic intermediate, in both nonpolar (solvent I) and polar solvents (solvent II) [92]. For the sake of simplicity no notice is taken of the different solvation of the initial reactants.

* Some definitions may be of importance for this and further discussions. A *two-step* reaction is one which takes place in two distinct kinetic steps, *via* a stable intermediate. A *one-step, concerted* reaction is one which takes place in a single kinetic step. A *synchronous* reaction is a concerted reaction in which all the bond-making and bond-breaking are parallel processes, having proceeded to the same extent in the activated complex. A *two-stage* reaction is a concerted but non-synchronous reaction. That is, some of the bond-making and bond-breaking processes take place mainly in the first half of the reaction (between reactants and activated complex), while the rest takes place in the second half of the reaction (between activated complex and products). *Cf.* [116] and the references cited therein.

reactions, the solvent effect depends on the relative size of $\Delta G_{\rm I}^{\neq}$ and $\Delta G_{\rm I}^{\neq\prime}$ resp. $\Delta G_{\rm II}^{\neq}$ and $\Delta G_{\rm II}^{\neq\prime}$ (*cf.* Fig. 5-6b). If the formation of the zwitterionic intermediate is irreversible, and $\Delta G^{\neq} \gg \Delta G^{\neq\prime}$, then the first step is rate-determining in all solvents. Consequently, there is a rate acceleration with increasing solvent polarity. When $\Delta G^{\neq} \ll \Delta G^{\neq\prime}$, this behaviour is reversed. If ever $\Delta G^{\neq} \approx \Delta G^{\neq\prime}$, then only relatively small, positive or negative, solvent effects will be observed. If the formation of the zwitterionic intermediate is reversible, then a rate increase with increasing solvent polarity will always be obtained. Under these circumstances, the distinction between a one-step and a two-step mechanism can be made only (if at all) when the solvent effects are large. Additionally, the activated complexes of the first stage of two-step reactions and of one-step reactions differ presumably only slightly in structure and dipolarity. Provided that the first step in the two-step reaction is rate-determining, the expected solvent effects for both reactions should be similar in direction and extent [92]. Three typical examples shall demonstrate solvent effects on [2+2] cycloadditions; for more examples see reference [92].

A prototype cycloaddition involving a zwitterionic intermediate is the reaction of *n*-butyl vinyl ether with tetracyanoethene as shown in Eq. (5-33) [94–98].

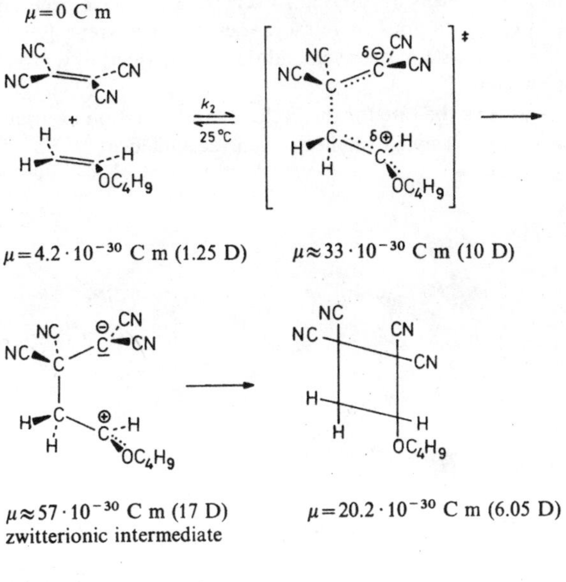

$\mu = 0$ C m

$\mu = 4.2 \cdot 10^{-30}$ C m (1.25 D) $\quad$ $\mu \approx 33 \cdot 10^{-30}$ C m (10 D)

$\mu \approx 57 \cdot 10^{-30}$ C m (17 D) zwitterionic intermediate $\quad$ $\mu = 20.2 \cdot 10^{-30}$ C m (6.05 D) (5-33)

The influence of the solvent on the cycloaddition rate constants for electron-rich alkenes is of an unusual magnitude as shown: $k_2(CH_3CN)/k_2(C_6H_{12}) = 2600$ for *n*-butyl vinyl ether, 10800 for 1-ethoxybutene, and 29000 for 1-methoxy-4-propenylbenzene (anethole) [94]. These ratios correspond to a $\Delta\Delta G^{\neq}$ of up to 23 kJ/mol (5.5 kcal/mol), for the increase in solvation during activation. The formation of the zwitterionic intermediate is assumed to be both reversible and rate-controlling. The dipole moments of reactants and adduct, as well as the calculated dipole moments of the activated complex and the zwitterion, reveal a notable increase in charge separation, corresponding to the large solvent effect obtained*). The dipole moment of the activated complex is larger than expected for the transition state of a concerted one-step cycloaddition reaction. The transition state dipole moment is two-thirds of the zwitterion dipole moment, which is reasonable because the charge separation is not yet completed in the transition state for zwitterion formation. Apart from solvent dependence of the rate constants, additional evidence for a two-step mechanism and formation of a zwitterionic intermediate in reaction (5-33) can be obtained by examining the structural variation of the enol ether on the rate [94], the activation parameter [94], and the solvent dependence of the activation volume [95, 96]. The intermediate 1,4-dipole can also be trapped using alcohols or 1,4-dipolarophiles [97, 98]. Further agreement with the two-step mechanism is the lack of solvent-dependent stereospecifity for the cycloaddition reaction [94]. Rotation of the assumed zwitterionic intermediate competes with the cyclization to give the cyclobutane product. – Similar large rate solvent effects have been observed for tetracyanoethene additions to ethyl propenyl thioether [518] and verbenene, a *trans*-fixed 1,3-diene [519]. The [2+2] cycloaddition of *cis*- and *trans*-1,2-bis(trifluoromethyl)-1,2-dicyanoethene to *tert*-butyl vinyl thioether also proceeds *via* a zwitterionic intermediate, corresponding to a rate acceleration of $k_2(CH_3CN)/k_2(CCl_4)=2160$ [99]. Further 1,4-dipolar zwitterionic intermediates, based on the rate changes by altering solvent polarity, have been postulated for the thermal [2+2] cycloadditions of tetracyanoethene to styrene [99a] and 1,1-diarylbutadienes [520], as well as for the cycloaddition of tris(methoxycarbonyl)-ethene to 4-(dimethylamino)styrene [521].

In contrast to reaction (5-33), the rate for the cycloaddition reaction of diphenylketene to *n*-butyl vinyl ether shows a much smaller solvent dependence; *cf.* Eq. (5-34) [100].

$\mu=5.9\cdot10^{-30}$ C m (1.76 D)

$$O{=}C{=}C(C_6H_5)_2 + H_2C{=}CH{-}OC_4H_9 \underset{31\,°C}{\xrightarrow{k_2}} \left[\overset{\delta\ominus}{O{\cdots}C{\cdots}}C(C_6H_5)_2 \;\cdots\; H_2C{\cdots}\overset{\delta\oplus}{C}H{-}OC_4H_9 \right]^{\ddagger} \longrightarrow \text{2,2-diphenyl-3-butoxycyclobutanone} \qquad (5\text{-}34)$$

$\mu=4.2\cdot10^{-30}$ C m (1.25 D) $\mu=10.1\cdot10^{-30}$ C m (3.20 D)

Solvent	c-C_6H_{12}	C_6H_5Cl	CH_3COCH_3	C_6H_5CN	CH_3CN
k_2^{rel}	1	13	43	63	163

The smaller rate increase with increasing solvent polarity, the large negative activation entropy as well as electronic and steric substituent effects and the stereospecifity of this reaction, are all in agreement with a concerted, but non-synchronous one-step

* The calculation of the dipole moment of the activated complex was based on an electrostatic model of Kirkwood, Laidler and Eyring (*cf.* Section 5.4.3) [2, 11]. The value for the dipole moment of the zwitterion was estimated as the vector sum of the partial moments [94].

mechanism. The unequal bond formation in the activated complex creates partial charges which are stabilized by increasingly polar solvents. Reaction (5-34) should be solvent-independent if the dipole moment of the activated complex approaches $10.1 \cdot 10^{-30}$ C m (*i.e.* sum of the dipole moments of the reactants), a figure which corresponds to the dipole moment of the product. The solvent dependence obtained shows, that there must be a larger charge separation in the activated complex than in the adduct [100]. – A similar solvent effect was reported for the dimerization reaction of dimethylketene ($k_2(CH_3CN)/k_2(CCl_4) = 29$), in agreement with a concerted one-step mechanism with charge separation in the activation step [101].

In the reaction of dimethylketene with enamines such as *N*-isobutenylpyrrolidine, the two-step process *via* a zwitterion (k_1) competes successfully with the concerted one-step mechanism (k_C), leading to cyclobutanones and 2:1 adducts, respectively [102]. In contrast to the activated complex formed from vinyl ethers, there is superior stabilization of the positive charge in the zwitterion, $>\overset{\delta\oplus}{C}\text{---}NR_2$ *vs.* $>\overset{\delta\oplus}{C}\text{---}OR$; *cf.* Eqs. (5-33) and (5-34). The experimental results for the cycloaddition of dimethylketene to *N*-isobutenyl-pyrrolidine show, that k_1 depends to a much higher extent on solvent polarity than k_C, the rate constant of the concerted reaction. In acetonitrile, the zwitterionic 1,4-dipole is produced 560 times faster than in cyclohexane, while the same solvent change accelerates the concerted reaction only 36-fold [102].

Graf [103] originally proposed a two-step mechanism for the cycloaddition reaction of chlorosulfonyl isocyanate to alkenes. This leads to the 1,4-dipole shown in Eq. (5-35), which can then ring close to give a β-lactam (and as by-product an unsaturated amide *via* a proton shift from R^1 or R^2 to $N^{\ominus}$). Moriconi [104], on the other hand, has proposed a nearly concerted, thermally allowed $[_{\pi}2_s + _{\pi}2_a]$ cycloaddition, probably initiated by a π-complex formation, and proceeding through the dipolar activated complex shown in the lower part of Eq. (5-35).

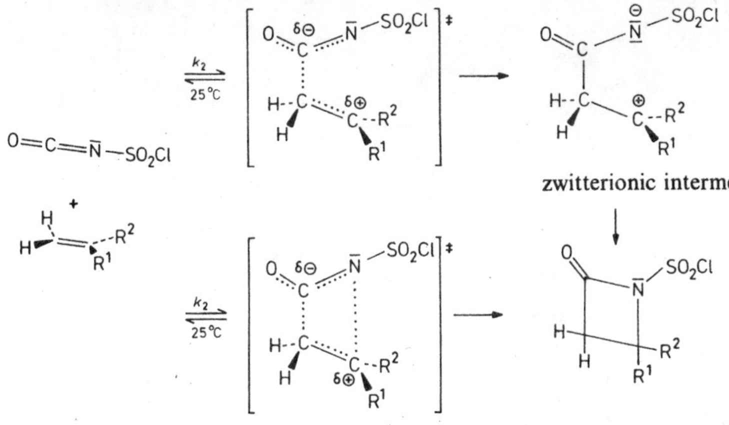

zwitterionic intermediate

(5-35)

Solvent	n-C_6H_{14}	$(C_2H_5)_2O$	$CHCl_3$	CH_2Cl_2	$C_6H_5NO_2$	CH_3NO_2
k_2^{rel}	1	31	250	1700	5000	18800

Among the evidence cited in favour of the two-step mechanism (for a review see reference [105]) is the marked increase in reaction rate produced by polar solvents for the reaction of chlorosulfonyl isocyanate with 2-ethyl-1-hexene ($R^1 = C_2H_5$; $R^2 = n\text{-}C_4H_9$) [106, 107]. Despite arguments favouring the one-step mechanism [104], the extraordinary solvent effects seem to support the two-step mechanism involving the zwitterionic 1,4-dipole intermediate [106, 107].

Not only thermal [2+2] cycloaddition reactions but also the corresponding [2+2] cycloreversions are subject to large solvent effects. A good example is the thermal decarboxylation of the *β*-lactone *trans*-3-*t*-butyl-4-phenyloxetan-2-one, as described in Eq. (5-36).

$$\xrightarrow[100\,^\circ\mathrm{C}]{k_1} \qquad [\ldots]^{\ddagger} \longrightarrow \qquad \xrightarrow{-CO_2} \qquad (5\text{-}36)$$

$\mu \approx 14 \cdot 10^{-30}$ Cm (4.2 D) $\qquad \mu \approx 27 \cdot 10^{-30}$ Cm (8.2 D) [524] $\qquad \mu \approx 0$ Cm

Solvent	Decalin	$C_6H_5OCH_3$	C_5H_5N	Propylene carbonate	$HCONHC_6H_5$
k_2^{rel}	1	4	14	98	438

The substantial rate increase with increasing solvent polarity is consistent with a two-step route *via* a highly dipolar activated complex [522]. This mechanism is also supported by the negative volume of activation, $\Delta V^{\neq} = -28$ cm³/mol, although the volume of reaction is $+52$ cm³/mol! [523]. MNDO calculations are more in favour of a one-step mechanism in which the two bonds O(1)–C(4) and C(2)–C(3) are successively broken [524].

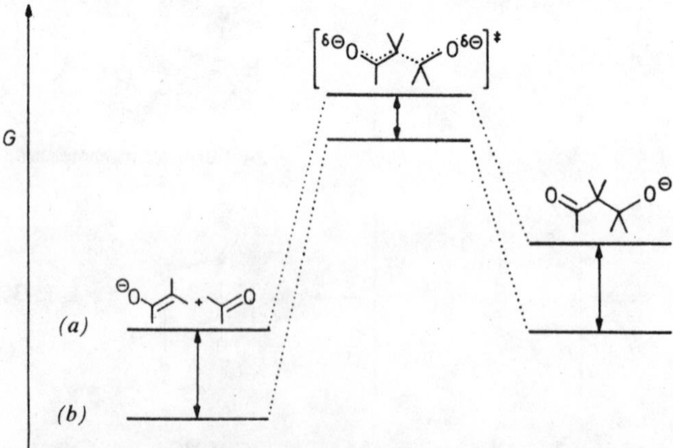

Fig. 5-7. Schematic Gibbs energy diagram for a general aldol addition of enolates to carbonyl compounds in both (a) nonpolar solvents, and (b) polar solvents, according to Heathcock [525].

One very important addition reaction is the aldol addition of enolate ions to carbonyl double bonds to give β-hydroxy aldehydes (aldols) or β-hydroxy ketones (ketols). According to Fig. 5-7, the negative charge is localized on one oxygen atome in both the reactants and products of the aldol addition. In the activated complex, however, the negative charge is shared between two oxygen atoms. In polar solvents, reactants and products are expected to be more stabilized by solvation than the activated complex with its dispersed charge. Thus, both aldol addition and retro-aldol reaction should be faster in nonpolar than in polar solvents [525]. Enolate, activated complex, and aldolate are considered as monomeric species, which is certainly an oversimplification. Nevertheless, the arguments regarding charge delocalization during the activation process should still apply for oligomers, and the conclusion that nonpolar solvents should accelerate aldol additions (and aldol reversals) is still valid.

Simple, clear-cut examples of aldol reactions exhibiting such solvent effects are scarce. Heathcock *et al.* [526] have reported that, the *erythro→threo* equilibration of lithium aldolates (*via* retro-aldol reaction) is much faster in pentane than in tetrahydrofuran or diethyl ether.

erythro-form ⇌ threo-form (5-37)

Thus, the *erythro* lithium aldolates given in Eq. (5-37) (R = Me, Et, *n*-Pr, *n*-Bu) equilibrate to their *threo* counterparts in less than two hours at 25 °C in pentane ($t_{1/2}$ = 45 min for the aldolate with R = CH_3), whereas in diethyl ether the rate of equilibration is much smaller ($t_{1/2}$ = 8 hours for the aldolate with R = CH_3) [525, 526].

We shall conclude this Section with four other examples of rearrangements, fragmentations, and isomerizations for solvent-dependent reactions involving dipolar activated complexes.

(*S*)-sulfoxide
$\mu \approx 13 \cdot 10^{-30}$ C m (4 D)

Achiral sulfenate
$\mu \approx 3 \ldots 7 \cdot 10^{-30}$ C m (1-2 D)

(*R*)-sulfoxide (5-38)

Solvent	Methylcyclohexane	C_6H_6	1,4-Dioxane	CH_3CN	C_2H_5OH	$CHF_2CF_2CH_2OH$
k_1^{rel}	332	121	73	25	11	1

A study of the solvent effects on the rate of thermal racemization of chiral allyl sulfoxides has revealed, that polar solvents significantly decelerate the racemization [108]. The reaction proceeds by way of a reversible and concerted rearrangement: achiral allyl sulfenates are formed as intermediates and an intramolecular α,γ-shift of the allyl group between the sulfoxide oxygen and sulfur termini occurs as shown in Eq. (5-38).

The proposed mechanism is in accordance with the observed solvent dependence of the reaction. Whereas the dipolar sulfoxide is expected to be more strongly solvated with an increase in solvent polarity, the less dipolar sulfenate should be relatively insensitive to such a solvent change. Stabilization of the sulfoxide, relative to the less dipolar activated complex (which should be similar to the sulfenate intermediate), increases the enthalpy of activation, $\Delta H^{\neq}$. This is reflected in the necessity of breaking increasingly strong solute-solvent interactions. On the other hand, because desolvation on activation is expected to increase the degrees of freedom in the system, a more positive $\Delta S^{\neq}$ is expected to work in the opposite direction and effect a compensating increase in k_1 with increasing solvent polarity. These suppositions are confirmed by experiment [108]. The sulfenate-sulfoxide equilibrium is shifted towards the more dipolar sulfoxide with increasing solvent polarity, as measured by the corresponding equilibrium constants [108]. In contrast to the previously mentioned examples of solvent-dependent reactions, this rearrangement belongs to those reactions in which the solvent effects originate mainly in changes in the initial-state enthalpy of the reactant.

The amine-catalyzed thermal decomposition of *tert*-butyl peroxyformate [110] respresents an example for a solvent-dependent heterolytic fragmentation reaction (for a review see reference [109]). This decomposition reaction occurs *via* a radial transition state in the absence of bases with $\Delta H^{\neq} = 159$ kJ/mol and $\Delta S^{\neq} = +63$ J/K · mol in chlorobenzene according to Eq. (5-39a). However, in the presence of pyridine, the reaction rate increases uniformly with increasing solvent polarity. This indicates a change in reaction mechanism from a radical to a heterolytic fragmentation pathway as shown in Eq. (5-39b).

The activation parameters obtained for reaction (5-39b) in chlorobenzene are in accordance with a dipolar activated complex involving considerable charge separation:

$$\mathrm{H{-}\overset{O}{\overset{\|}{C}}{-}O{-}O{-}C(CH_3)_3} \xrightarrow[140\,^{\circ}\mathrm{C}]{k_1} \left[\mathrm{H{-}\overset{O}{\overset{\|}{C}}{-}\overset{\delta\bullet}{O}\cdots\overset{\delta\bullet}{O}{-}C(CH_3)_3}\right]^{\neq} \longrightarrow \mathrm{H{-}\overset{O}{\overset{\|}{C}}{-}O\bullet + \bullet O{-}C(CH_3)_3} \qquad (5\text{-}39a)$$

$$\mathrm{H{-}\overset{O}{\overset{\|}{C}}{-}O{-}O{-}C(CH_3)_3} \xrightarrow[+\,\text{Pyridine},\ 90\,^{\circ}\mathrm{C}]{k_2} \left[\mathrm{C_5H_5\overset{\delta\oplus}{N}\cdots H\cdots\overset{O}{\overset{\|}{C}}\cdots O\cdots\overset{\delta\ominus}{O}{-}C(CH_3)_3}\right]^{\neq} \longrightarrow$$

$$\mathrm{C_5H_5\overset{\oplus}{N}{-}H + \overset{O}{\overset{\|}{C}}{=}O + \overset{\ominus}{O}{-}C(CH_3)_3} \qquad (5\text{-}39b)$$

Solvent	n-C_7H_{16}	CCl_4	C_6H_6	$CHCl_3$	CH_3NO_2	$C_6H_5NO_2$
k_2^{rel} for Eq. (5-39b)	1	4	14	30	110	140

$\Delta H^{\neq} = 64$ kJ/mol; $\Delta S^{\neq} = -96$ J/K · mol [110]. Thus, an increase in charge separation in the activation step should lead to more strongly orientated solvent molecules around the dipolar activated complex, as evidenced by the larger negative entropy of activation for reaction (5-39b). – The solvent-dependence of the amine-catalyzed fragmentation of 2-*tert*-butylperoxy-2-methylpropanoic acid can be explained in a similar manner [111].

Push-pull substituted azobenzenes are nice examples of solvent-dependent (*E*)/(*Z*)-isomerization reactions [527–529, 561, 729]. For instance, the thermal *cis*-to-*trans* isomerization of 4-(dimethylamino)-4′-nitroazobenzene exhibits a rate enhancement of *ca.* 10^5 on changing the solvent from *n*-hexane to formamide [528].

(5-40)

Solvent	*n*-C_6H_{14}	C_6H_6	CH_2Cl_2	CH_3COCH_3	CH_3OH	$HCONH_2$
k_1^{rel}	1	5	450	4760	$2.5 \cdot 10^4$	$8.2 \cdot 10^5$
$\Delta V^{\neq}/(cm^3 \cdot mol^{-1})$	-3[a]	-22[a]	-29	-30	-27	

[a] At 40 °C.

Two mechanisms have been proposed for this isomerization reaction; *cf.* Eq. (5-40). One involves rotation around the N=N double bond and includes complete π-bond rupture to give a highly dipolar activated complex. The other involves inversion at one (or both) of the azo-nitrogen atoms with simultaneous $sp^2 \rightarrow sp$ rehybridization during the activation process. The π-bond remains intact.

This huge solvent effect was first interpreted as an indication of the dipolar character of the activated complex, thus supporting the rotation mechanism [527]. The isomerization rate is also considerably accelerated by external pressure in benzene and in polar solvents, but is little affected in *n*-hexane; *cf.* the negative activation volumes, $\Delta V^{\neq}$, due to electrostriction during the activation process (see Section 5.5.11 for definitions). From this it was concluded that the isomerization mechanism changes from inversion in *n*-

hexane to rotation in benzene and polar solvents [528]. That is, in the absence of solvational stabilization of the rotational activated complex, the inversion route becomes dominant (*n*-hexane; gas-phase). Other observations, however, indicate that a modification of this dual mechanism seems to be necessary [529]. Taking into account that the inversional activated complex may also have some dipolar character due to resonance interactions of its push-pull substituents, the observed solvent, pressure, and substituent effects on rates can also be explained by means of a pure inversion mechanism [529, 561, 729]. – The rather small solvent influence on the analogous *cis*-to-*trans* isomerization of *N,N'*-diacetylindigo is best explained by a rotational mechanism *via* a biradical activated complex [530].

We shall conclude this Section with an example of a reaction which undergoes an extreme rate acceleration with an increase in solvent polarity. Thermolysis of α-chlorobenzyl methyl ether in a series of non-nucleophilic, non-HBD solvents shows rate variations up to 10^5, encompassing a $\Delta G^{\neq}$ range of 30 kJ/mol (7 kcal/mol) [112]. This dramatic solvent effect is best explained by a mechanism involving ionization of the C—Cl bond to form an ion pair, followed by a nucleophilic attack by $Cl^{\ominus}$ on CH_3 to give an aldehyde and chloromethane; *cf*. Eq. (5-41).

$$H_5C_6\text{-}CH(Cl)\text{-}O\text{-}CH_3 \underset{25\,°C}{\overset{k_1}{\rightleftharpoons}} \left[H_5C_6\text{-}CH(\overset{\delta\ominus}{\cdots Cl})\text{=}\overset{\delta\oplus}{O}\text{-}CH_3 \right]^{\ddagger} \longrightarrow H_5C_6\text{-}CH\text{=}\overset{\oplus}{O}\text{-}CH_3 \; Cl^{\ominus}$$

$$\longrightarrow H_5C_6\text{-}CH\text{=}O + Cl\text{-}CH_3 \qquad (5\text{-}41)$$

Solvent	CCl_4	C_6H_5Cl	$CHCl_3$	$C_6H_5NO_2$	CH_3CN
k_1^{rel}	1	36	750	22000	166000

The experimental activation entropies are all negative, showing a higher degree of solvent organization in the transition state than in the initial state. Surprisingly, the entropy decrease is greatest for solvents of higher polarity. Therefore, the degree of ground-state organization of the non-HBD solvents used must be considerably smaller than for the protic solvents, resulting in an extensive orientation of the non-HBD solvent molecules around the dipolar activated complex during activation [112].

5.3.3 Solvent Effects for Isopolar Transition State Reactions

Reactions involving isopolar activated complexes (neither dipolar nor radical in nature), normally exhibit only small solvent rate effects. This is because charge distribution in the activated complexes and the reactants is very similar. Owing to the different polarizability of the initial and transition state molecules, there are changes in the solute/solvent dispersion interactions leading to small rate changes. Many reactions involving isopolar transition states are electrocyclic, sigmatropic, cheletropic, or cycloaddition reactions, all of which are representative of a larger class of concerted reactions known as "pericyclic reactions". Woodward and Hoffmann have suggested that the course of these reactions is controlled by the symmetry of the molecular orbitals of the reactants and products [90, 113]. Changing either the substituent or the medium usually has little effect on the rates of pericyclic reactions. This fact has been used by some chemists as a criterion for establishing the mechanism of such reactions. Examples illustrating the lack of solvent sensitivity of pericyclic and related reactions are presented. Unfortunately, because no solvent effects are expected for these reactions thorough investigations in a large variety of solvents are lacking.

Some concerted $[m+n]$ cycloaddition reactions*), however, have been well examined [91–93], *e.g.* Diels-Alder reactions [114–116] and 1,3-dipolar cycloaddition reactions [117–120, 541–543]. Both of these reactions have been established as concerted [4+2] cycloaddition reactions.

The solvent effect on the bimolecular rate constant of Diels-Alder reactions is usually quite small. As a rule, in going from nonpolar to polar solvents, the rate constant increases only by a factor of about 3...15 [34, 35, 121–130, 531–537].

Typical examples are the cycloaddition of isoprene and maleic anhydride as shown in Eq. (5-42) [127], and the dimerization of cyclopentadiene (*cf.* Table 5-1 in Section 5.2) [33, 34].

H3C · + · $\xrightarrow[30\,°C]{k_2}$ [H3C ...]‡ → H3C ... (5-42)

Solvent	$(i\text{-}C_3H_7)_2O$	C_6H_6	C_6H_5Cl	CH_3NO_2	$C_6H_5NO_2$	*o*-Dichlorobenzene
k_2^{rel}	1	3.5	5.0	6.6	11	13

It is clear from the above figures that the activated complex is not much more dipolar than the initial state. Neutral reactants produce a neutral product in a single, concerted, often synchronous step. Although dienes and dienophiles may have a dipole moment, these dipols are usually incorporated unchanged into the product, and there is no reason to believe that they increase or decrease greatly during the reaction. The small solvent effects observed are in agreement with this picture of the Diels-Alder reaction.

Further recent examples of Diels-Alder cycloaddition reactions with small or negligible rate solvent effects can be found in the literature [531–535]. Quite weak solvent effects have been calculated for the Diels-Alder reaction between 1-hydroxybutadiene and acrolein by means of a cavity model [536]. The thermolysis of 7-oxabicyclo[2.2.1]hept-5-ene derivatives is an example of a solvent-independent *retro*-Diels-Alder reaction [537].

Nevertheless, there are some examples known with larger, although still moderate, solvent effects [124, 125, 129, 538–540]. Over a range of solvents from *o*-xylene to

* $[m+n]$-Cycloaddition reactions involve the addition of a system of m π-electrons to a system of n π-electrons to produce a new ring in a concerted, orbital-symmetry controlled process.

chloroform, the reaction rates for the addition of tetracyanoethene to anthracene have been found to increase by a factor of 70 [125]. In the case of the reaction between cyclopentadiene and acrolein, changing the solvent from ethyl acetate to acetic acid causes a 35-fold acceleration in rate [129]. A strongly dipolar activated complex is unlikely as reflected by this small sensitivity to solvent. These data are more consistent with the following mechanism: first the diene and dienophile form an EPD/EPA complex, this is then converted directly into adduct *via* an electron-rich, polarizable activated complex. Some Diels-Alder reactions experience a significant change in activation enthalpy when the solvent is changed. When the relative solvation enthalpies for the reactants are determined in EPD and EPA solvents by calorimetry, it is found that in EPD solvents the reactants are stabilized, whereas the more electronegative EPA solvents stabilize the electron-rich activated complex [128]. It appears therefore, that solvent effects on activation enthalpies of these Diels-Alder reactions (which include electron deficient maleic anhydride and tetracyanoethene as dienophiles) can be understood in terms of the electron donor-acceptor property of the solvent, dienophile solvation being increased by EPD solvents [128, 538–540] (*cf.* however reference [130]).

A very subtle solvent effect has been observed in the Diels-Alder addition of methyl acrylate to cyclopentadiene [124]. The polarity of the solvent determines the ratio of *endo* to *exo* product in this kinetically controlled cycloaddition reaction, as shown in Eq. (5-43). The more polar solvents favor *endo* addition.

$\mu = 1.4 \cdot 10^{-30}$ C m (0.42 D)

k_N, H_3CO, O, $\ddagger$ → CO_2CH_3 *endo*-Adduct

30 °C, H, H, H, CO_2CH_3 (5-43)

k_X, H_3CO, O, $\ddagger$ → CO_2CH_3 *exo*-Adduct

$\mu \approx 6 \cdot 10^{-30}$ C m (1.8 D)
(value for ethyl acrylate)

Solvent	$(C_2H_5)_3N$	$(CH_3OCH_2)_2$	CH_3CN	CH_3CO_2H	CH_3OH
$\lg \frac{[endo]}{[exo]}$	0.445	0.543	0.692	0.823	0.845

Consideration of the dipolarity of the two activated complexes can explain the observed trend. If the reactants are pictured as lying in roughly parallel planes, the dipole moments for the *exo* orientation are seen to be nearly opposite in direction, whereas for the *endo* orientation they are parallel. Therefore, the net dipole moment for the *endo* transition state is greater than that for the *exo*. Thus, the solvation of the *endo* activated complex will be more pronounced as the polarity of the solvent increases. This leads to a lowering of the activation enthalpy and preferential formation of *endo* adduct. The logarithm of the *endo*/*exo* product ratio in various solvents has been used to define an empirical solvent polarity scale [124] (*cf.* Section 7.3). The same explanation was adopted for the solvent-dependent *endo*/*exo* product ratio in 1,3-dipolar cycloadditions of phenylglyoxylonitrile oxide (C_6H_5–CO–$C^{\oplus}$=N–$O^{\ominus}$) to norbornadienes [124a]. Analogous solvent-dependent *endo*/*exo* product ratios have been obtained in [4+2]cycloadditions of cyclopentadiene to other acrylic acid derivatives [560].

As measured by the criteria of stereospecifity, regioselectivity, kinetic isotope effects, and solvent effects [117–120, 541–543], 1,3-dipolar cycloaddition reactions represent orbital symmetry-allowed [$_{\pi}4_s + _{\pi}2_s$] cycloadditions, which usually follow concerted pathways*). Diels-Alder reactions and 1,3-dipolar cycloadditions resemble each other as demonstrated by the small solvent effects on their bimolecular rate constants. In going from nonpolar to polar solvents, the rate constants of 1,3-dipolar cycloadditions change only by a factor of 2...10 [120, 131–134].

The cycloaddition of *N*-methyl-*C*-phenylnitrone to ethyl acrylate yielding a substituted isoxazolidine serves to illustrate this point [132]. In solvents of increasing polarity this reaction, shown in Eq. (5-44), exhibits only a 5.6-fold rate deceleration [132].

$\mu = 11.8 \cdot 10^{-30}$ C m (3.55 D)

$$\text{H(H}_5\text{C}_6\text{)C=N}^{\oplus}\text{(CH}_3\text{)–O}^{\ominus} + \text{H}_2\text{C=CH–CO}_2\text{C}_2\text{H}_5 \underset{85\,°\text{C}}{\xrightarrow{k_2}} [\ldots]^{\ddagger} \longrightarrow \text{isoxazolidine} \qquad (5\text{-}44)$$

$\mu = 5.9 \cdot 10^{-30}$ C m (1.76 D) $\qquad \mu = 8.3 \cdot 10^{-30}$ C m (2.48 D)

Solvent	$C_6H_5CH_3$	1,4-Dioxane	CH_3COCH_3	CH_3NO_2	CH_3CN	C_2H_5OH
k_2^{rel}	5.6	3.2	2.2	1.9	1.9	1

* Firestone has postulated that many 1,3-dipolar cycloadditions occur by a two-step mechanism with a discrete spin-paired diradical intermediate [118], but his arguments were criticized by Huisgen [119]. Distinction between concerted and stepwise-diradical mechanisms cannot be made on the basis of the negligible small solvent effects obtained for most 1,3-dipolar cycloaddition reactions. Reference [543] gives a fair discussion of this point. – Very recently, the first nonstereospecific, two-step 1,3-dipolar cycloaddition reaction via a zwitterionic intermediate has been reported by Huisgen *et al.*, using thiocarbonyl ylides, $R_2C{=}S^{\oplus}{-}CH_2^{\ominus}$, and dimethyl 2,3-dicyanofumarate as reaction partners; *cf.* R. Huisgen, G. Mloston, and E. Langhals, J. Am. Chem. Soc. *108*, 6401 (1986); J. Org. Chem. *51*, 4085 (1986).

Similar small rate factors were obtained for 1,3-dipolar cycloadditions between diphenyl diazomethane and dimethyl fumarate [131], 2,4,6-trimethylbenzenecarbonitrile oxide and tetracyanoethene or acrylonitrile [120], phenylazide and enamines [133], diazomethane and aromatic anils [134], azomethine imines and dimethyl acetylenedicarboxylate [134a], and diazo dimethylmalonate and diethylaminopropyne [544] or *N*-(1-cyclohexenyl)pyrrolidine [545]. Huisgen has recently produced a comprehensive review on solvent polarity and rates of 1,3-dipolar cycloaddition reactions [541, 542]. The observed small solvent effects can be easily explained by the fact, that the concerted, but non-synchronous bond formation in the activated complex may lead to the destruction or creation of partial charges, connected with the disappearance of the formal charges of the 1,3-dipole. A concerted mechanism, involving a small charge imbalance in the transition state, is characteristic of most 1,3-dipolar cycloaddition reactions. In the case of cycloadditions between diazomethane and certain aromatic anils to give substituted 1,2,3-triazolines, a specific protic/dipolar aprotic solvent effect was postulated, involving better solvation of the activated complex relative to the reactants in dipolar non-HBD solvents such as *N*,*N*-dimethylformamide [134].

1,3-Dipolar cycloadditions as well as 1,3-dipolar (and other) cycloreversions have been studied in a variety of solvents [546–549]. As expected, only rather small or negligible rate solvent effects have been observed.

Other examples of reactions closely related to the Diels-Alder cycloaddition reaction are: the ene reactions between alkenes with allylic hydrogen atoms (ene) and compounds with a double bond (enophile) [135, 136], and the dye-sensitized photooxygenation of allylic alkenes by singlet oxygen to give allylic hydroperoxides with a shifted double bond [137–139].

The observed range of solvent effects (less than a factor of four) for the ene reaction between 3-phenyl-1-*p*-tolylpropene-(1) and diethyl azodicarboxylate, given in Eq. (5-45), is best explained by a concerted mechanism involving an isopolar six-center transition state [136].

(5-45)

$R = 4\text{-}CH_3\text{-}C_6H_4$

Solvent	c-C_6H_{12}	1,4-Dioxane	$HCON(CH_3)_2$	CH_3CN	$(Cl{-}CH_2)_2$	$C_6H_5NO_2$
k_2^{rel}	1	1.1	2.0	2.2	3.1	3.9

The addition of singlet oxygen to 2-methyl-2-pentene occurs *via* a concerted "ene-type" mechanism as shown in Eq. (5-46)*. This is entirely consistent with the small solvent

* Since there exists a second group of allylic hydrogen atoms at the other end of the 2-methyl-2-pentene molecule, two different allylic hydroperoxides are obtained in the solvent-independent ratio of *ca.* 1 : 1 [138].

effect observed for this reaction [138]. When the solvent is changed from methanol to carbon disulfide the rate changes by a factor of seven*). Thus, it would appear that the activated complex does not involve much charge separation [138].

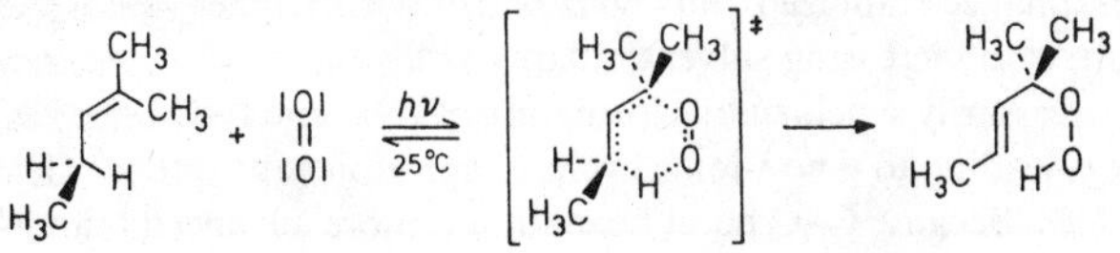

(5-46)

Solvent	CS_2	$CH_3CO_2C_2H_5$	CH_3SOCH_3	CH_3COCH_3	C_6H_6	CH_3OH
β-value*)	0.022	0.04	0.07	0.08	0.10	0.16

In the photo-sensitized oxygenation reactions of alkenes, not only the influence of the solvent on the reaction rate but also the effect of solvent on product distribution (*i.e.* from competing hydroperoxide, 1,2-dioxetane, and *endo*-peroxide formation) are rather small [550, 551].

In contrast to strongly solvent-dependent [2+2] cycloaddition reactions, which proceed through an 1,4-dipolar zwitterionic intermediate, by a two-step mechanism or through a dipolar activated complex by a one-step mechanism (*cf.* Section 5.3.2, and Eqs. (5-33) to (5-35) [92, 94–107], there are also [2+2] cycloadditions known, which exhibit concerted, nearly synchronous bond formation without significant charge separation on activation in the transition state. An example is given in Eq. (5-47). Since the rate constant for this diphenylketene/styrene addition is practically independent of solvent polarity [140], it can be classed as concerted.

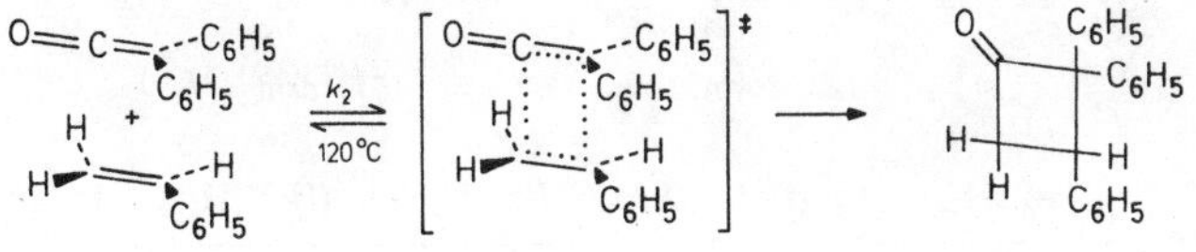

(5-47)

Solvent	C_6H_5Br	*o*-Dichlorobenzene	$(ClCH_2CH_2)_2O$	$HCON(CH_3)_2$
k_2^{rel}	1.2	1.1	1.1	1

The rates of [2+2]cycloaddition reactions between di-*tert*-butylthioketene and azomethines [141], thiobenzophenones and ketenimines [552] as well as [2+2+2]cycloadditions between quadricyclane and, for example, acrylonitrile [553] also reveal no significant solvent dependence.

* Only the ratio $\beta = k_d/k_A$ of the rate constant for decay of singlet oxygen to its ground-state triplet (designated k_d) and the specific rate constant for the product-forming step (designated k_A) is readily determined experimentally [137, 138]. Therefore, the β-values provide an inverse measure of the reactivity of the allylic alkene.

Another remarkable example of a [2+2] cycloaddition reaction is the Wittig reaction of alkylidenephosphoranes and carbonyl compounds [142, 143, 554]. The solvent dependence of some Wittig reactions has been studied [144–148]. The relative small inverse solvent effect found for the "salt-free" Wittig reaction between 4-nitrobenzaldehyde and the resonance-stabilized ylide benzoylmethylenetriphenylphosphorane – a 58-fold rate decrease with increasing solvent polarity [146–148] – is in agreement with a concerted, but not necessarily synchronous formation of C–C and P–O bonds in the rate-controlling first step, leading to a non-ionic cyclic oxaphosphetane intermediate such as described in Eq. (5-48). Because C–C bond formation is more advanced than P–O bond

$\mu = 8.0 \cdot 10^{-30}$ C m (2.41 D) [150]

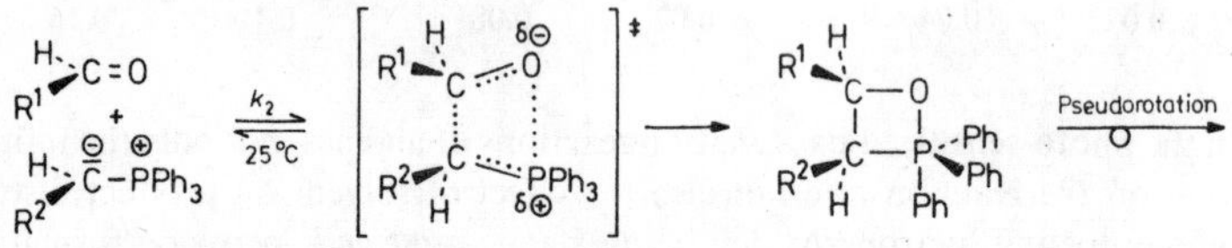

k_2, 25 °C; Pseudorotation; $-Ph_3P{=}O$

$\mu = 18.2 \cdot 10^{-30}$ C m (5.45 D) [149] $\mu^{\neq} < 26 \cdot 10^{-30}$ C m (7.9 D)

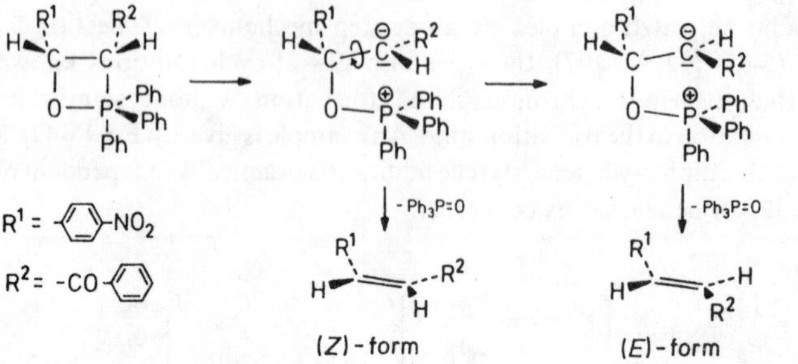

R^1 = 4-C_6H_4-NO_2

R^2 = $-CO-C_6H_5$

(Z)-form (E)-form (5-48)

Solvent	CCl_4	n-C_6H_{14}	CH_3OH	CH_3CN	C_6H_6	$(HOCH_2)_2$	$HCON(CH_3)_2$
k_2^{rel}	58	16	9.5	9	3.3	1.6	1

formation in the activation process [554], small partial charges are generated. However, according to the inverse solvent effect, the activated complex must be less dipolar than the reactants. That is, the dipole moment of the activated complex should definitely be less than the sum of the reactant dipole moments. This certainly rules out the formation of a zwitterionic phosphonium betaine in the first, rate-determining step. In addition, the somewhat solvent-dependent activation volumes are – with values of $\Delta V^{\neq} = -20 \ldots -30$ cm³/mol – insufficiently negative for a fully zwitterionogenic activation step [555].

According to Bestmann [554], the initially formed 1,2-oxaphosphetane with an apical-located oxygen atom undergoes a ligand rearrangement process (pseudorotation) thus bringing the bond necessary for alkene formation into the apical position. It is this step which may be rate-controlling in Wittig reactions with strongly basic, non-stabilized ylides (*e.g.* $R^2 = H$ or alkyl). After this conversion, cleavage of the C–P bond occurs to give a betaine. The substituent R^2 determines the lifetime of this zwitterion and consequently the configuration of the alkenes formed. If R^2 is an electron-donating group, fast phosphane oxide elimination occurs and the (*Z*)-alkene is formed. Electron-withdrawing groups R^2 [*e.g.* $R^2 = COC_6H_5$ as in Eq. (5-48)] stabilize the betaine which now isomerizes to the thermodynamically more stable conformer yielding (*E*)-alkenes *via* phosphane oxide elimination.

Protic solvents shift the alkene (*E*)/(*Z*) ratio in the direction of the (*E*)-form; *cf.* reference [554] for an explanation. The alkene (*E*)/(*Z*) ratio of salt-free Wittig reactions is thus influenced not only by the electronic character of R^2, but also by the solvent and the stereochemistry of the formation of the 1,2-oxaphosphetane in the first rate-determining step. According to Eq. (5-48), the thermodynamically less stable (*Z*)-1,2-oxaphosphetane is formed in the first activation step. A conformational analysis of the activated complex leading to the 1,2-oxaphosphetane intermediate provides a reasonable explanation for this unexpected *cis*-selectivity [556].

A complete analysis of the solvent influence on the Wittig reaction given in Eq. (5-48), based on Gibbs energies of transfer, $\Delta G_t^\circ(X, CH_3OH \rightarrow S)$, from methanol to thirteen other solvents for reactants, activated complex, and products has been given [148]. As confirmed by the kinetic results, this Wittig reaction is an example of a balanced reaction type [467] with similar solvation effects for both the initial and transition state, as well as for the final state. The activated complex appears to be reactant-like and product-like, as expected for an activated complex similar to a cyclic intermediate, without localized charges on definite ring atoms [148].

Sigmatropic reactions, *i.e.* reactions involving migration of a σ-bond, flanked by one or more conjugated systems, to a new position within the system, are also pericyclic reactions [90, 113]. Solvent studies of such reactions have been carried out, *e.g.* the effect on rate of solvent polarity has been examined for some [3, 3] sigmatropic reactions such as the Cope rearrangement of substituted 1,5-hexadienes [151, 152, 154] and the *ortho*-Claisen rearrangement of allyl aryl ethers [153, 154]. Cope and Claisen rearrangements show little response to variation to the polarity of the solvent, in accordane with an isopolar activated complex. Some [1,3] and [1,5] sigmatropic reactions, [557] and [558] resp., can also be considered as isopolar transition state reactions involving rather small rate solvent effects.

As an example, the rate of rearrangement of 2-allyl-2-(1-ethyl-1-propenyl)malononitrile changes by only a factor of 17, even when the solvent is changed drastically on going from cyclohexane to ethanol/water; *cf.* Eq. (5-49) [151].

NC, NC — $\xrightarrow[79\,°C]{k_1}$ [NC, NC]‡ ⟶ NC, NC (5-49)

Solvent	$c\text{-}C_6H_{12}$	1,4-Dioxane	C_2H_5OH	$C_2H_5OH/H_2O(1:1)$
k_1^{rel}	1	11	12	17

It can be concluded therefore, that there is negligible charge separation on activation, even when the compounds has two nitrile groups, which are ideal for stabilizing a potential carbanion in an ionic mechanism. Similar results were obtained for the Cope rearrangement of diethyl allylisopropenylmalonate [152]. It should be mentioned, however, that some 1,5-hexadienes, containing radical stabilizing substituents in 2,5-position, seem to react *via* a diradicaloid pathway [155]. The analogous azo Cope rearrangement of arylazo(α,α-dimethylallyl)malononitriles into N-(γ,γ-dimethylallyl)arylhydrazonomalononitriles, involving simultaneous carbon-carbon bond cleavage and carbon-nitrogen bond formation, exhibits a small but significant rate increase with increasing solvent polarity (a factor of 19 on changing from CCl_4 to Me_2SO). This result suggests a concerted mechanism *via* an activated complex of small dipolarity [593].

For the *ortho*-Claisen rearrangement of allyl *p*-tolyl ether shown in Eq. (5-50), the rate enhancement with increasing solvent polarity is modest, in agreement with a cyclic process involving concerted bond-making and -breaking on activation [153].

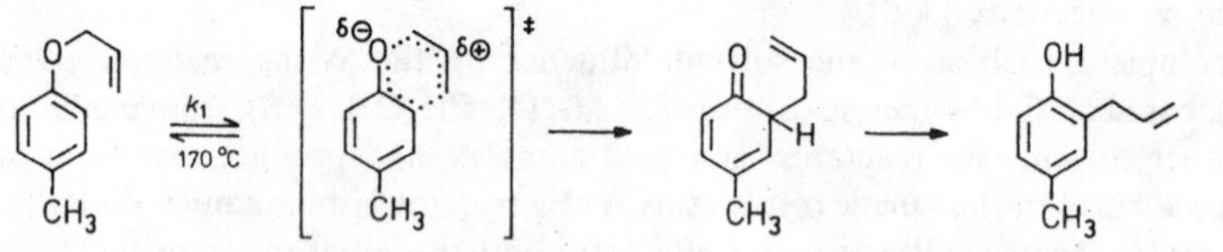

k_1, 170 °C (5-50)

Solvent	n-$C_{14}H_{30}$	$(n$-$C_4H_9)_2O$	Sulfolane	$(CH_2OH)_2$	C_2H_5OH/H_2O(28:72)
k_1^{rel}	1	1.1	3.6	22	36

For this reaction the rate was found to vary by a factor of 36 in going from the least polar (n-tetradecane) to the most polar solvent (ethanol/water), and a factor of 100 in going from the gas phase to the most polar solvent [153]. The small solvent and substituent effects observed suggest that a slightly dipolar activated complex must be formed during the Claisen rearrangement of this allyl aryl ether [153].

A similar slightly dipolar activated complex has been postulated for an aliphatic pendant to the aromatic Claisen rearrangement, that is for the [3,3]sigmatropic rearrangement of alkoxyallyl vinyl ethers [767]. For example, the rate of rearrangement of 6-methoxyallyl vinyl ether to 3-methoxy-4-pentenal is increased 3...68 fold upon changing the solvent from benzene to acetonitrile and methanol, respectively [767].

A reaction similar to the Claisen reaction is the [3,3]sigmatropic rearrangement of the allyl thionbenzoate into an allyl thiolbenzoate shown in Eq. (5-51) [156].

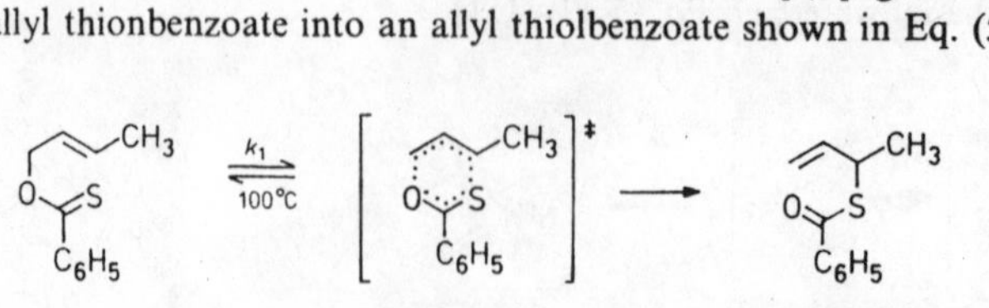

k_1, 100 °C (5-51)

Solvent	c-C_6H_{12}	C_6H_5Cl	CH_3COCH_3	CH_3CN	CH_3CO_2H
k_1^{rel}	1	1.9	2.3	4.9	4.8

From the first-order rate constants obtained in different solvents, it is apparent that this isomerization is not very sensitive to the polarity of the medium, in accordance with an isopolar, six-membered activated complex [156]. A similar small solvent effect has been observed for the [3,3]sigmatropic rearrangement of allyl S-methyl xanthate to allyl methyl dithiol carbonate [559].

Finally, the solvent dependence results for two electrocyclic reactions are mentioned. Electrocyclic reactions are generally defined as reactions involving the concerted cyclization of an n π-electron system to an (n-2) $\pi+2\sigma$-electron system, or the reverse process [90, 113].

The conrotatory cyclization of *all-cis*-deca-2,4,6,8-tetraene to *trans*-7,8-dimethylcycloocta-1,3,5-triene has been studied in solvents of different polarity [157]. In agreement with a synchronous conrotatory ring closure *via* an isopolar activated complex, the solvent effect is negligible as shown by the relative first-order rate constants in Eq. (5-52).

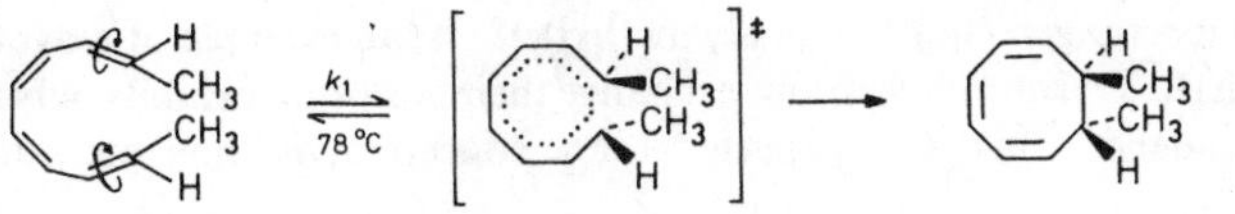

(5-52)

Solvent	$CDCl_3$	Pyridine	CH_3CN
k_1^{rel}	1.1	1.2	1

In the case of 2-methyl-4,4-diphenylcyclobutenone the reverse process, an electrocyclic ring opening, has also been examined in different solvents [158]; *cf.* Eq. (5-53).

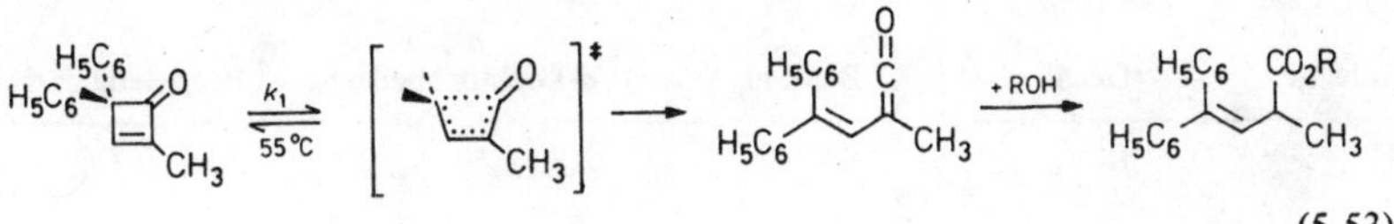

(5-53)

Solvent	c-C_6H_{12}	C_6H_6	$CH_3CO_2C_2H_5$	C_2H_5OH	CH_3OH
k_1^{rel}	3.0	2.3	1.4	1.1	1

The small inverse dependence of the first-order rate constant on solvent polarity is in agreement with a concerted electrocyclic ring cleavage through an isopolar activated complex to vinylketene, which is converted into the corresponding ester in alcoholic solvents [158].

The solvent-dependence of cheletropic reactions*) have also been investigated [158a]. The thermolysis of 3-methyl-2,5-dihydrothiophene 1,1-dioxide appears to involve

* Cheletropic reactions are defined as processes in which two σ-bonds that terminate at a single atom are made or broken in a concerted fashion.

a concerted fission of the two σ-bonds. This is in accordance with the very small solvent rate effect observed in six solvents of different polarity [158a].

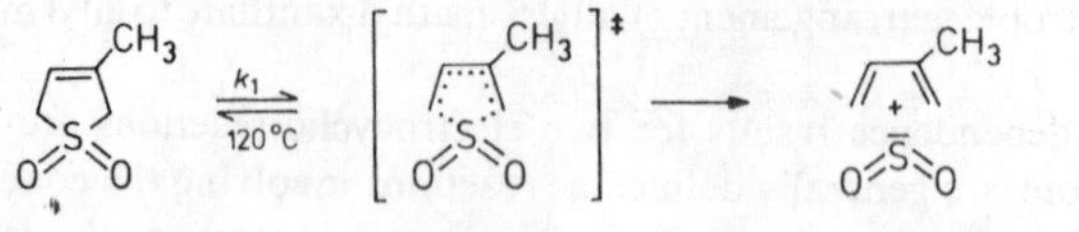

(5-54)

Solvent	n-$C_{10}H_{22}$	Triethylene glycol	$(H_5C_6)_2O$	H_5C_6—CO—C_6H_5	Sulfolane
k_1^{rel}	1	1.3	1.6	1.8	1.8

From these results it is clear that this reaction agrees with the requirements for a pericyclic reaction involving an isopolar activated complex.

The thermal rearrangement of allyl(silylmethyl)ethers, an example of a dyotropic reaction *), in which the silyl and allyl groups exchange their positions, exhibits only a very small solvent dependence; this is as expected for a concerted reaction according to Eq. (5-55) [158b]:

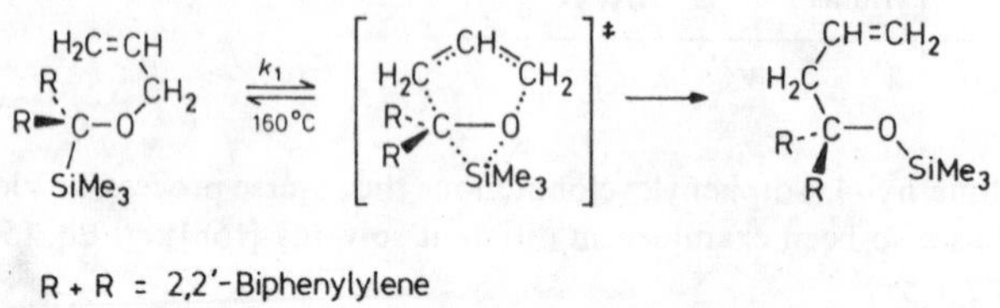

(5-55)

R + R = 2,2′-Biphenylylene

Solvent	Decalin	Benzene	o-Dichlorobenzene	Propylene carbonate
k_1^{rel}	1	1.1	2.8	5.1

Finally, it should be mentioned that no strict limit between reactions with dipolar and isopolar activated complexes exists. Some borderline cases with significant, but relatively small charge separation in going from the initial to the transition state, with corresponding small solvent rate effects, have been mentioned in this Section.

5.3.4 Solvent Effects for Free-Radical Transition State Reactions

According to Kosower, a third category of reactions involving free-radical activated complexes may be defined. These complexes are formed either through creation of unpaired electrons during radical pair formation or atom-transfer reactions [15, 468]

* Dyotropic rearrangements are uncatalyzed, intramolecular pericyclic reactions in which two σ-bonds simultaneously migrate [158b].

(*cf.* Section 5.3). These two different types of free-radical reactions have been investigated for solvent effects (for reviews see references [159–166]). Reactions in which the medium affects the reactivity of free radicals have received greater attention than radical-forming reactions. The solvent often affects the course of free-radical reactions by participating as a reactant. This is evidenced by the incorporation of the solvent into the products of the reaction. This Section, however, will be limited to those solvent effects which are truly medium effects.

Free radicals may be generated by oxidation, reduction, or by homolytic cleavage of one or more covalent bonds like C—C bonds (*e.g.* dimers of triarylmethyl radicals), N—N bonds (*e.g.* tetrasubstituted hydrazines), O—O bonds (*e.g.* hydroperoxides, dialkyl and diacyl peroxides, peroxycarboxylic esters), C—N bonds (*e.g.* dialkyl azo compounds), and N—O bonds (as in the thermolysis of nitrogen pentoxide O_2N—O—NO_2). Two typical examples, which have been investigated in different solvents, are given in Eqs. (5-56) and (5-57); *cf.* also reaction (5-39a) in Section 5.3.2.

$$(H_5C_6)_2C{=}C_6H_4{<}^{H}_{C(C_6H_5)_3} \underset{0\,°C}{\overset{k_1}{\rightleftharpoons}} \left[(H_5C_6)_2C^{\delta\odot}\cdots C_6H_4(H)\cdots C^{\delta\odot}(C_6H_5)_3\right]^{\ddagger} \longrightarrow 2\ \overset{\odot}{C}(C_6H_5)_3 \qquad (5\text{-}56)$$

$$(H_3C)_2C(CN){-}N{=}N{-}C(CN)(CH_3)_2 \underset{67\,°C}{\overset{k_1}{\rightleftharpoons}} \left[(H_3C)_2\overset{\delta\odot}{C}(CN)\cdots N{\equiv}N\cdots\overset{\delta\odot}{C}(CN)(CH_3)_2\right]^{\ddagger} \longrightarrow (H_3C)_2\overset{\odot}{C}CN + N{\equiv}N + \overset{\odot}{C}(CN)(CH_3)_2 \qquad (5\text{-}57)$$

The dissociation rate of the dimer of the triphenylmethyl radical*) in 28 solvents was studied by Ziegler *et al.* [167]. The decomposition rate of azobisisobutyronitrile in 36 solvents was measured by different authors [183–185, 562]. Despite the great variety of solvents the rate constants vary only by a factor of 2...4. This behaviour is typical for reactions involving isopolar transition states and often indicates, but does not prove, a radical-forming reaction. The lack of any marked solvent effects in most free-radical forming reactions will become more apparent after an examination of some further reactions presented in Table 5-8.

Generally it can be said that radical-forming reactions are usually not very sensitive to medium effects, because activated complexes which produce neutral radicals normally exhibit no charge separation. Even in the case of *p*-methoxyphenylazo-2-methylpropanedinitrile (*cf.* Table 5-8, last entry), which is predestinated for an ionic decomposition, there is only a modest 26-fold rate acceleration in going from *n*-decane to methanol as

* Ziegler *et al.* undertook their experiments with a compound which they believed to be hexaphenylethane [167]. In 1968 it has been shown that the dimer of the triphenylmethyl radical is not hexaphenylethane, but 1-diphenylmethylene-4-triphenylmethyl-2,5-cyclohexadiene [168, 169] in accordance with a proposal made by Jacobson in 1905 [170].

Table 5-8. Solvent influence on rates of monomolecular decomposition of various free-radical initiators [164].

Initiators	Number of solvents	Temperature °C	Range of variation of k_1/s^{-1}	$k_1^{rel.\,a)}$	References
$(H_5C_6)_2C{=}C_6H_4(H)C(C_6H_5)_3$	28	0	$(1.1\ldots4.2)\cdot10^{-3}$	3.9	[167]
bis(triphenylimidazolyl), R = C_6H_5	12	125	$(1.5\ldots3.5)\cdot10^{-5}$	2.3	[172]
$(H_3C)_3C{-}O{-}O{-}C(CH_3)_3$	6	75	$(1.9\ldots5.9)\cdot10^{-2}$	3.1	[171]
$(H_3C)_2C(C_6H_5){-}O{-}O{-}C(C_6H_5)(CH_3)_2$	5	140	$(2.2\ldots2.8)\cdot10^{-4}$	1.3	[173]
$H_3C{-}C(=O){-}O{-}O{-}C(=O){-}CH_3$	15	85.2	$(1.0\ldots1.7)\cdot10^{-4}$	1.7	[174–176]
$H_5C_6{-}C(=O){-}O{-}O{-}C(=O){-}C_6H_5$	40	80	$(1.8\ldots40)\cdot10^{-5}$	22	[177–179]
$(H_3C)_2CH{-}C(=O){-}O{-}O{-}C(=O){-}CH(CH_3)_2$	16	40	$(3.1\ldots68)\cdot10^{-5}$	22	[180, 181]
$H_5C_6{-}CH(CH_3){-}N{=}N{-}CH(CH_3){-}C_6H_5$	7	97.3	$(3.2\ldots4.1)\cdot10^{-5}$	1.3	[182]
$(H_3C)_2C(CN){-}N{=}N{-}C(CN)(CH_3)_2$	36	67	$(1.0\ldots2.0)\cdot10^{-4}$	2.0	[183–185, 562]
$(H_5C_6)_3C{-}N{=}N{-}C_6H_5$	7	25	$(1.9\ldots4.4)\cdot10^{-6}$	2.3	[186–188]
$(NC)_2C(CH_3){-}N{=}N{-}C_6H_4{-}p{-}OCH_3$	13	85	$(3.8\ldots100)\cdot10^{-5}$	26	[189]

a $k_1^{rel} = k_1$ (fastest solvent)/k_1 (slowest solvent).

solvent. This rate factor can be easily explained by assuming, that the cleavage of the two C—N bonds of the azo group is concerted, but non-synchronous [189].

The decomposition of dibenzoyl [177–179] and diisobutyryl peroxide [180, 181] (*cf.* Table 5-8) shows relatively large enhancing effects which may be explained by the fact that the diacyl peroxide molecule contains two mutually repelling $C^{\delta\oplus}{=}O^{\delta\ominus}$ dipoles. The molecule should therefore prefer the conformation shown in Eq. (5-58), with only a small net dipole moment – analogous to glyoxal, which possesses also a *transoid* conformation of the two carbonyl groups [190]. Dissociation of the diacyl peroxide leads to two

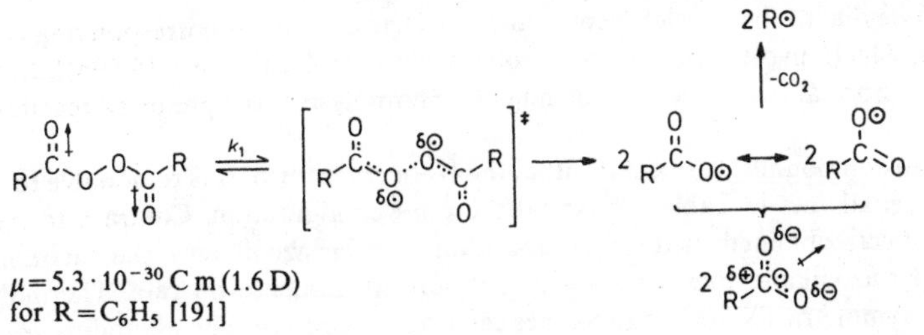

(5-58)

$\mu = 5.3 \cdot 10^{-30}$ C m (1.6 D) for $R = C_6H_5$ [191]

independent radical dipoles, with a comparatively greater net dipole moment (for comparison the dipole moment of methyl benzoate is $6.3 \cdot 10^{-30}$ C m = 1.9 D). Since this dipole separation occurs on activation, the slightly better solvation of the activated complex with increasing solvent polarity leads to the 22-fold rate acceleration experimentally observed on changing the solvent from *iso*-octane to acetonitrile [180, 181]. An additional influence of the solvent on the reaction of diisobutyryl peroxide is that it favors a heterolytic product in a subsequent, product-determining step. It is believed that a common intermediate, resulting from the rate-controlling step, then decomposes *via* homolytic and heterolytic pathways.

In the case of the thermolysis of unsymmetrical diacyl peroxides, R—CO—O_2—CO—Ar, with negatively substituted phenyl groups (*e.g.* Ar = *m*-chlorophenyl), there is a moderate increase in reaction rate with increasing solvent polarity. They are generally considered to involve ion-pair intermediates (*e.g.* $R^{\oplus}$ $^{\ominus}O_2C$—Ar), formed *via* dipolar activated complexes. A typical example is that of *endo*- and *exo*-(2-norbornyl)formyl *m*-chlorobenzoyl peroxide: $k_1(CH_3CN)/k_1(\text{cyclohexane}) = 320$ for the *exo*-reactant [563].

The thermal decomposition of symmetrical dialkyl peroxides such as diisopropyl peroxide in solution has been shown to consist of a competition between monomolecular homolysis (k_r) and an electrocyclic reaction yielding acetone and hydrogen (k_H); *cf.* Eq. (5-59) [564].

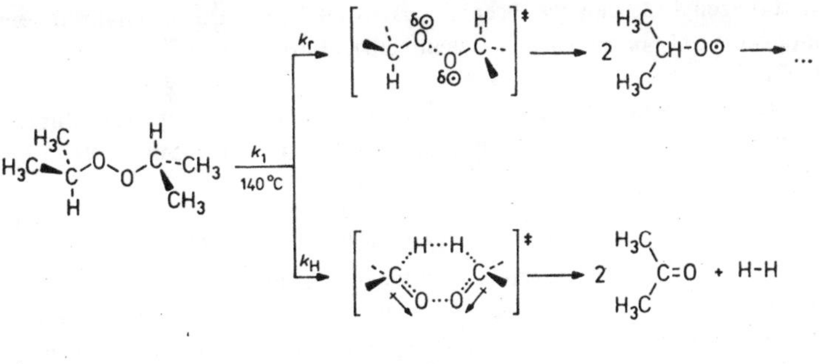

(5-59a)

(5-59b)

Whereas the activated complex for monomolecular homolysis [Eq. (5-59a)] has no dipolar character and k_r is nearly solvent-independent [$k_r(H_2O)/k_r(\text{toluene}) = 7$], k_H increases moderately with increasing solvent polarity [$k_H(H_2O)/k_H(\text{toluene}) = 59$]. This seems to be due to the development of dipolar character in the corresponding activated complex, which involves preformed dipolar acetone molecules [Eq. (5-59b)]. In the gas phase, the normal free-radical producing O–O homolysis is the preferred reaction route [564].

Azo compounds can exist in either the *cis* or *trans* form. It is reasonable to assume that the azoalkanes in Table 5-8 exhibit the *trans*-configuration. Contrary to the small solvent effects obtained in the decomposition of *trans*-azoalkanes, the thermolysis of definite *cis*-azoalkanes reveals a significant solvent influence on rate. Thermolysis of aliphatic symmetrical *cis*-*tert*-azoalkanes can lead either to the corresponding *trans*-*tert*-azoalkanes, presumably *via* an inversion mechanism, or to *tert*-alkyl radicals and nitrogen by decomposition *via* a free-radical transition state [192]. An example of the first type of reaction is the Z/E isomerization of [1,1′]azonorbornane. It has a reaction rate which is virtually solvent-independent thus agreeing with a simple inversion mechanism [565, 566]. The second reaction type is represented by the thermal decomposition of *cis*-2,2′-dimethyl-[2,2′]azopropane for which a substantial decrease in rate with increasing solvent polarity has been found [193]; *cf.* Eq. (5-60).

$$\underset{\text{N=N}}{(CH_3)_3C\ \ \ \ C(CH_3)_3} \xrightarrow[-22\,°C]{k_1} \left[\overset{\delta\ominus}{(CH_3)_3C} \cdots N{\equiv}N \cdots \overset{\delta\ominus}{C(CH_3)_3} \right]^{\ddagger} \longrightarrow 2\ (CH_3)_3C\odot + N{\equiv}N \qquad (5\text{-}60)$$

Solvent	n-C_5H_{12}	$(C_2H_5)_2O$	CH_3COCH_3	C_2H_5OH	CH_3OH
k_1^{rel}	65	31	8.4	2.1	1

Since *cis*-azoalkanes exhibit dipole moments of *ca.* $(7\ldots10)\cdot10^{-30}$ Cm (2...3 D) [194], this solvent effect is best rationalized by assuming a decrease and final loss of the dipole moment during activation. Due to their dipole moments, *cis*-azoalkanes are more stabilized by polar solvents than the less dipolar activated complexes. The activation process corresponds to a synchronous, two-bond cleavage, probably accompanied by widening of the C—N=N bond angles [193]. A two-step, one-bond cleavage process *via* short-lived diazenyl radicals has been discussed [567], but this mechanism seems to be more important only in the case of unsymmetrical azoalkanes, in particular arylazoalkanes [192].

Cyclic *cis*-azoalkanes (*e.g.* pyrazolines, pyridazines, and others) exhibit a similar rate solvent effect [193, 549], but angle expansion is of course impossible in the case of these compounds.

Another explanation of these solvent effects recognizes the fact that polar solvents have a higher internal pressure [*cf.* Eq. (3-6) in Section 3.2] [549]. Since the activation volume, $\Delta V^{\neq}$, for loss of nitrogen is positive, *cis*-azoalkanes should decompose more slowly in polar solvents. This should also be true for *trans*-azoalkanes. These, however, exhibit a negligible solvent dependence of the decomposition rate; see reference [192] for further examples.

Analogous to *cis*-azoalkanes (1,2-diazenes), the thermolysis of 1,1-diazenes is also solvent-sensitive. The monomolecular decomposition rate of *N*-(2,2,5,5-tetramethylpyrrolidyl)nitrene decreases with increasing solvent polarity [568].

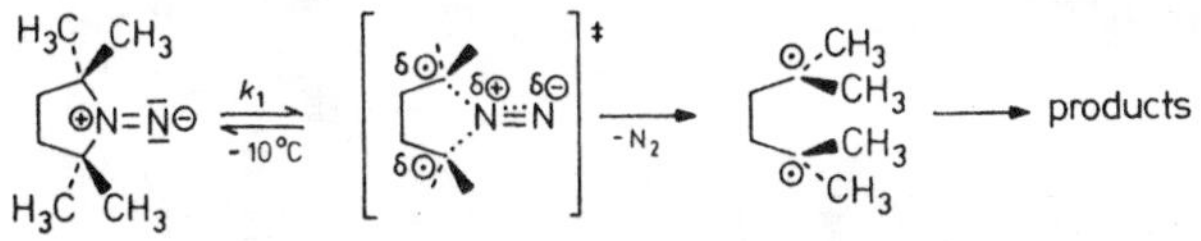

(5-61)

Solvent	$n\text{-}C_6H_{12}$	$(C_2H_5)_2O$	Tetrahydrofuran
k_1^{rel}	5.1	2.4	1.0

As shown by Eq. (5-61), the dipolarity of the solvated reactant molecule decreases during the activation process. Loss of nitrogen by simultaneous C—N bond breaking produces a 1,4-biradical.

The effect of the medium on the thermolysis of peroxycarboxylic esters deserves particular mention. Some examples are compiled in Table 5-9. An interesting aspect of this reaction is that the peresters, depending on structure, substituents, and medium, can decompose by two different mechanisms given in Eqs. (5-62a) and (5-62b) [195, 196].

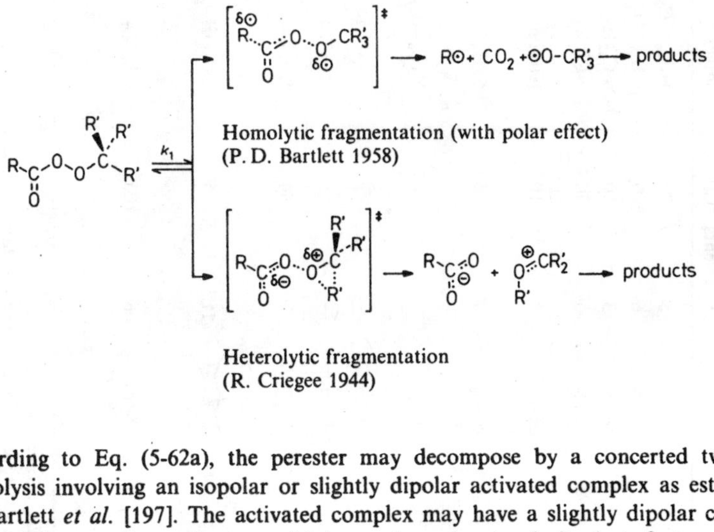

(5-62a)

Homolytic fragmentation (with polar effect)
(P. D. Bartlett 1958)

(5-62b)

Heterolytic fragmentation
(R. Criegee 1944)

According to Eq. (5-62a), the perester may decompose by a concerted two-bond homolysis involving an isopolar or slightly dipolar activated complex as established by Bartlett *et al.* [197]. The activated complex may have a slightly dipolar character resulting in a small rate increase with increasing solvent polarity (rate factors 2...11; *cf.* Table 5-9). This homolytic fragmentation, involving the so-called polar effect, is found in reactions no. 1...7 in Table 5-9. The products can be best explained by postulating a homolytic fragmentation of the peresters to form intermediate alkoxy and alkyl radicals.

A representative recent example of a homolytic perester fragmentation is given by the thermal decarboxylation of β-peroxylactones such as 4,4,5,5-tetramethyl-1,2-dioxolan-3-one [569]; *cf.* Eq. (5-63).

The observed rates vary only by a factor of three in the solvents used. This is consistent with a simple homolytic O—O-bond cleavage leading to a short-lived 1,5-

Table 5-9. Solvent influence on rates of monomolecular decomposition of various peroxycarboxylic esters R—CO—O—O—CR$'_3$ according to Eqs. (5-62a) and (5-62b).

No.	R	CR$'_3$	Number of solvents	Temperature °C	Slowest solvent	Fastest solvent	$k_1^{rel.}$ [a]	References
1	H_5C_6-O-CH_2-	-$C(CH_3)_3$	5	70	H_5C_6—C_2H_5	CH_3CN	1.6	[198]
2	n-C_5H_{11}-	-$C(CH_3)_3$	6	110	$CH_3(CH_2)_2CO_2H$	CH_3CO_2H	1.9	[199]
3	H_5C_6-$C(CH_3)_2$-	-$C(CH_3)_3$	3	50	n-$C_{12}H_{26}$	CH_3CN	3.5	[200]
4	H_3CO-C_6H_4-CH_2-	-$C(CH_3)_3$	3	70.4	c-C_6H_{12}	C_2H_5OH	3.8	[201]
5	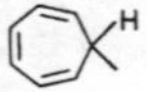 H	-$C(CH_3)_3$	3	10.3	tetralin	CH_3OH	6.2	[202]
6	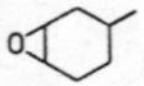 O	-$C(CH_3)_3$	9 [b]	95	C_6H_5Cl	n-C_4H_9OH	6.4	[203]
7	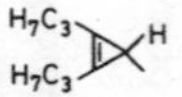 H_7C_3, H_7C_3, H	-$C(CH_3)_3$	6	30	H_5C_6—C_2H_5	H_5C_6—CH_2OH	11	[204]
8	o-$(H_5C_6)_2C$=CH-C_6H_4-	-$C(CH_3)_3$	2	90	c-C_6H_{12}	CH_3OH	62	[205]
9	o-H_5C_6S-C_6H_4-	-$C(CH_3)_3$	12	40	c-C_6H_{12}	CH_3OH	692	[206]
10	H_5C_6-CH_2-	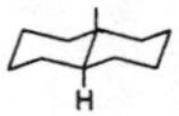 H	7	60	H_5C_6—C_2H_5	CH_3OH	31	[207]
11	H_5C_6-	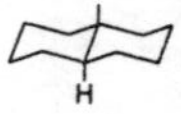 H	11 [c]	25	CH_3OH	CH_3OH/H_2O	3.5 [c]	[208, 209]
12	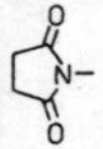 O, N-, O	-$C(CH_3)_3$	6	100	c-C_6H_{12}	CH_3OH	670	[210, 211]

[a] $k_1^{rel.} = k_1$ (fastest solvent)/k_1 (slowest solvent).

[b] Besides the two pure solvents seven chlorobenzene/n-butanol mixtures.

[c] The reaction rate increases in the order c-$C_6H_{12} < C_6H_6 < H_3C$—CO—$CH_3 < CHCl_3 < CH_3NO_2 < CH_3OH < CH_3CO_2H$ [208], but exact quantitative rate constants for this reaction are only available for eleven methanol/water-mixtures up to 2.8 mmol/mol of water [209].

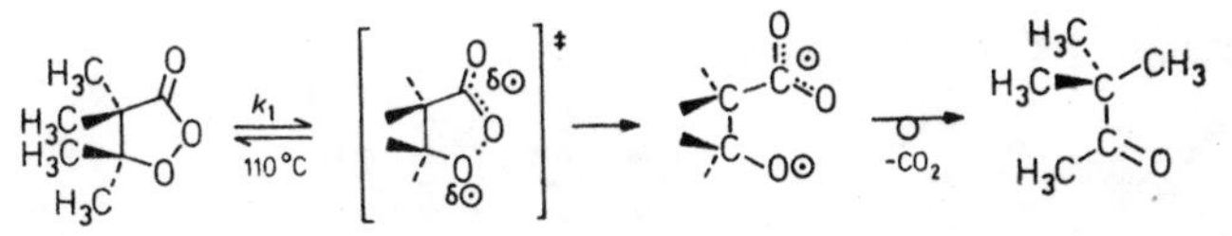

(5-63)

Solvent	c-C_6H_{12}	CCl_4	C_6H_6	CH_3CN
k_1^{rel}	1.0	1.2	1.8	2.8

biradical, which subsequently decarboxylates with concurrent β-alkyl 1,2-migration to afford the rearranged pinacolone as the major product [569].

Reactions no. 8 [205] and no. 9 [206] of Table 5-9 show special behaviour. These reactions which show larger medium effects, involve neighbouring group participation by the alkene or phenylthio groups in the homolytic cleavage of the O—O bond, as shown for the thermolysis of *tert*-butyl *o*-phenylthioperbenzoate in Eq. (5-64) [206]. The relatively

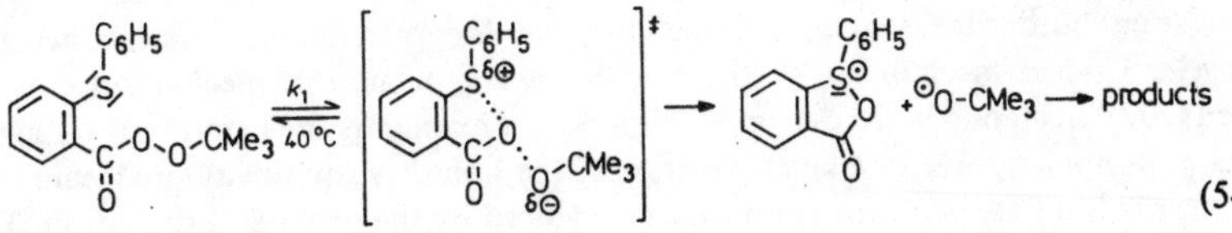

(5-64)

large solvent and substituent effects observed for this anchimerically assisted reaction (rate factor 692), indicate that a large contribution to the transition state is made by a dipolar structure involving a five-membered ring. There is a concentration of positive charge on the neighbouring group involved in the displacement from oxygen and of negative charge on the oxygen leaving group. The activated complex is best described as a dipolar singlet, represented by neutral and ionic canonical mesomeric structures. Products isolated from the perester decomposition are compatible with the proposed bonding interaction in the activated complex leading to free-radical fragments. The radical nature of the intermediate – and consequently the homolytic character of the reaction – is confirmed by the products observed and by the fact that galvinoxyl solutions (a radical scavenger) are decolorized. The fraction of radicals trapped does not decrease with increasing solvent polarity. The rates of reaction (5-64), measured in various solvents, correlate well with the rates of another anchimerically assisted reaction [206], the ionization of 4-methoxyneophyl tosylate, used to establish a scale of solvent ionizing power (*cf.* Section 7.3).

In contrast to reactions no. 1...9 in Table 5-9, the solvent effects and products observed in reactions no. 11 [208, 209] and no. 12 [210, 211] strongly suggest, that the predominant mode of decomposition of these two peresters involves heterolysis of the O—O bond and concurrent migration of the neighbouring alkyl group to the electron-deficient oxygen as described in Eq. (5-62b) [195, 196]. This ionic mechanism was first established by Criegee *et al.* [208]. Bartlett *et al.* [209] confirmed it using *trans*-9-decalyl-peroxybenzoate as substrate; *cf.* Eq. (5-65).

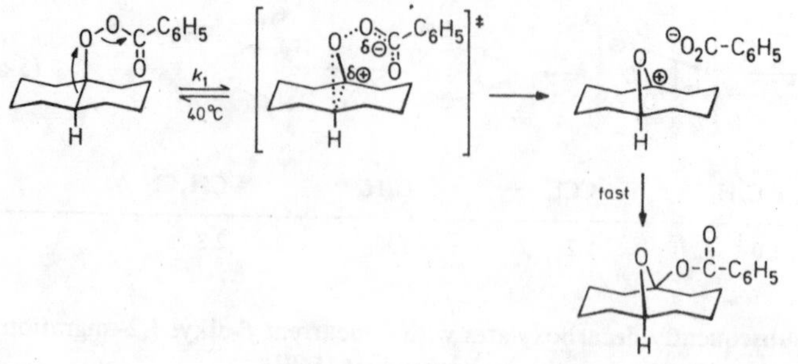

(5-65)

The strong rate acceleration observed in the thermal rearrangement of *trans*-9-decalyl-peroxybenzoate to 1-benzoyloxy-1,6-epoxycyclodecane with increasing solvent polarity (rate factor *ca.* 10^2) is in accordance with the postulated heterolytic O—O-bond cleavage leading to a dipolar activated complex.

The decomposition of *trans*-9-decalyl-peroxyphenylacetate (no. 10 in Table 5-9) is an interesting borderline reaction. Depending on the reaction medium, either the heterocyclic Criegee-mechanism or the homolytic fragmentation mechanism can be observed [207]. In alcohols, the decomposition occurs mainly by the heterolysis pathway whereas in nonpolar solvents like ethylbenzene, the homolysis pathway predominates. In acetonitrile both mechanisms compete as evidenced by the product distribution. The thermolysis of this particular perester is thus an impressive example of the strong influence which solvents may have, not only on reaction rates but also on the mechanism of a chemical reaction (*cf.* Section 5.5.7 for further examples). In addition it should be mentioned, that peresters such as *tert*-butyl cycloheptatrieneperoxycarboxylate undergo heterolytic fragmentation if the reaction is acid-catalyzed [195, 212]. The base-catalyzed ionic fragmentation of *tert*-butyl peroxyformate has already been mentioned [110]; *cf.* Eq. (5-39b).

Apart from the above described radical-forming reactions, the influence of solvent on the reactions of the radicals themselves has also been thoroughly investigated [159–166]. The most important elementary reactions of radicals are atom transfer, combination, addition, disproportionation, and electron transfer as listed in Table 5-10 [15, 213].

Most criteria for mechanisms depend upon intramolecular and extramolecular perturbation of the reacting system. A change in the medium is an extramolecular perturbation of the original system. The solvent effects produced by this perturbation can be predicted, as shown in Table 5-10, by taking into account whether or not the activated complex is dipolar or isopolar with the respect to the initial reactants. Only a small number of examples selected from a vast amount of solvent-dependence free-radical reactions [159–166] shall be used to illustrate this point.

The most extensively studied reactions are those in which a substitution reaction between radicals (like halogen atoms, alkoxy or peroxy radicals, *etc.*) and a neutral molecule A—X (*cf.* first reaction in Table 5-10) occurs. In this atom-transfer reaction the atom A, which is frequently a hydrogen atom, is transferred slowly from A—X to R·. In the isopolar activated complex of this reaction there is no appreciable charge separation.

Table 5-10. Expected substituent and solvent rate effects for elementary radical reactions [15, 213].

Reaction type	Reaction scheme	Intramolecular perturbation by substituents	Extramolecular perturbation by medium
Atom transfer	$R^{\bullet} + A\text{-}X \rightleftharpoons [R^{\delta\bullet} \cdots A \cdots X^{\delta\bullet}]^{\ddagger} \longrightarrow R\text{-}A + {}^{\bullet}X$	Modest	Modest
Combination	$R^{\bullet} + {}^{\bullet}R \rightleftharpoons [R^{\delta\bullet} \cdots R^{\delta\bullet}]^{\ddagger} \longrightarrow R\text{-}R$	Small	Small
Addition	$R^{\bullet} + CH_2{=}CH\text{-}R \rightleftharpoons [R^{\delta\bullet} \cdots CH_2 {=} \overset{\delta\bullet}{C}H\text{-}R]^{\ddagger} \longrightarrow R\text{-}CH_2\text{-}\overset{\bullet}{C}H\text{-}R$	Small	Small
Disproportionation	$R^{\bullet} + \underset{R}{C}H_2\text{-}\overset{\bullet}{C}H_2 \rightleftharpoons [R^{\delta\bullet} \cdots H \cdots \underset{R}{C}H {=} \overset{\delta\bullet}{C}H_2]^{\ddagger} \longrightarrow R\text{-}H + \underset{R}{C}H{=}CH_2$	Small	Small
Electron transfer	$R^{\bullet} \longrightarrow R^{\oplus} + e^{\ominus}$ or $R^{\bullet} + e^{\ominus} \longrightarrow R^{\ominus}$	Large	Large

In reactions of this type therefore only negligible solvent effects should be observed. However, there are also radical reactions known, in which a change in the polarity of the solvent can play an important role. For such processes, whose rate is clearly influenced by the medium, a certain charge separation during activation must be taken into account. The degree of separation of charges in an activated complex such as $[R^{\delta\ominus} \cdots A \cdots X^{\delta\oplus}]^{\neq}$ should depend on the electron affinity of the radical $R\cdot$ and the ionization potential of the molecule A—X.

The halogen abstraction by the stable free radical 1-ethyl-4-methoxycarbonylpyridinyl (Py·) proceeds by the mechanism shown in Eq. (5-66) [214, 570]. The first step which is rate-determining, is a transfer of the halogen atom to the pyridinyl radical.

$$\text{Py}^{\bullet} \text{ (1-ethyl-4-}CO_2CH_3\text{-pyridinyl)} + Br\text{-}CH_2Br \underset{25\,°C}{\overset{k_2}{\rightleftharpoons}} \left[\text{Py}^{\delta\bullet}(\text{H}) \cdots Br \cdots \overset{\delta\bullet}{C}H_2Br\right]^{\ddagger} \longrightarrow \text{Py(H)(Br)} + {}^{\bullet}CH_2Br \tag{5-66}$$

$$\text{Py(H)(Br)} \xrightarrow{\text{rapid}} \text{Py}^{\oplus}Br^{\ominus} \qquad {}^{\bullet}CH_2Br + \text{Py}^{\bullet} \xrightarrow{\text{rapid}} \text{Py-}CH_2Br$$

Products from attack at the 4-position are also observed

Solvent	CH_2Cl_2	CH_3CN	$(CH_3)_2CHOH$	C_2H_5OH
k_2^{rel}	1.0	2.1	2.0	3.5

The negligible solvent effect of this radical reaction with dibromomethane demonstrates that the activated complex for bromine atom-transfer has the same charge separation as the initial reactants. The dipole moment expected for a molecule like the pyridinyl radical is probably $(0...10) \cdot 10^{-30}$ C m (0...3 D). Dibromomethane has a modest dipole moment of $5 \cdot 10^{-30}$ C m (1.5 D). Consequently, in view of the negligible solvent effect upon rate, the activated complex must also have a dipole moment between $(0...10) \cdot 10^{-30}$ C m [214, 570].

In contrast to the preceding atom-transfer reaction, the solvent-induced rate change for the reaction between 1-ethyl-4-methoxycarbonylpyridinyl and 4-nitro-(halomethyl)benzenes is so large that a change in mechanism must be involved [215, 570]. In changing the solvent from 2-methyl-tetrahydrofuran to acetonitrile, the relative rate constant for 4-nitro-(bromomethyl)benzene increases by a factor of up to 14800. This is of the order expected for a reaction in which an ion pair is created from a pair of neutral molecules [*cf.* for example reaction (5-16)]. It has been confirmed therefore that, according to scheme (5-67), an electron-transfer process is involved [215, 570].

CO_2CH_3 CH_2X $\xrightleftharpoons[25\,°C]{k_2}$ [δ⊕ δ⊖]‡ → CO_2CH_3 CH_2X ⇌ Free Ions → products (5-67)

C_2H_5 NO_2 C_2H_5 $NO_2^{\ominus}$

Ion Pair

Solvent	2-Methyl tetra-hydrofuran	CH_2Cl_2	CH_3COCH_3	$HCON(CH_3)_2$	CH_3CN
k_2^{rel} for X=Br	1	46	278	7400	14800

It has been established, that for solvents in which specific solvation is not dominant, a small solvent effect implies an atom-transfer reaction and a large solvent effect suggests an electron-transfer reaction between neutral species. The high solvent sensitivity of electron-transfer reactions between neutral molecules should provide a useful test of their occurrence [215, 570]. From Table 5-11 it can be concluded, that atom-transfer, according to Eq. (5-66), is the rate-limiting step in the reaction of pyridinyl radical with (halomethyl)benzenes except for 4-nitro-(halomethyl)benzenes, for which the solvent effect is compatible only with an electron-transfer mechanism according to Eq. (5-67) (for a more detailed discussion see references [215, 570]).

Small but significant effects of solvent polarity were found in the autoxidation of a variety of alkenes and aralkyl hydrocarbons [216–220] (styrene [216, 218, 219], ethyl methyl ketone [217], cyclohexene [218], cumene [218, 219], tetralin [219], *etc.*). An extensive study on solvent effects in the azobisisobutyronitrile (AIBN)-initiated oxidation of tetralin in a great variety of solvents and binary solvent mixtures was made by Kamiya *et al.* [220].

Since solvent effects in radical oxidation (and autoxidation) reactions of organic compounds have been compiled and discussed in an excellent, recently published monograph by Emanuel, Zaikov, and Maizus [460], further discussion is not presented.

Table 5-11. Rate constant solvent effects for the reaction of haloalkanes with 1-ethyl-4-methoxycarbonylpyridinyl radicals at 25 °C [215, 570].

Haloalkane	k_2^{rel} [a]	$\Delta\Delta G^{\neq}/(kJ \cdot mol^{-1})$ [a]
Dibromomethane	*ca.* 0.5[b]	−1.3
(Bromomethyl)benzene	30	8.4
(Chloromethyl)benzene	63	10.3
4-Nitro-(chloromethyl)benzene	2900	19.7

[a] Rate constant for the reaction of pyridinyl radical in acetonitrile divided by that for the reaction in 1,2-dimethoxyethane and the corresponding change in Gibbs activation energy.
[b] $k_2^{rel} = k_2(i\text{-propanol})/k_2(\text{dichloromethane})$. This solvent polarity change is close to that used in the other cases.

Another example of a solvent-dependent atom-transfer reaction is hydrogen abstraction by chlorine atoms during the photochemical chlorination of hydrocarbons with molecular chlorine; for an excellent review see reference [571]. Russel reported that in the photochlorination of 2,3-dimethylbutane, according to reaction scheme (5-68), certain solvents do not have any effect on the selectivity of the reaction as measured by the rate ratio k_2^{tert}/k_2^{prim}, whereas other solvents increase this ratio significantly (*cf.* Table 5-12) [221]. The relative reactivity ratio k_2^{tert}/k_2^{prim} of the tertiary hydrogen atoms, with respect to the primary hydrogens in 2,3-dimethylbutane, can be determined from the relative amounts of 2-chloro-2,3-dimethylbutane and 1-chloro-2,3-dimethylbutane produced in

$$(H_3C)_2CH\text{-}CH(CH_3)_2 \xrightarrow[-HCl]{+\,\cdot Cl} \begin{cases} \xrightarrow{k_2^{tert}} (H_3C)_2CH\text{-}\dot{C}(CH_3)_2 \xrightarrow[-HCl]{+Cl_2} (H_3C)_2CH\text{-}CCl(CH_3)_2 \\ \xrightarrow{k_2^{prim}} (H_3C)_2CH\text{-}CH(CH_3)\dot{C}H_2 \xrightarrow[-HCl]{+Cl_2} (H_3C)_2CH\text{-}CH(CH_3)CH_2Cl \end{cases} \quad (5\text{-}68)$$

the photochlorination of this hydrocarbon*. In the absence of any solvent, and in several aliphatic and cycloaliphatic solvents, the ratio k_2^{tert}/k_2^{prim} varies from 3.3 to 9.1. In other solvents, particularly aromatic solvents and carbon disulfide, this ratio becomes comparatively large, ranging from 10 to 33 (*cf.* Table 5-12). Since both activated complexes leading to the primary and the tertiary haloalkane should not differ in their dipolarity, the solvation of the initial reactants must cause this selectivity. The latter group of solvents includes carbon disulfide and benzene derivates, containing substituents that increase the electron dènsity in the aromatic ring relative to that of benzene itself. These π-EPD solvents are able to form a loose π complex with the electrophilic chlorine radical

* It is necessary to multiply this product ratio by six to correct for the fact that there are twelve primary hydrogen atoms and two tertiary hydrogen atoms in the molecule.

Table 5-12. Solvent effect on the selectivity of the photochemical chlorination of 2,3-dimethylbutane at 55 °C (solvent concentration 4.0 mol/l) [221].

Solvent	Relative Reactivities k_2^{tert}/k_2^{prim} [a]
2,3-Dimethylbutane	3.7[b]
Nitromethane	3.3
Tetrachloromethane	3.5
Cyclohexene	3.6
Trichloroethene	3.6
Propionitrile	4.0
Methyl acetate	4.3
tert-Butanol	4.8
Nitrobenzene	4.9
1,4-Dioxane	5.6
Di-*n*-butyl ether	7.2
N,N-Dimethylformamide	9.1
Chlorobenzene	10.2
Fluorobenzene	10.3
Benzene	14.6
Toluene	15.4
Methoxybenzene	18.4
p-Xylene	18.6
1,3,5-Trimethylbenzene	25
Iodobenzene	31
Carbon disulfide	33[c]

[a] *Cf.* reaction scheme (5-68).
[b] Solvent concentration 7.6 mol/l.
[c] Value at 25 °C.

in a reversible reaction as shown in Eq. (5-69). As might be expected, the complexed and therefore less reactive chlorine atoms show greater selectivity as hydrogen abstractors. The differences in the relative amounts of tertiary and primary chloroalkanes formed become more pronounced when the chlorination is performed in such complexing solvents. On the other hand, electron-withdrawing substituents in the benzene ring (*e.g.* NO_2) decrease the selectivity relative to benzene. Even in the presence of benzene, a fairly good π-EPD solvent, there is still some attack on the primary hydrogen atoms of 2,3-dimethylbutane. More direct evidence for the existence of a chlorine atom/benzene π-complex in the photochlorination of 2,3-dimethylbutane in solution in the presence of benzene has been given recently by Ingold *et al.* [572]. The interaction of $Cl\cdot$ with carbon disulfide, which displays a remarkable high effectiveness in increasing the selectivity of the chlorine atom (*cf.* Table 5-12), probably leads to the formation of a σ-complexed radical according to Eq. (5-70), which certainly has less energy than free $Cl\cdot$ and act as more selective hydrogen abstractor. The comparatively small increase in selectivity caused by solvents lacking a π-electron system, but having non-bonding electrons (*n*-EPD solvents like alcohols, ethers, and *N,N*-dimethylformamide) might result from a complexation of the electrophilic chlorine atom by the oxygen of these solvents.

$$C_6H_6 + Cl^{\bullet} \rightleftharpoons \underset{\pi\text{-complex}}{[C_6H_6^{\delta\oplus} \rightarrow Cl^{\delta\ominus}]^{\bullet}} \xrightarrow[-HCl]{+R-H} C_6H_6 + R^{\bullet} \qquad (5\text{-}69)$$

$$S{=}C{=}S + Cl^{\bullet} \rightleftharpoons \underset{\sigma\text{-complex}}{S{=}C(Cl){-}S^{\bullet} \longleftrightarrow {}^{\bullet}S{-}C(Cl){=}S} \xrightarrow[-HCl]{+R-H} S{=}C{=}S + R^{\bullet} \qquad (5\text{-}70)$$

Recent studies of the influence of the solvent on relative selectivity in the photochlorination of 1,1-dichloroethane [573], 1-chlorobutane [574], and 2-chlorobutane [575] have shown that solvents can be divided into three classes. First, there are nonselective solvents which are approximately as selective as the neat liquid hydrocarbon reactants (*e.g.* CCl_4, CH_2Cl_2, CH_3CN). Secondly, there are moderately selective fluorohydrocarbon solvents (*e.g.* C_6F_{14}, C_8F_{18}, $C_{10}F_{20}O$) which exhibit selectivities similar to those in gas-phase chlorinations. Thirdly, there are solvents which lead to a greatly increased selectivity (*e.g.* C_6H_6, CS_2). The first group of solvents are relatively polarizable (as compared to the fluorohydrocarbon solvents) and are thus able to stabilize the activated complex by solvation. This lowers the Gibbs activation energy and tends to have a levelling effect. In contrast, the inert fluorohydrocarbons have no tendency to solvate the activated complex. Chlorinations in such solvents have a selectivity approaching that of the gas-phase reaction. The third group of solvents stabilize the chlorine atoms by specific solvation [*cf.* Eqs. (5-69) and (5-70)] and the resulting reactant/solvent-complex has to be broken open before hydrogen abstraction can occur [573–575].

The selectivity of free-radical side-chain bromination of toluene derivatives using *N*-bromosuccinimide and leading to mono- and dibromosubstituted toluenes has been studied in different solvents [577]. Surprisingly, yields and selectivities are much better in solvents such as methyl formate and dichloromethane than in the more commonly used tetrachloromethane.

Compared with chlorination, hydrogen abstraction reactions of alkoxy radicals are relatively insensitive to solvent effects [160, 222, 223]. The results of the AIBN-initiated radical chain chlorination of 2,3-dimethylbutane with *tert*-butyl hypochlorite, indicate a solvent effect for *tert*-butoxy radical reactions of much smaller magnitude, but greater selectivity in aromatic solvents [222, 223]. The reduced solvent effect for this hydrogen abstraction reaction has been attributed to steric effects. Due to the bulky methyl groups around the electrophilic oxygen atom, complex formation involving solvent molecules and activated complex is hindered.

tert-Alkoxy radicals involved in the reactions of dialkyl peroxides and alkyl hypochlorites are not only capable of abstracting a hydrogen from a hydrocarbon yielding an alkyl radical and *tert*-alkyl alcohol, they can also decompose into a ketone and an alkyl radical, which subsequently reacts with the hydrocarbon. Changes in the relative degree of hydrogen abstraction with respect to decomposition of the radical, can be determined from the relative amounts of the *tert*-alkyl alcohol and ketone produced as shown for the *tert*-butoxy radical in scheme (5-71) [160].

$$(CH_3)_3C\text{-}O^{\bullet} \xrightarrow[+H\text{-}R]{k_2^{Abstr}} (CH_3)_3C\text{-}OH + {}^{\bullet}R$$
$$(CH_3)_3C\text{-}O^{\bullet} \xrightarrow{k_1^{Dec}} H_3C^{\bullet} + (H_3C)_2C{=}O \quad (5\text{-}71)$$

The ratio k_1^{Dec}/k_2^{Abstr} may be solvent-dependent as has been shown for the reaction of *tert*-butyl hypochlorite with cyclohexane in different solvents [224].

One plausibe explanation is that the hydrogen abstraction reaction requires a radical without specific solvation of the oxygen atom, whereas a specifically solvated radical may be involved in the decomposition reaction. Complexation of the *tert*-alkoxy radical by solvent molecules, in the vicinity of the oxygen atom, causes a steric problem in the transition state of the hydrogen abstraction reaction. The solvent molecules must be released from the alkoxy radical before it can react as a hydrogen abstractor. On the other hand, the fragmentation of the *tert*-alkoxy radical in an unimolecular process is insensitive to any encumbering effect of the solvent molecules. Thus, the rate of hydrogen abstraction should be slightly retarded when the abstracting radical is specifically solvated, whereas the rate of fragmentation of the *tert*-alkoxy radical would not be affected by specific solvation. If the activated complex as a whole is unspecifically solvated, there may be an increase in rate for the fragmentation reaction (for a more detailed discussion see reference [160]).

A free-radical addition reaction, the solvent-dependence of which has been studied in thirty-nine solvents, is the addition of the 4-(dimethylamino)benzenthiyl radical to α-methylstyrene; *cf.* Eq. (5-72) [576].

$$Me_2N\text{-}C_6H_4\text{-}S^{\bullet} \leftrightarrow Me_2\overset{\oplus}{N}{=}C_6H_4{=}\overset{\ominus}{S}{}^{\bullet} \quad + \quad H_2C{=}C(C_6H_5)CH_3 \underset{23\,°C}{\overset{k_2}{\rightleftharpoons}} \left[Me_2N\text{-}C_6H_4\text{-}\overset{\delta\bullet}{S}\cdots CH_2{\cdots}\overset{\delta\bullet}{C}(C_6H_5)CH_3\right]^{\ddagger} \longrightarrow$$

$\mu \approx 13 \cdot 10^{-30}$ Cm (4.0 D) $\qquad \mu^{\ddagger} \approx 9 \cdot 10^{-30}$ Cm (2.6 D)

$$Me_2N\text{-}C_6H_4\text{-}S\text{-}CH_2\text{-}\overset{\bullet}{C}(C_6H_5)CH_3 \xrightarrow{+O_2} \text{peroxy radical} \longrightarrow \text{products} \quad (5\text{-}72)$$

Solvent	c-C_6H_{12}	CCl_4	C_6H_6	C_2H_5OH	CH_3CN	CH_3SOCH_3
k_2^{rel}	35	19	6.8	2.8	1.7	1.0

The rate of addition decreases moderately with increasing solvent polarity; there is a 35-fold rate decelaration in going from cyclohexane to dimethyl sulfoxide. In polar solvents, the dipolar reactant thiyl radical is more stabilized than the less dipolar activated complex. The stabilization of the thiyl radical by solvation has been proven by its strong positive solvatochromism (*i.e.* bathochromic shift of λ_{max} with increasing solvent polarity) [576]. Similar rate solvent effects have been observed in the addition of the 4-aminobenzenethiyl radical to styrene [577].

In conclusion, a solvent-dependent disproportionation reaction is discussed. The 2,6-di-*tert*-butyl-4-isopropylphenoxy radical disproportionates to the corresponding quinone methide and the parent phenol by a slightly solvent-dependent reaction, according to Eq. (5-73) [225]. The enthalpies of activation vary from 21 to 32 kJ/mol in going from cyclohexane to benzonitrile, increasing as the polarity of the solvent increases. Due to the compensating changes in the entropies of activation, the reaction rates are only to a small extent sensitive to changes in the medium. The formation of the activated complex may be regarded as the head to tail joining of two dipoles. Desolvation of one of the phenoxy radicals is essential for the formation of this activated complex. Thus, in a medium where radicals are highly solvated, the enthalpy of activation should be relatively high in order to provide the necessary desolvation energy. Such cases should be accompanied by the largest entropy increase. The linear relationship observed between $\Delta H^{\neq}$ and $(\varepsilon_r - 1)/\varepsilon_r$ strongly suggests, that the solvent effect in this reaction is mainly attributed to dipolar interactions between radical and solvent molecules [225].

k_2, 25 °C

(5-73)

+— = $(CH_3)_3C$-

Solvent	C_6H_6	c-C_6H_{12}	C_6H_5Cl	$C_6H_5OCH_3$	C_6H_5CN
k_2^{rel}	1	1.8	1.4	1.5	2.6
$\Delta H^{\neq}/(kJ \cdot mol^{-1})$	24	21	27	28	32

Apart from the selection of reactions involving dipolar, isopolar or free-radical activated complexes used to demonstrate the qualitative theory of solvent effects by Hughes and Ingold [16, 44] in the preceding Sections, further illustrative examples can be found in the literature (*e.g.* [14, 15, 18, 21, 23, 26, 29, 460, 468]).

5.3.5 Limitations of the Hughes-Ingold Rules

The qualitative theory of solvent effects introduced by Hughes and Ingold in 1935 [16, 44] is expressed as a set of rules that take into account the changes in charge magnitude and in charge distribution that occur between reactants and activated complex, as well as of the dielectric characteristics of solvents that enable them to solvate charges (*cf.* Section 5.3.1). Although the countless successful applications of this qualitative theory testifys to its widespread use, it does contain some inherent limitations.

One of these limitations is the assumption made by Hughes and Ingold that the contribution of entropy changes ($\Delta S^{\neq}$) to changes in Gibbs energy of activation ($\Delta G^{\neq}$) are negligible. This implies that enthalpy changes ($\Delta H^{\neq}$) dominate the Gibbs energy expression $\Delta G^{\neq} = \Delta H^{\neq} - \mathrm{T} \cdot \Delta S^{\neq}$. This assumption is necessary because an increase in solvation usually decreases the entropy of a given state. Decreases in the entropy of activation counteract increasing enthalpy changes, but fortunately these decreases are relatively small for most reactions. Therefore, most but not all chemical reactions are controlled by enthalpy changes. A number of reactions in which changes in the entropy of activation govern the Gibbs energy of activation were reported by Pearson [226]. As illustrated in Table 5-3 for the reaction of iodoethane and triethylamine (*cf.* Eq. (5-16) in Section 5.3.1 [59]) the rate constants increase by a factor of 52 with increasing solvent polarity, although the enthalpies of activation show only a relatively small change of $\Delta\Delta H^{\neq} = 9.2\ \mathrm{kJ \cdot mol^{-1}}$. The entropies of activation rise by $\Delta\Delta S^{\neq} = 26\ \mathrm{J \cdot mol^{-1} \cdot K^{-1}}$ for the same sequence of solvents. Polar solvents appear to reduce the loss of entropy on passing from the reactants to the activated complex of the reaction, and it is mainly this factor that causes the rates to increase with solvent polarity. An explanation in terms of solvent orientation around the dipolar activated complex of reaction (5-16) was given in Section 5.3.1.

Another example is the S_N1 solvolysis of 2-chloro-2-methylpropane (*cf.* Eq. (5-13) in Section 5.3.1 [40]) the rate data for which are given in Table 5-13. As the solvent is changed from ethanol to water, the rate of solvolysis increases by a factor of 335000, with a corresponding decrease in $\Delta G^{\neq}$. In going from ethanol to formic acid the values of $\Delta S^{\neq}$ vary only from -13 to $-7\ \mathrm{J \cdot mol^{-1} \cdot K^{-1}}$. In the first five solvents, the $\Delta S^{\neq}$ values are relatively constant (*ca.* $-12\ \mathrm{J \cdot mol^{-1} \cdot K^{-1}}$), in agreement with domination of the overall solvent effects by $\Delta H^{\neq}$. Since $\Delta S^{\neq}$ is constant in these solvents, $\Delta H^{\neq}$ and $\Delta \mathrm{G}^{\neq}$ must have a linear relationship. However, the large $\Delta S^{\neq}$ value for the reaction in water demonstrates that the $\Delta S^{\neq}$ changes are important in this case. This huge entropy effect is obviously due in part to the highly ordered structure of water.

Even in those cases where the rate constants, for a reaction in various solvents, are not significantly different, the activation parameters may indicate a significant amount of interaction between solute and solvent, as shown for the unimolecular decomposition of di-*tert*-butyl peroxide in Table 5-14 [172, 227]. The rate of decomposition of the peroxide does not vary by more than a factor of 2.3 in going from cyclohexane to acetonitrile, corresponding to a change in $\Delta G^{\neq}$ of only $4\ \mathrm{kJ \cdot mol^{-1}}$. However, the activation parameters $\Delta H^{\neq}$ and $\Delta S^{\neq}$ show that the solvent apparently have a marked effect on the decomposition reaction ($\Delta\Delta H^{\neq} = 41\ \mathrm{kJ \cdot mol^{-1}}$; $\Delta\Delta S^{\neq} = 94\ \mathrm{J \cdot mol^{-1} \cdot K^{-1}}$). An evaluation of these activation parameters indicates, that the compensating effects of the energy gained in solvating the radical-like activated complex, which would be expected to increase

Table 5-13. Rate constants and activation parameters for the S_N1 solvolysis of 2-chloro-2-methylpropane in six solvents at 25 °C [40].

Solvents	ε_r	$\frac{k_1}{(10^5 \cdot s^{-1})}$	$k_1^{rel.}$	$\frac{\Delta G^{\neq}}{(kJ \cdot mol^{-1})}$	$\frac{\Delta H^{\neq}}{(kJ \cdot mol^{-1})}$	$\frac{\Delta S^{\neq}}{(J \cdot mol^{-1} \cdot K^{-1})}$
Ethanol	24.55	0.00860	1	113	109	−13
Acetic acid	6.15	0.0213	2.5	111	108	−10
Methanol	32.70	0.0753	9	108	104	−13
Formamide	111.0	3.72	430	98.3	93.6	−16
Formic Acid	58.5	105	12200	90.0	87.9	− 7
Water	78.30	2880	335000	82.0	97.2	+51(!)

Table 5-14. Rate constants and activation parameters for the decomposition of di-*tert*-butyl peroxide at 125 °C [172, 227].

Solvents	ε_r	$\frac{k_1}{(10^{-5} \cdot s^{-1})}$	$k_1^{rel.}$	$\frac{\Delta G^{\neq}}{(kJ \cdot mol^{-1})}$	$\frac{\Delta H^{\neq}}{(kJ \cdot mol^{-1})}$	$\frac{\Delta S^{\neq}}{(J \cdot mol^{-1} \cdot K^{-1})}$
Gas phase	–	–	–	136	159	59
Cyclohexane	2.02	1.52	1	136	171	88
Tetrahydrofuran	7.58	1.84	1.2	135	155	51
Benzene	2.28	1.99	1.3	135	148	33
Nitrobenzene	34.82	2.39	1.6	134	149	38
Acetic acid	6.15	2.98	2.0	132	140	19
Acetonitrile	37.5	3.47	2.3	132	130	−6

the decomposition rate, is counterbalanced by the decrease in entropy resulting from the more highly ordered arrangement of solvent molecules around the activated complex. The result is that the reaction rate is in the same range as that observed for the gas phase reaction.

A second limitation of the Hughes-Ingold theory concerns the fact, that the solvent is treated as dielectric continuum, characterized by one of the following: its dielectric constant, ε_r, the dipole moment, μ, or by its electrostatic factor, *EF*, defined as the product of ε_r and μ [27]. The term "solvent polarity" refers then to the ability of a solvent to interact electrostatically with solute molecules. It should be remembered, however, that solvents can also interact with solute molecules through specific intermolecular forces like hydrogen bonding or EPD/EPA complexing (*cf.* Section 2.2). For example, specific solvation of anionic solutes by protic solvents may reduce their nucleophilic reactivity, whereas in dipolar aprotic solvents solvation of anions is less, resulting in enormous rate accelerations [6], which cannot be explained by the simple electrostatic theory (*cf.* Section 5.5.2).

A third limitation of the Hughes-Ingold concept of solvent effects on reaction rates arises from the fact that it is based on static equilibrium transition-state solvation (*cf.* Section 5.1). That is, the reorientational relaxation of solvent molecules during the activation process is considered to be sufficiently fast so that the activated complex will be in thermal equilibrium with the surrounding solvent shell. But this is not necessarily true in all cases, in particular not for very fast reactions. In such reactions, the rate will also depend on solvent reorientation rates and the standard transition-state theory will break down; *cf.* references [463, 465, 466] for further discussions.

It should be mentioned that an investigation of solvent rate effects is very often limited by the narrow range of solvents examined. When a particular reaction is studied in a small number of often very similar solvents, no far-reaching conclusions about the influence of solvent can be reached. As a rule, the reaction under consideration should be investigated in as many solvents of different polarity as possible. With a minimum of five different solvents a good general picture of the reaction should be obtained. Very often reactions have been carried out in a series of binary solvent mixtures such as alcohol/water. Solvents of such a similar structure will specifically solvate the solutes in a related manner throughout the whole solvent series, and a change of solvent mixture will mainly reflect the electrostatic influences of the solvents.

The Hughes-Ingold theory also neglects the changing solvent structure. Although solvent-solvent interactions are usually small compared to solute-solvent interactions, consideration should be given to solvent association when reactions are carried out in a highly structured solvent like water. The electrostatic theory ignores such solvent-solvent interactions.

The final limitation of the pure electrostatic theory is its inability to predict solvent effects for reactions involving isopolar transition states. Since no creation, destruction or distribution of charge occurs on passing from the reactants to the activated complex of these reactions, their rates are expected to be solvent-independent. But the observed rate constants usually vary with solvent, although the variations rarely exceed one order of magnitude (*cf.* Section 5.3.3). These solvent effects may be explained in terms of cohesive forces of a solvent acting on a solute, usually measured by the cohesive pressure of the solvent (*cf.* Section 5.4.2).

In spite of these limitations, the electrostatic Hughes-Ingold theory remains a good guide in predicting the solvent influence on chemical reactions, at least in a qualitative way. Exceptions can be safely assumed to involve strong specific solute-solvent interactions.

5.4 Quantitative Theories of Solvent Effects on Reaction Rates

5.4.1 General Remarks

Whereas the theoretical treatment of gas-phase reactions is comparatively simple, the calculation of rate constants for reactions in the liquid phase are very complicated. This is essentially due to the complexity of the many possible intermolecular solute-solvent interactions (*cf.* Section 2.2). When investigating solution-phase reaction kinetics, the problems to be faced include deciding which property of the solvent to use when setting up mathematical correlations with the reaction rates. Another problem is deciding which characteristics of the reacting molecules are to be considered when the effects of the solvent on their reactivity is determined. A quantitative allowance for the solvent effects on the rate constants k for elementary reactions consists in establishing the following functions:

$$k = f(a, b, c, \ldots m, n, o, \ldots) \qquad (5\text{-}74)$$

where a, b, and c are parameters characterizing the properties of the reactants, and m, n, and o are parameters characterizing the properties of the medium. The function thus obtained will then correctly describe the dependence of the rate constants on the medium. The problem then arises as to which reactant and solvent parameters are responsible for the observed dependences and must be included in Eq. (5-74).

When considering only pure electrostatic interactions leading to non-specific solvation, the solvent can be regarded as an isotropic continuum, with dielectric constant ε_r, and the reactants are characterized by both the magnitude and distribution of charge in the molecule. However, an analysis of solvent effects shows, that not only non-specific solvation caused by electrostatic and dispersion forces, but also specific solvation caused by hydrogen bonding and EPD/EPA complexation, must be considered. The reaction kinetics for any particular reaction in a given solvent will be determined by the predominating type of solvation for that reaction.

The theoretical treatment of liquid phase reaction kinetics is limited by the fact that no single universal theory on the liquid state exists at present. Problems which have yet to be sufficiently explained are: the precise character of interaction forces and energy transfer between reacting molecules, the changes in reactivity as a result of these interactions, and finally the role of the actual solvent structure. Despite some limitations, the absolute reaction rates theory is at present the only sufficiently developed theory for processing the kinetic patterns of chemical reactions in solution [2–5, 7, 8, 11, 24, 463–466]. According to this theory, the relative stabilization by solvation of the initial reactants and the activated complex must be considered (*cf.* Section 5.1).

Thus, the problem of making quantitative allowances for solvent-induced changes in the rate constant of a reaction $A+B \rightleftharpoons (AB)^{\neq} \rightarrow C+D$ (*cf.* Eq. (5-1) in Section 5.1) is reduced to the calculation of the difference between the partial Gibbs energies of solvation of the activated complex $(AB)^{\neq}$ and the initial reactants A and B as given in Eq. (5-75) [28].

$$\ln k = \ln k_0 - \frac{1}{R \cdot T}(\Delta G_{\mathrm{A,solv}} + \Delta G_{\mathrm{B,solv}} - \Delta G_{(\mathrm{AB})\neq,\mathrm{solv}}) \qquad (5\text{-}75)$$

k_0 is the rate constant of the reaction in a standard solvent or in the gas phase, k is the rate constant observed in the solvent under consideration. Some progress has been made in the calculation of initial-state and transition-state solvent effects on reaction rates using thermodynamic transfer functions [453, 467]; *cf.* also Section 5.5.3.

If all possible solute-solvent interactions are taken into account, then attempts to correlate the rate constant with the medium will generally lead to such complicated equations, that their experimental verification is impossible. Therefore, equations correlating rate constants with medium properties are usually derived on the basis of more or less theoretically justified models, allowing for only a limited number of dominating interaction factors. If the model adopted reflects correctly the dominating solute-solvent interactions, then a good quantitative description of the experimental data using the derived theoretical equation may be obtained. It is convenient, therefore, to distinguish between the following four reaction types:

(a) Reactions between neutral, apolar molecules (*via* isopolar activated complexes);
(b) Reactions between neutral, dipolar molecules (*via* dipolar activated complexes);

(c) Reactions between neutral, dipolar molecules and ions (*via* dipolar and charged activated complexes); and
(d) Reactions between ions (*via* dipolar and/or charged activated complexes).

The following chapters will deal only with the final results of these calculations based on suitable models, together with some examples, which shall illustrate the proposed dependences. More detailed discussions can be found in well-known monographs [2–5, 7, 8, 11, 12, 21, 24, 25, 28, 457, 459, 460]. Reviews covering both classical physical theories [5, 12, 21, 28, 457, 460] and quantum-mechanical theories [578, 579] of solvent effects on reaction rates have appeared.

5.4.2 Reactions between Neutral, Apolar Molecules

In any solution reaction cavities in the solvent must be created to accommodate reactants, activated complex, and products. Thus, the ease with which solvent molecules can be separated from each other to form these cavities, is an important factor in solute solubility (*cf.* Section 2.1). Furthermore, because solubility and reactivity are often related phenomena, the intermolecular forces between solvent molecules must also influence rates of reaction. The overall attraction forces between solvent molecules gives the solvent as a whole a cohesion which must be overcome before a cavity is created. The degree of cohesion may be estimated using the surface tension, but a more reliable estimate is obtained by considering the energy necessary to separate the solvent molecules. This is known as the *cohesive pressure c* (also called *cohesive energy density*) [228–232]. The cohesive pressure is defined as the energy of vaporization, ΔU_v, per unit molar volume, V_m, as shown in Eq. (5-76); *cf.* also Eq. (3-5) in Section 3.2.

$$c = \frac{\Delta U_v}{V_m} = \frac{\Delta H_v - R \cdot T}{M_r \cdot \varrho^{-1}} \tag{5-76}$$

Values of c are calculated from experimentally determined enthalpies (heats) of vaporization of the solvent to a gas of zero pressure, ΔH_v, at a temperature T, as well as from the molecular mass M_r, the density of the solvent ϱ, and the gas constant, R. The cohesive pressure characterizes the amount of energy needed to separate molecules of a liquid and is therefore a measure of the attractive forces between solvent molecules. The cohesive pressure c is related to the *internal pressure* π, because cohesion is related to the pressure within a liquid; *cf.* Eq. (3-6) in Section 3.2 for the precise definition of π*).

---

* Since the internal pressure is actually defined in a slightly different way, values of internal pressure approach those of the cohesive pressure only for nonpolar and nonassociated solvents (*cf.* Table 3-2 in Section 3.2) [228–232, 237]. Internal pressure is a measure of the instantaneous volume derivative of the cohesive pressure during isothermal expansion of a liquid (*cf.* Eq. (3-6) in Section 3.2). Because of the experimental difficulty in obtaining real internal pressures, it is usual to refer to $\Delta U_v / V_m$ as the internal pressure of a liquid.

In mixtures which are regular solutions*), the mutual solubility of the components depends on the cohesive pressure, hence Hildebrand termed the square root of c the *solubility parameter* δ, according to Eq. (5-77); *cf.* also Eq. (2-1) in Section 2.1 [228, 229, 231, 238].

$$\delta = \left(\frac{\Delta U_v}{V_m}\right)^{\frac{1}{2}} = \left(\frac{\Delta H_v - R \cdot T}{M_r \cdot \varrho^{-1}}\right)^{\frac{1}{2}} \tag{5-77}$$

A good solvent for a certain nonelectrolyte solute should have a δ-value close to that of the solute (*cf.* Section A.1). Extensive compilations of δ-values are given in references [231, 238]; a selection of δ-values for various organic solvents is given in Table 3-3 in Section 3.2.

Assuming that it is only van-der-Waals forces which are acting in the solute/solvent system, and that the heat of mixing is responsible for all deviations from ideal behaviour, as well as the fact that the solute/solvent interaction energy is the geometric mean of solute/solute and solvent/solvent interactions, Hildebrand [228, 229] and Scatchard [230] were able to develop the following expression for the activity coefficient f_i of the nonelectrolyte solute i dissolved in a solvent s (mole fraction basis), referred to a standard state of pure liquid solute (not infinite dilution):

$$RT \cdot \ln f_i = V_{m,i} \cdot \phi_s^2 \cdot \left[\left(\frac{\Delta U_{v,i}}{V_{m,i}}\right)^{\frac{1}{2}} - \left(\frac{\Delta U_{v,s}}{V_{m,s}}\right)^{\frac{1}{2}}\right]^2 \tag{5-78}$$

where $V_{m,i}$ is the molar volume of solute i (as a pure liquid); $\Delta U_{v,i}$ is the molar energy of vaporization of the solute (as a pure liquid); $V_{m,s}$ and $\Delta U_{v,s}$ are the same quantities for the solvent s; ϕ_s is the volume fraction of the solvent s, equal to unity for a dilute solution.

Since reactions are usually carried out in dilute solution, Eq. (5-78) can be simplified to Eq. (5-79),

$$RT \cdot \ln f_i = V_{m,i} \cdot (\delta_i - \delta_s)^2 \tag{5-79}$$

where δ_i and δ_s are the solubility parameters of the reactant solutes and the solvent, respectively, according to Eq. (5-77) [5, 228–230].

The rate constant k of a bimolecular reaction $A + B \rightleftharpoons (AB)^{\neq} \rightarrow C + D$ can be expressed either as in Eq. (5-75) or in terms of the activity coefficients as shown in Eq. (5-80),

$$\ln k = \ln k_0 + \ln f_A + \ln f_B - \ln f_{\neq} \tag{5-80}$$

where k_0 is the rate constant in an ideal solution [5, 28].

---

* Regular solutions are characterized by a disordered distribution of solute and solvent molecules which is the same as in an ideal solution. In going from an ideal to a regular solution there is no change in entropy ($\Delta S = 0$), whereas the change in the activity coefficient f_i of the solute i is determined only by the enthalpy component of the Gibbs energy: $RT \cdot \ln f_i = \Delta G = \Delta H$ [228, 229].

By substituting Eq. (5-79) into Eq. (5-80), the following relationship for the rate of reaction between the nonpolar reactants A and B is obtained [5, 28]:

$$\ln k = \ln k_0 + \frac{1}{RT}\left[V_A(\delta_A - \delta_s)^2 + V_B(\delta_B - \delta_s)^2 - V_{\neq}(\delta_{\neq} - \delta_s)^2\right] \tag{5-81}$$

where V_A, V_B, and $V_{\neq}$ are the molar volumes of A, B, and the activated complex, respectively. Thus, the rate constant depends not only on the difference in molar volumes between reactants and activated complex [called the volume of activation, $\Delta V^{\neq} = V_{\neq} - (V_A + V_B)$], but also on the relative cohesive pressure of reactants, activated complex, and solvent. If the reactants have a greater solubility in the solvent than the activated complex, the rate is lower compared to the rate in an ideal solution. The reverse is true if the activated complex is more soluble than the reactants.

Eq. (5-81) which is difficult to handle can be changed into the linear Eq. (5-82):

$$\begin{aligned} RT \cdot \ln (k/k_0) = {} & \delta_s^2 (V_A + V_B - V_{\neq}) \\ & + (V_A \cdot \delta_A^2 + V_B \cdot \delta_B^2 - V_{\neq} \cdot \delta_{\neq}^2) \\ & + 2\,\delta_s (V_{\neq} \cdot \delta_{\neq} - V_A \cdot \delta_A - V_B \cdot \delta_B) \end{aligned} \tag{5-82}$$

A reasonable assumption in some cases is that $V_{\neq} = V_A + V_B$, thus the first term in Eq. (5-82) becomes zero. The second term is constant for all solvents if molar volumes and cohesion pressures of reactants and activated complex are the same in these solvents. Thus, under certain conditions, the third term is the most important. In gas-phase reactions, only the second term is left ($\delta_s = 0$ for the gas phase).

Intuitively one would expect that the volume of the reactants, the volume of the activated complex, and the corresponding activation volume as well as the internal pressure of the solvent*) affect the reaction rate [27]. Already in 1929 Richardson and Soper [233] and later on Glasstone [234] put forward rules which consider the influence of cohesion of reactants, products, and the solvent on reaction rates. They observed that reactions in which the products possessed greater (or lower) cohesion than the reactants were generally accelerated (or decelerated) by solvents with high cohesion, whereas reactions in which reactants and products were of similar cohesion, the solvent had relatively little influence on the reaction velocity. Although the reaction products are not involved in the activation process, according to the transition state theory, the above observations are still valid, since the cohesion of the activated complex may be regarded as lying somewhere between the values for the reactants and the products [27]. It has been pointed out, however, that variations in rate, caused by the internal pressure or cohesive pressure of the solvents, should be small in most solvents and generally should not exceed an order of magnitude [235].

Use of Eqs. (5-81) or (5-82) to predict solvent effect on reactions rates between nonpolar solutes is limited by the fact, that there are no experimental data available for the

---

* How an external pressure may affect the rate of a reaction in solution is considered in Section 5.5.11. The activation volume of a reaction is usually obtained from external pressure measurements.

heats of vaporization and hence for δ_i of most reacting organic compounds. However, the δ_s-values are known for most organic solvents [231, 232, 236, 238]. A good approximation for the evaluation of $\delta_{\neq}$ is given by Eq. (5-83) [25, 28].:

$$\delta_{\neq} = \left(\frac{V_A\delta_A^2 + V_B\delta_B^2}{V_{\neq}}\right)^{\frac{1}{2}} \tag{5-83}$$

Nevertheless, it is possible to obtain correlations between reaction rates and δ_s^2 as shown in Figs. 5-8 and 5-9 (for further examples see references [25, 126]).

Fig. 5-8 shows that there is a rough correlation of lg (k/k_0) with δ_s^2 for the Diels-Alder dimerization of cyclopentadiene [35] as predicted by Eq. (5-81). The molar volumes of cyclopentadiene and the corresponding activated complex have been determined, 83.1 and 135 cm^3, respectively [126].

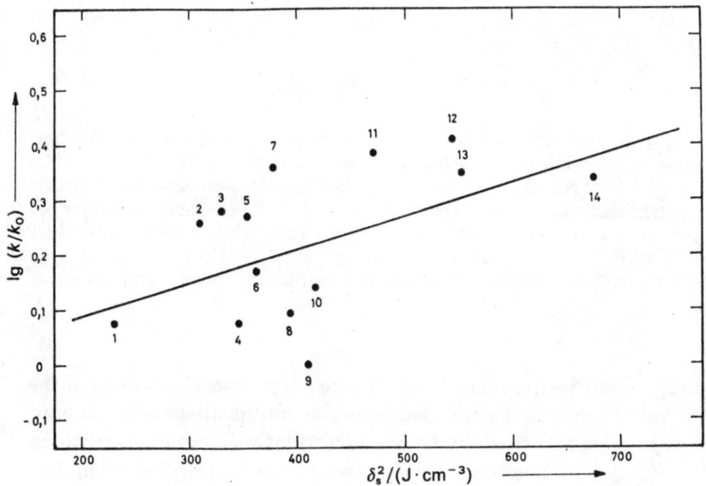

Fig. 5-8. Correlation of lg (k/k_0) [35] and the cohesive pressure δ_s^2 [238] in the Diels-Alder dimerization reaction of cyclopentadiene at 40 °C (rate constants relative to acetone as 'slowest' solvent):
1) diethyl ether, 2) tetrachloromethane, 3) toluene, 4) tetrahydrofuran, 5) benzene, 6) chloroform, 7) chlorobenzene, 8) dichloromethane, 9) acetone, 10) 1,4-dioxane, 11) *tert*-butanol, 12) 1-butanol, 13) 1-propanol, and 14) ethanol. The point for methanol shows a large deviation and is not included in this figure.

Fig. 5-9 shows that there is also a very rough, inverse correlation between lg (k/k_0) and δ_s^2 for the dissociation of the dimer of triphenylmethyl radical [167]. It can be safely assumed, that in this unimolecular reaction the molar volume of the activated complex is greater than the molar volume of the reactant, since a bond breaking must occur to some extent on activation.

The scattering of the points in the two figures is not surprising owing to the simplifications made in deriving Eq. (5-81). Furthermore, the solvent effects observed in

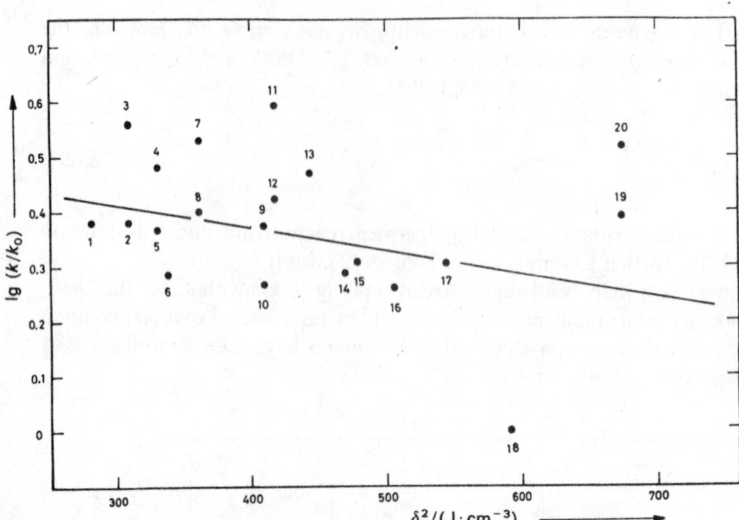

Fig. 5-9. Correlation of lg (k/k_0) [167] and the cohesive pressure δ_s^2 [238] in the dissociation reaction of 1-diphenylmethylene-4-triphenylmethyl-2,5-cyclohexadiene ("hexaphenylethane") at 0 °C; *cf.* Eq. (5-56) in Section 5.3.4 (rate constants relative to acetonitrile as 'slowest' standard solvent): 1) ethyl benzoate, 2) diethyl oxalate, 3) tetrachloromethane, 4) 1-bromopropane, 5) toluene, 6) 4-methyl-3-penten-2-one, 7) chloroform, 8) styrene, 9) propionic acid, 10) *N,N*-diethylacetamide, 11) carbon disulfide, 12) nitrobenzene, 13) aniline, 14) methyl salicylate, 15) pyridine, 16) ethyl cyanoacetate, 17) ethylene glycol monomethyl ether, 18) acetonitrile, 19) ethanol, and 20) nitromethane.

these reactions are very small with a comparatively large experimental error, and the solvents used include such dipolar and associated liquids as nitromethane and alcohols.

In principle it is possible to distinguish between reactions with negative activation volume $\Delta V^{\neq}$ $(= V_{\neq} - V_{\text{reactants}})$ and those with positive activation volume, indicating that the activated complex for the former occupies less volume than the reactants, according to the schematic Eqs. (5-84) and (5-85) [239].

Since in reaction (5-84) the reactants form a more compact activated complex, it is assumed that $(\delta_A + \delta_B) < \delta_{\neq}$. In case of Eq. (5-85) $(\delta_{A-B}) > \delta_{\neq}$ because a loosening

$$\text{(A+B)} \xrightleftharpoons[\text{negative}]{\Delta V^{\ddagger}} \left[\text{(A}\cdots\text{B)}\right]^{\ddagger} \longrightarrow \text{products} \qquad (5\text{-}84)$$

$$\text{(A-B)} \xrightleftharpoons[\text{positive}]{\Delta V^{\ddagger}} \left[\text{(A}\cdots\cdots\text{B)}\right]^{\ddagger} \longrightarrow \text{products} \qquad (5\text{-}85)$$

of bonds occurs on activation. If the nonpolar reaction is of type (5-84), *e.g.* dimerization of cyclopentadiene, ln k will increase as δ_s increases. Whereas for reactions of type (5-85), such as the unimolecular dissociation of the triphenylmethyl dimer, ln k will decrease as δ_s increases. Eq. (5-81) therefore predicts that the internal pressure (or the cohesive pressure) of solvents should influence reaction rates of nonpolar reactions in the same direction as external pressure (*cf.* also Section 5.5.11) [27, 232, 239].

The solvent influence on rates of bimolecular H-atom-transfer reactions $R\cdot + H{-}X \rightarrow R{-}H + \cdot X$ has been theoretically studied [580]. Rates for the model H-exchange reaction between $CH_3\cdot$ and CH_4 have been compared in the gas-phase and in rare gas solution (compressed Ar and Xe as solvents) over a wide range of internal pressure. Depending on the solvent and its internal pressure, relatively large rate enhancements have been calculated for this methyl/methane H-abstraction reaction. The major reason for this static solvent enhancement of rate in rare gas solutions seems to be the compact (tight) character of the activated complex, $[CH_3^{\delta\ominus} \cdots H \cdots {}^{\delta\ominus}CH_3]^{\neq}$ [580].

The concept of cohesive pressure (or internal pressure) is useful only for reactions between neutral, nonpolar molecules in nonpolar solvents, because in these cases other properties of the solvents such as the solvation capability or solvent polarity are neglected. For reactions between dipolar molecules or ions, the solvents interact with reactants and activated complex by unspecific and specific solvation so strongly, that the contribution of the cohesive pressure terms of Eq. (5-81) to ln k is a minor one. It should be mentioned, that cohesive pressure or internal pressure are not measures of solvent polarity. Solvent polarity reflects the ability of a solvent to interact with a solute, whereas cohesive pressure, as a structural parameter, represents the energy required to create a hole in a particular solvent to accommodate a solute molecule. Polarity and cohesive pressure are therefore complementary terms, and rates of reaction will depend on both of them [27, 232]. The influence of solvent polarity on reaction rates will be discussed in the following Sections.

5.4.3 Reactions between Neutral, Dipolar Molecules

In reactions between neutral, dipolar molecules, the electrostatic solute/solvent interactions such as dipole-dipole forces should above all determine the reaction rates in a manner as described qualitatively by the Hughes-Ingold theory [16, 44]. In order to describe electrostatic solvent effects, one considers the interaction between charged points separated by an isotropic dielectric continuum using functions of dielectric constant. In rate equations involving dipolar reactant molecules, Kirkwood's [240] equation is generally applicable. The Gibbs energy of an idealized dipole in a continuous medium of dielectric constant ε_r is compared to the Gibbs energy in a similar medium of unit dielectric constant ($\varepsilon_r = 1$):

$$\Delta G_{solv} = -\frac{N_A}{4\pi \cdot \varepsilon_0} \cdot \frac{\mu^2}{r^3} \cdot \frac{\varepsilon_r - 1}{2\varepsilon_r + 1} \qquad (5\text{-}86)$$

where μ is the permanent dipole moment and r the radius of the solute molecule containing the dipole, ε_0 is the permittivity of vacuum, and N_A is the Avogadro constant. This important formula relates the changes in Gibbs solvation energy of the dipolar solute both to the dielectric constant of the solvent and to the dipole moment and radius of the solute molecule, taking into account electrostatic forces only between solute and solvent molecules*).

Applying Kirkwood's formula to the transition state theory for the bimolecular reaction $A+B \rightleftharpoons (AB)^{\neq} \rightarrow C+D$ and combining Eq. (5-86) with Eq. (5-75), one obtains an expression for the rate constant of a reaction between two dipolar molecules A and B with moments μ_A and μ_B to form an activated complex with dipole moment $\mu_{\neq}$ [2]:

$$\ln k = \ln k_0 - \frac{1}{4\pi \cdot \varepsilon_0} \cdot \frac{N_A}{R \cdot T} \cdot \frac{\varepsilon_r - 1}{2\varepsilon_r + 1} \cdot \left(\frac{\mu_A^2}{r_A^3} + \frac{\mu_B^2}{r_B^3} - \frac{\mu_{\neq}^2}{r_{\neq}^3} \right) \tag{5-87}$$

where k is the rate constant in the medium of dielectric constant ε_r, and k_0 is the rate constant in a condensed medium of dielectric constant unity. Eq. (5-87) predicts that, if the activated complex is more dipolar than the reactants, the rate of the reaction increases with the dielectric constant of the medium.

Based on the same premises, but with some modification of the electrostatic model introduced by Kirkwood, Laidler and Landskroener obtained another similar expression for the reaction of two dipole molecules in a medium as given by Eq. (5-88) [11, 242]:

$$\ln k = \ln k_0 + \frac{1}{4\pi \cdot \varepsilon_0} \cdot \frac{3 \cdot N_A}{8 \cdot R \cdot T} \cdot \left(\frac{2}{\varepsilon_r} - 1 \right) \cdot \left(\frac{\mu_A^2}{r_A^3} + \frac{\mu_B^2}{r_B^3} - \frac{\mu_{\neq}^2}{r_{\neq}^3} \right) \tag{5-88}$$

This equation predicts that a plot of $\ln (k/k_0)$ versus $1/\varepsilon_r$ should give a straight line, and gives an explicit expression for the slope s of this line in terms of the radii and dipole moments as shown by Eq. (5-89):

$$s = \frac{1}{4\pi \cdot \varepsilon_0} \cdot \frac{3 \cdot N_A}{8 \cdot R \cdot T} \cdot \left(\frac{\mu_A^2}{r_A^3} + \frac{\mu_B^2}{r_B^3} - \frac{\mu_{\neq}^2}{r_{\neq}^3} \right) \tag{5-89}$$

If a reaction between neutral, dipolar molecules occurs with the formation of an activated complex with a dipole moment $\mu_{\neq}$ greater than either μ_A or μ_B, there will be an increase in the rate constant with increasing dielectric constant according to Eq. (5-88). This is because a higher dielectric constant favors the production of any highly dipolar species as, in this case, the activated complex. In applying Eqs. (5-87) and (5-88) to experimental data, a model for the activated complex has to be constructed in order to evaluate reasonable values for $\mu_{\neq}$ and $r_{\neq}$. This has been done, for example, for the acid and base hydrolysis of carboxylic esters [11, 242].

Another Coulombic energy approach for the calculation of electrostatic solvent effects on reactions between dipolar molecules was made by Amis [12, 21, 244]. He related the rate constant to the energy of activation by the well-known Arrhenius equation

---

* It should be noted that Kirkwood's formula does not appear in his first publication [240] in the form of Eq. (5-86). But it is this form of Kirkwood's formula which is widely known, representing only one of the terms of a more complex equation given in reference [240] with $n=1$. Kirkwood's theory was further developed in papers by Kirkwood, Westheimer, and Tanford [241], Laidler and Landskroener [242], and Hiromi [243].

$k = A \cdot \exp(-E_a/RT)$. It is assumed that the effect of the dielectric constant on the rate is given by Eq. (5-90):

$$\ln k = \ln k_{\infty} - \frac{1}{4\pi \cdot \varepsilon_0} \cdot \frac{2 \cdot \mu_A \cdot \mu_B \cdot N_A}{R \cdot T \cdot \varepsilon_r \cdot r^3} \tag{5-90}$$

where k is the rate constant in any medium of dielectric constant ε_r, k_{∞} is the rate constant in a medium of dielectric constant with a value of infinity, and μ_A and μ_B are the dipole moments of the two dipolar reactants A and B in vacuum. Eq. (5-90) predicts that a plot of $\ln k$ versus $1/\varepsilon_r$ will give a straight line, the slope of which should give a reasonable value of $r = r_A + r_B$, the distance of approach for the two dipolar molecules to react.

It should be mentioned, that there is no inconsistency in the fact that $\ln(k/k_0)$ is a function of $(\varepsilon_r - 1)/(2\varepsilon_r + 1)$ according to Eq. (5-87), and of $1/\varepsilon_r$ according to Eqs. (5-88) and (5-90). It can readily be seen, by carrying out explicitly the division of $(\varepsilon_r - 1)/(2\varepsilon_r + 1)$ according to Eq. (5-91), that $(\varepsilon_r - 1)/(2\varepsilon_r + 1)$ is linear in $1/\varepsilon_r$ to a good approximation [40].

$$\frac{\varepsilon_r - 1}{2\varepsilon_r + 1} = \frac{1}{2} - \frac{3}{4\varepsilon_r} + \frac{3}{8\varepsilon_r^2} - \frac{3}{16\varepsilon_r^3} + \ldots \tag{5-91}$$

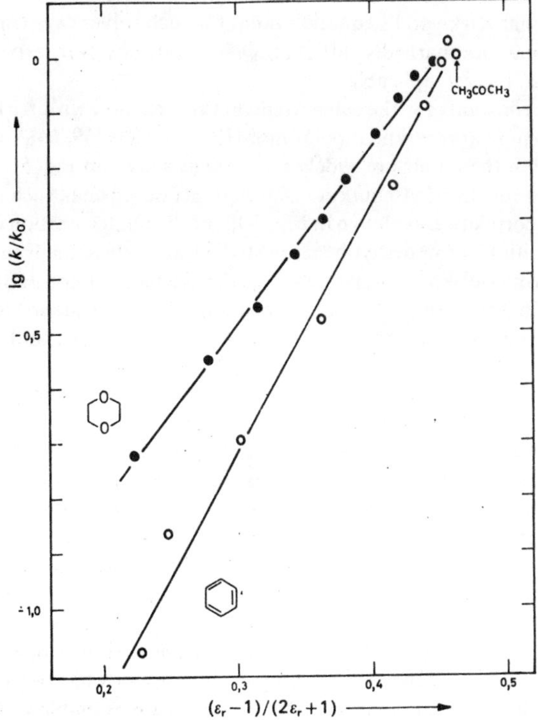

Fig. 5-10. Correlation between lg (k/k_0) [56] and the Kirkwood function $(\varepsilon_r - 1)/(2\varepsilon_r + 1)$ for the Menschutkin reaction between triethylamine and iodoethane at 40 °C in binary acetone/benzene and actone/1,4-dioxane mixtures (rate constants relative to acetone as common standard solvent).

For $\varepsilon_r = 8$, the error involved in neglecting terms in ε_r^2 and higher powers is less than one percent [40].

Laidler has pointed out [11, 242], that Eq. (5-88) is best considered as a semiquantitative formulation, which gives only a rough prediction of the effect of a change in dielectric constant on the rate of dipole-dipole reactions. This applies also to Eqs. (5-87) and (5-90). Nevertheless, in many cases a satisfactory correlation between rate constants and function of solvent dielectric constant has been obtained as, for instance, in the Menschutkin reaction between trialkylamines and haloalkanes forming quaternary tetraalkylammonium salts [2, 56, 58, 60, 61, 64, 65, 245–247].

Glasstone, Laidler, and Eyring [2] were the first to correlate rate data for some Menschutkin reactions according to Eq. (5-87) and they found in particular that a linear correlation between $\lg(k/k_0)$ and $(\varepsilon_r - 1)/(2\varepsilon_r + 1)$ is observed in the binary solvent mixture benzene/alcohol, while in benzene/nitrobenzene a monotonous deviation from the linear dependence was observed.

Fig. 5-10 shows typical dependences of $\lg(k/k_0)$ on the Kirkwood parameter $(\varepsilon_r - 1)/(2\varepsilon_r + 1)$ obtained for the Menschutkin reaction between triethylamine ($\mu = 2.9 \cdot 10^{-30}$ C m = 0.9 D) and iodoethane ($\mu = 6.3 \cdot 10^{-30}$ C m = 1.9 D) in binary solvent mixtures [56].

Binary solvent mixtures have the advantage that changing their composition will preferably change the electrostatic solute/solvent interactions, whereas nonelectrostatic and specific interactions remain the same within the whole solvent series and will be cancelled out. Fig. 5-10 shows that Kirkwood's equation holds for such solvent systems, but the slopes of the straight lines are markedly different. One might, however, expect them to be the same if Eq. (5-87) were applicable.

If one compares the rate constants for the same Menschutkin reaction with Kirkwood's parameter in thirty-two pure aprotic and dipolar non-HBD solvents [59, 64], one still finds a rough correlation, but the points are widely scattered as shown in Fig. 5-11.

Extending the media used for the Menschutkin reaction to protic solvents such as alcohols leads to an even worse correlation as shown in Fig. 5-12 for the quaternization of 1,4-diazabicyclo[2.2.2]octane with (2-bromoethyl)benzene studied in thirty-six solvents altogether [65]. The group of protic solvents is separated from the assembly of non-HBD solvents, each group showing a very rough distinct correlation with the function of dielectric constant. Such behaviour has been observed also for several other Menschutkin reactions [60, 61].

Although Eq. (5-87) is often qualitatively obeyed, as has been frequently mentioned, there is no exact linear correlation between the rate of Menschutkin reactions and the functions of dielectric constant as in the case of Fig. 5-12 [246, 247]. A complete absence of a regular effect of changes in the dielectric properties of the solvent on the reaction rate has also been observed [248, 249]. Sometimes a satisfactory correlation has been obtained because the reaction under consideration was studied in only a limited number of solvents. It can readily be seen from Fig. 5-11 that studying this Menschutkin reaction by chance in the six solvents no. 3, 11, 12, 26, 27, and 28 would simulate a very good correlation.

The fact that the curves of Figs. 5-10 to 5-12 (and others) have positive slopes is direct evidence for a partially ionized, dipolar activated complex. In cases for which Eq. (5-87) is reasonably well obeyed (especially for binary solvent mixtures), a reasonable value

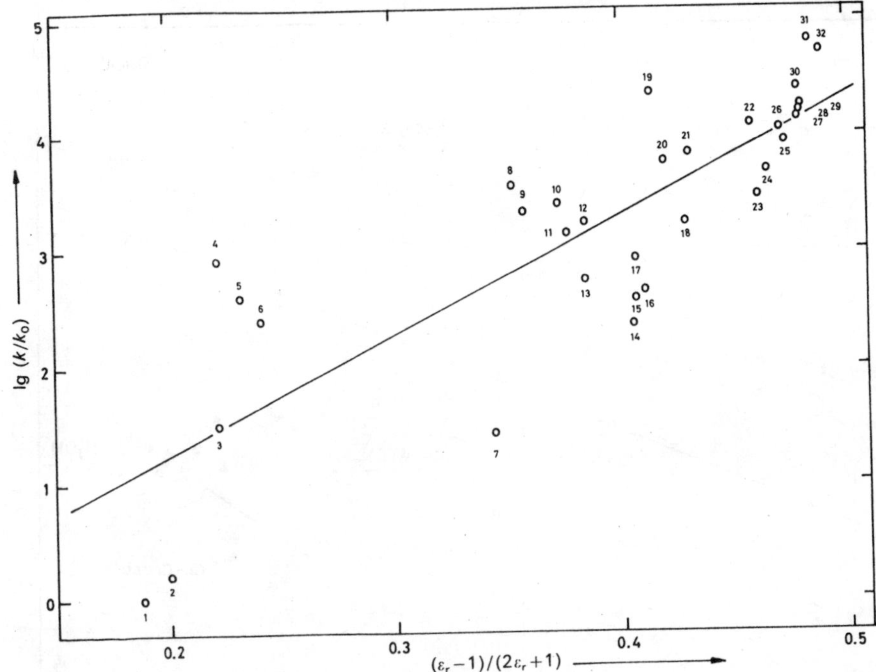

Fig. 5-11. Correlation between lg (k/k_0) [59, 64] and the Kirkwood function $(\varepsilon_r-1)/(2\varepsilon_r+1)$ for the Menschutkin reaction between triethylamine and iodoethane in 32 aprotic and dipolar aprotic solvents at 25 °C (rate constants relative to *n*-hexane as 'slowest' solvent): 1) *n*-hexane, 2) cyclohexane, 3) tetrachloromethane, 4) 1,4-dioxane, 5) benzene, 6) toluene, 7) diethyl ether, 8) iodobenzene, 9) chloroform, 10) bromobenzene, 11) chlorobenzene, 12) ethyl benzoate, 13) ethyl acetate, 14) 1,1,1-trichloroethane, 15) chlorocyclohexane, 16) bromocyclohexane, 17) tetrahydrofuran, 18) 1,1-dichloroethane, 19) 1,1,2,2-tetrachloroethane, 20) dichloromethane, 21) 1,2-dichloroethane, 22) acetophenone, 23) 2-butanone, 24) acetone, 25) propionitrile, 26) benzonitrile, 27) nitrobenzene, 28) *N,N*-dimethylformamide, 29) acetonitrile, 30) nitromethane, 31) dimethylsulfoxide, and 32) propylene carbonate.
The values of the second-order rate constants are taken from the compilation made by M. H. Abraham and P. L. Grellier, J. Chem. Soc., Perkin Trans. II *1976*, 1735.

for $\mu_{\neq}$ may be calculated from the slope according to Eq. (5-89) [2, 8] (for a compilation of such calculations see references [23, 64]). The calculated dipole moments $\mu_{\neq}$ range from *ca.* $17 \cdot 10^{-30}$ C m (*ca.* 5 D) to *ca.* $30 \cdot 10^{-30}$ C m (*ca.* 9 D) and show a considerable separation of charges for the activated complex of Menschutkin reactions, in agreement with the observed rate accelerations with increasing solvent polarity. However, it seems probable that if Eq. (5-87) is to hold over a substantial range of solvent composition, rather large-scale cancellations of nonelectrostatic effects must take place [23].

As the data for the Menschutkin reactions indicate, the character of the solute-solvent interactions is more complex than described by Eq. (5-87). It is evident that functions of dielectric constant alone, as given in Eq. (5-87), are not useful for describing

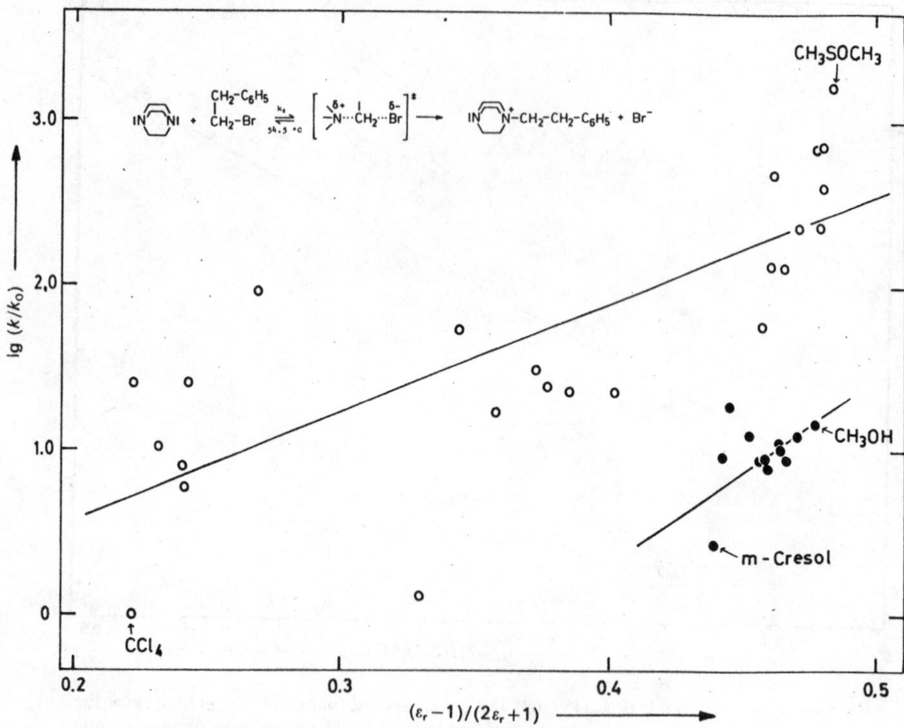

Fig. 5-12. Correlation between lg (k/k_0) [65] and the Kirkwood function $(\varepsilon_r-1)/(2\varepsilon_r+1)$ for the Menschutkin reaction between 1,4-diazabicyclo[2.2.2]octane and (2-bromoethyl)benzene in aprotic (○) and non-HBD (●) solvent at 54.5 °C (rate constants relative to tetrachloromethane as 'slowest' standard solvent).

the solvent effect on reactions between dipolar reactants except in certain special cases, such as when a mixture of two solvents is used. In addition to electrostatic forces, nonelectrostatic interactions, such as dispersion forces and hydrogen-bonding must be involved in Menschutkin reactions.

As seen from Fig. 5-11, although the three halobenzenes (points no. 8, 10, and 11) have similar dielectric constants, when used as solvents they lead to different reaction rates. Iodobenzene (no. 8), with the lowest dielectric constant, gives the largest rate. This observation strongly suggests that the polarizability of the solvent is an important factor in stabilizing the dipolar activated complex of this reaction. This was confirmed by Reinheimer *et al.* [57], who studied some Menschutkin reactions in benzene and its chlorine, bromine, and iodine derivates. They showed that the rate of the reaction increases with increasing polarizability of the solvent.

In case of alcoholic, protic solvents it should be particularly noted from Fig. 5-12, that the values of lg (k/k_0) are much lower than expected on the basis of their dielectric constants. It has been pointed out [64], that hydrogen bonding between the alcohols and the trialkylamine – thus reducing the Gibbs energy of the reactants – is not the only cause of this low accelerating power. This anomalous effect of aliphatic alcohols on the rates of Menschutkin reactions is due to the fact, that the activated complex for these reactions resembles relatively nonpolar solutes. Their structure lies somewhere between reactants and ion pair, probably nearer to the reactants, and compared to dipolar non-HBD solvents, which may effect stabilization through nonspecific interaction [64], is therefore destabilized by aliphatic alcohols. Within a series of aliphatic alcohols and in spite of the fact that the dielectric constant falls from 32.7 to 3.4, the reaction between bromoethane and *N,N*-dimethylaniline is accelerated by a factor of almost ten on going from methanol to octanol [245].

Eq. (5-87), which is a general rate equation predicting the effect of solvent on bimolecular reactions between dipolar molecules, can readily be modified to treat unimolecular reactions of dipolar molecules A by neglecting the term for the second reactant B. A classical example of this reaction type is the solvolysis of 2-chloro-2-methylpropane ($\mu = 7.1 \cdot 10^{-30}$ C m = 2.1 D). This is an example of an S_N1 reaction whose solvent dependence has been studied in many solvents and binary solvent mixtures (*cf.* Eq. (5-13) in Section 5.3.1) [40, 47, 250].

Assuming that the influence of nonelectrostatic terms and specific solute/solvent interactions are negligible, a linear relation is expected between $\lg(k/k_0)$ and the Kirkwood parameter $(\varepsilon_r - 1)/(2\varepsilon_r + 1)$. Such plots are shown in Figs. 5-13 and 5-14 for binary solvent mixtures and pure solvents. It can be seen that there is no simple relationship between $\lg(k/k_0)$ and $(\varepsilon_r - 1)/(2\varepsilon_r + 1)$ for all the solvents and solvent mixtures examined. As seen from Fig. 5-14, there is a large dispersion of the plot into separate lines for each solvent pair and curvature of the separate lines. Fig. 5-14 shows that there is a very poor correlation for pure solvents. In protic solvents, as a result of considerable specific solvation of the leaving anion through hydrogen bonding, the rate of solvolysis is higher than that expected from the dielectric constants of the solvents.

A somewhat better correlation using values of $\lg k_1$ at 120 °C was obtained by Koppel and Pal'm [250], but again protic and some non-HBD solvents do not conform to

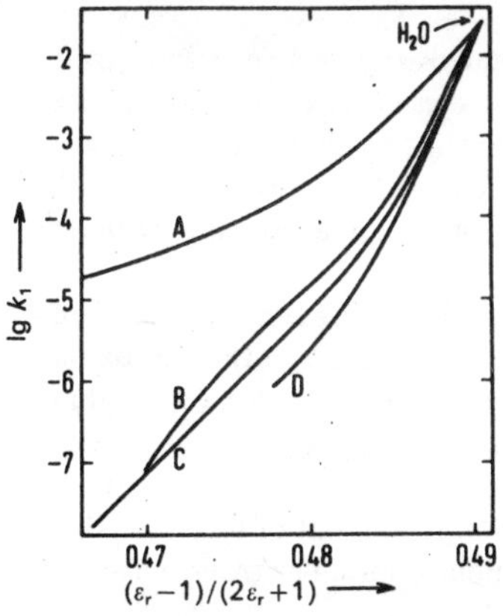

Fig. 5-13. Correlation between $\lg k_1$ [40] and the Kirkwood function $(\varepsilon_r - 1)/(2\varepsilon_r + 1)$ for the solvolysis of 2-chloro-2-methylpropane in binary 1,4-dioxane/water (A), ethanol/water (B), acetone/water (C), and methanol/water (D) mixtures at 25 °C; *cf.* Eq. (5-13) in Section 5.3.1.

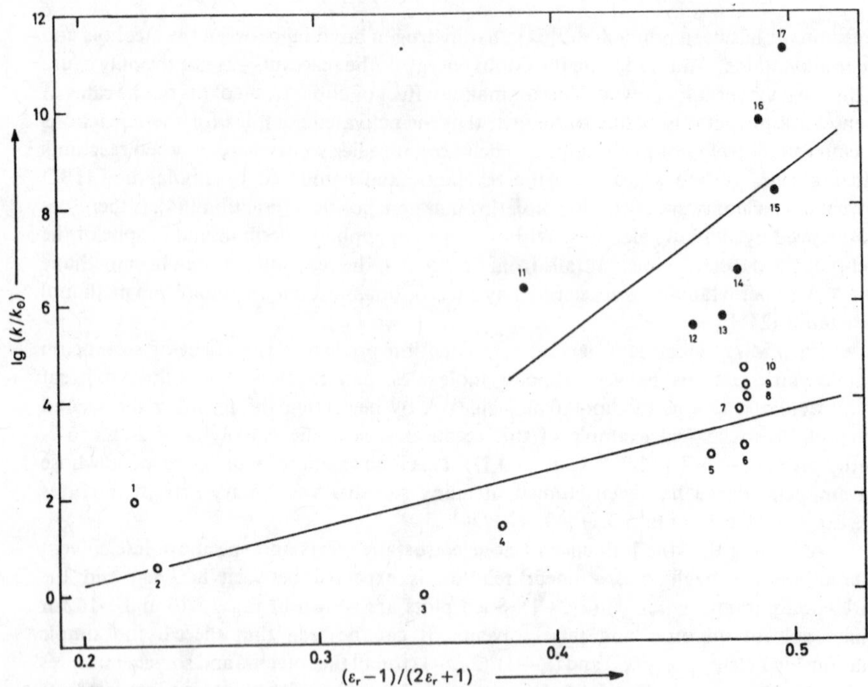

Fig. 5-14. Correlation between lg (k/k_0) [40, 47] and the Kirkwood function $(\varepsilon_r - 1)/(2\varepsilon_r + 1)$ for the solvolysis of 2-chloro-2-methylpropane in 17 pure aprotic (○) and non-HBD (●) solvents at 25 °C (rates relative to diethyl ether as 'slowest' standard solvent). Values of lg k are taken from the compilation in reference [47].
1) 1,4-dioxane, 2) benzene, 3) diethyl ether, 4) chlorobenzene, 5) acetone, 6) nitrobenzene, 7) 1-methyl-2-pyrrolidinone, 8) acetonitrile, 9) *N,N*-dimethylformamide, 10) nitromethane, 11) acetic acid, 12) 1-butanol, 13) ethanol, 14) methanol, 15) formamide, 16) formic acid, and 17) water.

the expected linear pattern. From the slope of the estimated straight line, a value of $\mu_{\neq} = 31 \cdot 10^{-30}$ C m = 9.2 D for the activated complex was calculated, which although reasonable must treated with caution, since the slope of the line depends largely on only a few values of lg k_1 in non-HBD solvents [250]. Nevertheless, it indicates that in the activated complex the charges are considerably separated, in agreement with the qualitative considerations made in Section 5.2.1.

The breakdown of the simple linear relationship between lg (k/k_0) and $(\varepsilon_r - 1)/(2\varepsilon_r + 1)$, required by Eq. (5-87), is obviously due to the failure of the approximations involved in deriving this equation, neglection of nonelectrostatic and specific solute-solvent interactions, and in the case of binary solvent mixtures due also to the selective solvation of the reactants and activated complex by one component of the mixture (*cf.* Section 2.4). It can be unequivocally concluded from a large amount of literature data, that the number of deviation for reactions between dipolar molecules from the simple electrostatic relationship, given in Eqs. (5-87), (5-88), and (5-90), greatly exceeds the number of cases where the relationships hold. Thus, allowances must be made also for electrostatic, nonelectrostatic, and specific solvation effects in the general effect of the medium.

It would appear from these observations, that the solvation capability might be better characterized using a linear Gibbs-energy relationship approach than functions of dielectric constant. There are now numerous examples known, for which the correlation between the rates of different reactions and the solvation capability of the solvent can be satisfactorily described with the help of semiempirical parameters of solvent polarity (*cf.* Chapter 7).

5.4.4 Reactions between Neutral Molecules and Ions

Numerous organic reactions are of the ion-dipole type, as for example the S_N2 reactions, given in Eqs. (5-17) and (5-18), Section 5.3.1. Considering the reaction between an ion A of charge $z_A \cdot e$ and a neutral, dipolar molecule B of dipole moment μ_B according to $A^{z_A \cdot e} + B \rightleftharpoons (AB)^{\neq z_A \cdot e} \rightarrow C + D$., Laidler and Eyring [2, 251] obtained Eq. (5-92) for the rate constant in a medium of ionic strength zero:

$$\ln k = \ln k_0 + \frac{1}{4\pi \cdot \varepsilon_0} \cdot \left[\frac{z_A^2 \cdot e^2 \cdot N_A}{2 \cdot R \cdot T} \cdot \left(\frac{1}{\varepsilon_r} - 1 \right) \cdot \left(\frac{1}{r_A} - \frac{1}{r_{\neq}} \right) - \frac{N_A}{R \cdot T} \cdot \frac{\mu_B^2}{r_B^3} \cdot \left(\frac{\varepsilon_r - 1}{2\varepsilon_r + 1} \right) \right] \tag{5-92}$$

where k and k_0 are the rate constants in media of dielectric constant ε_r and unity, respectively, and N_A is Avogadro's number. This equation predicts that $\ln k$ plotted against $1/\varepsilon_r$ for a molecule of zero dipole moment reacting with an ion of charge $z_A \cdot e$ should give a straight line, the slope of which would be $z_A^2 e^2 N_A / 2RT(1/r_A - 1/r_{\neq})$. This relationship should be especially true if the rate of the reaction is studied in mixtures of two solvents so that the dielectric constant can be varied by changing the proportions of each solvent. Since $r_{\neq}$ will be larger than r_A, the rate should be somewhat greater in a medium of lower dielectric constant. A quantitative test of Eq. (5-92) would require reasonable values of r_A and $r_{\neq}$. There seems to be no published data however which confirm Eq. (5-92).

A somewhat different equation relating the rate constant to the dielectric constant for a reaction $A^{z_A \cdot e} + B^{z_B \cdot e} \rightleftharpoons (AB)^{\neq (z_A + z_B) e} \rightarrow C + D$ in which the electrostatic interactions are more important than nonelectrostatic ones, was derived by Laidler and Landskroener [11, 242]:

$$\ln k = \ln k_0 + \frac{1}{4\pi \cdot \varepsilon_0} \cdot \left[\frac{e^2 \cdot N_A}{2 \cdot R \cdot T} \cdot \left(\frac{1}{\varepsilon_r} - 1 \right) \cdot \left(\frac{z_A^2}{r_A} + \frac{z_B^2}{r_B} - \frac{(z_A + z_B)^2}{r_{\neq}} \right) + \frac{3 \cdot N_A}{8 \cdot R \cdot T} \cdot \left(\frac{2}{\varepsilon_r} - 1 \right) \cdot \left(\frac{\mu_A^2}{r_A^3} + \frac{\mu_B^2}{r_B^3} - \frac{\mu_{\neq}^2}{r_{\neq}^3} \right) \right] \tag{5-93}$$

For a reaction between two dipoles having no net charge, the second term disappears, and the solvent effect is given entirely by the last term and Eq. (5-93) equals Eq. (5-88) in Section 5.4.3. For a reaction between an ion and a dipole (or between two charged dipoles) both terms must be included. The simplest case is for the reaction of a monovalent, structureless ion A of charge $z_A \cdot e(\mu_A = 0)$ with a neutral molecule B ($z_B \cdot e = 0$) of dipole moment μ_B. Eq. (5-93) has then the form:

$$\ln k = \ln k_0 + \frac{1}{4\pi \cdot \varepsilon_0} \cdot \left[\frac{e^2 \cdot N_A}{2 \cdot R \cdot T} \cdot \left(\frac{1}{\varepsilon_r} - 1 \right) \cdot \left(\frac{1}{r_A} - \frac{1}{r_{\neq}} \right) + \frac{3 \cdot N_A}{8 \cdot R \cdot T} \cdot \left(\frac{2}{\varepsilon_r} - 1 \right) \cdot \left(\frac{\mu_B^2}{r_B^3} - \frac{\mu_{\neq}^2}{r_{\neq}^3} \right) \right] \tag{5-94}$$

This equation predicts that $\ln k$ will vary linearly with the reciprocal of the dielectric constant, and gives an explicit expression for the slope in terms of radii and dipole moments. If data is available for a series of mixed solvents, one procedure for the application of this equation is to plot $\ln k$ against $1/\varepsilon_r$ and to determine the slope. Then it is possible to see whether its sign and magnitude may be predicted, using reasonable values for the radii and the dipole terms, in terms of Eq. (5-94). Of course, this procedure will depending on devising a suitable model for the activated complex. For Eq. (5-94) the same restrictions are valid as for Eq. (5-88). At best, Eq. (5-94) gives a semiquantitative formulation which allows only very rough predictions.

Another derivation of the dielectric constant dependence of $\ln k$ has been made by Amis [12, 21, 244] using a Coulomb energy approach for the ion-dipole interaction. Considering the mutual potential energy between an ion A of charge $z_A \cdot e$ and a dipole B of dipole moment μ_B in distance r_{AB} leads eventually to Eq. (5-95):

$$\ln k = \ln k_\infty + \frac{1}{4\pi \cdot \varepsilon_0} \cdot \frac{z_A \cdot e \cdot \mu_B \cdot N_A}{R \cdot T \cdot \varepsilon_r \cdot r_{AB}^2} \tag{5-95}$$

where k_∞ is the rate constant in a medium with dielectric constant of infinite magnitude. Taking the charge of the ion into account, Eq. (5-95) predicts that a plot of $\ln k$ versus $1/\varepsilon_r$ should be a straight line of positive slope if $z_A \cdot e$ is positive, and of negative slope if $z_A \cdot e$ is negative. This equation has been applied to both positive and negative ionic reactants reacting with dipole molecules [12, 21, 57, 252]. Illustrative examples of the applicability of Eqs. (5-94) or (5-95) are the alkaline hydrolysis of methyl propionate ($\mu = 6 \cdot 10^{-30}$ C m = 1.8 D) in acetone/water mixtures [252], and the S_N2 reaction between the azide anion and 1-bromobutane ($\mu = 7 \cdot 10^{-30}$ C m = 2.1 D) in pure dipolar non-HBD solvents [67]. In the case of alkaline hydrolysis, the rate determining step would be that between the negatively charged hydroxide ion and the dipolar ester. As suggested by the above mentioned equations, plots of lg (k/k_∞) versus $1/\varepsilon_r$ give the requisite straight lines as shown in Fig. 5-15 for both reactions. The positive slope of the line for the S_N2 reaction indicates, that in non-HBD solvents $r_{\neq} > r_A$, according to Eq. (5-94), whereas the negative slope of the line for the ester hydrolysis in protic solvents requires that $r_A > r_{\neq}$, which can be explained in terms of specific solvation of the hydroxide anion by hydrogen bonding [67, 252]. Further examples for application of Eq. (5-95) are given in references [12, 21].

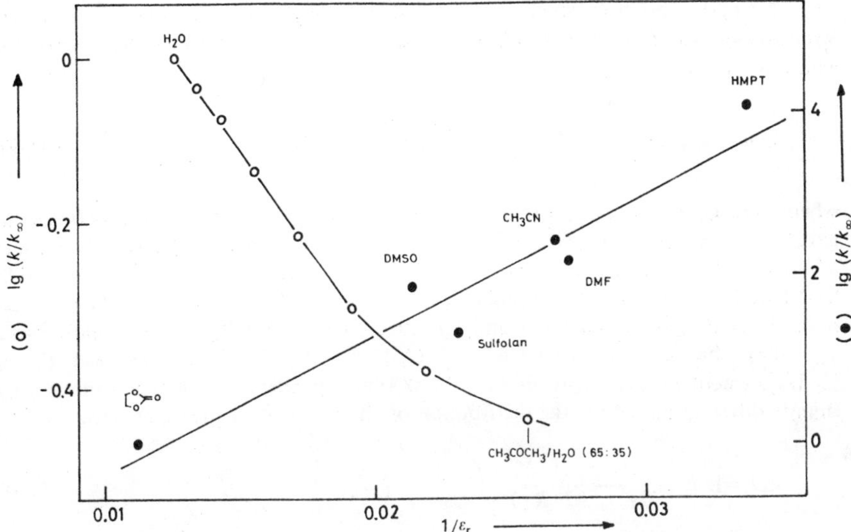

Fig. 5-15. Correlation between lg (k/k_∞) and $1/\varepsilon_r$ for the alkaline hydrolysis of methyl propionate in eight acetone/water mixtures at 25 °C (○) [252], and for the S_N2 reaction between the azide anion and 1-bromobutane in six pure dipolar non-HBD solvents at 25 °C (●) [67] (rate constants relative to the solvent with the largest dielectric constant).

5.4.5 Reactions between Ions

The combination of simple inorganic ions proceeds usually very fast and the rate of reaction is governed by the diffusion of the ions towards each other. There are, however, many reactions between ions which involve the making and breaking of covalent bonds or electron transfer, which may be as slow as reactions between neutral molecules (*cf.* for example reaction (5-19) in Section 5.3.1). Only reactions of this type will be considered in this Section.

Again the electrostatic theory will be useful in giving the general effect of the changes on the ionic reactants and the dielectric constant of the solvent. The reaction between ions was the first reaction to which the simple electrostatic model was applied for calculating the effect of the dielectric constant and the ionic strength of the medium on the reaction rate. The change in Gibbs energy during the formation of an ion pair from the ions A and B in a standard medium with dielectric constant ε_r, is equal to the electrostatic energy of the approach of two point charges $z_A \cdot e$ and $z_B \cdot e$ to a distance r_{AB} according to the following equation (N_A is Avogadro's number):

$$\Delta G_{\text{electrostatic}} = \frac{1}{4\pi \cdot \varepsilon_0} \cdot \frac{z_A \cdot z_B \cdot e^2 \cdot N_A}{\varepsilon_r \cdot r_{AB}} \tag{5-96}$$

From this equation, which applies to infinitely dilute solutions only, it is possible to write an expression for the dependence of the rate constant on the dielectric constant at zero ionic strength:

$$\ln k = \ln k_0 + \frac{1}{4\pi \cdot \varepsilon_0} \cdot \frac{z_A \cdot z_B \cdot e^2 \cdot N_A}{R \cdot T \cdot r_{AB}} \cdot \left(1 - \frac{1}{\varepsilon_r}\right) \tag{5-97}$$

where k and k_0 are the rate constants in media of dielectric constant ε_r and unity (*i.e.* in the gas phase), respectively. r_{AB} is the distance to which the two ionic reactants must approach in order to react and can be assumed to be equal to $r_A + r_B$. Eq. (5-97), first proposed by Scatchard [253], predicts a linear plot of $\ln k$ versus $1/\varepsilon_r$ with a negative slope if the charges of the ions have got the same sign, and a positive slope if the charges are of opposite sign.

For the reaction $A^{z_A \cdot e} + B^{z_B \cdot e} \rightleftharpoons (AB)^{\neq (z_A + z_B) e} \rightarrow C + D$ Laidler and Eyring [2, 251] derived another expression for the solvent dependence of rate constants using a slightly different model for the distribution of charges in the activated complex:

$$\ln k = \ln k_0 + \frac{1}{4\pi \cdot \varepsilon_0} \cdot \frac{e^2 \cdot N_A}{2 \cdot R \cdot T} \cdot \left(\frac{1}{\varepsilon_r} - 1\right) \cdot \left(\frac{z_A^2}{r_A} + \frac{z_B^2}{r_B} - \frac{(z_A + z_B)^2}{r_{\neq}}\right) \tag{5-98}$$

This equation is equivalent to the second term of Eq. (5-93). For reactions between two simple ions, the final term of Eq. (5-93) is less important and can generally be neglected. Eq. (5-98) is valid for reactions carried out in solutions of infinite dilution only, and reduces to Eq. (5-97) if $r_A = r_B = r_{\neq}$.

Some tests of Eqs. (5-97) and (5-98) have been made, particularly by carrying out a reaction between ions in a series of mixed solvents of varying dielectric constant. Examples are the reaction between bromoacetate and thiosulfate anions [11, 251], the reaction between negative divalent tetrabrómophenolsulfophthalein ions and hydroxide ions [254], and the ammonium cation/cyanate anion reaction yielding urea [255]. On the whole, the relationships (5-97) and (5-98) are a good approximation, although there are usually serious deviations especially at low dielectric constants (obviously due to association between unlike ions). Deviations from linearity may be explained as due to the failure of the simple approximations made in deriving these equations, due to selective solvation in binary solvent mixtures, and due to the neglection of such phenomena as the mutual polarization of ions or dipoles and the resulting inductive interactions, as well as specific solvation.

The entropy changes of reactions between ions deserve special mention. For reactions between ions of unlike sign there is an entropy increase going from reactants to activated complex. This is because the activated complex will have less charge than the reactants and will become partially desolvated. For ions of the same sign, however, the activated complex will be more charged than the reactants. This one would expect to be strongly solvated. There is a loss of entropy therefore, when the activated complex is formed. For examples which confirm these relationships see references [5, 11, 20].

Since the activity coefficients of the reactants vary with the ionic strength I of the solution, whose value is determined from the molar concentrations c_i of the ions i of charge z_i according to $I = 1/2 \cdot \sum_i c_i \cdot z_i^2$, the rate of reactions in solution must also depend on the ionic strength. This effect of the ionic strength of the solution on the rate constant described by Eq. (5-99) is termed the *Primary Salt Effect*.

$$\ln k = \ln k_0 + 2 \cdot z_A \cdot z_B \cdot A \cdot \sqrt{I} \qquad (5\text{-}99)$$

A is a constant for a given solvent and temperature ($A = 0.51 \cdot 2.303\, l^{1/2} \cdot mol^{-1/2}$ for aqueous solutions at 25 °C), and k_0 is the rate coefficient extrapolated to infinite dilution. Eq. (5-99), which was first derived by Brønsted [256], Bjerrum [257], and Christiansen [258] applying the Debye-Hückel theory [259] to the influence of neutral salts upon reaction rates in solution, predicts a linear relationship if $\ln k$ is plotted against the square root of the ionic strength. Since the simple Debye-Hückel relationship does not apply at those concentrations where ion association becomes important, Eq. (5-99) is only valid for solutions below *ca.* 10^{-2} mol/l for 1-1 electrolytes. Eq. (5-99) is perfectly accurate in dilute solutions.

Its conceptual significance is closely related to the Hughes-Ingold rules as given in Table 5-4 in Section 5.3.1: An increase in the ionic strength of a solution leads to an increase in its polarity. Thus, reactions between ions of like charge ($z_A \cdot z_B$ positive) are accelerated by an increase in the ionic strength, *i.e.* the addition of electrolytes, whereas reactions between ions of opposite charge ($z_A \cdot z_B$ negative) are retarded. If one of the reactants is a neutral molecule ($z_A \cdot z_B$ zero), the rate is expected to be independent of the ionic strength to a first approximation.

However, even in reactions involving neutral molecules ($z_A \cdot z_B$ zero) electrolyte effects can be found. Such reactions between neutral, nonpolar molecules can yield ionic products *via* a dipolar activated complex. This dipolar activated complex can be stabilized by an ionic medium, and a large positive salt effect is found. In reactions such as the solvolyses of 2-halo-2-methylpropanes, ions are produced and the ionic strength increases automatically during the reaction. This can lead to a kind of autocatalytical acceleration of the solvolyses. Small negative salt effects are frequently found in reactions between ions and neutral molecules because there is a greater charge dispersion during the activation process.

Eq. (5-99) was checked in detail in a large number of studies, but an extensive treatment of primary and secondary salt effects on reaction rates is outside the scope of this Section. The reader is referred to references [2–5, 11, 12, 19–21, 28], especially the complete and critical review of Davies [260].

In conclusion it can be said, that the electrostatic theory of solvent effects is a most useful tool for explaining and predicting many reaction patterns in solution. However, in spite of some improvements it still does not take into account a whole series of other solute/solvent interactions such as the mutual polarization of ions or dipoles, the specific solvation *etc.*, and the fact that the microscopic dielectric constant around the reactants may be different to the macroscopic dielectric constant of the bulk solvent. The deviations between observations and theory, and the fact that the dielectric constant cannot be considered as the only parameter responsible for the changes in reaction rates in solution, has led to the creation of different semiempirical correlation equations, which correlate the kinetic parameters to empirical parameters of solvent polarity (*cf.* Chapter 7).

5.5 Specific Solvation Effects on Reaction Rates

The intermolecular forces responsible for the association of dissolved ions or molecules with solvent molecules, consist of other forces as well as the non-specific directional, inductive, and dispersion forces which formed the essential basis for the quantitative relationships discussed in the preceding chapter. This other class of intermolecular forces consists of specific interactions such as hydrogen bonding and EPD/EPA forces (*cf.* Section 2.2). In the following discussion it will be shown, with the aid of some representative examples, how a consideration of the specific solvation of the reactants as well as the activated complexes can contribute to a better understanding of the rates, mechanisms, and sterical outcome of organic reactions. In this discussion, an even approximately complete treatment of the multitude of specific solvent effects cannot be attempted; for further reading see references [6, 10, 21, 26, 261–263, 452–454].

5.5.1 Influence of Specific Anion Solvation on Reaction Rates of S_N and other Reactions

In aliphatic nucleophilic substitution reactions the solvation of the departing anions is particularly important. In protic solvents this takes place preferably through hydrogen bonding, thus the activated complexes may be described as in Eqs. (5-100)*) and (5-101).

$$\text{R-X} + \text{H-S} \xrightleftharpoons{S_N1} [\text{R}^{\delta\oplus}\cdots\text{X}^{\delta\ominus}\cdots\text{H-S}]^{\ddagger} \longrightarrow \text{R}^{\oplus} + \text{X}^{\ominus}\cdots\text{H-S}$$

$$\xrightarrow{+\text{Y}^{\ominus}} \text{R-Y} + \text{X}^{\ominus}\cdots\text{H-S} \tag{5-100}$$

$$\text{Y}^{\ominus} + \text{R-X} + \text{H-S} \xrightleftharpoons{S_N2} [\text{Y}^{\delta\ominus}\cdots\text{R}\cdots\text{X}^{\delta\ominus}\cdots\text{H-S}]^{\ddagger}$$

$$\longrightarrow \text{Y-R} + \text{X}^{\delta\ominus}\cdots\text{H-S} \tag{5-101}$$

In these equations, the energy necessary for the ionization of the R–X bond will be lowered by the value of the X···H interaction.

Therefore, protic solvents usually have an accelerating effect on S_N reactions. This is one of the main reasons why S_N reactions of haloalkanes and sulfonate esters are carried out in solvents which consist wholly or partially of water, alcohols, or carboxylic acids. The relative strengths of the hydrogen bonding in the initial and transition state often overwhelm the electrostatic effect of solvents on the Gibbs energy of activation. On the other hand, the attacking nucleophile $Y^{\ominus}$ in the S_N2 reaction (5-101) can also be specifically solvated by protic solvents, so that its reactivity, and therefore the rate of the S_N2 reaction will be diminished. Examples of specific (electrophilic) solvation of anionic

* It should be pointed out that the formation of a carbenium ion in the course of an S_N1 reaction does not in fact occur as simply as described in Eq. (5-100), but takes place *via* intermediate contact- and solvent-separated ion pairs before the free solvated ions are formed; *cf.* Eqs. (2-19) and (2-20) in Section 2.6.

nucleophiles and leaving groups in S_N reactions can be found in references [264–269; 581–585]; see references [581, 582] for reviews.

Since the protic solvent usually is in large excess, its participation in the reaction cannot generally be established by means of kinetic measurements. However, if the reaction is carried out in a non-HBD solvent (*e.g.* C_6H_6, CCl_4) the effect of addition of small amounts of a protic solvent is easily observable. Thus, the S_N2 reaction between bromomethane and pyridine in benzene is accelerated by the addition of small amounts of alcohols or phenols. The more acidic the added solvent, the greater its ability to form hydrogen bonds, the greater the rate acceleration [264]. The electron-withdrawing effect of a protic solvent on the departing group X can be so pronounced that reactions of second-order in less strongly solvating media (*e.g.* ethanol), become first-order in acidic media (*e.g.* formic acid). The protic solvents exert an electrophilic pull on the departing anions in much the same way that heavy metal ions ($Ag^{\oplus}$, $Hg^{\oplus\oplus}$) catalyze nucleophilic substitution reactions of haloalkanes.

These results are supported by the observation that dipolar non-HBD solvents such as *N,N*-dimethylformamide or dimethyl sulfoxide, in spite of their high dielectric constants (36.7 and 46.7) and their high dipole moments ($12.3 \cdot 10^{-30}$ and $13.0 \cdot 10^{-30}$ C m) favour neither the ionization of haloalkanes nor S_N1 reactions (*cf.* Section 2.6). Dipolar non-HBD solvents cannot act as hydrogen bond donors and therefore badly solvate the departing anions. Thus, the anchimerically assisted ionization of 4-methoxyneophyl tosylate, shown in Eq. (5-102), is nine times faster in acetic acid than in dimethyl sulfoxide. This is in spite of the fact that the dielectric constant of acetic acid is eight times smaller than that of dimethyl sulfoxide; also the dipole moment of acetic acid is smaller than that of dimethyl sulfoxide [265].

→ products (5-102)

Solvent	$(C_2H_5)_2O$	$(CH_2)_4O$	CH_3COCH_3	$HCON(CH_3)_2$	CH_3SOCH_3
k_1^{rel}	1	17	169	980	3600

Solvent	C_2H_5OH	CH_3OH	CH_3CO_2H	H_2O	HCO_2H
k_1^{rel}	$1.2 \cdot 10^4$	$3.2 \cdot 10^4$	$3.3 \cdot 10^4$	$1.3 \cdot 10^6$	$5.1 \cdot 10^6$

It is furthermore remarkable that an approximately linear relationship between $(\varepsilon_r - 1)/(2\varepsilon_r + 1)$ and $\lg k_1$ values for reaction (5-102), measured in 19 solvents, is found only for non-HBD solvents (*cf.* Eq. (5-87) in Section 5.4.3), whereas protic solvents are much better ionizing media than their dielectric constant would suggest [265]. For example, acetic acid and tetrahydrofuran have very similar dielectric constants (6.2 and

7.6, respectively), and yet ionization in acetic acid exceeds that in tetrahydrofuran by a factor of $2 \cdot 10^3$! The reason for this extraordinary rate acceleration is that again the departing tosylate is better solvated due to hydrogen bonding in the protic solvent. The ability of the protic solvent to form hydrogen bonds is not reflected in its dielectric constant or in the dipole moment [265].

Another instructive example of electrophilic or H-bonding assistance of protic solvents (or co-solvents) in S_N1 reactions is the accelerated acetolysis rate of 2-bromo-2-methylpropane in the addition of phenols to the tetrachloromethane/acetic acid solution of the reactant [582]; see reference [582] for further examples. The usefulness of phenol as a solvent for S_N1 solvolysis reactions, in particular phenolysis of 1-halo-1-phenylethanes, has been stressed by Okamoto [582]. In spite of its low dielectric constant ($\varepsilon_r = 9.78$ at 60 °C), its low dipolarity ($\mu = 4.8 \cdot 10^{-30}$ C m = 1.45 D), and its low nucloeophilicity, it represents a solvent of high ionizing power due to its electrophilic driving force.

While increasing anion solvation by protic solvents has an accelerating effect on S_N1 reactions as described above, it is often a decelerating factor in S_N2 reactions. Thus, reaction (5-103) between (iodomethyl)benzene and radioactively labeled sodium iodide in acetone is clearly decelerated by the addition of protic solvents such as water, ethanol or phenol, as demonstrated in Fig. 5-16 [266].

$$\overset{*}{I}^{\ominus} + C_6H_5{-}CH_2{-}I \xrightarrow[0\,°C]{k_2} C_6H_5{-}CH_2{-}\overset{*}{I} + I^{\ominus} \qquad (5\text{-}103)$$

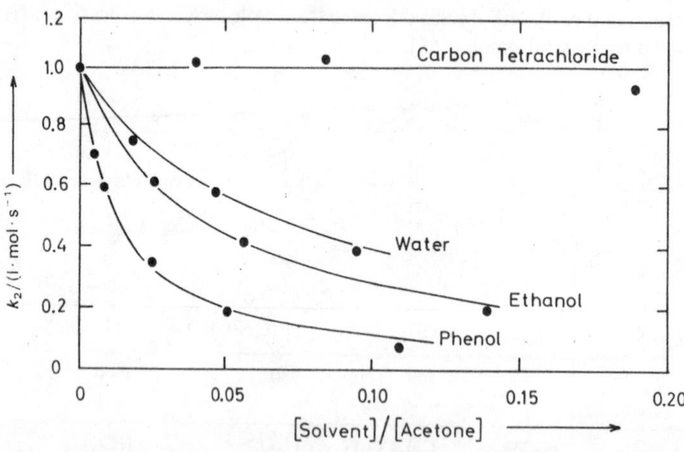

Fig. 5-16. The variation of k_2 for S_N2 reaction (5-103) with the ratio [solvent]/[acetone] [266].

The decrease in reaction rate is biggest for the addition of phenol, whereas the addition of the aprotic solvent CCl_4 leaves the reaction rate practically unchanged. The inhibitory effect of the protic solvents H–S (water < ethanol < phenol) correlates with their increasing tendency to form hydrogen bonds. This shows that the decrease in reaction rate is caused by specific solvation of the iodide anion, and depends on the position of the following equilibrium [266]:

$$(I^{\ominus})_{Acetone} + H\text{-}S \rightleftharpoons (I^{\ominus}\cdots H\text{-}S) + Acetone \tag{5-104}$$

Similar conclusions can be drawn from specific solvent effects on the following S_N2 reactions: (a) substitution of 1-halobutanes by iodide ion in acetone in the presence of small amounts of water [267]; (b) substitution by $N_3^{\ominus}$ and $SCN^{\ominus}$ in a series of *n*-octyl derivatives (*n*-$C_8H_{17}X$ with X = Cl, Br, I, OTs, OMs as leaving groups) in different solvents (CH_3OH in contrast to CH_3SOCH_3) [583]; (c) Menschutkin reaction between 1,4-diazabicyclo[2,2,2]octane and (2-haloethyl)benzenes in binary solvent mixtures, *i.e.* a main non-HBD solvent with low concentrations of protic co-solvents [584].

The nucleofugacity scales observed with the S_N2 reaction (b) in nonplar solvents (*e.g.* cyclohexane) are different from those obtained in protic solvents (*e.g.* methanol) and dipolar non-HBD solvents (*e.g.* dimethyl sulfoxide) [583]. For the Menschutkin reaction (c), specific rate inhibition has been observed on the addition of protic solvents which form hydrogen-bonds between the amine and the main solvent (*e.g.* C_6H_5OH, CH_3OH, $CHCl_3$). The diminution is proportional to the Brønsted acidity of the protic solvent added. Conversely, hydrogen-bond association between the halide nucleofuge and the protic solvent causes a specific catalytic effect which increases as the charge density of the leaving group increases. Both effects, deactivation of the amine and activation of the halide leaving group, are always in competition and the outcome depends on the chemical nature of the reactants [584].

In this connection it is noteworthy that the retarding effect of protic solvents on a given S_N2 reaction can also depend on the nature of the non-HBD solvent used [268, 269, 584]. Addition of a basic solvent may lead to a rate acceleration since the base can now compete with the nucleophile for hydrogen bond formation with the protic solvent. Thus, the S_N2 reaction (5-105) is accelerated by addition of 1,4-dioxane to a protic solvent such as methanol as reaction medium [268].

$$CH_3O^{\ominus} + CH_3\text{-}I \xrightarrow[25\,°C]{k_2} CH_3O\text{-}CH_3 + I^{\ominus} \tag{5-105}$$

Solvent	Methanol/1,4-Dioxane			
(1,4-Dioxane in cl/l)	0	20	40	60
k_2^{rel}	1	1.9	3.4	5.0

Most probably this rate increase is due to the fact that the specific solvation of the methoxide anion is decreased because of an association between 1,4-dioxane and methanol:

$$CH_3O^{\ominus} + H\text{-}OCH_3 \rightleftharpoons CH_3O^{\ominus}\cdots H\text{-}OCH_3 \tag{5-106}$$

$$CH_3O^{\ominus}\cdots H\text{-}OCH_3 + O(CH_2CH_2)_2O \rightleftharpoons CH_3O^{\ominus} + O(CH_2CH_2)_2O\cdots H\text{-}OCH_3 \tag{5-107}$$

In other words, by the addition of 1,4-dioxane, the reaction medium will become richer in free, not specifically solvated methoxide anions. Accordingly, the rate of the S_N2 reaction (5-105) increases [10, 268]. This interpretation is supported by the fact that reaction (5-105) is accelerated further by addition of the more highly basic solvent dimethyl sulfoxide, although admittedly this addition also leads to an increase in the dielectric constant of the reaction medium [269].

Similar results are found for the above-mentioned Menschutkin reaction (c). The added protic solvent can also combine with the main solvent. That solvent/solvent association leads to a diminution of the specific inhibitory and catalytic effect of protic solvents on the S_N2 reaction (c) [584]. The basicity of the main solvent determines the extent of deactivation of the protic solvent through H-bond association; this is anologous to Eq. (5-107).

There is also direct IR-spectroscopic evidence that, addition of a basic co-solvent to aqueous solutions reduces the percentage of non-associated "free" OH-groups which are necessary for hydrogen-bonding to S_N reactants [585]. When basic co-solvents are added, they scavenge the free OH-groups in water, thus lessening the specific protic solvent effect [585].

Nucleophiles with a heteroatom adjacent to the reaction center (*e.g.* $HOO^{\ominus}$, $ClO^{\ominus}$, $HONH_2$, NH_2NH_2) are found to be more reactive toward carbon electrophiles in S_N2 reactions in solution than would be expected from their basicity. This corresponds to a positive deviation from a Brønsted type plot (*i.e.* lg k_2 against pK_B). For instance, the hydroperoxide anion ($HOO^{\ominus}$) is 10^4 times less basic than the hydroxide anion ($HO^{\ominus}$), but it reacts *ca.* 50 times faster with (bromomethyl)benzene [459]. This behaviour is known as the *α-effect* [586] and various explanations for it have been given; *cf.* [587, 588] and references therein. If the α-effect is an intrinsic property of the anion, then it should manifest itself in the gas phase. Recently it has been reported that, in the gas phase $HOO^{\ominus}$ shows no evidence of enhanced nucleophilicity, compared to $HO^{\ominus}$, in its behaviour toward saturated (fluoromethane) and unsaturated carbon (methyl formate). If this is true, then the α-effect should arise because of a specific differential solvation of both the α- and normal nucleophiles ($HOO^{\ominus}/HO^{\ominus}$; $NH_2NH_2/R\text{-}NH_2$; *etc.*) which would lead to reduced solvation of the α-nucleophile. It is, however, not easy to understand why α-nucleophiles should be less solvated than normal nucleophiles; see the criticism given in reference [588].

Our final reaction is an instructive example of specific anion solvation in a non-S_N reaction. The decarboxylation rate of 6-nitrobenzisoxazol-3-carboxylate (in the presence of tetramethylguanidine as base) exhibits a dramatic rate increase in going from protic to non-HBD solvents as shown in Eq. (5-108) [589].

(5-108)

R = H

Solvent	H_2O	CH_3OH	C_2H_5OH	CH_2Cl_2	Et_2O	CH_3CN	CH_3SOCH_3	HMPT
k_1^{rel}	1	34	135	$6.4 \cdot 10^3$	$1.2 \cdot 10^4$	$3.9 \cdot 10^5$	$1.4 \cdot 10^6$	$9.5 \cdot 10^7$

The rate is slowest in aqueous solution and is enhanced in non-HBD solvents by a factor up to 10^7! Obviously, specific hydrogen-bonding between protic solvent molecules and the carboxylate group suppresses the decarboxylation. This is consistent with the fact that the decarboxylation rate of *4-hydroxy*benzisoxazol-3-carboxylate (R=OH), a compound with its own intramolecular protic environment, is very slow and does not show a substantial solvent effect. Furthermore, not only hydrogen-bonding but also ion-pairing seems to stabilize the carboxylate ion, at least in less polar solvents such as C_6H_6; CH_2Cl_2, and Et_2O [590]. Thus, the most reactive compound is the free carboxylate ion which lacks any specific stabilization.

In conclusion it can be stated, according to a general rule proposed by Palit already in 1947 [270], that solvents which block up the active center of a reactant through hydrogen bonding or by some other means, will suppress the reactivity of that reactant. In addition, solvents capable of promoting a favorable electron shift, necessary to the reaction, by specific solute/solvent interactions will enhance the reaction rate.

It should be mentioned that specific anion solvation is not only possible through hydrogen-bonding with protic solvents and ion-pairing, it is also possible through coordination with macrocyclic organic ligands, particularly with protonated cryptands [591, 592]. Recently, well-defined anion complexes, chloride ion cryptates, have been identified in aqueous solution using ^{35}Cl-NMR spectroscopy [591]. The anion is located inside the intramolecular cryptand cavity and is held by an array of hydrogen-bonds. An investigation of the nucleophilic reactivity of such specifically coordinated halide ions in various solvents has apparently not yet been reported.

5.5.2 Protic and Dipolar Aprotic Solvent Effects on Reaction Rates of S_N Reactions

In the course of a nucleophilic substitution reaction, a new bond is formed between the attacking nucleophile and the substrate. Since the nucleophile "attacks" with a free electron pair, a reagent should be all the more nucleophilic, the more readily the electron pair can be engaged in chemical bonding. Thus, the nucleophilicity of an anion is determined – among other things – by its basicity*) and its polarizability. One or the other property may dominate. While polarizability is hardly influenced by solvent, the basicity of an anion is strongly solvent dependent (*cf.* Section 4.2).

In the course of the activation process, the solvent shell of the nucleophile must be removed at the place of attack, while a new solvent shell around the activated complex is simultaneously forming. Thus, the Gibbs energy of activation will be higher, and the rate lower, the more strongly the molecules of the solvent shell are bound to the nucleophile. The less solvated a nucleophile, the more reactive it will be. Conversely, a nucleophile is stabilized by increasing solvation, its chemical potential falls, and its reactivity decreases.

---

* Basicity is measured in terms of a thermodynamic equilibrium involving coordination with H^+. Nucleophilicity is measured in terms of the rates of reaction with the most varied electrophiles. Hence, although a parallelism between basicity and nucleophilicity is often found (generally S_N reactions are faster with the stronger bases), it is by no means *a priori* necessary. A newer example for a direct simple relationship between nucleophilicity and basicity in S_N2 reactions is the reaction of 9-substituted fluorenide ions with (chloromethyl)benzene in dimethyl sulfoxide solution [595].

Since, according to Bunnett [271], nucleophilic reactivity depends on no less than seventeen factors (solvation effects included), it is very difficult to establish a general nucleophilicity order for all common nucleophiles in different solvents. Numerous attempts to establish such orders of nucleophilicities, which should be independent of any particular reaction electrophile, can be found in the literature [272–275].

When considering the solvation of anions, according to Parker [6] it is reasonable to distinguish between two classes of solvents: protic and dipolar aprotic solvents*) (*cf.* Section 3.4 and Fig. 3-3). The main difference between these two classes of solvents lies in their ability to solvate anions. Small anions with a large charge density (*i.e.* the ratio of charge to volume), which are strong hydrogen bond acceptors, are more strongly solvated in protic solvents. The hydrogen-bonding interaction is greatest for small anions (*e.g.* $F^{\ominus}$, $Cl^{\ominus}$, $HO^{\ominus}$), and least for large anions where the charge is dispersed (*e.g.* $SCN^{\ominus}$, $I^{\ominus}$, picrate$^{\ominus}$). Dipolar non-HBD solvents, which are polarizable, have mutual polarizability interactions with polarizable anions. This mutual polarizability interaction should be greatest for large, polarizable anions, and least for small weakly polarizable anions. Of course, mutual polarizability is not an exclusive prerogative of dipolar non-HBD solvents, some protic solvents may also be very polarizable. From the point of view of the "principle of hard and soft acids and bases" [275], hard anions are better solvated by hard (hydrogen-bonding) solvents, and soft anions have strong interactions with soft (dipolar non-HBD) solvents (*cf.* also Section 3.3.2). In other words, small ions lose in solvation and large, polarizable anions gain in solvation on transfer from protic to dipolar non-HBD solvents [6]. These considerations are essentially confirmed by the nucleophilic reactivities, observed for some S_N reactions in different media, presented in Table 5-15 [276–287].

It is apparent that the order of anion nucleophilicity is almost completely reversed on transfer from protic to dipolar non-HBD solvents. Especially for halide ions, the relative reactivity is completely reversed in both classes of solvents: whereas the order of reactivity is $I^{\ominus} > Br^{\ominus} > Cl^{\ominus} > F^{\ominus}$ in the protic solvent methanol (reactions no. 1 and no. 2 in Table 5-15), in dipolar non-HBD solvents such as *N,N*-dimethylformamide (no. 2), acetone (no. 3), dimethyl sulfoxide (no. 4), and acetonitrile (no. 5) the sequence of nucleophilicity is reversed. The traditional order of halide nucleophilicities, $I^{\ominus} > Br^{\ominus} > Cl^{\ominus}$ [261], applies only when the nucleophile is deactivated through solvation by protic solvents ($X^{\ominus} \cdots H{-}S$), whereas the natural order, $Cl^{\ominus} > Br^{\ominus} > I^{\ominus}$, is observed in dipolar non-HBD solvents. The sequences observed in dipolar non-HBD solvents give a much better picture of the true nucleophilicity of the weakly solvated anions, which also correlates better with their basicity.

With acetonitrile as solvent, there appears to be a general levelling of the nucleophilicities of anions as indicated by the relatively small variation of rate constants obtained for reaction no. 5 in Table 5-15 [282]. Quite surprisingly one of the best nucleo-

---

* Solvents referred to as *dipolar aprotic* are not in fact aprotic. In reactions where strong bases are employed their protic character can be recognized easily (*e.g.* $H_3C{-}SO{-}CH_3 + NH_2^{\ominus} \rightleftharpoons H_3C{-}SO{-}CH_2^{\ominus} + NH_3$). The term dipolar *aprotic* solvent is in fact rather misleading. Therefore, it has been recommended by Bordwell *et al.* [594] that the designation *dipolar aprotic* should be replaced by *dipolar nonhydroxylic* or better still by *dipolar non-HBD* solvents. The abbreviation *HBD* stands for *h*ydrogen-*b*ond *d*onor. *Cf.* also Chapter 3.4 and the footnote on page 69.

Table 5-15. Relative nucleophilic reactivities of free[a)] anions for S_N2 reactions in various protic and dipolar non-HBD solvents (no. 1...5) as well as in molten salts (no. 6 and no. 7) and in the gas phase (no. 8).

No.	Reaction	Solvent (temperature)	Sequence of decreasing reactivity (relative reaction rates)	References
(1)	$CH_3-I + Me^{\oplus}X^{\ominus}$	CH_3OH (25 °C)	$C_6H_5S^{\ominus} > S_2O_3^{2\ominus} \gg I^{\ominus} > SCN^{\ominus} \approx CN^{\ominus} > CH_3O^{\ominus} > Br^{\ominus} > N_3^{\ominus} > C_6H_5O^{\ominus} > Cl^{\ominus} > CH_3CO_2^{\ominus} > F^{\ominus}$	[276]
(2)	$CH_3-OTos + Me^{\oplus}X^{\ominus}$	CH_3OH (25 °C)	$C_6H_5S^{\ominus} > CH_3O^{\ominus} > I^{\ominus} > SCN^{\ominus} > Br^{\ominus} > Cl^{\ominus}$	[277]
		$HCON(CH_3)_2/H_2O$[b)]	$I^{\ominus} > Br^{\ominus} > Cl^{\ominus}$	[278]
		$HCON(CH_3)_2$ (0 °C)	$Cl^{\ominus} > Br^{\ominus} > I^{\ominus}$ (9 : 3 : 1)	[278]
(3)	$n\text{-}C_4H_9-OBros + (n\text{-}C_4H_9)_4N^{\oplus}X^{\ominus}$	CH_3COCH_3 (25 °C)	$Cl^{\ominus} > Br^{\ominus} > I^{\ominus}$ (18 : 5 : 1)	[279]
(4)	$n\text{-}C_3H_7-OTos + (n\text{-}C_4H_9)_4N^{\oplus}X^{\ominus}$	CH_3SOCH_3 (25 °C)	$S_2O_3^{2\ominus} > HO^{\ominus} \approx CH_3O^{\ominus} > F^{\ominus} > C_6H_5O^{\ominus} > N_3^{\ominus} > Cl^{\ominus} > Br^{\ominus} > I^{\ominus} > SCN^{\ominus}$	[280]
		CH_3SOCH_3 (50 °C)	$Cl^{\ominus} > Br^{\ominus} > I^{\ominus}$ (5.2 : 3.2 : 1)	[281]
		CH_3SOCH_3/H_2O[c)]	$I^{\ominus} > Br^{\ominus} > Cl^{\ominus}$ (6.7 : 1.3 : 1)	[281]
(5)	$C_6H_5CH_2-OTos + K^{\oplus}X^{\ominus}$[d)]	CH_3CN (30 °C)	$N_3^{\ominus} > CH_3CO_2^{\ominus} > CN^{\ominus} > F^{\ominus} > Cl^{\ominus} \approx Br^{\ominus} > I^{\ominus} > SCN^{\ominus}$ (10 : 9.6 : 2.4 : 1.4 : 1.3 : 1.3 : 1 : 0.3)	[282]
(6)	$CH_3-OTos + (C_2H_5)_4N^{\oplus} + X^{\ominus}$	$N_{2226}B_{2226}$[e)] (35 °C)	$Cl^{\ominus} > Br^{\ominus} > I^{\ominus}$ (2.1 : 1.2 : 1)	[283]
(7)	$(n\text{-}C_5H_{11})_4N^{\oplus} + X^{\ominus}$	$(n\text{-}C_5H_{11})_4N^{\oplus}X^{\ominus}$[f)] (180 °C)	$Cl^{\ominus} > Br^{\ominus} > I^{\ominus}$ (620 : 7.7 : 1)	[284]
(8)	$CH_3-Br + X^{\ominus}$	Gas phase	$HO^{\ominus} > CH_3O^{\ominus} \approx F^{\ominus} > CH_3S^{\ominus} \gg CN^{\ominus} > Cl^{\ominus} > Br^{\ominus}$	[285, 290]

[a] "Free" means without association with the corresponding cation, no ion pairing.
[b] 91 : 9 cl/l (0 °C).
[c] 70 : 30 cl/l (50 °C).
[d] The potassium salts are solubilized in acetonitrile with [18]crown-6, producing extremely reactive "naked" anions because of the weak anion solvation forces in acetonitrile solutions and the complete dissociation of the potassium salts [282].
[e] Molten triethyl-*n*-hexylammonium triethyl-*n*-hexyl-boride as solvent [283].
[f] Molten tetra-*n*-pentylammonium halides as solvent [284].

philes in acetonitrile is $CH_3CO_2^{\ominus}$, which is normally considered a very poor nucleophile in protic solvents, whereas $SCN^{\ominus}$, one of the more potent nucleophiles in aqueous solution, is approximately 30 times slower than $CH_3CO_2^{\ominus}$. Obviously, the so-called "naked", and therefore very reactive, anions in acetonitrile are solvated by much weaker forces than in protic solvents, and the variations in anion solvation in acetonitrile do not appreciably effect the relative nucleophilic reactivities of the anions [282].

Reactions no. 6 and no. 7 in Table 5-15 demonstrate that with molten quaternary ammonium salts as solvents, where deactivation by anion solvation is absent, the halide ions show the same nucleophilic order as in dipolar non-HBD solvents [283, 284]. This is in accordance with the theory of protic/dipolar non-HBD medium effects on $X^{\ominus}$

nucleophilicity [6]. It has been suggested, that fused-salts experiments should provide a good model for the determination of intrinsic relative nucleophilicities of anions towards saturated carbon atoms [284].

The results for reaction no. 8 in Table 5-15 indicate that nucleophilic reactivities of anions obtained in the gas phase are in principle in the same order as in molten salts and in dipolar non-HBD solvents [285, 290]. This again suggests, that specific solvation of the anions is responsible for the reversed order obtained in protic solvents relative to dipolar non-HBD solvents. Whereas the relative nucleophilicities in acetonitrile are similar to those found in the gas phase [282, 285, 290], the absolute gas-phase rates are some orders of magnitude greater than those in acetronitrile. The specific rates of displacement reactions of anions with halomethanes exceed those in solution by factors up to $\geq 10^{10}$ [285, 290]. These large differences in absolute rates demonstrate the moderating influence of the solvent on all the reactivities [282]. *Cf.* also Chapter 5.2.

In spite of evidence for increased stabilization of increasingly large anions in increasingly polarizable solvents due to dispersion interactions, there is some question as to whether the decreased reactivity of halide ions in dimethyl sulfoxide ($Cl^{\ominus} > Br^{\ominus} > I^{\ominus}$; *cf.* reaction no. 4 in Table 5-15) is due to increased solvation in the series $Cl^{\ominus} \rightarrow Br^{\ominus} \rightarrow I^{\ominus}$. Because the solvation order of these halide ions in dimethyl sulfoxide is $I^{\ominus} < Br^{\ominus} < Cl^{\ominus}$ (the same as in water!) as shown by solvation enthalpies, ΔH_{solv}, of the corresponding alkali halides in dimethyl sulfoxide and in water [281], Parker [6] suggested that the observed reactivity order $Cl^{\ominus} > Br^{\ominus} > I^{\ominus}$ in dipolar non-HBD solvents might be the order of increasing solvation in these solvents. Alternatively, this could also be the intrinsic order of nucleophilicity, which is only partially offset by decreasing solvation in the same order [281]. If the unsolvated $Cl^{\ominus}$ is a sufficiently stronger nucleophile than unsolvated $I^{\ominus}$, the more strongly solvated $Cl^{\ominus}$ in dimethyl sulfoxide is, nevertheless, more reactive than the comparatively less solvated $I^{\ominus}$. In protic solvents, however, the difference in solvation is much larger. Now the extremely strong solvation of $Cl^{\ominus}$ in protic solvents diminishes the reactivity of this anion to less than that of the more weakly solvated $I^{\ominus}$. Therefore, the reversed nucleophilic order of the halide ions in dimethyl sulfoxide relative to water must be attributed to the smaller differences in halide solvation in dimethyl sulfoxide than in water [281]. This [281] and other estimates [288, 289] are all in agreement with the observation that solvent stabilization of halide ions, although small compared to water, goes in the same direction, $Cl^{\ominus} > Br^{\ominus} > I^{\ominus}$, in dipolar non-HBD solvents.

Accordingly, recent gas-phase results for the clustering of $Cl^{\ominus}$, $Br^{\ominus}$, and $I^{\ominus}$ with Me_2SO and H_2O show that the bonding to dimethyl sulfoxide also decreases in the order $Cl^{\ominus} > Br^{\ominus} > I^{\ominus}$, *i.e.* with increasing ion radius [596]. Thus, anion solvation decreases with an increase in the ion radius for both protic and non-HBD solvents; however, the decrease is appreciable less in dipolar non-HBD solvents. This is the principle reason for the much higher rates of anion-molecule S_N2 reactions in dipolar non-HBD solvents [596].

An important contribution to a quantitative estimation of nucleophilic reactivity towards cations in different solvents was made by Ritchie *et al.* [274, 597]. According to Eq. (5-109),

$$\lg k - \lg k_0 = N_+ \qquad (5\text{-}109)$$

where k is the rate constant for the reaction of a cation (particularly triarylmethyl cations, aryldiazonium ions, and aryltropylium cations) with a given nucleophile/solvent-system, and k_0 is the rate constant for reaction of the same cation with a standard system (*e.g.* $HO^{\ominus}$ in H_2O), a parameter N_+ was defined, which represents the difference in nucleophilicity of the two systems. The N_+-values are measures of an inherent property of the (nucleophile *and* solvent)-system which is associated with the differential solvation of the nucleophile in different solvents. A selection of N_+-values is listed in Table 5-16 [597]. The higher the N_+-value, the less strongly solvated and the more reactive is the nucleophile in a given solvent. The large differences in N_+-values for the same nucleophile in protic and non-HBD solvents deserves special mention in this context; *cf.* for example entries (3), (13), and (17) for the cyanide ion, entries (15) and (19) for the monothioglycolate ion, and entries (6), (7), and (11) for hydrazine in Table 5-16.

Table 5-16. Nucleophilic parameters N_+ for various (nucleophile + solvent)-systems, based on reactions of malachite green or the tris-4-anisylmethyl cation [597].

No.	Nucleophile (solvent)	N_+
(1)	$H_2O(H_2O)$	0.73
(2)	$NH_3(H_2O)$	3.89
(3)	$CN^{\ominus}(H_2O)$	4.12
(4)	$HO^{\ominus}(H_2O)$	*4.75*[a]
(5)	$HO{-}NH_2(H_2O)$	5.05
(6)	$H_2N{-}NH_2(H_2O)$	6.01
(7)	$H_2N{-}NH_2(CH_3OH)$	6.89
(8)	$BH_4^{\ominus}(H_2O)$	6.95
(9)	$CH_3O^{\ominus}(CH_3OH)$	7.51
(10)	$N_3^{\ominus}(H_2O)$	7.54
(11)	$H_2N{-}NH_2(CH_3SOCH_3)$	8.17
(12)	$HOO^{\ominus}(H_2O)$	8.52
(13)	$CN^{\ominus}(CH_3SOCH_3)$	8.64
(14)	$N_3^{\ominus}(CH_3OH)$	8.78
(15)	$HOCH_2{-}CH_2S^{\ominus}(H_2O)$	8.87
(16)	$C_6H_5S^{\ominus}(H_2O)$	9.10
(17)	$CN^{\ominus}[HCON(CH_3)_2]$	9.44
(18)	$C_6H_5S^{\ominus}(CH_3OH)$	10.41
(19)	$HOCH_2{-}CH_2S^{\ominus}(CH_3SOCH_3)$	12.71

[a] All N_+-values are relative to $N_+ = 4.75$ for hydroxide ion in water.

The N_+-scale which is derived from nucleophile/electrophile combination reactions differs from the nucleophilicity scales generated from rates of S_N2 reactions (*cf.* Table 5-15). The N_+ equation has been tested for a wide variety of Lewis acid/base reactions and there is now a large set of data on nucleophile/electrophile combination reactions which are reasonably well correlated with N_+ [597].

The factors affecting the kinetic reactivities of nucleophiles in solution are still not well understood; *cf.* references [597, 598] for a sound discussion of this problem.

Since anions are solvated to a much lesser extent in dipolar non-HBD solvents than in protic solvents, they exist in these solvents as more or less "naked" and therefore extremely reactive ions. For S_N2 and S_NAr reactions involving anionic nucleophiles a change from protic to dipolar non-HBD solvent often cause a very dramatic acceleration of these reactions [6]*). Some typical examples are reactions (5-110) to (5-113), for which rate data in several solvents relative to solvent methanol are reported in Table 5-17 as the logarithms of rate ratios.

$$Cl^{\ominus} + CH_3\text{-}I \xrightarrow{k_2} Cl\text{-}CH_3 + I^{\ominus} \tag{5-110}$$

$$N_3^{\ominus} + CH_3CH_2CH_2CH_2\text{-}Br \xrightarrow{k_2} CH_3CH_2CH_2CH_2\text{-}N_3 + Br^{\ominus} \tag{5-111}$$

$$N_3^{\ominus} + F\text{-}C_6H_4\text{-}NO_2 \xrightarrow{k_2} [(F)(N_3)C_6H_4^{\ominus}\text{-}NO_2] \longrightarrow F^{\ominus} + N_3\text{-}C_6H_4\text{-}NO_2 \tag{5-112}$$

$$C_5H_{10}NH + F\text{-}C_6H_4\text{-}NO_2 \xrightarrow{k_2} [(F)(C_5H_{10}N^{\oplus}H)C_6H_4^{\ominus}\text{-}NO_2] \longrightarrow HF + C_5H_{10}N\text{-}C_6H_4\text{-}NO_2 \tag{5-113}$$

The rate increases (relative to methanol) involve factors in the range of 10^2 to 10^7 going from a protic to a dipolar non-HBD solvent. Even larger factors, as in the reaction 2,4-$(NO_2)_2C_6H_3I$ with $Cl^{\ominus}$ in hexamethylphosphoric triamide (ca. 10^9!) are not uncommon [291]. For reaction (5-112) the solvent change from methanol to hexamethylphosphoric triamide corresponds to a decrease in the Gibbs energy of activation, $\Delta\Delta G^{\neq}$, of 36.4 kJ/mol (8.7 kcal/mol) [292]. Hexamethylphosphoric triamide is usually the "fastest" of dipolar non-HBD solvents [291]. That these protic/dipolar non-HBD solvent effects are mainly anion-solvation phenomena is confirmed by the observation that similar large rate ratios are not observed for reaction (5-113), which involves a neutral nucleophile (piperidine). Also Menschutkin reactions between tertiary amines – neutral nucleophiles – and haloalkanes, are rather insensitive to protic/dipolar non-HBD solvent transfer [6, 23]. As expected for such an effect, attributed mainly to hydrogen-bonding interactions, protic/dipolar non-HBD solvent effects on the chemistry of anions are strongly temperature dependent; differences between protic and dipolar non-HBD solvents are greater at low temperatures than at high temperatures [291].

In addition, it should be mentioned that for protic/dipolar non-HBD solvent effects on S_N2 reactions $Y^{\ominus} + RX \rightleftharpoons [YRX^{\ominus}]^{\neq} \rightarrow YR + X^{\ominus}$ not only the solvation of the anion must be taken into account. An important factor in determining the rate of this reaction in different solvents may be the solvation of the reactant molecule RX and the activated complex $[YRX^{\ominus}]^{\neq}$. Although ions have much greater solvation enthalpies than nonelec-

---

* To speak of dipolar non-HBD solvents having an accelerating effect on the rates of bimolecular substitution reactions involving anionic nucleophiles seems to be looking at things in reverse order. The view that protic solvents have a retarding effect on such reactions seems to be much more consistent with the experimental data. But the above mentioned description of the protic/dipolar non-HBD solvent effect on reaction rates has been widely used in the literature.

Table 5-17. Relative rates of the S_N2 anion-molecule reactions (5-110) [291] and (5-111) [67], and of the S_NAr reactions (5-112) [291, 292] and (5-113) [293] in protic and dipolar non-HBD solvents at 25 °C.

Solvents	$\lg (k_2^{Solvent}/k_2^{MeOH})$ [a] for reaction			
	(5-110) [b]	(5-111) [c]	(5-112) [d]	(5-113) [e]
Protic Solvents:				
CH_3OH	0	0	0	0
H_2O	0.05	0.8	–	–
$CH_3CONHCH_3$	–	0.9	–	–
$HCONH_2$	1.2	1.1	0.8	–
$HCONHCH_3$	1.7	–	1.1	–
Dipolar non-HBD Solvents:				
$(CH_2)_4SO_2$ (Sulfolane)	–	2.6	4.5	–
CH_3NO_2	4.2	–	3.5	0.8
CH_3CN	4.6	3.7	3.9	0.9
CH_3SOCH_3	–	3.1	3.9	2.3
$HCON(CH_3)_2$	5.9	3.4	4.5	1.8
CH_3COCH_3	6.2	3.6	4.9	0.4
$CH_3CON(CH_3)_2$	6.4	3.9	5.0	1.7
N-methyl-2-pyrrolidinone (structure: O, N-CH$_3$)	6.9	–	5.3	–
$[(CH_3)_2N]_3PO$	–	5.3	7.3	–

[a] $k_2^{MeOH}/(l \cdot mol^{-1} \cdot s^{-1})$ is the second-order rate constant for the reaction in methanol; $k_2^{Solvent}$ is for reaction in the given solvent. $\lg k_2^{Solvent} = \lg (k_2^{Solvent}/k_2^{MeOH}) + \lg k_2^{MeOH}$.
[b] $\lg k_2^{MeOH} = -5.5$.
[c] $\lg k_2^{MeOH} = -5.1$.
[d] $\lg k_2^{MeOH} = -7.2$.
[e] $\lg k_2^{MeOH} = -3.85$ (50 °C).

trolytes, the solvation of nonelectrolytes should not be neglected, because important criteria are the differences in solvation. Many anionic activated complexes, especially those of S_NAr reactions, are expected to be highly polarizable and will have strong dispersion interactions in highly polarizable solvents such as dimethyl sulfoxide or *N,N*-dimethylformamide. On the other hand, due to the delocalization of the negative charge, they are weak hydrogen-bond acceptors and are not particularly well solvated by protic solvents. For example, solvation of large polarizable S_NAr transition state anions decreases in the solvent order $[(CH_3)_2N]_3PO$, $CH_3SOCH_3 > HCON(CH_3)_2 > CH_3OH > CH_3NO_2$, CH_3CN [291]. These differences in solvation of activated complexes may also contribute to the large rate accelerations obtained on protic/dipolar non-HBD solvent transfer for bimolecular substitution reactions*). However, in most cases S_N reactions are faster in dipolar non-HBD solvents because the reactant anion $Y^\ominus$ is much more solvated by protic than by dipolar non-HBD solvents and this outweighs any effects due to transition state anion or reactant molecule solvation [6].

Since small anions such as $F^\ominus$, $HO^\ominus$, and $CN^\ominus$ are often so poorly solvated by pure dipolar non-HBD solvents, it is difficult to find a soluble salt to act as the anion source.

* For an extensive discussion concerning the solvation of activated complexes of bimolecular substitution reactions for numerous variations of RX and $Y^\ominus$ as well as different mechanisms see references [6, 23, 291].

For example, electrolytes such as KF, KOH, and KCN are only slightly soluble in pure dimethyl sulfoxide. There are two possibilities in overcoming this experimental difficulty. First, the solubility of electrolytes in dipolar non-HBD solvents can be increased by using the corresponding tetraalkylammonium salts. Among these, tetraalkylammonium fluorides are the most prominent. In solutions of tetraalkylammonium fluorides in dipolar non-HBD solvents the fluoride ion is virtually unsolvated. These "naked" fluoride ions are both strong nucleophiles [599] and very effective bases [600]. They are more nucleophilic and basic than the fluoride ions of lower alkali metal fluorides (LiF, NaF, KF) in protic solvents. Secondly, protic/dipolar non-HBD solvent mixtures can be used as media for substitution reactions. Mixtures of dimethyl sulfoxide with water or alcohols have proven especially useful. This is because salts such as KOH, KF, KCN, CH_3ONa, and CH_3CO_2Na are reasonably soluble in solvent mixtures such as CH_3SOCH_3/H_2O (90 : 10) or CH_3SOCH_3/CH_3OH (90 : 10). Water or alcohols in combination with pure dipolar non-HBD solvents (*e.g.* 15 cl/l water in HMPT) provide a source of nucleophilic oxygen for the conversion of haloalkanes or sulfonate esters into the corresponding alcohols or ethers [601].

All the features discussed for protic/dipolar non-HBD solvent transfer are also observed for transfer from protic to dipolar non-HBD/protic mixture, but to a slightly lesser extent. In general, all S_N2 anion-molecule reactions are faster in mixtures than in pure protic solvents. They show a continuous, but not necessarily linear rate increase with an increase in the dipolar non-HBD component of the protic/dipolar non-HBD mixture [6]. Already small amounts of dipolar non-HBD component may cause a considerable acceleration in reaction rate.

The use of dipolar non-HBD instead of protic solvents as reaction media has often considerable practical synthetic advantages, which have been summarized by Parker [6], Madaule-Aubry [294], Liebig [295], and Illuminati [26]. A selection of common and less common dipolar non-HBD solvents is given in Table 5-18, together with some physical constants useful for their practical application. Reviews on particular dipolar non-HBD solvents have appeared; these are included in Table 5-18 (*cf.* also references [75–91] in Chapter 3).

Finally, some important examples emphasizing the versatility and synthetic utility of dipolar non-HBD solvents as reaction media will be given.

$RCO_2^{\ominus}$, an indifferent nucleophile in protic solvents, enjoys a large rate enhancement, permitting rapid alkylation with haloalkanes in hexamethylphosphoric triamide [301, 302]. When the Williamson ether synthesis is carried out in dimethyl sulfoxide [303], the yields are raised and the reaction time shortened. Displacements on unreactive haloarenes become possible [304] (conversion of bromobenzene to *tert*-butoxybenzene with *tert*-$C_4H_9O^{\ominus}$ in dimethyl sulfoxide in 86% yield at room temperature). The fluoride ion, a notoriously poor nucleophile or base in protic solvents, reveals its hidden capabilities in dipolar non-HBD solvents and is a powerful nucleophile in substitution reactions on carbon [305].

The observation, that bimolecular reactions of anions are often much faster in dipolar non-HBD solvents than in protic solvents of comparable dielectric constants is of great practical significance not only for substitution reactions but also for elimination, proton abstraction, and addition reactions [6].

Table 5-18. A selection of twenty-one organic dipolar non-HBD solvents in order of increasing dipole moment (*cf.* also Appendix, Table A-1).

Structure (⟷)[a)]	Name	$\mu/(10^{-30}$ C m)	t_{mp}/°C	t_{bp}/°C	References
$(H_3C)_2C{=}O$	Acetone	9.5	− 95.3	56.2	
$(Me_2N)_2C{=}O$	Tetramethylurea (Temur)	11.2	− 1.2	176.5	[602, 603]
$H_3C{-}N^{\oplus}(O^{\ominus}){=}O$	Nitromethane	11.9	− 28.6	101.2	[604]
cyclic $O{-}S(=O){-}O$	1,3,2-Dioxathiolane-2-oxide (Ethylene sulfite)	12.3		170–171 (dec.)	[605]
$Me_2N(H_3C)C{=}O$	*N,N*-Dimethylacetamide (DMAC)	12.4	− 20	166.1	[606]; *N,N*-Diethylacetamide: [633]
$Me_2N(H)C{=}O$	*N,N*-Dimethylformamide (DMF)	13.0	− 60.5	152.3	[606–608]
N-CH_3 pyrrolidinone	1-Methylpyrrolidin-2-one	13.6	− 16	202	[609]; 1-Ethylpyrrolidin-2-one: [633]

Table 5-18. (Continued)

Structure (+→)[a]	Name	$\mu/(10^{-30}$ C m)	t_{mp}/°C	t_{bp}/°C	References
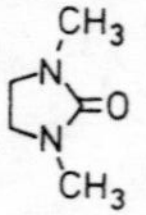	1,3-Dimethylimidazolidin-2-one (*N,N'*-Dimethylethylene urea, DMEU)	13.6	8.2	225.5	[603]
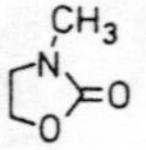	3-Methyloxazolidin-2-one	13.7	15.9	87–90 (1.3 mbar)	[610, 611]
$H_3C-C\equiv N$	Acetonitrile	13.7	− 45.7	81.6	[608, 612]
	Dimethyl sulfoxide	13.7	18.4	189.0	[608, 613–616]
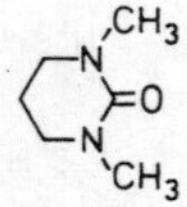	3,4,5,6-Tetrahydro-1,3-dimethyl-2(1H)-pyrimidinone (*N,N'*-Dimethylpropylene urea, DMPU)	14.1	< − 20	230	[603, 617–618]
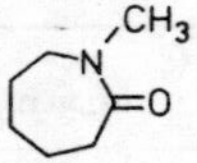	1-Methyl-hexahydroazepin-2-one, N-Methyl-ε-caprolactam	14.1		120 (25 mbar)	

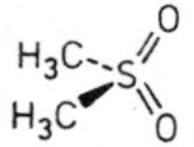	Dimethylsulfone	14.2	110	238	[619]
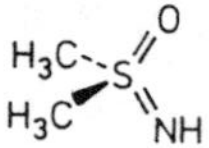	*S*,*S*-Dimethylsulfoximine (DMSOI)		52–53	100 (6.5 mbar)	[620]
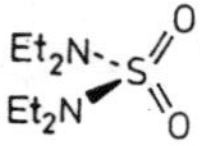	Tetraethylsulfamide (TES)	14.6[b)]		249–251	[621]
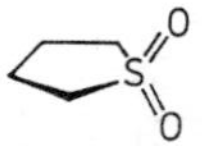	Tetrahydrothiophene-1,1-dioxide (Sulfolane)	15.7	28.4	285 (dec.)	[608, 622–624]
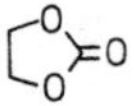	1,3-Dioxolan-2-one (Ethylene carbonate)	16.0	36	156	[625, 626]
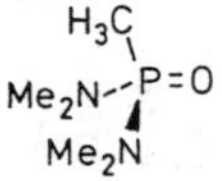	Methylphosphonic acid bis(dimethylamide)	16.0	– 3.2	62–63 (4 mbar)	[627, 628]
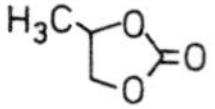	4-Methyl-1,3-dioxolan-2-one (Propylene carbonate)	16.5	– 49.2	241.7	[608, 625, 629, 630]
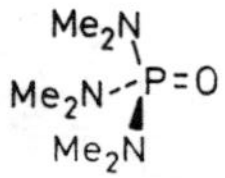	Hexamethylphosphoric acid triamide (HMPT)	18.5	7.2	235	[608, 617, 618, 631]

a) All formulae are oriented in such a way that the dipole moment of the solvent molecule is approximately in a parallel direction with this arrow.

b) Value for tetramethylsulfamide; H. Nöth, G. Mikulaschek, and W. Rambeck, Z. Anorg. Allg. Chem. *344*, 316 (1966).

Reduced solvation of commonly used E2 bases ($HO^{\ominus}$, $RO^{\ominus}$) in dipolar non-HBD solvents may elevate their reactivities to such an extent that E2 reactions of quite inert substrates occur [306]. Halide ions in dipolar non-HBD solvents are sufficiently strong bases to promote dehydrohalogenations of haloalkanes [73, 74]. Even the fluoride ion is the most efficient in this reaction [307, 308, 600]; the elimination rates decrease in the order $F^{\ominus} > Cl^{\ominus} > Br^{\ominus} > I^{\ominus}$.

A shift from $CH_3O^{\ominus}/CH_3OH$ to *tert*-$C_4H_9O^{\ominus}/CH_3SOCH_3$ can produce dramatic increases in rates (up to a factor of 10^9) of reactions which depend on proton abstraction from a C–H bond [31, 304]. The subsequent reaction of the carbanion may consist of isomerization, elimination, oxidation, and condensation, hence demonstrating the importance of the potassium *tert*-butoxide/dimethyl sulfoxide system in organic syntheses. The base-catalyzed alkene isomerization reaction (double bond migration) has been used to investigate efficacity of a wide variety of base/dipolar non-HBD solvent systems [309, 310].

Addition of anionic nucleophiles to alkenes and to heteronuclear double bond systems (C=O, C=S) also lie within the scope of this Section. Chloride and cyanide ions are efficient initiators of the polymerization and copolymerization of acrylonitrile in dipolar non-HBD solvents as reported by Parker [6]. Even some 1,3-dipolar cycloaddition reactions leading to heterocyclic compounds are often better carried out in dipolar non-HBD solvents in order to increase rates and yields [311]. The rate of alkaline hydrolysis of ethyl and *p*-nitrophenyl acetate in dimethyl sulfoxide/water mixtures, increases with increasing dimethyl sulfoxide concentration due to the increased activity of the hydroxide ion. This is presumably caused by its reduced solvation in the dipolar non-HBD solvent [312, 313]. Dimethyl sulfoxide greatly accelerates formation of oximes from carbonyl compounds and hydroxylamine as shown for substituted 9-oxofluorenes [314]. Nucleophilic attack on carbondisulfide by cyanide ion is possible only in *N,N*-dimethylformamide [315].

The superoxide ion, $O_2^{\ominus}\cdot$, produced by the electron-transfer reduction of dioxygen ($O_2 + e^{\ominus} \rightleftharpoons O_2^{\ominus}\cdot$), is a strong Brønsted base and an effective nucleophile. Because of rapid hydrolysis and disproportionation, the lifetime of $O_2^{\ominus}\cdot$ in aqueous solution is limited. This has led to investigations of its reaction chemistry in dipolar non-HBD solvents [632]. Under these conditions, the superoxide ion attacks haloalkanes via S_N2 displacement of the halides to eventually give dialkyl peroxides in a multi-step reaction [632].

5.5.3 Quantitative Separation of Protic and Dipolar Aprotic Solvent Effects for Reaction Rates by means of Solvent-Transfer Activity Coefficients

By considering the changes in ***standard molar Gibbs energy of solvation***, ΔG°_{solv}, of reactants and activated complexes, when a reaction is transferred from one solvent to another, it is possible to approach the problem from a quantitative point of view. The observed changes in Gibbs energy of solvation when an electrolyte $M^{\oplus}X^{\ominus}$ is transferred from a reference solvent O to another solvent S is commonly reported as the *solvent-transfer activity* *coefficient* $^{O}y^{S}$ *) [6, 291, 316–319, 454, 634, 635]. This coefficient makes it possible to relate the difference in chemical potential, μ, of a solute MX at infinite dilution on transferring it from an arbitrarily chosen reference solvent O to another solvent S at temperature T according to Eq. (5-114) **):

$$\mu_{MX}^{\infty}(\text{in S}) = \mu_{MX}^{\infty}(\text{in O}) + RT \cdot \ln {}^{O}y_{MX}^{S} \qquad (5\text{-}114)$$

A positive value of $\ln {}^{O}y_{MX}^{S}$ implies that the solute MX is better solvated by solvent O than by solvent S. A negative value indicates that solvation by solvent S is greater. Values of $\ln {}^{O}y_{MX}^{S}$ are obtained from various measurements, *e.g.* solubility, distribution, vapour pressure, and electrochemical measurements. Although solvent-transfer activity coefficients can be defined for individual ions $M^{\oplus}$ and $X^{\ominus}$ they cannot be independently measured. In order to divide the solvent-transfer activity coefficient ${}^{O}y_{MX}^{S}$ into separate contributions ${}^{O}y_{M^{\oplus}}^{S}$ and ${}^{O}y_{X^{\ominus}}^{S}$, an extrathermodynamic assumption must be considered; *cf.* Chapter 2.3 and [454, 634, 635].

The most convenient and generally accepted extrathermodynamic assumption is that ${}^{O}y_{Ph_4As^{\oplus}}^{S} = {}^{O}y_{Ph_4B^{\ominus}}^{S}$, using $Ph_4As^{\oplus}Ph_4B^{\ominus}$ as the reference electrolyte ($Ph = C_6H_5$). The tetrahedral ions of tetraphenylarsonium tetraphenylboride are of comparable structure and size. It is proposed, therefore, that the anion and the cation are similarly influenced on transfer from one solvent to another. This assumption makes it possible to calculate reasonable values for single-ion solvent-transfer activity coefficients (and single-ion Gibbs energies of transfer; *cf.* Table 2-9 in Chapter 2.3) between solvent pairs. Table 5-19 shows some selected values of $\lg {}^{O}y^{S}$ for anions and cations. These have been taken from Marcus's extensive critical compilation [634, 635]. All $\lg {}^{O}y^{S}$-values are expressed relative to water as the reference solvent O at 25 °C.

The striking increase in $\lg {}^{W}y_{ion}^{S}$-values for small anions with high charge density, on going from water to any of the dipolar non-HBD solvents, is consistent with the large rate increase observed in reactions of these anions and gives quantitative significance to the qualitative discussion in Section 5.5.2. The rather large increase in the solvent-transfer activity coefficient of $CH_3CO_2^{\ominus}$, compared with $Cl^{\ominus}$, for the isodielectric solvent change from CH_3OH to CH_3CN and $HCON(CH_3)_2$, reflects the considerable stabilization of this anion through hydrogen-bonding in both methanol and water. As expected, the larger and more polarizable anions show considerably smaller increases in $\lg {}^{W}y_{ion}^{S}$-values on transfer

---

* Also called the *medium effect*, *solvent activity coefficient*, or *transfer activity coefficient*, and also written as $y_t^{\infty}(MX, O \rightarrow S)$. It is a constant characteristic of the solute MX (or the solute ions $M^{\oplus}$ and $X^{\ominus}$) and the two solvents O and S.

** The change in standard molar Gibbs energy on transferring an electrolyte MX from one solvent to another may also be expressed as the *standard molar Gibbs energy of transfer*, ΔG_t° [6, 454, 634, 635]. The two quantities, solvent-transfer activity coefficient, ${}^{O}y^{S}$, and Gibbs energy of transfer, ΔG_t°, are related by the the following simple equation:

$$\Delta G_t^{\circ}(MX, O \rightarrow S) = \mu_{MX}^{\infty}(\text{in S}) - \mu_{MX}^{\infty}(\text{in O}) = RT \cdot \ln({}^{O}y_{M^{\oplus}}^{S} \cdot {}^{O}y_{X^{\ominus}}^{S}),$$

where $\mu_{MX}^{\infty}(\text{in S})$ and $\mu_{MX}^{\infty}(\text{in O})$ are the standard chemical potentials of MX referred to at infinite dilution in solvents S and O, respectively. In this Section, results are quoted as solvent-transfer activity coefficients using the molar concentration scale; *cf.* Eq. (2-12a) in Chapter 2.3.

Table 5-19. Some representative values of solvent-transfer activity coefficients for anions and cations at 25 °C with water (W) as reference solvent; molar concentration scale[a)].

Ions	lg $^{W}y^{S}$ for solvent S[b)]						
	S = CH_3OH	$HCONH_2$	CH_3CN	CH_3COCH_3	CH_3SOCH_3	$HCON(CH_3)_2$	$[(CH_3)_2N]_3PO$
$F^{\ominus}$	2.8	4.4	12.4	—	—	8.9	—
$Cl^{\ominus}$	2.3	2.4	7.4	10.0	7.0_5	8.5	10.2
$Br^{\ominus}$	1.9_5	1.9	5.5	7.4	4.8	6.3	8.0_5
$I^{\ominus}$	1.3	1.3	2.9	4.4	1.8	3.6	5.2_5
$N_3^{\ominus}$	1.6	1.9	6.5	7.5	4.5	6.3	8.6
$CN^{\ominus}$	1.5	2.3	6.1	8.4	6.1	7.0	—
$SCN^{\ominus}$	0.98	1.2	2.5	—	1.7	3.2	3.5
$ClO_4^{\ominus}$	1.1	−2.1	0.35	—	—	0.70	− 1.2
$CH_3CO_2^{\ominus}$	2.8	3.5	10.7	—	—	11.6	—
$(C_6H_5)_4B^{\ominus}$	−4.2	−4.2	− 5.7_5	− 5.6	-6.5_5	− 6.7_5	− 6.8
$Li^{\oplus}$	0.77	-1.7_5	4.4	—	−2.6	− 1.7_5	—
$Na^{\oplus}$	1.4	−1.4	2.6_5	—	-2.3_5	− 1.7	—
$K^{\oplus}$	1.7	−0.75	1.4	0.70	−2.3	− 1.8	− 2.8
$Ag^{\oplus}$	1.1_5	−2.7	− 4.0_5	1.6	−6.1	− 3.6_5	− 7.7
$(CH_3)_4N^{\oplus}$	1.0_5	—	0.53	0.53	−0.35	− 0.93	—
$(C_6H_5)_4As^{\oplus}$	−4.2	−4.2	− 5.7_5	− 5.6	-6.6_5	− 6.7_5	− 6.8

[a] Values recalculated from Table 2.4 in Y. Marcus, Pure Appl. Chem. *55*, 977 (1983) [634].

[b] A positive value of lg $^{W}y^{S}$ indicates that the ion is better solvated by water than by solvent S; a negative value means that the ion is more strongly solvated by transferring it to the solvent S. A difference in lg $^{W}y^{S}$ of unity corresponds to a difference in Gibbs energy of solvation of $\ln 10 \cdot RT \cdot \lg {}^{W}y^{S}$, *i.e.* 5.7 kJ/mol at 25 °C.

from water to dipolar non-HBD solvents (*e.g.* $I^{\ominus}$, $SCN^{\ominus}$). An extreme example of the difference between small, non-polarizable and large, polarizable anions is seen in the lg ${}^{W}y^{S}$-values of $F^{\ominus}$ and $(C_6H_5)_4B^{\ominus}$ for the solvent change $CH_3OH \rightarrow HCON(CH_3)_2$: lg ${}^{W}y^{DMF}_{F^{\ominus}}$ and lg ${}^{W}y^{DMF}_{Ph_4B^{\ominus}}$ differ by 8.65 units, which corresponds to a difference in Gibbs energy of transfer of 49 kJ/mol.

In addition, it is worth emphasising the contrasting behaviour of anions and cations on going from water to protic and to dipolar non-HBD solvents. Cations are generally smaller and less polarizable than anions and they are not hydrogen-bond acceptors. Some cations, however, are pronounced Lewis acids and are substantially better solvated in solvents which contain basic oxygen atoms and behave like EPD solvents (*cf.* Section 3.3.2). Therefore, small closed-shell cations such as $Li^{\oplus}$, $Na^{\oplus}$, and $K^{\oplus}$ are much more extensively solvated by solvents such as $[(CH_3)_2N]_3PO$, $HCON(CH_3)_2$, and CH_3SOCH_3, as well as by water than by methanol. The silver cation has a strong, apparently specific interaction with solvents such as $[(CH_3)_2N]_3PO$, $HCON(CH_3)_2$, CH_3SOCH_3, and CH_3CN. The interaction of $Ag^{\oplus}$ with CH_3COCH_3 and CH_3OH is much weaker. Large organic cations [*e.g.* $(CH_4)_4N^{\oplus}$, $(C_6H_5)_4As^{\oplus}$] are very poorly solvated by water but are much better solvated by dipolar non-HBD solvents.

When solvent-transfer activity coefficients are applied to reaction rates in terms of the absolute rate theory, where reactions are considered to involve an equilibrium between the reactants and the activated complex, Eq. (5-115) is obtained for a bimolecular reaction such as $Y^{\ominus} + RX \rightleftharpoons [YRX^{\ominus}]^{\neq} \rightarrow YR + X^{\ominus}$ [6, 291, 292].

$$\lg (k^{S}/k^{O}) = \lg {}^{O}y^{S}_{Y^{\ominus}} + \lg {}^{O}y^{S}_{RX} - \lg {}^{O}y^{S}_{[YRX^{\ominus}]^{\neq}} \qquad (5\text{-}115)^{*)}$$

The specific rate constant in a solvent S for this bimolecular reaction is related to the rate constant in the reference solvent O through the appropriate solvent-transfer activity coefficients. Eq. (5-115) shows to what extent solvent effects on the reaction rate are due to changes in the solvation of reactant anions, $Y^{\ominus}$, of reactant nonelectrolytes, RX, and of anionic activated complexes, $[YRX^{\ominus}]^{\neq}$. Anionic and uncharged activated complexes will behave in exactly the same way upon solvent transfer as "real" anions and non-electrolytes of comparable structure. Anionic acitvated complexes such as $[YRX^{\ominus}]^{\neq}$ should behave like large polarizable anions, and, therefore, should be better solvated in polarizable, dipolar solvents than in protic solvents.

The rate constants k in the reference solvent O and in solvent S can be readily measured. Values of lg ${}^{O}y^{S}$ for RX and $Y^{\ominus}$ are obtained as described above. Thus, by applying Eq. (5-115) lg ${}^{O}y^{S}$-values for the activated complex can be calculated, providing information about the structure and charge distribution of activated complexes. Values of lg ${}^{O}y^{S}_{[YRX^{\ominus}]^{\neq}}$ from Eq. (5-115) are compared with lg ${}^{O}y^{S}$-values for relevant solutes, or for

---

* An alternative way to analyze the solvent effect on this S_N2 reaction in terms of Gibbs energy of transfer, ΔG_t°, of reactants and activated complex from reference solvent O to another solvent S is given by the following equation [6, 23, 64, 292]:

$$\Delta G_t^{\neq} = \Delta G_t^{\circ}(Y^{\ominus}) + \Delta G_t^{\circ}(RX) - \Delta G_t^{\circ}([YRX^{\ominus}]^{\neq}).$$

For an attempt to correlate solvent transfer Gibbs energies of activation, ΔG_t°, of some S_N1 and S_N2 reactions, with the donor and acceptor numbers of the solvent, see reference [292a].

other well-established activated complexes, whose structure and charge distribution are such that they might act as models for the activated complex under consideration. The model with a lg ${}^{O}y^{S}$-value close to lg ${}^{O}y^{S}_{[YRX^{\ominus}]^{\neq}}$ may be used as a guide to the structure of $[YRX^{\ominus}]^{\neq}$. Parker and coworkers have estimated the lg ${}^{O}y^{S}$-values of activated complexes for numerous variations of RX, $Y^{\ominus}$ and mechanisms (*e.g.* S_N2, S_NAr) for substitution reactions of various charge types [6, 291, 292]. Using the "Gibbs energy of transfer" approach*), similar calculations of solvent effects on the Gibbs energy of reactants and activated complexes have been collected by Abraham [23], especially for Menschutkin reactions [64].

Table 5-20. Relative reaction rates and solvent-transfer activity coefficients[a)] for reactants and activated complex of S_N2 reaction (5-116) upon solvent transfer methanol (abbreviated to M) to *N,N*-dimethylformamide (DMF) at 25 °C [6, 291].

Leaving Group X	$\lg\left(\frac{k_2^{DMF}}{k_2^{M}}\right)$	$\lg {}^{M}y^{DMF}_{N_3^{\ominus}}$	$\lg {}^{M}y^{DMF}_{CH_3X}$	$\lg {}^{M}y^{DMF}_{[N_3CH_3X^{\ominus}]^{\neq}}$	$\lg {}^{M}y^{DMF}_{X^{\ominus}}$
$Cl^{\ominus}$	3.3	4.9	−0.4	1.2	6.5
$Br^{\ominus}$	3.9	4.9	−0.3	0.7	4.9
$I^{\ominus}$	4.6	4.9	−0.5	−0.2	2.6
$TsO^{\ominus}$	2.0	4.9	−0.6	2.3	3.5

[a] The more positive the value of lg ${}^{M}y^{DMF}$, the greater the solvation by methanol relative to *N,N*-dimethylformamide.

Examples of this procedure are shown in Table 5-20. These data for the S_N2 reaction (5-116) are taken from Parker *et al.* [6, 291], whose extensive compilation compares relative rates and solvent-transfer coefficients on transfer from the reference solvent methanol to the isodielectric solvent *N,N*-dimethylformamide at 25 °C.

$$N_3^{\ominus} + CH_3\text{-}X \xrightleftharpoons{k_2} \left[\overset{\delta\ominus}{N_3}\cdots CH_3\cdots\overset{\delta\ominus}{X}\right]^{\neq} \longrightarrow N_3\text{-}CH_3 + X^{\ominus} \qquad (5\text{-}116)$$

On transfer from methanol to *N,N*-dimethylformamide, the increase in the rate of these four analogous S_N2 reactions varies from $10^{2.0}$ to $10^{4.6}$, the rate being faster in the dipolar non-HBD solvent. Whereas the medium has only a small effect on the CH_3X Gibbs energy (these molecules being slightly better solvated in *N,N*-dimethylformamide almost independently of X), the anion $N_3^{\ominus}$ is substantially destabilized on transfer from CH_3OH to $HCON(CH_3)_2$. The increase in rate, superimposed by a relative increase in solvation of the activated complex in *N,N*-dimethylformamide in the order $[N_3CH_3OTs^{\ominus}]^{\neq} < [N_3CH_3Cl^{\ominus}]^{\neq} < [N_3CH_3Br^{\ominus}]^{\neq} < [N_3CH_3I^{\ominus}]^{\neq}$, clearly reflects this destabilization. It is, however, modified by a small destabilization of the activated complex when X = $TsO^{\ominus}$, $Cl^{\ominus}$, and $Br^{\ominus}$, thus showing the anionic nature of the activated complex. For X = halide this increase in stability correlates with the increasing polarizability of the activated complexes and corresponds to the increase in polarizability from $Cl^{\ominus} < Br^{\ominus} < I^{\ominus}$. Differences in solvation of activated complexes do not appear to contribute greatly to lg (k_2^{DMF}/k_2^{M}) when halide ions are the leaving groups. However, the comparatively low

* See footnote on page 227.

rate increase obtained for methyl tosylate must be due to the more positive value of lg $^{M}y^{DMF}_{[N_3CH_3OTs^{\ominus}]^{\neq}}$. Because the tosylate ion itself has a less positive lg $^{M}y^{DMF}$-value than does $Br^{\ominus}$ or $Cl^{\ominus}$, yet the activated complex containing tosylate has lg $^{M}y^{DMF}_{\neq}$ more positive than the activated complexes containing halide. This effect has been explained by assuming that a looser activated complex with a stronger localization of negative charge is present in the case of tosylate. Transition state anions with increased localization of negative charge at leaving and entering groups, become increasingly more solvated, for reasons already given. Departing tosylate is a better hydrogen-bond acceptor relative to free tosylate ion than is departing halide relative to free halide ion. Among other observations this leads to the generalization that, other things being equal, those ($Y^{\ominus}$ + RX)-reactions possessing the "tightest" activated complex (*i.e.* most covalently bound Y and X) give the greatest accelerations on transfer from protic to dipolar non-HBD solvents. For a more detailed description of "tight and loose S_N2 activated complexes" see the reviews of Parker [6, 291].

Assuming that lg $^{O}y^{S}_{RX}$ − lg $^{O}y^{S}_{[YRX^{\ominus}]^{\neq}}$ is roughly constant for closely related reactions, Parker obtained, from Eq. (5-115), the simple linear Gibbs energy relationship (5-117),

$$\lg (k^{S}/k^{O}) = \lg {}^{O}y^{S}_{Y^{\ominus}} + C \tag{5-117}$$

This involves a limited number of constants *C*, varying only with the nature of the substrates [6, 291]. It is possible to use Eq. (5-117) to estimate the rate constant for a given reaction in a new solvent S from its value in another reference solvent O, to within a factor of two. This is very good in view of the fact that rates can vary by a factor up to 10^{10} on switching solvents. Assuming the validity of Eq. (5-117), the easiest way of estimating lg $^{O}y^{S}_{Y^{\ominus}}$ is to measure a reaction rate for $Y^{\ominus}$ in two different solvents. A test of Eq. (5-117) for the S_N2 reaction between bromide ion and methyl tosylate in twelve solvents is given in reference [68].

Further examples of the dissection of initial state and transition state medium effects for reactions in protic and dipolar non-HBD solvents have been given by Buncel [467, 636], Abraham [23, 64, 637], Haberfield [638], and Blandamer *et al.* [639].

5.5.4 Acceleration of Base-Catalysed Reactions in Dipolar Aprotic Solvents

The specific solvation of anions in protic solvents, mediated mainly by hydrogen-bonding, diminishes not only their nucleophilic reactivity but also their basicity. All anions therefore are much stronger bases in dipolar non-HBD solvents than in protic solvents. This substantial increase in kinetic and thermodynamic basicity of anions in dipolar non-HBD solvents is very profitable in preparative organic chemistry in view of the activation of weakly acidic C—H, O—H, and N—H bonds. Especially useful for this purpose are solutions of potassium *tert*-butoxide in dimethyl sulfoxide, which is among the most basic media available for the organic chemist. The basicity of the *tert*-$C_4H_9O^{\ominus}/CH_3SOCH_3$ system is comparable to that of the $NH_2^{\ominus}$/liquid NH_3 system. Extremely basic solutions are also generated by the addition of dimethyl sulfoxide to aqueous or alcoholic solutions of alkali hydroxides or tetraalkylammonium hydroxides*). Substitution of dimethyl sulfoxide for alcohols or water, not only replaces the proton donor physically, but deactivates the remaining protic solvent by virtue of its own great H-bond acceptor capability according to Eq. (5-118).

$$RO^{\ominus}\cdots H{-}OR + (CH_3)_2\overset{\delta\oplus}{S}{=}\overset{\delta\ominus}{O} \rightleftharpoons RO^{\ominus} + (CH_3)_2\overset{\delta\oplus}{S}{=}\overset{\delta\ominus}{O}\cdots H{-}OR \qquad (5\text{-}118)$$

Many reactions become possible only in such superbasic solutions, or else they can be carried out under much milder conditions. Only some examples of preparative interest (which depend on the ionization of a C—H or N—H bond) will be mentioned here. The subsequent reaction of the resulting carbanion may involve electrophilic substitution, isomerization, elimination, or condensation [321, 322]. Systematic studies of solvent effects on intrinsic rate constants of proton-transfer reactions between carbon acids and carboxylate ions as well as amines as bases in various dimethyl sulfoxide/water mixtures have recently been carried out by Bernasconi *et al.* [769].

Rates of isotopic H/D exchange and racemization of optically active 2-methyl-3-phenylpropionitrile in the presence of potassium methoxide can be increased by a factor of $5 \cdot 10^7$ by substitution of dimethyl sulfoxide ($\varepsilon_r = 46.7$) for pure methanol ($\varepsilon_r = 32.7$) [31, 304, 231].

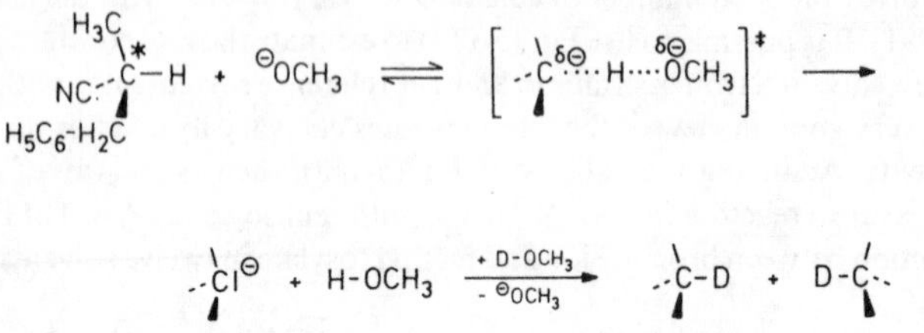

$$(H_3C)(NC)(H_5C_6{-}H_2C)\overset{*}{C}{-}H + {}^{\ominus}OCH_3 \rightleftharpoons \left[\ \overset{\delta\ominus}{C}\cdots H\cdots\overset{\delta\ominus}{O}CH_3\ \right]^{\ddagger} \longrightarrow$$

$$\overset{}{C}l^{\ominus} + H{-}OCH_3 \xrightarrow[-\,{}^{\ominus}OCH_3]{+\,D{-}OCH_3} C{-}D + D{-}C \qquad (5\text{-}119)$$

Solvent	CH_3OH/CH_3SOCH_3						
CH_3SOCH_3 in cg/g	0	25	50	76	90	98.5	100
k_2^{rel} (25 °C)	1	32	160	4900	$1.3 \cdot 10^5$	$5.0 \cdot 10^7$	ca. 10^9 [a]

[a] extrapolated

Much of the progress made in the area of base-catalysed alkene isomerization resulted from the use of dipolar aprotic solvents. Under homogeneous conditions this brings about high reactions rates at low temperatures [309, 310, 323, 324]. For example,

---

* The use of pure dimethyl sulfoxide with alkoxide ($CH_3O^{\ominus}$, *tert*-$C_4H_9O^{\ominus}$) or hydride ($H^{\ominus}$) ions leads to even more strongly basic systems which contain in part the very strongly basic dimethylsulfinyl ("dimsyl") anion, formed in the following equilibrium reaction:

$$RO^{\ominus} + CH_3{-}SO{-}CH_3 \rightleftharpoons RO{-}H + {}^{\ominus}|CH_2{-}SO{-}CH_3$$

the base-catalysed rearrangement of allyl to propenyl ethers [325], and of allyl to propenyl amines [326] is best carried out using *tert*-C_4H_9OK/CH_3SOCH_3 as reaction medium [323]. The alkenylation and aralkylation of a variety of alkylaromatic compounds in a homogeneous *tert*-C_4H_9OK/dipolar aprotic solvent system is greatly influenced by the solvent. In the case of the reaction between 4-isopropylpyridine (Ar = 4-pyridyl) and isoprene according to Eq. (5-120), the rate is faster in dimethyl sulfoxide ($t_{1/2} = 1.15$ min!) [324].

$$(H_3C)_2C(H)Ar + CH_2{=}C(CH_3){-}CH{=}CH_2 \xrightarrow[20\,^{\circ}C]{(t\text{-}C_4H_9O^{\ominus}K^{\oplus}/DMSO)} \underset{59\ \text{cmol/mol}}{Ar{-}C(CH_3)_2{-}CH_2{-}C(CH_3){=}CH{-}CH_3} + \underset{41\ \text{cmol/mol}}{(CH_3)_2C{=}CH{-}CH_2{-}C(CH_3)_2{-}Ar} \qquad (5\text{-}120)$$

The Wittig reaction proceeds more rapidly in dimethyl sulfoxide than in other customary solvents due to the enhanced deprotonation of the phosphonium salts to alkylidenephosphoranes [327]. Even camphor, a rather unreactive ketone, undergoes the Wittig reaction easily in base/CH_3SOCH_3 systems [327].

Similar spectacular increases in rate were found for the transformation of ketone hydrazones into hydrocarbons using the Wolff-Kishner reduction. This is mainly due to the enhanced ionization of the N—H bond. This reaction occurs even at room temperature in *tert*-C_4H_9OK/CH_3SOCH_3 and proceeds *via* solvent-mediated transfer of a proton from nitrogen to carbon [328, 329].

$$RR'C{=}N{-}NH_2 \xrightarrow[-ROH]{+RO^{\ominus}} [\,{>}C{\overset{\ominus}{\cdots}}N{\cdots}N{-}H\,] \xrightarrow{+ROH} [\text{cyclic transition state with } H{\cdots}O(R){\cdots}H]$$

$$\xrightarrow{-ROH} {>}CH{-}N{=}N^{\ominus} \xrightarrow{-N_2} {>}CH|^{\ominus} \xrightleftharpoons[-RO^{\ominus}]{+ROH} RR'CH_2 \qquad (5\text{-}121)$$

Since the hydrazone anion is stable in the absence of a proton donor, the overall reaction must depend on the presence of an alcohol molecule in the vicinity of the hydrazone anion. This is in agreement with the observation, that rate constants obtained in ROH/CH_3SOCH_3 mixtures, as a function of dimethyl sulfoxide concentration, exhibit a maximum at low alcohol concentration [330]. The rate-determining step of this reaction appears to be a concerted proton-transfer from the protic solvent to the hydrazone anion and a proton abstraction from the hydrazone anion by the basic solvent [329].

Since the Cope elimination reaction of amine oxides proceeds with intramolecular proton abstraction by the *N*-oxide group according to Eq. (5-122), it is not surprising, that with dimethyl sulfoxide as solvent the reaction occurs at room temperature [328, 331]. Protic solvents require temperatures of 120 to 150 °C because hydrogen bonding of the oxygen terminal makes the reaction more difficult.

$$\text{(cyclic amine oxide, } R_2N^{\oplus}\text{–}O^{\ominus}\text{)} \xrightarrow{\Delta} \text{alkene} + HO\text{-}NR_2 \tag{5-122}$$

Further examples of strong basic solvent systems are solutions of ionic fluorides in dipolar non-HBD solvents such as acetonitrile, *N,N*-dimethylformamide, dimethyl sulfoxide, and tetrahydrofuran. Tetraalkylammonium fluorides $R_4N^{\oplus}F^{\ominus}$ (R_4 = tetraethyl, tetra-*n*-butyl, benzyltrimethyl), which are soluble in dipolar non-HBD solvents, have received the most attention. Since they are highly hygroscopic they are usually associated with varying amounts of protic solvents (mainly water). The amount of these protic solvents influences the effective basicity of the fluoride ion by hydrogen bonding ($RO—H \cdot\cdot F^{\ominus}$). If only traces of the protic solvents are present then the fluoride ion can even abstract a proton from such weakly acidic solvents as acetonitrile, dimethyl sulfoxide, and nitromethane [640].

Among the fluoride ion promoted reactions which occur in dipolar non-HBD solvents are alkylations of alcohols and ketones, esterifications, Michael additions, aldol and Knoevenagel condensations as well as eliminations; for a review see reference [600]. In particular, ionic fluorides are useful in the dehydrohalogenation of haloalkanes and haloalkenes to give alkenes and alkynes (order of reactivity $R_4N^{\oplus}F^{\ominus} > K^{\oplus}$([18] crown-6)$F^{\ominus} > Cs^{\oplus}F^{\ominus} \approx K^{\oplus}F^{\ominus}$). For example, tetra-*n*-butylammonium fluoride in *N,N*-dimethylformamide is an effective base for the dehydrohalogenation of 2-bromo- and 2-iodobutane under mild conditions [641]; *cf.* Eq. (5-123).

$$H_3C\text{-}CH_2\text{-}\underset{\displaystyle I}{CH}\text{-}CH_3 \xrightarrow[\substack{50\,°C;\ 10\ \text{min} \\ \text{total yield 77 cmol/mol}}]{+\ n\text{-}Bu_4N^{\oplus}F^{\ominus}\ \text{in DMF}} \underset{89\ \text{cmol/mol}}{H_3C\text{-}CH{=}CH\text{-}CH_3} + \underset{11\ \text{cmol/mol}}{H_3C\text{-}CH_2\text{-}CH{=}CH_2} \tag{5-123}$$

Wittig reactions of phosphonium fluorides in dipolar non-HBD solvents can be carried out without an additional base: *e.g.* 4-nitrobenzaldehyde reacts slowly with (4-nitrobenzyl)triphenylphosphonium fluoride in refluxing acetonitrile to give 4,4'-nitrostilbene in good yield (84 cmol/mol) [642]. In this reaction, the fluoride ion attacks its own cation to give the corresponding Wittig ylide as intermediate.

Further examples of solvent effects on base-catalysed reactions can be found in references [297–300, 321, 322, 600].

5.5.5 Influence of Specific Cation Solvation on Reaction Rates of S_N Reactions

The nucleophilic reactivity of an anion depends not only on the extent of its specific solvation, but also on the degree of association with the corresponding cation. An ion-pair associated anion (or cation) is much less reactive than a free, non-associated ion*). As early

* A covalent compound dissociates into free ions in stages involving the formation of at least two types of ion pairs, contact and solvent-separated ion pairs; *cf.* Section 2.6 and Eqs. (2-19) to (2-21) [289, 333].

as 1912, Acree postulated that the reactivity of an anionic nucleophile should be depressed when its salt was incompletely dissociated [332]. Due to incomplete dissociation of the ionophore the reaction rate constant will fall as its concentration increases. The simple model given in Eq. (5-124) is consistent with the observation, that in all cases ion association deactivates the nucleophile [289].

$$M^{\oplus}Y^{\ominus} + \text{C–X} \underset{}{\overset{k_2^p}{\rightleftharpoons}} \left[M^{\oplus}\overset{\delta\ominus}{Y}\cdots\text{C}\cdots\overset{\delta\ominus}{X}\right]^{\ddagger} \longrightarrow \text{products} \qquad (5\text{-}124a)$$

Association ⇅ Dissociation

$$M^{\oplus} + Y^{\ominus} + \text{C–X} \underset{}{\overset{k_2^f}{\rightleftharpoons}} \left[\overset{\delta\ominus}{Y}\cdots\text{C}\cdots\overset{\delta\ominus}{X}\right]^{\ddagger} \longrightarrow \text{products} \qquad (5\text{-}124b)$$

Formation of the activated complex from an ion pair differs from formation from a free anion, in that in reaction (5-124a) the activation process suffers from a loss in coulombic interaction energy between the centres of opposite charges. Therefore, in the presence of $M^{\oplus}$ the activation energy is increased. This model also suggests that the ratio k_2^f/k_2^p should increase with increasing intensity of cation-anion interaction [289]. Ion pairs are more stable the higher the charge densities of the component ions are, and hence the stronger the electrostatic attraction is. Accordingly, the nucleophilicity of anions in weakly dissociating solvents (*i.e.* solvents with low dielectric constants; *cf.* Sections 2.6 and 3.2) depends on the nature of the binary salt, and therefore on the cation. The relative reactivity of free and associated ions has been examined in detail, not only for nucleophilic aliphatic substitution [333–335] but also for electrophilic aliphatic substitution reactions. In the latter, carbanionic ion pairs or compounds are involved [321, 322, 336–339]. The reactivity of carbanions is affected in the same way by the degree of interaction between the carbanion and the counter-ion.

According to Coulomb's law [*cf.* Eq. (2-17) in Section 2.6], the interaction between cation and anion can be minimized by increasing either the dielectric constant of the medium, or the interionic distance. There are several ways to separate anions from cations and thus increasing the anionic reactivity [643]:

(i) enlarge the size of the opposite cation by substituting bulky 'onium cations such as quaternary tetraalkylammonium ions for the small alkali cations;

(ii) use EPD solvents of high Lewis basicity as reaction media or additives because they strongly solvate small alkali cations, thus increasing the interionic distance; also they weakly solvate anions;

(iii) encage cations by means of macro(poly)cyclic ligands such as crown ethers or cryptands. Cation complexation leads to ion-pair dissociation via *ligand-separated ion pairs*.

Examples of each of these three types of anion activation*) are given.

---

* The term anion activation suggests that methods (i)...(iii) cause an increase in reaction rate. This does not hold for reactions in which the ion pair is more reactive than the free anion as a result of a cation-assisted reaction pathway. Methods (i)...(iii) exhibit cation *de*activation in conjunction with the anion activation!

An example of the S_N2 rate dependency on the nature of the counter-ion is given by the reaction of *n*-butyl *p*-bromobenzene sulfonate with lithium and tetra-*n*-butylammonium halides in the weakly dissociating solvent acetone ($\varepsilon_r = 20.7$) [279].

$$CH_3CH_2CH_2CH_2\text{-OBros} + X^{\ominus} \xrightarrow[25\,^{\circ}C]{k_2} CH_3CH_2CH_2CH_2\text{-X} + {}^{\ominus}\text{OBros} \qquad (5\text{-}125)$$

Anion $X^{\ominus}$		$I^{\ominus}$		$Br^{\ominus}$		$Cl^{\ominus}$
k_2^{rel}	$Li^{\oplus}$ salt[a]	6.2	>	5.7	>	1.0
		∨		∧		∧
	$(n\text{-}C_4H_9)_4N^{\oplus}$ salt[a]	3.7	<	18	<	68

[a] Salt concentration 0.04 mol/l

The nucleophilic reactivity of the lithium salts changes in the same order as in protic solvents ($I^{\ominus} > Br^{\ominus} > Cl^{\ominus}$; *cf.* Table 5-15). However, the order is completely reversed for the ammonium salts ($Cl^{\ominus} > Br^{\ominus} > I^{\ominus}$), and this latter order is the same as that found in dipolar non-HBD solvents such as *N,N*-dimethylformamide [278]. The small lithium cation with its high charge density has a strong tendency to form ion pairs with anions, whereas the electrostatic interaction between the large tetraalkylammonium ion and anions is comparatively weak. Quaternary ammonium salts, therefore, should be practically fully dissociated in acetone solution. Thus the reactivity order obtained with these salts corresponds to that of the free, non-associated halide ions. On the other hand, the sequence obtained with the lithium salts also reflects the dissociation equilibria of these salts in acetone solution [279].

Similarly, the reactivity of the phenoxide ions as the tetra-*n*-butylammonium salt is shown to be $3 \cdot 10^4$-fold higher than that of the corresponding potassium salt in the S_N2 alkylation reaction with 1-halobutanes, carried out in 1,4-dioxane [340]. Whereas the rate of alkylation of the potassium salt increases by a factor of *ca.* 10^5 on going from 1,4-dioxane ($\varepsilon_r = 2.2$) to *N,N*-dimethylformamide ($\varepsilon_r = 37.0$), the alkylation rate of the quaternary phenoxide is essentially insensitive to the same solvent change. Obviously, the phenoxide salt of the larger ammonium ion is already very reactive because of the relatively weak cation-anion interaction in the ion pair. In such cases dissociation to a truly free anion does not seem to be required in order to explain the high reactivity [340].

Examination of the S_N2 reaction between ethyl tosylate and halide ions in hexamethylphosphoric triamide ($\varepsilon_r = 29.6$) with a variety of counter-ions [$Li^{\oplus}$, $(n\text{-}C_4H_9)_4N^{\oplus}$] has shown, that the rates obtained with lithium salts are always higher than those of the corresponding tetra-*n*-butylammonium salts [341]. This is in contrast to the situation observed in acetone [279]. This means, that in this particular solvent, lithium salts are more dissociated than tetraalkylammonium salts. This has indeed been confirmed by conductivity measurements [341, 342]. The lithium cation apparently has specific interactions with strong EPD solvents such as $[(CH_3)_2N]_3PO$ (*cf.* Section 3.3.2).

Ion-pair dissociation can be achieved not only by using large cations with correspondingly low charge densities but also by employing EPD solvents as either reaction media or additives to salt solutions in other solvents. EPD solvents of high Lewis basicity are particularly good specific cation solvators and weak anion solvators – thus giving rise to highly reactive anions. The cation-solvating capacity of EPD solvents can be quantitatively described in terms of their high donor numbers *DN* (*cf.* Table 2-3 in Section 2.2.6) as well as their large negative solvent-transfer activity coefficients for cations (*cf.* Table 5-19 in Section 5.5.3). Good cation-solvating EPD solvents include most of the common dipolar non-HBD solvents (*cf.* Table 5-18 in Section 5.5.2) as well as open-chain polyethers such as oligoethylene glycol dialkyl ethers ("glymes") which contain the repeating unit $(\text{—CH}_2\text{—CH}_2\text{—O—})_n$ ($n \geq 2$) [345].

Alkali derivatives of CH-acidic compounds are usually highly aggregated in solvents of low dielectric constant such as benzene ($\varepsilon_r = 2.3$). The accelerating influences of a variety of EPD solvents used as additives in the alkylation of diethyl sodio-*n*-butylmalonate with 1-bromobutane in benzene according to Eq. (5-126) was discovered by Zaugg *et al.* [350]; *cf.* Table 5-21.

$$n\text{-C}_4\text{H}_9\text{-}\overset{\text{CO}_2\text{C}_2\text{H}_5}{\underset{\text{CO}_2\text{C}_2\text{H}_5}{\text{C}|^{\ominus}}}\ \text{Na}^{\oplus} + n\text{-C}_4\text{H}_9\text{-Br} \xrightarrow[25\,^{\circ}\text{C}]{k_2} (n\text{-C}_4\text{H}_9)_2\overset{\text{CO}_2\text{C}_2\text{H}_5}{\underset{\text{CO}_2\text{C}_2\text{H}_5}{\text{C}}} + \text{Na}^{\oplus}\,\text{Br}^{\ominus} \qquad (5\text{-}126)$$

Table 5-21. Relative rate constants of the alkylation of diethyl sodio-*n*-butylmalonate with 1-bromobutane in benzene at 25 °C with different additives [350].

Added Cation Solvator (0.648 mol/l)	$k_2^{\text{rel.}}$
No additive (benzene alone)	1
	1.1
CH_3COCH_3	1.3
C_2H_5OH	4.4
$CH_3O\text{—}CH_2CH_2\text{—}OCH_3$	6.4[a)]
	18
$HCON(CH_3)_2$	19
	30
$[(CH_3)_2N]_3PO$[b)]	54

[a] In pure 1,2-dimethoxyethane (8.7 mol/l) the reaction is 80 times faster than in benzene.
[b] Only 0.324 mol/l added.

Table 5-21 shows that the addition of even small proportions of EPD solvents affects the reaction rate markedly. The rate acceleration thus obtained is produced by a specific solvation of sodium ion which tends to dissociate the high-molecular weight ion-pair aggregate of the sodio-malonic ester that exists in benzene solution (degree of aggregation *n* is equal to 40...50 in benzene). This indicates that the kinetically active species is a lower aggregate of the free carbanion. Further evidence for a specific cation solvation is derived from the 6-fold rate difference observed in tetrahydrofuran ($\varepsilon_r = 7.6$)

and 1,2-dimethoxyethane ($\varepsilon_r = 7.2$) despite the fact that these two solvents possess nearly equal dielectric constants. The latter solvent is able to solvate sodium ions in the manner shown in Eq. (5-127). Especially noteworthy is the high reactivity exhibited on the addition of dicyclohexyl [18] crown-6. In benzene solution containing only 0.036 mol/l of crown ether, the alkylation rate is already equal to that observed in neat 1,2-dimethoxyethane [351].

The reactivity of sodio-butyrophenone toward 1-bromo-2-methylpropane is greatly enhanced in monoglyme (2560-fold) and diglyme (11400-fold) with respect to the rate in diethyl ether according to Zook *et al.* [352]. The ethylation of sodio-butyrophenone (0.13 mol/l) by bromoethane (1.6 mol/l) in diglyme at 30 °C is 75% complete in 152 sec, whereas the corresponding reaction time for a comparable ethylation in diethyl ether is 234 hours [353]. Again, specific solvation of the cation in the sodium enolate aggregates may explain these results. The reactivity toward bromoethane increases in the order $Li^{\oplus} < Na^{\oplus} < K^{\oplus}$-enolate corresponding to the decreasing charge density of the cations [353].

Finally, anion activation can also be achieved by the complexation of the cation using suitable macro(poly)cyclic ligand molecules [643]. Organic ligands that contain enforced cavities of dimensions at least equal to those of the smaller cations (and anions) have been called *cavitands* [644]. Cation complexation by such macro(poly)cyclic ligands leads to dissociation of the ion pairs as well as to salt solubilization. Complex formation with lipophilic organic ligands transforms small cations into voluminous lipophilic cationic species which are much more soluble in organic solvents of low polarity*). Therefore, the increase in reaction rates induced by specific cation-complexing ligands usually results from an increase in the anion reactivity *as well as* an increase in the reagent concentration.

Some general examples of anion activation by different types of cation complexation are given in Eqs. (5-127)...(5-131). Among the specific multidentate complexing agents for cations are:

(a) open-chain *podands* such as the oligoethylene glycol dialkyl ethers ("glymes"), investigated by Vögtle *et al.*; *cf.* Eq. (5-127) [345].

(b) cyclic *coronands* such as the crown ethers, first introduced by Pedersen in 1967 [343]; *cf.* Eq. (5-128) [348, 645–647];

(c) macrobi-/tricyclic spherical *cryptands*, first introduced by Lehn *et al.* in 1969 [344, 648]; *cf.* Eqs. (5-129) and (5-130). Eq. (5-130) illustrates that in the case of the ammonium ion the hydrogen-bond acceptor ability of the macrotricyclic ligand can also be important [349, 648];

(d) cyclic *spherands*, first introduced by Cram *et al.* in 1979 [644, 649]; *cf.* Eq. (5-131).

Whereas the outside of these cationic complexes is lipophilic, the cations are held inside the hydrophilic cavity of the organic ligands. In the case of Eq. (5-127), the podands

---

* Using water/organic solvent two-phase systems, the distribution of a salt in the organic layer is enhanced by lipophilic cation-complexing ligands. This is the basis of *liquid-liquid phase-transfer catalysis*. Phase-transfer catalysts help to transfer a water soluble reactant salt across the interface into the organic phase where a homogeneous reaction can occur, thus enhancing the rate of reaction. Phase-transfer catalysis is outside the scope of this book; the reader is referred to some excellent reviews [656–658].

$$H_3C-O\left[\diagup\diagdown O\right]_4\diagup\diagdown O-CH_3 + M^{\oplus}Y^{\ominus} \overset{K_S}{\rightleftharpoons} [\text{complex}]^{\oplus} + Y^{\ominus} \quad (5\text{-}127)$$

Pentaglyme, an open-chain *podand*

$$+ M^{\oplus}Y^{\ominus} \overset{K_S}{\rightleftharpoons} [\text{complex}]^{\oplus} + Y^{\ominus} \quad (5\text{-}128)$$

[18] Crown-6, a cyclic *coronand*

$$+ M^{\oplus}Y^{\ominus} \overset{K_S}{\rightleftharpoons} [\text{complex}]^{\oplus} + Y^{\ominus} \quad (5\text{-}129)$$

[2.2.2] Cryptand, a bicyclic spherical *cryptand*

$$+ NH_4^{\oplus}Y^{\ominus} \overset{K_S}{\rightleftharpoons} [\text{complex}]^{\oplus} + Y^{\ominus} \quad (5\text{-}130)$$

A tricyclic spherical *cryptand*

$$+ M^{\oplus}Y^{\ominus} \overset{K_S}{\rightleftharpoons} [\text{complex}]^{\oplus} + Y^{\ominus} \quad (5\text{-}131)$$

A cyclic *spherand*

cavity is built up during complexation by a template effect. The stability constants K_s increase from Eq. (5-127) to Eq. (5-131). Typical orders of magnitudes for K_s-values in methanol are $10^2 \ldots 10^4$ for podates, $10^4 \ldots 10^5$ for coronates, and $10^6 \ldots 10^8$ for cryptates [646]*⁾. The appropriate choice of ligand allows a selective complexation of a particular cation. For economic reasons, the simple and cheap open-chain glymes have had greater application in anion activation. Only a few representative examples of enhanced anion reactivity caused by cavitands will be mentioned here.

The effect of both podands and coronands on the S_N2 alkylation of potassium phenoxide with 1-bromobutane in 1,4-dioxane ($\varepsilon_r = 2.2$) has been investigated by Ugelstad *et al.* [354]. The addition of tetraethylene glycol dimethyl ether in concentrations equivalent to the phenoxide concentration leads to an alkylation rate 11 times faster than in pure 1,4-dioxane, whereas the same amount of dicyclohexyl [18]crown-6 causes an 8700-fold rate increase.

The base-catalysed hydrolysis of the sterically hindered ester methyl mesitoate can be carried out in KOH/DMSO at room temperature with 100% yield using a [2.2.2]cryptand as the potassium complexing ligand [652].

Liotta *et al.* [282, 355, 356, 359] investigated the chemistry of so-called "naked anions"**⁾, *i.e.* poorly solvated fluoride, cyanide, and acetate ions, solubilised as potassium salts in acetonitrile or benzene containing [18]crown-6 as additive. The reactivity of these naked anions in reactions with haloalkanes have been examined. According to the haloalkane structure, the reaction leads to substitution and/or elimination products. Solubilised fluoride ions are both potent nucleophiles and strong bases, whereas solubilised cyanide and acetate ions are good nucleophiles but rather weak bases. A variety of fluoroalkanes [282, 355] and nitriles [356, 361] can be prepared in good yields using crown ethers as cation solvators. The carboxylate ion is generally considered a bad nucleophile, but [18]crown-6 complexed potassium acetate (also called "bare acetate") reacts readily with haloalkanes in acetonitrile to yield the corresponding esters [359, 360]. The nitrite ion also displays a remarkable enhanced nucleophilicity in the presence of [18]crown-6. Under these conditions nitro compounds are formed from haloalkanes in good yields [361].

Other anions ($X^{\ominus} = Br^{\ominus}$, $I^{\ominus}$, $HO^{\ominus}$, $CH_3O^{\ominus}$) also exhibit enhanced nucleophilic reactivity in crown-ether mediated reactions of $K^{\oplus}X^{\ominus}$ [357, 358]. Montanari *et al.* [653] compared the effects of phosphonium salts, coronands, and cryptands on the S_N2 reaction of *n*-octylmethanesulfonate with various nucleophiles in chlorobenzene.

Apart from these examples of nucleophilic substitution reactions, there are many more applications of reactions which are enhanced by the addition of macro(poly)cyclic ligands; examples are found in recent reviews [346–348, 362, 643, 645–647].

---

* It should be mentioned that cation complexation by crown-type ligands can itself be solvent-dependent. For example, the dissociation rate of potassium [2.2.2]cryptate in EPD solvents increases with the donor number of the solvent [650]. In addition, coronands itself can interact with organic solvent molecules [651]. Such cation-solvent and ligand-solvent interactions can influence the formation of cation-ligand complexes.

** Completely unsolvated "naked anions" cannot be prepared in solution with coronands or even with cryptands as cation solvators. Even in this case ion-pairing still occurs leading to complexed ion pairs [646]. Totally unsolvated "naked anions" can exist only in the gas phase (*cf.* Section 5.2.).

Cation-assisted reactions, on the other hand, are hindered by complexation of the cation [648]. Reduction of carbonyl groups by metal hydrides [654] and the addition of organolithium compounds [655] are examples. This indicates that the coordination between carbonyl group and cation is an important step in these reactions.

In conclusion it can be said, that specific cation solvators – EPD solvents and particularly cavitands – have been demonstrated to be of proven utility for organic syntheses.

5.5.6 Solvent Influence on the Reactivity of Ambident Anions

Ambident anions are mesomeric, nucleophilic anions which have at least two reactive centers with a substantial fraction of the negative charge distributed over these centers*)**). Such ambident anions are capable of forming two types of products in nucleophilic substitution reactions with electrophilic reactants***). Examples of this kind of anion are the enolates of 1,3-dicarbonyl compounds, phenoxide, cyanide, thiocyanide, and nitrite ions, the anions of nitro compounds, oximes, amides, the anions of heterocyclic aromatic compounds (*e.g.* pyrrole, hydroxypyridines, hydroxypyrimidines) and others; *cf.* Fig. 5-17.

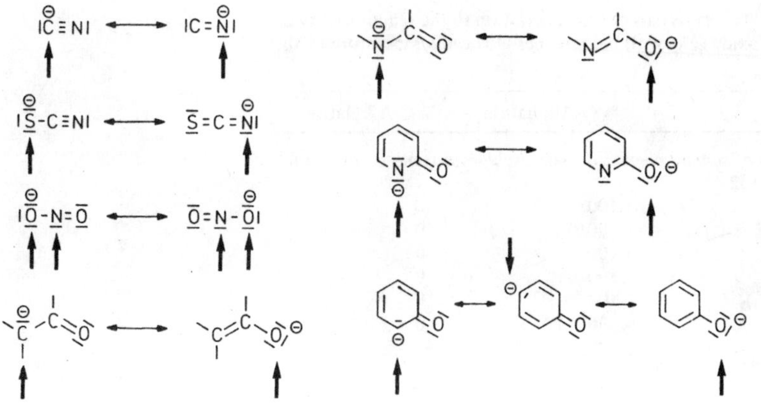

Fig. 5-17. Some ambident or ambifunctional anions. The arrows indicate the sites of dual reactivity towards electrophiles.

* The term *ambident anion* was proposed by Kornblum [363], whereas Gompper suggested that such anions should be called *ambifunctional* [364, 367].

** In principle, *ambident cations* may be defined in the same way, but the solvent dependence of their dual reactivity with nucleophiles has not yet been investigated thoroughly [368].

*** When a reaction can potentially give rise to two (or more) constitutional isomers but actually produces only one, the reaction is said to be *regioselective*. For example, the ambident nucleophile $NCO^{\ominus}$ usually gives only isocyanates R—NCO and not the isomeric cyanates, R—OCN.

If the substitution reaction is kinetically controlled, then the composition of the products is determined by the relative nucleophilicity of each of the donor atoms in the ambident anion in relation to the given electrophilic reactant. Among the factors influencing the mode of reaction (counterion, additives, concentration, temperature, pressure, leaving group, structure of the alkylating agent) the solvent plays a major role in the orientation of the electrophile, and this has been examined in several reviews [364–367, 367a, 367b]. Already in 1923 Claisen noticed that the O/C alkylation ratio observed in the reaction of phenol with 3-bromopropene in the presence of potassium carbonate depends strongly on the solvent used as reaction medium [369]. In acetone the major product formed was allyl phenyl ether, whereas in solvents like benzene or toluene *ortho*-allyl phenol was obtained as the main product [369].

Some representative data, taken from the extensive work of Kornblum *et al.* concerning the alkylation of phenoxides [*cf.* Eq. (5-132)] and β-naphthoxides, are collected in Table 5-22 [370, 371].

$$[C_6H_5O^{\ominus}]Na^{\oplus} + CH_2{=}CH{-}CH_2Cl \xrightarrow[-NaCl]{k_2} C_6H_5OCH_2CH{=}CH_2 + o\text{-}CH_2{=}CHCH_2C_6H_4OH + p\text{-}CH_2{=}CHCH_2C_6H_4OH \qquad (5\text{-}132)$$

Table 5-22. Oxygen versus carbon alkylation in the reaction of two sodium aryloxide salts with haloalkanes at room temperature [370, 371].

Solvents	% O-Alkylation	% C-Alkylation[a]
a) *Reaction of sodium phenoxide with 3-chloropropene according to equation* (5-132)		
1,4-dioxane	100	0
$(CH_3)_3COH$	100	0
C_2H_5OH	100	0
CH_3OH	100	0
$HCON(CH_3)_2$	100	0
$(CH_2)_4O$	96	0
H_2O	49	41
CF_3CH_2OH[b]	37	42
C_6H_5OH[c]	22	78
b) *Reaction of sodium β-naphthoxide with (bromomethyl)benzene*		
$HCON(CH_3)_2$	97	0
CH_3SOCH_3	95	0
$CH_3O{-}CH_2CH_2{-}OCH_3$	70	22
$(CH_2)_4O$	60	36
CH_3OH	57	34
C_2H_5OH	52	28
H_2O	10	84
CF_3CH_2OH	7	85

[a] Dialkylated products included.
[b] Reaction with 3-bromopropene.
[c] At 43 °C.

In protic solvents, the solvent effect on orientation has been interpreted in terms of preferential deactivation of the more electronegative atom by specific solvation through H-bonding*). As shown in Table 5-22, carbon alkylation competes more significantly with oxygen alkylation as the protic solvents become stronger proton-donors (*cf.* $CH_3CH_2OH \rightarrow H_2O \rightarrow CF_3CH_2OH \rightarrow C_6H_5OH$). In 2,2,2-trifluoroethanol, a very strong proton-donor, the yield of C-alkylated isomers approaches 42% and 85%, respectively. Owing to H-bonding, protic solvents selectively and effectively solvate the center with the maximum electron density. As a result of this, the accessibility of this center is reduced and substitution at the other donor-atom (in this case a carbon atom) may successfully compete with that at the oxygen atom. This protic solvent effect is also observed for other ambident anions. For example, in the alkylation of the enolate salt of acetoacetic ester, the yield of the O-alkylated isomer falls sharply on passing from dipolar non-HBD solvents to protic solvents such as alcohols [372]. Even relatively small amounts of added protic solvents influence the mode of reaction. Thus, in the alkylation of the enolate salt of acetoacetic ester with ethyl tosylate in hexamethylphosphoric triamide, the addition of equimolar amounts of protic solvents like water or *tert*-butanol significantly decreases the degree of O-alkylation [373]. High pressure promotes specific solvation of the more electronegative site in protic solvents, and increased alkylation occurs at the alternative position [378].

The tendency towards reaction at the center with the maximum electron density increases when dipolar non-HBD solvents are employed owing to the lack of specific solvation (*cf.* solvents $HCON(CH_3)_2$ and CH_3SOCH_3 in Table 5-22). Thus, in the alkylation of the enolate anions of 1,3-dicarbonyl compounds the greatest yields of the O-alkylated isomers are obtained in hexamethylphosphoric triamide, followed by dipolar non-HBD solvents of the amide type [372–375].

In poorly cation-solvating solvents there is also a marked dependence of reaction site on association with the corresponding cation. In non-dissociating solvents, it is likely that the cation will be preferentially coordinated to the atom in the ambident anion with the maximal electron density, this hinders reaction at this site. As shown in Table 5-22 by the results observed for the alkylation of sodium β-naphthoxide, screening of the electronegative oxygen atom due to association with the accompanying sodium cation leads to an increase of C-*versus* O-alkylation in aprotic, non-dissociating solvents such as 1,2-dimethoxyethane ($\varepsilon_r = 7.2$) and tetrahydrofuran ($\varepsilon_r = 7.6$), compared with dipolar non-HBD EPD-solvents. Alkylation of the less electronegative site can also be favoured by the addition of excess $M^{\oplus}$ to suppress dissociation of ambident ion pairs. The lithium enolates of 1,3-dicarbonyl compounds are an exception. They display a pronounced tendency towards O-alkylation (especially when compared to sodium enolates) which is not very sensitive to changes in the solvent (*e.g.* THF → DMF) [659]. It has been suggested that the tendency of lithium enolates towards O-alkylation is partly the result of a $Li^{\oplus}$-leaving group interaction and partly an intrinsic property of the enolate lithium ion pair [659].

---

* Kornblum termed this shielding of the center with maximal electron density in ambident anions by protic solvents, which prevents the reaction at this center, "selective solvation" [370]. In order to avoid confusion with selective solvation of ions or dipolar molecules in binary solvent mixtures (*cf.* Section 2.4) the designation "specific solvation" is preferred.

The ion-pair dissociation of ambident alkali enolates which results in increasing O/C alkylation ratios can be promoted not only by dissociating solvents but also by specific cation solvation. In the latter case, EPD solvents (*cf.* DMF and DMSO in Table 5-22b) or macro(poly)cyclic ligands such as coronands ("crown ethers") or cryptands are used [376, 377, 660]. For example, the alkylation of sodium β-naphthoxide with (bromomethyl)benzene or iodomethane in the presence of benzo-[18]crown-6 gives high O/C alkylation ratios when tetrahydrofuran or benzene are the solvents [660]. In dissociating solvents such as *N,N*-dimethylformamide or acetonitrile, however, so far no influence of added crown ether on the O/C product ratio has been observed [660]. – Recently, the influence of ion-pairing on the alkylation of preformed alkali 2,4-pentanedionates in dimethyl sulfoxide solution has been systematically studied [661]. High pressure results in a shift from paired to free ambident anions and increased O-alkylation in non-HBD solvents [378].

To summarize it can be stated, that the freer the ambident anion in every respect, the larger the O/C-alkylation ratio in the case of 1,3-dicarbonyl compounds [365]. Thus, if O-alkylation products are desired in the alkylation of enolates, dipolar non-HBD and dissociating solvents such as *N,N*-dimethylformamide, dimethyl sulfoxide, or, especially, hexamethylphosphoric triamide should be used. If C-alkylation is desired, protic solvents like water, fluorinated alcohols, or, in the case of phenols, the parent phenol will be the best choice [365].

Taking into account that the solvation of ambident anions in the activated complex may differ considerably from that of the free anion, another explanation for the solvent effect on orientation, based on the concept of hard and soft acids and bases (HSAB) [275] (*cf.* also Section 3.3.2), seems preferable [366]. In ambident anions the less electronegative and more polarizable donor atom is usually the softer base, whereas the more electronegative atom is a hard Lewis base. Thus, in enolate anions, the oxygen atom is hard and the carbon atom is soft, in the thiocyanate anion the nitrogen atom is hard and the sulfur atom is soft, *etc.* The mode of reaction can be predicted from the hardness or softness of the electrophile. In protic solvents the two nucleophilic sites in the ambident anion must interact with two electrophiles, the protic solvent and the substrate RX, of which the protic solvent is a hard*) and RX a soft acid. Therefore, in protic solvents it is to be expected that the softer of the two nucleophilic atoms (C *versus* O, N *versus* O, S *versus* N) should react with the softer acid RX.

In non-HBD, non-dissociating solvents a corresponding proposal can be made: hard counterions (alkali cations) should associate preferably with the hard site, and the substrate RX with the soft site in the activated complex composed of RX and ambident ion pair [366]. With increasing hardness of the counterion (increasing charge density) the fraction of C-alkylation should increase in non-HBD solvents and decrease on solvent insertion into the ion pair. Indeed, the C-ethylation of $M^{\oplus}$(ethyl acetoacetate)$^{\ominus}$ in dimethyl sulfoxide or hexamethylphosphoric triamide increases in the order $M^{\oplus} = R_4N^{\oplus} < Cs^{\oplus} < K^{\oplus} < Na^{\oplus} < Li^{\oplus}$ [373].

In the gas phase, the reaction of ethyl cations, $C_2H_5^{\oplus}$, with the ambident 2,4-pentanedione (which is 92% enolised at 25 °C in the gas phase) produces pre-

* Hydrogen atoms behave as hard Lewis acids in the formation of a hydrogen bond.

dominantly (>95%) alkylation on the hard oxygen site and not on the soft carbon atom, as predicted by the HSAB concept [662].

A theoretical explanation of the HSAB concept using quantummechanical perturbation theory was given by Klopman [379]. He showed that the softness of a particular Lewis acid (acceptor atom) is determined by the energy of the lowest unoccupied orbital and the softness of a Lewis base (donor atom) by the energy of the highest occupied orbital, while taking into account that the softness and hardness is influenced by the solvent. Limitations of the application of the HSAB concept on the reactivity of ambident anions have been recently discussed by Gompper *et al.* [367]. An alternative quantitative approach for the prediction of the dual reactivity of ambifunctional anions towards electrophiles has been proposed ("principle of allopolarization"), using so-called selectivity factors $S_f = Q_X/Q_Y$ (Q_X and Q_Y are the yields observed in reactions of ambident anions with two reaction sites X and Y) and polarity indices $P = l_X/l_Y$ (l_X and l_Y are the charge densities at reaction sites X and Y, taken from HMO-calculations) as determinating parameters.

In analogy with ambident anions, mesomeric *ambident cations* do exist, but the solvent influence on their dual reactivity with nucleophiles has not been thoroughly investigated; *cf.* reference [368] for a review.

A nice example of the solvent-dependent dual reactivity of an electrophilic crypto-cationic species has been given by Hünig *et al.* [663]. Ambident electrophilic α-enones react with nucleophiles such as the anion of the benzaldehyde O-(trimethylsilyl)cyanohydrine ($Nu|^{\ominus}$) in diethyl ether exclusively by 1,4-addition. In tetrahydrofuran (THF) or 1,2-dimethoxyethane (DME) the 1,2-adduct is formed predominantly; on the addition of HMPT or [12]-crown-4 it is formed exclusively; *cf.* Eq. (5-133).

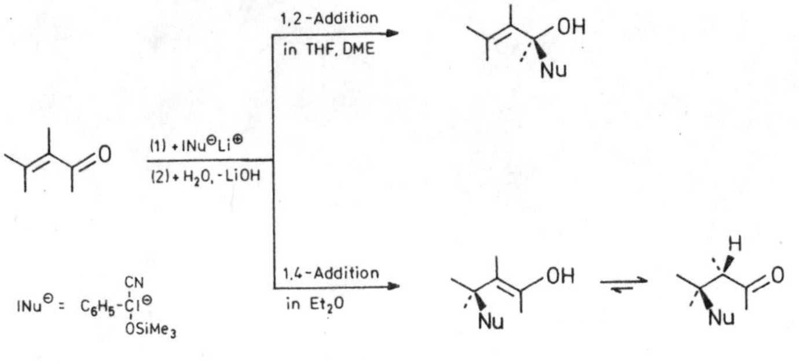

(5-133)

This dramatic solvent effect is a result of the difference in the extent of ion-pairing of $Li^{\oplus}Nu^{\ominus}$ in diethyl ether (→ contact ion pair) and THF or DME (→ solvent-separated ion pair), modified by the α-enone-$Li^{\oplus}$ complex [663].

5.5.7 Solvent Effects on Mechanisms and Stereochemistry of Organic Reactions

Elaborating on the preceding discussion on the dual reactivity of ambident anions, this Section contains a selection of reactions of various types which have in common the fact, that a change in solvent drastically changes the mechanism as well as the stereochemistry usually associated therewith. The examples are intended to demonstrate how a single, out of two or more alternatives, pathway for a reaction can be favoured through a proper choice of solvent. This, of course, is of considerable preparative interest. In this connection it should be emphasized, that alternative routes of a reaction may be separated by very small differences in Gibbs activation energy. It requires less than 12 kJ/mol (3 kcal/mol) of activation energy to change a 10:90 ratio to a 90:10 product mixture, and the sensitivity of this ratio can frequently be exploited by a proper choice of the reaction medium.

The stereochemical outcome of a nucleophilic substitution reaction at a saturated carbon is a function of the reaction mechanism (S_N1, S_N2, or S_Ni) which, in turn, can depend on the nature of the solvent, as already mentioned (*cf.* Sections 5.3.1 and 5.5.1). The fact that, in contrast to dipolar non-HBD solvents, the protic solvents diminish the nucleophilicity of anions and simultaneously favour the ionization of polarized bonds, makes it possible to displace a given reaction toward either the S_N1 or S_N2 type. A remarkable example is found in reaction (5-134) [380]. The nucleophilic substitution of cholesteryl tosylate in dipolar non-HBD solvents (*e.g.* $HCON(CH_3)_2$, CH_3SOCH_3) gives preferably the 3α-derivative with inversion of configuration. In contrast, in protic solvents (*e.g.* CH_3OH, $HCONHCH_3$) a mixture of the 3β- and 3,5-cyclo-6β-derivatives is formed, corresponding to an S_N1 mechanism with initial formation of a homoallyl carbenium ion. However, a change from a protic to a dipolar non-HBD solvent never results in the complete suppression of an S_N1 reaction [380].

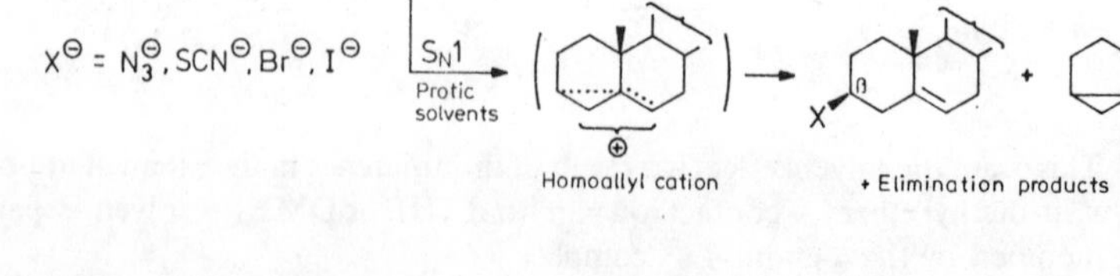

C_8H_{17} TosO β
$+X^{\ominus}$ $-TosO^{\ominus}$
S_N2 Dipolar non-HBD solvents → X α
S_N1 Protic solvents → Homoallyl cation → X β + X β + Elimination products
$X^{\ominus} = N_3^{\ominus}, SCN^{\ominus}, Br^{\ominus}, I^{\ominus}$

(5-134)

An "S_N2(intermediate)" mechanism has been proposed as a result of the consideration of the dependence of S_N1/S_N2 solvolyses rates on electrophilic (EPA) and nucleophilic (EPD) solvent assistance [664, 665].

S_N1 reactions of secondary and tertiary substrates R_3C—X (*e.g.* 2-adamantyl tosylate [665]; *t*-butyl heptafluorobutyrate [666]) proceed via activated complexes with high carbenium ion character to give ion-pair intermediates. They exhibit electrophilic solvent assistance in protic solvents SOH (H-bonding to leaving group) but they are insensitive to changes in solvent nucleophilicity; *cf.* Eq. (5-135).

$$R_3C\text{-}X \underset{\text{in SOH}}{\overset{S_N1}{\rightleftharpoons}} [R_3\overset{\delta\oplus}{C}\cdots\overset{\delta\ominus}{X}\cdots HOS]^{\ddagger} \longrightarrow R_3\overset{\oplus}{C} + \overset{\ominus}{X}\cdots HOS \longrightarrow \text{products} \qquad (5\text{-}135)$$

Electrophilic solvent assistance

S_N2 reactions of primary and secondary substrates R_3C—X are accelerated relative to S_N1 reactions by additional rearside nucleophilic attack on the substrate, which reduces the carbenium ion character in the activated complex. S_N2 reactions involving a concerted mechanism are also called "S_N2(one-stage)" reactions [665]. They span a huge spectrum of nucleophilic solvent assistance – varying from classical S_N2 processes involving strong nucleophilic solvent assistance (*e.g.* methyl tosylate) to weakly nucleophilic solvent-assisted processes with relatively high carbenium ion character in the activated complex (*e.g.* 2-propyl tosylate). This variability of S_N2 solvolysis reactions can be regarded as a solvent-dependent spectrum of intermediate mechanism rather than a varying mixture of only two distinct processes S_N1 and S_N2. To account for this gradation in mechanism, an "S_N2(intermediate)" mechanism has been proposed [664, 665]. This mechanism involves a pentacoordinated intermediate which is a nucleophilically solvated ion pair similar to the activated complex in the S_N2(one-stage) reaction. The solvent/carbocation interaction is covalent in character; *cf.* Eq. (5-136).

$$R_3C\text{-}X \underset{\text{in SOH}}{\overset{S_N2}{\rightleftharpoons}} \left[\begin{array}{c} \overset{H\backslash\delta\oplus}{\underset{S'}{O}}\cdots\overset{R}{\underset{R\ \ R}{C}}\cdots\overset{\delta\ominus}{X}\cdots HOS \end{array}\right]^{\ddagger} \longrightarrow \overset{H\backslash\oplus}{\underset{S'}{O}}\text{-}CR_3 + \overset{\ominus}{X}\cdots HOS \longrightarrow \text{products} \qquad (5\text{-}136)$$

Nucleophilic solvent assistance

The heterolysis of the bond between carbon and the leaving group in R_3C—X is supported by nucleophilic solvent assistance, the extent of which depends on the solvent and increases in the order $(CF_3)_2CHOH < CF_3CO_2H < CF_3CH_2OH < HCO_2H < H_2O < CH_3CO_2H < CH_3OH < CH_3CH_2OH$ [665]. If there is evidence for an intermediate as well as for nucleophilic solvent assistance, then the mechanism can be considered "S_N2(intermediate)". Examples for a gradual change of mechanism from S_N2(one-stage) through S_N2(intermediate) to S_N1 has been given [665].

The "S_N2(intermediate)" mechanism has not been without criticism; see for example [667, 668] and references cited therein. An alternative mechanism can be formulated, involving ion pairs formed by heterolysis of the R_3C—X bond *before* rate-limiting nucleophilic solvent attack occurs. Internal return to the educt would then occur more rapidly than attack by the nucleophile to give the products. For example, the solvolyses of secondary 1-arylethyl tosylates, $ArCH(OTs)CH_3$, can be also explained in terms of an ion-pair mechanism in which nucleophilic solvent attack on the ion pair plays a major role [667]. In less nucleophilic solvents this attack is rate-limiting, whereas attack of more nucleophilic solvents is fast, resulting in a rate-limiting initial ionization of $R_3C—X$. The ion pair generally interacts with the solvent although not by a specific interaction (*i.e.* covalent bonding with a single nucleophilic solvent molecule) as in the "S_N2(intermediate)" mechanism [667].

It is difficult to objectively prove or disprove either the S_N2(intermediate) or the ion-pair mechanism of S_N2 solvolysis reactions. According to Olah *et al.* [669], "the S_N2 intermediate can indeed be best characterized as a trivalent carbocationic centre solvated from both sides by the negatively charged nucleophile and the leaving group". According to Tidwell *et al.* [667], "this view deemphasizes the covalent character of the solvent-carbocation interaction, and is operationally equivalent to an ion pair".

The *ortho*/*para* alkylation ratio of the intramolecular nucleophilic substitution reaction of sodio-4-(3-hydroxyphenyl)butyl tosylate, according to Eq. (5-137), was shown to be solvent-dependent [381]. In general, an increase in the *ortho*/*para* alkylation ratio resulted when the polarity of the solvent was decreased. The orientation of the reaction was explained in terms of the nature of the association between the metal and phenoxide ions in different solvents [381].

OTos; $O^{\ominus}$; $Na^{\oplus}$ —(35 °C, $-TosO^{\ominus}Na^{\oplus}$)→ OH + OH (5-137)

	para-product	*ortho*-product
in $(CH_2)_4O$:	13 cmol/mol	87 cmol/mol
in CH_3OH:	49 cmol/mol	51 cmol/mol

The α/γ alkylation ratio of the intramolecular nucleophilic substitution reaction of an α,β-unsaturated cyclohexenone tosylate can also be controlled by the right choice of solvent. The desired γ-alkylation – the final step of the total synthesis of the sesquiterpene β-vetivone – is favoured by NaOH in CH_3SOCH_3/H_2O in contrast to $(CH_3)_3COK/(CH_3)_3COH$ which promotes α-alkylation [670].

That a single solvent molecule clustered to a nucleophile can drastically change the reaction pathway has been demonstrated by studying the reaction of phenyl acetate with methoxide ion in the gas phase [671, 672]. Alkaline hydrolysis of esters in solution are known to proceed by attack of the nucleophile at the carbonyl carbon atom to form a tetrahedral intermediate, followed by cleavage of the acyl-oxygen bond ($B_{AC}2$ mechanism); *cf.* Eq. (5-138a).

In the gas phase, however, nucleophilic aromatic substitution according to Eq. (5-138b) is the preferred pathway! Partially solvated nucleophiles such as $CH_3O^{\ominus}\ldots HOCH_3$, formed in the gas phase, react with phenyl acetate again via the $B_{AC}2$ mechanism. In the gas phase, charge-dispersed activated complexes such as formed in the S_NAr reaction obviously have lower activation barriers than charge-localized activated complexes leading to the tetrahedral intermediate of the $B_{AC}2$ mechanism. The latter can be stabilized by dispersing the localized negative charge over the carbonyl oxygen atom by means of H-bonding with protic solvent molecules. The attachment of even one HBD solvent molecule to the nucleophile is enough to change the reaction mechanism from S_NAr to $B_{AC}2$ [671].

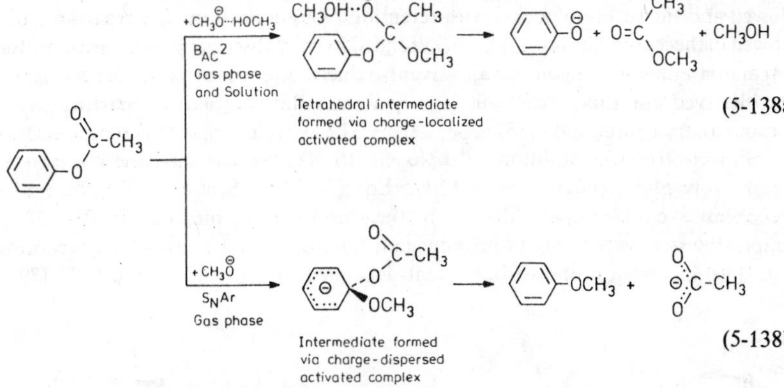

(5-138a)

(5-138b)

The mechanism and stereochemical course of electrophilic substitution reactions at saturated carbon atoms (S_E1, S_E2, or S_Ei) may also be affected by the medium [337, 382, 383]. This is especially true for S_E reactions of organomercury compounds as shown by Reutov *et al.* [384], Petrosyan [673], and Hughes and Ingold [385]. The isotopic exchange reaction of (α-ethoxycarbonyl)benzylmercuric bromide with radiomercuric bromide in anhydrous dimethyl sulfoxide, shown in Eq. (5-139), is first-order in substrate but zeroth-order in Hg^*Br_2 ($Hg^* = {}^{203}Hg$) and proceeds with complete racemization of the optically active substrate [384, 385]. These observations are in agreement with an S_E1 mechanism. This was the first time that such a mechanism has been established for organomercury compounds. The ionization of the C—Hg bond is facilitated by specific solvation of the leaving group as shown $(CH_3)_2S{=}O \rightarrow HgBr^{\oplus}$. However, when the same reaction is

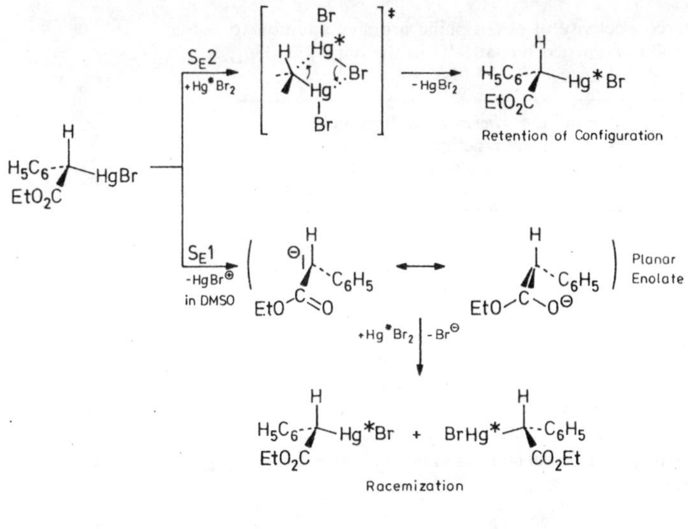

(5-139)

carried out in aqueous acetone, aqueous ethanol, or in pyridine, second-order kinetics are followed and the reaction occurs with retention of configuration. The reaction pathway followed in these solvents is presumably of the S_E2 type. Subsequent to the work of Reutov [384] and Hughes and Ingold [385], solvent-induced alterations of the mechanisms have been observed for other reactions of organometallic compounds, particularly other organomercury compounds [337, 338, 382, 383] as well as organotin compounds [674].

For electrophilic additions of halogens to alkenes, not only the reaction rate is strongly solvent-dependent [79–81] (*cf.* Eq. (5-29) in Section 5.3.2), but also its stereochemical course may be affected by the polarity of the medium [79, 386–388]. For example, the stereoselectivity of bromine addition to *cis*- and *trans*-stilbene according to Eq. (5-140) has been found to be solvent-dependent as shown in Table 5-23 [79, 386].

cis-Form · *trans*-Form · polar solvents · A · B · Racemic Form (threo) · $R = C_6H_5$ · *meso*-Form (erythro) (5-140)

Table 5-23. Stereoselectivity of electrophilic bromine addition to *cis*- and *trans*-stilbene, carried out at 0 °C in the dark [79, 386]; *cf.* Eqs. (5-29) and (5-140).

Solvents	cmol/mol *racemic* 1,2-dibromo-1,2-diphenylethane[a]	
	from *cis*-stilbene	from *trans*-stilbene
CS_2	81.4	5.5
CCl_4	77.0	11.0
	67.4	—
C_2H_5OH	52.0	—
$CCl_3CO_2CH_3$	51.0	21.0
CCl_3CN	34.0	18.5
$C_6H_5NO_2$	29.5	16.5

[a] The difference to 100 cmol/mol consists of *meso*-1,2-dibromo-1,2-diphenylethane.

Whereas in nonpolar solvents such as CS_2 and CCl_4, bromide adds to *cis*-stilbene in a highly stereoselective manner to give 81.4 cmol/mol *racemic* stilbene dibromide, and to *trans*-stilbene to give 94.5 cmol/mol *meso*-stilbene dibromide, in polar solvents, the degree of stereoselectivity is considerably reduced in both cases. Similar results have been observed for bromine addition to 1-phenylpropene [387]. These results can be interpreted in terms of reaction scheme (5-140) which involves both ethene bromonium ions (symmetrical or unsymmetrical) and α-halogeno carbenium ions as intermediates. The observed solvent dependence of this polar bromination reaction may be the result of differential solvation requirements of the bromonium ion and the carbenium ion in their equilibria, leading to more carbenium ion character, as opposed to bromonium ion character, as the solvent polarity increases. In nonpolar solvents, the increased relative bromonium ion stability may be attributed to "internal solvation" of the carbenium ion by the neighbouring bromine atom. In polar solvents, however, the carbenium ion is more stabilized, enabling it to rotate about the C—C single bond (A $\rightleftharpoons$ B), particularly at higher temperatures. In other words, the cyclic halonium ion becomes relatively more stable than the carbenium ion as the solvent becomes less capable of nucleophilic solvation. The observation that there is an increase in stereoselectivity for electrophilic alkene halogenation in nonpolar solvents at low temperatures, may serve as a qualitative example of solvent and temperature effects on halonium $\rightleftharpoons$ carbenium ion equilibria. These effects have been quantitatively determined by McManus and Peterson [388]. The fact that in polar solvents *racemic* and *meso*-forms are not obtained in equal amounts, although the reaction seems to proceed through an α-halogeno carbenium ion in these solvents, is simply explained by the different conformational energies of A and B. The conformer B in scheme (5-140), leading to *meso*-form, is sterically more favored than A. Generally, solvent polarity plays a significant role in determining the stereochemistry of addition reactions only in borderline cases such as stilbene where there is stabilization of the carbenium ion by the neighbouring phenyl group. Halogen addition to other alkenes such as diethyl fumarate and diethyl maleate also exhibits high stereoselectivity in polar solvents [79].

In cycloaddition and cycloreversion reactions, the mechanism and product distribution may be also strongly influenced by the reaction medium. Two examples have already been briefly mentioned in Sections 5.3.2 [102] and 5.3.3 [124]. In the reaction of dimethylketene with enamines such as *N*-isobutenylpyrrolidine, a two-step process *via* a zwitterionic intermediate, leading preferably to a δ-methylene-δ-lactone as 2:1 adduct, competes with a concerted mechanism to give a cyclobutanone derivative according to Eq. (5-141) [102]. In cyclohexane 92% of the enamine reacts concertedly to give the cyclobutanone derivative, whereas 8% goes through the zwitterionic intermediate. The path *via* the zwitterion increases in importance with increasing solvent polarity and reaches 57% in acetonitrile [102].

An interesting example of the solvent-dependent cycloreversion of a substituted $3H$-Δ^1-pyrazoline (formed by the 1,3-dipolar cycloaddition of diazomethane to an allylic bromide) has been given by Kolsaker *et al.* [675].

The stereoselectivity of the Diels-Alder reaction between methyl acrylate and cyclopentadiene depends to some extent on solvent polarity [124]; *cf.* Eq. (5-43) in Section 5.3.3. The *endo*/*exo* ratio for the cycloaddition product has been found to vary from 74:26 cmol/mol in triethylamine, to 88:12 cmol/mol in methanol at about 30 °C. It has been

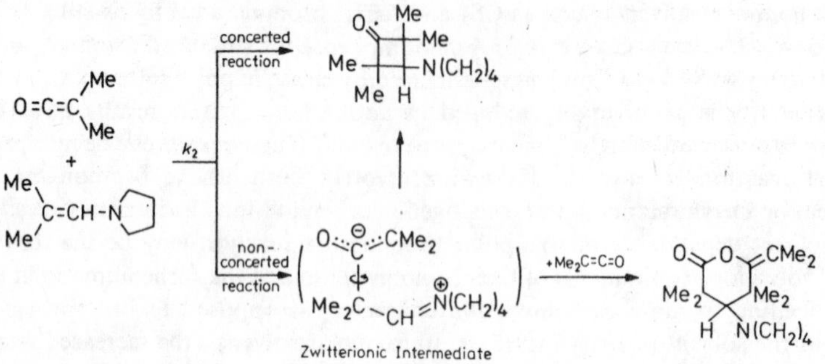

(5-141)

Solvent	c-C_6H_{12}	C_6H_6	C_6H_5Cl	$CHCl_3$	CH_3COCH_3	CH_3CN
% concerted reaction	92.0	79.4	72.8	62.3	52.1	43.0

observed, that for the same reaction not only the *exo*/*endo* stereoisomer ratio is affected by solvent polarity, but also the product ratio of *exo* (or *endo*) adduct to *endo*-dicyclopentadiene [389]. The latter is formed as a by-product in the Diels-Alder reaction. As expected, the *exo* (or *endo*) adduct/*endo*-dicyclopentadiene ratio increases with increasing solvent polarity [389]. – The rate and regioselectivity of Diels-Alder reactions with inverse electron demand between the ketene aminal $CH_2{=}C(NMe_2)_2$ and 3-phenyl-1,2,4,5-tetrazine as well as substituted 1,2,4-triazines is strongly solvent-dependent [676]. The "*ortho*"/"*meta*" product ratio of regioisomers decreases with increasing solvent polarity in non-HBD solvents [676].

Examples of the solvent-influenced competition between concerted [4+2]Diels-Alder type cycloaddition reactions and 1,4-dipolar reaction pathways with zwitterionic intermediates can be found in references [677–679]. For example, in solvents of low polarity ($CHCl_3$, CH_2Cl_2), homofuran reacts with tetracyanoethene to form the seven-membered [4+2]cycloadduct **A** in quantitative yield. In solvents of high polarity (CH_3CN), however, it is the [2+2]cycloadduct **B** which is formed mainly *via* an 1,4-dipolar activated complex and a zwitterionic intermediate [679]; *cf*. Eq. (5-142).

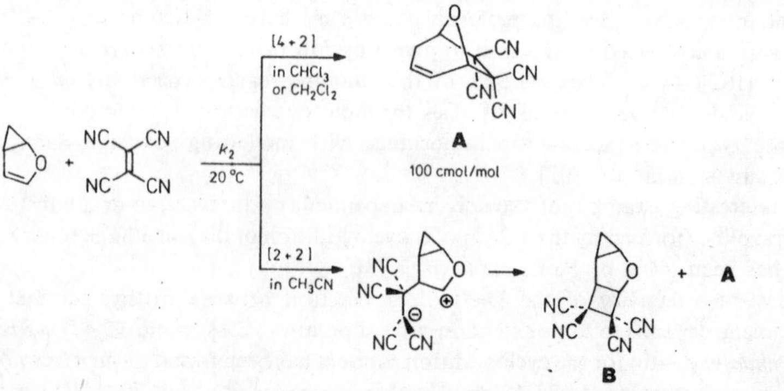

(5-142)

Appreciable solvent effects are also obtained in photochemically induced [2+2]cycloadditions and cycloreversions [390–392]. Examples are given in Eqs. (5-143) [390] and (5-144) [680].

The [2+2]photodimerization of 2-cyclopenten-1-one leads to the regioisomeric cyclobutane derivatives shown in Eq. (5-143) [390]. The *anti*/*syn* product ratio decreases as the solvent polarity increases. This is due to the more dipolar, and hence more strongly solvated, activated complex which leads to the dipolar *syn*-cycloadduct.

2 (2-cyclopenten-1-one) $\xrightarrow[20\,°C]{h\nu}$ *anti*-form ($\mu = 0$) + *syn*-form ($\mu \neq 0$) (5-143)

Solvent	C_6H_6	C_6H_5Cl	CH_2Cl_2	CH_3COCH_3	CH_3CN
$\lg \frac{[anti]}{[syn]}$	0.61	0.39	0.28	0.16	−0.14

When 1,2-bis(methoxycarbonyl)-3,4-bis(2-naphthyl)cyclobutane is irradiated in solution (cyclohexane/tetrahydrofuran and tetrahydrofuran/acetonitrile mixtures) in the presence of triethylamine, it undergoes [2+2]cycloreversion *via* an exciplex with triethylamine [680]. Interestingly, as shown in Eq. (5-144), the mode of cycloreversion changes with the solvent polarity, thus reflecting the different dipolar electronic structure of the exciplex. In nonpolar solvent mixtures, a 'horizontal' cleavage of the cyclobutane ring takes place, in highly polar solvent mixtures a 'vertical' one. It has been assumed that in the latter case the intermediate exciplex exhibits increased charge-transfer character [680].

Cyclobutane ($2\times$ 2-naphthyl, $2\times CO_2CH_3$) $\xrightarrow[+Et_3N]{h\nu}$ Nonpolar Solvents: 2 (2-naphthyl)CH=CH(CO_2CH_3); Polar Solvents: (2-naphthyl)CH=CH(2-naphthyl) + H_3CO_2C–CH=CH–CO_2CH_3 (5-144)

Photo-sensitized oxygenation of olefins with singlet oxygen can in principle proceed *via* three competitive reaction pathways: [4+2]cycloaddition to *endo*-peroxides, ene reaction to allylic hydroperoxides, and [2+2]cycloaddition to 1,2-dioxetanes (see reference [681] for a review). With suitable olefinic substrates, the chemical outcome of such photo-oxygenation reactions can be strongly influenced by the solvent. This is shown in the somewhat simplified Eq. (5-145).

[4+2] Nonpolar solvents — endo-Peroxide
Ene R. Nonpolar solvents — Hydroperoxide
[2+2] Polar protic solvents — Zwitterionic intermediate (1,4-dipol) — 1,2-Dioxetane

(5-145)

Since [4+2]cycloaddition and ene reactions are generally assumed to proceed in a concerted manner *via* isopolar activated complexes, they should exhibit virtually the same small often negligible response to changes in solvent polarity. This is what in fact has been found; *cf.* for example [138, 682, 683]. However, two-step [2+2]cycloaddition reactions of singlet oxygen to suitably substituted electron-rich olefins proceed *via* dipolar activated complexes to zwitterionic intermediates (1,4-dipoles or perepoxides). In this case, the relative amounts of 1,2-dioxetane and allylic hydroperoxides or *endo*-peroxides should vary markedly with solvent polarity if two or even all three of the reaction pathways shown in Eq. (5-145) are operative [681, 683, 684].

In the photo-oxygenation of enol ethers where the ene reaction and [2+2]cycloaddition compete, polar solvents favour cycloaddition whereas nonpolar solvents favour ene product formation [681, 683–685]. For example, 2,3-dihydro-4-methyl-4*H*-pyrane reacts with singlet oxygen to yield both a 1,2-dioxetane and an allylic hydroperoxide as primary products in nonpolar cyclohexane and dipolar acetonitrile, with product ratios of 13:87 and 84:16 cmol/mol, respectively, as shown in Eq. (5-146) [685].

$+ {}^1O_2$ → products (5-146)

Cyclohexane:	13 cmol/mol	87 cmol/mol
Acetonitrile:	84 cmol/mol	16 cmol/mol

Further evidence in support of zwitterionic intermediates in the [2+2]cycloaddition of singlet oxygen to electron-rich alkenes has been obtained by Jefford *et al.* [684]. The photo-oxygenation of 2-(methoxymethylidene)adamantane creates a zwitterionic intermediate (peroxide or perepoxide) which can be captured by acetaldehyde to give 1,2,4-trioxanes in addition to 1,2-dioxetanes; *cf.* Eq. (5-147).

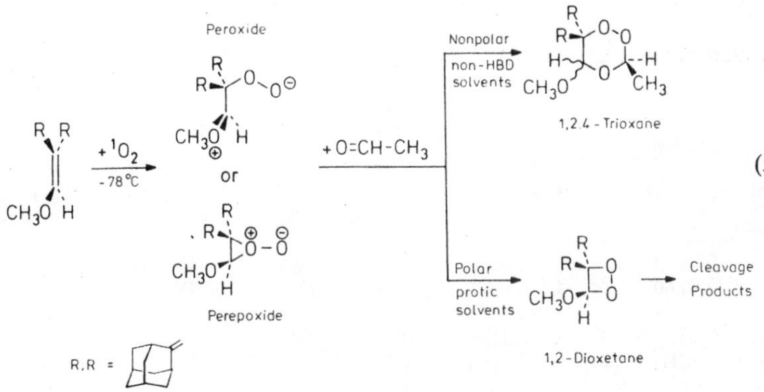

Peroxide

$+{}^1O_2$, $-78\,°C$

or

Perepoxide

$+ O{=}CH{-}CH_3$

Nonpolar non-HBD solvents → 1,2,4-Trioxane

Polar protic solvents → 1,2-Dioxetane → Cleavage Products

R,R =

(5-147)

The ease of capture itself is also solvent-dependent: polar protic solvents (methanol, ethanol) stabilize the zwitterion by H-bonding, thus rendering it unreactive towards external electrophiles such as acetaldehyde, and facilitating its closure to a 1,2-dioxetane. On the other hand, nonpolar solvents (toluene, tetrahydrofuran, ethyl acetate) prevent the dispersion of charge, thus enhancing the reactivity towards aldehydes and favouring 1,2,4-trioxane formation [684]. The zwitterionic intermediate in Eq. (5-147) can be formulated as either peroxide or perepoxide (peroxirane). A distinction between these two species is difficult to make. In both cases, the positive charge is stabilized as an oxonium ion while the negative charge remains localized on the terminal oxygen atom. According to calculations, a peroxide would have a permanent dipole moment of *ca.* $34 \cdot 10^{-30}$ C m (10.1 D) and a perepoxide a dipole moment of *ca.* $19 \cdot 10^{-30}$ C m (5.6 D) [686].

Further recent examples of solvent effects on photo-oxygenation reactions with singlet oxygen can be found in references [687–689].

Similar solvent effects can be found in fragmentation reactions [109]. Such reactions may proceed heterolytically or homolytically depending on the solvent used as reaction medium. Thus, the observed solvent effect for the thermolysis of p-CH_3—C_6H_4—N=N—S—C_6H_4—p—$C(CH_3)_3$ indicates that this benzenediazoaryl sulfide decomposes in apolar non-HBD solvents homolytically according to Eq. (5-148a) [393].

The marked increase in rate observed in protic solvents such as acetic acid and aqueous ethanol, however, may be due to heterolytic decomposition according to Eq. (5-148b). Suitably substituted benzenediazoaryl sulfides such as p-CH_3O—C_6H_4—N=N—S—C_6H_4—p—NO_2 should be more inclined to decompose heterolytically because of mesomeric stabilization of the diazonium and thiolate ions formed. Indeed, the thermolysis of this compound is extremely sensitive to solvent polarity, in polar solvents heterolysis being presumably predominant [393]. Later on, it was found that the thermolysis of (Z)-benzenediazoalkyl sulfides leads to isomerization to the (*E*)-form as well as to simultaneous decomposition [394]. Only the decomposition is strongly accelerated with increasing solvent polarity, whereas the rate of (*Z*)/(*E*)-isomerization is almost independent of the solvent.

$$Ar{-}N{=}N{-}SR \xrightarrow{k_1} \xrightarrow{\text{Homolysis}} [Ar{-}N{=}\overset{\delta\odot}{N}\cdots\overset{\delta\odot}{SR}]^{\ddagger} \longrightarrow Ar{-}N{=}\overset{\odot}{N} + \overset{\odot}{S}R \longrightarrow \ldots \quad (5\text{-}148a)$$

$$\xrightarrow{\text{Heterolysis}} [Ar{-}\overset{\delta\oplus}{N}{\equiv}N\cdots\overset{\delta\ominus}{SR}]^{\ddagger} \longrightarrow Ar{-}\overset{\oplus}{N}{\equiv}N + {}^{\ominus}SR \longrightarrow \ldots \quad (5\text{-}148b)$$

Solvent	i-C_8H_{18}	C_6H_6	C_5H_5N	CH_3SOCH_3	CH_3CO_2H	C_2H_5OH/H_2O (96:4 cl/l)
$k_1^{\text{rel a)}}$	0.32	1	0.47	0.91	47	44
$k_1^{\text{rel b)}}$	0.60	1	25	100	45	190

[a] p-$CH_3{-}C_6H_4{-}N{=}N{-}S{-}C_6H_4{-}p{-}C(CH_3)_3$
[b] p-$CH_3O{-}C_6H_4{-}N{=}N{-}S{-}C_6H_4{-}p{-}NO_2$

According to Zollinger *et al.* [690], depending on the solvent used the thermal dediazoniation of arenediazonium ions in solution can proceed *via* the competitive heterolytic and homolytic reaction pathways; *cf.* the somewhat simplified Eq. (5-149).

$$Ar{-}\overset{\oplus}{N}{\equiv}N \xrightarrow[25\,°C]{k_1} \xrightarrow[\text{low nucleophilicity}]{\text{Solvents of}} [Ar\cdots\overset{\delta\oplus}{N}{\equiv}\overset{\delta\oplus}{N}]^{\ddagger} \longrightarrow Ar^{\oplus} + N{\equiv}N \longrightarrow \text{products} \quad \text{(Heterolysis)}$$

$$\xrightarrow[\text{Solvents of high nucleophilicity}]{+\,:S} [Ar\cdots\overset{\delta\oplus}{N}{\equiv}N\cdots\overset{\delta\oplus}{S}]^{\ddagger} \longrightarrow Ar\odot + N{\equiv}N + \odot S^{\oplus} \longrightarrow \text{products} \quad \text{(Homolysis)} \quad (5\text{-}149)$$

In solvents of low nucleophilicity (*e.g.* $(CF_3)_2CHOH$, CF_3CH_2OH, CH_3CO_2H, H_2O with pH < 1), the familiar heterolytic dediazoniation takes place to give an aryl cation and products derived therefrom. The solvent effect on heterolytic dediazoniation rates is small, the slowest (in 1,4-dioxane) and fastest rate (in 2-propanol) differing by a factor of only 9 [690].

However, in solvents of high nucleophilicity (*e.g.* DMSO, HMPT, pyridine) the formation of products derived from aryl radicals is favoured. Therefore, in these solvents a homolytic dediazoniation must be taking place. This mechanism involves addition of a nucleophilic solvent molecule to the electrophilic β-nitrogen atom of the diazonio group, followed by homolysis into a radical pair and nitrogen. The homolytic dediazoniation is not only enhanced by sufficiently nucleophilic solvent molecules but also by the addition of other nucleophiles which form relatively stable radicals through electron transfer. Under comparable reaction conditions, the rates of homolytic dediazoniation are higher than the corresponding rates of heterolysis [690].

Dichotomous homolytic and heterolytic decomposition reactions can also be found in the thermolysis of peroxycarboxylic esters [195, 207–209]; *cf.* Eq. (5-62) and Table 5-9 in Section 5.3.4. An already classical example, given in Eq. (5-150), is exhibited by the decomposition of *trans*-9-decalyl peroxyphenylacetate [207].

in C_6H_5-C_2H_5, 90 °C: + CO_2 + ⊙R ⟶ products

k_1

R = C_6H_5-CH_2

in CH_3CN, 50 °C

(5-150)

Product analysis and reaction-rate medium effects reveal such a change in mechanism on passing from nonpolar to polar solvents [207, 209]. The decomposition of this perester occurs by heterolysis in alcohols and mainly by homolysis in ethylbenzene, whereas in acetonitrile both mechanisms appear to compete with each other [207].

For further examples of dichotomous solvent-influenced radical/ionic perester decompositions see the base-catalyzed perester fragmentation shown in Eq. (5-39) in Section 5.3.2 [110] as well as the decomposition of *t*-butyl heptafluoroperoxybutyrate, C_3F_7—CO—O—O—C(CH_3)$_3$ [691]. – The relative extent of monomolecular and induced thermal decomposition of disubstituted dibenzyl peroxydicarbonate, ArCH_2—O—CO—O—O—CO—O—CH_2Ar, is also substantially influenced by the reaction medium [692]. – The thermolysis of suitable dialkyl peroxides can also proceed via two solvent-dependent competitive reaction pathways (homolytic and electrocyclic reaction) as already shown by Eq. (5-59) in Section 5.3.4 [564].

The solvent-influenced *syn/anti* dichotomy for bimolecular eliminations of acyclic and medium-ring bromides, tosylates, and 'onium salts has been reviewed [395, 693] and will be mentioned only briefly. As a rule, the *syn*-elimination pathway gains importance in non-dissociating solvents, while dissociating solvents facilitate the more common *anti*-elimination reaction. The more unusual *syn*-elimination is favoured in non-dissociating solvents because of ion-pair association which favours a cyclic six-membered activated complex as shown in Eq. (5-151a); see reference [395].

The advantage of *syn*-elimination in solvents of low dielectric constants lies also in the formation of contact ion pair in the product, whereas *anti*-elimination produces a product-separated ion pair according to Eq. (5-151b). Thus, reaction of free ions proceeds in standard *anti* fashion, while ion pairs (or higher aggregates) tend to react by a *syn* pathway. Solvent separation of the $RO^{\ominus}M^{\oplus}$ ion pair in the reactant state removes the driving force for *syn*-elimination (*i.e.* dissociating solvents, increase in cation size, addition of crown ethers, *etc.*).

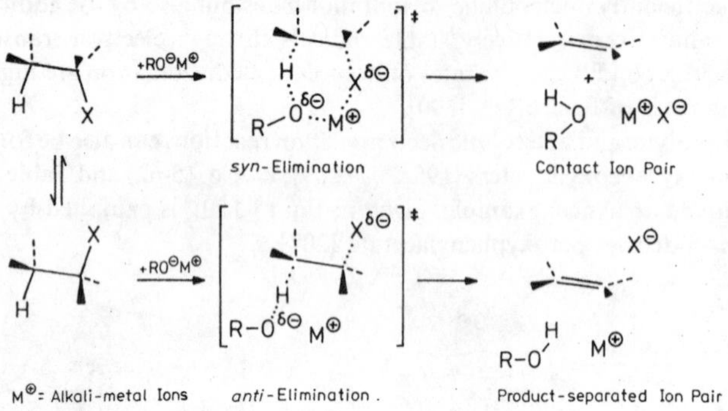

(5-151a)

(5-151b)

For example, the *syn/anti* product ratio of the base-promoted dehydrochlorination of *meso*-3,4-dichloro-2,2,5,5-tetramethylhexane with potassium *t*-butoxide to give (*Z*)- and (*E*)-3-chloro-2,2,5,5-tetramethyl-3-hexene changes from 92:8 in tetrahydrofuran ($\varepsilon_r = 7.6$) and 86:14 in *t*-butanol ($\varepsilon_r = 12.5$) to 7:93 in dimethyl sulfoxide ($\varepsilon_r = 46.7$), respectively [694].

An example of extreme solvent dependence is seen in the competition between dehydrobromination (believed to be E_2C) and debromination (designated E_2Br) as shown in Eq. (5-152) [396].

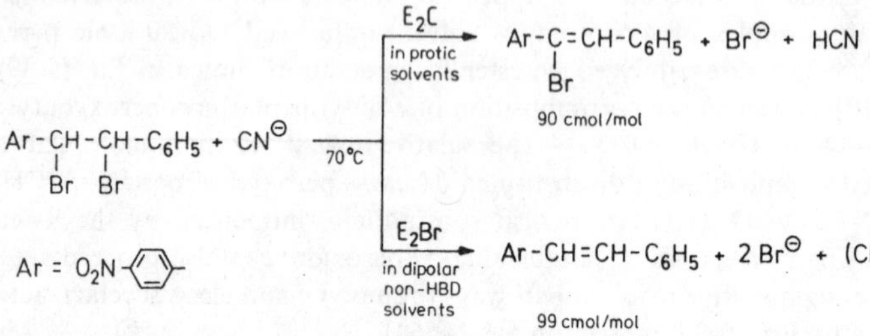

(5-152)

In *N,N*-dimethylformamide, *anti*-debromination of *erythro*-1,2-dibromo-1-(4-nitrophenyl)-2-phenylethane with $(n\text{-}C_4H_9)_4N^{\oplus}CN^{\ominus}$ gives 99 cmol/mol 4-nitro-*trans*-stilbene. As the solvent composition approached pure ethanol, the proportion of α-bromo-4-nitro-*cis*-stilbene, the product of *anti*-dehydrobromination, increases to 90 cmol/mol. The E_2C reaction involves a looser activated complex, better solvated by protic solvents than its tighter E_2Br-like counterpart [396].

The ability of reagents to differentiate between functional groups is called *chemoselectivity* and can be solvent-dependent. A nice clear-cut example is the reduction of the bifunctional compound 11-bromoundecyl tosylate with lithium aluminium hydride in different solvents as shown in Eq. (5-153) [695].

With diethyl ether as solvent, the reaction proceeds with the selective reduction of the tosyl group to give 1-bromo-*n*-undecane, whereas in diethylene glycol dimethyl ether ("diglyme") the bromo substituent is selectively reduced to yield *n*-undecyl tosylate. In diethyl ether, the lithium cation is poorly solvated and $LiAlH_4$ reacts as an ion pair. In

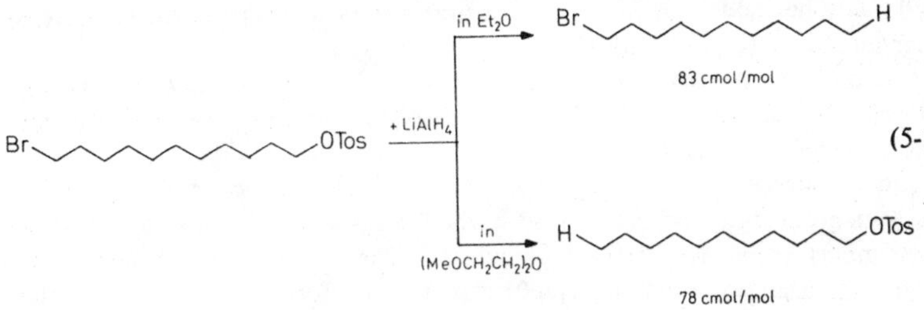

Br–(CH₂)–OTos + LiAlH$_4$ → in Et$_2$O: Br–…–H (83 cmol/mol); in (MeOCH$_2$CH$_2$)$_2$O: H–…–OTos (78 cmol/mol)

(5-153)

diglyme, the lithium cation is better solvated, forming solvent-separated ion pairs, thereby enhancing the nucleophilicity of the $AlH_4^{\ominus}$ ion which results in rapid reduction of the bromo substituent. The high reactivity of $LiAlH_4$ toward alkyl tosylates in the weakly cation-solvating diethyl ether is presumably due to the complexation of the lithium cation to the tosyl group thereby increasing its leaving-group ability [695]. – Other solvent-dependent chemoselective reductions of bifunctional compounds with lithium aluminium hydride [696, 697] and with lithium borohydride [709] have been reported.

The dissociation equilibrium of associated $Li^{\oplus}AlH_4^{\ominus}$ ion pairs in a variety of ethereal solvents has been investigated by ^{7}Li- and ^{27}Al-NMR spectroscopy [768], $\delta(^{27}Al)$ is nearly independent of the solvent as is $^1J_{Al-H}$, whereas $\delta(^7Li)$ depends on the $LiAlH_4$ concentration and the EPD-character of the ethereal solvent. According to this result, diglyme seems to be the solvent of choice for reactions in which the $AlH_4^{\ominus}$ should be least affected by ion-pair formation [768].

Another interesting example of solvent-dependent control of chemoselectivity is the reaction of the bromo α-enone with an organo-cupper reagent as seen in Eq. (5-154) [698]. The lithium dimethylcuprate displacement reaction of haloalkanes and the conjugate addition of cuprate reagents to α-enones exhibit opposite responses on the addition of EPD solvents. An appropriate choice of substrate and reaction medium should therefore make it possible to select either the $LiCuMe_2$ coupling reaction with a haloalkane or its conjugate addition.

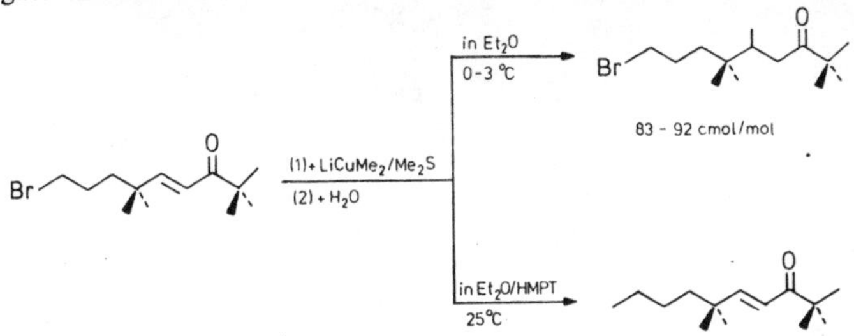

(1) + $LiCuMe_2/Me_2S$; (2) + H_2O → in Et$_2$O, 0-3 °C: 83 - 92 cmol/mol; in Et$_2$O/HMPT, 25 °C

(5-154)

Indeed, usually lithium dimethylcuprate reacts in diethyl ether with the α-enone group to give a methylsubstituted bromo ketone. Addition of hexamethylphosphoric triamide (HMPT), however, slows down this reaction to such an extent that displacement of the bromo substituent takes place [698]. – Another remarkable example of the influence of HMPT on chemoselectivity is the reaction of an arsonium ylide, $Ph_3As{=}CH{-}CH{=}$

CH—Ph, with benzaldehyde in tetrahydrofuran solution, yielding either an epoxide (in THF) or an alkene (in THF/HMPT) [699].

Not only chemoselectivity but also *enantioselectivity* and *diastereoselectivity* of organic reactions can be controlled by choice of solvent; for recent reviews on asymmetric induction see references [700, 701].

Some examples of enantioselective syntheses, carried out in chiral media, have already been given at the end of Section 3.2 which deals with chiral solvents; *cf.* also Table A-2 (Appendix). In general, asymmetric inductions as a result of chiral solvents or chiral cosolvents are disappointingly low [700]. The reasons for this are that the differential solvation by the chiral solvent of the two enantiomorphic activated complexes which lead to either the (*R*)- or (*S*)-product is not sufficient. That is, the difference in Gibbs energy of activation, $\Delta\Delta G^{\neq} = \Delta G^{\neq}_{(R)} - \Delta G^{\neq}_{(S)}$, is not large enough to favour only one of the two enantiomeric products. It should be remembered that a difference of $\Delta\Delta G^{\neq} = 10.8$ kJ/mol (2.6 kcal/mol) at 20 °C would be sufficient to get an (*S*)/(*R*) product ratio of 99:1 (*i.e.* an enantiomeric excess *e.e.* = 98%).

Therefore, a more direct covalent linking of the chiral information with the reactants is necessary in order to obtain diastereomorphic activated complexes with greater differential solvation, thus leading to better diastereoselectivity.

The Grignard reaction of (±)-3-phenylbutanone with phenylmagnesium bromide to give stereoisomeric 2,3-diphenylbutan-2-ols, shown in Eq. (5-155), represents such an example [702].

(*R*) + (*S*) —[(1) + C_6H_5MgBr at 30 °C; (2) + H_2O, − Mg(OH)Br; 20 - 98 cmol/mol]→ (*R,R*) + (*S,R*) + (*S,S*) + (*R,S*) (5-155)

Solvent	Et_3N	Et_2O	1,4-dioxane	$(CH_2)_4O$	diglyme	$(MeOCH_2)_2$
% [(*S,R*)+(*R,S*)]	26	36	49	61	66	73

In this particular case, the observed diastereoselectivity, *ds* = % [(*S,R*)+(*R,S*)], increases with increasing solvent polarity. The (*SR, RS*)-carbinol is preferably formed in more polar solvents, whereas the (*RR, SS*)-carbinol dominates in less polar solvents. This result can be understood by a careful analysis of the two diastereomorphic activated complexes in terms of steric and polar effects. The activated complex which leads to the (*SR, RS*)-carbinol appears to be more dipolar and hence more strongly solvated; see reference [702] for details.

Another nice example is the addition of lithium dimethylcuprate to a chiral oxazolidine (prepared from (*E*)-cinnamaldehyde and (−)-ephedrine), followed by hydrolysis, to give 3-phenylbutanal; *cf.* Eq. (5-156) [703].

$$\text{oxazolidine} \xrightarrow[\text{(2) + } H_3O^{\oplus}]{\text{(1) + LiCuMe}_2 \text{ at } -42\,^{\circ}\text{C}} \text{(S)-Aldehyde} + \text{(R)-Aldehyde} \qquad (5\text{-}156)$$

50-80 cmol/mol

(*S*)-Aldehyde (*R*)-Aldehyde

Diastereoselective addition in Et_2O/HMPT (1:1) leads to the (*S*)-aldehyde with an enantiomeric excess of 40%, whereas in *n*-hexane the (*R*)-aldehyde is formed with 80% (!) enantiomeric excess. The (*R*)-configurated aldehyde is also obtained in benzene and in dichloromethane but with lower *e.e.*-values, 50% and 25%, respectively. Inverse results were obtained with a chiral oxazolidine prepared from (*E*)-cinnamaldehyde and (+)-ephedrine. Here the (*S*)-aldehyde with *e.e.* = 79% is formed in *n*-hexane, and the (*R*)-aldehyde in Et_2O/HMPT (1:1) with *e.e.* = 50% [703]. This result may be due to different structures of the organocopper reagent, and hence the diastereomorphic activated complexes, in nonpolar solvents (*n*-hexane, benzene, dichloromethane) and in EPD solvents (Et_2O/HMPT) [703].

A striking solvent effect was observed in the reduction of a chiral α-keto amide, C_6H_5—CO—CO—NR_2 (NR_2 = (*S*)-proline methyl ester), with sodium tetrahydridoborate, leading to mandelic acid after hydrolysis [704]. When the α-keto amide was reduced in pure tetrahydrofuran or methanol, the resulting enantiomeric excess of (*S*)-mandelic acid produced was 36% and 4%, respectively. However, when a tetrahydrofuran/methanol (99:1 cl/l) solvent mixture was used, the enantiomeric excess increased to 64% (!). In either solvent mixtures, a catalytic amount of a protic solvent (CH_3OH or H_2O) was found to be necessary for good asymmetric inductions [704].

Extremely diastereoselective solvent-influenced alkylations were obtained with propionates of chiral alcohols derived from (+)-camphor, as shown in Eq. (5-157) [705].

Metallation of the propionate with lithium cyclohexylisopropylamide (LiCA) in a solvent-controlled reaction gives either the (*Z*)- (in THF) or (*E*)-lithium enolate (in THF/HMPT 4:1). Rearside shielding of both diastereomeric enolates, provided by the 3,5-dimethylphenyl group attached to the sulfonamide moiety at C-2, prevents alkylation of these enolates from the backface. Hence, frontface attack with iodo-*n*-tetradecane leads to diastereomeric esters with either (*R*)- or (*S*)-configuration at the α-carbon atom of the propionate group. Using LiCA/HMPT mixtures with increasing HMPT molar ratio in the metallation step leads to a smooth changeover from diastereoselective formation of the (*R*)-ester to diastereoselective formation of the (*S*)-ester. The turning point is at an LiCA/HMPT molar ratio of 1:1. Obviously, this configurational conversion is caused by the kinetically controlled formation of the diastereomeric enolates by deprotonation with either LiCA or the LiCA/HMPT complex. The EPD solvent HMPT acts as an lithium-solvating agent in this reaction, causing preferential formation of the (*E*)-lithium enolate. An explanation for this dramatic solvent effect has been given by analysis of the steric requirements for the enolization of esters [706]. The better solvation of $Li^{\oplus}$ in the presence of HMPT and the enhanced reactivity of the amide base favours the activated complex leading to the (*E*)-enolate. In the less coordinating solvent THF, the interaction between $Li^{\oplus}$ and the estercarbonyl oxygen atom must be strong, thus favouring the activated complex leading to the (*Z*)-enolate; see reference [706] for details. – Addition of cation-solvating HMPT *after* the deprotonation step in reaction (5-157) also leads to an increase in diastereoselectivity, *i.e.* preferential formation of the (*R*)-ester. This solvent effect may be caused by disruption of conformationally unfavourable interactions between $Li^{\oplus}$ and the 3,5-dimethylphenyl group of the sulfonamide moiety; *cf.* Eq. (5-157) [705].

+ LiCA, −78 °C; in THF → (*Z*)-Enolate; + $n\text{-}C_{14}H_{29}I$, −40 °C, 74 cmol/mol → (*R*)-Ester, *d.e.* = 95 %

in THF/HMPT 4 : 1 → (*E*)-Enolate; + $n\text{-}C_{14}H_{29}I$, −40 °C, 77 cmol/mol → (*S*)-Ester, *d.e.* = 96 %

(5-157)

Solvent-controlled diastereoselectivities have also been observed in Diels-Alder cycloaddition reactions of cyclopentadiene with bis-(−)-menthyl fumarate [707] and with the acrylate of (*S*)-ethyl lactate, $CH_2{=}CH{-}CO{-}OCH(CH_3){-}CO_2Et$ [708]. In the latter reaction, giving four diastereomeric cycloadducts, diastereoselectivities of up to 85:15 have been obtained in *n*-hexane [708]. The diastereoselectivities decrease with increasing solvent polarity, while the *endo*/*exo* selectivity increases. This is in agreement with the pattern found for simple achiral acrylates [124]; *cf.* Eq. (5-43) in Section 5.3.3.

We shall conclude this Section with reference to the (*E*)/(*Z*)-isomerization of imino compounds $R_2C{=}N{-}R'$, which in principle may proceed by a rotation mechanism or an inversion mechanism. Determination of the solvent dependence of this isomerization has been used in deciding between the two possible reaction mechanisms [397, 398]. The small solvent effects usually observed are in agreement with an inversion mechanism [398]. – An inversion mechanism was also postulated for the (*E*)/(*Z*)-isomerization of push-pull substituted azobenzenes [529, 561]; *cf.* Eq. (5-40) in Section 5.3.2.

Only a few representative particularly interesting examples of solvent effects on the mechanisms and stereochemistry of organic reactions were mentioned in this Section. The significance of these, often very specific solvent effects are well recognized by organic chemists, but their detailed understanding is frequently insufficient and the subject of current research.

5.5.8 Influence of Solvophobic Interactions on Reaction Rates and Mechanisms

The modification of chemical reactions through the incorporation of the reactant molecules into aqueous micelles or other organized assemblies has received considerable attention in recent years. Reactions are known where rates, mechanisms, and even the stereochemistry have been significantly affected by the addition of so-called *amphiphiles* to the reaction medium.

Solvophobic and especially hydrophobic interactions provide the driving force for aggregation of organic ions known as amphiphiles in dilute aqueous solutions (*cf.* Sections 2.2.7 and 2.5) [399]. Amphiphilic ions possessing long normal hydrocarbon chains have both pronounced hydrophobic and hydrophilic properties and exhibit, therefore, the important property of forming, over a narrow concentration range, termed the *critical micelle concentration* (*cmc*), molecular aggregates in solution, called *micelles* (*cf.* Fig. 2-12 in Section 2.5)*). These micelles, rather than individual amphiphilic ions, may cause alterations of rates and mechanisms of organic reactions in aqueous solutions of surfactants. A suitable choice of surfactant can lead to rate increases of 5- to 1000-fold compared to the same reaction in the absence of surfactant. The catalysis of organic reactions by ionic micelles can be explained in terms of electrostatic and hydrophobic interactions of the reactants and activated complexes with the micelle. The reaction substrate is partitioned between micellar and bulk aqueous phases by hydrophobic binding of the substrate to a micelle. Then, by simple electrostatics, this complex either attracts (rate acceleration) or repels (rate retardation) an incoming ionic reactant. The field of micellar catalysis has recently been the subject of comprehensive reviews [289, 400–403, 711–713]. Therefore in this Section, only a few typical examples shall demonstrate the influence of hydrophobic interactions on organic reactions.

According to Eq. (5-158), the reaction of 1-fluoro-2,4-dinitrobenzene with phenoxide or thiophenoxide ions is accelerated by micelles of cetyltrimethylammonium bromide (CTABr) in aqueous solution by factors of 230 and 1100, respectively [404].

This enhancement of the reaction of bulk-phase anions $C_6H_5Y^{\ominus}$ with the organic nonelectrolyte 2,4-$(NO_2)_2C_6H_3F$, the latter partitioned between bulk and micellar phase, is expected in the presence of cationic amphiphiles such as CTABr from pure electrostatic

$$2,4\text{-}(NO_2)_2C_6H_3\text{-}F + {}^{\ominus}Y\text{-}C_6H_5 \xrightarrow[25\,°C]{k_2} 2,4\text{-}(NO_2)_2C_6H_3\text{-}Y\text{-}C_6H_5 + F^{\ominus} \quad (5\text{-}158)$$

Surfactant	no CTABr added	with cationic CTABr
$k_2^{rel}(Y=O)$	1	230
$k_2^{rel}(Y=S)$	1	1100

* Depending on their chemical structure, surfactants capable of forming micelles are usually classified into cationic (*e.g.* ammonium salts), anionic (*e.g.* sulfates, carboxylates), ampholytic (*e.g.* zwitterionic salts), and non-ionic surfactants (usually containing polyoxyethylene chains); *cf.* Table 2-10 in Section 2.5.

considerations. Amphiphiles with opposite charge to that of the reactant ion accelerate, whereas amphiphiles of like charge inhibit the reactions of neutral substrates. The catalytic efficiency increases with increasing distribution of neutral substrate into the micelles, that is, with increasing lipophilicity of the reactant molecules. A compilation of micellar catalytic effects expected for organic reactions of different charge type in the presence of cationic or anionic micelles can be found in reference [289].

Even at reactant concentrations well below critical micelle concentrations, hydrophobic interactions may result in marked rate accelerations. Thus, the rates of bimolecular aminolysis of *p*-nitrophenyl decanoate and acetate by *n*-decylamine and ethylamine have been determined in aqueous solution and a distinct rate enhancement in the aminolysis of the long chain ester by the long chain amine has been reported [405, 406]. As shown in Table 5-24, the ratio $k_2^{decylamine}/k_2^{ethylamine}$ for *p*-nitrophenyl

Table 5-24. Second-order rate constants for the aminolysis of *p*-nitrophenyl esters in water/acetone (99:1) at 35 °C according to $CH_3(CH_2)_nCO{-}OC_6H_4{-}p{-}NO_2 + C_2H_5(CH_2)_nNH_2 \rightarrow CH_3(CH_2)_nCO{-}NH(CH_2)_nC_2H_5 + HOC_6H_4{-}p{-}NO_2$, with $n=0$ and 8 [405].

Ester	$k_2^{ethylamine}$	$k_2^{decylamine}$	$k_2^{decylamine}/k_2^{ethylamine}$
p-Nitrophenyl acetate	5.74	39.1	6.8
p-Nitrophenyl decanoate	0.42	133	317
$k_2^{decanoate}/k_2^{acetate}$	0.073	3.4	46.6

decanoate is 317*). That is, *n*-decylamine attacks the long-chain ester nearly 47 times faster than expected on the basis of its reactivity toward *p*-nitrophenyl acetate. This rise in the observed rate constant is best explained by the association of the reactant molecules to form a kind of "micro-micelle" prior to reaction. This is due to hydrophobic interactions between the long hydrocarbon chains of both reactants in aqueous solution. In accordance with this interpretation, the ratios $k_2^{decylamine}/k_2^{ethylamine}$ fall to values near unity for both esters, when the reactions are carried out in aqueous 1,4-dioxane (50:50) – a medium, in which hydrophobic interactions are seriously disrupted [405].

Diels-Alder cycloaddition reactions have undergone impressive improvements recently, taking advantage of hydrophobic interactions existing between the essentially nonpolar reactants in the aqueos medium. The use of water as a solvent in Diels-Alder reactions leads to greatly enhanced reaction rates and selectivities. This remarkable result has been pioneered by Breslow *et al.* [714] and further explored by Grieco *et al.* [715–717]; for a recent review *cf.* [718].

For example, the rate of the Diels-Alder cycloaddition reaction between 9-hydroxymethylanthracene and *N*-ethylmaleimide, as shown in Eq. (5-159), is only slightly altered changing the solvent from dipolar acetonitrile to nonpolar isooctane [714] – as expected for an isopolar transition state reaction; *cf.* Chapter 5.3.3. – In water, however,

* Even larger rate increases (up to 10^7-fold) were observed for the same, but amine-catalyzed termolecular reaction of *p*-nitrophenyl ester with alkylamines, carried out in water or aqueous ethanol at 25 °C [406].

(5-159)

Solvent	i-C_8H_{18}	n-C_4H_9OH	CH_3OH	CH_3CN	H_2O	H_2O + 4.86 M LiCl
k_2^{rel}	7.4	6.2	3.2	1	211	528

the reaction is 211 times faster than in acetonitrile! The addition of lithium chloride, known as a solute which increases hydrophobic effects, the rate in water is increased by a further 2.5 fold. This exceptional behaviour of the solvent water is best explained by means of hydrophobic interactions which promote the association of the diene and the dienophile during the activation process [714]. For some Diels-Alder reactions of cyclopentadiene, which can give both *endo* and *exo* addition of dienophiles, the *endo*/*exo* product ratio increases when water is the solvent [714]. This is related to the well known influence of polar solvents on such ratios [124] as well as to the need to minimize the activated complex surface area in aqueous solution.

The Diels-Alder cycloaddition reaction of 2,6-dimethyl-*p*-benzoquinone with methyl (*E*)-3,5-hexadienoate, carried out in toluene as solvent, even after seven days gives only traces of the cycloadduct shown in Eq. (5-160). However, when the solvent is changed to water and sodium (*E*)-3,5-hexadienoate is the diene, 77 cmol/mol of the desired cycloadduct have been obtained after one hour and esterification with diazomethane [715, 716]*). Again, hydrophobic interactions between diene and dienophile in the aqueous medium seem to be responsible for this remarkable and synthetically useful rate acceleration. – This work has been extended to aza Diels-Alder cycloaddition reactions between dienes and imonium ion dienophiles in aqueous solution [717].

(5-160)

R = CH_3: in Toluene, 7d → traces only

R = $Na^{\oplus}$: in Water, 1 h → 77 cmol/mol (after esterification with CH_2N_2)

A sugar-assisted solubilization of the diene in aqueous Diels-Alder reactions has been recently proposed, using glucose as the hydrophilic moiety [744]. – Based on the aqueous Diels-Alder cycloaddition of cyclopentadiene with diethyl fumarate, an increase in reaction rate with the increasing solvophobicity of the medium, characterized by

* Under the conditions of the aqueous Diels-Alder reaction the initially formed *cis*-cycloadduct equilibrates to the more stable *trans*-fused ring system [716].

Abraham's solvophobicity parameter *Sp* (*cf.* Section 7.3), has been found [745]. Addition of β-cyclodextrin to the reaction medium promotes this cycloaddition to an additional degree. Increasing medium solvophobicity and the addition of β-cyclodextrin also influences the diastereoselectivity (*i.e.* the *exo/endo* product ratio) of this Diels-Alder reaction: the amount of *endo*-product increases on addition of β-cyclodextrin and with increasing medium solvophibicity, due to the more compact activated complex leading to the *endo*-adduct [745].

An interesting phenomenon related to the acceleration of Diels-Alder reactions in aqueous media is found when clays such as montmorillonite are suspended in an organic solvent (*e.g.* ethanol or dichloromethane) [719]. The layered structure of the clay enables it to trap pools of internal water which obviously are capable of exerting hydrophobic effects.

The hydrophobicity-driven association of reactant molecules in aqueous solution has been found even in aldol reactions. The trimethylsilyl ether of cyclohexanone adds to benzaldehyde in aqueous solution at 20 °C without a catalyst to give aldol addition products with a *syn/anti* stereoselectivity that is the reverse of the acid-catalayzed reaction carried out in nonaqueous dichloromethane [746].

The useful ability of micelles in producing high local concentrations of bound organic reactants at low bulk concentrations has been used in photodimerization reactions; see references [712, 713] for recent reviews. The increase in reactant concentration per unit volume of the micelle promotes the probability of encounters between two molecules resulting in up to 1000 fold rate enhancements. The photodimerization of acenaphthylene illustrates this effect nicely; *cf.* Eq. (5-161) [720, 721].

2 [acenaphthylene] $\xrightarrow[\text{in } C_6H_6 \text{ + surfactant}]{h\nu}$ *cis*-Adduct + *trans*-Adduct (up to 97 cmol/mol) (5-161)

At acenaphthylene concentrations as low as $2 \cdot 10^{-3}$ mol/l, facile photodimerization takes place in benzene in the presence of nonionic or anionic surfactants, whereas in pure benzene at these concentrations no detectable amount of the two dimers were found [720]. Furthermore, the *cis/trans* product ratio is slightly dependent on the type of surfactant used [720, 721]. The product ratio during the course of photodimerization of 1-substituted acenaphthylenes is influenced by both solvent polarity and the addition of surfactants. The photodimerization reaction in micelles yields stereoisomer ratios similar to those obtained in polar solvents [721].

Most chemical studies in this field have concentrated on the effect of micellar surfactants on reaction rates and only few attempts have been made to investigate the effect micelles might have in altering the relative extent of competing reactions. For example, in studying the competitive hydrolysis and aminolysis of aryl sulfates in aqueous solution, Fendler *et al.* have found [407], that cationic micelles such as cetyltrimethylammonium bromide (CTABr) are able to alter the balance between S—O bond fission and C—O bond fission as shown in Eq. (5-162).

$$2,4\text{-}(O_2N)_2C_6H_3\text{-}O\text{-}SO_2\text{-}O^{\ominus} + R\text{-}NH_2 \xrightarrow{39\,°C} \begin{cases} \xrightarrow[\text{with CTABr}]{k_2^{\text{hydrolysis}}} Ar\text{-}OH + R\text{-}NH\text{-}SO_2\text{-}O^{\ominus} & \text{S-O bond fission} \\ \xrightarrow[\text{without CTABr}]{k_2^{\text{nucleophilic attack}}} Ar\text{-}NH\text{-}R + {}^{\ominus}O\text{-}SO_2\text{-}OH & \text{C-O bond fission} \end{cases} \tag{5-162}$$

The reaction of amines with 2,4-dinitrophenyl sulfate can result in the formation of phenol and sulfate ion (*via* S—O bond fission), or alternatively in the production of *N*-substituted anilines and hydrogen sulfate ions (*via* C—O bond fission). Under nonmicellar conditions, C—O bond cleavage is the dominating reaction, while cationic micelles are able to induce complete suppression of aniline formation. This dramatic effect has been explained in terms of change in the micro-environment of both the reactants and activated complexes through contribution of hydrophobic and electrostatic interactions [407].

To summarize, solubilization of reactants in micelles can lead to the following effects: local concentration, cage, pre-orientational, microviscosity, and polarity effects [713]. *Local concentration effects* are attributable to the increase in substrate concentration per unit volume of micelle due to the tendency of hydrophobic organic substrates to be solubilized in micelles. The ability of micelles to hold reactive intermediates together long enough for intramicellar reactions between them to occur is called the *cage effect*; *cf.* Section 5.5.10. The *pre-orientational effect* is the capability of micelles to solubilize substrates in a specific orientation, thus determining the regioselectivity of their reactions. Since the viscosity inside the micelles is generally much higher compared to the surronding aqueous solution, a substrate molecule incorporated into a micelle has less translational and rotational freedom. This could be reflected in its chemical reactivity and is called the *microviscosity effect*. Finally, the micropolarity of the hydrophobic interior of a micelle and of the micelle/solution interface is different from the polarity of the bulk aqueous solution. This change in substrate reactivity due to the variation in micropolarity is called the *polarity effect*. Solvatochromic dyes can be used as molecular probes (*e.g.* pyridinium-*N*-phenoxide betaine dyes) to determine directly the actual micropolarity experienced by a substrate in the region of the micelle; *cf.* [722–724]. Space limits us to only a few representative examples of this type of hydrophobic interaction in this Section.

5.5.9 Liquid Crystals as Reaction Media

Liquid crystals possess physical properties which lie somewhere between those of solids and liquids; *cf.* Section 3.1 and [725]. The rigidity which is present in a solid matrix is absent in liquid crystals, thus permitting molecular motion as well as conformational flexibility of the dissolved solute molecules. At the same time, due to the order in the liquid crystalline phase, the randomness in motion and conformational flexibility of the dissolved solute molecules is to some extent restricted. If the structures of the solute and solvent molecules are compatible, then solute molecules can be incorporated into the liquid-crystalline phase without disrupting its order. Thus, the reactivity of substrate molecules incorporated into liquid crystals without destroying their order should be different from that in isotropic solvents. Apart from the first report on the influence of liquid crystals on chemical reactions by Svedberg in 1916 [726], the use of liquid crystals as solvents for chemical reactions has become a subject of further research only in recent years [713]*⁾.

For example, reactions which would lead to a product with steric demands that do not fit in the liquid-crystalline order can be retarded. That is, the rigidity of the solvent molecules can prevent uni- or bimolecular reactions which would be feasable in isotropic solvents. Alternatively, the ability of liquid crystals to orient dissolved solute molecules act as a driving force in bimolecular reactions, particularly in entropy-controlled reactions which are subject to severe orientational constraints in the activation process. Two factors seem to be primarily responsible in influencing the reactivity of solutes ordered in liquid crystals: (*a*) the solvation ability of the liquid-crystalline solvent, *i.e.* the efficiency with which the solute molecules are solvated, and (*b*) the degree of distortion which the reacting system must undergo during the activation process. – Only a few examples showing the

k_1, 24–48 °C (5-163)

cis-N,N'-Diacylindigo $R = n\text{-}C_{17}H_{35}$ → *trans-N,N'*-Diacylindigo

Solvent	C_6H_6 (48 °C)	$C_6H_5CH_3$ (48 °C)	isotropic BS (46 °C)	smectic BS**⁾ (24 °C)
k_1^{rel}	68	53	31	1 (!)
$\Delta H^{\neq}/(kJ \cdot mol^{-1})$	93		91	130
$\Delta S^{\neq}/(J \cdot K^{-1} \cdot mol^{-1})$	−25		−33	+88

* A recent *Tetrahedron Symposia-in-Print* (No. 29) on *Organic Chemistry in Anisotropic Media* collects 43 topical papers devoted to the study of various organic reactions in crystalline solids, *liquid crystals*, micelles, vesicles, monolayers, zeolites, host/guest assemblies, polymer matrices, and surfaces: J. Scheffer (ed.), Tetrahedron *43*, 1197–1745 (1987).

** *n*-Butyl stearate has an enantiotropic smectic B phase from 14...27 °C [727].

influence of solvent order on chemical reactions can be presented here; for a recent review of organic photochemical reactions in organized media see [713].

The rates and activation parameters for the thermal *cis* → *trans* isomerization of *N,N'*-distearoylindigo have been determined in both isotropic and liquid-crystalline solvents [727].

In isotropic nonpolar solvents such as benzene, toluene and *n*-butyl stearate (BS; $t > 27\,°C$) the long alkyl chains have no influence on the rate of the *cis* → *trans* isomerization reaction. However, in the smectic phase of the liquid-crystalline solvent *n*-butyl stearate, the isomerization rate is considerably lower than in isotropic solvents. The correspondingly higher activation enthalpy and the increased positive activation entropy are obviously caused by the migration of the two long stearoyl chains involved in the *cis* → *trans* isomerization. The solute alkyl chains are intertwined with ordered solvent molecules and their translocation will be resisted by nearby *n*-butyl stearate molecules. This anchoring effect and not the shape changes of the isomerizing indigoid moiety is what is responsible for the observed rate decrease in smectic *n*-butyl stearate [727].

Similar rate decelerations in liquid-crystalline solvents have been observed for the thermal *cis* → *trans* isomerization of a bulky tetrasubstituted ethene in cholesteric phases [728]. – On the other hand, the activation parameters for the thermal *cis* → *trans* isomerization of less-dipolar substituted azobenzenes show no dependence on the solvent order. This indicates that the *cis*-isomers and their corresponding activated complexes present a similar steric appearance to the solvent environment [729]. This result is more consistent with an isomerization mechanism which proceeds via inversion rather than rotation; *cf.* Eq. (5-40) in Section 5.3.2 and [527–529, 561]. The latter reaction represents a nice example of the use of liquid-crystalline solvents as mechanistic probes [729].

As a consequence of the alignment of solute molecules in liquid-crystalline solvents the ratio of products formed in competitive reaction pathways can be different from that observed in isotropic liquids. This is illustrated by the Norrish type II photolysis of alkyl phenyl ketones with varying alkyl chain length in the isotropic, smectic, and solid phase of

(5-164)

Solvent	$n\text{-}C_7H_{16}$	$n\text{-}C_{17}H_{36}$ *)	$CH_3CO_2\text{—}n\text{-}C_4H_9$	$n\text{-}C_{17}H_{35}CO_2\text{—}n\text{-}C_4H_9$ **)
ketone / *cyclobutanol*	1.1	1.2	2.2	21 (!)

* Solid at 20 °C; $t_{mp} = 22.5\,°C$ [730].

** *n*-Butyl stearate has an enantiotropic smectic B phase from 14...27 °C [727].

n-butyl stearate (BS) [730]. For example, the ratio of elimination-to-cyclization products, formed in the photolysis of *n*-heptadecyl phenyl ketone, was shown to exhibit a strong phase dependence with a 19-fold increase in the smectic *n*-butyl stearate phase relative to the isotropic solvents *n*-heptane and *n*-butyl acetate; *cf*. Eq. (5-164).

The small increase observed on changing the solvent from *n*-heptane to *n*-butyl acetate is caused by the increase in solvent polarity. When the photolysis was carried out in solid *n*-heptadecane, no specific change was observed in the ketone/cyclobutanol product ratio.

These results have been explained as follows [730]; the cyclization of the intermediate 1,4-biradical, formed after γ-H-abstraction, requires rotation of the phenyl-substituted radical centre away from its equilibrium position. That is, the phenyl group must be placed in a direction perpendicular to the long axis of the alkyl phenyl ketone and the surrounding *n*-butyl stearate molecules. This should cause a disruption of the smectic solvent order and, as a consequence, cyclobutanol formation would be unfavourable in *n*-butyl stearate. For the competitive elimination reaction the intramolecular motions required in the activation process should cause only a small distortion of the liquid-crystalline solvent order. Thus, the formation of elimination products is enhanced in the ordered solvent.

For alkyl phenyl ketones with shorter alkyl chains (*e.g. n*-butyl and *n*-decyl phenyl ketone), the elimination-to-cyclization ratio is virtually unaltered in the smectic solvent [730]. These ketones are not as rigidly incorporated into the liquid-crystalline lattice as *n*-heptadecyl phenyl ketone and thus are able to disrupt their local smectic environment to a greater extent. Hence, the corresponding conformations of the 1,4-biradical which lead to the elimination products are not more favourable than those which lead to cyclization products in the liquid-crystalline solvent.

In general, the greater the similarity in size and shape between the solute and liquid-crystalline solvent molecules, the easier it is for the solute to incorporate itself into the liquid-crystalline phase. If the result of a chemical reaction depends on the solvent order, then the largest effects can be expected for those solute molecules which fit best into the liquid crystal structure. Since the structure of *n*-heptadecyl phenyl ketone is identical to *n*-butyl stearate except for the terminal phenyl and butoxy groups, it exhibits the largest solvent effect of the ketones studied [730].

Another illustrative example is the photodimerization of *n*-octadecyl *trans*-cinnamate which has been studied in the isotropic, smectic, and crystalline phases of *n*-butyl stearate, BS [731]; *cf*. Eq. (5-165).

Irradiation of cinnamate esters give *trans/cis*-isomerization, dimerization, and ester cleavage reactions. The latter represents only less than 5% of the total reaction pathway in this particular case. As expected, dimerization is favoured over isomerization with increasing initial solute concentrations. The *trans/cis*-isomerization involves only a relatively small perturbation of the surrounding solvent molecules. Therefore, it occurs with almost equal facility in the isotropic, smectic, and solid phases of *n*-butyl stearate. However, the regioselectivity of the photochemical [2+2]cycloaddition reaction is strongly phase-dependent. Of the possible dimers, only the two stereoisomers shown in Eq. (5-165) were detected in the irradiated samples. A strong preference for *head-to-tail* dimerization has been found in the smectic and solid phases. This result has been explained by assuming that not only solvent-mediated solute alignments but also dipole-dipole

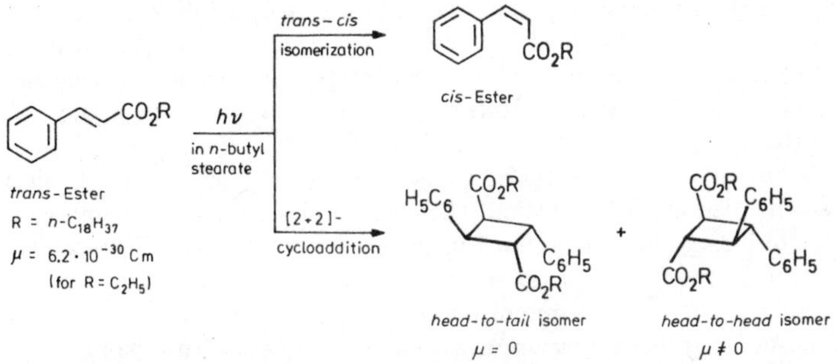

(5-165)

Solvent	BS at 32 °C (isotropic)	BS at 18 °C*) (smectic)	BS at 8 °C (solid)
*head-to-tail***) / *head-to-head*	3.3	14.2	15.1

induced interactions between dipolar cinnamate molecules must exist in the mesophase. The dipole-dipole interaction leads to pairwise, antiparallel associations of the cinnamate molecules with interdigitation among neighbouring solute molecules. Irradiation of these antiparallel oriented solute associates preferentially yields the *head-to-tail* dimers. In summary, the regioselectivity of the photodimerisation of *n*-octadecyl *trans*-cinnamate is controlled by the combination of two factors: solute alignment by ordered solvents and dipole-dipole interactions between solute molecules [731].

Similar results have been observed in the photodimerization reaction of acenaphthylene [732, 733]; *cf.* Eq. (5-161) in Section 5.5.8. A considerable increase in the production of the *trans*-adduct was reported in cholesteric liquid-crystalline media compared to the isotropic solvent benzene in which the *cis*-adduct of acenaphthylene is the dominant product [733]. That is, the ordered structure of the cholesteric mesophase affects the formation of *trans*-adduct advantageously. Furthermore, the *trans*/*cis* product ratio depends significantly on the initial acenaphthylene concentration. In isotropic solutions, the dimerization of singlet-excited acenaphthylene molecules is known to yield exclusively the *cis*-adduct whereas a mixture of *cis*- and *trans*-adducts result from triplet-excited solute molecules. The lowering of *cis*-adduct production in the mesophase has been attributed to the enhanced efficiency of the triplet reaction in comparison with the singlet reaction as shown by quantum yield measurements [732]. The increase in triplet reaction efficiencies has been ascribed to the increase in the fraction of acenaphthylene-acenaphthylene collisions which have coplanar or parallel-plane orientations with respect to the surrounding solvent molecules, and not to the increase in the total number of collisions per unit time [732]. See references [713, 732, 733] for a more detailed discussion of this photodimerization reaction.

Surprisingly, some Diels-Alder cycloaddition reactions show no variation in *endo*/*exo* product ratio with changes in solvent phase. Ordered liquid-crystalline solvents

* *n*-Butyl stearate exists as an enantiotropic smectic B phase from 14...27 °C [727].

** Dimer ratios.

are not able to differentiate between *endo*- and *exo*-activated complexes in the Diels-Alder reaction of 2,5-dimethyl-3,4-diphenylcyclopentadienone with dienophiles of varying size (cyclopentene, cycloheptene, indene, and acenaphthylene), when it is carried out in isotropic (benzene), cholesteric (cholesteryl propionate), and smectic liquid-crystalline solvents at 105 °C [734].

Cholesteric liquid crystals are optically active nematic phases as a result of their gradual twist in orientational alignment. Therefore, cholesteric liquid-crystalline solvents are expected to induce enantioselectivity in chemical reactions; see reference [713] for a review on photoasymmetric induction by chiral mesophases. The existing results are not very promising. So far the maximum photoasymmetric induction reported has been described for the hexahelicene formation in cholesteric phases with an enantiomeric excess of the right-handed helix of about 1% [735]. – The decarboxylation of ethylphenylmalonic acid in cholesteryl benzoate as liquid-crystalline solvent gave 2-phenylbutyric acid (overall yield 80 cmol/mol) which was shown to be optically active with an enantiomeric excess of 18% of the (*R*)-(−) enantiomer; *cf.* Eq. (5-166) [765]. In contrast to the 18% *e.e.*

$$H_5C_2\text{-}C(CO_2H)(C_6H_5)\text{-}CO_2H \xrightarrow[-CO_2]{160\,°C,\ 2\,h} \underset{(S)\text{-}(+)}{H_5C_2\text{-}CH(CO_2H)(C_6H_5)} + \underset{(R)\text{-}(-)}{H_5C_2\text{-}CH(CO_2H)(C_6H_5)} \qquad (5\text{-}166)$$

found in the cholesteric solvent, decarboxylation of ethylphenylmalonic acid in bornyl acetate, an isotropic chiral solvent, yielded 2-phenylbutyric acid which was essentially racemic (*e.e.* = 0%) [765]. – Using cholesteryl 4-nitrobenzoate as liquid-crystalline solvent, the *ortho*-Claisen rearrangement of γ-methylallyl 4-tolyl ether yielded optically active 2-(α-methylallyl)-4-methylphenol but the absolute configuration and optical purity of the rearrangement product were unknown [766]. – Various explanations for the failure of asymmetric induction in thermal reactions in liquid-crystalline media have been given in the literature [736, 737].

Obviously, liquid-crystalline media are not generally useful solvents in controlling the rate and stereochemistry of chemical reactions. In each case, careful consideration of the fine details regarding the structure of educts and activated complex, their preferred orientations in a liquid-crystalline solvent matrix, and the disruptive effects that each solute has on the solvent order has to be made. A mesophase effect can only be expected when substantial changes in the overall shape of the reactant molecule(s) occur during the activation process [734].

5.5.10 Solvent Cage Effects

When two chemical species react together at every encounter*), chemical change can take place only as fast as the reactants can diffuse together. In contrast to the gas phase, there may be many collisions per encounter in solution. Considering a pair of molecules that

* The whole process of two species coming together and remaining together for a number of subsequent collisions in solution has been called an "encounter".

have just encountered each other or, what in this regard is equivalent, a pair of molecules that have just arisen from the decay of a parent molecule, these molecules can become separated from each other only as a result of their diffusional motion through the inert solvent. If this diffusion is comparatively slow in solution, diffusion may be answerable for the observable rates of many reactions, since the slowest step in a sequence determines the net rate of change. Such reactions are called "encounter-" or "diffusion-controlled" [408, 409, 409a]. Second-order rate contants for such reactions can, in principle, be calculated from the laws of diffusion.

Reactions that occur with an activation energy of less than about 20 kJ/mol (5 kcal/mol) depend upon the rate at which the reactants move together through the solution. Since the movement of a molecule through an inert solvent has itself an activation energy of diffusion of about 20 kJ/mol, this movement becomes the slowest step for diffusion-controlled reactions. Almost all radical-radical recombination reactions require so little activation energy that they are usually diffusion-controlled in solvents of normal viscosity. Thus, the recombination rates for simple alkyl radicals in inert solvents lie in the range $10^8 \ldots 10^{10}\ \mathrm{l \cdot mol^{-1}\, s^{-1}}$ at 25 °C [410]. Ion combination reactions such as the acid-base reaction between the solvated proton and solvated hydroxide ion are also known to be diffusion-controlled. This reaction has one of the largest rate constants known for liquid-phase reactions, $1.4 \cdot 10^{11}\ \mathrm{l \cdot mol^{-1}\, s^{-1}}$ in water at 25 °C [411].

When a pair of radicals (or ions) have reached adjacent positions through diffusion, or it is generated from a single parent molecule as in the decomposition of radical initiators, it is hemmed in by a cage of solvent molecules – an effect, which has been called "solvent cage effect"*). The two species must diffuse to become statistically distributed in the solvent, but, because of the activation energy of diffusion, they collide with each other several times before they separate, and recombination may occur during these collisions**). This type of reaction is called a "cage reaction" [413].

With azoalkanes, peroxides, and other initiators, varying percentages of the radicals formed react together as a geminate pair in the solvent cage as shown, for instance, for the photolysis of azomethane according to Eq. (5-167) [415–417]. The efficiency of free-radical production, F, in the decomposition of azomethane can be described in terms of an effective rate constant, k_d, which approximately describes the diffusional separation of caged radical pairs to give free radicals, a process competing with radical-radical reactions within the solvent cage, the rate constant of which is designated as k_c. The combination reaction must obviously be very fast, in order to compete successfully with diffusion. It is usually assumed that k_c is nearly independent of the nature of the solvent, and that all the variation in cage products may be attributed to k_d. Very clear evidence for the existence of a cage effect in reaction (5-167) was found from the

---

* The concept of a *solvent cage* restricting the separation of two reactants was first introduced by Frank and Rabinowitsch [412] in predicting a reduction in the quantum yield for photodissociation processes in solution compared with the gas phase. Thus, when iodine in solution is dissociated by a flash, the quantum yield is much less than unity indicating that most of the iodine atoms recombine before escaping from the solvent cage. The *solvent cage* is an aggregate of solvent molecules that surrounds the fragments formed by thermal or photochemical bond cleavage of a precursor species.

** Other radical reactions which can compete with recombination inside the solvent cage are those of disproportionation [414] and of reaction with the adjacent solvent molecules.

$$H_3C{-}N{=}N{-}CH_3 \xrightarrow[-N_2]{h\nu} (H_3C\odot\odot CH_3)_{Solv} \begin{cases} \xrightarrow{k_c} (H_3C{-}CH_3)_{Solv} \quad \text{Cage product} \\ \xrightarrow{k_d} (CH_3\odot)_{Solv} + (\odot CH_3)_{Solv} \xrightarrow{+\,\text{Scavenger}} \text{Scavenged products} \end{cases} \tag{5-167}$$

Geminate radical pair

results of crossover experiments [416, 417]. When a mixture of azomethane and perdeuterioazomethane is photolyzed in the gas phase, the $CH_3\cdot$ and $CD_3\cdot$ radicals formed recombine rapidly to yield CH_3CH_3, CD_3CD_3, and CH_3CD_3 in such proportions (1:1:2) that it is clear that $CH_3\cdot$ and $CD_3\cdot$ are randomly mixed before recombination. When, however, the same experiment is repeated in the inert solvent *iso*-octane, no CH_3CD_3 is detected when CH_3CH_3 and CD_3CD_3 are formed along with N_2. Evidently, solvent molecules keep methyl radicals formed from the same parent molecule together until they recombine. The cage effect can occur in the gas phase at high pressures, as has been demonstrated for the photolysis of azomethane in propane at 49 bar. Here propane and intact reactant molecules act as a quasi-solvent and prevent formation of CH_3CD_3 [416].

Cage effects account also for the fact that not all the radicals produced from the decomposition of initiators such as azobisisobutyronitrile (AIBN) are effective in initiating radical polymerizations. In the somewhat simplified reaction scheme (5-168) depicting the thermolysis of AIBN, two types of cyanopropyl radicals are shown, one still within the solvent cage, whereas the others reached their statistical separation in the solution. Adding certain reactive compounds called scavengers to the solution, should divert the free radicals away from dimer formation. In other words, radical scavengers

$$H_3C{-}C(CH_3)(CN){-}N{=}N{-}C(CH_3)(CN){-}CH_3 \xrightarrow[-N_2]{\Delta} \left(H_3C{-}\overset{CH_3}{\underset{CN}{C}}\odot \;\; \odot\overset{CH_3}{\underset{CN}{C}}{-}CH_3\right)_{Solv} \rightleftharpoons \left(H_3C{-}\overset{CH_3}{\underset{CN}{C}}\odot\right)_{Solv} + \left(\odot\overset{CH_3}{\underset{CN}{C}}{-}CH_3\right)_{Solv} \tag{5-168}$$

Free Radicals $\xrightarrow{+\,\text{Scavenger}}$ Scavenged products

Cage pair and free radicals $\longrightarrow \left(H_3C{-}\overset{H_3C}{\underset{NC}{C}}{-}\overset{CH_3}{\underset{CN}{C}}{-}CH_3\right)_{Solv}$

should reduce the yield of tetramethylsuccinonitrile derived from the free radicals (but not that formed from the caged ones) to a limiting value. This then represents the amount of product formed within the solvent cage. Indeed, the yield of tetramethylsuccinonitrile from the thermolysis of AIBN in tetrachloromethane at 80 °C, falls from about 96 cmol/mol when no scavenger is present to a constant value of about 19 cmol/mol using *n*-butanethiol, a good hydrogen donor, as scavenger [418]. This means, that 19 cmol/mol of the initiator radicals are lost in solvent cage recombinations and only 81 cmol/mol of AIBN molecules are efficient in free-radical production under these reaction conditions. The initiator efficiency of AIBN is in fact lower under most circumstances, since only few substrates approach the reactivity of *n*-butanethiol as scavenger. The fraction of AIBN molecules leading to initiation of styrene polymerization increases as the concentration of the monomer increases, reaching a limiting value of 68 cmol/mol above 2 mol/l styrene in an inert solvent [419]. The remaining 32 cmol/mol may be attributed to the reactions of the geminate radical pairs.

Solvent cage effects have been observed not only in the decomposition of azoalkanes, but also in the thermolysis of other compounds such as peroxides (*e.g.* diacyl peroxides [420], peresters [421]) which generate two radicals simultaneously. For example, diacetyl peroxide initially labeled with ^{18}O in the carbonyl oxygen was partially thermolysed in *iso*-octane at 80 °C and it was found, that *ca.* 38 cmol/mol of all geminate radical pairs formed in the decomposition recombine to give diacetyl peroxide with scrambling of the label [420]. Solvent cage effects have been found even for ion-producing reactions, only however at high external pressures. The solvolysis of 2-bromopropane in a 4:1 methanol/ethanol mixture at 46 °C becomes diffusion-controlled at pressures higher than 40 kbar [738]. – Liquid-crystalline solvents and micellar systems can also provide solvent cages. Liquid crystals and micelles have the capacity to hold two reactive species together for a longer period of time compared to isotropic, homogeneous solutions. Micellar aggregates are particularly effective in imposing cage constraints on chemical reactions. Photochemical examples of the consequences of restricting solute diffusion in liquid crystals and in micelles can be found in a recent review [713]; *cf.* also Sections 5.5.8 and 5.5.9.

The prediction of solvent effects on k_d, as reflected by the efficiency of free-radical production, F, has been attempted by considering the correlation between F and macroscopic solvent parameters such as solvent viscosity [413]. That the proportion of cage reactions depends on the viscosity of the solvent used, was shown by Kochi [422], who decomposed a series of diacyl peroxides in *n*-pentane and decalin. In decalin, the more viscous solvent, photolysis of two different diacyl peroxide gave a high yield of the symmetrical dimers and only a minor yield of cross-dimer, thus indicating the strong solvent cage effect. In the more fluid solvent *n*-pentane, the cage effect is less important and a higher proportion of the cross-products were formed. Usually, however, no simple relationship with solvent viscosity was found. Studying the photolysis of azomethane in a wide range of different media, Martin *et al.* [423] were able to correlate the observed amount of cage product ethane to macroscopic solvent parameters other than viscosity, that is solvent internal pressure and cohesive energy density (*cf.* Section 5.4.2).

We shall conclude this Section with an example of solvent cage effects of ion-molecule recombination reactions as found in the ozonolysis of alkenes in nonpolar solvents [739, 740]. According to the Criegee mechanism [424], unsymmetrically *cis*-Alkene unsymmetrical — Primary ozonide — Carbonyl oxide + Carbonyl compound — Each ozonide as *cis/trans* pair

(5-169)

substituted alkenes ought to give two zwitterions and two carbonyl compounds after decomposition of the unstable primary ozonide, as shown in reaction scheme (5-169). If both scission and recombination are statistical, the three possible final ozonides *A*, *B* and *C* should be formed in a 1 :2 :1 ratio, provided there is no preferred breakdown of the initial ozonide and no solvent cage. The experimentally determined values of this molar ratio differ from the statistical value, the yield of the symmetrical cross-ozonides always being lower as observed for 2-pentene [425] and 2-hexene [426]. This means that part of the recombination reaction must occur in a solvent cage. With an increase in the initial concentration of alkene, the relative proportions of "normal" and "cross" ozonides approach the statistical values. When he failed to find cross-ozonides in the ozonolysis of 3-heptene, Criegee postulated that the zwitterion and carbonyl fragments were formed and recombined in a solvent cage, thus preventing cross-recombination [427]. The earlier failure of Criegee to obtain any cross-ozonides in the case of 3-heptene can now probably be attributed to the alkene concentration used [428]. At the concentration of alkene normally used in ozonolysis the predominant product is the normal parent ozonide of the unsymmetrically substituted alkene. A plot of the ratio of normal ozonides to cross-ozonides *versus* alkene concentration shows that cross-ozonide formation decreases with dilution [428]; *cf.* also [428a].

Reduced solvent cage effects have however been observed for ozonations in polar solvents such as dichloromethane and ethyl acetate, in contrast to nonpolar hydrocarbons; *cf.* [739, 740]. Increasing the polarity of the ozonation solvent leads to an increase in the ratio of cross to normal ozonides formed from unsymmetrical alkenes. Obviously, the increased solvent polarity increases the separation of the carbonyl oxide and the carbonyl compound, allowing them to react more independently rather than in a solvent cage. Accordingly, in contrast to nonpolar solvents, an increase in reaction temperature has no effect on the ozonide ratio. This is due to the fact that in polar solvents the solvent cage effect is minimal. For the ozonolysis of an equimolar mixture of ethene and tetradeuterio-ethene, leading to two normal and one cross-ozonide, it has been estimated that in polar solvents such as CH_2F_2 at least 90% of the final ozonide formation occurs outside the solvent cage in which the primary ozonide decomposition took place. This percentage decreases to about 10% out-of-cage recombination in a nonpolar solvent such as *iso*-butane [741].

In conclusion, the relative yields of the normal and cross ozonides are a function of initial alkene concentration, solvent polarity, and temperature. High alkene concentration, polar solvents, and high temperatures maximize the cross-ozonide yields, with proportions close to statistical amounts in favourable cases [739, 740].

For recent comprehensive reviews on the ozonation reaction, dealing not only with solvent effects on the relative yields of normal and cross ozonides but also with solvent effects on the total ozonide yield and the *cis/trans* ozonide ratio from *cis-* and *trans-*alkenes, see references [739, 740]. Supplementary solvent effects have been reported recently for the ozonolysis of 3-aryl-1-methylindenes yielding solvent-dependent mixtures of final *exo-* and *endo-*ozonides [742].

5.5.11 External Pressure and Solvent Effects on Reaction Rates

Not only the internal pressure of a solvent can affect chemical reactions (see Section 5.4.2 [231, 232]), but also the application of external pressure can exert large effects on reaction rates and equilibrium constants [239, 429–433, 747–750]. According to Le Chatelier's principle of least restraint, the rate of a reaction should be increased by an increase in external pressure, if the volume of the activated complex is less than the sum of the volumes of the reactant molecules, whereas the rate of reaction should be decreased by an increase in external pressure if the reverse is true. The fundamental equation for the effect of external pressure on a reaction rate constant k was deduced by Evans and Polanyi on the basis of transition-state theory [434]:

$$\left(\frac{\partial \ln k}{\partial p}\right)_T = -\frac{\Delta V^{\neq}}{RT} \tag{5-170}$$

provided that k is expressed in pressure-independent concentration units (mole fraction or molality scale) at a fixed temperature and pressure*). The volume of activation, $\Delta V^{\neq}$, is interpreted as the difference between the partial molar volume of the activated complex including its molecules of solvation, $V^{\neq}$, and the sum of the partial molar volumes of the reactants with their associated solvent molecules, $\sum r \cdot V^{\mathrm{R}}$, at the same temperature and pressure: $\Delta V^{\neq} = V^{\neq} - \sum r \cdot V^{\mathrm{R}}$, where r is the stoichiometric number of the reactant R and V^{R} its partial molar volume.

* If k is expressed in the pressure-dependent molarity scale (*i.e.* $\mathrm{mol \cdot l^{-1}}$), its pressure-dependence is given by

$$\left(\frac{\partial \ln k}{\partial p}\right)_T = -\frac{\Delta V^{\neq}}{RT} - \kappa(n-1)$$

where κ is the compressibility of the solution (usually that of the solvent) and n is the kinetic order of the reaction. For first-order reactions ($n=1$), the equations for the pressure-dependence of k are identical. For reactions of higher order, however, the additional compressibility term must be taken into account.

To produce significant changes in $\ln k$, pressures of several hundred bars are commonly used. While a change in reaction rate by a factor of 2 to 4 can be anticipated for a change in temperature of 10 °C, a pressure change of about 700 bar is required to bring about the same effect even for reactions with the relatively large activation volume of $\pm 40\ \text{cm}^3 \cdot \text{mol}^{-1}$ [435]. Normally, the magnitude of $\Delta V^{\neq}$ lies somewhere between $+25$ and $-25\ \text{cm}^3 \cdot \text{mol}^{-1}$, which requires pressures of *ca.* 1200 bar to produce rate changes of a factor of 3. An extensive table showing the effects of pressure on the reaction rate for some negative $\Delta V^{\neq}$-values can be found in reference [749]. The minus sign in Eq. (5-170) means that pressure accelerates reactions which are characterized by a volume shrinkage in passing from reactants through activated complex (negative $\Delta V^{\neq}$) and retards those with a volume expansion (positive $\Delta V^{\neq}$). In many cases a plot of $\ln k$ against pressure is linear, indicating that for these reactions the quantity $\Delta V^{\neq}$ is independent of pressure. However, above *ca.* 10 kbar reactions do not obey the ideal rate equation (5-170) since activation volumes are pressure-dependent, the values of $\Delta V^{\neq}$ generally decrease as pressure increases.

The activation volume changes arise from two sources: (a) making and breaking of chemical bonds, and (b) interaction of reactants and activated complex with the surrounding solvent molecules. Therefore, $\Delta V^{\neq}$ can be considered as being composed of $\Delta V_i^{\neq}$, the *intrinsic* change in molar volume of the reactant molecules themselves in forming the activated complex, as well as $\Delta V_{solv}^{\neq}$, the change in molar volume of the solvating solvent molecules during the activation process, according to Eq. (5-171):

$$\Delta V^{\neq} = \Delta V_i^{\neq} + \Delta V_e^{\neq} \qquad (5\text{-}171)$$

$\Delta V_e^{\neq}$ is related to the well-known volume contraction normally observed on dissolution of electrolytes and called *electrostriction*. The contraction of the solvent surrounding an ion is best expressed by the Drude-Nernst equation (5-172):

$$\Delta V_e = -\frac{N_A \cdot z^2 \cdot e^2}{2r \cdot \varepsilon_r^2} \frac{\partial \varepsilon_r}{\partial p} \qquad (5\text{-}172)$$

The solvent is assumed to be a continuum of relative permittivity ε_r, and the ion to be a hard sphere of radius r with charge $z \cdot e$; N_A is the Avogadro number. Eq. (5-172) represents the electrostatic contraction change in volume for a mole of ions. According to the Drude-Nernst equation, electrostriction should be proportional to the square of charge on the ions, inversely proportional to the ionic radii, and should increase in proportion to the value of $(1/\varepsilon_r^2)(\partial \varepsilon_r / \partial p)$, which depends on the nature of the solvent. It follows that ΔV_e will vary strongly in reactions where charges are created or neutralized, which, in turn, should show up as a solvent dependence of $\Delta V^{\neq}$.

$\Delta V_i^{\neq}$ represents the change in volume due to changes in bond lengths and angles. It is this contribution to $\Delta V^{\neq}$ that is connected to the reaction mechanism in terms of the relative positions of the atoms in reactants and the activated complex. The absolute size of $\Delta V_i^{\neq}$ has been concluded to be approximately $+10\ \text{cm}^3 \cdot \text{mol}^{-1}$ for bond cleavage and approximately $-10\ \text{cm}^3 \cdot \text{mol}^{-1}$ for bond formation in reactions of organic molecules [430].

In a reaction, $A \rightleftharpoons [X]^{\neq} \rightarrow C+D$, bonds will be stretched to form $[X]^{\neq}$, so that $\Delta V_i^{\neq}$ would be expected to be positive for unimolecular dissociative reactions and negative for the reverse bimolecular associative reaction of C and D. The $\Delta V_e^{\neq}$ contribution is caused by the rearrangement of the solvent molecules due to steric requirements of the reaction and to change in charge density on activation. The latter effect can be predicted on the basis of the qualitative theory of solvent effects introduced by Hughes and Ingold [16, 44] (*cf.* Section 5.3.1), representing the reacting ions or dipolar molecules by a sphere and the solvent by a dielectric. Creation or concentration of charge on the reacting species, on passing from initial to transition state, will increase the intermolecular electrostatic forces between the solute and the permanent or induced dipoles in the solvating molecules. This leads to a reduction in volume, called *electrostriction*, of the solvate complex. Since the extent of electrostriction varies as the square of the charge on the sphere according to Eq. (5-172), the association of ions with like charges will increase the electrostriction in the solvent and $\Delta V_e^{\neq}$ will be large and negative, whereas association of two oppositely charged reactants will be accompanied by a large positive $\Delta V_e^{\neq}$. Because the degree of solvation of ionic or strongly dipolar species may be extensive, the $\Delta V_e^{\neq}$ term often predominates over the $\Delta V_i^{\neq}$ term. The activation volume, therefore, reflects not only the intrinsic differences in molecular dimensions of reactants and activated complex, but also the difference in their degree of solvation. Pressure is therefore a probe uniquely suited for the study of solvation changes during a reaction.

The importance of the solvent in determining the effects of pressure on reaction rates has been recognized in general terms for a long time, but the first satisfactory discussion was given by Buchanan and Hamann in 1953 [436]. A schematic compilation of pressure and solvent effects on reactions of different charge type, established by Dack [27, 239], is given in Table 5-25. The entire basis of the effect of solvent polarity on $\Delta V^{\neq}$, as shown in Table 5-25 is that the less polar solvents have higher compressibilities and are therefore more constricted by ionic or dipolar solutes than the more polar solvents, which exhibit smaller compressibilities owing to the strong intermolecular interactions already present in the absence of a solute. This consideration would also suggest a correlation between $\Delta V^{\neq}$ and the entropy of activation, $\Delta S^{\neq}$, because an increase in electrostriction due to an intensification of the electric field around the solute corresponds to a decrease in both volume and entropy owing to loss in freedom of motion within the solvent complex. That is, a large negative $\Delta V^{\neq}$ value should, in principle, correspond to a large negative $\Delta S^{\neq}$ value for reactions, whose temperature and pressure dependence has been investigated in solvents of different polarity. Indeed, a linear relationship of $\Delta V^{\neq}$ with $\Delta S^{\neq}$ has been reported for many reactions.

Two typical examples shall illustrate the predictions made by Table 5-25. Clear-cut examples of reaction type 2 are the Diels-Alder cycloaddition reactions*). The solvent effect on the activation volumes for such a cycloaddition reaction is reported in Table 5-26 [437]. In agreement with an isopolar cyclic activated complex being intrinsically smaller than the reactants, large negative values of $\Delta V^{\neq}$ have been found for this reaction. As expected, solvent polarity has comparatively little influence on $\Delta V^{\neq}$ ($\Delta\Delta V^{\neq} = 7.3$

---

* The first Diels-Alder reaction studied under high pressure was the dimerization of cyclopentadiene [751].

Table 5-25. Effect of external pressure and solvent polarity on the rate of reactions of different charge type and on their volume of activation, $\Delta V^{\neq}$ [27, 239].

No.	Reactants	Activated Complex	Effect of increased pressure on reaction rate	Effect of increased solvent polarity on $\Delta V^{\neq}$
1	R	$\xrightarrow{+\Delta V^{\neq}}$ R	Decrease	None
2	$R_1 + R_2$	$\xrightarrow{-\Delta V^{\neq}}$ $R_1 \cdots R_2$	Increase	None
3	RX	$\xrightarrow{-\Delta V^{\neq}}$ $\overset{\delta+}{R} \cdots \overset{\delta-}{X}$	Increase	More negative
4	$R_1R_2^+$	$\xrightarrow{+\Delta V^{\neq}}$ $\overset{\delta+}{R_1} \cdots \overset{\delta+}{R_2}$	Decrease	More positive
5	$R_1 + R_2$	$\xrightarrow{-\Delta V^{\neq}}$ $\overset{\delta+}{R_1} \cdots \overset{\delta-}{R_2}$	Increase	More negative
6	$R_1^+ + R_2^-$	$\xrightarrow{+\Delta V^{\neq}}$ $\overset{\delta+}{R_1} \cdots \overset{\delta-}{R_2}$	Decrease	More positive
7	$RX + Y^-$	$\xrightarrow{-\Delta V^{\neq}}$ $\overset{\delta-}{X} \cdots R \cdots \overset{\delta-}{Y}$	Decrease	More positive
8	$RX + Y^+$	$\xrightarrow{-\Delta V^{\neq}}$ $\overset{\delta+}{X} \cdots R \cdots \overset{\delta+}{Y}$	Decrease	More positive

Table 5-26. Effect of external pressure and solvent polarity on reaction rate and activation volume of the Diels-Alder reaction between isoprene and maleic anhydride at 35 °C [437]; *cf.* Eq. (5-42) in Section 5.3.3.

Solvents	$k_2 \cdot 10^4/s^{-1}$ [a]		$k_2^{rel.}$ [b]	$\Delta V^{\neq}/(cm^3 \cdot mol^{-1})$ at 1 bar [c]
	at 1 bar	at 1336 bar		
Dichloromethane	5.28	26.1	4.9	−39.8
Dimethyl carbonate	1.82	—	—	−39.3
Acetone	2.18	11.8	5.4	−39.0
Diisopropyl ether	0.597	3.56	6.0	−38.5
1-Chlorobutane	1.59	9.31	5.9	−38.0
Acetonitrile	6.25	33.9	5.4	−37.5
Ethyl acetate	1.22	6.33	5.2	−37.4
1,2-Dichloroethane	5.50	32.2	5.9	−37.0
Nitromethane	9.86	44.4	4.5	−32.5

[a] Rate constants based on mol fraction scale. $\Delta\Delta V^{\neq} = 7.3\ cm^3 \cdot mol^{-1}$

[b] $k_2^{rel.} = k_2(1336\ bar)/k_2(1\ bar)$.

[c] Limit of error $\pm 0.8\ cm^3 \cdot mol^{-1}$.

$cm^3 \cdot mol^{-1}$ between nitromethane and dichloromethane). This suggests that the solute-solvent interactions of the activated complex are small and similar to those of the reactants. The small solvent effects obtained nevertheless, might be explained in terms of solvent internal pressure [27, 438], where solvent internal pressure acts on the rates of nonpolar reactions in the same direction as external pressures*).

Similar pressure effects have been observed in the 1,3-dipolar cycloaddition of diazodiphenylmethane to various alkenes. This is in agreement with a concerted mechanism involving an isopolar activated complex [752, 753]; *cf.* Section 5.3.3. For the 1,3-dipolar cycloaddition of diazodiphenylmethane to dimethyl acetylenedicarboxylate at 25 °C in *n*-hexane ($\Delta V^{\neq} = -24$ $cm^3 \cdot mol^{-1}$) and in acetonitrile ($\Delta V^{\neq} = -15$ $cm^3 \cdot mol^{-1}$), the solvent-induced difference $\Delta\Delta V^{\neq}$ is only 9 $cm^3 \cdot mol$, this corresponds to a very small rate constant solvent effect; $k_2(CH_3CN)/k_2(n\text{-}C_6H_{14}) = 3.4$ [752]. – Relatively little is known about the pressure dependence of other pericyclic reactions; *cf.* [754] and references cited therein. For a thorough investigation of pressure- and solvent-dependent degenerative sigmatropic rearrangements [*e.g.* 5-(trimethylsilyl)cyclopentadiene, bullvalene] see reference [754].

Contrary to reactions going through isopolar transition states, reactions of type 3 to 8 in Table 5-25, which involve formation, dispersal or destruction of charge, should exhibit large solvent effects on activation volume. This is shown in Table 5-27 for the S_N2 substitution reaction between triethylamine and iodoethane [441], an example of the well-known Menschutkin reaction, the pressure dependence of which has been investigated thoroughly [439–445, 755].

Table 5-27. Effect of external pressure and solvent polarity on reaction rate and activation volume of the Menschutkin reaction between triethylamine and iodoethane at 50 °C [441]; *cf.* also Table 5-5 in Section 5.3.1 [59].

Solvents	$k_2 \cdot 10^6/(l \cdot mol^{-1} \cdot s^{-1})$		$k_2^{rel.}$ [a]	$\Delta V^{\neq}/(cm^3 \cdot mol^{-1})$ at 1 bar [b]
	at 1 bar	at 1961 bar		
n-Hexane	0.123	1.35	11.0	−51.5
Acetone	318	2600	8.2	−48.5
Benzene	27.8	–	–	−46.2
Chlorobenzene	92.0	627	6.8	−44.6
Methanol	50.6	208	4.1	−40.6
Nitrobenzene	934	4450	4.8	−37.0

[a] $k_2^{rel.} = k_2(1961\ bar)/k_2(1\ bar)$. $\Delta\Delta V^{\neq} = 15\ cm^3 \cdot mol^{-1}$

[b] The limit of error for $\Delta V^{\neq}$ lies between ± 1.6 and ± 2.6 $cm^3 \cdot mol^{-1}$.

Corresponding to reaction type 5 in Table 5-25, increasing pressure leads to an increase in reaction rate, which is more pronounced in less polar solvents. The $\Delta V^{\neq}$ values observed in solvents of different polarity demonstrate clearly that nonpolar solvents undergo more electrostriction than polar media ($\Delta\Delta V^{\neq} = 15$ $cm^3 \cdot mol^{-1}$ between

* For common organic solvents the internal pressures range from 1800 to 5000 bar at 25 °C [438].

n-hexane and nitrobenzene). – Somewhat more positive activation volumes have been reported for the Menschutkin-type reaction of triphenylphosphane with iodomethane at 30 °C: $\Delta V^{\neq} = -28.0\ cm^3 \cdot mol^{-1}$ in acetonitrile and $\Delta V^{\neq} = -17.6\ cm^3 \cdot mol^{-1}$ in the slightly more polar propylene carbonate; the reaction being 245 times faster in propylene carbonate than it is in nonpolar diisopropyl ether [755].

Reaction type 3 in Table 5-25 is best represented by the solvolysis of 2-chloro-2-methylpropane; *cf.* Eq. (5-13) in Section 5.3.1. According to the heterolysis of the C—Cl bond one would expect the activation volume to be positive because of the C—Cl stretching during the activation process. However, a negative activation volume of $\Delta V^{\neq} = -22.2\ cm^3 \cdot mol^{-1}$ has been found for this solvolysis at 30 °C in ethanol/water (80:20 cl/l), indicating a strong volume contraction due to solvation of the dipolar activated complex (electrostriction) [756]. Typical values of activation volume of haloalkane solvolyses in protic solvents are in the range of $-15 \ldots -30\ cm^3 \cdot mol^{-1}$.

The solvolysis rate of 2-bromopropane in a mixture of methanol/ethanol (4:1 cl/l) increases strongly up to *ca.* 40 kbar. Above 40 kbar, however, the reaction is retarded by applied pressure [738]. Consequently, the reaction is considered to be diffusion-controlled at very high pressure, as frequently encountered in the case of free-radical reactions; *cf.* Section 5.5.10.

In summarizing it can be stated that solvents which lower the value of $\Delta V^{\neq}$ of a reaction by electrostriction accelerate the rate of that reaction, whereas those solvents capable of raising $\Delta V^{\neq}$ cause the rates to fall [27, 239]. For a more detailed discussion and other examples of the reaction types given in Table 5-25 see references [27, 239, 430, 433, 435, 439, 747–750]. In general, the effect of solvent polarity and external pressure can be used to draw conclusions about whether the activated complex is more dipolar (that is, interacts more strongly with the solvent) than the initial reactants.

Finally, it should be mentioned that external pressure has recently found more and more application in organic synthesis, not least because of the increasing commercial availability of high pressure devices. For the following types of organic reactions a rate enhancement with increasing pressure is expected [749]: (a) associative reactions in which the number of molecules decreases in forming the products; *e.g.* cycloaddition and condensation reactions; (b) reactions which proceed via cyclic isopolar activated complexes; *e.g.* Cope and Claisen rearrangements; (c) reactions which proceed *via* dipolar activated complexes; *e.g.* Menschutkin-type S_N2 reactions, aromatic electrophilic substitution reactions; and (d) reactions with steric hindrance. A comprehensive review on organic synthesis under high external pressure can be found in reference [749].

A very unusual case is the use of supercritical-fluid (SCF) solvents as reaction media. Whereas the application of SCF solvents in extraction processes has been an active area of research and technical development [757], the use of SCF solvents as reaction media has been considerably less explored. An important property of supercritical fluids as solvents is the possibility to manipulate the physicochemical properties of these solvents through small changes in *pressure* and temperature. This can influence solubilities, mass transfer (diffusivity), and rate constants of the reacting systems which are dissolved in SCF solvents. The effect of external pressure on reaction kinetics measured in SCF solvents can occur through the pressure-dependence of concentrations of the reactants or through the pressure-dependence of the rate constants; the latter being substantial for inert SCF solvents at conditions approaching the critical point of the solvent [758].

For instance, the critical point of carbon dioxide is at $t_c = 31\,°C$ and $p_c = 74$ bar. Below this point, CO_2 is a normal liquid, easily maintained under modest pressures (*ca.* 65 bar at 25 °C). Above 31 °C, no amount of pressure will be sufficient to liquify CO_2. There only exists the supercritical fluid phase which behaves as a gas, although when highly compressed, this fluid can be denser than liquid CO_2. In comparison to the subcritical liquid phase of CO_2, supercritical-fluid CO_2 behaves like another solvent: it has higher compressibility, higher diffusivity, lower viscosity, and lower surface tension than does the subcritical liquid phase. By means of some empirical parameters of solvent polarity (*cf.* Chapter 7), it has been found that supercritical-fluid CO_2 behaves very much like a hydrocarbon solvent with very low polarizability [759].

The Diels-Alder cycloaddition reaction of maleic anhydride with isoprene has been studied in supercritical-fluid CO_2 at conditions near the critical point of CO_2 [758]. The rate constants obtained for supercritical-fluid CO_2 as solvent at 35 °C and *high* pressures (> 200 bar) are similar to those obtained when normal liquid ethyl acetate has been used as solvent. However, at 35 °C and pressures approaching the critical pressure of CO_2 (74 bar), the effect of pressure on the rate constant becomes substantial. Obviously, $\Delta V^{\neq}$ takes on large negative values at temperatures and pressures near the critical point of CO_2. Thus, pressure can be used to manipulate reaction rates in SCF solvents at near-critical conditions. This effect of pressure on reacting systems in SCF solvents appears to be unique. A discussion of fundamental aspects of reaction kinetics at near-critical reaction conditions within the framework of transition-state theory can be found in reference [758].

5.5.12 Solvent Isotope Effects

Solvent isotope effect (SIE) is a term frequently used to describe changes in kinetic and equilibrium processes produced by replacing a normal solvent by its isotopically substituted counterpart. Since replacement of hydrogen by deuterium gives the relatively largest mass change and easily measurable results from a perturbation of those molecular properties which are sensitive to mass, the ratio of a measurement X in light water (H_2O) to the corresponding value in heavy water (D_2O), X_{H_2O}/X_{D_2O}, is usually what is meant by the term solvent isotope effect of that property [446–449, 760, 761]. Comparatively little is known of solvent isotope effects in solvents other than water (*e.g.* CH_3OH/CH_3OD, CH_3CO_2H/CH_3CO_2D) [447]. Kinetic solvent isotope effects (KSIE) k_{H_2O}/k_{D_2O} range from 0.5 to about 6, with the most common values falling between 1.5 and 2.8 [447]. The isotope effect for reactions carried out in an isotopically substituted solvent can be used to indicate direct or indirect solvent participation in the reaction. Unfortunately, the observed effects are a combination of three factors: (a) The solvent can be a reactant. For instance, if an O—H or O—D bond of the solvent is broken in the rate-determining step, there will be a *primary isotope effect*; (b) The reactant molecules may become labelled with deuterium by a fast H/D exchange reaction, and then the newly labeled molcule can cleave in the rate-determining step; (c) The intermolecular solute/solvent interactions (*i.e.* the solute solvation) may be different in the labelled and nonlabelled solvent. This can change the differential solvation of educts and activated complex and hence the Gibbs energy of activation of the reaction. This is called a *secondary isotope effect*. – In many cases not only

Table 5-28. Some important physical properties of light and heavy water [446, 451, 635].

Property	Unit	Value for H_2O	Value for D_2O
Relative molecular mass, M_r	$g \cdot mol^{-1}$	18.015	20.028
Freezing point, t_{fp}	°C	0.00	3.81
Temperature of maximal density, t	°C	3.98	11.23
Boiling point (at 1013 mbar), t_{bp}	°C	100.00	101.42
Density, ϱ [a]	$g \cdot cm^{-3}$	0.997047	1.10448
Molar volume, V_m [a]	$cm^3 \cdot mol^{-1}$	18.069	18.133
Viscosity, η [a]	Pa · s	$8.903 \cdot 10^{-4}$	$11.03 \cdot 10^{-4}$
Vapour pressure, p [a]	Pa	3166	2737
Heat of vaporization, ΔH_v [a]	$kJ \cdot mol^{-1}$	44.04	45.46
Refractive index, n_D^{25} [a]		1.33250	1.32841
Dielectric constant, ε_r [a]		78.46	77.94
Ionization constant, K_w [a]	$mol \cdot l^{-1}$	$1.81 \cdot 10^{-16}$	$3.54 \cdot 10^{-17}$
Dipole moment, μ	C · m	$6.12 \cdot 10^{-30}$	$6.14 \cdot 10^{-30}$
Solubility parameter, δ [a]	$MPa^{1/2}$	47.9 [b]	48.7
Solubility of NaCl in moles of salt per 55.5 moles of solvent [a]		6.1	5.8

[a] At 25 °C. – [b] *Cf.* Table 3-3 in Section 3.2.

the first and third factors but often the second are operating simultaneously and it is difficult to separate them [762].

In Table 5-28 a number of the physical properties of light and heavy water are compared with each other [446, 451, 635].

Whereas light and heavy water have nearly identical dielectric constants and dipole moments, it can be concluded from the greater boiling point, heat of vaporization, density, and viscosity of heavy water, that liquid D_2O is more structured than the already highly structured H_2O at room temperature (*cf.* Fig. 2-1 in Section 2.1). This corresponds also to the fact that salts acting as structure-breakers are generally less soluble in D_2O than H_2O [446].

Bearing this in mind, and assuming that the direction and magnitude of medium-influenced kinetic solvent isotope effects of type (c) are determined by the different solvation of reactants and activated complex, Swain and Bader draw the following conclusion [450]*: A reaction which destroys the structure of water during formation of the activated complex (by creation of charge) will proceed more rapidly in light water. A reaction which returns structure to the solvent when the activated complex is reached (by neutralisation or dispersal of charge) will exhibit an enhanced rate in heavy water. In addition to these non-specific effects, there also can exist differences in specific interactions such as hydrogen-bonding between reactants or activated complex and the solvent. The rather scarce experimental results available confirm these simple predictions only in part. Most of the systems studied involve acid or base catalysis and consequently participation of water as reactant. Thus, the observed solvent isotope effects often include a composite of primary and secondary isotope effects, because any protons which are

* Another view of the origin of solvent isotope effects has been given by Bunton and Shiner [763]; see also [764].

exchangeable with solvent protons or deuterons give rise to primary and secondary isotope effects. For example, D_3O^+ is a stronger acid than H_3O^+, and DO^- is a stronger base than HO^-. Since proton-transfer is an integral part of kinetic solvent isotope effects in acid- and base-catalyzed reactions, and the solvent is involved in the reaction as reaction partner, further treatment of this topic lies beyond the scope of this book. The reader therefore is referred to some excellent and comprehensive reviews relating to solvent isotope effects [20, 446–449, 760, 761].

6 Solvent Effects on Absorption Spectra of Organic Compounds

6.1 General Remarks

When absorption spectra are measured in solvents of different polarity it is found that the position, intensity, and shape of absorption bands are usually modified by these solvents [1–4]. These changes are a result of physical intermolecular solute-solvent interaction forces (such as ion-dipole, dipole-dipole, dipole-induced dipole, hydrogen bonding, *etc.*), which above all tend to alter the energy difference between ground and excited state of the absorbing species containing the chromophore*). The medium influence on absorption spectra can be considered by comparing the spectral change observed (a) on going from the gas phase to solution, or (b) simply by changing the nature of the solvent. Because in most cases it is not possible to measure the absorption spectrum in the gas phase, the treatment of this topic will be restricted to approach (b) in this chapter. This is possible, because there is increasing evidence that there is no lack of continuity between the magnitude of spectral changes, in going from an isolated molecule in the gas phase to a weakly interacting or to a strongly interacting liquid medium, provided there are no specific interactions like hydrogen bonding or EPD/EPA complexation [3].

All those spectral changes which arise from alteration of the chemical nature of the chromophore-containing molecules by the medium, such as proton or electron transfer between solvent and solute, solvent-dependent aggregation, ionization, complexation, or isomerization equilibria lie outside the scope of this Chapter. Theories of solvent effects on absorption spectra assume principally that the chemical states of the isolated and solvated chromophore-containing molecules are the same and treat these effects only as a physical perturbation of the relevant molecular states of the chromophores.

Thus, solvent effects on absorption spectra can be used to provide information about solute-solvent interactions [1, 4]. On the other hand, in order to minimize these effects, it would be preferable to record absorption spectra in less interacting nonpolar solvents, such as hydrocarbons, whenever solubility permits. Suitable choice of a spectral solvent may be facilitated by consulting Tables A-4 (UV/Vis), A-5 (IR), A-6 (^{1}H-NMR), and A-7 (^{13}C-NMR) in the Appendix, which list some of the more common solvents and their absorption properties.

6.2 Solvent Effects on UV/Vis Spectra [5-17]

6.2.1 Solvatochromic Compounds

The term *solvatochromism* is used to describe the pronounced change in position (and sometimes intensity) of an UV/Vis absorption band, accompanying a change in the polarity of the medium. A hypsochromic (or blue) shift, with increasing solvent polarity, is

* A *chromophore* is generally regarded as any grouping of an organic molecule (sometimes the whole molecule itself) which is responsible for the light absorption under consideration. O. N. Witt introduced this term, although it had a different meaning from that accepted today; *cf.* Ber. Dtsch. Chem. Ges. *9*, 522 (1876). For example, the C=C group of ethene which is responsible for the $\pi \rightarrow \pi^*$ absorption is its chromophore.

usually called *negative solvatochromism*. The corresponding bathochromic (or red) shift is termed *positive solvatochromism*. What kind of compounds exhibit this attitude towards changes in solvent polarity?

To begin with, the solvent effect on spectra, resulting from electronic transitions, is primarily dependent on the chromophore and the nature of the transition ($\sigma \rightarrow \sigma^*$, $n \rightarrow \sigma^*$, $\pi \rightarrow \pi^*$, $n \rightarrow \pi^*$, and charge-transfer absorption). The electronic transitions of particular interest in this respect are $\pi \rightarrow \pi^*$ and $n \rightarrow \pi^*$, as well as charge-transfer absorptions. Organic compounds with chromophores containing π-electrons can be classified into three different groups according to their idealized π-electronic structure: *aromatic compounds*, *polyenes* (and *polyines*), and *polymethines* (*cf.* Fig. 6-1 [18, 19])*).

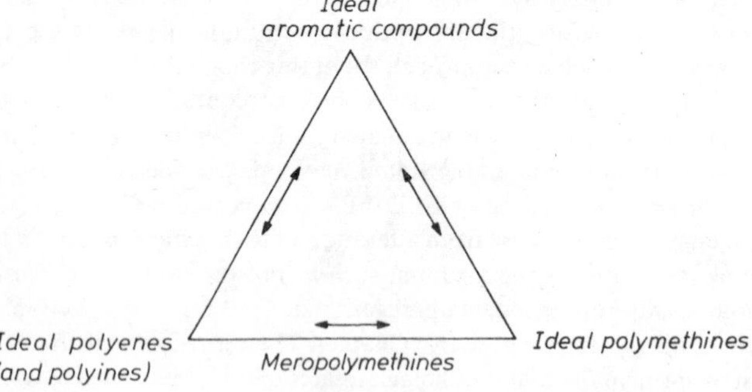

Fig. 6-1. Classification of organic compounds with π-electron system according to Dähne [18, 19]*).

In contrast with compounds of aromatic and polyene-like electronic structure, polymethines are conjugated chain molecules with equal bond-lengths and charge alternation along the methine chain [18, 19]. They exhibit the following common structural features:

$\ldots(n+3)\pi\ldots$	$n=1, 3, 5, 7, \ldots$
$X-(CR)_n-X'$	$R=H$ or other substituents
	X, X' = terminal chain atoms (N, O, P, S) or atom groups

$X=X'$ polymethine dyes ($X=X'=N$: cyanines; $X=X'=O$: oxonoles)
$X \neq X'$ meropolymethine dyes ($X=N$ and $X'=O$: merocyanines)

* A slight extension of Dähne's triad system of basic chromophores has been recently given by J. Fabian, in D. Fassler, K.-H. Feller, and B. Wilhelmi (eds.): *Progress and Trends in Applied Optical Spectroscopy*. Teubner-Texte zur Physik, Vol. 13, p. 151–177, Teubner, Leipzig 1987. In addition to the three classical chromophores of polyenic, aromatic, and polymethinic type, two non-classical chromophores of anti-aromatic (*e.g.* [4n]annulenes) and non-Kekulé type (*e.g.* biradicaloid dyes) have been included in Dähne's scheme, thus constituting a squared pyramid instead of the triangle given in Fig. 6-1.

Of particular interest are the intramolecularly ionic *meropolymethine dyes* (especially the *merocyanines*), whose electronic structure lies somewhere between that of polyenes and that of polymethines depending on the nature of X and X′ as well as on solvent polarity [20]. These are systems in which an electron-donating group, D, is linked by a conjugated system, R, to an electron-accepting group, A. Their intermediate π-electronic structure can be described in terms of two mesomeric structures, D—R—A↔D$^{\oplus}$—R—A$^{\ominus}$, as, for example, this special vinylogous merocyanine dye (n=0, 1, 2, . . .):

Its electronic transition is associated with an intramolecular charge-transfer between donor and acceptor group, producing an excited state with a dipole moment (μ_e) appreciably different from that in the ground state (μ_g).

It has been established experimentally, that only those molecules with π-electrons, for which the charge distribution (and consequently the dipole moment) in the electronic ground state is considerably different from that in the excited state exhibit a pronounced solvatochromism. Thus, for the following organic compounds only a comparatively small solvent dependence of their UV/Vis absorption spectra is observed: aromatic compounds (without electron donor and/or acceptor groups, *e.g.* benzene [21, 22]), polyenes (*e.g.* lycopene [23], carotinoids [24]), polyines (*e.g.* polyacetylenes [25]), and symmetrical polymethine dyes [26–28, 292, 293], *e.g.* the heptamethine cyanine dye shown below [293]:

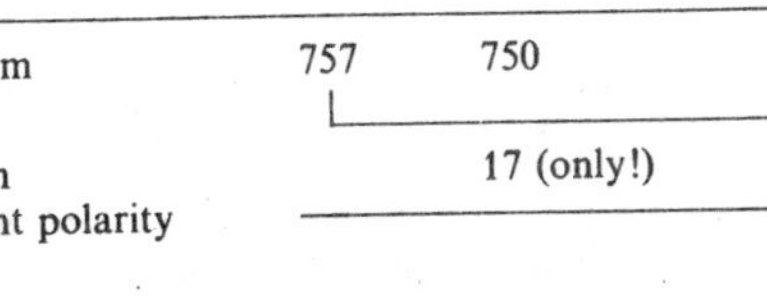

Solvent	$CHCl_3$	CH_3SOCH_3	CH_3OH
λ_{max}/nm	757	750	740
$\Delta\lambda$/nm	17 (only!)		
Solvent polarity	→		

A solvent change from methanol to chloroform causes a bathochromic shift of only *ca.* 300 cm^{-1} (17 nm) for its longest wavelength $\pi \rightarrow \pi^*$ absorption band [293].

In contrast with these nonpolar compounds, very dramatic solvent effects on UV/Vis spectra have been observed for dipolar meropolymethine dyes, especially merocyanines, due mainly to the change in their dipole moments on electronic transition. An example is the following negatively solvatochromic pyridinium-*N*-phenoxide betaine, which exhibits one of the largest solvatochromic effects ever observed (*cf.* Fig. 6-2 [29]):

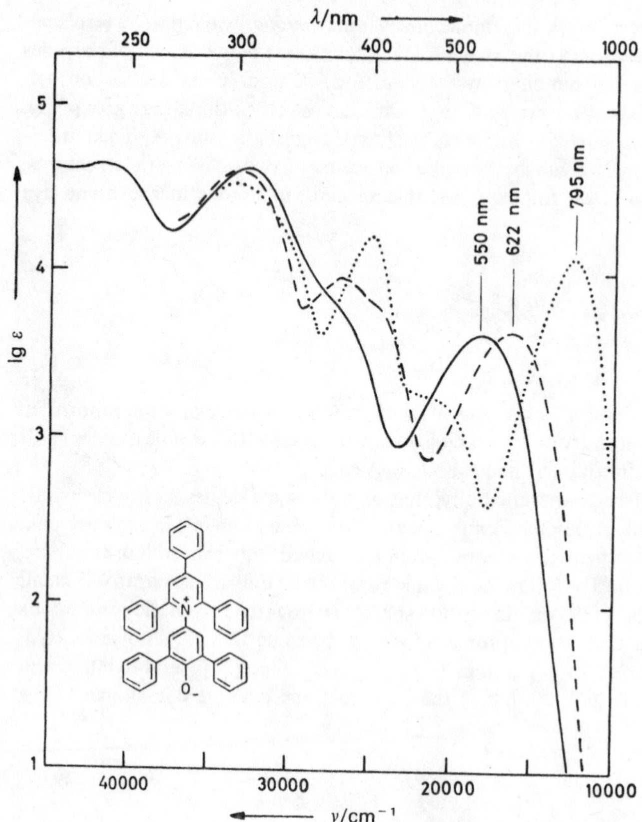

Fig. 6-2. UV/Vis absorption spectrum of 2,6-diphenyl-4-(2,4,6-triphenyl-1-pyridinio)phenoxide in ethanol (———), acetonitrile (----), and 1,4-dioxane (·····) at 25 °C [29].

Its long wavelength band is shifted by 9730 cm^{-1} (357 nm) on going from diphenylether to water as solvent. Solutions of this betaine dye are red-colored in methanol, violet in ethanol, blue in isoamylalcohol, green in acetone, and yellow in anisole, thus comprising the whole visible region. This extraordinary large solvent-induced shift of the visible $\pi \rightarrow \pi^*$ absorption band of intramolecular charge-transfer character has been used to introduce an empirical parameter of solvent polarity, called the $E_T(30)$-value [10, 29, 294]; *cf.* Section 7.4. – It can also be used for the UV/Vis-spectroscopic determination of water or other polar solvents in binary mixtures of solvents of different polarity [30, 31, 295, 296]. Applications of the solvatochromism of the pyridinium-*N*-phenoxide betaine dye in analytical chemistry have been reviewed [297].

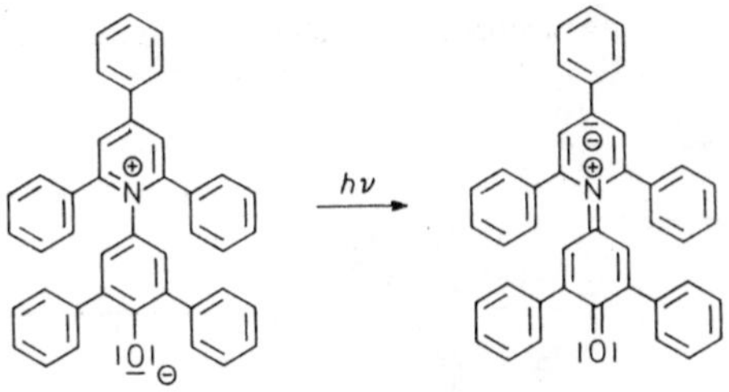

Solvent	$(C_6H_5)_2O$[a]	$C_6H_5OCH_3$	CH_3COCH_3	i-$C_5H_{11}OH$	C_2H_5OH	CH_3OH	H_2O[a]
λ_{max}/nm	810	769	677	608	550	515	453
Solution colour	—	yellow	green	blue	violet	red	—
Solvent polarity	⟶						

[a] Solubility very low.

The extreme sensitivity of the visible absorption spectrum to small changes in the surrounding medium has made this betaine dye a useful molecular probe in the study of micelle/solution interfaces [298, 299], microemulsions and phospholipid bilayers [299], model liquid membranes [300], rigid rod-like isocyanide polymers [301], and the retention behaviour in reverse-phase liquid chromatography [302].

Not only the position of the long-wavelength absorption band of the pyridinium-*N*-phenoxide betaine dye is strongly solvent-dependent but also its band width and band shape. A thorough band-shape analysis of a betaine dye with *tert*-butyl groups instead of the phenyl groups in *o*,*o*′-position to the phenolic oxygen atom has been given recently [431]. Bandshape features are dominated by electronic-vibrational coupling parameters such as solvent reorganization Gibbs energy, molecular nuclear displacements, and vibrational frequencies. In non-HBD solvents, the notably asymmetric charge-transfer absorption band with a slower fall-off on the high-wavenumber site indicates substantial coupling to a molecular mode with a vibrational wavenumber of about 1600 cm^{-1}. This value is close to those expected for C—C, C—N, and C—O stretching modes likely to be the ones displaced on photoexcitation. For protic solvents, a different kind of bandshape pattern is observed, most likely caused by specific betaine/solvent interactions such as hydrogen-bonding [431].

A representative selection of some thoroughly investigated positive and negative solvatochromic compounds is given in Table 6-1. Further interesting recent examples of solvatochromic dyes can be found in references [311–314].

Table 6-1 reveals that the long wavelength absorption band undergoes a bathochromic shift as the solvent polarity increases (positive solvatochromism), if the excited state is more dipolar than the ground state ($\mu_g < \mu_e$; dyes no. 1...11). If the ground state is more dipolar than the excited state ($\mu_g > \mu_e$), the opposite behaviour, a hypsochromic shift, occurs (negative solvatochromism; dyes no. 12...22). In valence bond theory language, the extent and direction of solvatochromism depends on whether the zwitterionic mesomeric structure is more important in the ground state or in the excited state. The quadrupole-merocyanines no. 8 and 10 represent special cases for which the dipole

Table 6-1. A selection of 22 representative solvatochromic compounds comprising their dipole moments in the ground (μ_g) and excited state (μ_e) [32, 33] and their long wavelength $\pi-\pi^*$ absorption maxima in two solvents of widely different polarity.

No.	Formula	$\frac{\mu_g \cdot 10^{30}}{\mathrm{C\,m}}$ a)	$\frac{\mu_e \cdot 10^{30}}{\mathrm{C\,m}}$ a)	$\tilde{\nu}/\mathrm{cm}^{-1}$ (nonpolar solvent)	$\tilde{\nu}/\mathrm{cm}^{-1}$ (polar solvent)	$\Delta\tilde{\nu}/\mathrm{cm}^{-1}$ b)	References
(a) *Positive solvatochromic compounds*							
1	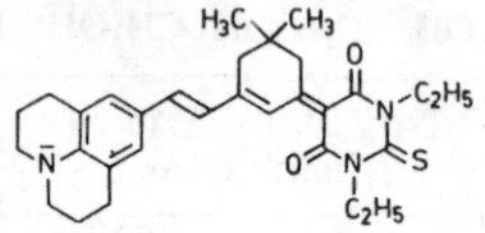	–	–	17790 (*iso*-octane)	13390 (lutidine-water 20:80)	4400	[34]
2	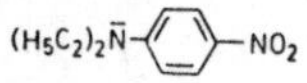 $(H_5C_2)_2\bar{N}-C_6H_4-NO_2$	17 c)	43 c)	27400 (cyclohexane)	23230 (water)	4170	[35]
3	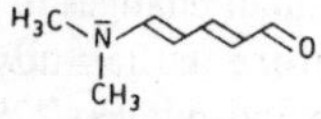	26	–	30170 (*n*-hexane)	26140 (water)	4030	[37]
4	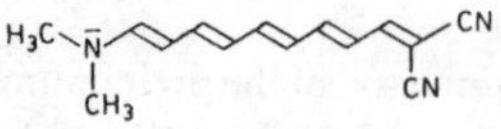	46	99	18800 (methylcyclohexane)	14810 (ethanol)	3990	[36]
5	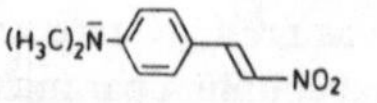	26	60	25000 (cyclohexane)	21280 (ethanol-water 1:9)	3720	[38, 432]
6	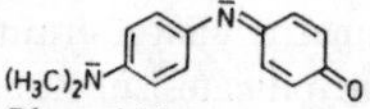 Phenol blue	19	25–30	18120 (cyclohexane)	14970 (water)	3150	[23, 39–41] e)
7	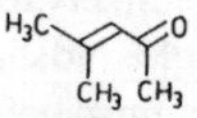	9	–	43370 (*iso*-octane)	41220 (water)	2150	[42]
8	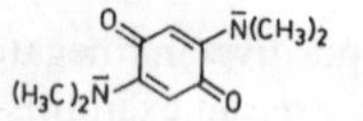	0	0	20830 (*n*-hexane)	18830 (water)	2000	[43]
9	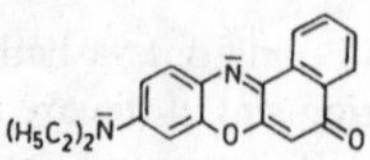	–	–	19190 (benzene)	18180 (methanol)	1010	[44, 45]
10	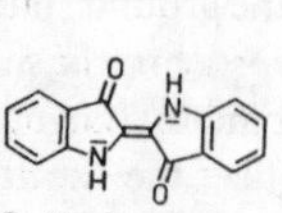 Indigo	0	0	17010 tetrachloromethane	16130 (dimethyl sulfoxide)	880	[46]

Table 6-1. (Continued)

No.	Formula	$\frac{\mu_g \cdot 10^{30}}{\mathrm{C\,m}}$ a)	$\frac{\mu_e \cdot 10^{30}}{\mathrm{C\,m}}$ a)	$\tilde{\nu}/\mathrm{cm}^{-1}$ (nonpolar solvent)	$\tilde{\nu}/\mathrm{cm}^{-1}$ (polar solvent)	$\Delta\tilde{\nu}/\mathrm{cm}^{-1}$ b)	References
11		28	63	24100 (cyclohexane)	23500 (methanol)	600	[47, 47a]
(b) *Negative solvatochromic compounds*							
12		49 d)	20 d)	12350 (diphenyl ether)	22080 (water)	−9730	[29, 30]
13		–	–	14600 (toluene)	24100 (water)	−9500	[34]
14		87	–	16130 (chloroform)	22620 (water)	−6490	[48–50, 50a, 304–309]
15		–	–	16080 (chloroform)	22170 (water)	−6090	[51]
16		97	63	16390 (pyridine)	21280 (water)	−4890	[52, 53]
17		–	–	18520 (benzene)	23150 (methanol)	−4630	[54]
18		–	–	17000 (benzene)	21550 (water)	−4550	[55]
19		–	–	16640 (toluene)	19760 (methanol)	−3120	[56, 310]

Table 6-1. (Continued)

No.	Formula	$\frac{\mu_g \cdot 10^{30}}{\mathrm{C\,m}}$ [a]	$\frac{\mu_e \cdot 10^{30}}{\mathrm{C\,m}}$ [a]	$\tilde{\nu}/\mathrm{cm}^{-1}$ (nonpolar solvent)	$\tilde{\nu}/\mathrm{cm}^{-1}$ (polar solvent)	$\Delta\tilde{\nu}/\mathrm{cm}^{-1}$ [b]	References
20	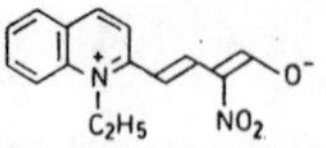 C_2H_5, NO_2, O^-	35	–	19010 (benzene)	22080 (water)	−3070	[57]
21	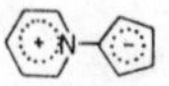	44	–	19560 (chloroform)	22060 (water)	−2500	[58, 59]
22	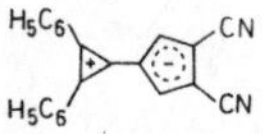 H_5C_6, H_5C_6, CN, CN	48	–	27250 (cyclohexane)	28820 (acetonitrile)	−1570	[60]

[a] Dipole moments in Coulombmeter (C m). 1 Debye $= 3.336 \cdot 10^{-30}$ C m. Methods of determination of μ_e and results have been reviewed [32, 33, 303].
[b] $\Delta\tilde{\nu} = \tilde{\nu}$ (nonpolar solvent) $- \tilde{\nu}$ (polar solvent). [c] Value for 1-dimethylamino-4-nitrobenzene.
[d] Value for 2,6-di-*tert*-butyl-4-(2,4,6-triphenyl-1-pyridinio)phenoxide [61, 62]. [e] The solvatochromism of phenol blue in supercritical-fluid solvents (C_2H_4, $CFCl_3$, CHF_3) has recently been studied by S. Kim and K. P. Johnston, Ind. Eng. Chem. Res. *26*, 1206 (1987).

moments μ_g and μ_e must be zero due to the presence of a center of symmetry. Some of the dyes included in Table 6-1 have been used to derive empirical scales of solvent polarity (no. 1, 2, 4, 12, and 13; *cf.* Section 7.4) [10, 294].

Not only intramolecularly ionic compounds such as dipolar meropolymethine dyes, but also EPD/EPA complexes (*cf.* Section 2.2.6) with an intermolecular charge-transfer (CT) absorption can exhibit a pronounced solvatochromism. The CT transition also involves ground and excited states with very different dipole moments. This suggests that the CT absorption band should exhibit marked solvent polarity effects [7c, 17, 63, 64].

A striking example is the negatively solvatochromic effect observed for 1-ethyl-4-methoxycarbonylpyridinium iodide, whose UV/Vis spectrum in a variety of solvents is shown in Fig. 6-3 [65–67]. The longest wavelength band of the ground-state ion-pair complex corresponds to an intermolecular transfer of an electron from the iodide to the pyridinium ion with annihilation of charge during the transition. The large dipole moment

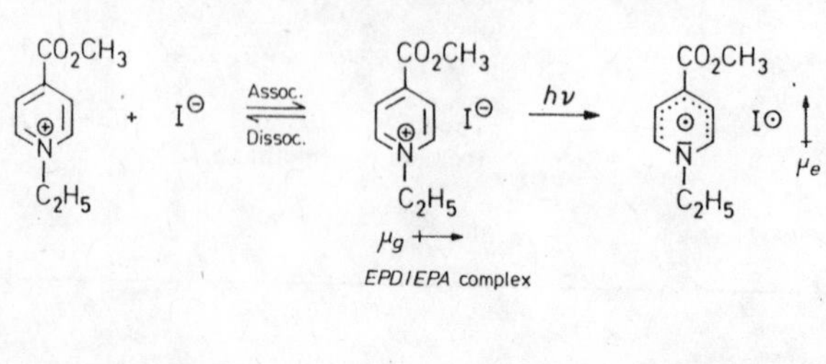

CO_2CH_3 C_2H_5 + $I^{\ominus}$ Assoc. / Dissoc. CO_2CH_3 C_2H_5 $I^{\ominus}$ $\xrightarrow{h\nu}$ CO_2CH_3 C_2H_5 $I\odot$ μ_e

μ_g

EPD/EPA complex

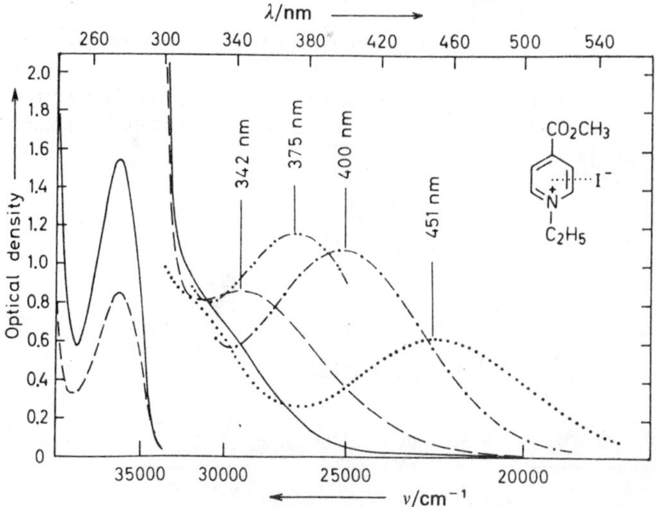

Fig. 6-3. The first charge-transfer band in the UV/Vis absorption spectrum of 1-ethyl-4-methoxycarbonylpyridinium iodide in water (———), methanol (----), 2-propanol (–··–··–··–), acetonitrile (–·–·–·–), and *cis*-1,2-dichloroethene (·····) at 25 °C [65–67].

of the ground state is at right angles to the pyridinium ring. In the excited state, however, it is much smaller and will lie, at least that part resulting from the pyridinyl radical, in the plane of the ring ($\mu_g > \mu_e$; $\mu_g \perp \mu_e$). The corresponding dipole moments for the ground and excited state have been calculated as $\mu_g = 46 \cdot 10^{-30}$ C m (13.9 D) and $\mu_e = 29 \cdot 10^{-30}$ C m (8.6 D) [315]. The large negative solvatochromism of this pyridinium iodide has been explained by the stabilization of the more dipolar ground state and destabilization of the less dipolar excited state on transfer of the ion pair to more polar solvents [65–67]. Alternatively, the negative solvatochromism can result if both states are destabilized when transferred to more polar solvents, resulting in a relatively greater destabilization of the excited state [315]. The pronounced negative solvatochromism of this pyridinium iodide has been used to establish an empirical scale of solvent polarity, the so-called *Z*-scale [65–67]; *cf.* Section 7.4.

Less pronounced, but significant solvent shifts of the CT band are observed for EPD/EPA complexes, if the ground state is not ionic and the excited state is ionic ($\mu_g < \mu_e$). An example is the CT absorption band of the acenaphthene/3,5-dinitrophthalic anhydride complex which shows a bathochromic shift with increasing solvent polarity [64].

Organometal complexes composed of a central metal atom and organic ligands containing a π-electron system exhibit two kinds of solvent-dependent charge-transfer absorptions, depending on the relative electron-donor/electron-acceptor properties of metal and ligand: (i) metal-to-ligand charge-transfer absorption (MLCT), and (ii) ligand-to-metal charge-transfer absorption (LMCT). The strong solvatochromism observed for both types of charge-transfer absorptions has been thoroughly investigated for a variety of group 6 metal complexes with different organic ligands (mainly diimine derivatives,

—N=C—C=N—, and their heterocyclic counterparts); *cf.* references [423, 424] for recent reviews. Only two representative examples can be mentioned here: the negative solvatochromic 2,2′-bipyridine tetracarbonylwolfram(0) complex with MLCT absorption [425, 426], and the positive solvatochromic imidazole pentacyanoferrate(III) complex with LMCT absorption [427].

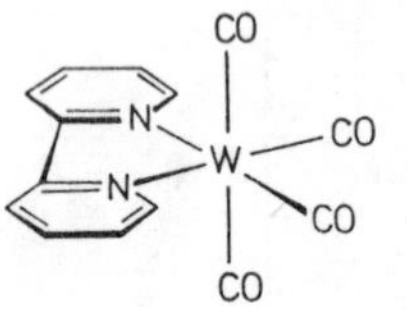

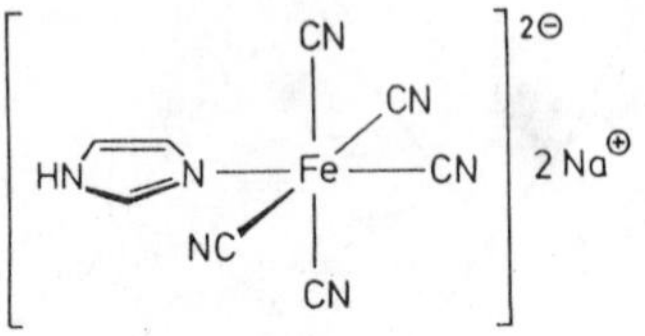

Solvent	i-C_8H_{17}	CH_3OH
λ_{max}/nm	574	474
$\Delta\lambda$/nm	−100	

Solvent	$HCONMe_2$	H_2O
λ_{max}/nm	427	500
$\Delta\lambda$/nm	+73	

In the MLCT-case, upon excitation an electron is transferred from the electron-rich tetracarbonylwolfram(0) moiety to the electron-accepting 2,2′-bipyridine system ($d \rightarrow \pi^*$ transition), with a simultaneous change in polarizability and dipole moment, leading to an excited state with reduced dipole moment. In the LMCT-case, upon excitation an electron jumps from the electron-rich imidazole ring to the electron-accepting pentacyano-ferrate(III) fragment ($\pi \rightarrow d$ transition) with similar but opposite changes in polarizability and dipole moment. – The MLCT absorption of centrosymmetric binuclear coordination compounds without permanent dipole moment such as bis(pentacarbonylwolfram)pyrazine, $(CO)_5W$-pyrazine-$W(CO)_5$, experiences an even stronger solvatochromism than their mononuclear, non-centrosymmetric, dipolar analogues, thus demonstrating the importance of solvent dipole/solute induced-dipole and dispersion interactions in solutions of such complexes [426]; *cf.* the analogous solvatochromism of the centrosymmetric organic dyes no. 8 and 9 in Table 6-1.

Recently, a consistent model permitting rationalization and prediction of the solvatochromic behaviour of coordination compounds with MLCT transition has been described [428]. According to this qualitative model, the changing relation between the metal-ligand bond dipolarities in the ground and MLCT excited state determines whether the complex is negative, positive, or not solvatochromic [428].

Special cases of charge-transfer spectra are the so-called charge-transfer-to-solvent (CTTS) spectra [17, 68]. In this type of CT transitions solute anions may act as electron-donor and the surrounding solvent shell plays the role of the electron-acceptor. A classical example of this kind of CTTS excitation is the UV/Vis absorption of the iodide ion in solution, which shows an extreme solvent sensitivity [68, 316]. Solvent-dependent CTTS absorptions have also been obtained for solutions of alkali metal anions in ether or amine solvents [317].

6.2.2 Theory of Solvent Effects on UV/Vis Absorption Spectra

A qualitative interpretation of solvent shifts is possible by considering (a) the momentary transition dipole moment present during the optical absorption, (b) the difference in permanent dipole moment between the ground and excited state of the solute, (c) the change in ground-state dipole moment of the solute induced by the solvent, and (d) the Franck-Condon principle [69]. According to Bayliss and McRae, four limiting cases can be distinguished for intramolecular electronic transitions in solution [69, 318]:

(1) *Nonpolar solute in a nonpolar solvent.* In this case only dispersion forces contribute to the solvation of the solute. Dispersion forces, operative in any solution, invariably cause a small bathochromic shift, the magnitude of which is a function of the solvent refractive index n, the transition intensity, and the size of the solute molecule. The function $(n^2-1)/(2n^2+1)$ has been proposed to account for this general red shift [69, 70]. Corresponding linear correlations between this function of n and $\Delta\tilde{\nu}$ have been observed for aromatic compounds (*e.g.* benzene [22], phenanthrene [71]), polyenes (*e.g.* lycopene [23]), and symmetrical polymethine dyes (*e.g.* cyanines [26, 27, 292, 293].

(2) *Nonpolar solute in a polar solvent.* In the absence of a solute dipole moment there is no significant orientation of solvent molecules around the solute molecules, and again a general red shift, depending on the solvent refractive index n, is expected.

(3) *Dipolar solute in a nonpolar solvent.* In this case, the forces contributing to solvation are dipole-induced dipole and dispersion forces. If the solute dipole moment increases during the electronic transition, the Franck-Condon excited state*) is more solvated by dipole-solvent polarization, and a red shift, depending on solvent refractive index n and the change in solute dipole moment, is expected. The Franck-Condon excited state is less solvated, if the solute dipole moment decreases during the electronic transition, and a blue shift, again proportional to the above mentioned two factors, is expected. In the latter case, the resultant shift may be red or blue depending on the relative magnitude of the red shift caused by polarization and the blue shift.

(4) *Dipolar solute in a polar solvent.* Since the ground-state solvation results largely from dipole-dipole forces in this case, there is an oriented solvent cage around the dipolar solute

---

* The Franck-Condon principle states that since the time required for a molecule to execute a vibration (ca. 10^{-12} s) is much longer than that required for an electronic transition (*ca.* 10^{-15} s), the nuclei of the chromophore (and of the surrounding solvent molecules) do not appreciably alter their positions during an electronic transition; *cf.* J. Franck, Trans. Faraday Soc. *21*, 536 (1926), and C. U. Condon, Phys. Rev. *32*, 858 (1928). Therefore, at the instant of its formation, the excited solute molecule is momentarily surrounded by a solvent cage whose size and orientation are those suitable to the ground state – a situation, which is usually called the Franck-Condon state [69]. The equilibrium excited state is subsequently reached by a process of relaxation. The Franck-Condon excited molecule and its solvent cage are in a strained state, whose energy is necessarily greater than that of the equilibrium state. Because the Franck-Condon excited state is directly reached, the magnitude of the solvent effect does not correspond exactly to the extent of charge-transfer on electronic activation, as was proposed for reaction rates, where the activated complex is supposed to be in thermal equilibrium with its environment.

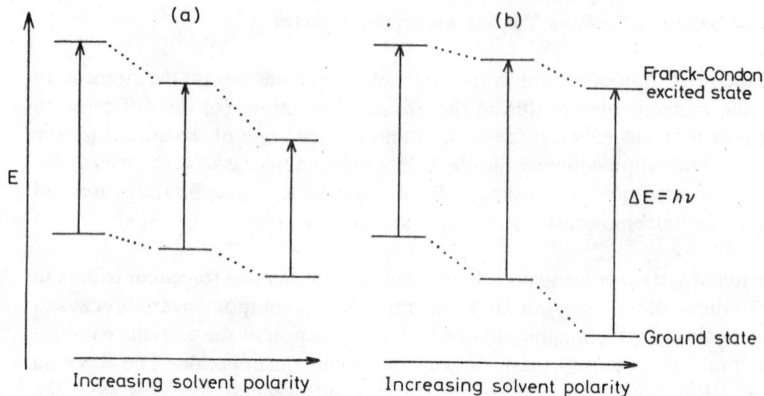

Fig. 6-4. Schematic qualitative representation of solvent effects on the electronic transition energy of dipolar solutes in polar solvents [2, 69]. (a) $\mu_g < \mu_e$, *i.e.* the dipole moment of the Franck-Condon excited state of the solute is larger than those of its ground state; (b) $\mu_g > \mu_e$.

molecules, resulting in a net stabilization of the ground state of the solute molecules. If the solute dipole moment increases during the electronic transition ($\mu_g < \mu_e$), the Franck-Condon excited state*) is formed in a solvent cage of already partly oriented solvent dipoles. The better stabilization of the excited state relative to the ground state, with increasing solvent polarity, will result in a bathochromic shift. Its magnitude will depend on the extent of the change in solute dipole moment during the transition, the value of the solvent dipole moment, and the extent of interaction between solute and solvent molecules. This situation is schematically illustrated in Fig. 6-4a.

If the dipole moment of the solute decreases during the electronic transition, the Franck-Condon excited state is in a strained solvent cage of oriented dipoles not correctly disposed to efficiently stabilise the excited state. Thus, with increasing solvent polarity, the energy of the ground state is lowered more than that of the excited state, and this produces a hypsochromic shift (*cf.* Fig. 6-4b). The superimposed bathochromic shift due to polarization will usually be less, resulting in a net hypsochromic shift. For intermolecular charge-transfer transitions (except charge-transfer-to-solvent) the direction of the solvent-induced wavelength shift may be determined in a similar manner.

For strongly solvatochromic compounds (*cf.* Table 6-1) the observed solvent-induced wavelength shifts cannot be explained only in terms of a change in the permanent dipole moment on electronic transition ($\mu_g \neq \mu_e$). The change in ground-state dipole moment of the solute, induced by the surrounding solvent cage ($\mu_g \rightarrow \mu_g'$) must also be taken into account [20, 36, 72–79]. The dipolar solute molecules cause an electronic polarization of the surrounding solvent molecules, creating a so-called reaction field**),

---

* see Footnote p. 295.

** According to Onsager, a reaction field is the electric field arising from an interaction between an ideal nonpolarizable point dipole and a homogeneous polarizable dielectric continuum in which the dipole is immersed [80]. The reaction field is the electric field felt by the solute molecule due to the orientation and/or electronic polarization of the solvent molecules by the solute dipole.

which affects the solute's ground-state dipole moment μ_g. That is, the interaction of the dipolar solute molecules with this induced reaction field, due to the total dipole moment (permanent and induced) of the solvent molecules, may cause an alteration of the electronic structure of the chromophore*). For meropolymethine dyes such as the positively solvatochromic merocyanine described below, this implies that increasing solvent polarity should shift the electronic structure from a polyene-like state (*a*) to a more polymethine-like state (*b*) [20, 74–79].

$(2n+4)\pi$

(*a*) Polyene-like state ⟷ (*b*) Polymethine-like state ⟷ (*c*) Polyene-like state

Increasing solvent polarity

The consequences of such a change in electronic ground-state structure on the position of absorption have been calculated on the basis of valence-bond theory by Förster already in 1939 [72]. According to this calculations, an intermediate meropolymethine (*b*) with equal contribution of both mesomeric structures (*a*) and (*c*) will have the longest wavelength absorption. Therefore, a chromophore with a polyene-like electronic structure (*a*) will exhibit a bathochromic shift with increasing solvent polarity (positive solvatochromism), whereas a chromophore with a polymethine-like electronic structure (*b*) will show a hypsochromic shift on the same solvent change (negative solvatochromism) [72].

That the electronic ground-state structure of a dipolar solute is indeed affected by solvent polarity has been independently shown by ^{1}H-NMR- [20, 50, 73, 75, 78], ^{13}C-NMR- [77], and IR-measurements [20] of merocyanines. Some of these results observed with the positively solvatochromic 3-dimethylamino-acrolein are presented in Table 6-2.

The transition from polyene-like state (*a*) (with a balanced π-electron density) to polymethine-like state (*b*) (with strongly alternating π-electron density along the methine chain) with increasing solvent polarity can be seen clearly from the ^{13}C- and ^{15}N-chemical shifts and from the increasing equalization of the $^3J_{HH}$ coupling constants, as well as from the decreasing wave-number of the C=O stretching vibration. Furthermore, it has been shown that the electronic polarizability of this merocyanine also depends strongly on solvent polarity [79, 83]. In non-polar solvents, the polarizability is similar to that of polyenes, whereas in polar solvents, it very nearly reaches that of ideal polymethines [79].

As shown by the ^{1}H-NMR chemical shifts of negatively solvatochromic meropolymethine dyes (*e.g.* phenol blue), the electronic ground-state structure of these dyes

* This reaction field, caused by the solvent molecules surrounding the dipolar solute molecule, is of the order of 10^7 V/cm and can influence an absorption spectrum in the same manner as an externally applied electric field. The spectral changes produced by means of a homogeneous external electric field have been termed *electrochromism*. Thus, solvatochromism is closely related to electrochromism [13, 81, 82].

Table 6-2. ^{13}C- and ^{15}N-Chemical shifts [77], $^3J_{HH}$ coupling constants [73], and C=O stretching vibrations [20] of 3-dimethylamino-acrolein in solvents of increasing polarity.

$$(CH_3)_2\bar{N}-\overset{3}{C}H=\overset{2}{C}H-\overset{1}{C}H=O \leftrightarrow (CH_3)_2\overset{\delta\oplus}{N}\cdots CH\cdots\overset{6\pi}{C}H\cdots CH-\overset{\delta\ominus}{O} \qquad \mu_g = 21\cdot10^{-30}\ \text{C m}^{a)}$$

(a) (b) (benzene; 25 °C)

Solvents	CCl_4	CS_2	$CDCl_3$	CD_3SOCD_3	CD_3CN	D_2O
$\delta_{^{13}C}$/ppm C-1	187.6	–	–	188.4	–	190.5
C-2	101.2	–	–	101.3	–	100.3
C-3	160.1	–	–	161.6	–	164.4
$\delta_{^{15}N}$/ppm[b,c]	56.8	–	62.4	64.9	–	85.4
$^3J_{1-H/2-H}$/Hz	–	7.95	8.22	–	8.25	9.32
$^3J_{2-H/3-H}$/Hz	–	12.7	12.5	–	12.5	12.2
$\tilde{\nu}_{C=O}$/cm^{-1}[b]	1633[d]	–	1620	1620	1624	1592

[a] M. H. Hutchinson and L. E. Sutton, J. Chem. Soc. *1958*, 4382.
[b] Values for non-deuterated solvents.
[c] R. Radeglia, R. Wolff, B. Bornowski, and S. Dähne, Z. Phys. Chem. (Leipzig) *261*, 502 (1980).
[d] Value for c-C_6H_{12}.

changes from a polymethine-like state (*b*) in non-polar solvents to a polyene-like state (*c*) in polar solvents [50, 78].

A particularly interesting solvatochromic merocyanine dye is 4-[2-(1-methyl-4-pyridinio)ethenyl]phenolate. First it exhibits a bathochromic and then a hypsochromic shift of the long-wavelength $\pi \rightarrow \pi^*$ absorption band as the solvent polarity increases [309]. This is in agreement with semiempirical MO calculations [308]; *cf.* also entry 14 in Table 6-1.

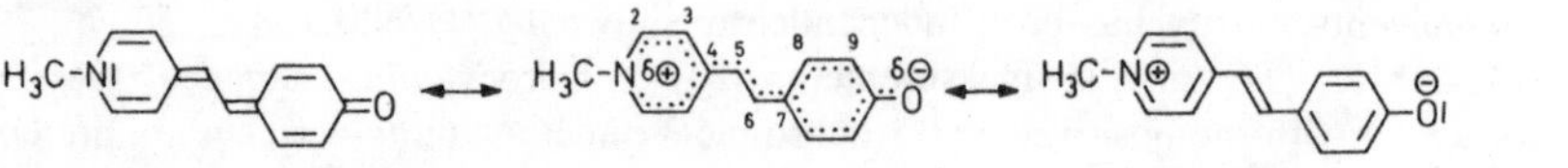

(a) (b) (c)

Solvent	c-C_6H_{12}	$CHCl_3$	H_2O
λ_{max}/nm	592	620	442
$\Delta\lambda$/nm	+28	−178	
Solvent polarity	→	→	→

This surprising inverted solvatochromism indicates clearly that the ground-state electronic structure of this stilbazolium merocyanine dye changes, with increasing solvent polarity, from the quinonoid structure (*a*) to the benzenoid structure (*c*) via the intermediate polymethine-like structure (*b*). The longest-wavelength absorption is near the balanced valence structure (*b*), that is in chloroform as solvent. In agreement with this inverted solvatochromism, the calculated dipole moments of the ground and excited state show the following evolution: $\mu_g < \mu_e$ in nonpolar and $\mu_g > \mu_e$ in polar solvents for the $\pi \rightarrow \pi^*$ transition [308]. Increasing solvent polarity also causes a dramatic change in molecular geometry as shown by both the calculated bond lengths and the net π-electron charges [308]. In a solvent of medium polarity, the merocyanine dye takes on an ideal polymethine valence structure with intermediate bond lengths of *ca.* 140 pm along the methine chain C-4/C-5/C-6/C-7. The experimentally not observed *trans* → *cis* photoisomerization of this transoid stilbazolium merocyanine dye can also be explained in terms of the contribution of both the benzenoid and the quinonoid mesomeric structure to its ground-state electronic structure [430]. – These experimental and theoretical results are at variance with models of solvatochromism based on the Onsager theory of dielectrics, in which the solute is simply treated as a point dipole whose moment is assumed to be solvent-independent.

In addition to changes in ground-state electronic structure and changes in dipole moment on electronic transition, a third possibility responsible for solvatochromism has been discussed [84]: if the potential curves (Morse curves) of ground and excited states of a meropolymethine are shifted towards one another by solvent interaction, then, according to the Franck-Condon principle, the relative intensity of the vibronic transitions should be altered, with a net shift or change of shape of the absorption band as a consequence [84]. It has been shown, however, that this effect contributes to solvatochromism only to a minor extent, if at all [37]. A remarkable solvent effect influencing the shape of $\pi \rightarrow \pi^*$ absorption bands, attributed to solvent-induced changes in vibronic interactions, has been reported for a unsymmetrical pyrylothiacyanine dye [319].

Sometimes, external solvent polarization interactions can lift internal symmetry restrictions in the solute molecule and can induce new bands not observable in the gas-phase spectrum. A well-known example is the vibrationally forbidden $0 \rightarrow 0$ vibrational component of the long-wavelength $\pi \rightarrow \pi^*$ absorption of benzene (at 262 nm in *n*-hexane) which appears when benzene is dissolved in organic solvents, but not in the gas phase [320].

It should be mentioned, that solvent effects on the intensity of UV/Vis absorption bands cannot be interpreted in a simple qualitative fashion as is the case for the band position shifts [85, 308, 309, 321–323].

Quantitative calculations of the solvent dependence of UV/Vis absorption spectra, based on different models, have been carried out by Bayliss and McRae [69], Oshika [86], McRae [70], Lippert [47], Bakhshiev [87], Bilot and Kawski [88], Weigang and Wild [71], Abe [89], Liptay [90, 94], Kuhn and Schweig [59, 91], Nicol *et al.* [16, 92], Suppan [93], Kampas [95], Germer [95a], Nolte and Dähne [95b], Bekárek [324], and Mazurenko [325]. The underlying theory has been repeatedly reviewed [1, 4, 13–17], and critical discussions of the differences in the results of previous calculations can be found in references [1, 15, 16, 90, 318].

According to the McRae-Bayliss model of solvatochromism [69, 70] which is directly evolved from Onsager's reaction field theory [80], the electronic transition from ground (*g*) to excited state (*e*) of a solvatochromic solute is given by Eq. (6-1) [318]:

$$\tilde{\nu}_{\mathrm{eg}}^{\mathrm{sol}} = \tilde{\nu}_{\mathrm{eg}}^{0} - [(\mu_{\mathrm{g}} \cdot \mu_{\mathrm{e}} \cdot \cos\varphi - \mu_{\mathrm{g}}^{2}) \cdot 1/a_{\mathrm{w}}^{3}] \cdot [L(\varepsilon_{\mathrm{r}}) - L(n^{2})] - [(\mu_{\mathrm{e}}^{2} - \mu_{\mathrm{g}}^{2}) \cdot 1/a_{\mathrm{w}}^{3}] \cdot [L(n^{2})] \qquad (6\text{-}1)$$

where $\tilde{\nu}$ is the solute transition energy in the gas phase ($\tilde{\nu}_{\mathrm{eg}}^{0}$) and in solution ($\tilde{\nu}_{\mathrm{eg}}^{\mathrm{sol}}$), a_{W} the solute cavity radius, μ_{g} and μ_{e} the permanent dipole moments of the ground and excited-state molecule, and φ the angle between the ground and excited-state dipoles. $L(\varepsilon_r)$ and $L(n^2)$ are reaction field functions defined by $L(x)=2(x-1)/(2x+1)$ with $x=\varepsilon_r$ or n^2; ε_r and n are the bulk static relative permittivity and refractive index of the solvent, respectively. $[L(\varepsilon_r)-L(n^2)]$ and $L(n^2)$ are the relevant solvent variables, whereas the remaining terms $(\mu_g, \mu_e, \varphi, a_W)$ are assumed to be solute-dependent constants and independent of the solvent. As already mentioned, this is not strictly correct since any real solute dipole is polarizable and therefore affected by the reaction field.

This McRae-Bayliss model of solvatochromism as well as modifications thereof have been recently carefully reexamined and tested by Brady and Carr [318] and Ehrenson [318]. The inadequacies of the McRae-Bayliss approach to solvatochromism are outlined. Alternate reaction field models have been tested, however, with limited success only [318].

A more rigorous approach to general UV/Vis-absorption/solvent correlations has been given by Liptay [33, 90, 94]. According to Liptay, the solvent-dependent wave number shift for an absorption corresponding to an electronic transition from ground- (g) to excited-state (e) molecules can be described by Eq. (6-2)*):

$$\begin{aligned} hca\Delta\tilde{\nu}_{eg}^{sol} &= hca(\tilde{\nu}_{eg}^{sol}-\tilde{\nu}_{eg}^{0}) \\ &= (W_{Ce}^{FC}-W_{Cg})+(W_{De}^{FC}-W_{Dg}) \\ &\quad -\tfrac{1}{2}(\tilde{\boldsymbol{\mu}}_e-\tilde{\boldsymbol{\mu}}_g)\mathbf{f}'(\mathbf{1}-\mathbf{f}'\boldsymbol{\alpha}_e)^{-1}(\boldsymbol{\mu}_e-\boldsymbol{\mu}_g) \\ &\quad -(\tilde{\boldsymbol{\mu}}_e-\tilde{\boldsymbol{\mu}}_g)\mathbf{f}(\mathbf{1}-\mathbf{f}\boldsymbol{\alpha}_g)^{-1}\boldsymbol{\mu}_g \\ &\quad -\tilde{\boldsymbol{\mu}}_g(\mathbf{1}-\mathbf{f}'\boldsymbol{\alpha}_e)^{-1}(\mathbf{1}-\mathbf{f}\boldsymbol{\alpha}_g)^{-2}\mathbf{f}(\boldsymbol{\alpha}_e-\boldsymbol{\alpha}_g) \\ &\quad \times[\tfrac{1}{2}(\mathbf{1}-\mathbf{f}'\boldsymbol{\alpha}_g)\mathbf{f}\boldsymbol{\mu}_g+(\mathbf{1}-\mathbf{f}\boldsymbol{\alpha}_g)\mathbf{f}'(\boldsymbol{\mu}_e-\boldsymbol{\mu}_g)] \end{aligned} \tag{6-2}$$

$\tilde{\nu}_{eg}^{0}$ and $\tilde{\nu}_{eg}^{sol}$ are the absorption wave numbers of the transition considered in the gas phase and in solution, respectively, h is Planck's constant, c speed of light in vacuum, and a $(=10^2\ \mathrm{m}^{-1}\cdot\mathrm{cm})$ a conversion constant (thus the unit of $\tilde{\nu}_{eg}$ is cm^{-1}).

The first term on the right-hand side of Eq. (6-2) represents the differences of the energies required to form a cavity in the solvent for the ground state and Franck-Condon excited state molecules. Since, for most transitions its change in size during the excitation process is small, it is usually assumed that $W_{Ce}^{FC}-W_{Cg}=0$. The second term represents the dispersion interaction between the solute molecule and the surrounding solvent molecules, approximated by $W_{De}^{FC}-W_{Dg}=-hcaD_{eg}f'$, where D_{eg} is a quantity almost independent of the solvent. According to Eq. (6-4) the dispersion interactions are dependent on the

* The tensors $\mathbf{f}$ and $\mathbf{f}'$ in Eq. (6-2) may be represented by Eqs. (6-3) and (6-4) [33, 90], assuming that the solvent can be approximated to a homogeneous and isotropic dielectric, where the solute molecules are localized in a spherical cavity (then the tensor $\mathbf{f}_e$ is reduced to the scalar f_e), and approximating the dipole moment of the solute molecule by a point dipole localized in the center of this sphere:

$$\mathbf{f}=f\mathbf{1}=[2(\varepsilon_r-1)/4\pi\varepsilon_0 a_w^3(2\varepsilon_r+1)]\mathbf{1} \tag{6-3}$$

$$\mathbf{f}'=f'\mathbf{1}=[2(n^2-1)/4\pi\varepsilon_0 a_w^3(2n^2+1)]\mathbf{1} \tag{6-4}$$

In these Eqs. ε_r represents the relative permittivity of the solution, ε_0 the permittivity of vacuum, n the refractive index of the solution (for $\tilde{\nu}\rightarrow 0$), and a_W is the radius of the sphere.

refraction index n of the solvent. The third and fourth terms of Eq. (6-2) represent the energy change due to the change in the permanent dipole moment of the solute molecule during excitation. The third term depends on the change in the dipole moment and essentially on the refractive index of the solvent used; the fourth term depends on the ground-state dipole moment, the change of dipole moment, and the dielectric constant of the solvent. The fifth and last term depends on the change of the polarizability $(\alpha_e - \alpha_g)$ of the solute molecule on excitation*).

In the case of nonpolar solute molecules, the third and fourth as well as the fifth term in Eq. (6-2) is zero, thus the solvent dependence will be determined by dispersion interactions and only the second term is essential. The solvent shift, compared to the vapour state, will be approximately 70 to 3000 cm^{-1} to lower wave numbers (general red shift) depending on the function **f**′ only (*cf.* Eq. (6-4)). If, on excitation, there is a sufficiently large change in dipole moment, the third and fourth term have to be taken into account. In the case of an increase in the dipole moment, these terms cause a bathochromic shift, and in the case of a decrease, a hypsochromic shift of the absorption band. It has been calculated, that for a molecule with an interaction radius $a_W = 6 \cdot 10^{-8}$ cm, a dipole moment $\mu_g = 20 \cdot 10^{-30}$ C m (6 D), and a dipole change $(\mu_e - \mu_g) = 100 \cdot 10^{-30}$ C m (30 D), the shift between the vapour state and a nonpolar solvent ($\varepsilon_r = 2$) will be *ca.* 4000 cm^{-1}, and the shift from a nonpolar to a medium polar solvent ($\varepsilon_r = 30$) will be *ca.* 12000 cm^{-1} [90]. The last term in Eq. (6-2) will be important only if μ_g or $(\mu_e - \mu_g)$ is also large. In this case the two preceding terms are usually much larger and will essentially determine the solvent dependence. Therefore, in an approximation, the last term may be usually neglected [33, 90]. Finally, it should be mentioned that because of the approximations made in deriving Eq. (6-2), one cannot expect that this equation for the solvent dependence of UV/Vis absorptions is fully accurate [33, 90].

Combining the idea of solvent-induced changes in molecular structure with the conception of a solvent continuum around the solvatochromic molecule, a microstructural model of solvatochromism has been developed by Dähne *et al.*, which reproduces qualitatively correct and quantitatively satisfactory the solvatochromic behavior of simple merocyanine dyes [95b]. The results, obtained with this model for 5-dimethylamino-2,4-pentadien-1-al are in good agreement with the solvent-dependent experimental data such as transition energies, oscillator strengths, π-electron densities, and π-bond energies [95b]; *cf.* also [326, 327].

Very recently, a more general approach to the calculation of solvent-induced shifts of absorption (and emission) band maxima has been given. It is based upon the use of a thermodynamic cycle to calculate the energy of a charged sphere in a solvent with a non-equilibrium polarization. Interestingly, the new expressions for the shift in absorption with solvent changes, derived from this reversible-work approach, reduce to equations previously derived with the reaction-field method by McRae, Ooshika, Lippert, Liptay, and others, when point-dipole approximations for the solute are made; *cf.* B. S. Brunschwig, S. Ehrenson, and N. Sutin, J. Phys. Chem. *91*, 4714 (1987).

* For the addition of a sixth and a seventh term depending on the fluctuation of the reaction field see references [33, 94].

6.2.3 Specific Solvent Effects on UV/Vis Absorption Spectra

In principle, the general rules for solvent effects on the position of $\pi \rightarrow \pi^*$ absorption bands are also valid for $n \rightarrow \pi^*$ (and $n \rightarrow \sigma^*$) absorption bands of *N*-heterocycles and of compounds with heteronuclear double bonds such as $>$C=X or —N=X (with X=O, S, N, *etc*). For instance, the $n \rightarrow \pi^*$ excited state of a carbonyl group*) is less dipolar than the ground state. During the process of excitation, one *n*-electron is promoted from a nonbonding orbital on the oxygen atom to an antibonding π^* orbital which is delocalized over the carbonyl group. Removal of an electron from the oxygen atom implies a considerable contribution by the $>\bar{C}^{\ominus}-\bar{O}^{\oplus}$ mesomeric structure, with a decrease or even reversal in direction of the excited-state dipole moment. Indeed, it has been found, that the dipole moment of the lowest singlet $n \rightarrow \pi^*$ excited state of benzophenone, with $\mu_e(n \rightarrow \pi^*) = 5 \cdot 10^{-30}$ C m (1.5 D), is only half as large as the dipole moment of the ground state, $\mu_g = 10 \cdot 10^{-30}$ C m (3 D) [32, 33, 96]. This dipole diminution should correspond to a hypsochromic shift of the $n \rightarrow \pi^*$ absorption band with increasing solvent polarity (negative solvatochromism). In addition, protic solvents are capable of hydrogen-bond formation with oxygen lone pairs, thus lowering the energy of the *n*-state further, whereas to a first approximation the energy of the π^*-state is not modified by hydrogen bonding. Thus, the blue shift observed for $n \rightarrow \pi^*$ absorption bands of carbonyl compounds with increasing solvent polarity can be interpreted as the result of cooperating effects of both electrostatic and hydrogen bonding interactions on solute molecules [97–106]. This well-known blue shift can be attributed to superior general and specific solvation of the dipolar ground state and/or inferior solvation of the less dipolar $n \rightarrow \pi^*$ excited state by the polar solvent. The results observed for pyridazine as example, shown in Fig. 6-5, clearly implicate hydrogen bonding as the principle cause of the hypsochromic shift of the $n \rightarrow \pi^*$ transition which occurs when ethanol is added to *n*-hexane solutions of pyridazine [98].

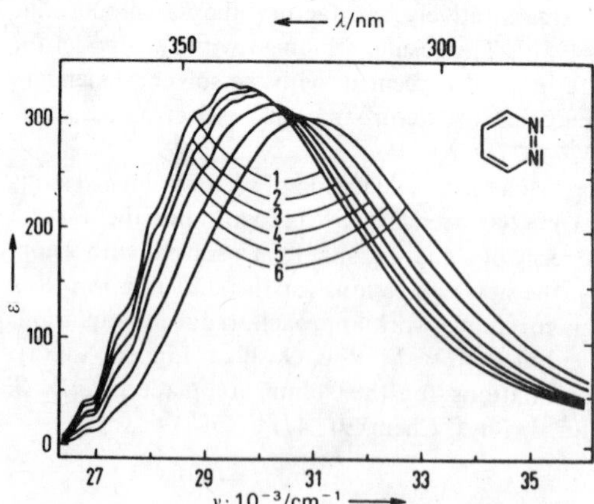

Fig. 6-5. The $n \rightarrow \pi^*$ band in the UV/Vis absorption spectrum of pyridazine in *n*-hexane/ethanol mixtures from zero to 3.2 cl/l ethanol ($c = 1.01 \cdot 10^{-2}$ mol/l pyridazine) [98]. Ethanol concentrations: curve (1) zero, (2) 0.0343, (3) 0.0686, (4) 0.137, (5) 0.274, and (6) 0.549 mol/l.

* Again, the excited state referred to here is the Franck-Condon excited state, which has a solvent shell identical with that of the ground state.

A different interpretation of the blue shift of the carbonyl $n \rightarrow \pi^*$ transition in HBD solvents has been given [328, 329]. According to MO calculations of the formaldehyde/water system, the major contribution to the blue shift arises from changes induced in the internal geometry of the solute/solvent components by hydrogen-bonding [329]. Such changes in the geometry will, as a result of the Franck-Condon principle, alter the vibrational band structure of the $n \rightarrow \pi^*$ transition relative to the non-hydrogen-bonded case. This would mean that the blue shift arises from intensity redistributions among the vibrational sub-bands of the $n \rightarrow \pi^*$ absorption band, and not from different stabilization of the ground and excited states by hydrogen bonding which is assumed to exist in both states [329].

The study of the $n \rightarrow \pi^*$ absorption of benzophenone in non-HBD solvents like 1,2-dichloroethane and acetonitrile, where hydrogen bonding is unimportant, reveals that these polar solvents also induce a blue shift (*cf.* Table 6-3) [102, 104]. This shift, however, is smaller than the shift caused by hydrogen bonding. For the $n \rightarrow \pi^*$ transition of benzophenone, a shift of 680 cm^{-1} between the non-HBD solvents *n*-hexane and acetonitrile is observed, compared to a shift of 2200 cm^{-1} between *n*-hexane and water. The larger wavenumber shift is mainly due to hydrogen bonding (*cf.* Table 6-3). The $n \rightarrow \pi^*$ transition energies of carbonyl groups in different solvents are found to vary linearly with the infrared stretching frequencies in the same solvents, indicating the importance of ground state stabilization by solvents [102].

A combination of spectroscopic and calorimetric measurements have been used to separate solvent effects on the $n \rightarrow \pi^*$ ground and excited states of some ketones (acetone, acetophenone, and benzophenone) [106]. It was found that the enthalpies of transfer of the Franck-Condon excited states*) were endothermic on going from a dipolar non-HBD

Table 6-3. The $n \rightarrow \pi^*$ and $\pi \rightarrow \pi^*$ band maxima in the UV/Vis absorption spectrum of benzophenone in solvents of increasing polarity [104].

Solvents	$\tilde{\nu}(n \rightarrow \pi^*)/cm^{-1}$	$\tilde{\nu}(\pi \rightarrow \pi^*)/cm^{-1}$
n-Hexane	28860	40400
Cyclohexane	28860	40240
Diethyl ether	29070	40160
1,2-Dichloroethane	29370	39600
Dimethyl sulfoxide	29370	–
N,N-Dimethylformamide	29330	–
Acetonitrile	29540	39920
1-Butanol	29990	39600
1-Propanol	29900	39600
Ethanol	30080	39680
Methanol	30170	39600
Water	ca. 31060 (sh)	38830
	$\Delta\tilde{\nu} = -2200$ cm^{-1}	$\Delta\tilde{\nu} = 1570$ cm^{-1}

* The enthalpy of transfer of the Franck-Condon excited-state molecule from one solvent to another has been calculated from the corresponding difference in the excitation energies and the difference between the calorimetrically determined heats of solution of the ground-state molecule in the two solvents of interest [106].

solvent (*e.g.* dimethylformamide) to a polar protic solvent (*e.g.* methanol) and larger than the corresponding endothermic ground-state transfer enthalpies. That is, contrary to accepted views, the solvent effect on the $n \rightarrow \pi^*$ Franck-Condon *excited* state of carbonyl compounds makes the major contribution to the blue shift observed on going to polar protic solvents. This has been explained by the assumption that the orientation strain in the Franck-Condon excited state should be greater in protic HBD solvents than in dipolar non-HBD solvents [106]. However, this interpretation of the carbonyl blue shift has been critisized for not taking into account so-called cavity terms, *i.e.* terms which account for solvent/solvent interactions which have to be overcome in order to create a cavity for ground- and excited-state molecules [303a, 330]. The enthalpy of transfer of a solute from one solvent to another is the sum of a solute/solvent interaction term and a solvent/solvent interaction term, usually called a cavity term. See references [106], [303a, 330] and [329] for further discussions.

In contrast to the $n \rightarrow \pi^*$ absorption of carbonyl compounds, for the $\pi \rightarrow \pi^*$ absorption of $>$C=O, $\mu_e(\pi \rightarrow \pi^*)$ is not only colinear with μ_g but also has increased magnitude with respect to μ_g(*e.g.* benzophenone: $\mu_g = 10 \cdot 10^{-30}$ C m; $\mu_e(\pi \rightarrow \pi^*) = 16 \cdot 10^{-30}$ C m [107]). Consequently, in going from a nonpolar to a polar solvent, the $\pi \rightarrow \pi^*$ absorption should undergo a bathochromic shift, while the $n \rightarrow \pi^*$ absorption undergoes a hypsochromic shift. This contradictory behaviour of absorption bands with changes in solvent polarity, illustrated for benzophenone in Fig. 6-6 and Table 6-3, is of diagnostic importance in order to distinguish between the $n \rightarrow \pi^*$ and $\pi \rightarrow \pi^*$ transitions of carbonyl compounds [98, 109]. – For the distinction of $n \rightarrow \pi^*$ and $\pi \rightarrow \pi^*$ transitions of substituted azobenzenes see for example reference [331].

The solvent-dependence of the $n \rightarrow \pi^*$ transition of chromophores other than the carbonyl group such as $>$C=S in thiocarbonyl compounds [332], $>\overset{\oplus}{N}-\overset{\ominus}{O}$ in *N*-oxides [65, 333], $>\overset{\odot\oplus}{N}-\overset{\ominus}{O}$ in aminyloxide radicals [334–336], and $>\overset{\oplus}{N}=\overset{\ominus}{N}$ in 1,1-diazenes [337] has also been investigated. One example, the 2,2,6,6-tetramethylpiperidine-1-oxyl

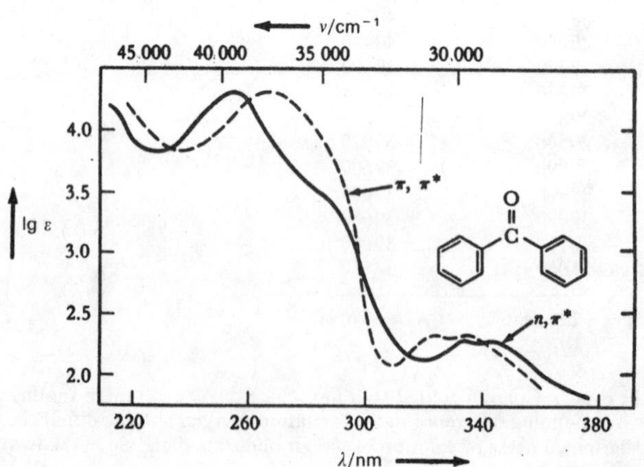

Fig. 6-6. UV/Vis absorption spectrum of benzophenone in cyclohexane (——) and ethanol (- - -) at 25 °C [104, 108].

radical, exhibits a comparatively large negative solvatochromism of the weak $n \rightarrow \pi^*$ absorption band, whereas the intense $\pi \rightarrow \pi^*$ transition in the ultraviolet region is not sensitive to solvent changes [334, 336].

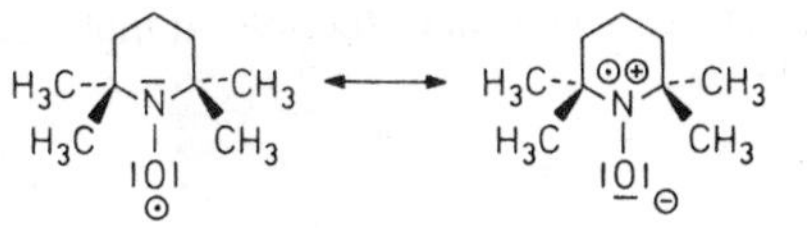

Solvent	n-C_6H_{14}	CH_3CN	CH_3OH	H_2O
$\lambda_{max}^{n \rightarrow \pi^*}$/nm	477	461	446	424
$\Delta\lambda$/nm		−53		

The visible $n \rightarrow \pi^*$ absorption band is shifted hypsochromically by 53 nm as the medium changes from n-hexane to water. Based on the negative solvatochromism of this aminyloxide radical, a further spectroscopic solvent polarity scale, called the E_B-scale of Lewis acidity, has been recently proposed [336]; *cf.* Section 7.4.

The marked solvent-dependence of the $n \rightarrow \pi^*$ transition of benzophenone [110] and of 2,2,6,6-tetramethylpiperidine-1-oxyl [338] when solubilized by micellar surfactants, has also been used to investigate the molecular-microscopic polarity of the micellar environment in which the transition takes place.

Specific hydrogen bonding is also involved in the solvation of saturated compounds with heteroatoms carrying lone pairs of electrons, which give rise to $n \rightarrow \sigma^*$ transitions. These transitions also undergo a significant blue shift with increasing solvent polarity, especially in protic solvents [111]. Direct evidence of the general validity of the concept of a blue shift of $n \rightarrow \sigma^*$ absorption bands is provided by a comparison of the vacuum UV bands of water, ammonia, hydrogen sulfide, and phosphane in aqueous solution relative to the location of these bands in the gas phase or in non-HBD nonpolar solvent solution spectra [111].

6.2.4 Solvent Effects on Fluorescence Spectra

When excited states of a molecule are created in solution by continuous or flash excitation, the excited-state molecule interacts to a varying degree with the surrounding solvent molecules, depending on their polarity, before returning to the ground state. These excited-state solute/solvent interactions found in fluorescent molecules are often reflected in the spectral position and shape of the emission bands as well as in the lifetimes of the excited-state molecules.

When considering the solvent dependence of the position of emission bands, the finite relaxation time τ_R for the rearrangement of the solvent molecules surrounding the solute molecule in the Franck-Condon excited state and the finite liftime τ_e of the molecule in the excited state have to be taken into account [112–116, 339, 340].

In the case $\tau_R \gg \tau_e$*) the emission will occur before any rearrangement of solvent molecules in the solvation shell takes place. The initial state of the emission process is the Franck-Condon excited state and the final state is the equilibrium ground state. Hence the wavenumber of emission will be equal to the wavenumber of the corresponding absorption. In the case $\tau_R \ll \tau_e$ (*cf.* Fig. 6-7)*) reorientation of the solvent molecules can

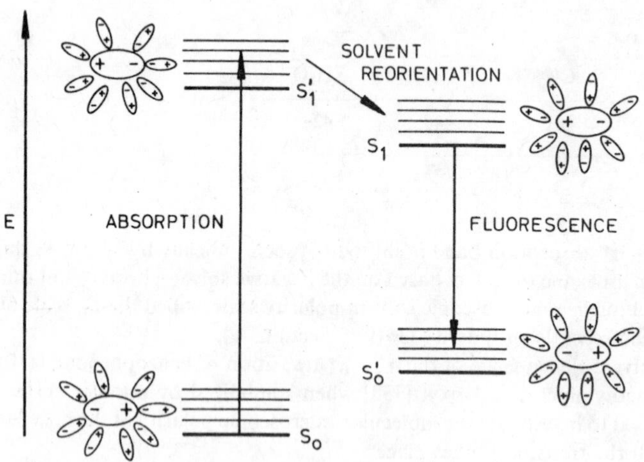

Fig. 6-7. Effect of solvent reorientation in the excited state on the fluorescence band of a dipolar molecule with dipole flip on excitation. S_1' and S_0' are the Franck-Condon excited and ground states, respectively; S_1 and S_0 are the corresponding equilibrium states; $\tau_R \ll \tau_e$ [116].

take place after electronic excitation and a relaxed excited state is obtained in which another solvation equilibrium has been established. It is from this equilibrium state that fluorescence occurs at room temperature. By analogy, there is a Franck-Condon ground state after emission which persists momentarily until the solvent molecules reorganize to the equilibrium arrangement for the ground state.

A general explanation of solvent effects on emission spectra has been given by Lippert [7b, 47]. If, for example, the fluorescent molecule has a higher dipole moment in the excited state than it has in the ground state, the emitted photon will have much less energy than the absorbed photon with increasing solute/solvent interaction, consequently there will be an anomalously large Stokes red shift of the fluorescence band. Hence,

* In liquid solutions, the rotational relaxation time τ_R for the solvent molecules is in the range 10^{-12} to 10^{-10} s at room temperature, the lifetime τ_e of an excited singlet state is of the order of 10^{-8} s. Hence, under these circumstances $\tau_R \ll \tau_e$ will be generally valid. τ_R increases strongly with decreasing temperature while τ_e is only slightly temperature dependent. Therefore, at lower temperatures the case where $\tau_R \approx \tau_e$ can occur. In solid solutions $\tau_R \gg \tau_e$ is the norm.

emission bands are found to undergo bathochromic shifts with increasing solvent polarity in case of $\mu_g < \mu_e$ (*cf.* as example dye no. 11 in Table 6-1) [7b, 47]*).

An illustrative example of solvent effects on absorption and emission spectra is the positive solvatochromic laser dye 7-amino-4-trifluoromethylcoumarin. Structural restrictions on the rotatory motion of the amino group are inherent [341].

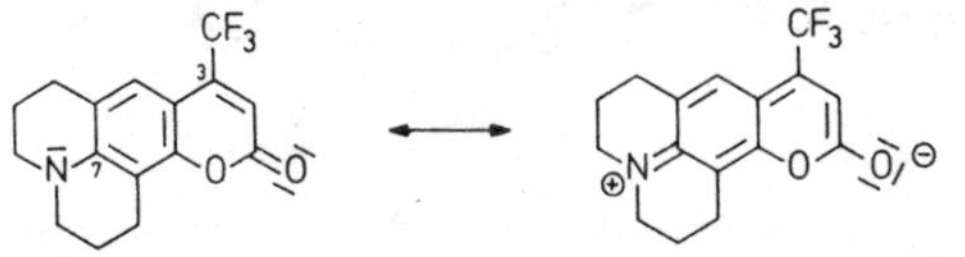

(a) ⟷ (b)

Solvent	c-C_6H_{12}	$CH_3CO_2C_2H_5$	CH_3CN	C_2H_5OH	H_2O	$\Delta\lambda$/nm
$\lambda_{max}^{Abs.}$/nm	393	409	418	421	430	+37
$\lambda_{max}^{Fluor.}$/nm	455	501	521	531	549	+94
$\tau_e \cdot 10^9$/s	4.3	5.4	5.6	3.4	–	

With increasing solvent polarity, both absorption and emission bands undergo a bathochromic shift, the latter more pronounced than the former. This indicates an *i*ntramolecular *c*harge-*t*ransfer (ICT) absorption of the less dipolar ground-state molecule with (*a*) as the dominant mesomeric structure, leading to highly dipolar excited-state molecule with (*b*) as the prominent structure. It is this planar ICT excited-state molecule (with a lifetime τ_e) from which the emission occurs. Increasing solvent polarity stabilizes the ICT excited-state molecule relative to the ground-state molecule with the observed red shift of the fluorescence maximum as the experimentally observed result [341].

In the case of 7-diethylamino-4-trifluoromethylcoumarin, which has an amino group which is free to rotate, another competitive solvent-dependent decay path has been observed: rotation of the amino group of the planar ICT excited-state molecule leads to a *t*wisted *i*ntramolecular *c*harge-*t*ransfer (TICT) excited-state molecule from which a radiationless decay to the ground-state molecule occurs [341]. Solvent-dependent rate constants for both the radiative and nonradiative decay of excited-state coumarin dyes have been determined [341].

The existence of two interconvertible ICT and TICT excited-state molecules can lead to a dual, variable solvent-dependent fluorescence. This dual fluorescence was first discovered by Lippert *et al.* [342] using 4-(*N,N*-dimethylamino)benzonitrile as the fluorescent compound, and then correctly interpreted by Grabowski *et al.* [343]. They identified the TICT excited state as the origin of the long-wavelength solvent-dependent fluorescence of this nitrile. The increasing number of bichromophoric organic molecules with two fluorescent states and the underlying concepts of the TICT excited state have been extensively reviewed recently [344].

* These solvent-induced shifts in absorption and emission bands have been used to calculate the dipole moments of electronically excited molecules [32, 33, 47, 303]. Excited-state dipole moments have also been obtained by the measurement of fluorescence polarization caused by external electric fields [32, 33].

Particularly well studied examples of another solvent-dependent dual fluorescence are 6,2-(arylamino)napththalene sulfonates (ANS) whose absorption and emission characteristics are as follows (np = nonplanar; ct = charge transfer) [119, 120, 340]:

In contrast to the absorption maxima of ANS, the fluorescence maxima are sensitive to solvent changes, but to a different extent in solvents of varying polarity. The fluorescence of ANS arises from two different excited-state molecules, the apolar locally excited state $S_{1,np}$ and the zwitterionic excited state $S_{1,ct}$ which emerges from $S_{1,np}$ by intramolecular electron or charge transfer. The first emission, predominant in nonpolar solvents, varies only modestly with solvent polarity, as expected for a transition $S_{1,np} \rightarrow S_0$ between two states of similar charge separation. The second emission, observed in more polar solvents, is quite sensitive to solvent polarity, as anticipated for a transition $S_{1,ct} \rightarrow S_0$ between two states of very different charge separation [119, 120, 340].

The conversion of the initially formed $S_{1,np}$ state to the $S_{1,ct}$ state by intramolecular electron transfer is very fast and varies in a way parallel but not identical to the dielectric relaxation time for the solvent used. This is because the local environment around the excited-state molecule is different from that surrounding a solvent molecule [120, 340]. That is, the ICT process is to a large extent determined by the dielectric relaxation processes of the solvent surrounding the ANS molecule. Thus, solvent motion seems to be the controlling factor in the formation and decay of the ICT excited state of ANS and other bichromophoric organic molecules [120, 340]. A detailed mechanism for fast intramolecular electron-transfer reactions of ANS and 4-(*N*,*N*-dimethylamino)benzonitrile, using two simplified molecular-microscopic models for the role of the solvent molecules, has been given recently by Kosower [340].

Dielectric friction is the measure of the dynamic interaction of a charged or dipolar solute molecule with the surrounding polar solvent molecules. This concept has been applied, by Hynes *et al.* [339], to solvent- and time-dependent fluorescence shifts resulting from the electronic absorption by a solute in polar solvents. If the solvent molecules are strongly coupled to the charge distribution in ground- and excited-state molecules, the relatively slow solvent reorientation can lead to an observable time evolution of the fluorescence spectrum in the nano- to picosecond range. This time-dependent fluorescence (TDF) has been theoretically analysed in terms of dynamic non-equilibrium solvation of excited-state molecules in polar solvents. It was shown that the TDF shift is proportional to the time-dependent dielectric friction on the absorbing dipolar molecule. That is, the relaxation rate of the fluorescence spectrum directly reflects a dynamic solvent property, namely the reorientation rate of the surrounding polar solvent molecules closest to the fluorescent solute molecule. This is analogous to studies of dynamic polar solvent effects of chemical reactions; *cf.* Chapter 5.1 and reference [463] cited therein. In fast chemical reactions dielectric friction leads to deviations from the rate constants predicted by the transition-state theory. This assumes that equilibrium solvation holds throughout the passage from educts via the activated complex to the products. The closest possible connection between time-dependent fluorescence and fast solvent-dependent chemical reactions should occur for electron-transfer reactions which involve a large charge separation similar to those encountered in absorbing molecules with subsequent ICT processes as mentioned above [339, 340].

An extreme example of solvent-dependent fluorescence of an ICT excited-state molecule is shown by 1-phenyl-4-(4-cyano-1-naphthylmethylene)piperidine, a bichromophoric rod-shaped molecule containing electron donor (D) and acceptor groups (A) separated by an elongated cycloalkyl spacer [345]. The absorption spectrum of this piperidine derivative is virtually independent of solvent polarity and closely resembles the expected sum of the spectra of the two separate chromophores, that is *N*,*N*-dialkylaniline (D) and 1-vinylnaphthalene (A). The fluorescence spectrum does not exhibit any emission caused by the two separate, locally excited chromophores D and A. Instead a single, rather broad emission band is observed the maximum of which undergoes a dramatic bathochromic shift with increasing solvent polarity ($\Delta\lambda = 282$ nm for *n*-hexane → acetonitrile). The position of the fluorescence maximum covers almost the entire visible spectral region if the solvent polarity is varied, thus making this compound a visual probe of solvent polarity. This behaviour is characteristic of an emission originating from a highly dipolar or zwitterionic ICT excited-state molecule. The dipole moment of this excited-state molecule, $\mu_e = 83 \cdot 10^{-30}$ C m, corresponds to a full charge separation over the distance between the centers of the electron donor and acceptor group. This indicates an almost complete electron transfer on excitation despite the lack of either direct D/A contact or mesomeric coupling via a π-electron system between D and A [345]. The combination of strong solvatochromism with high fluorescence quantum yield makes this bichromophoric piperidine derivative attractive as a fluorescent solvent polarity probe [345].

H C N CN $\xrightarrow[\text{ICT}]{h\nu_A}$ H C N CN

$\mu_g \leq 5 \cdot 10^{-30}$ C m (1.6 D) $\mu_e \approx 83 \cdot 10^{-30}$ Cm (25 D)

Solvent	n-C_6H_{14}	C_6H_6	$CHCl_3$	CH_2Cl_2	C_5H_5N	CH_3CN	$\Delta\lambda$/nm
$\lambda_{max}^{Fluor.}$/nm	412	478	531	579	627	694	+282 (!)
$\tau_e \cdot 10^9$/s	1.2	15	13	9	1	–	

Further interesting examples showing pronounced solvent effects on fluorescence spectra have been studied by Davis [17, 117] (EPD/EPA complex of hexamethylbenzene/tetrachlorophthalic anhydride), Eastman [118] (aromatics and *N*-heterocycles),

Kosower *et al.* [346] (8-phenylamino-1-napththalenesulfonates) and [347] (*syn*-bimanes), Kreysig *et al.* [348] (benzo[g]quinoline) and [349] (1,3,4-oxadiazoline-2-imines), Werner *et al.* [350] (methyl anthroates), Capomacchia *et al.* [351] (benzo[a]pyrene), Dogra *et al.* [352] (5-aminoindazole), Johnson *et al.* [353] (*N*-ethyl-3-acetylcarbazole), Wang [354] (2-substituted 1-indanones), Halpern *et al.* [355] (1-aza-bicyclo[2.2.2]octane), Staab *et al.* [429] (4,5,7,8-tetracyano[2.2]paracyclophane), and Langhals [433] (4-aminophthalimide); *cf.* also [344] for further examples.

Fluorescence measurements on short-lived (<1 ns) and long-lived (>10 ns) electronically excited organic molecules in binary solvent mixtures have been used by Suppan to study the phenomenon of *selective* or *preferential solvation* of dipolar solute molecules [394]. Even in ideal solvent mixtures, the $\pi^* \rightarrow \pi$ emission energy is often highly non-linear with the mole fraction of polar solvent. This non-linearity results from both (i) the nonspecific solute/solvent association described as dielectric enrichment in the solvent shell of dipolar solute molecules, and (ii) specific solute/solvent associations such as hydrogen-bonding or EPD/EPA interactions. Both solute/solvent interactions lead to molecular-microscopic local solute-induced non-homogeneities in the binary solvent mixture, generally called selective or preferential solvation; *cf.* Section 2.4.

Differential solvent interactions with ground- and excited state molecules not only lead to shifts in the fluorescence maxima but also to perturbation of the relative intensities of the vibrational fine structure of emission bands. For instance, symmetry-forbidden vibronic bands in weak electronic transitions can exhibit marked intensity enhancements with increasing solute/solvent interaction [320, 359]. A particularly well-studied case is the solvent-influenced fluorescence spectrum of pyrene, first reported by Nakajima [356] and later used by Winnik *et al.* [357] for the introduction of an empirical solvent polarity parameter, the so-called *Py*-scale; *cf.* Section 7.4.

The $\pi^* \rightarrow \pi$ emission spectrum of monomer pyrene exhibits five well-resolved major vibronic bands between 370 and 400 nm, labelled I...V in progressive order (the $0 \rightarrow 0$ band being labelled I, *etc.*). Peak I ($0 \rightarrow 0$ band) shows significant intensity enhancement with increasing solvent polarity compared with peak III ($0 \rightarrow 2$ band). Thus, the ratio of emission intensities of the vibronic bands I and III serves as a quantitative measure of solvent polarity ($Py = I_{I}/I_{III}$) although it is difficult to determine exactly [358]. The exact mechanism of how polar solvents enhance the intensity of symmetry-forbidden vibronic transitions by means of a reduction in local symmetry is not fully understood [357, 359].

Fluorescent organic compounds have been widely used as molecular-microscopic probes in biophysical studies of the local environment in micelle-forming surfactant solutions, in phospholipid dispersions, and in membranes. It is assumed that the nature of the probe environment is reflected in its emission characteristics (*i.e.* position and intensity of emission maxima, vibrational fine structure, quantum yields, excited-state lifetime, polarization of the fluorescence); *cf.* [360] for a review.

Extensive theoretical and experimental studies of the solvent effects on emission spectra of organic molecules led to quantitative expressions very similar to Eq. (6-2) [4, 13–16, 90]. Liptay has developed Eq. (6-5) for the solvent dependence of the difference between the wavenumbers $\tilde{\nu}_{eg}^{sol(a)}$ of the absorption ($g \rightarrow e$) and $\tilde{\nu}_{eg}^{sol(f)}$ of the corresponding emission ($e \rightarrow g$) for the limiting case when $\tau_R \ll \tau_e$ [33, 90]. All terms depending on polarizability change are neglected.

$$
\begin{aligned}
hca\Delta\tilde{\nu} &= hca(\tilde{\nu}_{\mathrm{eg}}^{\mathrm{sol}(a)} - \tilde{\nu}_{\mathrm{eg}}^{\mathrm{sol}(f)}) \\
&= hca(\tilde{\nu}_{\mathrm{eg}}^{0(a)} - \tilde{\nu}_{\mathrm{eg}}^{0(f)}) \\
&\quad + \frac{2}{4\pi\varepsilon_0 a_{\mathrm{W}}^3}\left[\frac{\varepsilon_{\mathrm{r}}-1}{2\varepsilon_{\mathrm{r}}+1} - \frac{n^2-1}{2n^2+1}\right]|\mu_{\mathrm{e}} - \mu_{\mathrm{g}}|^2 \qquad (6\text{-}5)
\end{aligned}
$$

$\tilde{\nu}_{\mathrm{eg}}^{0(a)}$ and $\tilde{\nu}_{\mathrm{eg}}^{0(f)}$ are the wavenumbers of the electronic transitions in absorption and fluorescence in the gas phase, respectively; the other terms are as in Eq. (6-2). Eq. (6-5) has been widely used for the determination of dipole moment changes from the solvent dependence of spectra. The main source of error is the limited accuracy of the estimated value for the interaction radius a_{W} of the solute molecule since $\Delta\tilde{\nu}$ is a cubic function of a_{W}.

6.2.5 Solvent Effects on ORD and CD Spectra

Optical rotatory dispersion (ORD) involves the measurement of the rotation of plane-polarized light by a chiral compound as a function of the wavenumber. Circular dichroism (CD) is the unequal absorption of right and left circularly polarized light as a function of its wavenumber. CD and anomalous ORD curves observed for chiral solute molecules are different manifestations of the so-called *Cotton effect* [121–124]. A necessary condition for the appearance of a Cotton effect is the absorption of light in the UV/Vis spectral range by the chiral solute. The position of the maximum of the UV/Vis absorption corresponds fairly well to the position of the CD maximum and to the wavenumber where the ORD curve crosses the zero rotation line. Therefore, all solute/solvent interactions which influence position and intensity of UV/Vis absorption bands will also affect the ORD and CD spectra [121–124].

Most of the research on ORD and CD has involved chiral ketones, because the carbonyl chromophore has a convenient weak $n \rightarrow \pi^*$ absorption band in the 33300 cm^{-1} (300 nm) region*⁾. The Cotton effect as observed in either ORD or CD curves of molecules containing a carbonyl chromophore varies with a change of solvent. This variation occurs both in the wavenumber of the ORD extrema or of CD maxima, and in the intensity of the Cotton effect, as measured by the rotational strength (R), ellipticity (θ), differential absorption ($\Delta\varepsilon$), or the ORD amplitude (a) [361]. Thus, the hypsochromic shift observed for $n \rightarrow \pi^*$ transitions of carbonyl chromophores with increasing solvent polarity and increasing solvent capacity for hydrogen-bond formation (*cf.* Section 6.2.3), gives rise to a corresponding blue shift of the CD and ORD curves. Typical wavelengths for the maximum of the $n \rightarrow \pi^*$ CD band are: 297 nm in *n*-hexane, 295 nm in 1,4-dioxane, 293 nm in acetonitrile, 290 nm in ethanol or methanol, and 283 nm in 2,2,2-trifluoroethanol [361]. This hypsochromic shift of the $n \rightarrow \pi^*$ band of carbonyl chromophores with increasing solvent polarity is largely a result of the stabilization of the solute *n*-orbital by solvation, particularly hydrogen bonding in protic solvents. However, intensity redistributions among the vibrational sub-bands of the $n \rightarrow \pi^*$ absorption band with increasing

---

* Chiral compounds with chromophores that absorb light strongly in the UV/Vis region are usually unsatisfactory for ORD and CD measurements as insufficient incident light is transmitted to permit the measurements.

solute/solvent interaction can also be responsible for the observed blue shifts [328, 329]; *cf.* Section 6.2.3 for a discussion of this point. – In addition, an enhanced Cotton effect is usually observed with increasing solvent polarity. This change in $\Delta\varepsilon$ can be reasonably explained by the assumption that the total symmetry of a tightly bonded solute/solvent complex should be greater than that of a weakly solvated solute molecule [361]. – Typical examples of solvent-dependent Cotton effects for carbonyl and other heteronuclear X=O compounds can be found in references [125] (camphor derivatives), [126] and [328] (ketosteroids), [127] (α-silyl ketones), [128] (cyclic lactones and lactams), [129] (uracil nucleosides), and [130] (α-chloro sulfoxides).

For some carbonyl compounds a reverse in optical rotation sign has been observed when the ORD spectra were measured in solvents of different polarity. For example, (*S*)-5-hydroxy-1,7-diphenyl-3-heptanone is dextrorotatory in chloroform (molar rotation $[M]_D^{25} = +39.6°$) and levorotatory in methanol ($[M]_D^{25} = -7.6°$), whereas its acetate is dextrorotatory in both solvents [363]. This observation suggests that the interaction between the β-ketol moiety and the solvent is responsible for the change and reversal in optical rotation with increasing solvent polarity.

H_5C_6 … C_6H_5 (intramolecular O–H···O) $\underset{25°C}{\overset{+EPD}{\rightleftharpoons}}$ H_5C_6 … C_6H_5 (EPD···H–O)

$[M]_D^{25} > 0°$ $[M]_D^{25} < 0°$

Solvent	CCl_4	$CHCl_3$	C_6H_6	1,4-Dioxane	CH_3COCH_3	CH_3CN	CH_3OH
$[M]_D^{25}$ [a)]	+48.8°	+39.6°	+39.6°	+13.5°	−2.5°	−7.6°	−7.6°

[a] $c = 0.025$ mol/l.

The solvent-induced variations in the optical rotation of the β-ketol moiety are presumably caused by conformational changes, associated with the formation and breaking of the intramolecular hydrogen bond. In non-EPD solvents, the intramolecularly hydrogen-bonded conformer with coplanar hydroxyl and carbonyl groups dominates. Increasing EPD-character of the solvents causes breaking of the intramolecular hydrogen bond, followed by competitive intermolecular hydrogen bonding and nonspecific solvation of the β-ketol moiety; *cf.* references [363, 364] for further discussions.

Another reason for the solvent-dependent Cotton effects of $n \rightarrow \pi^*$ absorption bands is the solvent-induced alteration of equilibria between distinct conformers of the carbonyl compound; *cf.* Section 4.4.3. Differential solvation of equilibrating conformers can change the position of the equilibrium and thus the intensity and even the sign of the Cotton effect. An illustrative example is (+)-*trans*-2-chloro-5-methylcyclohexanone, the Cotton effect of which reverses its sign on transfer from water to *n*-hexane [362]; *cf.* equilibrium (*32a*) $\rightleftharpoons$ (*32b*) in Section 4.4.3. The molar rotations of this conformationally mobile α-chloro ketone, taken from its ORD spectrum measured in twenty-eight solvents, manifests huge changes with increasing solvent polarity: $[M]_{330}^{25}$ at 330 nm equals −1819° in cyclohexane, −428° in diethyl ether, +382° in water, and +680° in dimethyl sulfoxide [362]. This is obviously due to a shift of the diaxial/diequatorial equilibrium in favour of the more dipolar diequatorial conformer with increasing solvent polarity; *cf.* Section 4.4.3 and reference [364] for further discussions.

Structurally symmetric achiral compounds may show optical activity in the presence of chiral solvent molecules owing to asymmetry induced by the chiral solvent. For example, the achiral carbonyl compounds benzil and benzophenone surprisingly show optical activity in the region of the $n \rightarrow \pi^*$ absorption in the CD spectrum when dissolved in the chiral solvent (*R*,*R*)-(−)-2,3-butanediol [131, 365]. This phenomenon, first recorded for organic molecules by Bosnich [131], has been named *induced optical activity* [365]. Obviously, the optical activity will be induced in the carbonyl chromophore by the surrounding chiral and protic solvent molecules, leading to a dissymmetric environment, even if there is a completely random distribution of the chiral solvent molecules in the solvation shell [365].

When 2-benzoylbenzoic acid and chiral (*R*)-(−)-amphetamine were dissolved together in equimolar proportions in nonpolar solvents, a strong CD was induced in the region of the $n \rightarrow \pi^*$ carbonyl absorption [132]. The sign of the CD was positive and its magnitude was reduced with increasing solvent polarity: the molar ellipticity $[\theta]$ at *ca.* 320 nm equals +1320 in tetrachloromethane, +229 in acetonitrile, and nearly zero in methanol.

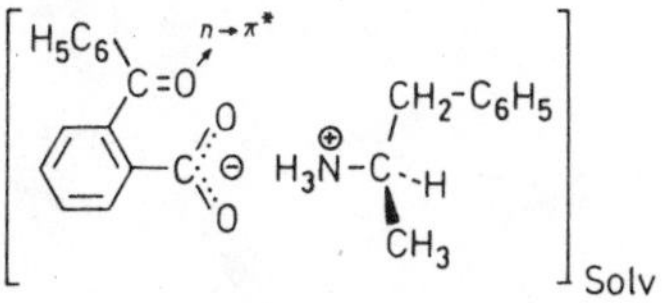

$[H_5C_6-C(=O)-C_6H_4-CO_2^{\ominus} \; H_3\overset{\oplus}{N}-CH(CH_3)-CH_2-C_6H_5]_{Solv}$ ($n \rightarrow \pi^*$)

Contact ion pair in non-dissociating solvents

Contact ion pairs derived from salt formation between the keto acid and the chiral amine appear to be responsible for this observation. The proximity of the counter ions of the tight ion pair in non-dissociating solvents is the origin of the induced CD. Increasing dissociation of the ion pair in more polar solvents should then reduce the strength of the induced CD, in agreement with the experimental observations [132].

6.3 Solvent Effects on Infrared Spectra

The vibrational spectrum of a molecule A—B depends not only on the strength of the bond between A and B, but may also be markedly affected by environmental factors. Such intermolecular interactions modify the infrared spectra in a number of ways: the wavenumber of the normal vibrational modes of a molecule may be shifted to higher or lower values, the intensities can be altered, and the half-width of bands may be greatly increased. A typical example, exhibiting two parts of the infrared spectrum of 1,1-dichloroethene, is illustrated in Fig. 6-8. With increasing solvent polarity, the two absorption bands are shifted to lower $[\tilde{\nu}_{as}(CH_2)]$ and higher wavenumbers $[\gamma(CH_2)]$,

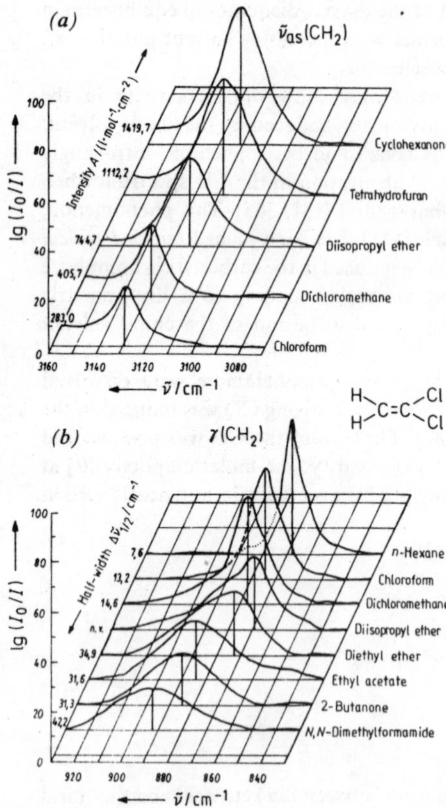

Fig. 6-8. Infrared spectrum of 1,1-dichloroethene, $H_2C{=}CCl_2$. Solvent effect on (a) the wavenumber $\tilde{\nu}$ and intensity A of the antisymmetrical stretching vibration $\tilde{\nu}_{as}(CH_2)$; and (b) the wavenumber $\tilde{\nu}$ and half-width $\Delta\tilde{\nu}_{1/2}$ of the out-of-plane bending vibration $\gamma(CH_2)$ [366].

respectively. At the same time, both the absorption intensity and the half-width of the two bands increase steadily with increasing solute/solvent interaction. Obviously, both infrared vibrations of 1,1-dichloroethene are influenced by nonspecific and specific solute/solvent interactions to a different extent.

The measurement of such solvent-induced IR spectral changes have been extensively used in spectroscopic studies of solute/solvent interactions [1–4, 367], especially hydrogen-bond (HBD/HBA) interactions [134, 135, 367]. Experimental solvent effect studies and solvent shift theories have been previously reviewed [1–4, 136–144]. The IR transmission characteristics of solvents commonly used in IR spectroscopy can be found in Table A-5 (Appendix).

The wavenumber shift, $\Delta\tilde{\nu}$, is generally represented as the difference between the absorption in the vapor phase ($\tilde{\nu}^0$) and in the solvent under consideration ($\tilde{\nu}^s$). When measured in solution the band maxima of all simple stretching vibrations are displaced to lower wavenumbers (*e.g.* stretching vibration $\tilde{\nu}_{C=O}$ of carbonyl compounds) whereas those of bending vibrations are shifted to higher wavenumbers (*e.g.* out-of-plane deformation γ_{C-X} of halobenzenes). The most solvent-sensitive infrared stretching

Table 6-4. Absorption band maxima in the infrared spectra of acetone [145–148, 368] (C=O stretch), methan[^{2}H]ol [149] (O—D stretch), and 1-chloropropane [150] (C—Cl stretch) in a selection of solvents of increasing polarity.

Solvents	$\tilde{\nu}_{C=O}/cm^{-1}$ of $Me_2C{=}O$ [a]	$\tilde{\nu}_{O-D}/cm^{-1}$ of $CH_3O{-}D$ [b]	$\tilde{\nu}_{C-Cl}/cm^{-1}$ of n-C_3H_7Cl [a, c]
Vapour ($\tilde{\nu}^0$)	1738	2720	743
n-Hexane	1721.5	2696	735
Triethylamine	1719.5	2406 (!)	–
Tetrachloromethane	1718	2689	–
Carbon disulfide	1718	–	730
Benzene	1717	2668	725
Tetrahydrofuran	1716.5	2575	–
Ethyl acetate	–	2631	723.5
1,4-Dioxane	1715	2592	722.5
Acetone	1715	2597	721
Acetonitrile	1713.5	2617	720
N,N-Dimethylformamide	1713	2554	–
Nitromethane	1711.5	2661	719 (!)
Dichloromethane	1711.5	2677	–
Dimethyl sulfoxide	1709	2528	–
Ethanol	1708.5 [d]	–	–
Methanol	1707.5 [e]	–	–
Ethane-1,2-diol	1703.5	–	–
Aniline	1703	2511	–
Water (D_2O)	1697.5 (!)	–	–
	$\Delta\tilde{\nu}=\tilde{\nu}^0-\tilde{\nu}^{H_2O}=40.5\ cm^{-1}$	$\Delta\tilde{\nu}=\tilde{\nu}^0-\tilde{\nu}^{Et_3N}=314\ cm^{-1}$	$\Delta\tilde{\nu}=\tilde{\nu}^0-\tilde{\nu}^{CH_3NO_2}=24\ cm^{-1}$

[a] Dilute solution.
[b] 0.04 mol/l CH_3OD.
[c] *trans*-Conformer.
[d] With shoulder at 1718 cm^{-1}.
[e] With shoulder at 1717 cm^{-1}.

vibrations are those of $X^{\delta\oplus}{=}O^{\delta\ominus}$ bonds (X = C, N, P, S), $X^{\delta\ominus}{-}H^{\delta\oplus}$ bonds (X = C, N, O, S, halogens), and $C^{\delta\oplus}{-}X^{\delta\ominus}$ bonds (X = halogens). Three typical examples are given in Table 6-4 [145–150, 368]. In all three cases, the C=O, O—D, and C—Cl stretching vibrations are shifted to lower wavenumbers with increasing solvent polarity. Excessive shifts of $\tilde{\nu}_{C=O}$ in HBD solvents (*e.g.* H_2O), and of $\tilde{\nu}_{O-D}$ in HBA solvents (*e.g.* Et_3N) caused by hydrogen-bonding are observed. When the solute/solvent interaction is nonspecific as in non-HBD and non-HBA solvents, the bands shift monotonically from one extreme to the other.

Solvents effects of the infrared stretching vibrations of the following organic compounds have been studied in depth: $\tilde{\nu}_{C=O}$ of 4-pyridones [151], tropone and tropolones [152], benzophenone and *N,N*-dimethylformamide [154], acetophenone [155], aliphatic aldehydes [157], *N*-methylacetamide [369], esters and dialkyl carbonates [370]; $\tilde{\nu}_{N=O}$ for a nitrosyl protoporphyrine derivative [371]; $\tilde{\nu}_{P=O}$ for triarylphosphane oxides [153] and triethylphosphane oxide [372]; $\tilde{\nu}_{S=O}$ for dimethyl sulfoxide [154, 373]; $\tilde{\nu}_{C-H}$ for chloroalkanes [160], chloroform [374], and *n*-octane [375]; $\tilde{\nu}_{\equiv C-H}$ of 1-alkynes [133, 138]; $\tilde{\nu}_{C-Cl}$ for haloalkanes [150, 161]; $\tilde{\nu}_{C\equiv N}$ for nitriles [156]; $\tilde{\nu}_{Si-H}$ for silanes [159]; $\tilde{\nu}_{N-H}$ of pyrrol [158], *N*-methylacetamide [369], and *N*-methylanilines [376]; as well as $\tilde{\nu}_{O-H}$ of *tert*-butyl hydroperoxide [377]. A comprehensive list of earlier publications on solvent effects on infrared absorptions is given by Hallam (*cf.* page 420 of reference [134]).

The wavenumber displacement of a solute vibration is a complex function of both solute and solvent properties and can be explained in terms of weak nonspecific electrostatic interactions (dipole-dipole, dipole-induced dipole, *etc.*) and of strong specific association of solute with solvent molecules, usually of the hydrogen-bond type [140]. It should be realized that the duration of vibrational transitions is very short with respect to motion of the solvent molecules (*e.g.* for an O—H stretching vibration, the frequency is *ca.* $10^{14}\ s^{-1}$). Thus it is possible to observe such transitions even for short-lived entities such as may arise after a collision in the liquid phase (collision complexes) [140].

To a first approximation the bathochromic shift observed for the C=O stretching vibration of acetone (*cf.* Table 6-4) in non-HBD solvents may be explained by the degree of C=O dipolarity as determined by the relative contribution of the two mesomeric forms in Eq. (6-6).

$$\boxed{\overset{\delta\oplus}{X}=\overset{\delta\ominus}{O}}\quad X = C,N,P,S \qquad \left(\begin{matrix}R\\R\end{matrix}\!\!>C=\overline{\underline{O}} \longleftrightarrow \begin{matrix}R\\R\end{matrix}\!\!>\overset{\oplus}{C}-\overline{\underline{O}}|^{\ominus} \right) + H{-}S \rightleftharpoons \begin{matrix}R\\R\end{matrix}\!\!>\overset{\delta\oplus}{C}\cdots\overset{\delta\ominus}{O}\cdots H{-}S \tag{6-6}$$

$$\boxed{\overset{\delta\ominus}{X}-\overset{\delta\oplus}{H}}\quad X = B,C,N,O,S,\text{Halogens} \qquad R-\overset{\delta\ominus}{O}-\overset{\delta\oplus}{H} + |S{-}H \rightleftharpoons R-\overset{\delta\ominus}{O}-\overset{\delta\oplus}{H}\cdots S{-}H \tag{6-7}$$

A change in the external environment produces small alterations in the relative contribution of the two mesomeric structures, and affects the wavenumbers of absorption in much the same way as do changes in the internal chemical environment [146]. Accordingly, the $\tilde{\nu}_{C=O}$ absorption band of acetone is displaced from 1721.5 cm^{-1} in *n*-hexane to 1709 cm^{-1} in the dipolar non-HBD solvent dimethyl sulfoxide [368]; *cf.* Table 6-4. In protic solvents, however, hydrogen-bonding superimposes this nonspecific solvent effect according to Eq. (6-6). Even the less dipolar HBD solvent aniline causes a larger bathochromic shift of the $\tilde{\nu}_{C=O}$ band than dimethyl sulfoxide, up to 1703 cm^{-1}; this is only surpassed by water [368].

In this respect, camphor is a particularly well-studied example [378]; *cf.* Fig. 6-9. The IR absorption band of the C=O stretching vibration of camphor shows that the carbonyl group exists free in the gas-phase and experiences nonspecific solute/solvent interactions in *n*-heptane, tetrachloromethane, and pyridine, as shown by the sharp, single bands in Figs. (6-9b)...(6-9d). However, in protic solvents such as methanol, two absorption bands are present; *cf.* Fig. (6-9e). These bands are attributed to the presence of both nonspecifically solvated camphor ($\tilde{\nu}_{C=O} = 1745.4\ cm^{-1}$) and a specific equimolar 1:1 camphor/methanol complex ($\tilde{\nu}_{C=O} = 1732.2\ cm^{-1}$), according to the equilibrium given in Eq. (6-6). – In the strong HBD solvent hexafluoro-2-propanol (HFIP), two carbonyl bands are observed as well. That at higher wavenumbers ($\tilde{\nu}_{C=O} = 1723.2\ cm^{-1}$) is assigned to the 1:1 camphor/HFIP complex, whereas the band at lower wavenumbers is due to complexes of higher order [378].

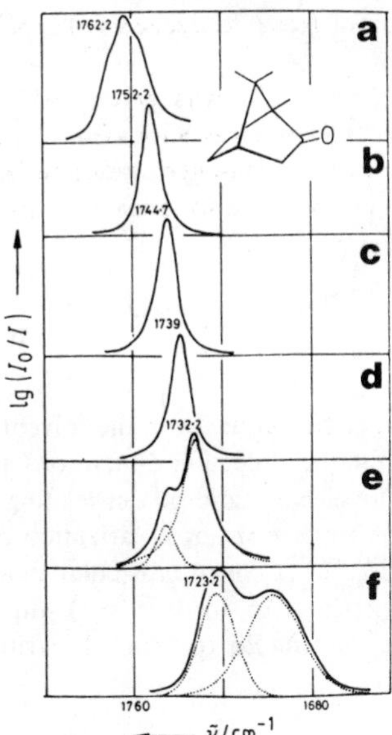

Fig. 6-9. The $\tilde{\nu}_{C=O}$ stretching absorption band in the infrared spectrum of camphor (a) in the gas phase, (b) in *n*-heptane, (c) in tetrachloromethane, (d) in pyridine, (e) in methanol, and (f) in hexafluoro-2-propanol [378].

The differentiation between effects due to specific solute/solvent interactions and bulk dielectric solvent effects is not easy to visualize and often a matter of debate [367]. The experimental data indicate that the solvent sensitivites of $\tilde{\nu}_{X=O}$ vibrations are complex functions of several factors including contributions from bulk dielectric effects, non-specific dispersion and induction forces, specific HBD/HBA interactions as well as steric effects [134]. – Solvent effects on the $\tilde{\nu}_{C=O}$ IR stretching absorption have been linearly correlated with the solvent shifts of $n \rightarrow \pi^*$ UV/Vis absorptions [102] and with ^{13}C-NMR chemical shifts (^{13}C=O) [367, 368] of some carbonyl compounds.

Contrary to $X^{\delta\oplus}{=}O^{\delta\ominus}$, in $X^{\delta\ominus}{-}H^{\delta\oplus}$ groups, the free end of the dipolar X—H bond available for specific solvent association is of the opposite sign. Consequently, the largest solvent-induced $\tilde{\nu}_{X-H}$ shifts do not occur in HBD solvents but in HBA (=EPD) solvents of Lewis-base type [162, 163]. Thus, in accordance with Eq. (6-7), the $\tilde{\nu}_{O-D}$ IR stretching absorption band of CH_3OD undergoes the largest bathochromic shift in the less dipolar HBA solvent triethylamine, as compared to the shifts observed in dipolar non-HBA solvents such as nitromethane and acetonitrile [149]; *cf.* Table 6-4. Since $\tilde{\nu}_{O-D}$ measures the strength of the HBD solute/EPD solvent interaction, empirical solvent scales of Lewis basicity have been introduced, based on the solvent dependence of $\tilde{\nu}_{O-D}$ for CH_3OD [149] or $\tilde{\nu}_{O-H}$ for phenol; *cf.* Section 7.4.

In order to test whether the factors responsible for solvent shifts are the same for different compounds, the so-called Bellamy-Hallam-Williams (BHW) plot is often used [162]. If the dipoles in a given family such as X=O, X—H, or C-halogen exhibit a common pattern of solvent effects, then the $\Delta\tilde{\nu}/\tilde{\nu}^0$ values of any one compound measured in a range of solvents can be plotted against the corresponding values of any other compound containing the same grouping to give a straight line. Consequently, the method of solvent variation is often useful for the identification of group frequencies of dipolar links (for examples see references [164, 165]).

The first theoretical treatment of infrared solvent shifts was given 1937 by Kirkwood [166] and by Bauer and Magat [167]. Eq. (6-8) – known as Kirkwood-Bauer-Magat (KBM) relationship – has been derived on the basis of Onsager's reaction field theory [80] using the simple model of a diatomic oscillator within a spherical cavity in a continuous solvent medium of macroscopic dielectric constant ε_r.

$$\frac{\Delta\nu}{\nu^0} = \frac{\nu^0 - \nu^s}{\nu^0} = C \cdot \frac{\varepsilon_r - 1}{2\varepsilon_r + 1} \tag{6-8}$$

ν^0 is the vibrational frequency in the gas phase, ν^s is the frequency in the solvent of dielectric constant ε_r, and C is a constant depending upon the molecular dimensions and electrical properties of the vibrating solute dipole. The electrostatic model leading to Eq. (6-8) assumes that only the electronic contribution to the solvent polarization can follow the vibrational frequencies of the solute (*ca.* 10^{14} s^{-1}). Since molecular dipole relaxations are characterized by much lower frequencies (10^{10} to 10^{12} s^{-1}), dipole orientation cannot be involved in the vibrational interaction, and Eq. (6-8) may be written in the following modified form [158, 168]:

$$\frac{\Delta\nu}{\nu^0} = \frac{\nu^0 - \nu^s}{\nu^0} = C \cdot \frac{n^2 - 1}{2n^2 + 1} \tag{6-9}$$

Here n is the refractive index of the solvent. Both equations have been widely used and tested for a large number of compounds [136–144] and found to be valid only in a very limited range in dilute solutions of nonpolar solvents, where specific interactions can be neglected. In polar solvents, the point of the KBM plot are usually shifted toward higher values of $\Delta\nu/\nu^0$. The deviations from the KBM equation have been attributed to hydrogen bonding and formation of molecular complexes and such interactions are not taken into account by the KBM equation.

The relationships (6-8) and (6-9) have subsequently been modified and improved by a number of workers [169–172]; see [1] for a review. Eqs. (6-10) and (6-11), derived by Buckingham [170],

$$\frac{\Delta\nu}{\nu^0} = C_1 + C_2 \cdot \frac{\varepsilon_r - 1}{2\varepsilon_r + 1} + C_3 \cdot \frac{n^2 - 1}{2n^2 + 1} \quad \text{(for polar solvents)} \tag{6-10}$$

$$\frac{\Delta\nu}{\nu^0} = C_1 + \tfrac{1}{2}(C_2 + C_3) \cdot \frac{\varepsilon_r - 1}{2\varepsilon_r + 1} \quad \text{(for nonpolar solvents)} \tag{6-11}$$

give better agreements than Eqs. (6-8) and (6-9) for solvent shifts in infrared spectra in the absence of specific interactions. C_1, C_2, and C_3 are constants for the solute under consideration and may be evaluated by applying Eq. (6-11). Plotting the observed relative shifts in nonpolar solvents against the dielectric function $(\varepsilon_r - 1)/(2\varepsilon_r + 1)$ yields values of C_1 from the intercept and $(C_2 + C_3)$ from the slope. C_2 and C_3 can then be evaluated from Eq. (6-10) using shifts in polar solvents.

The Buckingham Eq. (6-10) takes into account the fact that the influence of solvent dipolarity [characterized by $f(\varepsilon_r)=(\varepsilon_r-1)/(2\varepsilon_r+1)$] and solvent polarizability [characterized by $f(n^2)=(n^2-1)/(2n^2+1)$] on the solute IR vibrations are two independent effects. Based on the assumption that solute/solvent collision complexes are formed in solution, which should lead to a mutual correlation in dipolarity/polarizability changes, Bekárek *et al.* have added a third cross-term $f(\varepsilon_r)\cdot f(n^2)$ to the two terms of Eq. (6-10) [379]. Indeed, using the modified three-term Eq. (6-12),

$$\frac{\Delta\nu}{\nu^0}=C_1+C_2\cdot f(\varepsilon_r)+C_3\cdot f(n^2)+C_4\cdot f(\varepsilon_r)\cdot f(n^2) \qquad (6\text{-}12)$$

surprisingly better correlations have been obtained for a variety of experimental solvent-dependent IR stretching vibrations [379]. Even with the cross-term $f(\varepsilon_r)\cdot f(n^2)$ alone, the correlations were better than those obtained with Eq. (6-10), thus demonstrating its predominant influence on solvent-induced IR band shifts [380].

Other more sophisticated approaches to the calculation of medium effects on IR absorption bands can be found in the comprehensive, excellent review of Lutskii *et al.* [1]. According to Lutskii, even for quite simple molecules acceptable precise calculations of $\Delta\nu/\nu^0$ still present insuperable difficulties. This explains the growing practice of correlating $\Delta\nu/\nu^0$ with empirical parameters of solvent polarity within the framework of linear Gibbs energy relationships. Some of these empirical parameters are even derived from solvent-dependent IR absorptions as reference processes as, for example, the *G*-values of Schleyer *et al.* [154]; *cf.* Section 7.4.

Attention has also been given to the change in intensities of infrared absorption bands in going from the gas phase to solution from theoretical [168, 170, 173–175] and experimental points of view [176–181]. In general, infrared absorption intensities are relatively little altered by a solvent change. Usually, an increase of the integrated band intensity *A* is obtained in passing from vapour to solution in a nonpolar solvent and further to a polar solvent; *cf.* Fig. (6-8a) [366]. A fair parallelism is often found between solvent effects on intensity *A* and on relative absorption shifts $\Delta\nu/\nu^0$. A striking example is the C=O stretching absorption of acetone [179]. When specific interactions occur, the solvent effect on intensity *A* is more pronounced. For instance, the C—D stretching vibration of deuterochloroform increases thirtysixfold in going from tetrachloromethane to triethylamine solution [182]. Many equations have been proposed to predict the ratio of the intensities observed in the gas and liquid phase [168, 170, 173–175]. All theories predict an increase in intensity but none is able to give the precise magnitude of the solvent effect.

6.4 Solvent Effects on Electron Spin Resonance Spectra

From the point of view of the solvent influence, there are three features of an electron spin resonance (ESR) spectrum of interest for an organic radical measured in solution: the *g*-factor of the radical, the isotropic hyperfine splitting (HFS) constant *A* of any nuclei with nonzero spin in the molecule, and the widths of the various lines in the spectrum [2, 183–186, 390]. The *g*-factor determines the magnetic field at which the unpaired

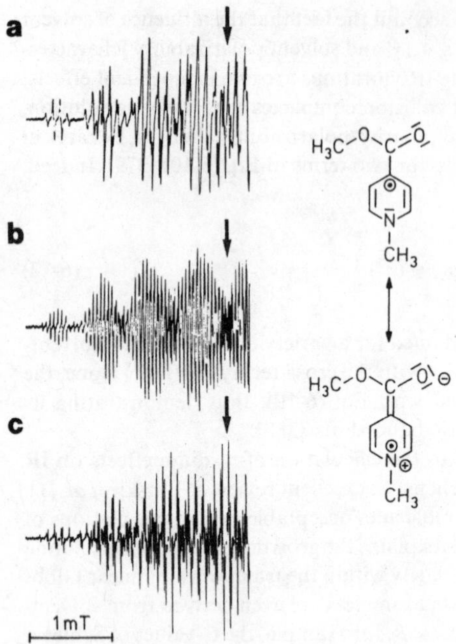

Fig. 6-10. Low-field parts of the ESR spectra of 4-methoxycarbonyl-1-methylpyridinyl radical in (a) 2-methylpentane, (b) benzene, and (c) *N,N*-dimethylformamide. The arrows indicate the center of the ESR spectra [381].

electron of the free radical will resonate with the fixed frequency of the ESR spectrometer (usually 9.5 GHz). The isotropic HFS constants are related to the distribution of the π-electron spin density (also called spin population) of π-radicals. Line-width effects are correlated with temperature-dependent dynamic processes such as internal rotations and electron transfer reactions. Some up-to-date reviews on organic radicals in solution are given in reference [390].

An illustrative example of a strongly solvent-dependent ESR spectrum is given in Fig. 6-10. It shows low-field portions of the ESR spectra of the stable, distillable (!) 4-methoxycarbonyl-1-methylpyridinyl radical, measured in three solvents of increasing polarity [381]. The different patterns of the ESR lines observed in different media arises from the solvent influence on the spin distribution, as measured by the HFS constants $A(^{1}H)$ and $A(^{14}N)$. This change in spin distribution results from solvent-induced polarization of the carbonyl group which is coplanar and conjugated with the pyridinyl ring.

Whereas solvent effects on line-widths and *g*-factors have been studied only in a small number of cases, there are numerous data on the solvent dependence of the HFS constants in the ESR spectra of organic free radicals. The sensitivity of nuclear HFS constants of free radicals to changes in solvent polarity have been noted for ketyls

Table 6-5. g-Factors and isotropic nitrogen and oxygen hyperfine splitting constants, $A(^{14}N)$ and $A(^{17}O)$, of 2,2,6,6-tetramethyl-4-piperidone-1-oxyl in ten solvents of increasing polarity [216].

Solvents	g-Factor	$A(^{14}N)/mT$ [a)]	$A(^{17}O)/mT$ [a)]
Benzene	2.0062	1.445	1.929
Toluene	2.0063	1.447	1.952
N,N-Dimethylformamide	2.0062	1.466	1.940
Dichloromethane	2.0061	1.477	1.936
Dimethyl sulfoxide	2.0060	1.490	1.974
1-Butanol	2.0060	1.501	1.911
Formamide	2.0060	1.528	1.888
1,2-Ethanediol	2.0061	1.540	1.872
Diethyl ether	–	1.594	1.784
Water	2.0058	1.601	1.786
[a] mT = milliTesla	$\Delta g = 0.0005$ ≙ 0.02%	$\Delta A(^{14}N) = -0.156$ mT ≙ 11%	$\Delta A(^{17}O) = +0.143$ mT ≙ 7%

[187–188], *o*- and *p*-benzosemiquinone anion radicals [183, 189–195, 382], semidiones*) [196], phenoxyl radicals [197–200], nitroaromatic anion radicals [201–205, 383], dialkyl and diaryl aminyloxides (nitroxides) [206–218, 335, 384–388], acyl aminyloxides [219], azomethine aminyloxides [220], thioindigo radicals [221], and pyridinyl radicals [222, 381, 389].

The solvent dependence of g-factors has been studied for *p*-benzosemiquinone anion radical [191], di-*tert*-butyl aminyloxide [212], 2,2,6,6-tetramethyl-4-piperidone-1-oxyl [216], and nitrosyl protoporphyrine derivatives [371]. Apparently, changes in distribution of the unpaired electron and changes in spin-orbit coupling in going from one solvent to another are reflected in the g-factor changes. A dramatic line width alteration has been observed for the nitrobenzene anion radical in hexamethylphosphoric triamide upon addition of methanol [205].

Table 6-5 illustrates the influence of the medium on the g-factors and HFS constants for the following ^{17}O-labelled aminyloxide radical [216]:

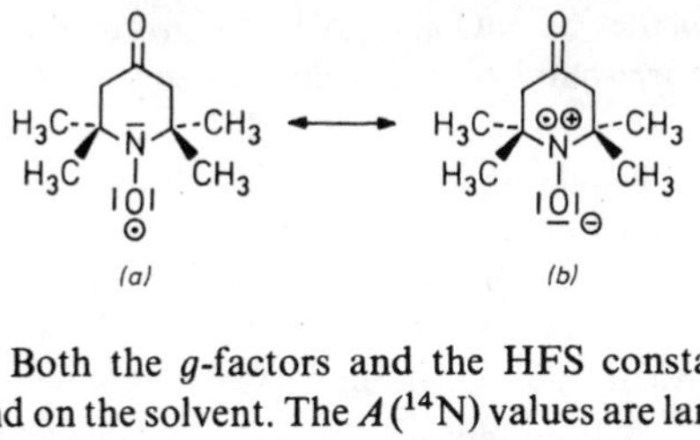

(a) (b)

Both the g-factors and the HFS constants of this neutral aminyloxide radical depend on the solvent. The $A(^{14}N)$ values are largest in polar protic solvents such as water and smallest in nonpolar non-HBD solvents such as benzene. The reverse is true for the $A(^{17}O)$ values. That is, solvents which reduce $A(^{14}N)$ increase $A(^{17}O)$. This opposing

* Semidiones are vinylogs of the superoxide anion radical, ⊙Ō—Ō|⊖ ↔ ⊖|Ō—Ō⊙, in the same way that benzosemiquinones are phenylogs of the same radical anion. Semidiones may be obtained by a one-electron reduction of conjugated aliphatic or cycloaliphatic diketones [196].

behaviour of the HFS constants with increasing solvent polarity may be rationalized in terms of valence bond theory by considering the relative contribution of the two main mesomeric structures, (*a*) and (*b*) in Eq. (6-13) to the actual electronic structure.

$$\left(\underset{(a)}{R_2\bar{N}-\bar{\underline{O}}\cdot} \longleftrightarrow \underset{(b)}{R_2\overset{\oplus}{N}(\overset{\ominus}{\cdot})-\bar{\underline{O}}|^{\ominus}} \right) + H{-}S \rightleftharpoons R_2\overset{\delta\oplus}{N}\overset{\cdot}{\cdots}\overset{\delta\ominus}{O}\cdots H{-}S \qquad (6\text{-}13)$$

Since $A(^{14}N)$ and $A(^{17}O)$ are determined by the unpaired π-electron spin density at the nitrogen and oxygen nuclei respectively, effects that favor the dipolar structure (*b*) relative to (*a*) will be associated with an increase in the magnitude of $A(^{14}N)$ and with a decrease in the magnitude of $A(^{17}O)$. Both solvent polarity and hydrogen bonding capability, will exert a similar influence on the relative contributions of (*a*) and (*b*) because hydrogen bonding is expected to occur predominantly with an oxygen lone pair in structure (*b*), according to Eq. (6-13). Thus, the greater the polarity and hydrogen bonding capability of the solvent, the greater the increase in the electron density on oxygen and the spin density on nitrogen, and the more favoured structure (*b*). The corresponding very small decrease in the *g*-factors in polar solvents is also attributed to the redistribution of the spin density, as well as to the decrease in spin-orbit coupling which would accompany a decrease in unpaired electron density on the oxygen atom [216]. As spin density diminishes around the oxygen atom with increasing solvent polarity, the *g*-factor of the aminyloxide is shifted towards the free-spin value of $g = 2.00232$ (*cf.* Table 6-5).

Similar solvent effects on HFS constants have been observed for other aminyloxides such as diphenyl aminyloxide [207, 212], di-*tert*-butyl aminyloxide [218, 385], *tert*-butyl aminyloxide [213, 217], and 2,2,6,6-tetramethylpiperidyl-1-oxide [384, 387, 388]. The $A(^{14}N)$ constants of di-*tert*-butyl and two other aminyloxides have been proposed as an empirical solvent polarity parameter because $A(^{14}N)$ is easily measured in most solvents [218, 389]; *cf.* Section 7.4.

A comparison of the $A(^{14}N)$-values, the $\tilde{\nu}$-values of the N—O IR stretching vibration, and the $\tilde{\nu}_{max}$-values of the $n \rightarrow \pi^*$ UV/Vis-absorption of di-*tert*-butyl aminyloxide reveals that the mode of solvation for the $>$C=O and $>$N—O· group is similar [335]. For example, the $A(^{14}N)$-values of di-*tert*-butyl aminyloxide in a range of solvents are linearly correlated to the $\tilde{\nu}_{max}$-values of its $n \rightarrow \pi^*$ absorption [335].

Semiquinones were among the first radicals to be studied in solution by ESR spectroscopy. Semiquinones have been observed as anion radicals [189, 191, 193–195, 382] and as protonated neutral radicals [192] in solution. *p*-Benzosemiquinone is of particular interest because it is possible to label the carbonyl carbon atom with ^{13}C and the carbonyl oxygen atom with ^{17}O. Together with the four remaining carbon atoms, which all have a hydrogen bonded to them, it is possible to obtain three solvent-dependent HFS constants [$A(^{1}H)$, $A(^{13}C)$, and $A(^{17}O)$]. It has been shown, for example, that the carbon atoms of the *p*-benzosemiquinone anion radical which are adjacent to the oxygen atoms exhibit marked changes in $A(^{13}C)$ constants with changes in solvent polarity [189]. The latter are 0.213 mT for both acetonitrile and dimethyl sulfoxide, but in water the splitting is only 0.04 mT [189]. Just as for aminyloxides, the redestribution of spin density within the semiquinone leads to a modification of the *g*-factor [191]. For the *p*-benzosemiquinone anion radical in water solution, $g = 2.00469$, and in dimethyl sulfoxide $g = 2.00541$ [191].

A particularly well-studied anion radical is the sodium 9-fluorenone ketyl radical shown in Eq. (6-14). The solvent influence on the $A(^{13}C)$ HFS constant of the ^{13}C-enriched 9-fluorenone ketyl group was examined in dipolar non-HBD solvents and their binary mixtures with toluene and tetrahydrofuran [391].

(a) ⟷ (b) $Na^{\oplus}$ (6-14)

Solvents	$(Me_2N)_3PO$	N-Methyl-pyrrolidone	CH_3CONMe_2	$HCONMe_2$	CH_3SOCH_3
$A(^{13}C)$/mT	0.176	0.224	0.230	0.275	0.320

With the increasing anion solvation power of the pure solvents, the $A(^{13}C)$-value of the free anion radical increases from 0.176 mT in HMPT to 0.320 mT in DMSO. The negatively charged oxygen atom of the carbonyl group attracts the positive end of the dipolar solvent molecules, which on their part inductively reduce the electron density around the carbon nucleus. This corresponds to a greater contribution of the mesomeric structure (*b*) to the electronic ground state; consequently there is an increase in the ^{13}C splitting.

Addition of small amounts of dipolar non-HBD solvents to solutions of the sodium 9-fluorenone ketyl radical in toluene, in which diamagnetic dimers or higher aggregates are present, gives rise to well-resolved HFS patterns. The ^{13}C splitting first decreases with an increase in the mole fraction of the dipolar non-HBD solvent. A limiting $A(^{13}C)$-value is reached at mole fraction $x = 0.2 \ldots 0.3$, due to dissociation into monomeric contact ion-pairs. The $A(^{13}C)$-value keeps constant until the ion pairs finally dissociate into free ions at a mole fraction of greater than 0.6 (the concentration of the anion radical being ca. 10^{-4} mol/l). Then, the $A(^{13}C)$-values vary with the solvent as given in Eq. (6-14) [391]. Thus, the solvent dependence of the ^{13}C splitting can be used to follow the formation and dissociation of ion pairs between cations and anion radicals; *cf.* Section 2.6.

Gendell, Freed, and Fraenkel [183] proposed a theory (GFF theory) to account for these solvent effects. It is based on the assumption that the solvent forms localized complexes with the oxygen atoms (or other heteroatoms), altering their electronegativity and consequently redistributing the spin density within the π-system of the free radical. In constructing a model for these complexes, Fraenkel *et al.* focused their attention on radicals containing dipolar substituents (or heteroatoms) and postulated, that each substituent was able to form a localized complex with one solvent molecule. This theory, together with HMO calculations of spin densities applied to the solvent effects on proton HFS constants of *p*-benzosemiquinone anion radical, gave good quantitative agreement [183]. The GFF theory has been tested, however, with limited success by Luckhurst and

Orgel [187], who examined the ESR spectrum of 9-fluorenone ketyl in mixtures of *N,N*-dimethylformamide and methanol.

Various solvent effect theories concerning HFS constants in ESR spectra using various reaction field approaches have been developed by Reddoch *et al.* [385] and Abe *et al.* [392]. According to Reddoch *et al.*, none of the continuum reaction field models is entirely satisfactory. Therefore, a dipole-dipole model using a field due only to a dipole moment of one solvent molecule instead of various reaction fields was proposed, and applied to di-*tert*-butyl aminyloxide [385]. However, Abe *et al.* found that the HFS constants are proportional to the reaction field of Block and Walker [393] when protic solvents are excluded [392]. This relationship has been successfully applied to di-*tert*-butyl and diaryl aminyloxides, to the 4-methoxycarbonyl-1-methylpyridinyl radical (*cf.* Fig. 6-10), and to the 4-acetyl-1-methylpyridinyl radical (see below) [392].

Finally, an example which exhibits extraordinarily large variations of the HFS constants with solvent should be mentioned. The neutral 4-acetyl-1-methyl-pyridinyl radical shown in Eq. (6-15) exhibits variations of the $A(^1H)$ constants, caused by the protons of the heteroaromatic ring and the acetyl group, of the order of 200 to 300% [222, 381, 389]. The greatest changes of HFS constants with solvent obtained so far with aminyloxides [216, 218] and some nitroaromatic anions [204] are at most of the order of 50%.

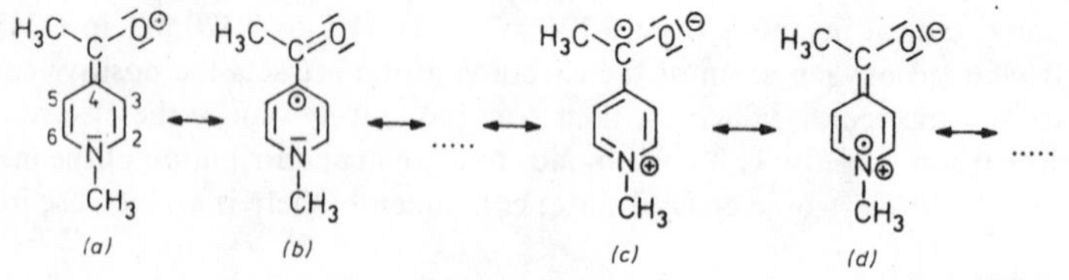

(a) ⟷ (b) ⟷ ..... ⟷ (c) ⟷ (d) ⟷ ..... (6-15)

Solvents	$(CH_2)_4O$	$(Me_2N)_3PO$	CH_3SOCH_3	CH_3OH	H_2O
$A(^1H)$/mT of C-2	0.294	0.270	0.252	0.210	0.164
$A(^1H)$/mT of C-3	0.048	0.073	0.086	0.138	0.185
$A(^1H)$/mT of $COCH_3$	0.195	0.231	0.261	0.410	0.541

It is also worthwhile noticing that the magnitude of the $A(^1H)$ constants caused by the 2-H and 3-H hydrogen atoms is reversed on going from tetrahydrofuran to water. Again, this unusual solvent dependence of the HFS constants may be best explained in terms of valence bond theory, considering the relative contribution of the nonpolar mesomeric structures (*a*), (*b*), *etc.*, and the dipolar mesomeric structures (*c*), (*d*), *etc.*, to the actual electronic structure of the radical. In solvents of low polarity, large spin densities are observed at the C-2, C-4, and C-6 positions, while the methyl of the acetyl group and the protons at C-3 and C-5 cause relatively small splitting constants. However, in solvents of higher polarity, which strongly favour the dipolar mesomeric structures, an increase of the acetyl and 3-H and 5-H splittings and a parallel decrease of the 2-H and 6-H couplings is observed [222]. An additional stabilization of the dipolar structures in protic solvents may come from hydrogen bonding to the negatively charged carbonyl oxygen atom.

A completely different medium influence on ESR spectra has been observed for radical anions dissolved in non-dissociating solvents. In non-dissociating solvents of low dielectric constant, ion pairing can lead to the appearance of hyperfine features in the ESR spectra of radical anions. These are caused by interactions between the unpaired electron and the nuclei of the diamagnetic cationic gegenion [204, 223–225, 391]. For example, in the ESR spectrum of the ion pair $Na^{\oplus}A^{\ominus}$, every absorption line in the ESR spectrum of $A^{\ominus}$ will be split into a quartett of lines resulting from coupling to the ^{23}Na nuclei, which have a nuclear spin of $I = 3/2$. Generally, the stronger the solvation of the cation and accordingly the dissociation of the ion pair, the smaller will be the HFS constants. Strong coordination by the solvent will decrease the effective electron affinity of the cation. Ultimately, there may be an insertion of solvent molecules between cation and anion, leading to solvent-separated ion pairs (*cf.* Fig. 2-14 in Section 2.6). This process may be quantized or may appear as a more continuous pulling-off process. Both processes are especially favoured by solvents which are good cation solvators such as the oligoethylene glycol dialkyl ethers (glymes) (*cf.* Section 5.5.5).

With respect to these solvent effects, it is unfortunate, that in most cases ethers have been used as solvents for determining ESR spectra of radical anions. In these solvents, the metal coupling due to ion pair formation is almost always obtained. In general, in coordinating solvents of relatively high dielectric constant such as acetone. *N,N*-dimethylformamide, or acetonitrile, metal ion hyperfine splitting has not been detected. In binary solvent mixtures, gradual changes in the HFS constants resulting from increased proportion of the better cation solvating solvent were reported for naphthalenides [204] and 9-fluorenides [391]. The use of tetraalkylammonium salts in ESR studies of radical anions often serves to overcome the difficulties involved in ion-pair formation. Lowering the temperature corresponds to an increase in ionic solvation. Therefore, on cooling, the ESR spectra of free radical anions often appear, in addition to those of the ion pairs. A compilation of solvent and cation dependence of metal hyperfine splittings for ion pairs has been given by Sharp and Symons [204].

6.5 Solvent Effects on Nuclear Magnetic Resonance Spectra

6.5.1 Nonspecific Solvent Effects on NMR Chemical Shifts

Position and line-width of nuclear magnetic resonance (NMR) signals, as well as the magnitude of the spin-spin coupling constants in the high resolution NMR spectrum of a particular molecule in solution, are affected by the surrounding molecules of the same or different species. In solution, the position of the resonance signal is often found to be concentration dependent. This effect of neighbouring solute molecules can be easily eliminated by carrying out the measurements at different concentrations and extrapolating to infinite dilution. The solvent-dependence of NMR chemical shifts is most conveniently referred to the shift values, extrapolated to infinite dilution in a nonpolar solvent having as nearly isotropic molecular properties (in particular shape, polarizability, and magnetic susceptibility) as possible. Tetrachloromethane, *n*-hexane, and cyclohexane are commonly used as inert reference solvents. The characteristic ^{1}H- and ^{13}C-NMR resonance signals of commonly used NMR solvents are compiled in Tables A-6 and A-7 (Appendix).

In general, two different solvent effects on NMR spectra can be distinguished: (a) shifts due to a difference in the bulk magnetic susceptibility χ_m (sometimes written as χ_v, for volume susceptibility) of the solute and the solvent; (b) shifts arising from intermolecular interactions between solute and solvent molecules. Since the bulk susceptibility effect depends on the shape of the sample and, therefore, is not of chemical interest, some form of correction for it is applied. For two coaxial cylindrical samples with axis perpendicular to the applied magnetic field and differing in magnetic susceptibility by $\Delta\chi_m$, the bulk susceptibility correction is given by Eq. (6-16):

$$\delta_{corr} = \delta_{exp} + \frac{2\pi}{3} \cdot \Delta\chi_m \cdot 10^6$$

$$= \delta_{exp} + \frac{2\pi}{3} \cdot (\chi_m^{reference} - \chi_m^{solution}) \cdot 10^6 \qquad (6\text{-}16)$$

Only shifts observed in excess of this amount may be then attributed to intermolecular interaction effects. Use of an internal standard provides an automatic compensation for the bulk susceptibility effect, but for comparison of shifts measured in this way in different solvents it must be kept in mind that the standard itself may be subject to solvent effects. These are minimized in ^{1}H- and ^{13}C-NMR spectroscopy by use of tetramethylsilane (TMS) as an internal standard.

The intermolecular solute/solvent interactions may arise from nonspecific interaction forces such as dispersion, dipole-dipole, dipole-induced dipole, *etc.*, as well as from specific interactions found in protic and aromatic solvents. Solvent effects on NMR spectra were first observed by Bothner-By and Glick [226] and independently by Reeves and Schneider [227] in 1957. Since then, the influence of solvent on chemical shifts (and coupling constants) has been extensively studied by scores of workers and has been thoroughly reviewed by several specialists [1–4, 288–237].

Some illustrative examples of solvent-dependent NMR chemical shifts for several different nuclei in a cation [238, 239], two dipolar molecules [240, 241], and an apolar molecule [242, 243] are given in Table 6-6.

Whereas the solvent influence on the ^{1}H and ^{13}C chemical shifts of the apolar tetramethylsilane is comparatively small ($\Delta\delta$ *ca.* 0.5...1.5 ppm), much greater effects are observed in the case of dipolar molecules such as 4-fluoro-nitrosobenzene and triethylphosphane oxide ($\Delta\delta$ *ca.* 3...25 ppm) as well as for the thallous ion ($\Delta\delta > 2000$ ppm!).

Amongst the cations, $^{205}Tl^{\oplus}$ is exceptionally sensitive to its solvent environment. The solvent-dependent chemical shift observed for this cation is over 2600 ppm [238, 239]. In comparison, the known solvent chemical shift for $^7Li^{\oplus}$ is only 6 ppm [244], for $^{23}Na^{\oplus}$ 20 ppm [245, 396], and for $^{133}Cs^{\oplus}$ it is only 130 ppm [246]. The remarkable solvent sensitivity of the chemical shift of $^{205}Tl^{\oplus}$ makes it an exceptionally useful probe for the study of common and preferential solvation [247]; *cf.* Section 2.4. Similar huge solvent-induced chemical shifts have been found for ^{59}Co salts and complexes [395]. No single bulk property of the solvent has been shown responsible for these large shifts. A correlation between cation chemical shifts and Gutmann's donor number, which is a measure of

Table 6-6. A selection of solvent-dependent NMR chemical shifts [a)] of thallous ion ($^{205}Tl^{\oplus}$) [238, 239], triethylphosphane oxide (^{31}P) [240, 367], 4-fluoro-nitrosobenzene (^{19}F) [241], and tetramethylsilane (^{13}C and ^{1}H) [242, 243].

Solvents	$\delta(^{205}Tl)$ of $Tl^{\oplus}X^{\ominus}$ [b)]	$\delta(^{31}P)$ of $(C_2H_5)_3PO$ [c)]	$\Delta\delta(^{19}F)$ of 4-FC_6H_4NO [d)]	$\delta(^{13}C)$ of $(CH_3)_4Si$ [e)]	$\delta(^{1}H)$ of $(CH_3)_4Si$ [f)]
n-Hexane	–	0.00	–	0.19	–
Cyclohexane	–	–	10.50	0.16	−0.04
1,2-Diaminoethane	2147	–	–	–	–
1-Aminobutane	1896	–	–	–	–
Benzene	–	3.49	11.50	−0.26	0.30
Tetrachloromethane	–	3.64	11.10	−0.82	−0.16
1,4-Dioxane	–	4.59	11.60	−0.07	–
Acetone	–	5.33	12.45	0.09	0.03
Pyridine	644	6.04	12.55	−0.34	0.34
N,*N*-Dimethylformamide	126	6.82	12.85	−0.11	–
Dimethyl sulfoxide	369	8.22	13.20	–	–
Dichloromethane	–	8.67	12.90	−0.18	−0.08
Chloroform	–	9.83	12.95	−0.53	−0.14
Formamide	96	16.95	13.05	–	–
Methanol	–	17.60	12.45	0.51	–
Water	0	23.35	–	–	–
	$\Delta\delta =$ 2147 ppm	$\Delta\delta =$ 23.35 ppm	$\Delta\Delta\delta =$ 2.70 ppm	$\Delta\delta =$ 1.33 ppm	$\Delta\delta =$ 0.50 ppm

[a] Shifts in ppm; a positive sign corresponds to paramagnetic downfield shifts.
[b] Shifts are extrapolated to zero anion concentration and are in units of ppm from water.
[c] Shifts extrapolated to infinite dilution, referred to *n*-hexane and corrected for the difference in volume susceptibilities between *n*-hexane and the respective solvent.
[d] Shifts relative to fluorobenzene as internal standard, at high dilution ("shielding parameters" [241]).
[e] Apparent ^{13}C shift of TMS (20 ml/100 ml) with respect to the resonance of ^{13}C in neat TMS.
[f] Intrinsic ^{1}H shift of TMS (20 ml/100 ml) with respect to the resonance of ^{1}H in neat TMS.

solvent Lewis basicity (*cf.* Table 2-3 in Section 2.2.6), has often been found [245, 246]. The solvent-induced chemical shifts can be viewed as a measure of the strength of cation/solvent interaction, with the solvent acting as Lewis base and interacting electrostatically and covalently with the cation. – Relatively large solvent-induced ^{1}H-chemical shifts have been also observed for organic cations such as 1-methylpyridinium [248], 1,4-diethylpyridinium [249], 1-methylquinolinium [250], and tropylium ion [250].

In the case of the dipolar molecules included in Table 6-6, parts of the solvent-induced chemical shifts may be qualitatively explained in terms of valence bond theory as shown for 4-fluoro-nitrosobenzene [241]. Its ^{19}F-signal is increasingly shifted to lower field strengths with increasing solvent polarity. The greater the solvent polarity, the greater the charge separation in the dipolar molecule due to an increasing contribution of the dipolar mesomeric structure to the electronic ground state, and the greater the

$$\overline{O}=\overline{N}-C_6H_4-\overline{F}| \longleftrightarrow |\overline{\underline{O}}^{\ominus}-\overline{N}=C_6H_4=\overline{F}^{\oplus}$$

deshielding of the fluorine atom. This corresponds to the downfield ^{19}F-chemical shifts observed in polar solvents.

An analogous interpretation explaining part of the large downfield ^{31}P-chemical shift of triethylphosphane oxide, $Et_3P{=}O \leftrightarrow Et_3P^{\oplus}{-}O^{\ominus}$, which is observed with increasing solvent polarity, has also been given [240, 367]. The high ^{31}P-NMR shift sensitivity of triethylphosphane oxide makes this dipolar compound particularly useful as probe molecule in the study of solute/solvent interactions. Indeed, Gutmann *et al.* have used ^{31}P-chemical shifts to measure solvent Lewis acidity, also called the acceptor number [240]; *cf.* Table 2-5 in Section 2.2.6, and Section 7.4. The ^{31}P-chemical shifts of triethylphosphane oxide, measured in a variety of solvents, are linearly correlated to the corresponding IR wavenumber shifts $\tilde{\nu}(P{=}O)$ of its P=O stretching vibration [367].

The solvent influence on X = O dipolarity (X = C, N, P, S) has been also studied, particularly for carbonyl compounds such as aliphatic ketones and esters [251–254, 367, 397, 398]. For example, the ^{13}C and ^{17}O chemical shifts of the carbonyl atoms are very sensitive to solvents, especially to protic solvents capable of hydrogen-bonding to the carbonyl oxygen atom, as shown for acetone in Eq. (6-17) [251, 253, 367, 397].

$$\left((H_3C)_2{}^{13}C{=}^{17}O \longleftrightarrow (H_3C)_2\overset{\oplus}{C}{-}\overset{\ominus}{O} \right) + H{-}S \rightleftharpoons (H_3C)_2\overset{\delta\oplus}{C}{\cdots}\overset{\delta\ominus}{O}{\cdots}H{-}S \qquad (6\text{-}17)$$

Solvents	CCl_4	C_6H_6	CH_3COCH_3	$HCONMe_2$	CH_3CN	CH_3OH	H_2O	HCO_2H
$\Delta\delta(^{13}C{=}O)$/ppm[a,b]	−1.3	−0.8	0	+0.7	+2.1	+3.7	+ 9.1	+ 9.1
$\Delta\delta(C{=}^{17}O)$/ppm[a,c]	+5	0	0	−2	−4	−12	−37	−40

[a] Chemical shifts relative to neat acetone; a negative sign corresponds to diamagnetic highfield shifts.
[b] At 15.1 MHz [251].
[c] At 7.65 MHz [254].

The shift data given in Eq. (6-17) demonstrate that, as the ^{13}C shielding decreases with increasing solvent polarity, the corresponding ^{17}O shielding increases [254]. A monotonic relationship is seen to exist between the ^{13}C and ^{17}O solvent-induced chemical shifts of acetone. This behaviour has been interpreted qualitatively in terms of the altered dipolarity of the carbonyl group as represented by the relative importance of the two mesomeric structures in the valence bond description of acetone. Furthermore, these solvent effects on ^{13}C-chemical shifts can be linearly correlated with the corresponding infrared C=O stretching frequencies measured in the same variety of solvents, as has been shown for acetone, acetophenone, and ethyl acetate [253, 367].

Other representative dipolar compounds for which solvent-induced chemical shifts have been thoroughly studied are dipolar aliphatic compounds (^{13}C) [399], substituted aromatic hydrocarbons (^{1}H) [252, 255, 400], heterocyclic aromatic compounds (^{1}H) [255, 256], *N*,*N*-dimethylbenzamide (^{13}C) [401], meropolymethine dyes (^{1}H, ^{13}C, and ^{15}N) [20, 50, 73, 75, 77, 78, 402], pyridinium-*N*-phenoxide betaine dyes (^{13}C) [403, 404], chloroform (^{13}C) [257], pyridine-1-oxide (^{13}C) [258], fluoropyridines (^{19}F) [259], 4′-substituted 4-fluorophenylphenyl compounds (^{19}F) [260], and triphenylphosphane oxide (^{31}P) [261]. Leading references for further examples are [248–261]; *cf.* also [1–4, 228–237].

A particularly well-studied dipolar example of solvent-induced NMR chemical shifts is 3-dimethylaminoacrolein. As shown by ^{1}H- [20, 50, 73, 75, 78], ^{13}C- [77, 402], and ^{15}N-NMR investigations [402], this vinylogous formamide exhibits a polyene-like π-electron structure (*a*) in nonpolar solvents, and a dipolar polymethine-like structure (*b*) in polar solvents; *cf.* Eq. (6-18) and Table 6-2 in Section 6.2.2.

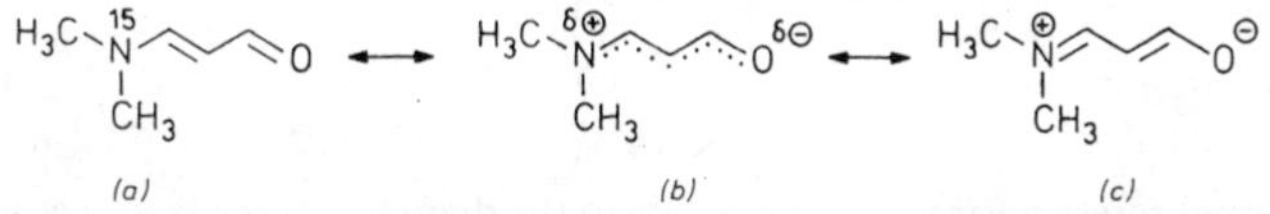

(a) ⟷ (b) ⟷ (c) (6-18)

Solvents	C_6H_6	1,4-Dioxane	CH_3COCH_3	$HCONMe_2$	CH_3SOCH_3	CH_3OH	H_2O
$\delta(^{15}N)/ppm^{a)}$	55.6	57.0	59.2	62.2	64.9	72.9	85.4

[a] $c = 0.5$ mol/l; aqueous saturated $^{15}NH_4Cl$ solution as external reference.

Accordingly, the ^{15}N signal of the ^{15}N-labelled merocyanine exhibits a large downfield shift of ca. 30 ppm with increasing solvent polarity, corresponding to increased deshielding of the ^{15}N atom. These solvent-induced ^{15}N-chemical shifts correlate linearly with the Gibbs energy of activation, $\Delta G^{\neq}_{rot}$, for the rotation of the dimethylamino group around the ^{15}N—C bond [402]. This can be rationalized in terms of the π-electron distribution along the N—C fragment of 3-dimethylaminoacrolein: increasing solvent polarity favours nitrogen lone-pair delocalization with downfield ^{15}N-chemical shifts and increasing $\Delta G^{\neq}_{rot}$-values as consequence.

Compared with the pronounced solvent-induced chemical shifts observed with ionic and dipolar solutes, the corresponding shifts of nonpolar solutes such as tetramethylsilane are rather small; *cf.* Table 6-6. A careful investigation of ^{13}C-chemical shifts of unsubstituted aromatic, alternant and nonalternant, hydrocarbons in aliphatic and aromatic non-HBD solvents by Abboud *et al.* has shown that, the differential solvent-induced chemical shift range (relative to benzene as reference) is of the order of only $-1.4 \ldots +1.0$ ppm (positive values representing downfield shifts) [405]. The ^{13}C-NMR spectra of these aromatic compounds have been shown to be sensitive to solvent dipolarity and polarizability except for aromatic solvents, for which an additional specific aromatic solvent-induced shift (ASIS; see later) has been found. There is no simple relationship between the solvent-induced chemical shifts and the calculated charge distribution of the aromatic solute molecules. This demonstrates the importance of quadrupoles and higher multipoles in solute/solvent interactions involving aromatic solutes [405].

In conclusion, the screening (or shielding) constant of a nucleus in a particular molecule is not only determined by the electronic distribution within the molecule, but also by the nature of the surrounding medium. The observed screening constant, σ_{obsd}, is the sum of the screening constant for the isolated molecule, σ_o, and a contribution σ_{medium}, arising from the surrounding medium, according to

$$\sigma_{obsd} = \sigma_o + \sigma_{medium} \quad (6\text{-}19)$$

The screening constant σ_{medium} is given by Eq. (6-20),

$$\sigma_{medium} = (H_0 - H)/H_0 \tag{6-20}$$

where H_0 is the applied field strength producing resonance in an isolated gaseous molecule, and H is the field required to produce resonance in the medium. Buckingham, Schaefer, and Schneider [262] have suggested that σ_{medium} contains contributions from five different sources according to the generally accepted Eq. (6-21).

$$\sigma_{medium} = \sigma_b + \sigma_a + \sigma_w + \sigma_e + \sigma_s \tag{6-21}$$

σ_b arises if an external reference is used, and is due to the diamagnetic bulk susceptibility differences between solution and reference sample. The magnitude of σ_b depends on the shape of the sample; *cf.* Eq. (6-16). It is zero for a spherical sample or by use of an internal reference. σ_a is derived from the anisotropy of the magnetic susceptibility of some solvent molecules. It is particularly important for disc-shaped molecules of aromatic solvents, and rod-like molecules such as carbon disulfide. σ_a has been detected experimentally by major deviations from the expected behaviour of methane based on σ_b and σ_w. σ_w is a downfield shift, thought to arise from weak dispersion interactions between solute and solvent molecules (van der Waals forces) [263]. This effect is measured by use of nonpolar, isotropic solutes (*e.g.* methane) in nonpolar, isotropic solvents (*e.g.* tetrachloromethane) and an external reference, followed by a susceptibility correction. Its magnitude increases with increasing molecular polarizability of the solvent molecules. σ_e represents the contribution of a polar effect, caused by the charge distribution in the dipolar solute molecule [262, 264, 265]. Dipolar molecules induce a dipole moment in the surrounding solvent molecules. The electric field E thus created (the reaction field according to Onsager [80]) produces a small change in the solute chemical shift. This effect should therefore depend on the dipole moment and polarizability of the solute and the dielectric constant of the solvent, *i.e.* $(\varepsilon_r - 1)/(2\varepsilon_r + 1)$ [262, 264]. The solvent shifts experienced by dipolar molecules include all the terms described so far. They are commonly reported as the solvent-induced difference in the chemical shifts of the solute of interest and the internal reference compound (usually TMS) in a nonpolar dilute reference phase such as cyclohexane. Finally, a fifth term, σ_s, may be added on the right side of Eq. (6-21) to account for specific solute/solvent interactions such as hydrogen-bonding or EPD/EPA interactions. With EPD solutes in protic solvents (or *vice versa*) σ_s is usually by far the largest term.

Each of the contributions to σ_{medium} has been the subject of several separate investigations. It is usually found, that σ_b and σ_w cause paramagnetic shifts (that is, they lead to resonance at lower applied field). σ_a leads to diamagnetic shifts for disc-shaped solvent molecules such as benzene, and to paramagnetic shifts for cylindrically symmetric, rod-shaped ones such as carbon disulfide, whereas σ_e can be responsible for either diamagnetic or paramagnetic shifts, depending on the position of the nucleus relative to the polar group in the solute molecule [262].

Fig. 6-11 illustrates the dissection of the solvent-induced ^{1}H-NMR shift into various terms using the rather elementary example of methane [3, 266].

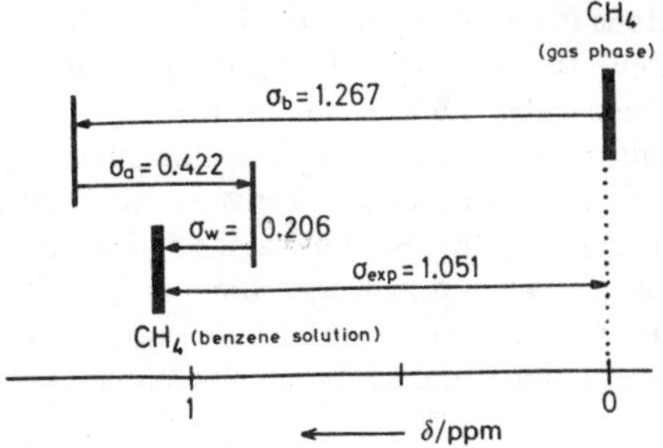

Fig. 6-11. Solvent-induced ^{1}H-NMR chemical shift for methane on going from the gas phase into benzene solution [3, 266].

Since screening constant σ and chemical shift δ are related to each other, in practice a modified Eq. (6-21), Eq. (6-22), is often used,

$$\Delta\delta_{obsd} = \delta_b + \delta_a + \delta_w + \delta_e + \delta_s \tag{6-22}$$

where $\Delta\delta_{obsd}$ represents the chemical shift increment caused by the medium in going from the gaseous to the condensed state.

The numerous attempts and several proposed models for calculating and describing the first four quantities of Eq. (6-21) or Eq. (6-22) in a quantitative way have been thoroughly reviewed [1, 3, 232, 233, 235, 267, 268, 398, 406, 407], and will therefore not be mentioned here. It has been stated by Webb *et al.* [406], that, since precise mathematical expressions are not available for most of the non-bonded solute/solvent interactions considered, from a quantum chemical point of view the separation of the various contributions is not very meaningful; *cf.* also [1]. Comparison of experimental and calculated solvent-induced chemical shifts is often hampered by the fact that most of the experimental results are obtained with liquid samples. It would be preferable to have chemical shifts (and spin-spin coupling constants) quoted relative to the isolated probe molecule in the gas-phase [406]. The crux of the matter is that, by definition, chemical shifts are differential quantities and may contain contributions from several different solvent effects which cannot be separated *a priori* [235]. According to Webb *et al.*, it is hardly surprising that, with the approximations in the solute/solvent interaction models, the overall agreement between observed and predicted solvent effects on NMR parameters is less than quantitatively satisfactory in many cases [406].

6.5.2 Specific Solvent Effects on NMR Chemical Shifts

Specific solvent effects on the resonance positions of the nuclei of dissolved compounds consist mainly of hydrogen-bonding effects and aromatic solvent-induced shifts (ASIS effects). The interactions between the solute and the surrounding solvent molecules lead, in these cases, to molecular species which are more or less definable entities. If the lifetime of a given nucleus in each of the distinct species present in the solution is sufficiently short, a single averaged resonance signal is observed.

Since the discovery, in 1951, that protons involved in hydrogen bonds experience large shifts in their resonance signals [268, 269], the behaviour of ^{1}H-chemical shifts, as affected by hydrogen-bond interactions, has been widely studied and thoroughly reviewed [232–235, 270–272]. Hydrogen bonding usually results in a paramagnetic downfield shift of the resonance signal of the proton involved; the solvents most effective in causing this are those with strong EPD properties. There is a general observation that the magnitude of the change in the ^{1}H-chemical shift which occurs when a hydrogen bond is formed is related to its strength. Since hydrogen bonds are usually made and broken very rapidly relative to the chemical shift difference between bonded and nonbonded forms, separate lines are not observed for different species.

Proton-donors possessing OH, NH, SH, and CH functional groups have been employed. Processes ranging from self association of chloroform to inter- and intramolecular hydrogen bonding of ambifunctional species (*e.g.* carboxylic acids, *o*-aminophenols) have been examined. A great variety of examples is given in recent reviews [270–272]. In particular, the behaviour of the hydroxy-proton shift, because of its extreme sensitivity to hydrogen bonding interaction, has been widely studied. In almost all cases, formation of the hydrogen bond causes the resonance signal of the bonded OH-proton to move downfield by as much as 10 ppm. Intramolecularly hydrogen bonded enols and phenols display resonances at especially low field. In solutions of dipolar compounds in protic solvents such as water, hydrogen bonding is the most important kind of intermolecular interaction.

Another illustrative example is that of phenylacetylene. Table 6-7 summarizes the ^{1}H-chemical shifts of its alkyne proton in a variety of solvents [273]. Most solvents (except aromatic solvents) cause the acetylenic hydrogen nuclei to undergo decreased shielding. The corresponding low-field shift has been interpreted in terms of weak specific association between the alkyne as proton-donor and electron pair-donor groupings of the solvent [273]. The high-field shifts in aromatic solvents arise from the magnetic anisotropy of the solvent molecules (see below). The order of effectiveness of the solvent in shifting the $\equiv$C—H signal is qualitatively the same as that observed for the shifts of the infrared NH stretching frequencies of pyrrole in dilute solutions [162]. The basicity of a variety of EPD solvents has been measured by the ^{1}H-chemical shift of the chloroform proton at

Table 6-7. ^{1}H-NMR chemical shifts of the alkyne proton of phenylacetylene in various solvents at infinite dilution [273].

Solvents	$\delta(\equiv C-H)$/ppm	$\Delta\delta$/ppm[a]
Toluene	2.64	−0.10
Benzene	2.71	−0.03
Cyclohexane	2.74	0.00
Nitromethane	3.26	0.52
Acetonitrile	3.35	0.61
Nitrobenzene	3.44	0.70
1,4-Dioxane	3.50	0.76
Acetone	3.61	0.87
Pyridine	4.10	1.36
N,N-Dimethylformamide	4.35	1.61

[a] $\Delta\delta = \delta(\text{solvent}) - \delta(\text{cyclohexane})$.

infinite dilution in the solvent of interest and in cyclohexane as inert reference solvent (*cf.* Table 3-5 in Section 3.3.1) [143, 274, 275].

The reason for the pronounced shifts of hydrogen-bonded protons cannot be completely explained by simple electrostatic considerations. First, it is evident that the electron distribution in the X—H covalent bond in the hydrogen-bonded system X—H···|Y is altered by the electric field of|Y in such a way that the proton is deshielded. But, in addition, the proton may experience an anisotropy effect of the neighbouring group |Y. When the proton is bonded more or less to the center of the π-electron cloud of an aromatic solvent, the ring current effect leads to a large upfield shift (*cf.* Table 6-7), which predominates over any deshielding due to other factors. A hydrogen bond to π-electrons of an aromatic or heteroaromatic ring is the only type of hydrogen bond that results in an upfield, rather than downfield, shift.

The effect of hydrogen bonding on the chemical shift of the proton-acceptor atom |Y (and of X) has not been studied in any depth. For example, hydrogen bonding affects the ^{17}O-chemical shift of hydroxy groups. The effect of solvent on the ^{17}O-chemical shift of water, methanol, and acetic acid has been investigated [276]. It appears that the hydroxylic oxygen experiences greater downfield shifts when it acts as proton-donor than when it serves as proton-acceptor [276]. The ^{13}C- and ^{17}O-chemical shifts of acetone [251, 254] (*cf.* Eq. (6-17) in the preceding Section), and the ^{31}P-chemical shift of triethylphosphane oxide [240] (*cf.* Table 6-6) reveal also a comparatively large medium effect in protic solvents, obviously due to hydrogen bonding. – The large solvent-induced ^{59}Co-chemical shift of $(n\text{-Bu})_4N^{\oplus}$ $Co(CN)_6^{\ominus}$ observed in protic solvents has been recommended as a measure of the HBD capacity of protic solvents [395]. In protic solvents, a downfield shift of the ^{59}Co resonance occurs upon hydrogen-bonding to the cyanide ligand, according to $^{59}Co—C{\equiv}N\cdots H—S$. A ^{59}Co-chemical shift difference of *ca.* 1 ppm is even observed between the solvents H_2O and D_2O; D_2O being the weaker HBD solvent [395].

Solvent-induced ^{1}H-chemical shifts of hydrogen-bonded protons are not only found for hydrogen-bonded *solute/solvent* complexes but also for hydrogen-bonded *solute/solute* complexes. A well-studied example is the intermolecularly hydrogen-bonded 1:1 complex between trifluoroacetic acid (as HBD) and 2,4,6-trimethylpyridine (as HBA), as shown in Eq. (6-23) [408]; *cf.* also Eq. (4-29) in Section 4.4.1.

$$\cdots \rightleftharpoons CF_3\text{-}\overset{O}{\overset{\|}{C}}\text{-}O\text{-}H\cdots N(C_5H_2(CH_3)_3) \rightleftharpoons CF_3\text{-}\overset{O}{\overset{\|}{C}}\text{-}\overset{\ominus}{O}\cdots H\text{-}N^{\oplus}(C_5H_2(CH_3)_3) \rightleftharpoons \cdots \qquad (6\text{-}23)$$

(a) (b)

Solvents	C_6H_6	C_6H_5Cl	CH_2Cl_2	$ClCH_2CH_2Cl$	CH_3CN
$\delta(OH)$/ppm[a]	19.02	18.92	18.10	18.00	16.95

[a] Equimolar solution with $c = 0.15$ mol/l [408].

In this case, the solvent-induced ^{1}H-chemical shift of *ca.* 2 ppm is best explained by assuming a double-minimum potential model of the hydrogen-bond, *i.e.* the existence of a rapid solvent-dependent proton-transfer equilibrium between (*a*) the covalent and (*b*) the ionic hydrogen-bonded complex. With increasing solvent polarity, the proton-transfer equilibrium is shifted in favour of the ionic complex (*b*).

In conclusion, proton NMR has proven to be one of the most sensitive spectroscopic methods both qualitatively and quantitatively of studying hydrogen bonds [270–272, 277]; *cf.* also Section 2.2.5.

When a dipolar molecule is dissolved in a magnetically anisotropic solvent consisting of disc-shaped molecules, *e.g.* benzene, the NMR signals of the solute protons are usually shifted upfield with respect to their positions in an isotropic solvent such as 2,2-dimethylpropane (neopentane) or tetrachloromethane; *cf.* the preceding discussion of σ_a in Eq. (6-21). The specific solvent-induced ^{1}H-chemical shift of a solute proton signal when the solvent is changed from a reference aliphatic solvent to an aromatic solvent is called *aromatic solvent-induced shift* (ASIS)*) and is defined according to Eq. (6-24) [278, 279],

$$\Delta\delta(\text{ASIS}) = \delta_{\text{AS}} - \delta_{\text{S}} \qquad (6\text{-}24)$$

where δ_{AS} is the position of a proton signal in the aromatic solvent (*e.g.* C_6H_6 or C_6D_6) and δ_S is the value for the same signal in the aliphatic solvent (*e.g.* CCl_4 or $CDCl_3$). A positive value of $\Delta\delta$(ASIS) indicates a downfield shift relative to the signal position in the aliphatic solvent. $\Delta\delta(^1$H-ASIS)-Values can be as large as ± 1.5 ppm and have proven to be a powerful tool in the elucidation of structural, stereochemical, and conformational problems [3, 232–234, 408]. The utility of $\Delta\delta(^{13}$C-ASIS) is still unknown since ^{13}C-chemical shifts include a far greater range than ^{1}H-chemical shifts and, therefore, contributions due to solvent anisotropies become relatively insignificant in ^{13}C-NMR spectroscopy [409].

The pronounced magnetic anisotropy of benzene helps to reveal subtle solute-solvent interactions which otherwise could not be detected. For example, ASIS's can be used to differentiate between axial and equatorial protons or methyl groups adjacent to carbonyl groups. A typical shift is that of an axial 2-methyl group in a cyclohexanone ring, 0.2...0.3 ppm upfield, and that of the corresponding equatorial 2-methyl group, 0.05...0.10 ppm downfield, in benzene relative to tetrachloromethane as solvent [3]. This can be used to determine the configuration at the 2-position and to assess the position of the conformational equilibrium in 2-methylcyclohexanone.

The so-called carbonyl plane rule permits the locating of protons behind or in front of a keto group [408, 410, 411]. Assuming a reference plane perpendicular to the C_α—CO—$C_{\alpha'}$ plane and passing through the carbonyl carbon atom, the value of $\Delta\delta$(ASIS) is positive for protons lying in front of this perpendicular plane (toward the oxygen atom), and negative for those protons lying behind the perpendicular plane. For example, the $\Delta\delta$(ASIS)-values for the terpene ketone pulegone are positive for the 10-methyl group and negative for the 7- and 9-methyl group as the former lies in front of

---

* Somewhat related to the ASIS's are the lanthanide-induced NMR chemical shifts. These involve addition of a lanthanide salt to form a complex, rather than a solvent effect. In fact, aromatic solvents are of the same class as the diamagnetic lanthanide shift reagents; *cf.* reference [415].

the carbonyl reference plane and the latter behind the reference plane [412]. Analogous results have been observed for camphor whose 10-methyl group signal experiences a downfield shift whereas the signals of the 8- and 9-methyl groups are shifted upfield in benzene relative to tetrachloromethane [412]; see reference [408] for further examples of the carbonyl plane rule. Interestingly, when the aromatic solvent is hexafluorobenzene instead of benzene, the carbonyl plane rule is exactly reversed [412, 413].

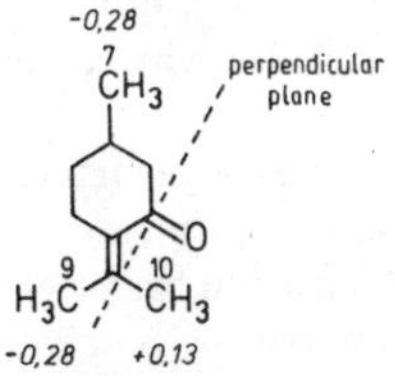

$\Delta\delta$(^{1}H-ASIS) of pulegone, in ppm

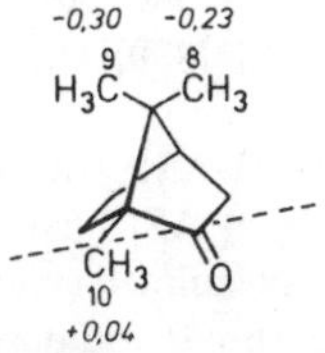

$\Delta\delta$(^{1}H-ASIS) of camphor in ppm

For both terpene ketones, pulegone and camphor, the $\Delta\delta$(^{13}C-ASIS) has been estimated by a chemical shift comparison method using *tert*-butylcyclohexane as an additional reference compound [409]. By comparing the $\Delta\delta$(ASIS) of both ^{1}H and ^{13}C nuclei it seems that the carbonyl plane rule is also valid for aromatic solvent-induced ^{13}C-chemical shifts of carbonyl compounds [409]. Obviously, the particular geometrical arrangement of the aromatic solvent molecules around the carbonyl dipole influences both the ^{1}H and ^{13}C nuclei in the same way.

The ASIS for many other classes of organic compounds have also been studied and plane rules for lactones, lactams, and acid anhydrides have been suggested; *cf.* reference [3] for a comprehensive selection of successful applications of ASIS.

In an investigation of the origin of the ASIS phenomenon, camphor has been studied in a variety of solvents [279]. It was found that there is an excellent linear correlation between the $\Delta\delta$(^{1}H-ASIS)-values of the 8- and 9-methyl group for more than fifty different aromatic solvents. This is in accordance with a formulation of the type

$$\Delta\delta(\text{ASIS}) = (\text{solute property}) \cdot (\text{solvent parameter}) \qquad (6\text{-}25)$$

The solute property was identified with a site factor which depends only on the geometry of the solute. The solvent parameter is simply proportional to the concentration of benzene rings in the medium [279]. Based on these results, Laszlo *et al.* proposed a solute/solvent cluster model which at present is the most accepted ASIS approach [279]. This model considers ASIS as resulting from a slight organization of aromatic solvent molecules around the dipolar site of the solute molecule due to weak intermolecular interactions between solute dipoles and solvent quadrupoles [413]. The lifetime of such transient solute/solvent complexes must be very short on the NMR time scale and the resultant NMR spectrum will be a weighted average among all solute/solvent species involved. The exact stoichiometry and geometry of these transient complexes are not known, but the flat surface of benzene molecules is likely to face the positive end of the solute dipole. Thus, benzene molecules induce upfield shifts due to their magnetic anisotropy for protons near the positive dipole end. The peripheric part of benzene molecules is likely to face the negative end of the solute dipole. Thus, downfield shifts are induced for protons near the negatively charged end of the solute dipole; *cf.* reference [413] for a picture of the most stable orientation of benzene molecules around a solute dipole. – The opposite ASIS observed with hexafluorobenzene as aromatic solvent can then easily be explained in terms of the different charge distribution in hexafluorobenzene as compared to benzene [413]. The strong H—F dipoles cause the fluorosubstituted solvent to present its negatively charged fluorine atoms to the positive end of the solute dipole. That is, hexafluorobenzene would take up the edgeway stance whereas benzene would be face on [413].

A multivariate data analysis of $\Delta\delta$(^{1}H-ASIS) of ethereal solutes, using tetramethylsilane as internal reference, has shown that ASIS can be described by a single solvent parameter model [414]. This statistically calculated single parameter can be correlated to Laszlo's solvent parameter, given in Eq. (6-25) and derived from the ASIS of camphor. These results are in favour of the time-averaged transient cluster model where ASIS is a product of a solute specific site parameter and a solvent parameter; *cf.* Eq. (6-25). This solvent parameter has been interpreted as being composed of solvent molar volumes and electronic effects caused by substituents in the aromatic solvent molecules [414, 417]. Since the intermolecular solute/solvent interactions causing ASIS are weak, solvent/solvent interactions can compete with them. Thus, it seems reasonable that solvent packing effects as expressed by molar volumes can influence the value of $\Delta\delta$(ASIS).

The solvent clustering model as well as other attempts at explaining the ASIS phenomenon (not given here) have been reviewed and criticized, but no descriptive alternative model has been given [416].

Liquid crystals provide another kind of magnetically anisotropic solvents which have been used in NMR spectroscopy [281–283]. Liquid crystals are known to form partially ordered structures; *cf.* Sections 3.1 and 5.5.9 [280]. Small anisotropic solute molecules dissolved in liquid-crystalline solvents experience partial orientation. Thus, rapid tumbling of the solute molecule about only two of the three axes is possible. This results in some averaging but still allows coupling between the magnetic dipoles of the nuclei as well as chemical shift anisotropies. If the solute molecules are not free to tumble rapidly enough for dipole/dipole averaging, as they usually do in the gas or liquid phase, rather complex NMR spectra with line broadening are observed. However, from the positions and the number of lines observed in the NMR spectra of solutes dissolved in a liquid-crystalline solvent, it is possible to determine their bond angles, relative bond lengths, and the sign of spin-spin coupling constants. For example, this restricted tumbling results in magnetic nonequivalence of the ^{1}H-chemical shifts of benzene and normal coupling constants between the *ortho*-, *meta*-, and *para*-protons can be obtained.

Finally, an interesting application of chiral solvents in the determination of the optical purity and the absolute configuration of solutes by NMR spectroscopy should be mentioned. Experimental observations indicate that the NMR spectra of enantiomeric mixtures in certain optically active solvents show small splittings of some of the peaks (*cf.* Table A-2 in the Appendix for chiral solvents). For example, Pirkle *et al.* [284] have examined the ^{1}H and ^{19}F-NMR spectra of enantiomeric 2,2,2-trifluoro-1-phenylethanol in optically active 1-(1-naphthyl)ethylamine. In the chiral solvent, the solute gives rise to distinct signals for each enantiomer. These observations are explained as the result of strong specific and non-specific interactions which produce labile diastereomeric solvates. In these solvates each isomer is sufficiently different for some of the enantiomeric nuclei to be in magnetically different environments. – Another nice example of the formation of ^{1}H-NMR spectroscopically different diastereomeric solvates are solutions of (−)-cocaine in (*R*)- and (*S*)-methyl phenylcarbinol [418].

6.5.3 Solvent Effects on Spin-Spin Coupling Constants

Variation of solvent not only affects chemical shifts in NMR spectra of dissolved compounds, but also the spin-spin coupling or splitting constants. Splitting occurs when there is isotropic coupling of nuclear spins through the bonding electrons. When the signals of two nuclei A and B are split by one another, the magnetic field experienced by A is modified *via* the effects on the bonding electrons of the $(I+1)$ nuclear spin orientations of B, and *vice versa*. Therefore, solvent-induced changes in a coupling constant must reflect changes in the electronic structure of the solute molecule. The best way to change the electronic structure of a ground state molecule is to subject the molecule to an external electric field. Some of the known intermolecular solute/solvent interaction mechanisms (*cf.* Section 2.2) are expected to produce such electric fields in the solute cavity (reaction fields). Thus, solvent effects on coupling constants have been observed frequently, but attempts to derive widespread correlations and generalizations have not been very successful. Reviews by Smith [236], Barfield and Johnston [237], and Hansen [419] give comprehensive information in solvent effects on coupling constants as well as on the underlying theories used to explain the observed results.

In one of the earliest publications devoted primarily to solvent effects on coupling constants, Evans [285] reported a 9.6 Hz increase for the one-bond coupling constant $^1J_{^{13}C-H}$ of chloroform in thirteen solvents ranging from cyclohexane ($J=208.1$ Hz) to dimethyl sulfoxide ($J=217.7$ Hz). Additional values of $^1J_{^{13}C-H}$ for chloroform in N and O containing heterocyclic solvents have been provided by Laszlo [286] ($J=215.0$ Hz in pyridine; $J=213.0$ Hz in 1,4-dioxane). The changes in $^1J_{^{13}C-H}$ observed in solvents of increasing basicity and polarity have been attributed to hydrogen bonding and reaction field effects. Hydrogen bonding might lengthen the C—H bond of chloroform which ought to result in a decrease in $^1J_{^{13}C-H}$, but *via* an electrostatic repulsion mechanism the carbon 2*s* contribution is increased resulting in an increase in $^1J_{^{13}C-H}$; the result actually observed [285, 286]. A shift of electrons away from hydrogen towards carbon produces an increase in the contribution of the carbon 2*s*-orbital to the C—H bond and a corresponding increase in $^1J_{^{13}C-H}$. Further support for the primacy of hydrogen bonding effects on $^1J_{^{13}C-H}$ have been provided by examining bromoform in thirty solvents [287]. Solvent effects on $^1J_{^{13}C-H}$ coupling constants have been calculated by the finite perturbation theory [420]. According to these calculations, solvent effects on $^1J_{^{13}C-H}$ arise mainly from electronic changes in the solute molecule caused by intermolecular solute/solvent interactions [420]. – $^1J_{^{13}C-H}$ for pyridine-1-oxide is also solvent-dependent [258]. – In summary, although the solvent effects are rather small, $^1J_{^{13}C-H}$ and most other one-bond coupling constants always increase in solvents of increasing polarity and basicity. Solvent-induced changes in ^1J of maximally 7% have been observed, and for long-range couplings even less than that [236, 419].

Geminal two-bond couplings between hydrogens, ${}^2J_{H-H}$, may change from 2 to 80%, always decreasing (in the absolute sense) in solvents of increasing polarity [236, 288]. A typical example is α-chloroacrylonitrile, the ${}^2J_{H-H}$ solvent dependence of which has been studied in various solvents [289, 290]. For fluoroform, ${}^2J_{H-F}$ coupling constant varies only by 1% in a range of solvents, decreasing algebraically with increasing van der Waals solute/solvent interactions [421]. Geminal ${}^2J_{H-H}$ coupling constants may be either positive or negative. Therefore, it must be recognized that positive geminal coupling constants which apparently decrease and negative geminal coupling constants which apparently increase are, in fact, showing exactly the same behaviour in the absolute sense: the coupling constants become more negative or decrease algebraically.

Vicinal three-bond couplings, ${}^3J_{A-B}$, pesent rather ambiguous results [236, 237]. When changes do occur, ${}^3J_{A-B}$ often increases as the polarity of the solvent increases. An example of this behaviour, ${}^3J_{H^1-H^2}$ of 3-dimethylaminoacrolein [20, 73], has been already mentioned in Table 6-2 (*cf.* discussion in Section 6.2.2). A detailed study of H—H, H—F, and F—F three-bond coupling constants of difluoroalkenes has been given [291]. The dielectric solvent effects on ${}^3J_{F-H}$ of mono-, di-, and trifluoroethanes have been calculated and satisfactorily compared with the experimental results [422]. Observed vicinal coupling constants in some ethane derivatives are also found to vary with solvent. In such cases, however, vicinal couplings may be altered as a result of conformational changes in an adjacent part of the molecule. Three-bond coupling constants are known to vary with the dihedral angle between the C—H bonds.

7 Empirical Parameters of Solvent Polarity

7.1 Linear Gibbs Energy Relationships

Solvent effects on organic reactivity (*cf.* Chap. 4 and 5) and on absorption spectra (*cf.* Chap. 6) have been studied for more than a century (*cf.* Chap. 1). Organic chemists have usually attempted to understand these solvent effects in terms of the *polarity* of the solvent. Solvent polarity is a commonly used term related to the capacity of a solvent for solvating dissolved charged or dipolar species. This concept of solvent polarity is easily grasped qualitatively, but it is difficult to define precisely and even more difficult to express quantitatively. Attempts to express it quantitatively involve mainly physical solvent properties such as dielectric constant, dipole moment, or refractive index (*cf.* Section 3.2.). From idealized theories, the solvent dielectric constant (*i.e.* the relative permittivity ε_r) is often predicted to serve as a quantitative measure of solvent polarity. However, this approach is often inadequate since these theories regard solvents as a non-structured continuum, not composed of individual solvent molecules with their own solvent/solvent interactions, and they do not take into account specific solute/solvent interactions such as hydrogen-bonding and EPD/EPA interactions which often play a dominating role in solute/solvent interactions. Similarly, solvent dipole moments are inadequate measures of solvent polarity since the charge distribution of a solvent molecule may not only be given by its dipole moment but also by its quadrupole or higher multipole moments, leading to dipolar, quadrupolar, octupolar *etc.* solvent molecules [121]. Therefore, a more general definition of the commonly used term *solvent polarity* would be useful.

The author stated in 1965 [1, 3] that the polarity of a solvent is determined by its solvation capability (or solvation power) for reactants and activated complexes as well as for molecules in the ground and excited states. This in turn depends on the action of *all* possible, specific and nonspecific, intermolecular forces between solvent and solute molecules. These intermolecular forces include Coulomb interactions between ions, directional interactions between dipoles, inductive, dispersion, hydrogen-bonding, and charge-transfer forces, as well as solvophobic interactions (*cf.* Chap. 2). Only those interactions leading to definite chemical alterations of the solute molecules through protonation, oxidation, reduction, complex formation, or other chemical processes are excluded. It is obvious that such a definition of solvent polarity cannot be measured by an individual physical quantity such as the dielectric constant. Indeed, very often it has been found that no correlation between the dielectric constant (or its different functions such as $1/\varepsilon_r$, $(\varepsilon_r-1)/(2\varepsilon_r+1)$, *etc.*) and the logarithms of rate or equilibrium constants of solvent-dependent chemical reactions exists. No single macroscopic physical parameter could possibly account for the multitude of solute/solvent interactions on the molecular-microscopic level. Until now the complexity of solute/solvent interactions has also prevented the derivation of generally applicable mathematical expressions, which allow the calculation of reaction rates or equilibrium constants of reactions carried out in solvents of different polarity.

In such a situation other indices of solvent polarity are sought. The lack of reliable theoretical expressions for calculating solvent effects and the inadequacy of defining "solvent polarity" in terms of simple physical constants has stimulated attempts to introduce empirical scales of solvent polarity, based on convenient, well-known, solvent-sensitive reference processes. A common approach is to assume that some particular reaction rate, equilibrium, or spectral absorption is a suitable model for a large class of other solvent-dependent processes. If one carefully selects an appropriate, sufficiently solvent-sensitive reference process, one can assume that this process reflects all possible solute/solvent interactions which are also present in related solvent-influenced processes. It should therefore give an empirical measure of solvent polarity – or, more precisely, an empirical measure of the solvation capability of a particular solvent for the given reference process. This reference process can be considered as a probe in the solvation shell of the standard solute – a probe that sums up a wide variety of possible intermolecular interactions such as ion-dipole, dipole-dipole, dipole-induced dipole, hydrogen-bonding, *etc.* Naturally, the most useful model processes should be those best understood on a molecular basis. Model processes used to establish empirical scales of solvent polarity have been reviewed [1–9, 122–124].

At best, this approach provides a quantitative index to solvent polarity from which absolute or relative values of rate or equilibrium constants for many reactions, as well as absorption maxima in various solvents, can be derived. Since they reflect the complete picture of all the intermolecular forces acting in solution, these empirical parameters constitute a more comprehensive measure of the polarity of a solvent than any other single physical constant. In applying these solvent polarity parameters, however, it is tacitly assumed that the contribution of intermolecular forces in the interaction between the solvent and the standard substrate, is the same as in the interaction between the solvent and the substrate of interest. This is obviously true only for closely related solvent-sensitive processes. Therefore, it is not expected that an empirical solvent scale based on a particular reference process is universal and useful for all kinds of reactions and absorptions. Any comparison of the effect of solvent on a process of interest with a solvent polarity parameter is, in fact, a comparison with a reference process.

This kind of procedure, *i.e.* empirical estimation of solvent polarity with the aid of actual chemical or physical reference processes, is very common in chemistry. The well-known Hammett equation for the calculation of substituent effects on reaction rates and chemical equilibria, was introduced by Hammett using the ionization of *meta*- or *para*-substitued benzoic acids in water at 25 °C as a reference process in 1937 in the same way [10]. Usually, the functional relationships between substituent or solvent parameters and various substituent- or solvent-dependent processes take the form of a *linear Gibbs energy relationship*, frequently still known as *linear free-energy* (LFE) *relationship* [11–15, 125–127].

Let us consider a certain reaction series, *i.e.* only small changes are involved in going from one reaction to another. These changes may be structural, such as a series of differently substituted compounds, or may involve a single reaction carried out in a series of different solvents or solvent mixtures. It has been found, that the changes in rate and in equilibrium constant occurring in one reaction series can be often related to those in another, closely related series. Thus, plotting the logarithms of rate or equilibrium constants for one reaction series against the corresponding constants for a second, related series frequently gives a straight line, which can be expressed by Eq. (7-1).

$$\lg k_i^B = m \cdot \lg k_i^A + c \tag{7-1}$$

k_i^A and k_i^B are rate or equilibrium constants of two reaction series *A* and *B* which are subject to the same changes in the structure or the surrounding medium.

Since the relationship between the equilibrium constant, *K*, for a reaction and the difference between the standard molar Gibbs energies of the products and reactants, ΔG°, is given by Eq. (7-2),

$$\lg K = -\frac{\Delta G^\circ}{\ln 10 \cdot R \cdot T} \tag{7-2}$$

and a similar expression using the standard molar Gibbs energy of activation, $\Delta G^{\neq}$, for the rate constant *k* of a reaction can be written as in Eq. (7-3),

$$\lg\left(\frac{k}{R \cdot T/N_A \cdot h}\right) = -\frac{\Delta G^{\neq}}{\ln 10 \cdot R \cdot T} \tag{7-3}$$

Eq. (7-1) essentially describes a relationship between standard molar Gibbs energies*⁾. It is often convenient to express linear Gibbs energy relationships in terms of ratios of constants by referring all members of a reaction series to a reference member of the series; thus, the correlation in Eq. (7-1) can also be expressed by Eq. (7-4).

$$\lg (k_i^B/k_0^B) = m \cdot \lg (k_i^A/k_0^A) \tag{7-4}$$

where k_0^A and k_0^B are the constants for the reference substituent or the reference solvent.

Such relationships are useful in two ways. The first application is in the study of reaction mechanisms. The correlation of data for a new reaction series by means of a linear Gibbs energy relationship establishes a similarity between the new series and the reference series. The second use of linear Gibbs energy equations is in the prediction of reaction rates or equilibrium constants dependent on substituent or solvent changes. Let us consider a reaction between a substrate and a reagent in a medium *M*, which leads, *via* an activated complex, to the products**⁾.

$$(\text{Substrate})_M + (\text{Reagent})_M \rightleftharpoons [\text{S} \cdots \text{R}]_M^{\neq} \rightarrow (\text{Products})_M \tag{7-5}$$

There are three ways of introducing small changes in order to establish a reaction series:

---

* Linear Gibbs energy relationships are manifestations of so-called extrathermodynamic relationships. Extrathermodynamic approaches are combinations of detailed models with the concepts of thermodynamics. Since it involves model building, this kind of approach lacks the rigor of thermodynamics, but it can provide information not otherwise accessible. Although linear Gibbs energy relationships are not a necessary consequence of thermodynamics, their occurrence suggests the presence of a real connection between the correlated quantities, and the nature of this connection can be explored.

** The designation of one reactant as the *substrate* and another as the *reagent* is arbitrary but useful in considering chemical reactivity. The substrate always undergoes some change in the reaction series, the reagent does not. A catalyst is always considered to be a reagent.

(a) First we can change the substrate by introducing different substituents. In the case of *meta-* and *para-*substituted benzene derivatives this leads to the well-known Hammett equation (7-6), where k_X is a rate or equilibrium constant for *meta-* or *para-*substituted

$$\lg k_X - \lg k_O = \lg (k_X/k_O) = \sigma \cdot \varrho \tag{7-6}$$

substrates, σ is the substituent constant, and ϱ is the reaction constant ($\varrho = 1$ for the standard reaction by definition) [10]. A typical Hammett correlation is shown in Fig. 7-1 for the S_N2 alkylation reaction of substituted pyridinium-*N*-phenoxide betaines with iodomethane [16]. As expected for this S_N2 reaction (negative ϱ-value), electron-releasing substituents increase the reaction rate, whereas electron-withdrawing groups decelerate.

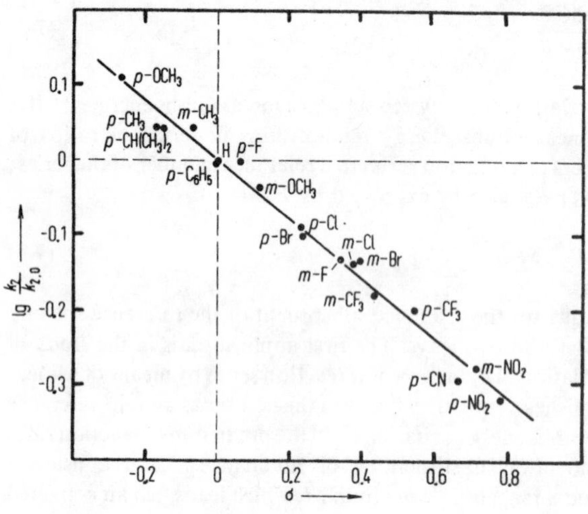

Fig. 7-1. Hammett correlation between σ-values and the logarithms of the relative rate constants of the S_N2 alkylation reaction of substituted pyridinium-*N*-phenoxide betaines with iodomethane in chloroform at 25 °C [16].

(b) Second we can change the reagent to give, for example, a Gibbs energy relationship called the Brønsted-Pedersen catalysis equation, developed as a result of studies on the base-catalyzed decomposition of nitramide [17]. This equation establishes a quantitative relationship between acid and base strengths and their effectiveness as catalysts in reactions subject to general acid or base catalysis: the stronger the acid (or base), the better it is as a catalyst. This Brønsted catalysis law, introduced in 1924, was the first linear Gibbs energy relationship.

(c) Finally, we can change the surrounding medium *M* while leaving the other reaction parameters unchanged. Provided the selected reaction is sufficiently solvent-sensitive, this gives us the desired empirical solvent parameters [1–9, 122–124]. Thermodynamically, solvation may be considered in the same general terms as the modification of the properties of the substrate molecule by substituent changes, the solvating molecules being equivalent to loosely attached substituents [18]. One important difference between substituent and solvent effects on chemical reactivity is that substituents may change the chemical reactivity of a given substrate in a discontinuous manner only, whereas solvents, especially solvent mixtures, allow a continuous modification of the substrate reactivity. Empirical parameters of solvent polarity based on solvent effects on chemical equilibria and reaction rates will be described in Sections 7.2 and 7.3.

In principle, the same considerations as in Eq. (7-5) can be made for the spectral excitation of a substrate, dissolved in a medium M, with photons $h \cdot \nu$. Although linear Gibbs energy relationships usually deal with relative reactivities only, in the form of reaction-rate and equilibrium data, this approach can be extended to various physical

$$\underset{\text{Ground state}}{(\text{Substrate})_M} + \text{h} \cdot \nu \rightarrow \underset{\text{Excited state}}{(\text{Substrate})_M^*} \tag{7-7}$$

measurements such as spectroscopic investigations of the members of a reaction series in various spectral ranges (UV/Vis [19], IR [19], NMR [20], *etc.*). Spectroscopic measurements very often can be obtained under conditions of greater precision and variety than available in reactivity measurements.

In order to establish a reaction series, the substrate in Eq. (7-7) can be altered in two ways:

(a) Variation by substituents leading to spectroscopic Hammett equations in the form presented in Eq. (7-8),

$$\frac{E_{\text{T,X}} - E_{\text{T,O}}}{\ln 10 \cdot R \cdot T} = \sigma \cdot \varrho_{\text{A}} \tag{7-8}$$

and first introduced by Kosower, Hofmann, and Wallenfels [21]. $E_{\text{T,X}}$ and $E_{\text{T,O}}$ are the transition energies in kJ/mol (or kcal/mol) of the substituted and unsubstituted substrate, respectively; σ is the Hammett substituent constant, and ϱ_{A} has been denoted as absorption constant [22, 23]. The transition energies are divided by ($\ln 10 \cdot RT$) in order to convert them into an appropriate form for Hammett substituent constants, which are commonly derived form equilibrium or rate constants. A representative example of this kind of Hammett correlation for the long-wavelength $\pi - \pi^*$ absorption of substituted pyridinium-*N*-phenoxide betaines is shown in Fig. 7-2 [23].

As expected, electron-withdrawing substituents cause a bathochromic shift of this absorption band, which exhibits a strong intramolecular charge-transfer character (*cf.* Section 6.2.1), whereas electron-releasing substituents give rise to a corresponding hypsochromic shift.

(b) Provided the position of the absorption band is sufficiently solvent-sensitive, changing the medium M, in which the substrate is to be dissolved, permits the

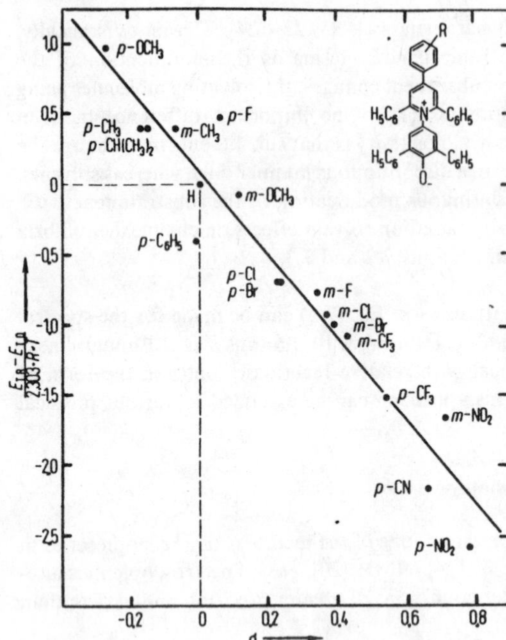

Fig. 7-2. Hammett correlation between σ-values and the modified transition energies of the long-wavelength charge-transfer $\pi \rightarrow \pi^*$ absorption of substituted pyridinium-*N*-phenoxide betaines in methanol at 25 °C [23]; *cf.* Fig. 6-2 in Chapter 6.

introduction of spectroscopic solvent polarity scales. Empirical parameters of solvent polarity based on spectroscopic measurements will be described in Section 7.4. The most comprehensive solvent polarity scale is based on solvatochromic pyridinium-*N*-phenoxide betaines, already mentioned in Section 6.2.1 and in Figs. 7-1 and 7-2.

In the early use of linear Gibbs energy relationships, simple single-term equations such as the Hammett equation were considered sufficient to fit given sets of experimental data from reaction series. Later on, more complicated multi-term equations with more than one product term were formulated in order to model the simultaneous influence of several effects on chemical reactions or optical excitations, one product term per effect [15]. The connection between such multiparameter relationships and solvent effects will be described in Section 7.7.

Simple and multiple linear Gibbs energy relationships can be generally interpreted in two distinct ways [126, 127]:

(a) Traditionally, as relations expressing combinations of fundamental chemical or physical effects, universally present in chemical reactions. Deviations from such relations have to be explained by new effects in addition to the old, known effects. If all such effects are discovered and quantified, retrospective rationalization and prediction of chemical processes (*i.e.* chemical reactions and equilibria) is fully possible and it is conceivable that at some time in the future basic research will no longer be needed. The main problems are to find reaction series which are best suited to discovering all these fundamental effects, and to establish how large a deviation can be tolerated before another fundamental effect has to be postulated. –

(b) Chemometrically*, as locally valid linearizations of very complicated unknown functional relationships using empirical models of similarity. That is, linear Gibbs energy relationships are considered as approximate models with local validity for similar reactions series only, *e.g.* one model for reactions of substituted benzene derivatives, another model for reactions of substituted aliphatic compounds. Or in the case of solvent effects, one model would apply to solvatochromic non-HBD, another to solvatochromic HBD reference dyes. In contrast to the traditional interpretation (a), this means that a given linear Gibbs energy relationship is not necessarily universally valid. In any new chemical reaction series there will be regularities in the observed data that *cannot* be predicted from the behaviour of previously investigated chemical model processes. In other words, the data observed in a chemical reaction series can be divided into two parts: one part which will coincide with other known reaction series, and another part which will be system-specific. The latter part is often not negligible and sometimes substantial. The larger the common part, the closer the similarity between a new reaction series and the reference reaction series previously investigated. The mathematics of fitting PC and FA models to a matrix of chemical data permits one to judge objectively the applicability of a given linear Gibbs energy relationship, without making assumptions about a single term model fitting part of the data matrix.

A fascinating discussion on the problem whether linear Gibbs energy relationships are (a) fundamental laws of chemistry, or (b) only local empirical rules, can be found in references [126, 127]. Attempts at discriminating between these two kinds of interpretation of linear Gibbs energy relationships based on an evaluation of the practicality of models derived from either view are described. It seems that from a more rigorous mathematical point of view, interpretation (b), stressed by Wold and Sjöström [126], appears to be the correct one, whereas from a more practical, descriptive point of view, interpretation (a), stressed by Taft and Kamlet [127], will be the one preferred by experimental chemists. The final answer is ambiguous.

7.2 Empirical Parameters of Solvent Polarity from Equilibrium Measurements

The first attempt to introduce an empirical relationship between an equilibrium constant and solvent polarity was made in 1914 already by K. H. Meyer [24]. Studying the solvent-dependent keto-enol tautomerism of β-dicarbonyl compounds he found a proportionality between the equilibrium constants of various tautomeric compounds in the same set of

---

* Chemometrics stands in this context for analysis of multivariate chemical data by means of statistical methods such as principal component analysis (PCA) or factor analysis (FA); *cf.* Section 3.5.

different solvents (*cf.* Table 4-2 in Section 4.3.1). He therefore split the tautomeric equilibrium constant K_T into two independent factors according to Eq. (7-9).

$$K_T = \frac{[\text{enol}]}{[\text{diketo}]} \equiv L \cdot E \tag{7-9}$$

E is the so-called enol constant and measures the enolization capability of the keto form ($E=1$ for ethyl acetoacetate by definition). Thus, the so-called *desmotropic constant* L is a measure of the enolization power of the solvent. By definition, the values of L are equal to the equilibrium constants of ethyl acetoacetate ($E=1$), determined in different solvents [24]. This desmotropic constant seems to be the first empirical solvent parameter. It describes the relative solvation power of a solvent for diketo and enol forms of β-dicarbonyl compounds. It was measured for a few solvents only and was soon forgotten.

In contrast to this early empirical solvent scale, one of the latest, introduced by Eliel and Hofer in 1973 [25] and based on the solvent-dependent conformational equilibrium of 2-isopropyl-5-methoxy-1,3-dioxane, should be mentioned (*cf.* Table 4-9 in Section 4.4.3). In general, polar solvents shift this conformational equilibrium towards the more dipolar *axial cis* isomer. The authors propose calling the standard molar Gibbs energy changes associated with this equilibrium, $-\Delta G^{\circ}_{OCH_3}$, the D_1-scale (D for dioxane, 1 for the number of carbons in the alkoxy group). They also recommend using this solvent scale, known for seventeen solvents, in the estimation of other solvent-dependent equilibria and reaction rates [25].

Another approach to a new solvent scale was introduced by Gutmann in 1966 [26, 27]. Based on the fact that many chemical reactions are to be influenced primarily by coordination interactions between an electron-pair acceptor (EPA) substrate and electron-pair donor (EPD) solvents (*cf.* Sections 2.2.6 and 2.6), he provided an empirical measure of the Lewis basicity of solvents using the so-called *donor number DN* (or *donicity*). Antimony pentachloride was chosen as reference compound in order to obtain a measure of the nucleophilic properties of a solvent since the antimony becomes readily coordinated on the acceptance of an electron-pair from an EPD solvent. According to Eq. (7-10), the donor number is defined as the negative value of the molar enthalpy for the

$$\text{D:} + \text{SbCl}_5 \underset{\text{in Cl-CH}_2\text{CH}_2\text{-Cl}}{\overset{\text{room temp.}}{\rightleftharpoons}} \overset{\oplus}{\text{D}}-\overset{\ominus}{\text{Sb}}\text{Cl}_5$$

$$\text{Solvent Donor Number } DN \equiv -\Delta H_{\text{D-SbCl}_5}/(\text{kcal}\cdot\text{mol}^{-1}) \tag{7-10}$$

adduct formation between antimony pentachloride and EPD solvents, D, measured calorimetrically in highly diluted solutions of 1,2-dichloroethane as inert solvent at room temperature [26]. Corresponding enthalpy measurements in the reference solvent tetrachloromethane have led to analogous results [28].

Donor numbers vary from 2.7 for nitromethane which is a weak EPD solvent, to 38.8 for hexamethylphosphoric triamide, a strong EPD solvent; *cf.* Table 2-3 in Section 2.2.6. The donor number was measured directly, *i.e.* calorimetrically, for *ca.* fifty solvents [26–28, 128]. It was also estimated by other means such as ^{23}Na- [29, 129], ^{27}Al- [130], and

^{1}H-NMR [131] spectroscopy. Visual estimates of the varying donor numbers of EPD solvents can easily be made using the colour reactions with copper(II), nickel(II), or vanadyl(IV) complexes as acceptor solutes instead of $SbCl_5$ [132]. A selection of donor numbers has already been given in Table 2-3 in Section 2.2.6. A critical compilation and discussion of donor numbers has been given by Marcus [133].

Since donor numbers are defined in the non-SI unit kcal · mol^{-1}, Marcus has recommended the use of dimensionless normalized donor numbers DN^N, which are defined as $DN^N = DN/(38.8 \text{ kcal} \cdot \text{mol}^{-1})$ [133]. The non-donor solvent 1,2-dichloroethane ($DN = DN^N = 0.0$) and the strong donor solvent hexamethylphosphoric triamide (HMPT: $DN = 38.8$ kcal · mol^{-1}; $DN^N = 1.0$) have been used to fix the DN^N-scale. Although solvents with higher donicity than HMPT are known, it is expedient to choose this EPD solvent with the highest directly, *i.e.* calorimetrically, determined *DN*-value as the second reference solvent [133]*). The normalized DN^N-values are also included in Table 2-3 in Section 2.2.6.

Since the donor numbers were measured in dilute solution in inert 1,2-dichloroethane**), they reflect the donicity of the *isolated* EPD solvent molecules. However, in neat, associated EPD solvents an increase in the donicity occurs [134]. For such highly-structured neat EPD solvents (*e.g.* water, alcohols, amines) the term *bulk donicity* has been introduced [135] in order to rationalize the deviations of these solvents in plots of $^{23}Na^{\oplus}$-NMR chemical shifts [136] and ESR-parameter [137] *vs.* the donor number. Unfortunately, great discrepancies exist between the DN_{bulk}-values given in the literature when estimated by different methods. For this reason they are not included in Table 2-3 in Section 2.2.6; *cf.* reference [133] for a collection and discussion of bulk donicities, DN_{bulk}.

Donor numbers are considered semiquantitative measures of solute/EPD-solvent interactions, particularly useful in the prediction of other EPD/EPA interactions in coordination chemistry. Numerous examples of the application of donor numbers have been given by Gutmann [26, 27, 30]; *cf.* also [113, 133]. It has been shown that donor number correlations are parallel with correlations based on the highest occupied molecular orbital (HOMO) eigenvalues of EPD solvent molecules [139]. For non-HBD solvents a fair correlation has been obtained between their donor numbers and their gas-phase proton affinities *PA*, indicating that the *DN*-values do indeed reflect the intrinsic molecular properties of EPD solvents [140].

The donor number approach has been criticized for both conceptual [141] and experimental reasons [28, 133, 138, 265]. Therefore, the search for other empirical Lewis basicity parameters has continued.

---

* Although a donor number of 38.8 kcal · mol^{-1} for HMPT was given by Gutmann [26, 27], a much higher *DN*-value of 50.3 kcal · mol^{-1} has been recently measured by Bollinger *et al.* [128]. This shows the serious problems which arise when measuring the Lewis basicity of this EPD solvent towards $SbCl_5$.

** 1,2-Dichloroethane is obviously not a chemically inert solvent under all circumstances. For example, the EPD solvent triethylamine is rapidly quaternized by 1,2-dichloroethane under the catalytic action of $SbCl_5$, leading to an overevaluated donor number for triethylamine [128]. Even solutions of HMPT and $SbCl_5$ in 1,2-dichloroethane contain non-negligible amounts of charged species, the formation of which contributes to the measured enthalpy [138].

Another remarkable Lewis basicity scale for 75 non-HBD solvents has been established recently by Gal and Maria [138, 142]. This involved very precise calorimetric measurements of the standard molar enthalpies of 1:1 adduct formation of EPD solvents with gaseous boron trifluoride, $\Delta H^{\circ}_{D-BF_3}$, in dilute dichloromethane solutions at 25 °C, according to Eq. (7-11).

$$\text{D:} + BF_3 \underset{\text{in } CH_2Cl_2}{\overset{25\,°C}{\rightleftharpoons}} \overset{\oplus}{D}-\overset{\ominus}{B}F_3 \tag{7-11}$$

A selection of $\Delta H^{\circ}_{D-BF_3}$-values has already been given in Table 2-4 in Section 2.2.6. This new Lewis basicity scale is more comprehensive and seems to be more reliable than the donor number scale. Analogously, a Lewis basicity scale for 88 carbonyl compounds (esters, carbonates, aldehydes, ketones, amides, ureas, carbamates) has been derived from their standard molar enthalpies of complexation with gaseous boron trifluoride in dichloromethane solution [143]. The corresponding $\Delta H^{\circ}_{CO-BF_3}$-values range from 33 kJ·mol^{-1} for di-*tert*-butyl ketone to 135 kJ·mol^{-1} for 3-diethylamino-5,5-dimethylcyclohexen-2-one.

A comparison of various Lewis basicity scales has recently been given by Persson [144].

Equilibrium and rate constants for the reaction of donor solvents with the coordinately unsaturated pentacarbonyl chromium, produced by flash photolysis of hexacarbonyl chromium, have been determined in various solvents. They have been used to obtain empirical solvent parameters for organometallic reactions such as the catalytic oligomerization of butadiene [31].

Another, more general approach for the estimation of EPD/EPA interactions between Lewis acids and bases, not restricted to the solvent as a reaction partner, was given by Drago [32]; *cf.* Eq. (2-12) and Table 2-6 in Section 2.2.6.

The solvent-dependent tautomerization of a pyridoxal 5′-phosphate Schiff's base has been shown to be another appropriate model process for the measurement of solvent polarities [32a]. This model process seems to be particularly useful for the determination of the polarities of sites of proteins at which pyridoxal 5′-phosphate is bound [32a].

Finally, an empirical hydrophobicity parameter derived from measurements of the distribution of a solute between two immiscible liquids should be mentioned; *cf.* Section 2.2.7 dealing with hydrophobic interactions. The hydrophobic or lipophilic character of organic compounds plays an important role in their ability to interfere with biochemical systems. Therefore, systematic efforts have been made to obtain numerically defined constants to assess the hydrophobic character of organic compounds. A hydrophobicity parameter which has proven quite valuable in the fields of toxicology, pharmacology, and environmental science is the ***Hansch-Leo 1-octanol/water partition coefficient*** $K_{o/w}$ or $P_{o/w}$ as defined in Eq. (7-12),

$$K_{o/w} = \frac{c_i(\text{1-octanol})}{c_i(\text{water})} \equiv P_{o/w} \tag{7-12}$$

where c_i(1-octanol) and c_i(water) are the molar equilibrium concentrations of the solute i in the two immiscible phases 1-octanol and water, respectively [145–148]. The 1-octanol/water partition coefficients are often used as hydrophobicity parameters in the Gibbs energy-based form of lg $P_{o/w}$. They are known for a vast amount of organic compounds, particularly for those compounds normally used as organic solvents. A lg $P_{o/w}$-value of 3.90 for *n*-hexane means that this hydrophobic solvent is preferably found in the 1-octanol phase, whereas a lg $P_{o/w}$-value of −1.35 for dimethyi sulfoxide reveals the hydrophilicity of this particular solvent. A compilation of lg $P_{o/w}$-values for 102 organic solvents can be found in reference [149] which deals also with correlations of lg $P_{o/w}$ with some solvatochromic solvent parameters. 1-Octanol/water partition coefficients have been served as the basis for many papers in a field of biochemistry known as *q*uantitative *s*tructure/*a*ctivity *r*elationship (QSAR), inspired by Hansch's original observation that distribution of a solute between these two solvents provide an accurate measure of lipophilic/hydrophilic interactions. For example, lg $P_{o/w}$ is a good measure of the ease with which drugs penetrate membranes and bind to hydrophobic surfaces. The interrelation between the narcotic potencies of various substances and the partition coefficient has long been known. Many applications of this hydrophobicity parameter in the framework of linear Gibbs energy relationships have been reported [145–148].

Using 1-octanol/water partition coefficients for aromatic solutes, Hansch also defined a hydrophobicity parameter π_X for organic substituents X according to Eq. (7-12a) [146–148].

$$\pi_X = \lg P_{o/w}^{C_6H_5X} - \lg P_{o/w}^{C_6H_6} \tag{7-12a}$$

$P_{o/w}^{C_6H_5X}$ and $P_{o/w}^{C_6H_6}$ are the partition coefficients between 1-octanol and water for C_6H_5X and C_6H_6, respectively. A positive π_X such as +0.56 for $X = CH_3$ means that these substituents favour the organic phase relative to the hydrogen atom (X=H). If substituents X have negative values of π_X such as −0.67 for OH, then they cause partitioning into water. The π_X substituent parameters have also been widely used in quantitative structure/activity relationships, particularly in pharmacology.

Another *m*icroscopic *h*ydrophobicity substituent parameter *MH* was evaluated by Menger *et al.* [150]. Addition of ammonium salts $R\text{-}NMe_3^{\oplus}\ X^{\ominus}$ (R=alkyl, aryl; $X^{\ominus}$ = halide ions) to 10^{-5} M 4-nitrophenyl laurate in water destroys or disrupts the ester aggregates formed in aqueous solution. The corresponding shift of the aggregation equilibrium deshields the laurate ester group, resulting in enhanced rate constants for the basic hydrolysis of 4-nitrophenyl laurate. The more hydrophobic the R group of the ammonium salt, the greater its disaggregation power, and the greater the rate enhancement [150].

Recently, a quantitative measure of the solvent solvophobic effect*) has been introduced by Abraham *et al.* [282]. It has been shown that the standard molar Gibbs energies of transfer of nonpolar, hydrophobic solutes X (X = argon, alkanes, and alkane-like compounds) from water (W) to another solvents (S) can be linearly correlated through

* The solvophobic or hydrophobic effect is simply regarded as the experimental observation of the relative insolubility in water (or other highly structured liquids) of certain organic solutes, as compared to their solubility in nonaqueous solvents. The so-defined solvophobic (hydrophobic) *effect* should be distinguished from the solvophobic (hydrophobic) *interaction* between two or more solute molecules in solution; *cf.* Section 2.2.7. The solvophobic (hydrophobic) *effect* is only related to solute/solvent interactions [282].

a set of equations such as Eq. (7-12b),

$$\Delta G_t^\circ(\mathrm{X, W \rightarrow S}) = M \cdot R_\mathrm{T} + D \qquad (7\text{-}12b)$$

where R_T is a solute parameter, and M and D characterize the solvent. M-values are referred to water, *i.e.* $M=0$ for water by definition. With another fixed point, *i.e.* the M-value of the most hydrophobic solvent *n*-hexadecane ($M = -4.2024$), a scale of solvophobic power Sp has been defined according to Eq. (7-12c),

$$Sp = 1 - \frac{M\,(\text{solvent})}{M\,(n\text{-hexadecane})} = 1 + \frac{M\,(\text{solvent})}{4.2024} \qquad (7\text{-}12c)$$

where the Sp-values of water and *n*-hexadecane are arbitrarily taken as unity and zero. The Sp-values provide a simple quantitative measure of the solvophobic power of solvents, relative to the two reference solvents water ($Sp=1$) and *n*-hexadecane ($Sp=0$) at 298 K. For pure solvents, the following order of decreasing solvophobic power has been obtained [282]: water > formamide > 1,2-ethanediol > methanol > ethanol > 1-propanol > 2-propanol > 1-butanol ≫ *n*-hexadecane. First linear correlations between Sp-values and rate constants of Diels-Alder reactions between cyclopentadiene and diethyl fumarate in hydrophobic media have been reported by Schneider *et al.* [283]; *cf.* also Section 5.5.8.

Finally, a parameter describing the *softness* of solvents should be mentioned. In terms of Pearson's principle of hard and soft acids and bases the hardness of an ion or molecule is understood as resistance to a change or deformation in the electronic cloud (*cf.* Section 3.3.2). This property, or its complement, the softness of solvents is expected to play a role in the solvation of hard and soft solutes. Marcus defined a μ-scale of solvent softness (from *malakos* = soft in Greek) according to Eq. (7-12d) as the difference between the mean of the standard molar Gibbs energies of transfer of sodium and potassium ions from water (W) to a given solvent (S) and the corresponding transfer energy for silver ions, divided by 100 [285].

$$\mu = \left\{ \frac{1}{2}\,[\Delta G_t^\circ(\mathrm{Na}^\oplus, \mathrm{W \rightarrow S}) + \Delta G_t^\circ(\mathrm{K}^\oplus, \mathrm{W \rightarrow S})] - \Delta G_t^\circ(\mathrm{Ag}^\oplus, \mathrm{W \rightarrow S}) \right\} \cdot \frac{1}{100} \qquad (7\text{-}12d)$$

Since water is a hard solvent, the Gibbs energy of transfer of ions from water as a reference solvent to other solvents should depend on the softness of these solvents in a different manner for hard and soft ions. For ions of equal charge and size, hard ions should prefer water and soft ions the softer solvents. The definition of μ by Eq. (7-12d) has been given because the size of the soft $\mathrm{Ag}^\oplus$ ion is intermediate between those of hard $\mathrm{Na}^\oplus$ and $\mathrm{K}^\oplus$. μ-Values for 34 organic solvents have been determined; *e.g.* $\mu = -0.12$ for 2,2,2-trifluoroethanol, 0.00 for water (by definition), 0.64 for pyridine, and 1.35 for *N,N*-dimethylformamide. The degree of softness among solvents with oxygen, nitrogen, and sulfur donor atoms increases in the series O-donor (alcohols, ketones, amides) < N-donor (nitriles, pyridines, amines) < S-donor solvents (thioethers, thioamides). Applications of this softness scale of solvents have been given [285]. For other recently introduced solvent scales ranking the donor strength of EPD solvents towards soft acceptors (*e.g.* $HgBr_2$) and hard acceptors (*e.g.* $\mathrm{Na}^\oplus$) see reference [287].

7.3 Empirical Parameters of Solvent Polarity from Kinetic Measurements

Since reaction rates can be strongly affected by solvent polarity (*cf.* Chap. 5) the introduction of solvent scales using suitable solvent-sensitive chemical reactions was obvious [33, 34]. One of the most ambitious attempts to correlate reaction rates with empirical parameters of solvent polarity is that of Winstein and his coworkers [35–37]. They found that the S_N1 solvolysis of 2-chloro-2-methylpropane (*tert*-butyl chloride) is strongly accelerated by polar, especially protic solvents; *cf.* Eq. (5-13) in Section 5.3.1. Grunwald and Winstein [35] defined a solvent "ionizing power" parameter Y using Eq. (7-13),

$$Y = \lg k_A^{t-BuCl} - \lg k_0^{t-BuCl} \tag{7-13}$$

where k_0^{t-BuCl} is the first-order rate constant for the solvolysis of *tert*-butyl chloride at 25 °C in aqueous ethanol (80 cl/l ethanol and 20 cl/l water; $Y=0$) as reference solvent, and k_A^{t-BuCl} is the corresponding rate constant in another solvent. This reaction was selected as the model process because it was believed to occur by an essentially pure S_N1 mechanism, with ionization of the C—Cl bond as the rate-determining step. Choosing a standard reaction and a reference solvent, a linear Gibbs energy relation is then written in the familiar form of Eq. (7-14):

$$\lg k_A - \lg k_0 = \lg (k_A/k_0) = m \cdot Y \tag{7-14}$$

where m is a substrate parameter measuring the substrate sensitivity to changes in the "ionizing power" of the medium, and Y is a parameter characteristic of the given solvent. Scales of Y and m were established by taking $Y=0$ for 80 : 20 cl/l ethanol/water and $m=1$ for the solvolysis of *tert*-butyl chloride. It is expected that Eq. (7-14) can be applied to reactions very similar to the standard reaction, that is, S_N1 substitutions. The similarity between Y and m of Eq. (7-14), and σ and ϱ of the Hammett equation (7-6) is obvious. Y-values are known for some pure, mainly protic solvents and for various binary mixtures of organic solvents with water or a second organic solvent [35, 36]. Typical Y-values are indicated in Table 7-1. It should be noted that the Y-value of the standard solvent lies about midway between the extremes. Y-values of binary mixtures are not related to solvent composition in a simple manner.

The rate constant for the solvolysis of *tert*-butyl chloride at 120 °C was also chosen as a solvent-dependent model reaction by Koppel and Pal'm [38] in order to calculate a solvent polarity scale which is restricted to nonspecific interactions only. In order to confirm the limiting S_N1 mechanism for the solvolysis of *tert*-butyl chloride, Schleyer and his coworkers [39] have compared the solvolysis rates of *tert*-butyl chloride and 1-bromoadamantane in a large series of solvents. They reasoned that any rate-determining elimination or nucleophilic solvent assistence in *tert*-butyl chloride solvolysis will result in a failure to correlate with 1-bromoadamantane. In 1-bromoadamantane backside nucleophilic solvent attack and elimination are both impossible. An excellent correlation between solvolysis data for *tert*-butyl chloride and 1-bromoadamantane has been found,

Table 7-1. Empirical parameters based on kinetic measurements for solvents of decreasing polarity.

Solvents	Y [35, 36]	lg k_1 [a]	X [51]	$\mathscr{S}$ [b]	Ω [52]
Water	3.493	−1.180	–	–	0.869[k]
Formic Acid	2.054	−0.929	–	–	–
2,2,2-Trifluoroethanol	1.045[c,d]	–	–	–	–
Formamide	0.604	–	–	–	0.825[k]
Methanol/Water (80 cl/l + 20 cl/l)	0.381	–	–	–	–
Ethanol/Water (80 cl/l + 20 cl/l)	0.000	−2.505	–	–	–
Tetra-*n*-hexylammonium benzoate	−0.39[e]	–	–	–	–
Acetone/Water (80 cl/l + 20 cl/l)	−0.673	–	–	–	–
1,4-Dioxane/Water (80 cl/l + 20 cl/l)	−0.833	–	–	–	–
Methanol	−1.090	−2.796	0.91	−1.89	0.845
Acetic acid	−1.675	−2.772	0.00	–	0.823
Ethanol	−2.033	−3.204	–	−2.02	0.718
2-Propanol	−2.73	–	–	–	–
tert-Butanol	−3.26	–	–	–	–
Dimethyl sulfoxide	–	−3.738	1.6	–	–
Nitromethane	–	−3.921	–	0.041	0.680
Acetonitrile	–	−4.221	0.04	−0.328	0.692
N,N-Dimethylformamide	−3.5[f]	−4.298	0.8	−0.222	0.620
Acetic anhydride	–	−4.467	–	–	–
1,2-Dichloroethane	–	–	–	−0.420	0.600
Dichloromethane	–	–	–	−0.553	–
Pyridine	–	−4.670	–	–	0.595
Acetone	–	−5.067	–	−0.824	0.619
Chloroform	–	–	–	−0.886	–
Chlorobenzene	–	–	−1.9	−1.15	–
1,2-Dimethoxyethane	–	–	–	–	0.543
1,4-Dioxane	–	–	–	−1.43	–
Ethyl acetate	–	−5.947	–	−1.66	–
Tetrahydrofuran	–	−6.073	–	−1.54	–
Benzene	–	–	–	−1.74	0.497[g]
Tetrachloromethane	–	–	−4.8	−2.85	–
Diethyl ether	–	−7.3	–	−2.92	0.466[i]
Cyclohexane	–	–	–	−4.15	0.595[g]
Decalin	–	–	–	–	0.537[h]
Triethylamine	–	–	–	–	0.445
n-Hexane	–	–	–	*ca.* −5	–

[a] Ionization of 4-methoxyneophyl tosylate at 75 °C [37].
[b] S_N2 reaction of tris(*n*-propyl)amine with iodomethane at 20 °C [49, 50].
[c] V. J. Shiner, W. Dowd, R. D. Fisher, S. R. Hartshorn, M. A. Kessick, L. Milakofsky, and M. W. Rapp, J. Am. Chem. Soc. *91*, 4838 (1969).
[d] D. E. Sunko, I. Szele, and M. Tomič, Tetrahedron Lett. *1972*, 1827, 3617.
[e] C. G. Swain, A. Ohno, D. K. Roe, R. Brown, and T. Maugh, J. Am. Chem. Soc. *89*, 2648 (1967).
[f] S. D. Ross and M. M. Labes, J. Am. Chem. Soc. *79*, 4155 (1957).
[g] W. M. Jones and J. M. Walbrick, J. Org. Chem. *34*, 2217 (1969).
[h] J. F. King and R. G. Pews, Can. J. Chem. *43*, 847 (1965).
[i] R. Braun and J. Sauer, Chem. Ber. *119*, 1269 (1986); value at 20 °C.
[k] A. Lattes, I. Rico, A. de Savignac, and A. A. Samii, Tetrahedron *43*, 1725 (1987).

indicating that *tert*-butyl chloride seems to solvolyze *via* a limiting S_N1 mechanism, free from nucleophilic solvent participation and from rate-determining elimination. Similar good correlations have been observed between the solvolysis rates of *tert*-butyl chloride and 1-adamantyl tosylate [40] as well as 2-adamantyl tosylate [41]. The absence of any S_N2 character in the solvolysis of *tert*-butyl heptafluorobutyrates in aqueous ethanol and similar solvents has recently been confirmed [151]. In most cases, the solvolysis rate data of *tert*-butyl substrates are best explained by an S_N1 process with electrophilic solvent assistance of the leaving group in protic solvents forming strong hydrogen bonds. That is, the ionization rates of *tert*-butyl halides should be mainly dependent on solvent polarity and solvent electrophilicity, but not on solvent nucleophilicity [152]; *cf.* Eq. (5-135) in Section 5.5.7. However, according to Bentley *et al.* [153], Kevill *et al.* [155], and Bunton *et al.* [156] it is now thought that *tert*-butyl chloride reacts solvolytically with weak but significant nucleophilic solvent assistance. Only 1- and 2-adamantyl substrates seem to react by an S_N1(limiting) mechanism due to their cage structure which absolutely precludes nucleophilic attack by solvents from the rear. This has been concluded from the observation that the relative reactivities of *tert*-butyl chloride and 1-chloroadamantane are different in nucleophilic solvents and in weakly nucleophilic solvents such as trifluoroacetic acid, hexafluoroisopropanol, or trifluoroethanol where *tert*-butyl chloride is abnormally unreactive [153]. See also reference [157] for a nice picture for the solvolysis of 1-bromoadamantane as compared to *tert*-butyl bromide with and without backside solvent assistance.

The Grunwald-Winstein equation (7-14) is fairly successful in a large number of cases. Good linear relationships between $\lg k_1$ and Y are shown for the solvolysis of various tertiary haloalkanes and secondary alkyl sulfonates, *i.e.* reactions which proceed *via* an S_N1 mechanism such as found for the standard reaction. For reactions involving borderline mechanisms (*e.g.* solvolysis of secondary haloalkanes) and for S_N2 reactions (*e.g.* solvolysis of primary haloalkanes) the application of Eq. (7-14) is less satisfactory. When solvolysis rates for different binary mixtures of solvents are correlated using Eq. (7-14), the well-known phenomenon of dispersion is often observed, that is, lines or curves of slightly different slopes are observed for each solvent system [35, 36]. In other words, the substrate parameter m is solvent dependent in these cases. If Eq. (7-14) were strictly obeyed, all the points should lie on a single straight line. These observations indicate that the reaction rate does depend not only on the ionizing power of the solvent (for which Y is a measure) but also on the nucleophilicity of the solvent. Obviously, nucleophilic solvent assistance may be relevant in reactions with increasing S_N2 characteristics. Therefore, the trends in m-values suggest their usefulness in determining the extent of solvent nucleophilic participation. Those reactions which are S_N1, exhibit m-values near 1.00, while values for S_N2 substrates range from 0.25 to 0.35. Values of m between these extremes are typical for secondary substrates lying in the borderline area between limiting S_N1 and pure S_N2 mechanisms. The term "S_N2(intermediate) mechanism" has been suggested for reactions where there is evidence of nucleophilic solvent assistance and evidence for a nucleophilically solvated ion pair reaction intermediate; *cf.* Eq. (5-136) in Section 5.5.7. Thus, a gradual change in mechanism going from S_N2(one-stage) through S_N2(intermediate) to S_N1(limiting) mechanism has been postulated [41, 153]; *cf.* Section 5.5.7 for a discussion of this solvent-influenced S_N2/S_N1 mechanistic spectrum.

To account for nucleophilically solvent assisted processes, Winstein *et al.* [42] later provided a four-parameter equation of the type shown in Eq. (7-15)*),

$$\lg (k_A/k_O) = m \cdot Y + l \cdot N \tag{7-15}$$

where m and l are substrate parameters, Y is the solvent ionizing power, N represents the solvent nucleophilicity, and l measures the substrate response to changes in solvent nucleophilicity. Instead of this extended Winstein equation (7-15), the simple Winstein equation (7-14) can only be used when the $l \cdot N$ term makes either a negligible or constant contribution. The extended Winstein equation (7-15) was explicitly evaluated by both Peterson *et al.* [43] and Schleyer *et al.* [44], using two reference substrates, one with a high sensitivity to solvent nucleophilicity (*e.g.* bromomethane [43] or methyl tosylate [44]; $l = 1.00$), and the other with a low sensitivity to solvent nucleophilicity (*e.g. tert*-butyl chloride [43] or 2-adamantyl tosylate [44]; $l = 0.00$ and $m = 1.00$)**). Eq. (7-15) correlates solvolysis data for such widely varying substrates as methyl and 2-adamantyl tosylates [44].

Another variation of Eq. (7-15), introduced by Bentley *et al.* [44, 154], is the three-parameter Eq. (7-16) which contains the adjustable parameter Q:

$$\lg (k_A/k_O) = (1 - Q) \cdot \lg (k_A/k_O)_{CH_3OTs} + Q \cdot \lg (k_A/k_O)_{2-AdOTs} \tag{7-16}$$

---

* The original equation was expressed as the partial differential equation [42]:

$$d \lg k = \left(\frac{\partial \lg k}{\partial Y}\right)_N dY + \left(\frac{\partial \lg k}{\partial N}\right)_Y dN$$

If the partial derivatives are constant and equal to m and l, respectively, this expression leads to Eq. (7-15).

** For example, solvent nucleophilicity constants N for Eq. (7-15) have been calculated by rearrangement of Eq. (7-15) and solving for N, and using the solvolysis of methyl tosylate to provide appropriate data, according to

$$N = \lg (k_A/k_O)_{CH_3OTs} - 0.3\ Y_{OTs}$$

$\lg (k_A/k_O)_{CH_3OTs}$ refers to solvolytic data for methyl tosylate (at 50 °C) in any solvent (k_A) relative to ethanol/water 80 cl/l (k_O). The sensitivity m of methyl tosylate solvolysis to the ionizing power Y is obtained from a solvent series in which Y varies but not N: since acetic acid and formic acid seem to be equally nucleophilic, the corresponding m-value of 0.3 for methyl tosylate provides a reasonably good estimate of m for substitution into Eq. (7-15). Eventually, l has been defined to 1.00 (l = 1.00) in view of the extremely high sensitivity of methyl tosylate to changes in solvent nucleophilicity. Values of Y in this equation are based on 2-adamantyl tosylate instead of *tert*-butyl chloride solvolysis in order to retain the same leaving group (OTs). N-values derived in this way are also designated as N_{OTs}-values [158]; they range from −5.56 (CF_3CO_2H) to 0.2 (i-C_3H_7OH) [44, 158].

Another scale of solvent nucleophilicity N for Eq. (7-15) has been based on the solvolysis of triethyloxonium hexafluorophosphate which is a positively charged substrate minimizing the electrophilicity term Y of Eq. (7-15) [160]. This N-scale was developed using the equation

$$N = \lg (k_A/k_O)_{Et_3O^{\oplus}} - 0.55\ Y^+$$

where Y^+ is the value for $\lg (k_A/k_O)_{t-Bu\overset{\oplus}{S}Me_2}$ at 50 °C [160].

Methyl and 2-adamantyl tosylates model the mechanistic extremes, S_N2(one-stage) and S_N1(limiting), within the range of compounds examined. This equation allows one to determine to what extent a substrate's reactivity resembles methyl tosylate (S_N2: $Q=0$) or 2-adamantyl tosylate (S_N1: $Q=1$).

For a completely limiting case, with no nucleophilic solvent assistance and no anchimeric assistance, Eq. (7-16) reduces to Eq. (7-17),

$$\lg (k_A/k_O) = Q \cdot \lg (k_A/k_O)_{2-\text{AdOTs}} = m \cdot Y_{OTs} \qquad (7\text{-}17)$$

which has the same form as the original Winstein Eq. (7-14) where Q and m are equivalent. However, Eq. (7-17) uses a new scale of solvent ionizing power, Y_{OTs}, which is based on 2-adamantyl tosylate, with $m=1$ for solvolysis rates in any solvent (k_A) relative to solvolysis in 80:20 cl/l ethanol/water (k_O) at 25 °C [44, 154, 158]. Although it is a secondary substrate, 2-adamantyl tosylate solvolyses through a S_N1(limiting) or k_c mechanism*) *via* the rate-determining formation of contact ion pairs without detectable solvent assistance or ion-pair return [154]. The Y_{OTs}-scale of solvent ionizing power spans a rate range of over eight orders of magnitude, from $Y_{OTs}=4.57$ for trifluoroacetic acid to $Y_{OTs}=-3.74$ *tert*-butanol [158], and can even be extended to aqueous sulfuric acid as the ionizing medium [159]. The Y_{OTs}-scale has proven very useful for the correlation of other solvolysis rates [153, 154]. For example, Y_{OTs}-values correlate linearly with the logarithms of solvolysis rate constants for 1-adamantyl tosylate as well as with the spectroscopically derived solvent polarity parameter $E_T(30)$ [158]; *cf.* Section 7.4. Y_{OTs}-values correlate *non*linearly with Winstein's original Y-values [158].

Recent very precise experimental studies clearly show that Winstein's original Y-values as defined by Eq. (7-14) do not apply to leaving groups other than the chloride ion [153, 158–166], most probably due to the variable amount of specific solvation of different leaving groups by solvent molecules. If allowance is made for different leaving groups X, the solvolytic rate data of substrates RX can be correlated with data for a reference compound according to the general Eq. (7-18).

$$\lg (k_A/k_O)_{RX} = m \cdot Y_X \qquad (7\text{-}18)$$

Y_X-values for different leaving groups X have recently been defined using $m=1.00$ for the solvolyses of 1-adamantyl substrates (X=Cl [153], Br [153], I [161], picrate [162], and $^{\oplus}SMe_2$ [163]) and 2-adamantyl substrates (X=OTs [44, 158], $OClO_3$ [162, 166], OSO_2CF_3=OTf [164, 166], and picrate [166]). Additional Y_{OTs}-values have been obtained from solvolyses of 1-adamantyl tosylate (X=OTs [158, 162]). Further Y_{OTf}-values for trifluoromethane sulfonates (triflates) were found from the solvolyses of 7-norbornyl triflate [165]. This range of Y_X-values ensures that specific leaving group solvation effects

* According to Winstein, real solvolyses are either nucleophilically solvent assisted (designated k_s and including both substitution and elimination processes) or anchimerically assisted (designated k_Δ), with k_c representing the hypothetical limit which is reached when nucleophilic solvent assistance and anchimeric assistance approach zero. *Anchimeric* assistance (from the Greek *anchi+meros*, neighbouring parts) means neighbouring group participation during the ionization step; *cf.* Eq. (7-20) for an example.

do not interfere with mechanistic deductions from the solvolyses rate data. – Sufficient data is now available to evaluate trends in the various Y_X-scales. A comparison of the Y_X-scales and a discussion of the correlations between them can be found in reference [166]. For example, the relative rates for the S_N1 solvolyses of different bridgehead substrates *having the same leaving group* are found to be almost solvent-independent [167].

Quite recently, an analysis of Winstein's *Y*- and *N*-values and their modifications in terms of Kamlet and Taft's linear solvation energy relationship (*cf.* Section 7.7) has been given by Abraham *et al.* [288].

A four-parameter relationship (7-19) similar to Eq. (7-15) was proposed by Swain *et al.* [45].

$$\lg (k_A/k_0) = c_1 \cdot d_1 + c_2 \cdot d_2 \tag{7-19}$$

d_1 and d_2 are measures of solvent nucleophilicity and electrophilicity, respectively, and c_1 and c_2 measure the substrate's sensitivity to these solvent properties. The rate constant k_0 refers to the reaction in the standard solvent, 80:20 cl/l ethanol/water. This approach is essentially statistical since in contrast to Eq. (7-15) four rather than two parameters (*m* and *l*) are varied. Numerous values of d_1, d_2, c_1, and c_2 were calculated, using scales based on certain standard systems, in such a way to fit a large number of experimental rate constants [45]. Although a satisfactory correlation was achieved for a wide range of solvents and substrates, this approach has been criticized for a lack of connection with chemical reality [42, 46]. Thus, the substrate factors c_1 and c_2 are of little mechanistic significance, *e.g.* *tert*-butyl chloride has a higher c_1 value than bromomethane, suggesting that bromomethane (S_N2 substrate) should be less sensitive to solvent nucleophilicity than *tert*-butyl chloride (S_N1 substrate).

An alternative model system, potentially superior to *tert*-butyl chloride in the evaluation of solvent ionizing power, is based on the specific rate of anchimeric, β-aryl assisted solvolysis or decomposition of 4-methoxyneophyl tosylate at 75 °C according to Eq. (7-20) [37]. Use of this compound avoids some practical disadvantages of *tert*-butyl

(7-20)

chloride: *tert*-butyl choride being rather volatile is troublesome to weigh accurately; it reacts extremely rapidly in highly polar solvents, and its solvolysis rate constant usually can not be determined in nonhydroxylic solvents. Therefore, Winstein *et al.* [37] put forward the values of $\lg k_1$ for the ionization of 4-methoxyneophyl tosylate as a reference reaction suitable for measuring solvent polarity even in fairly nonpolar and non-HBD solvents (*cf.* Table 7-1)*). The $\log k_1$ (or $\lg k_{ion}$) scale is equivalent to the *Y* scale for protic

* In his appreciation of Winstein's scientific work Bartlett [47] proposed the term *W* for "Winstein parameter" for these values commonly referred to as "$\lg k_1$" or "$\lg k_{ion}$".

solvents and has been extended to non-HBD solvents, for which *Y* values are not available. – Although a primary substrate, 4-methoxyneophyl tosylate is believed to solvolyze *via* concerted anchimeric β-aryl participation without nucleophilic solvent interference and internal ion-pair return [37]. A more sophisticated picture of solvent effects on β-arylalkyl tosylate solvolyses has been given by Schleyer *et al.* He demonstrated that primary and secondary β-arylalkyl substrates solvolyze through competition between discrete β-aryl assisted (k_Δ) and β-aryl nonassisted reaction pathways (k_s), each of which leads to distinct sets of products [168]. This is due to the fact that the neighbouring aryl group and the solvent molecules are in competition to displace the leaving group. By decreasing the solvent nucleophilicity as well as introducing electron-releasing substituents in the aryl group, the kinetic influence of the β-aryl group becomes more noticeable. Therefore, the $\lg k_1$ (or $\lg k_{ion}$) values as measures of the solvent ionizing power are subject to the same limitations as found for the *Y* values; *cf.* the preceding discussion of Eq. (7-14).

In addition to the application of S_N1 reactions as model reactions for the evaluation of solvent polarity, Drougard and Decroocq [48] suggested that the value of $\lg k_2$ for the S_N2 Menschutkin reaction of tri-*n*-propylamine and iodomethane at 20 °C – termed "$\mathscr{S}$" according to Eq. (7-21) – should also be used as a general measure of solvent polarity.

$$\mathscr{S} = \lg k_2 \; [(CH_3CH_2CH_2)_3N + CH_3I] \tag{7-21}$$

The second-order rate constants of this quaternization reaction have been determined for seventy-eight solvents by Lassau and Jungers [49, 50]. A selection from this relatively extensive solvent scale is given in Table 7-1.

Analogous to the *Y* scale and based on a nucleophilic substitution reaction, Gielen and Nasielski [51] suggested that a solvent polarity scale could be based on electrophilic aliphatic substitution reactions such as the reaction of bromine with tetramethyltin shown in Eq. (7-22).

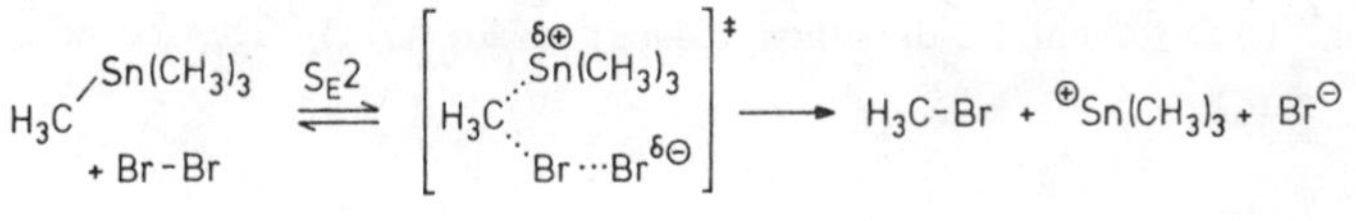

$$H_3C{-}Sn(CH_3)_3 + Br{-}Br \;\underset{}{\overset{S_E2}{\rightleftharpoons}}\; \left[H_3C\cdots\overset{\delta\oplus}{Sn}(CH_3)_3 \;\; \cdots Br\cdots Br^{\delta\ominus} \right]^{\ddagger} \longrightarrow H_3C{-}Br + {}^{\oplus}Sn(CH_3)_3 + Br^{\ominus} \tag{7-22}$$

This model reaction probably passes through a non-cyclic dipolar activated complex (S_E2 mechanism) in polar solvents, while in less polar solvents a cyclic activated complex gains importance (S_F2 mechanism) [51]. The linear Gibbs energy relationship given by these authors is as follows:

$$\lg (k_A/k_0) = p \cdot X \tag{7-23}$$

k_A and k_0 are the rate constants of reaction (7-22) in a given solvent A and in glacial acetic acid, respectively. The latter is used as the reference solvent ($X=0$ by definition). *X* is a parameter characteristic of the given solvent, and *p* is a parameter characteristic of the given reaction; $p=1$ for Eq. (7-22) by definition. *X*-values are known for only seven solvents to date (*cf.* Table 7-1). Both solvent scales, *X* and *Y*, do not quite agree, since the reaction rate is influenced by both the electrophilic and nucleophilic character of the solvent to a different degree. The electrophilic substitution reaction (7-22) is probably supported by a nucleophilic pull of the solvent on the leaving group ($-SnMe_3$). The coupling constants $^2J_{^{117}Sn-^1H}$ obtained from ^{1}H-NMR spectra of $(CH_3)_3SnBr$ in various solvents, which can be taken as a measure of the nucleophilic character of the solvent with respect to tin, show a correlation with the *X*-values, in agreement with this assumption [51].

The fact that the rate of some Diels-Alder [4+2] cycloaddition reactions is affected, although only slightly, by the solvent was used by Berson *et al.* [52] in establishing an empirical polarity parameter called Ω. These authors found that, in the Diels-Alder addition of cyclopentadiene to methyl acrylate, the ratio of the *endo*-product to the *exo*-product depends on the reaction solvent. The *endo*-addition is favored with increasing solvent polarity; *cf.* Eq. (5-43) in Section 5.3.3. Later on Pritzkow *et al.* [53] found, that not only the *endo*/*exo* product ratio but also the absolute rate of the Diels-Alder addition of cyclopentadiene to acrylic acid derivatives increases slightly with increasing solvent polarity. The reasons for this behaviour have been already discussed in Section 5.3.3. Since reaction (5-43) is kinetically controlled, the product ratio [*endo*]/[*exo*] equals the ratio of the specific rate constants, and Berson *et al.* [52] define

$$\lg\,(k_{endo}/k_{exo}) = \lg\,[endo]/[exo] \equiv \Omega \tag{7-24}$$

The values of Ω lie between 0.445 (triethylamine) and 0.869 (water), and are known for fourteen solvents (*cf.* Table 7-1). Owing to the low solubility of the reactants in polar media, an extension of this scale is limited.

Related to Diels-Alder [2+2]cycloadditions are 1,3-dipolar cycloadditions which are known to be far less solvent-dependent; *cf.* Eq. (5-44) in Section 5.3.3. Nagai *et al.* [169] found that the 1,3-dipolar cycloaddition reaction of diazodiphenylmethane to tetracyanoethene (TCNE) is an exception; it is 180 times faster in nonbasic chloroform than in the EPD solvent 1,2-dimethoxyethane; *cf.* Eq. (7-25). The second-order rate

$R_2C{=}N^+{=}N^-$ + $(NC)_2C{=}C(CN)_2$ (EPD-solvents) $\underset{30\,°C}{\xrightarrow{k_2}}$ $[\ldots]^{\ddagger}$ $\rightarrow$ pyrazoline $\xrightarrow{-N_2}$ tetracyanocyclopropane; $R = C_6H_5$ (7-25)

constant decreases with increasing solvent basicity. This solvent effect can be interpreted in terms of the solute/solvent interaction between the soft Lewis-acidic π-acceptor TCNE and basic EPD solvents. Thus, one of the educts is stabilized through specific solvation resulting in a corresponding rate deceleration. Using reaction (7-25) as the reference process, an empirical parameter D_π of solvent Lewis basicity for interactions with soft π-acceptors has been proposed according to Eq. (7-26):

$$\lg\,(k_0/k_A) = D_\pi \tag{7-26}$$

k_0 and k_A are the rate constants of reaction (7-25) at 30 °C in benzene (reference solvent with $D_\pi = 0$) and other solvents respectively [169]. D_π-values are known for 34 solvents [169]. They are relatively large for aromatic solvents (*i.e.* soft EPD solvents) compared to other empirical Lewis basicity parameters which have been determined by employing rather hard Lewis-acid probes; *e.g.* $SbCl_5$-derived DN values [*cf.* Eq. (7-10)], CH_3OD-derived *B* values [*cf.* Eq. (7-34)], and 4-nitrophenol-derived β values [*cf.* Eq. (7-51)]. The D_π-values have been applied successfully to reactions of diazodiphenylmethane with various soft π-acceptors such as other cyano-substituted alkenes as well as quinones [170].

7.4 Empirical Parameters of Solvent Polarity from Spectroscopic Measurements

Spectroscopic parameters of solvent polarity have been derived from solvent-sensitive standard compounds absorbing light in spectral ranges corresponding to UV/Vis-, IR-, ESR-, and NMR-spectra (*cf.* Chap. 6) [1–9]. The first suggestion that solvatochromic dyes should be used as indicators of solvent polarity was made by Brooker *et al.* [54] in 1951, but Kosower [5, 55] in 1958 was the first to set up a comprehensive solvent scale.

Kosower [5, 55] has taken the longest-wavelength intermolecular charge-transfer (CT) transition of 1-ethyl-4-methoxycarbonylpyridinium iodide as a model process. It exhibits a marked negative solvatochromism; *cf.* the formula of this dye and its UV/Vis absorption spectra in Fig. 6-3 in Section 6.2.1. A solvent change from pyridine to methanol causes a hypsochromic shift of the longest-wavelength CT band of 105 nm. This is due to stabilization of the electronic ground state, which is an ion pair, relative to the first excited state, which is a radical pair, with increasing solvent polarity. The general conditions for the appearance of solvatochromism have already been discussed in Section 6.2.2. Kosower defined his polarity parameter, *Z*, as the molar transition energy, E_T, expressed in kcal/mol, for the CT absorption band of 1-ethyl-4-methoxycarbonylpyridinium iodide in the appropriate solvent according to Eq. (7-27).

$$E_T/(\text{kcal}\cdot\text{mol}^{-1}) = h\cdot c\cdot\tilde{\nu}\cdot N_A = 2.859\cdot 10^{-3}\cdot\tilde{\nu}/\text{cm}^{-1} \equiv Z \qquad (7\text{-}27)$$

h is Planck's constant, *c* is the velocity of light, $\tilde{\nu}$ is the wavenumber of the photon which produces the electronic excitation, and N_A is Avogadro's number. A *Z*-value of 83.6 for methanol means that a transition energy of 83.6 kcal is necessary to bring one mole of the standard dye, dissolved in methanol, from electronic ground state to first excited state*). The stronger the stabilizing effect of the solvent on the ground-state ion-pair as compared

* For conversion according to the International System of Units the following equations can be used:

$$1\ \text{kcal}\cdot\text{mol}^{-1} = 4.184\ \text{kJ}\cdot\text{mol}^{-1};$$

$$E_T/(\text{kJ}\cdot\text{mol}^{-1})\ \text{or}\ Z/(\text{kJ}\cdot\text{mol}^{-1}) = 1.196\cdot 10^{-2}\cdot\tilde{\nu}/\text{cm}^{-1}.$$

The choice of $\text{kcal}\cdot\text{mol}^{-1}$ as unit for the *Z* scale is related to the usefulness of this quantity for comparison to chemical reactions and their activation energies. Since this and other parameters have been widely used in the literature, conversion of the units into $\text{kJ}\cdot\text{mol}^{-1}$ has been avoided in order to avert confusion.

with the less dipolar radical pair in the excited state, the higher this transition energy and, thus, the *Z*-value. A high *Z*-value corresponds to high solvent polarity. The basis for the use of *Z* as a measure of solvent polarity has already been given in Section 6.2.1; *cf.* also [5, 55, 171, 172]. An alternative explanation for the negative solvatochromism of substituted pyridinium iodides has been given [171].

Z-values cover the range 94.6 (water) to about 60 kcal/mol (*i*-octane) and have originally been measured for 21 pure solvents and 35 binary solvent mixtures [5, 56], as well as some electrolyte [57] and surfactant solutions [58]. Various authors have since gradually extended this to include 45 pure solvents. *Z*-values for a further 41 pure solvents have been determined by Griffiths and Pugh [172], who also compiled all available *Z*-values and their relationships with other solvent polarity scales. A selection of *Z*-values together with some other spectroscopic solvent polarity parameters are given in Table 7-2.

There are some serious limitations in the determination of *Z*-values. First, *Z*-values can be obtained by direct mesurement only over the solvent range chloroform ($Z = 63.2$ kcal/mol) to 70 : 30 cl/l ethanol/water ($Z = 86.4$ kcal/mol). In highly polar solvents, the long-wavelength charge-transfer band moves to such short wavelengths that it cannot be observed underneath the much stronger $\pi \rightarrow \pi^*$ absorption band of the pyridinium ion. Therefore, the *Z*-value for water was obtained by extrapolating the *Z*-values measured for acetone/water, ethanol/water, and methanol/water mixtures to zero organic component in a plot against Winstein's *Y*-values. Because the lines were extrapolated a considerable distance, the original *Z*-value for water (94.6 kcal/mol) has been reexamined and a lower *Z*-value (91.8 kcal/mol) is advocated by Griffiths and Pugh [172]. Secondly, the standard pyridinium iodide is not soluble in many nonpolar solvents. By using the more soluble 4-*tert*-butoxycarbonyl-1-ethylpyridinium iodide and pyridine-1-oxide as secondary standards it was possible to calculate the *Z*-values of nonpolar solvents [5, 55].

Z-values are both temperature- and pressure-dependent. The CT absorption band of substituted pyridinium halides is shifted hypsochromically as the temperature of the solution is decreased [59]. Thus, *Z*-values decreases with increasing temperature due to a lowering of the solute/solvent interactions at the higher temperature. Furthermore, it has been shown that the CT absorption band of 1-ethyl-4-methoxycarbonylpyridinium iodide is shifted bathochromically for solutions in methanol and ethanol with increasing pressure (up to 1920 bar), while for other solvents such as acetone and *N,N*-dimethylformamide it is shifted hypsochromically [60]. Except for the lower alcohols, the bulk solvent polarity generally increases with pressure [60].

Z-values have now been widely used empirically to correlate other solvent-sensitive processes with solvent polarity, *e.g.* the $n \rightarrow \sigma^*$ absorption of haloalkanes [61], the $n \rightarrow \pi^*$ and $\pi \rightarrow \pi^*$ absorption of 4-methyl-3-penten-2-one [62], the $\pi \rightarrow \pi^*$ absorption of phenol blue [62], the CT absorption of tropylium iodide [63], as well as many kinetic data (Menschutkin reactions, Finkelstein reactions, *etc.* [62]). Copolymerized pyridinium iodides, embedded in the polymer chain, have also been used as solvatochromic reporter molecules for the determination of microenvironment polarities in synthetic polymers [173]. No correlation was observed between *Z*-values and the dielectric constant. Measurement of solvent polarities using empirical parameters such as *Z*-values has already found favour in textbooks for practical courses in physical organic chemistry [64].

Brownstein suggested Eq. (7-28) analogous to the Hammett equation (7-6), for the general description of solvent effects [65].

Table 7-2. Empirical parameters based on spectroscopic measurements for solvents of decreasing polarity (25 °C; 1013 mbar).

Solvents	Z/(kcal · mol^{-1}) [5, 55, 172]	S [65]	χ_R/(kcal · mol^{-1}) [77]	χ_B/(kcal · mol^{-1}) [77]	Φ [79]	G [85]	$A(^{14}N)$/mT [89]
Water	94.6	0.1540	–	68.9	0.545	–	1.7175
2,2,2-Trifluoroethanol	–	–	–	–	0.475	–	–
2,2,3,3-Tetrafluoro-1-propanol	86.3	–	–	–	–	–	–
1,2-Ethanediol	85.1	0.0679	40.4	–	0.295	–	1.6364
Ethanol/Water 80:20 cl/l	84.8	0.0650	–	–	–	–	–
2-Aminoethanol	84.4[a)]	–	–	–	–	–	–
Methanol	83.6	0.0499	43.1	63.0	0.285	–	1.6210
Formamide	83.3	0.0463	–	–	0.245	–	–
Ethanol	79.6	0.0000	43.9	60.4	0.255	–	1.6030
Acetic acid	79.2	0.0050	–	–	0.37	–	1.6420
2-Octanol	78.6[a)]	–	–	–	–	–	–
1-Propanol	78.3	−0.0158	44.1	–	–	–	–
Diethyleneglycol monomethylether	78.1[a)]	–	–	–	–	–	–
N-Methylacetamide	77.9[b)]	–	–	–	–	–	–
1-Butanol	77.7	−0.0240	44.5	56.8	–	–	1.6018
Tetrahydrothiophene-1,1-dioxide	77.5[b)]	–	–	–	–	–	–
2-Propanol	76.3	−0.0413	44.5	56.1	0.19	–	1.5973
Propylene carbonate	72.4[a)]	–	–	–	–	–	–
2-Methyl-2-propanol, *tert*-Butanol	71.3	−0.1047	–	–	0.130	–	1.5860
Acetonitrile	71.3	−0.1039	45.7	53.7	0.135	93	1.5666
Nitromethane	71.2[c)]	−0.134	44.0	–	–	99	1.5759
Dimethyl sulfoxide	71.1	–	42.0	–	0.115	–	1.5692
N,N-Dimethylformamide	68.4	−0.1416	43.7	51.5	0.135	–	1.5635
N,N-Dimethylacetamide	66.9[b)]	–	43.0	–	–	–	–
Tri-*n*-hexyl-*n*-heptylammonium iodide (mp. 95.7 °C; in molten form)	66.4[d)]	–	–	–	–	–	–
Acetone	65.5	−0.1748	45.7	50.1	–	–	1.5527
Nitrobenzene	–	−0.218	42.6	–	–	100[c)]	–
Diethyleneglycol diethylether	65.2[a)]	–	–	–	–	–	–
Benzonitrile	65.0[c)]	–	43.3	–	–	–	–
Dichloromethane	64.7	−0.1890	44.9	47.5	–	100	1.5752
1,1,2,2-Tetrachloroethane	64.3[f)]	−0.083	–	–	–	–	–

Table 7-2. (Continued)

Solvents	$Z/(kcal \cdot mol^{-1})$ [5, 55, 172]	S [65]	$\chi_R/(kcal \cdot mol^{-1})$ [77]	$\chi_B/(kcal \cdot mol^{-1})$ [77]	Φ [79]	G [85]	$A(^{14}N)/mT$ [89]
Pyridine	64.0	−0.1970	43.9	50.0	—	94	1.5608
cis-1,2-Dichloroethene	63.9[e]	—	—	—	—	—	—
1,2-Dichloroethane	63.4[f]	−0.151	—	—	—	95	1.5655
Chloroform	63.2	−0.2000	44.2	—	0.095	106	1.5863
Hexamethylphosphoric triamide	62.8[g]	—	—	—	0.115	—	—
1,1-Dichloroethane	62.1[f]	—	—	—	—	—	—
Triethyleneglycol dimethylether	61.3[c]	—	—	—	—	90[c]	—
Fluorobenzene	60.2[e]	—	—	—	—	—	—
Di-*n*-butylether	60.1[a]	−0.286	48.6	—	—	61	—
o-Dichlorobenzene	60.0[e]	—	—	—	—	—	—
Ethyl acetate	59.4[f]	−0.210	47.2	—	—	—	—
Bromobenzene	59.2[e]	−0.164	44.6	—	—	—	1.5479
1,2-Dimethoxyethane	59.1	—	—	—	—	—	1.5424
Ethoxybenzene, Phenetole	58.9[c]	—	—	—	—	89[c]	—
Tetrahydrofuran	58.8[c]	—	46.6	—	0.10	—	1.5373
Chlorobenzene	58.0[e]	−0.182	45.2	—	—	—	1.5472
2-Methyltetrahydrofuran	55.3	—	—	—	—	—	—
1,4-Dioxane	—	−0.179	48.4	—	0.115	86	1.5452
Diethyl ether	—	−0.277	48.3	—	0.08	64	1.5334
Benzene	54	−0.215	46.9	—	0.02	80	1.5404
Toluene	—	−0.237	47.2	41.7	—	74	1.5347
Carbon disulfide	—	−0.240	—	—	—	74	1.5289
Tetrachloromethane	—	−0.245	48.7	—	−0.01	69	1.5331
Triethylamine	—	−0.285	49.3	—	—	62	—
n-Hexane	—	−0.337	50.9	—	0.07	44	1.5134
Cyclohexane	—	−0.324	50.0	—	0.035	49	—
Gas phase	—	−0.556	—	—	—	0	—

[a] W. N. White and E. F. Wolfarth, J. Org. Chem. *35*, 2196 (1970).
[b] R. S. Drago, D. M. Hart, and R. L. Carlson, J. Am. Chem. Soc. *87*, 1900 (1965); R. S. Drago and K. F. Purcell, in T. C. Waddington (ed.): *Non-Aqueous Solvent Systems*. Academic Press, London, New York 1965, p. 211 ff.
[c] J. F. King and R. G. Pews, Can. J. Chem. *43*, 847 (1965).
[d] J. E. Gordon, J. Am. Chem. Soc. *87*, 4347 (1965).
[e] C. Walling and P. J. Wagner, J. Am. Chem. Soc. *86*, 3368 (1964).
[f] P. H. Emslie and R. Foster, Rec. Trav. Chim. Pays-Bas *84*, 255 (1965).
[g] J. E. Dubois and A. Bienvenüe, Tetrahedron Lett. *1966*, 1809.

$$\lg k_A - \lg k_0 = \lg (k_A/k_0) = S \cdot R \tag{7-28}$$

k_A is the rate constant, equilibrium constant, or the function of a spectral shift for a reaction or absorption in various solvents, and k_0 is the corresponding quantity for dry ethanol as reference solvent ($S = 0.00$ for dry ethanol by definition). S is characteristic of the solvent and R gives the susceptibility of the given property towards a change of solvent. As a standard process, Brownstein [65] chose the CT absorption of Kosower's dye [55], and assigned to it an R-value of 1.00. Having chosen a standard solvent and a standard reaction, it was then possible to calculate R and S-values for other reactions and solvents, respectively. From Kosower's work, 58 S-values were used to determine R-values for 9 reactions. In a continuation of this process, 158 S-values and 78 R-values were deduced, including R-values for solvent-dependent UV/Vis, IR, and NMR absorption, rates of reactions, and positions of equilibria [65]. The S-values represent statistical averages of a variety of different solvent polarity parameters, including Z and Y, and therefore, cannot be related to a specific model process. In principle, this is an interesting attempt at generalization but many of the correlations used to calculate R and S-values are rather poor. It would seem that too many different solvent-dependent processes are being mixed-up and treated in an oversimplified way.

The practical limitations in the Z-value approach can be overcome by using pyridinium-N-phenoxide betaine dyes such as *(44)* as the standard probe molecule. They exhibit a negatively solvatochromic $\pi \rightarrow \pi^*$ absorption band with intramolecular charge-transfer character; *cf.* discussion of this dye in Section 6.2.1, its UV/Vis spectrum in Fig. 6-2, and its dipole moment in electronic ground and excited state mentioned in Table 6-1, dye no. 12.

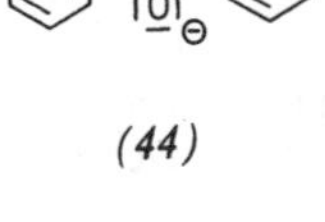

(44) 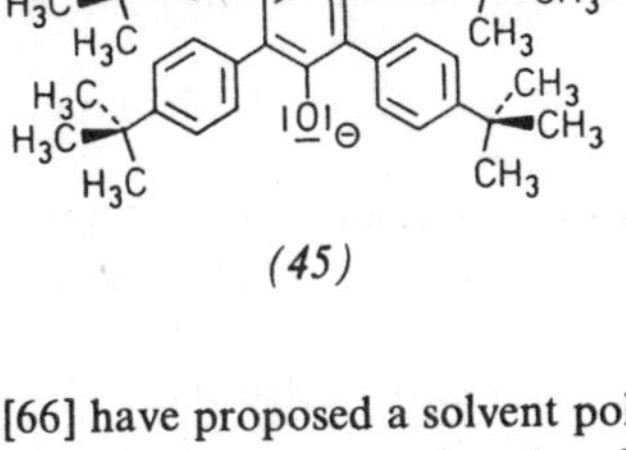 *(45)*

Dimroth and Reichardt [66] have proposed a solvent polarity parameter, $E_T(30)$, based on the transition energy for the longest-wavelength solvatochromic absorption band of the pyridinium-N-phenoxide betaine dye *(44)* (dye no. 30 in reference [66]). According to Eq. (7-27), the $E_T(30)$-value for a solvent is simply defined as the transition energy of the dissolved betaine dye *(44)* measured in kcal/mol [2, 66–68] (for conversion into SI units see footnote on page 359). The major advantage of this approach is that the solvatochromic absorption band is at longer wavelengths for *(44)* than for Kosower's dye, generating an extraordinarily large range for the solvatochromic behaviour: from $\lambda = 810$ nm, $E_T(30) = 35.3$ kcal/mol, for diphenyl ether, to $\lambda = 453$ nm, $E_T(30) = 63.1$ kcal/mol, for water. Since the greater part of this solvatochromic range lies within the visible region of the spectrum, it is even possible to make a visual estimation of solvent polarity. For example, the solution colour of *(44)* is red in methanol, violet in ethanol, green in acetone, blue in isoamyl alcohol, and greenish-yellow in anisole [66]. A remarkable feature of these solution colour changes is that nearly every colour of the visible spectrum can be obtained by applying suitable binary mixtures of solvents of different polarity. To date, the betaine dye *(44)* holds the world record in solvatochromism with a direct experimentally observed hypsochromic shift of more than 350 nm (ΔE_T *ca.* 28 kcal/mol = 117 kJ/mol) in case of a solvent change from diphenyl ether to water. Owing to this exceptionally large displacement of the solvatochromic absorption band, the $E_T(30)$-values provide an excellent and very sensitive characterization of the polarity of solvents, high $E_T(30)$-values corresponding to high solvent polarity. $E_T(30)$-values have been determined for more than 270 pure organic solvents [66, 174–176] and for a great number of binary solvent mixtures [68–72, 124, 177–192]. A collection of $E_T(30)$-values for pure organic solvents, representing the most comprehensive empirical solvent polarity scale so far known, is given in Table 7-3*⁾. Some discrepancies between the $E_T(30)$-values given in Table 7-3 and those published in earlier papers [66, 174] can be explained by improved techniques for the purification of organic solvents and obtaining UV/Vis spectra of highly diluted betaine solutions [175, 176]. $E_T(30)$-values are also known for various binary solvent mixtures: alcohols/water [68, 124, 177–181], ethers/water [68, 70, 72, 180, 182], dipolar non-HBD solvents/water [68, 70, 178, 180, 181, 183, 184], alcohols/alcohols [180, 184–186], dipolar non-HBD solvents/alcohols [183–187], and numerous other solvent/solvent combinations [68–72, 124, 177–190]. $E_T(30)$-values of binary solvent mixtures with limited mutual miscibility have also been investigated [191]. The polarity of binary solvent mixtures has been reviewed by Langhals [192].

The primary standard betaine dye *(44)* is only sparingly soluble in water and less polar solvents; it is insoluble in nonpolar solvents such as aliphatic hydrocarbons. In order to overcome the solubility problems in nonpolar solvents, the more lipophilic penta-*tert*-butylsubstituted betaine dye *(45)* has additionally been used as a secondary reference probe [174]. The excellent linear correlation between the E_T-values of both dyes allows the calculation of $E_T(30)$-values for solvents in which the solvatochromic indicator dye *(44)* is not soluble.

Unfortunately, $E_T(30)$-values have by definition the dimension of kcal/mol, a unit which should be abandoned in the framework of SI units. Therefore, the use of so-called normalised E_T^N-values has been recommended [174]. They are defined according to Eq. (7-29), using water and tetramethylsilane (TMS) as extreme reference solvents.

$$E_T^N = \frac{E_T(\text{solvent}) - E_T(\text{TMS})}{E_T(\text{water}) - E_T(\text{TMS})} = \frac{E_T(\text{solvent}) - 30.7}{32.4} \qquad (7\text{-}29)$$

* A compilation of the 100 most important organic solvents together with their physical constants and ordered by decreasing E_T^N-values can be found in the Appendix (Table A-1).

Table 7-3. Empirical parameters of solvent polarity $E_T(30)$ [*cf.* Eq. (7-27)] and normalized E_T^N-values [*cf.* Eq. (7-29)], derived from the long-wavelength UV/Vis absorption band of the negative solvatochromic pyridinium-*N*-phenoxide betaine dyes *(44)* and *(45)*, for 271 organic solvents, measured at 25 °C and 1 bar [66, 174–176].

Solvents	$E_T(30)/(\text{kcal} \cdot \text{mol}^{-1})$[a]	E_T^N
Tetramethylsilane, TMS	(30.7)[b,e,f]	*0.000*
Alkanes and Cycloalkanes		
3-Methylbutane	(30.9)[b,f]	(0.006)[b]
n-Pentane	(31.0)[b,f]	(0.009)[b]
n-Hexane	(31.0)[b,f]	(0.009)[b]
n-Heptane	(31.1)[b,f]	(0.012)[b]
n-Octane	(31.1)[b,f]	(0.012)[b]
n-Nonane	(31.0)[b,f]	(0.009)[b]
n-Decane	(31.0)[b,f]	(0.009)[b]
n-Dodecane	(31.1)[b,f]	(0.012)[b]
Cyclohexane	(30.9)[b,f]	(0.006)[b]
cis-Decahydronaphthalene, *cis*-Decalin	(31.2)[b,f]	(0.015)[b]
Haloalkanes and Haloalkenes		
Dichloromethane	40.7[f]	0.309
Dibromomethane	39.4[h]	0.269
Diiodomethane	36.5[h]	0.179
Chloroform	39.1[d,e,f]	0.259
Deuteriochloroform	39.0[d,e]	0.256
Bromoform	37.7[e]	0.216
Tetrachloromethane	32.4[f]	0.052
Bromoethane	37.6[d,e]	0.213
1,1-Dichloroethane	39.4[d,e]	0.269
1,2-Dichloroethane	41.3[j]	0.327
1,2-Dibromoethane	38.3[i]	0.235
1,1,1-Trichloroethane	36.2[d,e]	0.170
1,1,2,2-Tetrachloroethane	39.4[j]	0.269
1-Chloropropane	37.4[d,e]	0.207
1,3-Dichloropropane	40.2[j]	0.293
1,2-Dibromopropane	39.1[i]	0.259
Chlorocyclohexane	36.9[j]	0.191
1,1-Dichloroethene	37.0[e]	0.194
Trichloroethene	35.9[d,e]	0.160
Alkylarenes and Haloarenes		
Benzene	34.3[f,g]	0.111
Methylbenzene, Toluene	33.9[f,g]	0.099
(Trifluoromethyl)benzene	38.5[g]	0.241
1,4-Dimethylbenzene, *p*-Xylene	33.1[f]	0.074
1,3,5-Trimethylbenzene, Mesitylene	32.9[g]	0.068
1,2,3,4-Tetramethylbenzene, Prehnitene	33.0[g]	0.071
Ethinylbenzene, Phenylacetylene	37.2[f]	0.201
Cyclohexylbenzene	34.2[e]	0.108
1,2,3,4-Tetrahydronaphthalene, Tetralin	33.7[j]	0.093
Fluorobenzene	37.0[g]	0.194
1,2-Difluorobenzene	39.3[g]	0.265
1,3-Difluorobenzene	37.3[g]	0.204
1,4-Difluorobenzene	36.4[g]	0.176
1,3,5-Trifluorobenzene	33.2[g]	0.077

Table 7-3. (Continued)

Solvents	$E_T(30)/(kcal \cdot mol^{-1})$[a]	E_T^N
Pentafluorobenzene	38.4[g]	0.238
Hexafluorobenzene	34.2[g]	0.108
Chlorobenzene	36.8[g]	0.188
1,2-Dichlorobenzene	38.0[g]	0.225
1,3-Dichlorobenzene	36.7[g]	0.185
1,2,4-Trichlorobenzene	36.2[g]	0.170
Bromobenzene	36.6[g]	0.182
1,2-Dibromobenzene	37.6[g]	0.213
1,3-Dibromobenzene	36.5[g]	0.179
2,5-Dibromo-1-methylbenzene	34.7[g]	0.123
Iodobenzene	36.2[g]	0.170
1-Methylnaphthalene	35.3[g]	0.142
1-Chloronaphthalene	37.0[e]	0.194
1-Iodonaphthalene	36.9[g]	0.191
Aliphatic and Cycloaliphatic Alcohols		
Methanol	55.4[f]	0.762
Benzyl alcohol	50.4[k]	0.608
Ethanol	51.9[d,e]	0.654
Ethanol/water (80:20 cl/l)	53.7[d,e]	0.710
1-Phenylethanol	46.7[k]	0.494
2-Phenylethanol	49.5[k]	0.580
2-Cyanoethanol	59.6[l]	0.892
2-Chloroethanol	55.5[l]	0.765
2,2,2-Trichloroethanol	54.1[j]	0.722
2,2,2-Trifluoroethanol	59.8[f]	0.898
1-Propanol	50.7[d,e]	0.617
2,2,3,3-Tetrafluoro-1-propanol	59.4[d,e]	0.886
3-Phenyl-1-propanol	48.5[k]	0.549
2-Propanol	48.4[f]	0.546
1,1,1,3,3,3-Hexafluoro-2-propanol	65.3[f]	1.068
1-Butanol	50.2[d,e]	0.602
2-Butanol	47.1[d,e]	0.506
2-Methyl-1-propanol, Isobutyl alcohol	48.6[k]	0.552
2-Methyl-2-propanol, *tert*-Butyl alcohol	43.3 (30 °C)[f]	0.389
1-Pentanol	49.1[e,k]	0.568
2-Pentanol	46.5[d,e]	0.488
3-Pentanol	45.7[d,e,l]	0.463
3-Methyl-1-butanol, Isoamyl alcohol	49.0[e]	0.565
2-Methyl-2-butanol, *tert*-Pentyl alcohol	41.1[l]	0.321
Cyclopentanol	47.0[k]	0.503
1-Hexanol	48.8[e]	0.559
Cyclohexanol	46.9[e]	0.500
1-Heptanol	48.5[k]	0.549
3-Ethyl-3-pentanol	38.5[l]	0.241
2,4-Dimethyl-3-pentanol	40.1[l]	0.290
1-Octanol	48.3[e]	0.543
3-Ethyl-2,4-dimethyl-3-pentanol	37.9[l]	0.222
1-Decanol	47.7[k]	0.525
1-Dodecanol	46.7[k]	0.494
2-Propene-1-ol, Allyl alcohol	52.1[e]	0.660

Table 7-3. (Continued)

Solvents	$E_T(30)/(\text{kcal} \cdot \text{mol}^{-1})$[a]	E_T^N
2-Propyne-1-ol, Propargyl alcohol	(53.0)[c,e]	(0.688)[c]
2-Aminoethanol	51.8[d,e]	0.651
2-Amino-1-butanol	50.2[e]	0.602
1-Amino-2-propanol	50.1[j]	0.599
2-Methoxyethanol	52.3[d,e]	0.667
2-Ethoxyethanol	51.0[e,j]	0.627
2-(*n*-Butoxy)ethanol	50.2[e]	0.602
1,2-Ethanediol, Glycol	56.3[d,e]	0.790
1,2-Propanediol	54.1[m]	0.722
1,3-Propanediol	54.9[n]	0.747
1,2,3-Propanetriol, Glycerol	57.0[m]	0.812
1,2-Butanediol	52.6[j]	0.676
1,3-Butanediol	52.8[m]	0.682
1,4-Butanediol	53.5[j]	0.704
2,3-Butanediol	51.8[j]	0.651
Diethylene glycol	53.8[d,e]	0.713
Triethylene glycol	53.5[d,e]	0.704
Tetraethylene glycol	52.2[e]	0.664
2-(Hydroxymethyl)furan, Furfuryl alcohol	(50.3)[c,e]	(0.605)[c]
2-(Hydroxymethyl)tetrahydrofuran	50.3[e]	0.605
Aromatic alcohols (Phenols)		
Phenol	61.4 (40 °C)[o]	0.948
2-Methylphenol, *o*-Cresol	52.5[o]	0.673
3-Methylphenol, *m*-Cresol	53.4[o]	0.701
4-Methylphenol, *p*-Cresol	60.8 (40 °C)[o]	0.929
2-Ethylphenol	51.0[o]	0.627
4-Ethylphenol	61.2 (40 °C)[o]	0.941
2-Isopropylphenol	50.1[o]	0.599
2-(*tert*-Butyl)phenol	49.0[o]	0.565
4-(*n*-Butyl)phenol	59.5[o]	0.889
2,4-Dimethylphenol, 2,4-Xylenol	50.8[o]	0.620
2-Isopropyl-5-methylphenol, Thymol	48.9 (40 °C)[o]	0.562
5-Isopropyl-2-methylphenol, Carvacrol	51.6[o]	0.645
2-Chlorophenol	55.4[o]	0.762
3-Chlorophenol	60.8[p]	0.929
2-(Methoxycarbonyl)phenol, Methyl salicylate	45.4[o]	0.454
2-(Ethoxycarbonyl)phenol, Ethyl salicylate	44.7[o]	0.432
Aliphatic, Cycloaliphatic, and Aromatic Ethers and Acetals		
Diethyl ether	34.5[f]	0.117
Di-*n*-propyl ether	34.0[q]	0.102
Di-*n*-butyl ether	(33.0)[b,f]	(0.071)[b]
tert-Butyl methyl ether	35.5[e]	0.148
Dibenzyl ether	36.3[g]	0.173
Diphenyl ether	35.3 (30 °C)[d,e]	0.142
Ethyl vinyl ether	36.2[e]	0.170
Dimethoxymethane	35.8[e]	0.157
1,2-Dimethoxyethane	38.2[d,e]	0.231
Diethylene glycol dimethyl ether	38.6[d,e]	0.244
Diethylene glycol diethyl ether	37.5[d,e]	0.210

Table 7-3. (Continued)

Solvents	$E_T(30)/(\text{kcal} \cdot \text{mol}^{-1})$[a]	E_T^N
Triethylene glycol dimethyl ether	38.9[d,e]	0.253
Methyloxirane, Propylene oxide	39.8[e]	0.281
2-(Chloromethyl)oxirane, Epichlorohydrin	44.5[e]	0.426
Furan	36.0[g]	0.164
Tetrahydrofuran	37.4[d,e]	0.207
Tetrahydro-2-methylfuran	36.5[d,e]	0.179
Tetrahydro-2,5-dimethylfuran (*cis*/*trans*-isomers)	35.5[j]	0.148
1,3-Dioxolane	43.1[e]	0.383
Tetrahydropyran	36.6[e]	0.182
3-Methyl-tetrahydropyran	36.1[j]	0.167
5-Acetyl-5-methyl-1,3-dioxane	41.5[d,e]	0.333
1,4-Dioxane	36.0[d,e,f]	0.164
Methoxybenzene, Anisol	37.1[g]	0.198
Methyl phenyl sulfane, Thioanisole	37.0[g]	0.194
1,2-Dimethoxybenzene, Veratrole	38.4[e]	0.238
Ethoxybenzene, Phenetole	36.6[g]	0.182
Aliphatic, Cycloaliphatic, and Aromatic Ketones		
Acetone	42.2[d,e]	0.355
2-Butanone	41.3[d,e]	0.327
2-Pentanone	41.1[q]	0.321
3-Pentanone	39.3[d,e]	0.265
3-Methyl-2-butanone, Isopropyl methyl ketone	40.9[r]	0.315
2-Hexanone	40.1[r]	0.290
4-Methyl-2-pentanone, Isobutyl methyl ketone	39.4[d,e]	0.269
3,3-Dimethyl-2-butanone, *tert*-Butyl methyl ketone	39.0[e]	0.256
4-Heptanone	38.9[s]	0.253
2,4-Dimethyl-3-pentanone, Diisopropyl ketone	38.7[e]	0.247
5-Nonanone, Di-*n*-butyl ketone	37.5[q]	0.210
2,6-Dimethyl-4-heptanone, Diisobutyl ketone	38.0[d,e]	0.225
Cyclopentanone	39.4[k]	0.269
Cyclohexanone	39.8[r]	0.281
Phenylethanone, Acetophenone	40.6[g]	0.306
2,4-Pentanedione, Acetylacetone	49.2[e]	0.571
Carboxylic Acids and Anhydrides		
Formic acid	(54.3)[c,e]	(0.728)[c]
Acetic acid	(51.7)[c,e]	(0.648)[c]
Propanoic acid	(50.5)[c,e]	(0.611)[c]
Acetic anhydride	(43.9)[c,e,t]	(0.407)[c]
Aliphatic and Aromatic Esters		
Methyl formate	45.0[j]	0.441
Ethyl formate	40.9[e]	0.315
Methyl acetate	40.0[d,e]	0.287
Ethyl acetate	38.1[d,e]	0.228
Ethyl chloroacetate	39.4[j]	0.269
Vinyl acetate	38.0[e]	0.225
n-Propyl acetate	37.5[e]	0.210
n-Butyl acetate	38.5[j]	0.241
Glycerol triacetate, Triacetin	41.6[d]	0.336
Methyl acrylate	44.5[e]	0.426

Table 7-3. (Continued)

Solvents	$E_T(30)/(\text{kcal}\cdot\text{mol}^{-1})$[a]	E_T^N
Ethyl acetoacetate	49.4[e]	0.577
4-Butyrolactone	44.3[e]	0.420
Dimethyl carbonate	41.1[e]	0.321
Diethyl carbonate	37.0[k]	0.194
1,3-Dioxolan-2-one, Ethylene carbonate	48.6 (40 °C)[e]	0.552
4-Methyl-1,3-dioxolan-2-one, Propylene carbonate	46.6[e]	0.491
Ethyl benzoate	38.1[e]	0.228
Dimethyl phthalate	40.7[e]	0.309
Amides and Cyanamides		
Formamide	56.6[d,e]	0.799
N-Methylformamide	54.1[d,e]	0.722
N,N-Dimethylformamide, DMF	43.8[d,e]	0.404
N,N-Dimethylthioformamide	44.0[j]	0.410
N-Methylacetamide	52.0 (30 °C)[d,e]	0.657
N,N-Dimethylacetamide, DMAC	43.7[d,e]	0.401
N,N-Diethylacetamide	41.4[j]	0.330
Pyrrolidin-2-one	48.3[e]	0.543
1-Methylpyrrolidin-2-one	42.2[d,e]	0.355
1-Methyl-hexahydroazepin-2-one, N-Methyl-ε-caprolactam	41.6[j]	0.336
N,N-Dimethylcyanamide	43.8[j]	0.404
N,N-Diethylcyanamide	43.3[j]	0.389
N,N-Diisopropylcyanamide	42.0[j]	0.349
1-Pyrrolidinecarbonitrile, *N*-Cyanopyrrolidine	42.6[j]	0.367
1-Piperidinecarbonitrile, *N*-Cyanopiperidine	42.1[j]	0.352
4-Morpholinecarbonitrile, *N*-Cyanomorpholine	42.8[j]	0.373
Ureas		
Tetramethylurea	41.0[d,e]	0.318
Tetraethylurea	43.1[e]	0.383
N,N,N',N'-Tetramethylguanidine	39.3[d,e]	0.265
1,3-Dimethylimidazolidin-2-one, DMEU	42.5[e]	0.364
1,3-Dimethyl-2-oxo-hexahydropyrimidine, DMPU	42.1[e]	0.352
Aliphatic and Aromatic Nitriles		
Ethanenitrile, Acetonitrile	45.6[f]	0.460
Propanenitrile	43.7[d,e]	0.401
Acrylonitrile	46.7[e]	0.494
Butanenitrile	43.1[e]	0.383
n-Pentanenitrile, Valeronitrile	42.4[j]	0.361
(Cyanomethyl)benzene, Phenylacetonitrile	42.7[g]	0.370
Cyanobenzene, Benzonitrile	41.5[g]	0.333
Nitroalkanes and Nitroarenes		
Nitromethane	46.3[d,e]	0.481
Nitroethane	43.6[d,e]	0.398
2-Nitropropane (with 3.5 cl/l 1-Nitropropane)	42.8[e]	0.373
Nitrobenzene	41.2[g]	0.324
Aliphatic, Cycloaliphatic, and Aromatic Amines		
1,1-Dimethylethaneamine, *tert*-Butylamine	36.8[e]	0.188
1,2-Diaminoethane	42.0[d,e]	0.349

Table 7-3. (Continued)

Solvents	$E_T(30)/(\text{kcal} \cdot \text{mol}^{-1})$[a]	E_T^N
Diethylamine	35.4[d,e]	0.145
Triethylamine	(32.1)[b,f]	(0.043)[b]
Tri-*n*-butylamine	(32.1)[b,f]	(0.043)[b]
Pyrrolidine	39.1[j]	0.259
Piperidine	35.5[d,e]	0.148
Morpholine	41.0[e]	0.318
Aniline	44.3[d,e]	0.420
2-Chloroaniline	45.5[o]	0.457
N-Methylaniline	42.5[e]	0.364
N,N-Dimethylaniline	36.5[g]	0.179
Heteroarenes		
Pyridine	40.5[f,g]	0.302
2-Methylpyridine, 2-Picoline	38.3[d,e]	0.235
3,4-Dimethylpyridine, 3,4-Lutidine	38.9[f,g]	0.253
2,6-Dimethylpyridine, 2,6-Lutidine	36.9[g]	0.191
2,6-Di-*tert*-butylpyridine	34.0[g]	0.102
2,4,6-Trimethylpyridine, Collidine	36.4[g]	0.176
2-Fluoropyridine	42.4[f,g]	0.361
2-Chloropyridine	41.9[u]	0.346
3-Bromopyridine	39.7[f,g]	0.278
2,6-Difluoropyridine	43.3[g]	0.389
Pentafluoropyridine	36.3[g]	0.173
2-Cyanopyridine	44.2 (30 °C)[j]	0.417
Quinoline	39.4[d,e]	0.269
Phosphorus Compounds		
Trimethyl phosphate	43.6[s]	0.398
Triethyl phosphate	41.7[s]	0.340
Tri-*n*-propyl phosphate	40.5[s]	0.302
Tri-*n*-butyl phosphate	39.6[s]	0.275
Hexamethylphosphoric acid triamide, HMPT	40.9[d,e]	0.315
Hexamethylphosphorothioic triamide, HMPTS	39.5 (30 °C)[j]	0.272
Methylphosphonic acid bis(dimethylamide)	42.3[e]	0.358
Sulfur Compounds		
Carbon disulfide	32.8[f]	0.065
Dimethyl sulfoxide	45.1[f]	0.444
N,N,N',N'-Tetraethylsulfamide, TES	41.0[j]	0.318
Thiophene	35.4[g]	0.145
Tetrahydrothiophene	36.8[j]	0.188
Tetrahydrothiophene-1,1-dioxide, Sulfolane	44.0[e]	0.410
3-Methyl-tetrahydrothiophene-1,1-dioxide, 3-Methylsulfolane	43.0[j]	0.380
1,3,2-Dioxathiolane-2-oxide, Ethylene sulfite	50.0[j]	0.596
Bis(2-hydroxyethyl)sulfane	54.5[j]	0.735
1.Methylpyrrolidin-2-thione	42.8[j]	0.373
Miscellaneous		
Carbon dioxide (in its liquid and supercritical state at 69 bar)	(33.8) (24 °C)[v]	0.096
tert-Butyl hydroperoxide	49.7[w]	0.586
Isopropyl nitrate	43.1[e]	0.383

Table 7-3. (Continued)

Solvents	$E_T(30)$/(kcal · mol^{-1})[a)]	E_T^N
(2*S*, 3*S*)-(+)-Bis(dimethylamino)-2,3-dimethoxybutane, DDB	36.6[e)]	0.182
3-Methyloxazolidin-2-one	45.0[j)]	0.441
Tetra-*n*-hexylammonium benzoate	44.3[d, e)]	0.420
Water and Heavy Water		
Deuterium oxide (99.75 cg/g)	62.8[e)]	0.991
Water	63.1[d, e)]	*1.000*

ᵃ Since the standard betaine dye *(44)* was numbered *30* in the first publication by Dimroth, Reichardt *et al.* [66][d)], its molar *t*ranstition *e*nergies were designated as $E_T(30)$-values.

ᵇ These $E_T(30)$-values in parentheses are secondary values, determined by means of the more lipophilic penta-*tert*-butylsubstituted betaine dye *(45)*; *cf.* references [174][e)] and [175, 176][f, g)].

ᶜ These $E_T(30)$-values in parentheses are secondary values, calculated from Kosower's *Z*-values [55] by means of the correlation equation $E_T(30)/(\text{kcal} \cdot \text{mol}^{-1}) = 0.752 \cdot Z/(\text{kcal} \cdot \text{mol}^{-1}) - 7.87$ ($n = 15$; $r = 0.998$). They can be used in correlations but they are still subject to revision. The correlation equation has been calculated by Griffiths and Pugh [172] with 15 carefully selected solvents common to both scales.

ᵈ K. Dimroth, C. Reichardt, T. Siepmann, and F. Bohlmann, Liebigs Ann. Chem. *661*, 1 (1963); K. Dimroth and C. Reichardt, ibid. *727*, 93 (1969); C. Reichardt, ibid. *752*, 64 (1971); K. Dimroth and C. Reichardt, Fortschr. Chem. Forsch. *11*, 1 (1968).

ᵉ C. Reichardt and E. Harbusch-Görnert, Liebigs Ann. Chem. *1983*, 721.

ᶠ C. Laurence, P. Nicolet, and C. Reichardt, Bull. Soc. Chim. Fr. *1987*, 125.

ᵍ C. Laurence, P. Nicolet, M. Lucon, and C. Reichardt, Bull. Soc. Chim. Fr. *1987*, 1001.

ʰ V. Bekárek and J. Juřina, Coll. Czech. Chem. Commun. *47*, 1060 (1982).

ⁱ S. Balakrishnan and A. J. Easteal, Austr. J. Chem. *34*, 933 (1981).

ʲ C. Reichardt *et al.*, unpublished results.

ᵏ M. H. Aslam, G. Collier, and J. Shorter, J. Chem. Soc., Perkin Trans. II *1981*, 1572.

ˡ H. Elias, M. Dreher, S. Neitzel, and H. Volz, Z. Naturforsch., Part B *37*, 684 (1982).

ᵐ E. M. Kosower and H. Dodiuk, J. Am. Chem. Soc. *98*, 924 (1976).

ⁿ E. M. Kosower, H. Dodiuk, K. Tanizawa, M. Ottolenghi, and N. Orbach, J. Am. Chem. Soc. *97*, 2167 (1975).

ᵒ J. Hormadaly and Y. Marcus, J. Phys. Chem. *83*, 2843 (1979).

ᵖ K. Hesse and S. Hünig, Liebigs Ann. Chem. *1985*, 715.

q Z. Ilić, Z. Maksimović, and C. Reichardt, Glasnik Hem. Društva Beograd [Bull. Soc. Chim. Beograd] *49*, 17 (1984).

ʳ A. G. Burden, N. B. Chapman, H. F. Duggua, and J. Shorter, J. Chem. Soc., Perkin Trans. II *1978*, 296.

ˢ Z. B. Maksimović, C. Reichardt, and A. Spirić, Z. Anal. Chem. *270*, 100 (1974).

ᵗ T. G. Beaumont and K. M. C. Davis, J. Chem. Soc., Part B *1968*, 1010.

ᵘ W. Jeblick and K. Schank, Liebigs Ann. Chem. *1977*, 1096.

ᵛ J. A. Hyatt, J. Org. Chem. *49*, 5097 (1984).

ʷ H. Langhals, E. Fritz, and I. Mergelsberg, Chem. Ber. *113*, 3662 (1980).

Hence, the corresponding E_T^N-scale ranges from 0.000 for TMS, the least polar solvent, to 1.000 for water, the most polar solvent. These E_T^N-values are dimensionless numbers which are included in Table 7-3. For example, an E_T^N-value of 0.500 for cyclohexanol means that this solvent exhibits 50% of the solvent polarity of water, as empirically measured by means of the standard dye *(44)*.

The major limitation of the $E_T(30)$-values is the fact that no $E_T(30)$-values can be measured for acidic solvents such as carboxylic acids. Addition of traces of an acid to solutions of *(44)* or *(45)* immediately changes the colour to pale yellow due to protonation at the phenolic oxygen atom of the dye. The protonated form no longer exhibits the long-wavelength solvatochromic absorption band. The excellent linear correlation between $E_T(30)$ and Kosower's *Z*-values, which are available for acidic solvents, allows the calculation of $E_T(30)$-values for such solvents [174]. – A further limitation is the fact that it was not possible to measure the absorption maximum of the standard betaine dye *(44)* in the gas phase as reference state.

The solvents in Table 7-3 can be roughly divided into three groups according to their $E_T(30)$ or E_T^N-values depending on their specific solvent/solute interactions: (i) protic (HBD) solvents (E_T^N *ca.* 0.5...1.0), (ii) dipolar non-HBD solvents (E_T^N *ca.* 0.3...0.5), and (iii) apolar non-HBD solvents (E_T^N *ca.* 0.0...0.3). This solvent classification is confirmed by other solvent characteristics; *cf.* Fig. 3-3 in Section 3.4.

Because of the rather localized negative charge at the phenolic oxygen atom*), the standard dye *(44)* is capable of specific Lewis acid/base interactions. Therefore, in addition to the nonspecific dye/solvent interactions, the betaine dye *(44)* preferably measures the specific Lewis acidity of organic solvents. On the other hand, the positive charge of the pyridinium moiety of *(44)* is delocalized. Therefore, the solvent Lewis basicity will not be registered by the probe molecule *(44)*. If this solvent property is relevant for the system under study, other empirical measures of Lewis basicity should be used; *cf.* Section 7.7.

An analysis of the $E_T(30)$-values, using multivariate statistical methods, has been carried out by Chastrette *et al.* [193]. According to this analysis, the $E_T(30)$-values of non-HBD solvents are measures of the dipolarity and polarizability as well as the cohesion of the solvents. Another analysis of $E_T(30)$-values in terms of functions of the dielectric constant (ε_r) and refractive index (n_D) of forty non-HBD solvents has been given by Bekárek *et al.*; he emphasises the predominant influence of the f(ε_r) term on the $E_T(30)$-parameter of those solvents [194].

The $E_T(30)$-values of binary solvent mixtures are not related to their composition in a simple linear manner [68–72, 124, 177–192]. Most binaries behave as more or less non-ideal solvent mixtures. A monotonous, but not always linear change in $E_T(30)$ with mole fraction of one solvent component (x_s) is obtained for some alcohol/water mixtures such as CH_3OH/H_2O [68, 124, 177–178, 181], for alcohol/alcohol mixtures such as CH_3OH/C_2H_5OH [185], for dipolar non-HBD/dipolar non-HBD mixtures such as CH_3CN/DMSO [185], and for apolar non-HBD/apolar non-HBD mixtures such as C_6H_6/C_5H_5N [185].

---

* The X-ray analysis of the bromo-substituted betaine dye *(44)* shows that it is not planar. Not only the five peripheral phenyl groups are all twisted but also the pyridinium and phenoxide rings, the latter with a mutual interplanar angle of about 65° [75].

However, addition of a small amount of a polar solvent to the betaine dye *(44)* in nonpolar solvents causes a disproportionately large hypsochromic band shift which corresponds to an excessively large increase in $E_T(30)$. This phenomenon can easily be explained by strong preferential or selective solvation of the dipolar betaine dye by the more polar component of the binary solvent mixture; *cf.* Section 2.4. Typical examples of solvent mixtures with preferential solvation of *(44)* are 1,4-dioxane/water [68, 124, 177–178, 180], C_5H_5N/H_2O [68, 124, 177], *t*-BuOH/H_2O [180], 2-(*n*-butoxy)ethanol/H_2O [180], CH_3CN/H_2O [183], *c*-C_6H_{12}/C_2H_5OH [190] and C_6H_6/C_2H_5OH [190]. In these cases, the $E_T(30)$-values do not in fact measure the polarity of the bulk solvent mixture but rather the micropolarity of the solvation shell on the molecular-microscopic level. Solvatochromic dyes such as *(44)* are thus a simple means for the study of the phenomenon of preferential solvation [192].

A particular interesting synergistic solvent effect has been found for some binary solvent mixtures composed of HBA and HBD solvents. For example, a graph of $E_T(30)$ against composition for the binaries $(RO)_3PO/CHCl_3$ [68], $CH_3COCH_3/CHCl_3$ [68, 183], CH_3COCH_3/CH_2Cl_2 [183], DMSO/ROH (R = *i*-Pr, *t*-Bu) [185] and CH_3CN/ROH (R = Et, *i*-Pr, *t*-Bu) [185] shows a pronounced maximum. This means that the binary solvent mixtures behave as a more polar medium than either of its two components! The deviations in $E_T(30)$ in these synergistic solvent mixtures are attributable to specific intermolecular solvent/solvent hydrogen bond interactions (*e.g.* $P{=}O\cdots HCCl_3$, $C{=}O\cdots HCCl_3$, $S{=}O\cdots HOR$, *etc.*) which create a new, more polar solvent system [68, 184, 187]. Relative strong interactions between acetone and chloroform have been proposed many times on the basis of a variety of experimental evidence. This synergistic polarity behaviour of HBA/HBD solvent mixtures can be of particular interest in the acceleration of solvent-dependent chemical reactions.

The polarity of binary solvent mixtures with limited mutual miscibility such as *n*-BuOH/H_2O and *c*-C_6H_{12}/DMF [191] as well as solid polymer mixtures (organic glasses) [195] have also been studied using solvatochromic probes such as the betaine dye *(44)*.

According to Langhals [184, 191, 192, 196], the $E_T(30)$-values of binary solvent mixtures can be quantitatively described by the two-parameter equation (7-30).

$$E_T(30) = E_D \cdot \ln (c_p/c^* + 1) + E_T^\circ(30) \qquad (7\text{-}30)$$

c_p is the molar concentration of the more polar component (*i.e.* the solvent with the higher $E_T(30)$-value) and $E_T^\circ(30)$ is the $E_T(30)$-value of the pure component with lower polarity. E_D ("Energiedurchgriff") and c^* ("Erscheinungskonzentration") are adjustable parameters specific for the binary solvent system under study. c^* is a threshold value defining a transition between two regions. For low concentrations of the more polar solvent ($c^* \gg c_p$), a good approximation of Eq. (7-30) is $E_T(30) = E_D \cdot (c_p/c^*) + E_T^\circ(30)$, and the $E_T(30)$-values increase linearly with the molar concentration of the more polar solvent c_p. For high concentrations of the more polar solvent ($c^* \ll c_p$), Eq. (7-30) can be written as $E_T(30) = E_D \cdot \ln (c_p/c^*) + E_T^\circ(30)$, and values of $E_T(30)$ correlate linearly with the logarithm of the molar concentration of the more polar solvent $\ln c_p$. Eq. (7-30) is valid with high precision for about 80 solvent mixtures investigated and can even be applied to more complicated solvent systems as mentioned above [191, 192].

It should be mentioned that an equation analogous to Eq. (7-30) has been successfully applied to salt effects on reaction rates arising from variations in solvent polarity on the addition of electrolytes (ionophores) [197]; *cf.* also Eq. (5-99) in Section 5.4.5. For electrolyte solutions, the added salt can be treated as a more polar "co-solvent" [197].

The strong dependence of $E_T(30)$-values on the composition of binary mixtures of solvents with different polarity can be used for the quantitative UV/Vis spectroscopic determination of water in organic solvents [68, 192, 198, 199].

In general, the $E_T(30)$-values exhibit a good, mostly linear correlation with a large number of other solvent-sensitive processes such as light absorption, reaction rates, and chemical equilibria [2]. Applications of $E_T(30)$-values to chemical reactivity [124, 200–202] and analytical chemistry [203] have been reviewed. Their application to photochemical processes has been discussed [204, 205]. Its extreme sensitivity to small changes in the surrounding medium has made the betaine dye *(44)* a useful molecular probe in the study of micelle/solution interfaces, microemulsions and phospholipid bilayers, model lipid membranes, rigid rod-like isocyanide polymers, and the retention behaviour in reverse-phase chromatography; *cf.* page 289 and references [298–302] of Chapter 6. Even standard molar Gibbs energies of transfer of chloride ions, $\Delta G_t^\circ(Cl^\ominus, H_2O \rightarrow S)$, could be linearly correlated to the $E_T(30)$-values. $E_T(30)$-values have also been used to calculate acceptor numbers, AN_E, which are not otherwise available by direct measurements [207]; *cf.* Table 2-5 in Section 2.2.6.

It should be mentioned that the pyridinium-*N*-phenoxide betaine dye *(44)* is not only very sensitive to changes in solvent polarity, but in addition its longest-wavelength solvatochromic absorption band also depends on changes in temperature [73, 175, 180, 208], pressure [74, 182, 208], on the addition of electrolytes (ionophores) [209–213], as well as on the introduction of substituents in the peripheral phenyl groups; *cf.* Fig. 7-2 in Section 7.1.

The *thermo-solvatochromism* of *(44)* can be easily observed by means of a betaine solution in ethanol: at −75 °C the solution is red-colored, and at +75 °C the solution is blue-violet, corresponding to absorption maxima of 513 nm and 568 nm, respectively [73]. The reason for this thermo-solvatochromism is the increased stabilization of the dipolar electronic ground state of *(44)* relative to the less dipolar excited state with decreasing temperature, due to better solute/solvent interactions at low temperature. It can be stated that, the lower the temperature, the higher the $E_T(30)$-value.

Tamura and Imoto [74] and Kelm *et al.* [182] observed pressure effects on the solution spectra of *(44)*. In all solvents used they found a hypsochromic shift of the longest-wavelength absorption band with increasing pressure. The observed hypsochromic shift of a betaine solution in acetone runs up to 9 nm by raising the pressure from 1 to 1920 bar (674 nm → 665 nm; $\Delta E_T(30) = 0.6$ kcal/mol) [74]. On the supposition that this *piezo-solvatochromism* results from better solute/solvent interactions with increasing pressure, it can be stated that, the higher the pressure, the more polar the solvent, and the higher the $E_T(30)$-value. However, this conclusion must be considered carefully, because other factors such as conformational changes may also contribute to the observed spectral changes.

The addition of electrolytes (ionophores) to solutions of *(44)* causes hypsochromic shifts of its solvatochromic absorption band [209–213]. This phenomenon can be designated as *halo-solvatochromism**⁾. For example, the addition of KI, NaI, LiI, BaI_2, $Ca(SCN)_2$, and $Mg(ClO_4)_2$ to solutions of *(44)* in acetonitrile leads to a differential hypsochromic band shift which increases with this ionophore order, *i.e.* with increasing charge density of the cation [211]. Obviously, salts act similarily to other polar compounds (solvents) when added to solutions of *(44)*. The polarity of binary ionophore/solvent mixtures as a function of composition can be quantitatively described in a manner similar to other binary solvent/solvent mixtures [197, 213].

It should be noted that the polarity of the medium also influences the ^{1}H- and ^{13}C-NMR chemical shifts of dye *(44)* [215, 216]; *cf.* Section 6.5.1. The sites in the betaine molecule most influenced by the solvent are those nearest to the positive and negative charges within the dye molecule, and this is reflected in the NMR chemical shifts.

The solvent, temperature, pressure, ionophore, and substituent effects on the UV/Vis spectra of the pyridinium-*N*-phenoxide betaine dyes indicate the extreme sensitivity of this class of compounds to small changes in the environment. Their behaviour may be compared to that of the Princess and the Pea in one of H. C. Andersen's fairy-tales [76]**⁾. Their utility for setting up linear Gibbs energy relationships is demonstrated by the fact that the same betaine dye can be used for establishing kinetic and spectroscopic scales of substituents (*cf.* Figs. 7-1 and 7-2) as well as a spectroscopic scale of solvent polarity (*cf.* Table 7-3).

Further solvent polarity scales based on UV/Vis absorption as well as fluorescence spectra have been proposed by Brooker *et al.* [77], Dähne *et al.* [78], de Mayo *et al.* [217], Dubois *et al.* [79], Mukerjee *et al.* [218] and Wrona *et al.* [219], Walter *et al.* [220], Walther [81] and Lees *et al.* [82], Zelinskii *et al.* [80], Winnik *et al.* [222], and Kamlet and Taft [84, 84a, 224, 226].

The solvent dependence of the $\pi \rightarrow \pi^*$ transition energies of two meropolymethine dyes was used by Brooker *et al.* [77] to establish the solvent polarity parameters χ_R and χ_B (*cf.* Table 7-2). χ_R is based on the positive solvatochromic merocyanine dye no. 1 in Table 6-1 of Section 6.2.1 (red shift with increasing solvent polarity), while χ_B represents the transition energies of the negative solvatochromic merocyanine dye no. 13 in Table 6-1.

Dähne *et al.* [78] proposed the positive solvatochromic 5-dimethylaminopentadien-2,4-al-1 (dye no. 3 in Table 6-1) as a solvent polarity indicator and recommended a relative polarity function *RPM* (from the German "Relatives Polaritätsmaß").

The $\pi \rightarrow \pi^*$ transition energy E_{sp} of a spiropyran zwitterion of the type described in Section 4.4.2 [*(27a)* $\rightleftharpoons$ *(27b)*] has been used by de Mayo *et al.* to characterize the polarity of solid oxide surfaces such as that of silica gel [217].

---

* The *halo-solvatochromism* of *(44)* can be considered as the only genuine halochromism, *i.e.* a colour change with increasing ionic strength of the medium *without* chemically changing the chromophore. The term *halochromism*, as introduced by Baeyer *et al.* [214], denotes the trivial colour change of a dye on addition of acids or bases. This is simply caused by the creation of a new chromophore in an acid/base reaction whereby a colourless compound is rendered coloured on salt formation, *e.g.*

$$(C_6H_5)_3C{-}Cl \text{ (colourless)} + AlCl_3 \rightleftharpoons (C_6H_5)_3C^{\oplus} \text{ (yellow)} + AlCl_4^{\ominus}.$$

** The princess was so sensitive to her surroundings that she was able to feel a pea through 20 mattresses and 20 eider-down quilts in her bed. This extreme sensitivity corresponds in some way to the sensitivity of the pyridinium-*N*-phenoxide betaine dye *(44)*.

Dubois *et al.* [79] formulated Φ-values (formerly F-values) as solvent polarity parameters based on the position of the solvent-sensitive $n \rightarrow \pi^*$ transition of certain saturated aliphatic ketones as shown in Eq. (7-31)

$$\Delta\tilde{\nu}_{\mathrm{H}}^{\mathrm{S}} = \tilde{\nu}^{\mathrm{S}} - \tilde{\nu}^{\mathrm{H}} = \Phi(\tilde{\nu}^{\mathrm{H}} - 32637) - 174 \tag{7-31}$$

Solvents are characterized by deviations from unity of the slope of $\tilde{\nu}^{\mathrm{S}}$-values (wavenumber of absorption for various ketones in solvent S) against $\tilde{\nu}^{\mathrm{H}}$ (for various ketones in *n*-hexane as reference solvent); *cf.* Table 7-2.

Mukerjee *et al.* [218] and Wrona *et al.* [219] have used the highly solvatochromic $n \rightarrow \pi^*$ transition energy of the stable 2,2,6,6-tetramethylpiperidine-1-oxide radical (TMPNO) for the development of a solvent polarity scale. So-called $E_{\mathrm{B}}^{\mathrm{N}}$-values as empirical measures of solvent Lewis acidity have been determined for 53 pure organic solvents and some binary solvent/water mixtures [219].

Using the negative solvatochromism of the $n \rightarrow \pi^*$ absorption of *N,N*-(dimethyl)-thiobenzamide-*S*-oxide, an $E_{\mathrm{T}}^{\mathrm{SO}}$ solvent scale has been proposed by Walter *et al.* [220]. This scale comprises of 36 solvents and three binary solvent/water mixtures and is thought to be particularly useful for characterizing protic solvents.

Furthermore, the E_{K}-scale of Walther [81] and the E_{MLCT}^*-scale of Lees *et al.* [82] should be mentioned. Both rather comprehensive solvent scales are based on the negative solvatochromic metal-to-ligand charge-transfer absorption (MLCT; $d \rightarrow \pi^*$) of the two zero valent group 6 metal complexes *(46)* and *(47)* of the common formula $M(CO)_4$ (diimine). A consistent explanation for the solvatochromic behaviour of such coordination compounds with MLCT-absorption has recently been given by Kaim *et al.* [83]; *cf.* also Section 6.2.1.

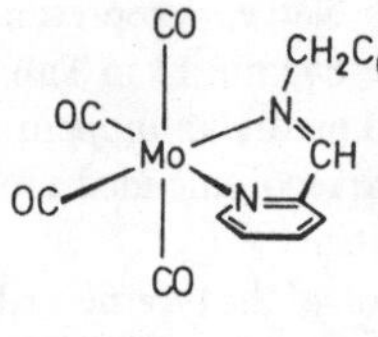

(46) 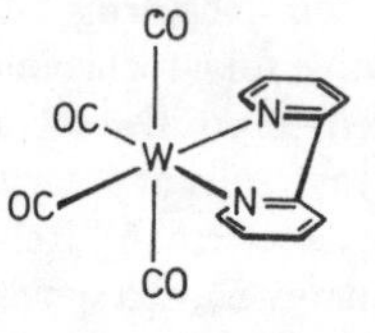 *(47)*

For a review of the use of solvatochromic metal complexes as visual indicators of solvent polarity see reference [221].

Solvatochromic fluorescent probe molecules have also been used to establish solvent polarity scales. The solvent-dependent fluorescence maximum of 4-amino-*N*-methylphthalimide was used by Zelinskii *et al.* to establish a "universal scale for the effect of solvents on the electronic spectra of organic compounds" [80, 213]. More recently, a comprehensive *Py*-scale of solvent polarity including 95 solvents has been proposed by Winnik *et al.* [222]. This is based on the relative band intensities of the vibronic bands I and III of the $\pi^* \rightarrow \pi$ emission spectrum of monomer pyrene; *cf.* Section 6.2.4. A significant enhancement is observed in the $0 \rightarrow 0$ vibronic band intensity I_{I} relative to the

0 → 2 vibronic band intensity I_{III} with increasing solvent polarity. The ratio of emission intensities for bands I and III serves as an empirical measure of solvent polarity: $Py = I_I/I_{III}$ [222]. However, there seems to be some difficulty in determining precise *Py*-values shown by the varying *Py*-values from different laboratories; the reasons for these deviations have been investigated [223].

An interesting approach, called the *solvatochromic comparison method*, used to evaluate a β-scale of solvent hydrogen-bond acceptor (HBA) basicities, an α-scale of solvent hydrogen-bond donor (HBD) acidities, and a π*-scale of solvent dipolarity/polarizability using UV/Vis spectral data of solvatochromic compounds, was employed and further developed by Kamlet, Taft *et al.* [84, 84a, 224, 226]. Magnitudes of enhanced solvatochromic shifts, $\Delta\Delta\tilde{\nu}$, in HBA solvents*) have been determined for 4-nitroaniline relative to homomorphic**) *N,N*-diethyl-4-nitroaniline (compound no. 2, Table 6-1, Section 6.2.1). Both standard compounds are capable of acting as HBA substrates (at the nitro oxygens) in HBD solvents, but only 4-nitroaniline can act as a HBD substrate in HBA solvents. Taking the $\Delta\Delta\tilde{\nu}$-value of 2800 cm^{-1} for hexamethylphosphoric triamide (a strong HBA solvent) as a single fixed reference point ($\beta_1 = 1.00$), a β-scale of solvent HBA basicities for HBA solvents was developed [84]. Using the same solvatochromic comparison method, *i.e.* the enhanced solvatochromic shifts $\Delta\Delta\tilde{\nu}$, in HBD solvents for 4-nitroanisole and the pyridinium-*N*-phenoxide betaine dye *(44)*, an α-scale of HBD acidities was evaluated. Taking the $\Delta\Delta\tilde{\nu}$-value of 6240 cm^{-1} for methanol (a strong HBD solvent) as a fixed reference point ($\alpha_1 = 1.00$), an α-scale of solvent HBD acidities for HBD solvents was established [84].

The same authors also introduced a π*-scale of solvent dipolarity/polarizability [84a]. This π*-scale is so named because it is derived from solvent effects on the $\pi \rightarrow \pi^*$ electronic transitions of a variety of nitroaromatics (4-nitroanisole, *N,N*-diethyl-3-nitroaniline, 4-methoxy-β-nitrostyrene, 1-ethyl-4-nitrobenzene, *N*-methyl-2-nitro-*p*-toluidine, *N,N*-diethyl-4-nitroaniline, and of 4-dimethylaminobenzophenone). Solvent effects on the $\tilde{\nu}_{max}$-values of these seven primary solvatochromic indicators have been employed in the initial construction of the π*-scale, which was then expanded and refined by multiple least-squares correlations with additional solvatochromic indicators. In this way, an averaged π*-scale of solvent dipolarity/polarizability has been established which measures the ability of the solvent to stabilize a charge or a dipole by virtue of its dielectric effect [84a]. A normalized range of 0.00 (for cyclohexane) to 1.00 (for dimethyl sulfoxide) for the π*-values of common solvents has been chosen so that, taken with the α-scale of solvent HBD acidities and the β-scale of solvent HBA basicities (which have also been scaled to range from 0.00 to 1.00), these parameters can be used together in a multiparameter equation (*cf.* Section 7.7).

---

* The term hydrogen-bond acceptor (HBA) refers to the acceptance of the proton of a hydrogen-bond. Therefore, HBA solvents are also electron-pair donor (EPD) solvents. Hydrogen-bond donor (HBD) refers to the donation of the proton. Therefore, HBD solvents behave as protic solvents.

** The term *homomorphic* molecules was introduced by Brown *et al.* [225]. Molecules having the same or closely similar molecular geometry are *homomorphs*, *e.g.* ethane is a homomorph of methanol, toluene of phenol, 4-aminobenzenesulfonamide of 4-aminobenzoic acid, and *N,N*-diethyl-4-nitroaniline of 4-nitroaniline.

A selection of Kamlet-Taft's solvatochromic parameters β, α, and π^* for 33 organic solvents, taken from a more recent comprehensive and improved collection [226], is given in Table 7-4.

The π^*-scale has even been extended to perfluorinated hydrocarbons [227] and to the gas phase [228]. The parameters in Table 7-4 were arrived at by averaging the multiple normalized solvent effects on a variety of solvent-dependent properties involving various types of solvatochromic indicator dyes. Therefore, the solvatochromic parameters of Table 7-4 are no longer directly based on the solvent effects indicated by a distinct single solvatochromic indicator dye. Rather they are statistically averaged values resulting from

Table 7-4. A selection of Kamlet-Taft's solvatochromic parameters β, α, and π^* for 33 organic solvents[a)], taken from reference [226].

Solvents	β	α	π^*
Gas phase			−1.1
n-Hexane, *n*-Heptane	0.00	0.00	−0.08
Cyclohexane	0.00	0.00	*0.00*
Dichloromethane	0.00	(0.30)	0.82
Chloroform	0.00	(0.44)	0.58
Tetrachloromethane	0.00	0.00	0.28
Benzene	0.10	0.00	0.59
Methylbenzene, Toluene	0.11	0.00	0.54
Chlorobenzene	0.07	0.00	0.71
Nitrobenzene	0.39	0.00	1.01
Pyridine	0.64	0.00	0.87
Diethyl ether	0.47	0.00	0.27
1,4-Dioxane	0.37	0.00	0.55
Tetrahydrofuran	0.55	0.00	0.58
Triethylamine	0.71	0.00	0.14
Acetone	0.48	0.08	0.71
Cyclohexanone	0.53		0.76
Ethyl acetate	0.45	0.00	0.55
Ethyl benzoate	0.41	0.00	0.74
Propylene carbonate	0.40	0.00	0.83
Dimethyl sulfoxide	0.76	0.00	*1.00*
Sulfolane		0.00	0.98
Acetonitrile	0.31	0.19	0.75
Hexamethylphosphoric acid triamide	1.05	0.00	0.87
N,N-Dimethylformamide	0.69	0.00	0.88
Formamide	(0.55)	0.71	0.97
tert-Butanol	(1.01)	0.68	0.41
2-Propanol	(0.95)	0.76	0.48
1-Propanol		0.78	0.52
Ethanol	(0.77)	0.83	0.54
2,2,2-Trifluoroethanol	0.00	1.51	0.73
Methanol	(0.62)	0.93	0.60
Acetic acid		1.12	0.64
Water	(0.18)[b)]	1.17	1.09

[a] Values in parentheses are relatively less certain.

[b] Value for monomeric water molecules. For the bulk liquid, a revised β-value of 0.47 has been used; *cf.* Y. Marcus, J. Phys. Chem. *91*, 4422 (1987) and reference 4 cited therein.

a series of successive approximations [226]*. Kamlet and Taft's solvatochromic parameters have been used in one-, two-, and three-parameter correlations involving different combinations of these parameters which are called *linear solvation energy relationships* (LSER's). An impressive series of more than 40 articles entitled "Linear solvation energy relationships" has been so far published (May 1987): Part 1 [229]... Part 41 [230]; *cf.* Section 7.7 for a further discussion of such multiparameter correlations.

An extension of the β-scale of solvent HBA basicity, using only 4-nitroaniline and *N,N*-diethyl-4-nitroaniline as solvatochromic indicators (B_{KT}-scale), has been given by Krygowski *et al.* [231, 232]. Some further improvements of Kamlet and Taft's solvatochromic parameters have been proposed by Kolling [233] and Bekárek [234]. According to Bekárek, better correlations are obtained using modified β_n-, α_n-, and π^*_n-values which are derived from the original β-, α-, and π^*-values by dividing them by the refractive index function $(n^2-1)/(2n^2+1)$. This is in order to eliminate the polarization contribution of the solvent molecules in the cybotactic solvation sphere during the electronic excitation of the solvatochromic indicators. According to Bekárek, the original β-, α-, and π^*-values are adequate for correlating certain types of spectral properties (*e.g.* HFS constants in ESR spectra), but their applicability to other types of solvent effects could be improved by using the modified β_n-, α_n-, and π^*_n-values [234]. However, this modification has strongly been criticized by Kamlet, Taft, and Abboud for conceptual and computational reasons. This leads to the conclusion that in *all* correlations the original solvatochromic parameter are in fact best [235].

Interestingly, a statistical principal component analysis of the solvatochromic shift data sets previously used by Kamlet and Taft in defining the π^*-scale has shown that, instead of one (π^*), two solvent parameters (θ_{1k} and θ_{2k}) are necessary to describe the solvent-induced band shifts of the studied solvatochromic indicators [236]. This is not unexpected since the π^* parameters are assumed to consist of a blend of dipolarity and polarizability contributions to the solute/solvent interactions.

Laurence *et al.* [237–239] have tried to improve Kamlet and Taft's solvatochromic comparison method [224] by introducing a new *thermosolvatochromic comparison method*. In doing this they tried to eliminate various shortcomings including sometimes insufficiently precise determined solvatochromic parameters. According to the solvatochromic comparison method, the π^*-value of a solvent S is measured by the bathochromic shift relative to cyclohexane, $-\Delta\tilde{\nu}_S$, of the $\pi \rightarrow \pi^*$ transition of a non-HBD indicator dye (*e.g.* 4-nitroanisole). The basicity parameter β of the same solvent is measured by the

* There is a discussion in the literature concerning the use of averaged and statistically optimized "constant" solvent polarity parameters (*e.g.* β, α, π^*) instead of experimentally derived parameters which are based on a distinct, single, and well-understood solvent-dependent reference process [*e.g.* Y_{OTs}, Z, E_T (30)]. There are some practical reasons in favour of the experimentally derived solvent parameters: They are related to a distinct chemical or physical reference process and the corresponding solvent scale can easily be enlarged by new precise measurements. Averaged and statistically optimized solvent parameters are no longer directly related to a distinct reference process and are, thus, ill-defined. New data may be difficult to incorporate into the existing framework, and even if they are, then this can lead to a modification of the already calculated "constant" solvent parameters. It has been pointed out "that it is better to study one good model with precision than to take the average of results obtained from many poor models" [237]; *cf.* also [153].

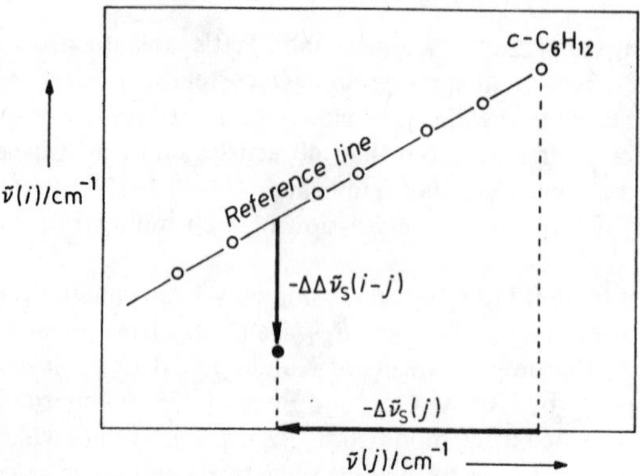

Fig. 7-3. Illustration of the solvatochromic comparison method according to [237]. Reference line equation: $\tilde{\nu}(i) = a_{ij} \cdot \tilde{\nu}(j) + b_{ij}$. Reference solvent: cyclohexane. ○ non-HBA and non-HBD solvents; ● an HBA solvent.

supplementary bathochromic shift $-\Delta\Delta\tilde{\nu}_S$ obtained using a second HBD indicator dye (*e.g.* 4-nitrophenol), which is a homomorph of the first dye. Plotting the absorption wavenumbers of the HBD indicator dye *i* against the wavenumbers of its non-HBD homomorph *j*, for non-HBD and non-HBA solvents, leads to the linear reference line given in Fig. 7-3.

Because of the specific HBD-solute/HBA-solvent interaction, all HBA solvents fall below this line, namely the higher the basicity the lower their position. The Kamlet-Taft π^*- and β-values are the means of the normalised values $-\Delta\tilde{\nu}_S(j)$ and $-\Delta\Delta\tilde{\nu}_S(i-j)$ for several indicator dyes *j* and various homomorph pairs *i*/*j*, according to Eqs. (7-32) and (7-33), respectively (CH refers to cyclohexane) [237].

$$-\Delta\tilde{\nu}_S(j) = \tilde{\nu}_S(j) - \tilde{\nu}_{CH}(j) \tag{7-32}$$

$$-\Delta\Delta\tilde{\nu}_S(i-j) = [a_{ij} \cdot \tilde{\nu}_S(j) + b_{ij}] - \tilde{\nu}_S(i) \tag{7-33}$$

It is obvious, that the accuracy of the determination of $-\Delta\Delta\tilde{\nu}_S(i-j)$ depends strongly on the precision with which the reference line of Fig. 7-3 is established.

Laurence *et al.* [237, 238] have shown that some of the absorption bands of the solvatochromic indicator dyes previously used by Kamlet and Taft to fix the reference lines exhibit a solvent- and temperature-dependent vibrational fine structure which makes the determination of the band maxima, $\tilde{\nu}_{max}$, difficult. Furthermore, the indicator dyes used are thermo-solvatochromic, *i.e.* $\tilde{\nu}_{max}$ is temperature-dependent. No allowance for this temperature-dependence of the β-, α-, and π^*-values has been made previously.

Finally, the right choice of solvent on going from the gas phase to the most polar non-HBA and non-HBD solvents, has to be made in order to fix the reference line as given in Fig. 7-3. Taking all these difficulties into account, and using 4-nitrophenol/4-nitroanisol and 4-nitroaniline/*N*,*N*-dimethyl-4-nitroaniline as distinct solvatochromic indicator couples *i* and *j*, as well as a carefully selected set of non-HBA and non-HBD solvents,

Laurence *et al.* [237, 238] have determined more precise reference lines. From these it was possible to derive new solvent dipolarity/polarizability and solvent HBA basicity parameters which they believe to be more correct than the original π^* and β parameters. In particular, the β-scale seems to be mainly a scale of solvent HBA basicity against NH-donor solutes and does not have the degree of general validity as originally claimed [84].

Other spectroscopic measurements of solvent polarity use as standard processes solvent-sensitive infrared stretching absorptions of groups such as X=O and X—H···B, where X may be C, S, N, O, or P, and B is a hydrogen-bond acceptor (HBA) [*cf.* Eqs. (6-6) and (6-7) in Section 6.3]. Schleyer *et al.* [85] proposed the relationship (7-34) for the correlation of solvent-sensitive IR vibrations. If X—H···B vibrations are being

$$\frac{\tilde{\nu}^0 - \tilde{\nu}^s}{\tilde{\nu}^0} = a \cdot G \tag{7-34}$$

examined, $\tilde{\nu}^0$ refers to its gas-phase value. The value for a is a measure of solvent susceptibility of a particular IR vibration, and G*) is a function of the solvent only. Since solvent shifts of $\tilde{\nu}_{C=O}$ and $\tilde{\nu}_{S=O}$ are proportional to solvent shifts of $\tilde{\nu}_{X-H\cdots B}$, G-values are calculated from the solvent shifts of the carbonyl bands of *N,N*-dimethylformamide and benzophenone and the sulfonyl band of dimethyl sulfoxide. An arbitrary value of 100 was assigned to dichloromethane to fix the scale ($G=0$ for the gas phase) [85]. Values of G are given in Table 7-2. Further G-values have recently been determined by Somolinos *et al.* [240] and their relationships to other solvent polarity parameters have been investigated [241].

Kagiya *et al.* [86] have measured the IR wavenumber of the O—D and C=O stretching vibration of monodeuteromethanol (CH_3OD) and acetophenone, respectively, in various solvents. Taking benzene as the reference solvent, they have used the wavenumber shift relative to the maximum found in benzene as a measure of the electron-pair donating power and the electron-pair accepting power of a given solvent. Measurements of the O—D stretching band of CH_3OD, $\tilde{\nu}_{MeOD}$, have been greatly extended (up to 55 aprotic solvents) and improved by Shorter *et al.* [87]. Based on the measurements of Kagiya *et al.* [86], Koppel and Pal'm [6] have defined a Lewis basicity parameter B_{MeOD} using Eq. (7-35),

$$B_{MeOD}/\text{cm}^{-1} = \Delta\tilde{\nu}_{OD} = \tilde{\nu}^0_{MeOD} - \tilde{\nu}_{MeOD\cdots B} \tag{7-35}$$

where $\tilde{\nu}^0_{MeOD}$ and $\tilde{\nu}_{MeOD\cdots B}$ refer to the O—D stretching vibration of CH_3OD measured in the gas phase ($\tilde{\nu}^0_{MeOD} = 2720$ cm^{-1} [87]) and in a given HBA solvent B. A comprehensive collection of newly determined B_{MeOD}-values can be found in reference [87].

In a similar way, another set of solvent Lewis basicity parameters B_{PhOH} based on band shifts of the O—H stretching vibration of phenol in tetrachloromethane induced by hydrogen-bond formation with added HBA solvents B, was used by Koppel and Paju [88]

* The abbreviation G was chosen from the name of a close aquaintance of one of the authors [85]; private communication to C. R.

to classify 198 solvents, according to Eq. (7-36) ($B_{PhOH}=0$ for CCl_4):

$$B_{PhOH}/cm^{-1}=\Delta\tilde{\nu}_{OH}^{CCl_4}=\tilde{\nu}_{PhOH}^{CCl_4}-\tilde{\nu}_{PhOH\cdots B}^{CCl_4} \tag{7-36}$$

This B_{PhOH}-scale was further extended by Makitra *et al.* [242]. It has been shown that the spectroscopically determined B_{PhOH}-values correlate well with Gutmann's calorimetrically measured donor numbers *DN* [243]; *cf.* Eq. (7-10).

Another remarkable IR spectroscopic parameter of solvent Lewis basicity has recently been introduced by Laurence *et al.* [239], using the so-called *infrared comparison method* analogous to Kamlet and Taft's solvatochromic comparison method [224]. The band maxima wavenumbers of the C=O stretching vibration of the two homomorphs CCl_3CO_2H and $CCl_3CO_2CH_3$ have been measured in the gas phase as well as in non-HBD and non-HBA solvents in order to establish a reference line which follows Eq. (7-37); *cf.* Fig. 7-3.

$$\tilde{\nu}_{C=O}^{CCl_3CO_2H}=1.1715\cdot\tilde{\nu}_{C=O}^{CCl_3CO_2CH_3}-292.8 \tag{7-37}$$

This linear relationship demonstrates the similarity of the polarity effects on both homomorphic vibrators, trichloroacetic acid and its methyl ester. For HBA solvents B, however, the $\tilde{\nu}_{C=O}$ data points are displaced below the reference line of Eq. (7-37). These deviations are caused by the formation of solute/solvent hydrogen bonds $CCl_3CO_2H\cdots B$, resulting in a decrease in the C=O vibration wavenumber. The hydrogen-bond induced wavenumber shift $\Delta\tilde{\nu}_{C=O}$ for a HBA solvent is then calculated by Eq. (7-38); *cf.* Eq. (7-33).

$$\begin{aligned}\Delta\tilde{\nu}_{C=O}&=\tilde{\nu}_{C=O}^{CCl_3CO_2H}\text{ [calc. from Eq. (7-37)]}-\tilde{\nu}_{C=O}^{CCl_3CO_2H}\text{ [observed]}\\&=[1.1715\cdot\tilde{\nu}_{C=O}^{CCl_3CO_2CH_3}-292.8]-\tilde{\nu}_{C=O}^{CCl_3CO_2H}\end{aligned} \tag{7-38}$$

Due to this subtraction the nonspecific polarity effect of the solvent is disentangled from its basicity effect on the C=O vibrator of CCl_3CO_2H in HBA solvents. $\Delta\tilde{\nu}_{C=O}$-values range from 1.4 cm^{-1} for benzene to 27.2 cm^{-1} for trimethyl phosphate [239]. The $\Delta\tilde{\nu}_{C=O}$ displacements correlate well with the $\Delta\tilde{\nu}_{OD}$-values of Eq. (7-35). However, this $\Delta\tilde{\nu}_{C=O}/\Delta\tilde{\nu}_{OD}$ relationship is family-dependent, *i.e.* different linear correlations lines are obtained for different classes (families) of organic HBA solvents such as π, S, O, and N bases [239]. Unfortunately, $\Delta\tilde{\nu}_{C=O}$-values are not available for carbonyl-containing solvents because of insufficient solvent transparency, and for more basic solvents such as water, DMSO and HMPT because of decarboxylation and proton-transfer reactions.

Electron spin resonance (ESR) and nuclear magnetic resonance (NMR) measurements have also been used to establish solvent polarity scales. Knauer *et al.* [89] determined the nitrogen hyperfine splitting constants, $A(^{14}N)$, of several aminyloxide radicals (nitroxides) such as di-*tert*-butyl aminyloxide in 31 solvents. It has been known for some time that the ^{14}N isotropic HFS constant in the ESR spectrum of aminyloxide free radicals is sensitive to the polarity of the solvents in which they are dissolved (*cf.* Table 6-5 and the discussion in Section 6.4). Since $A(^{14}N)$ is easily measured in almost all solvents, it may prove to be useful as an empirical solvent polarity parameter, especially in cases where the other parameters are difficult to determine owing to limited solubility. Only very low radical concentrations are usually needed to obtain an ESR spectrum. The $A(^{14}N)$ values of di-*tert*-butyl aminyloxide are included in Table 7-2. Further $A(^{14}N)$-values of di-*tert*-butyl aminyloxide have been determined by Kolling [244], Reddoch *et al.* [245], and Symons *et al.* [246], who also discussed their solvent-dependence, their applicability as solvent parameters, and their correlations with other solvent parameters in more detail. The $A(^{14}N)$-values of 2,2,6,6-tetramethylpiperidine-1-oxide (TMPNO) [247, 248] and 4-acetyl-1-methylpyridinyl free radicals [244] have also been discussed as potential solvent polarity probes. A critical comparison of the TMPNO $A(^{14}N)$-values with other physical and empirical solvent parameters has been given by Kecki *et al.* [248]. For example, the $A(^{14}N)$-values of aminyl oxides correlate well with the $E_T(30)$- and *AN*-values, but not with the solvent dielectric constants or dipole moments or functions thereof [245, 248].

Interesting solvent scales based on NMR measurements have been proposed by Taft *et al.* [90] and by Gutmann *et al.* [91]. A solvent polarity parameter, designated as *P*, has been defined by Taft *et al.* [90] as the ^{19}F chemical shift (in ppm) of 4-fluoronitrosobenzene in a given solvent, relative to the same quantity in the reference solvent cyclohexane (*cf.* Table 6-6 and the discussion in Section 6.5.1). These parameters define a scale ranging from $P=0.0$ in cyclohexane to $P=2.7$ in sulfolane, and can easily be measured in a wide variety of solvents. The *P*-values appear to be related to the ability of the solvents to form specific 1:1 complexes with the nitroso group of the standard compound. A compilation of *P*-values is found in reference [92]. In addition, ^{13}C chemical shifts of (trifluoromethyl)benzene and phenylsulfur pentafluoride have been used by Taft *et al.* to study nonspecific dipolar interactions with HBD solvents and utilized to define π^*-values of solvent dipolarity/polarizability for protic solvents [249].

Complementary to the donor numbers *DN* [*cf.* Eq. (7-10)], Gutmann *et al.* introduced so-called acceptor numbers *AN* as measures of the Lewis acidity or EPA property of organic solvents [91, 134, 207, 251]. Acceptor numbers are derived from the relative ^{31}P-NMR chemical shift values, δ_{corr}, of triethylphosphane oxide, related to those of the 1:1 adduct Et_3PO—$SbCl_5$. δ_{corr} is the observed ^{31}P chemical shift corrected for concentration effects and for differences in magnetic volume susceptibility.

$$(Et_3P{=}O \longleftrightarrow Et_3\overset{\oplus}{P}{-}\overset{\ominus}{O}) + A \rightleftharpoons Et_3\overset{\delta\oplus}{P}{\cdots}\overset{\delta\ominus}{O}{-}A$$

With *n*-hexane as the reference solvent, values of *AN* for a given solvent A are calculated according to Eq. (7-39).

$$AN=\frac{\delta_{corr}(A)-\delta_{corr}(n\text{-}C_6H_{14})}{\delta_{corr}(Et_3PO\text{—}SbCl_5)-\delta_{corr}(n\text{-}C_6H_{14})}\cdot 100 \qquad (7\text{-}39)$$

The *AN*-scale was set up by defining $AN=0$ for *n*-hexane, and $AN=100$ for Et_3PO—$SbCl_5$ dissolved in 1,2-dichloroethane. The term $[\delta_{corr}(Et_3PO\text{—}SbCl_5)-\delta_{corr}(n\text{-}C_6H_{14})]$ was experimentally found to be 42.58 ppm; Eq. (7-39) reduces therefore to Eq. (7-40).

$$AN=\frac{\delta_{corr}(A)-\delta_{corr}(n\text{-}C_6H_{14})}{42.58\ \text{ppm}}\cdot 100=\Delta\delta_{corr}\cdot 2.348/\text{ppm} \qquad (7\text{-}40)$$

A selection of *AN*-values has already been given in Table 2-5 of Section 2.2.6; *cf.* also Table 6-6 in Section 6.5.1. The observed solvent-dependent ^{31}P chemical shifts result mainly from the polarization of the dipolar P=O group, induced by the interaction with electrophilic solvents A, particularly HBD solvents. The decrease in electron density at the phosphorus atom results in a deshielding proportional to the strength of the probe/solvent interaction. In solutions of protonic acids, the ^{31}P chemical shift of the O-protonated triethyl hydroxyphosphonium salt is observed. Since Et_3PO is very hygroscopic and therefore not very suitable from an experimental point of view, the use of $(n\text{-Bu})_3PO$ instead of Et_3PO as probe molecule has been recommended [250].

The *AN*-values represent dimensionless numbers expressing the Lewis acidity properties of a given solvent A relative to those of $SbCl_5$, which is also the standard compound for setting up the donor number scale. Owing to the good solubility properties of triethylphosphane oxide, acceptor numbers are available for many types of coordinating and non-coordinating solvents. They are particularly useful in characterizing the Lewis acidities of protic solvents and protonic acids. Relationships have been found between the acceptor number and other empirical parameters such as Z, $E_T(30)$, and Y-values, as well as many thermodynamic solvation quantities and other solvent-dependent processes [91, 139, 140, 207, 250–253]. The fairly good linear correlations of AN with Z and $E_T(30)$ reveal that the latter represent, to a large extent, measures of the electrophilic properties of the solvents, and therefore of necessity fail when applied to reactions which are mainly influenced by the nucleophilic properties of the solvents. The linear correlation between AN and $E_T(30)$-values for 21 non-HBD solvents is so satisfactory that an AN_E-scale has been established using the $E_T(30)$-values available for a greater number of solvents (*cf.* Table 7-3), using the correlation equation $AN_E = 1.29 \cdot E_T(30) - 40.52$ [207].

As a further measure of the solvent electrophilicity the ^{59}Co NMR chemical shift values of tetra-*n*-butylammonium hexacyanocobaltate(III), $(n\text{-Bu}_4N)_3Co(CN)_6$, have been recommended [254]. Hydrogen bonding of the type $^{59}Co{-}CN\cdots HS$ brings about a large shift of the ^{59}Co resonance signal in direct proportion to the strengths of this specific solute/solvent interaction. Except for water, a good linear correlation exists between $\delta(^{59}Co)$ and AN-values as well as for Y_{OTs}-values. From the observed water deviation it has been concluded that spectroscopic solvent parameters such as AN and $\delta(^{59}Co)$ characterize the electrophilicity of "monomeric" solvents, whereas kinetic parameters such as Y_{OTs} characterize the electrophilicity of "polymeric" solvents. Plots of Y_{OTs} against $\delta(^{59}Co)$ and AN show that the electrophilicity of "polymeric" water, $(H_2O)_n$, is greater than that of "monomeric" water [254].

7.5 Empirical Parameters of Solvent Polarity from Other Measurements

An important measure of the total molecular cohesion per unit volume of liquid, is the *cohesive pressure c* (also called *cohesive energy density*), which characterizes the energy associated with all the intermolecular solvent/solvent interactions in a mole of the solvent. The cohesive pressure is defined as the molar energy of vaporization to a gas at zero pressure, ΔU_v, per molar volume of the solvent, V_m, according to Eqs. (3-5) and (5-76) in

Sections 3.2 and 5.4.2, respectively [93, 94]. The cohesive pressure c is related to the *internal pressure* π; *cf.* Eq. (3-6) and Table 3-2 in Section 3.2.

The square root of the cohesive pressure has been termed by Hildebrand and Scott as the solubility parameter δ because of its value in correlating and predicting solubility behaviour of nonelectrolytes (ionogens) [93, 94]; *cf.* Eqs. (2-1) and (5-77) in Sections 2.1 and 5.4.2, respectively. Comprehensive lists of δ-values are given in references [94] and [95]; *cf.* Table 3-3 in Section 3.2 for a selection. Correlations of δ-values with other solvent polarity parameters have been attempted, but in most cases only relatively poor mutual relationships were found [96, 97, 255, 256]. Since δ-values which are characterized by the energy needed to separate molecules of the liquid measure the attractive forces between the solvent molecules only, it need not necessarily be a measure of the solute/solvent interaction forces. It could be that some of the solvent/solvent interaction forces are also of relevance in particular solute/solvent interactions. Rather, the δ-values are related to the energy necessary to form a cavity in the solvent which can then accommodate the solute molecule. In accordance with this is the experimental observation that a good solvent for a certain nonelectrolyte solute should have a δ-value close to that of the solute [93, 94]. Depending on the solvent, creation of a cavity in an organic solvent requires about 20...40 kJ/mol (5...10 kcal/mol).

Sometimes, Hildebrand's solubility parameter δ has been incorrectly used in linear Gibbs energy relationships; *cf.* for example [96, 97, 226, 255]. Since in linear Gibbs energy relationships the correlated solvent-dependent solute properties (*e.g.* lg K, lg k, $h \cdot \nu$) are proportional to Gibbs energy changes of reaction or activation (ΔG, $\Delta G^{\neq}$) and excitation energies (E_T), all the terms of a regression equation should include an energy dimension; *cf.* Eqs. (7-2) and (7-3) in Section 7.1. Therefore, it is unjustified to use δ-values in such regression equations because they have the dimension of the square root of the energy ($J^{1/2} \cdot cm^{-3/2}$). Instead of δ, the cohesive pressure c which is equal to δ^2 should be used, as has been demonstrated in some more recent work dealing with the inclusion of cavity terms in multiparameter regression equations [256–258].

It has been shown, that gas-liquid chromatographic methods are particularly suitable for a quantitative characterization of the polarity of solvents. In gas-liquid chromatography it is possible to determine the solvent power of the stationary liquid phase very accurately for a large number of substances [98, 99, 259, 260]. Many groups of substances exhibit a certain dependence of their relative retention parameters on the solvation characteristics of the stationary phase or of the separable components. In determining universal gas-chromatographic characteristics the so-called *retention index*, I, introduced by Kováts [100], is frequently used. The elution maxima of individual members of the homologous series of n-alkanes (C_nH_{2n+1}) form the fixed points of the system of retention indices. The retention index is defined by means of Eq. (7-41),

$$I = 100 \cdot \frac{\lg V_x - \lg V_n}{\lg V_{n+1} - \lg V_n} + 100 \cdot n \qquad (7\text{-}41)$$

where V_x is the specific retention volume of solute x and V_n and V_{n+1} are the retention volumes of two n-alkanes with n and $n+1$ carbon atoms, respectively; hence $V_n < V_x < V_{n+1}$. The retention index is independent of the gas-chromatographic equipment used, and depends only on the solute x, temperature, and stationary phase. The retention index specifies with which *n*-alkane a solute leaves the separation column, whereby the number of carbon atoms of the respective *n*-alkane is multiplied by 100. In other words, a retention index of 800 or 1100 means that the solute leaves the column simultaneously with either *n*-octane or *n*-undecane, respectively. A retention index of 732 implies that the solute exhibits the same retention time as a hypothetical *n*-alkane with 7.32 carbon atoms.

Kováts and Weiß [101] used these retention indices to examine the polarity of stationary liquid phases. If the total Gibbs energy of dissolution is determined by the work of separating the solvent molecules to form a hole, by a dispersion term, a polar term, and the energy of the donor/acceptor interaction, the sum of the last two terms is proportional to the difference between the retention indices of the solute on the column with a given liquid phase X (I_T^X) and with a standard nonpolar stationary phase (*i.e.* a hydrocarbon; $I_T^{St\,Ap}$) at temperature T. This difference, $\Delta I_T^X = I_T^X - I_T^{St\,Ap}$, is then proposed as the polarity parameter. Using 1-chloro- and 1-bromo-*n*-hexadecane as the standard dipolar stationary phase, and *n*-hexadecane as the standard nonpolar stationary phase, $\Delta I_{50}^{Cl/Br}$ was defined as the new polarity parameter according to Eq. (7-42) [101].

$$\Delta I_{50}^{Cl/Br} = I_{50}^{1-\text{chloro or } 1-\text{bromo}-n-\text{hexadecane}} - I_{50}^{n-\text{hexadecane}} \qquad (7\text{-}42)$$

The $\Delta I_{50}^{Cl/Br}$-values show correlations with other empirical solvent polarity parameters such as Z, $E_T(30)$, and $\lg k_1$ of 4-methoxyneophyl tosylate solvolysis [101].

Another, more rigorous approach, based on retention indices and taking into account that the polarity of the column depends not only on the nature of the stationary phase but also on the type of substance analyzed, has been suggested by Rohrschneider [98, 102]. The polarity of a stationary phase must therefore be assessed simultaneously with respect to a whole group of compounds possessing varying donor/acceptor properties. Rohrschneider proposed the five-term equation (7-43) for the solute retention which has found wide use in gas-chromatography [98, 99, 259, 260].

$$\Delta I = a \cdot x + b \cdot y + c \cdot z + d \cdot u + e \cdot s \qquad (7\text{-}43)$$

a, b, c, d, and e are polarity factors of the stationary liquid phase, and x, y, z, u, and s are polarity factors of the substances undergoing analysis. Having determined the polarity factors of 22 different stationary phases with respect to chosen standard substances (benzene, ethanol, 2-butanone, nitromethane, and pyridine), and those of 30 other solutes, Rohrschneider calculated 660 retention indices with only small limits of error.

Using a different set of standard substances, *i.e.* substituting 1-butanol, 2-pentanone, and 1-nitropropane for the rather volatile ethanol, 2-butanone, and nitromethane, McReynolds developed an analogous approach [103]. Altogether, he characterized over 200 liquid stationary phases using a total of 10 probes. A statistical analysis of the McReynolds retention index matrix using the principal component analysis method has shown, that only three components are necessary to reproduce the experimental data matrix [262]. The first component is related to the polarity of the liquid phase, the second depends almost solely on the solute, and the third is connected to specific interactions with solute hydroxy groups [262].

Both of these approaches used in the characterization of stationary liquid phase polarities by means of retention indices have been further explored and expanded [104, 259–261]. Similar methods for the characterization of solvent polarity in liquid-liquid and liquid-solid chromatography are found in references [105–107]; *cf.* also Chapter A-7 and Tables A-10 and A-11 in the Appendix.

7.6 Interrelation and Application of Solvent Polarity Parameters

From the previous Sections we can conclude that there are many empirical solvent scales, the most comprehensive of which are the solvatochromic ones; *cf.* for example Table 7-3. Unfortunately, almost every year new solvent scales are proposed. At present, more than 30 different solvent polarity scales are known*). However, only about eight of them have found wider application in the correlation analysis of solvent effects, *i.e.* Y, Z, $E_T(30)$ α, β, π^*, DN, and AN. The application of most solvent scales is restricted by the fact that they are known only for an insufficient number of solvents. The repertoire of common organic solvents available to the chemist consists of about 300 solvents, not to speak of the infinite number of solvent mixtures! The extension of most solvent scales is restricted by the inherent properties of the selected reference process which exclude the determination of solvent parameters for certain, often important solvents (*e.g.* chemical reactions between solute and solvent; solubility problems, *etc.*). For this reason, the most comprehensive solvent polarity scales are those derived from spectroscopic reference processes, which are the most easily measured for a large set of organic solvents.

In general, all these parameters constitute more comprehensive measures of solvent polarity than the dielectric constant or any other single physical characteristic, since they reflect more reliably the complete picture of all intermolecular forces acting between solute and solvent molecules. The solvent-dependent processes used to define solvent polarity parameters may be regarded as probes which permit a purely empirical investigation of solvent effects. In applying these parameters, however, it is tacitly assumed that the intermolecular interactions in the reference system used to develop a particular solvent scale are similar to those in the system the prediction of whose solvent effects is being raised. This is obviously true only for strongly related solvent-dependent processes. Therefore, the use of single solvent parameters to predict solvent effects on equilibria, reaction rates, and spectral absorptions should be very restricted. One cannot expect a parameter to be universally useful for all kinds of solvent-sensitive processes, since any correlation of solvent effects of a particular process with a solvent polarity parameter is, in fact, a comparison with the effect of solvent on a reference process. But, if one compares the various empirical solvent scales (*cf.* Tables 7-1 to 7-3), based on very different solvent-sensitive reference processes, varying strongly in the energies involved, one finds surprisingly, that most of the existing empirical solvent scales agree with each other very well qualitatively and even sometimes quantitatively. In spite of the large

* At present, the situation is not quite as bad as in the correlation analysis of substituent effects where even more substituent parameters than common substituents seem to be known. It has been suggested that new solvent polarity scales should only be introduced into the literature if they exhibit significant advantages over existing solvent scales [235].

energy changes connected with a solvent change from methanol to hydrocarbon solvents for Z (*ca.* 100 kJ/mol), $E_T(30)$ (*ca.* 105 kJ/mol), $\lg k_1$ of 4-methoxyneophyl tosylate solvolysis ($-\ln 10 \cdot RT \cdot \lg k_1$ *ca.* 29 kJ/mol), and Ω values ($-\ln 10 \cdot RT \cdot \Omega$ *ca.* 2 kJ/mol), in all four cases one obtains similar polarity orders for the solvents. This led Berson to observe that "... in this respect a set of solvents behaves like an elephant, which can lift a log or a peanut with equal dexterity" [52].

In particular, there are good linear correlations between the $E_T(30)$-values and some other empirical solvent polarity parameters according to Eq. (7-44),

$$y = a \cdot E_T(30) + b \tag{7-44}$$

where a and b were determined by the method of least squares. A compilation of such linear correlations between $E_T(30)$-values and twelve other solvent polarity parameters can be found in reference [124]. For example, there is a satisfactory linear correlation between the values of $E_T(30)$ and Z according to Eq. (7-45) ($n = 54$ solvents; correlation coefficient $r = 0.978$).

$$Z/(\text{kcal} \cdot \text{mol}^{-1}) = 1.337 \cdot E_T(30)/(\text{kcal} \cdot \text{mol}^{-1}) + 9.80 \tag{7-45}$$

The excellent $Z/E_T(30)$ correlation for a selected set of 15 solvents common to both scales have been used to calculate $E_T(30)$-values from Z-values for acidic solvents for which $E_T(30)$-values are not available [172]; *cf.* footnote[c)] of Table 7-3. A similar satisfactory linear correlation between $E_T(30)$-values and acceptor numbers allows the calculation of AN-values which are not directly available [207].

Of particular interest is the correlation between $E_T(30)$-values and Kamlet-Taft's solvatochromic parameters α, β, and π^* [84a, 226, 235]. For 32 non-HBD solvents, a satisfactory linear correlation according to Eq. (7-46) is obtained ($n = 32$; $r = 0.972$) [226],

$$E_T(30)/(\text{kcal} \cdot \text{mol}^{-1}) = 14.6\,(\pi^* - 0.23\,\delta) + 30.31 \tag{7-46}$$

where δ, a polarizability correction term, is 0.0 for nonchlorinated aliphatic solvents, 0.5 for polychlorinated aliphatics, and 1.0 for aromatic solvents; *cf.* Eq. (7-53) in Section 7.7. For a group of twelve so-called *select solvents*, *i.e.* non-HBD aliphatic solvents with a single dominant group dipole moment, the δ-term disappears ($\delta = 0$) and a simplified and improved linear correlation is obtained ($n = 12$; $r = 0.987$) [84a]. – If protic solvents are included in the overall solvent set, then Eq. (7-46) has to be enlarged by the introduction of the α-parameter of solvent HBD acidity to give the two-parameter Eq. (7-47) ($n = 44$ solvents); *cf.* Section 7.7 for a further discussion of multiparameter correlation equations.

$$E_T(30)/(\text{kcal} \cdot \text{mol}^{-1}) = 14.6\,(\pi^* - 0.23\,\delta) + 16.5\,\alpha + 30.31 \tag{7-47}$$

Eq. (7-47) is in agreement with the assumption that $E_T(30)$-values measure not only a blend of solvent dipolarity and polarizability but also the solvent HBD acidity (Lewis acidity) of protic solvents. For attempts to improve these $E_T(30)/\pi^*$ correlations using modified Kamlet-Taft parameters (π^*_n and α_n) see references [234, 235].

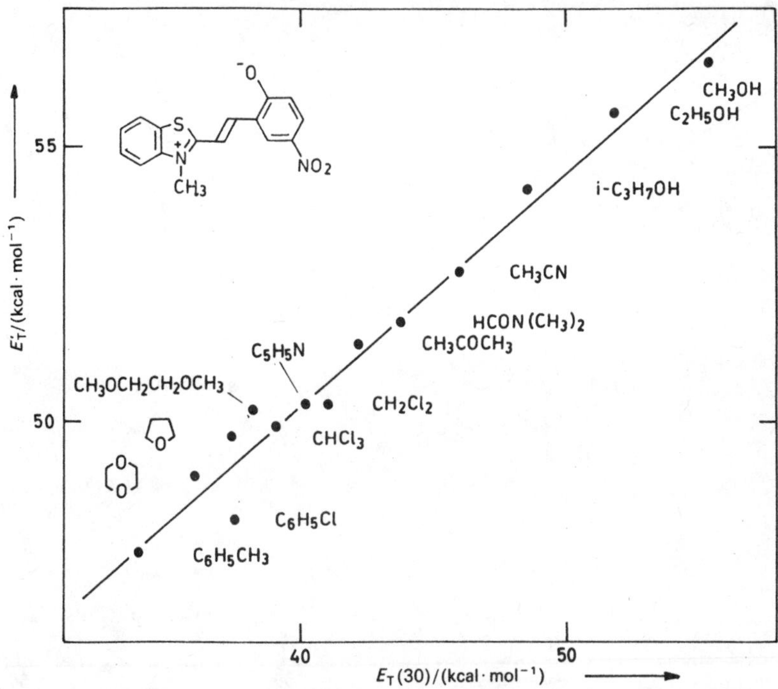

Fig. 7-4. Correlation between $E_T(30)$ [66] and the $\pi-\pi^*$ transition energy, E_T', of a merocyanine dye [108]. $E_T' = 0.42 \cdot E_T(30) + 33.5$ ($n = 14$; $r = 0.986$; $s = 0.482$).

According to Bekárek *et al.* [194], the $E_T(30)$-values can be correlated with solvent dielectric constants ε_r and solvent refraction indices n of 40 non-HBD aliphatic solvents according to Eq. (7-48) ($n = 44$; $r = 0.958$),

$$E_T(30)/(\text{kcal} \cdot \text{mol}^{-1}) = 72.02\, f^2(\varepsilon_r) - 29.16\, f(\varepsilon_r, n^2) + 29.87 \tag{7-48}$$

where $f(\varepsilon_r)$ is equal to the Kirkwood function $(\varepsilon_r - 1)/(2\varepsilon_r + 1)$ and $f(\varepsilon_r, n^2)$ represents the cross term $(\varepsilon_r - 1)(n^2 + 1)/(2\varepsilon_r + 1)(2n^2 + 1)$. From the magnitude of the regression coefficients it can be concluded that the $f(\varepsilon_r)$ term is the predominant influence on the $E_T(30)$-parameter when only non-specific solute/solvent interactions are considered [194].

The application of single solvent parameters such as the $E_T(30)$-values in correlating other solvent-dependent processes has been shown to be surprisingly successful. This means that the blend of different solute/solvent interaction forces, as measured by the solvatochromic pyridinium-*N*-phenoxide betaine dye *(44)*, is nearly the same as in numerous other solvent-dependent chemical reactions and spectral absorptions. It seems, therefore, that the blend of solute/solvent interactions as measured by the $E_T(30)$-values represents a kind of "mean solvent polarity", representative for many solvent-sensitive

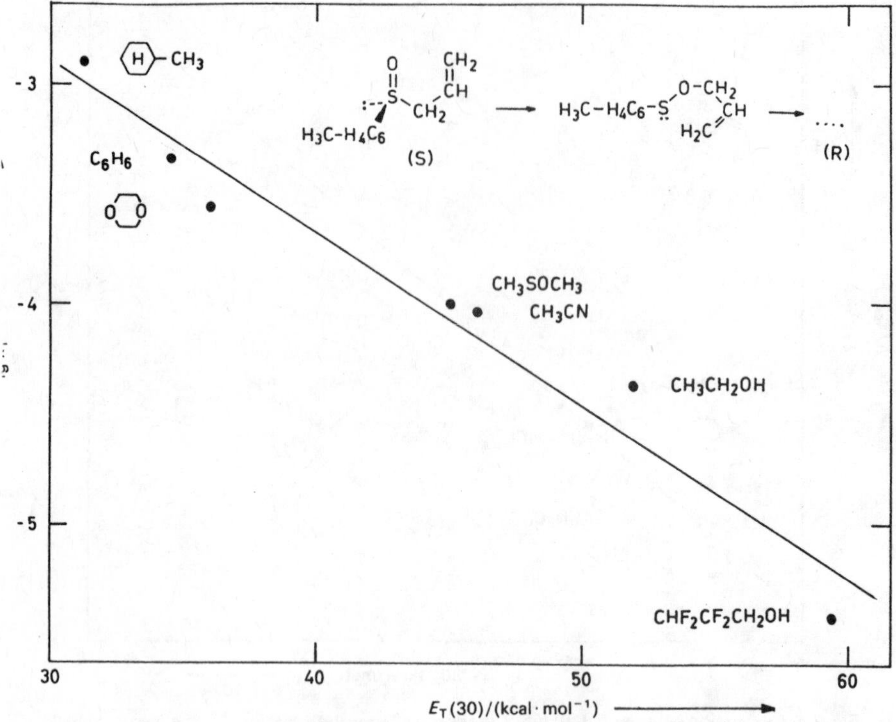

Fig. 7-5. Correlation between $E_T(30)$ [66] and lg k_1 of the thermal racemization of chiral allyl *p*-tolyl sulfoxide at 60.7 °C [109]; *cf.* Eq. (5-38). $\lg k_1 = -0.078 \cdot E_T(30) - 0.55$ ($n = 7$; $r = -0.976$; $s = 0.222$).

processes. As Eq. (7-47) exhibits, $E_T(30)$-values represent a combined measure of solvent dipolarity/polarizability and solvent HBD acidity (Lewis acidity), the latter property being significant only for protic (HBD) solvents. $E_T(30)$-correlations for more than 100 solvent-sensitive processes have been collected in references [2, 124, 200–203]. Three typical examples are given in Figs. 7-4 to 7-6.

In Fig. 7-4 an excellent correlation between $E_T(30)$-values and the $\pi - \pi^*$ transition energies of the open-chain form of a merocyanine dye obtained by irradiation of the corresponding photochromic benzthiazolospiropyrane is shown [108] (*cf.* the related spiropyrane *(27a)*/merocyanine *(27b)* equilibrium in Section 4.4.2).

The structure of the open-chain form was assigned on the basis of its negative solvatochromic behaviour, similar to that of other meropolymethines such as the pyridinium-*N*-phenoxide betaines [108]. The correlation shown in Fig. 7-4 allows one to calculate absorption maxima of the merocyanine dye in other solvents for which $E_T(30)$-values are known.

A correlation between $E_T(30)$-values and the rate of thermal racemisation of chiral allyl *p*-tolyl sulfoxide according to Eq. (5-38) in Section 5.3.2 is shown in Fig. 7-5 [109].

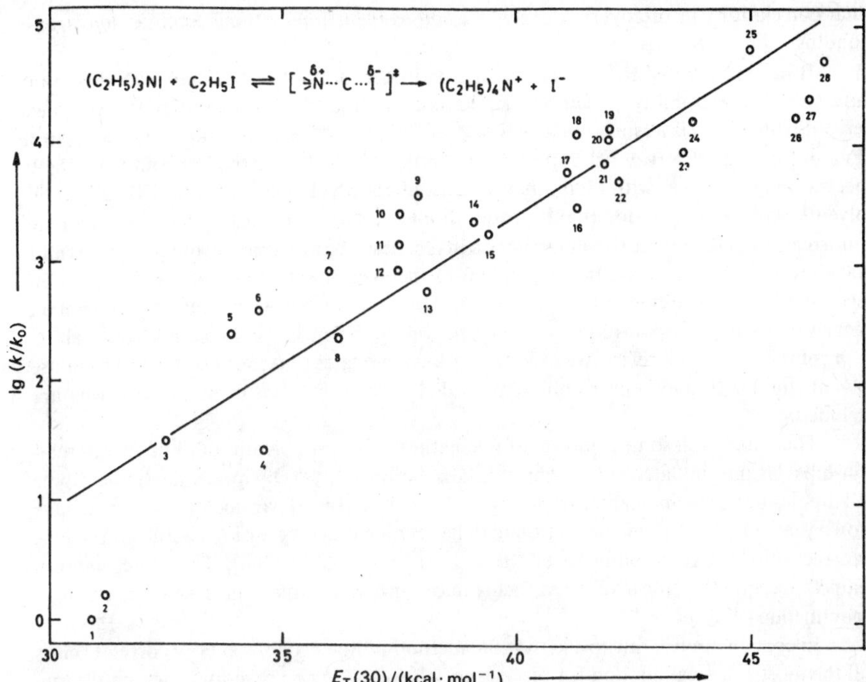

Fig. 7-6. Correlation between $E_T(30)$ [66] and lg (k/k_0) for the reaction between triethylamine and iodoethane in aprotic and dipolar aprotic solvents at 25 °C (rate constants relative to *n*-hexane as 'slowest' solvent) [110].
lg $(k/k_0) = 0.248 \cdot E_T(30) - 6.54$ ($n = 28$; $r = 0.920$). (1) *n*-hexane, (2) cyclohexane, (3) tetrachloromethane, (4) diethyl ether, (5) toluene, (6) benzene, (7) 1,4-dioxane, (8) 1,1,1-trichloroethane, (9) iodobenzene, (10) bromobenzene, (11) chlorobenzene, (12) tetrahydrofuran, (13) ethyl acetate, (14) chloroform, (15) 1,1-dichloroethane, (16) 2-butanone, (17) dichloromethane, (18) acetophenone, (19) nitrobenzene, (20) benzonitrile, (21) 1,2-dichloroethane, (22) acetone, (23) propionitrile, (24) *N*,*N*-dimethylformamide, (25) dimethyl sulfoxide, (26) acetonitrile, (27) nitromethane, and (28) propylene carbonate.

The linear decrease in reaction rate with increasing solvent polarity has been considered as evidence in support of the proposed reaction mechanism, involving a less dipolar cyclic activated complex (*cf.* discussion of this reaction in Section 5.3.2).

Fig. 7-6 demonstrates the correlation between $E_T(30)$ and the relative rates for the S_N2 Menschutkin reaction between a tertiary amine and a haloalkane in non-HBD solvents. The values of the second-order rate constants are taken from the compilation made by Abraham and Grellier [110].

A comparison of Fig. 7-6 with Fig. 5-11 in Section 5.4.3 reveals some improvement of the poor correlation between lg (k/k_0) and the Kirkwood function $(\varepsilon_r - 1)/(2\varepsilon_r + 1)$. If protic solvents are included in the $E_T(30)$/lg (k/k_0) correlation, two lines are obtained as in Fig. 5-12, one for non-HBD and one for protic solvents. This demonstrates the utility of such correlations in discovering specific solute/solvent interactions such as hydrogen-bonding.

It has been stated that, when specific hydrogen-bonding effects are excluded, and differential polarizability effects are similar or minimized, the solvent polarity scales derived from UV/Vis absorption spectra (Z, S, $E_T(30)$, π^*, χ_R, E_K), fluorescence spectra (Py), infrared spectra (G), ESR spectra [$A(^{14}N)$], ^{19}F-NMR spectra (P), and ^{31}P-NMR spectra (AN) are linear with each other for a set of select solvents, *i.e.* non-HBD aliphatic solvents with a single dominant group dipole [263]. This result can be taken as confirmation that all these solvent scales do in fact describe intrinsic solvent properties and that they are to a great extent independent of the experimental methods and indicators used in their measurement [263]. That these empirical solvent parameters correlate linearly with solvent dipole moments and functions of the dielectric constants (either alone or in combination with refractive index functions) indicates that they are a measure of the solvent dipolarity and polarizability, provided specific solute/solvent interactions are excluded.

There has been some criticism of the method of linear solvation Gibbs energy relationships. Using simulated solvatochromic correlation analyses by modeling dipole/dipole and dipole/induced-dipole interactions with a single combined parameter, Carr *et al.* have shown that, although good correlations can be obtained, the regression coefficients can be incorrect and not representative of the system under study [264]. Therefore, caution against overinterpretation of solvatochromic regression equations has been strongly recommended [264].

In conclusion it is fair to say that the method of linear Gibbs energy correlations is still the most practical method for predicting solvent effects on reaction rates, equilibria, and spectral absorptions, as well as for predicting substituent effects for reactions in solution. In so far as one understands why the model process responds to a solvent change, something can be learnt about the particular process under study. This kind of procedure has been criticized as being too empirical. But one should take into account that not only the basic postulates of linear Gibbs energy relationships (*i.e.* additivity and separability) are of theoretically acceptable form, but the choice of a suitable reference process also requires intensive application of theory since the solute/solvent interaction of the model process and the process under investigation must be closely related.

7.7 Multiparameter Approaches

In spite of the observation that single empirical parameters, such as those mentioned in the foregoing Sections, may serve as good approximations of solvent polarity in the sense defined in Section 7.1 [1,3], there are many examples of solvent-sensitive processes known which cannot be correlated to only one empirical solvent parameter. It has been repeatedly found, that the simple concept of "polarity" as a universally determinable and applicable solvent characteristic is a gross oversimplification. The solvation capability or solvation power of a solvent, which has been roughly divided into non-specific and specific solute/solvent interactions, is the result of many different kinds of interaction mechanisms between the molecules of the solute and the solvent (*cf.* Section 2.2). Solvent effects are basically more complicated and often more specific than substituent effects. In the latter case, linear Gibbs energy relationships such as the Hammett equation are well-established and known to work very well. In order to take into account two or more aspects of solvation, a multiparameter approach of the general form

$$A = A_0 + b \cdot B + c \cdot C + d \cdot D + \ldots \quad (7\text{-}49)$$

has been tried where A is the value of a solvent-dependent physico-chemical property (lg K, lg k, $h \cdot \nu$, *etc.*) in a given solvent, and A_0 is the statistical quantity corresponding to the value of this property in the gas phase or in an inert solvent. B, C, D,... represent independent but complementary solvent parameters which account for the different solute/solvent interaction mechanisms; b, c, d,... are the regression coefficients describing the sensitivity of property A to the different solute/solvent interaction mechanisms. Such an equation can be applied only to data for a large number of well-chosen solvents, and its success must be examined by proper statistical methods [14, 15]. The separation of solvent polarity into various solute/solvent interaction mechanisms is purely formal and may not be even theoretically valid as the interactions could be coupled, *i.e.* not operate independently of each other. But, if this separation can be reasonably done, the resultant parameters may be used to interpret solvent effects through such multiple correlations, thus providing information about the type and magnitude of interactions with the solvent. On the basis of Eq. (7-49) the often observed failure of single solvent parameters can readily be understood. Any single empirical solvent parameter must have a fixed relative sensitivity to each of the various interaction mechanisms implied in Eq. (7-49). Thus, only application to processes that have the same relative sensitivity to various interaction mechanisms as the single solvent parameter will give a good correlation. Studies of multiparameter approaches to solvent effects on physical and chemical properties based on the general Eq. (7-49) have been given by Katritzky *et al.* [111], Koppel and Palm [6, 112], Kamlet and Taft [84a, 224, 226], Krygowski and Fawcett [113], Swain *et al.* [265], Mayer [266, 251], and Dougherty [114].

Katritzky *et al.* [111] tested various multiparameter equations using linear combinations of existing empirical solvent parameters. The most successful treatment combines the $E_T(30)$-values (*cf.* Table 7-3) with functions of dielectric constant and index of refraction. Using $E_T(30)$ and the Kirkwood function $(\varepsilon_r - 1)/(2\varepsilon_r + 1)$, a two-parameter equation was constructed which allows independent variation of dipole/dipole and hydrogen-bonding forces. This equation is based on the assumptions, that the Kirkwood function adequately represents dipole/dipole interactions, and that $E_T(30)$-values are sensitive to both dipolar interactions and the interaction between solute and hydrogen-bond donor (HBD) solvents. It could be shown that correlations of rates, equilibria, and spectral properties are indeed significantly improved by such multi-parameter treatment which implicitly allows for various independent interaction mechanisms between solvent and solute ground, transition, and excited states [111].

A more rigorous approach has been suggested by Koppel and Palm [6, 112], who argue that a complete description of all solute/solvent interactions must include both non-specific and specific effects. They proposed the general four-parameter equation (7-50) which relates the variation of a given property A to two non-specific (Y and P) and two specific characteristics of the solvent (E and B).

$$A = A_0 + y \cdot Y + p \cdot P + e \cdot E + b \cdot B \quad (7\text{-}50)$$

A and A_0 are defined as in Eq. (7-49); by definition, A is equal to A_0 for the gas phase. The non-specific parameters Y*) and P measure solvent polarization and polarizability, respectively, according to classical dielectric theory. E and B are specific parameters measuring the Lewis acidity (electrophilic solvating power) and Lewis basicity (nucleophilic solvating power) of the solvent; and y, p, e, and b are the corresponding regression coefficients indicating the sensitivity of A to the four different solvent parameters. Dielectric constants, ε_r, are the basis of Y, and were used in the form of the Kirkwood function, $(\varepsilon_r - 1)/(2\varepsilon_r + 1)$, or as $(\varepsilon_r - 1)/(\varepsilon_r + 2)$, a function based on the expression for molar polarization. The functions $(n^2 - 1)/(2n^2 + 1)$ or $(n^2 - 1)/(n^2 + 2)$ of the refractive index for sodium light were used for the polarizability parameter, P. In the effective numerical range, these pairs of functions are approximately co-linear, and the choice between the two functions $f(\varepsilon_r)$ and the two functions $f(n^2)$, respectively, is largely arbitrary.

A scale of Lewis acidity, E, was based on the $E_T(30)$-values discussed in Section 7.4 (*cf.* Table 7-3), but these were corrected for the influence of non-specific effects, and adjusted to an origin $E = 0$ for the gas phase by means of Eq. (7-51) [6, 112, 115].

$$\begin{aligned} E &= E_T(30) - E_T^0(30) - y \cdot Y - p \cdot P \\ &= E_T(30) - 25.57 - 14.39 \cdot Y - 9.08 \cdot P \end{aligned} \quad (7\text{-}51)^{**)}$$

The subtraction of the polarization $(y \cdot Y)$ and polarizability $(p \cdot P)$ contributions from the total solvent effect allows an estimation of the contribution from specific solute-solvent interactions. This correction of $E_T(30)$-values was made using least-squares regression analysis by correlating the data for suitably selected non-specifically and specifically interacting solvents. E-values derived in this way from $E_T(30)$-values are presented in references [6, 115]; they range from zero (gas phase, saturated hydrocarbons) to about 22 kcal/mol for water. By definition, $e = 1$ in Eq. (7-50) for the reference process, *i.e.* the $\pi \rightarrow \pi^*$ transition of the pyridinium-*N*-phenoxide betaine dye *(44)*. The reasons for assuming that $E_T(30)$ and, thus, E largely relates to Lewis acidity in protic solvents was already mentioned***).

A scale of Lewis basicity, B was based on the O—D infrared stretching band of CH_3OD, according to Eq. (7-35) in Section 7.4 [6, 87, 112]. EPD solvents reduce $\tilde{\nu}_{O-D}$

---

* Not to be confused with the Y-values of Grunwald and Winstein in Eq. (7-14).

** The E scale has been extended and improved by Koppel and Paju [115], based on new published data for $E_T(30)$-values [66]. Eq. (7-51) thus becomes

$$E = E_T(30) - 25.10 - 14.84 \cdot Y - 9.59 \cdot P$$

*** The procedure for parametrization of solvent electrophilicity has been criticized, mainly because it was found that the use of $E_T(30)$ instead of E in the multiple regression treatment of solvent effects is often quite successfull; see reference [15, 116] for examples. It has been shown that values of $E_T(30)$ and E are linearly correlated, at least for solvents with an $E_T(30)$-value of greater than ca. 40 kcal/mol [178]. This calls into question the value of Koppel and Palm's division of $E_T(30)$ into pure electrophilicity effects and non-specific effects by means of Eq. (7-51).

through hydrogen-bonding and the wavenumber shift measures the strength of the HBD-solute/EPD solvent interaction. For use in Eq. (7-50) the shifts were adjusted to an origin $B=0$ for the gas phase. By definition, $b=1$ in Eq. (7-50) for the reference process, the IR wavenumber shifts of CH_3OD.

A detailed analysis of solvent effects on various solvent-sensitive processes by means of Eq. (7-50) has been given by Koppel and Palm [6, 112]. If Eq. (7-50) quantitatively reflects the influence of all basic types of solute/solvent interactions, it should be possible to correlate any solvent-dependent kinetic or spectral data in terms of Y, P, E, and B solvent parameters. This has been done for more than sixty solvent-sensitive processes [6]. Surprisingly, the greater part of the processes investigated by Koppel and Palm [6] depends only on a single specific solvent parameter: in fifty cases only electrophilic solvation is important (thus confirming the wide applicability of single $E_T(30)$-values), while in seven cases, nucleophilic solute/solvent interaction is the predominant solvation mechanism. Only one representative example will be mentioned in more detail in order to demonstrate what conclusions for any particular case can be drawn from Eq. (7-50). For a critical discussion of the Koppel-Palm treatment of solvent effects and valuable comments on its applicability see reference [15].

The rate of reaction between benzoic acid and diazodiphenylmethane in 44 non-HBD solvents correlates significantly with all four Koppel-Palm parameters of Eq. (7-50), according to Eq. (7-52) with $n=44$, $r=0.976$, and standard deviation $s=0.188$ [15, 116]; *cf.* also [117].

$$\lg k_2 = -3.13 + 4.58\, f(\varepsilon_r) + 11.96\, f(n^2) + 0.195\, E - 0.018\, B \qquad (7\text{-}52)$$

The regression coefficients e and b are of opposite sign, and the order of decreasing significance of the four terms is $B > f(\varepsilon_r) > E > f(n^2)$ as shown by the stepwise regression. This result supports the mechanism given in the following reaction scheme:

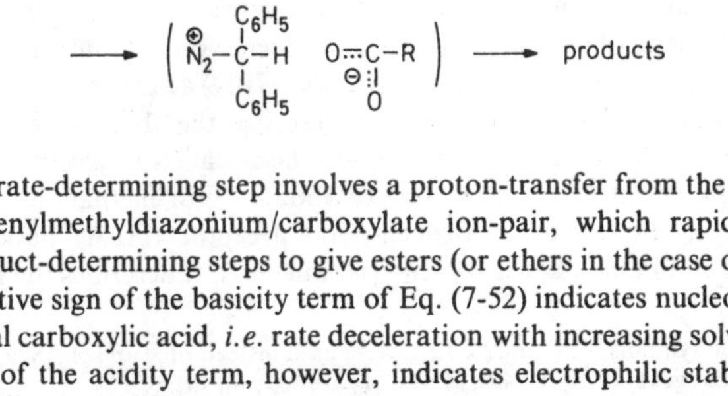

$$N_2{=}C(C_6H_5)_2 + \overset{\delta\oplus}{H}{-}\overset{\delta\ominus}{O}{-}\underset{\|}{C}(=O){-}R \;\underset{37^\circ C}{\overset{k_2}{\rightleftharpoons}}\; \left[\overset{\delta\oplus}{N_2}{\cdots}C(C_6H_5)_2{\cdots}H{\cdots}O{\cdots}C(\overset{\delta\ominus}{O}){-}R\right]^{\ddagger}$$

Nucleophilic solvation

Electrophilic solvation

$$\longrightarrow \left(\overset{\oplus}{N_2}{-}CH(C_6H_5)_2 \quad O{\cdots}C(\overset{\ominus}{O}){-}R\right) \longrightarrow \text{products}$$

The rate-determining step involves a proton-transfer from the carboxylic acid to form a diphenylmethyldiazonium/carboxylate ion-pair, which rapidly reacts in subsequent product-determining steps to give esters (or ethers in the case of alcoholic solvents). The negative sign of the basicity term of Eq. (7-52) indicates nucleophilic stabilization of the initial carboxylic acid, *i.e.* rate deceleration with increasing solvent basicity. The positive sign of the acidity term, however, indicates electrophilic stabilization of the activated complex, resulting in rate acceleration with increasing solvent acidity. Thus, the two counteracting solvent effects are nicely unravelled by this kind of treatment.

Another important treatment of multiple interacting solvent effects, in principle analogous to Eq. (7-50) but more precisely elaborated and more generally applicable, has been proposed by Kamlet and Taft [84a, 224, 226]. Both theirs and Koppel and Palm's approaches have much in common, *i.e.* that it is necessary to consider non-specific and specific solute/solvent interactions separately, and that the latter should be subdivided into solvent Lewis acidity interactions (HBA solute/HBD solvent) and solvent Lewis basicity interactions (HBD solute/HBA solvent). Using the solvatochromic solvent parameters α, β, and π^*, which have already been introduced in Section 7.4 (*cf.* Table 7-4), the multiparameter equation (7-53) has been proposed for use in so-called *linear solvation energy relationships* (LSER).

$$A = A_0 + s(\pi^* + d\delta) + a\alpha + b\beta \tag{7-53}$$

The solute property A can represent, for example, the logarithm of a rate or equilibrium constant, as well as a position of maximal absorption in an UV/Vis, IR, NMR, or ESR spectrum*); A_0 is the regression value of this solute property in cyclohexane as reference solvent.

π^* is an index of solvent dipolarity/polarizability which measures the ability of the solvent to stabilize a charge or a dipole by virtue of its dielectric effect. For a set of *select solvents*, *i.e.* non-HBD aliphatic solvents with a single dominant group dipole, the π^*-values are proportional to the dipole moment of the solvent molecule. The π^* scale was selected to run from 0.00 for cyclohexane to 1.00 for dimethyl sulfoxide. The π^*-values correspond to the use of dielectric constant and refractive index in the Koppel-Palm equation (7-50) as measures of polarization and polarizability interactions. Therefore, functions of ε_r and n^2 are not included in the Kamlet-Taft equation (7-53). The advantage of using π^*-values instead of the functions $f(\varepsilon_r)$ and $f(n^2)$ is that the latter terms are ground-state properties of the *bulk* solvent, whereas the π^*-values are derived from electronic transitions occuring on a molecular-microscopic level in solute-organized cybotactic regions, *i.e.* within the solvation shell of the solute [84a].

δ is a discontinuous polarizability correction term equal to 0.0 for non-chlorinesubstituted aliphatic solvents, 0.5 for poly-chlorinesubstituted aliphatics, and 1.0 for aromatic solvents. The δ-values reflect the observation that differences in solvent polarizability are significantly greater between the three solvent classes than within the individual classes. Thus, the sign and magnitude of the δ-term is related to the variable dipolarity/polarizability blend observed in the solvent influence on the solute property A.

α is a measure of the solvent hydrogen-bond donor (HBD) acidity and corresponds to Koppel and Palm's Lewis acidity parameter E. It describes the ability of a solvent to donate a proton in a solvent-to-solute hydrogen bond. The α-scale was selected to extend from zero for non-HBD solvents (*e.g. n*-hexane) to about 1.0 for methanol.

β is a measure of the solvent hydrogen-bond acceptor (HBA) basicity and corresponds to Koppel and Palm's Lewis basicity parameter B. It describes the solvent's

* In the original papers the symbols XYZ and $(XYZ)_0$ were used instead of A and A_0 [84a]. For the sake of consistency with the general Eq. (7-49), A and A_0 are used in Eq. (7-53).

ability to accept a proton (or, *vice versa*, donate an electron pair) in a solute-to-solvent hydrogen bond. The β-scale was selected to extend from zero for non-HBD solvents (*e.g.* n-hexane) to about 1.0 for hexamethylphosphoric acid triamide (HMPT).

The regression coefficients s, d, a, and b in Eq. (7-53) measure the relative susceptibilities of the solvent-dependent solute property A to the indicated solvent parameters. Due to the normalization of the α, β, and π^* scale (from *ca.* 0.0 to about 1.0), the a/s, b/s, and a/b ratios are assumed to provide quantitative measures of the relative contribution of the indicated solvent parameters.

Eq. (7-53) has been used in the correlation analysis by multiple regression of numerous reaction rates and equilibria, spectroscopic data, and various other solvent-dependent processes. An impressive series of more than 40 articles entitled "Linear Solvation Energy Relationships" has been so far published (May 1987): Part 1 [229]. . . Part 41 [230]; *cf.* also the summarizing articles [127, 224]. Later on, Eq. (7-53) has been extended to Eq. (7-54) by the introduction of two further solvent parameters, the δ_H^2 term*) and the ξ term [226]

$$A = A_0 + s(\pi^* + d\delta) + a\alpha + b\beta + h\delta_H^2 + e\xi \tag{7-54}$$

δ_H^2 represents Hildebrand's solubility parameter squared and corresponds to the cohesive pressure c, which characterizes the energy associated with the intermolecular solvent/solvent interactions; *cf.* Eqs. (3-5) and (5-77) in Sections 3.2 and 5.4.2, respectively, for the definition of c and δ. Thus, δ_H^2 is considered as a measure of the enthalpy or Gibbs energy input required to separate solvent molecules to provide a suitably sized cavity for the solute. This solvent cavity term is important for multiple correlations dealing with enthalpies or Gibbs energies of solution, Gibbs energies of transfer between two solvents, or gas-liquid chromatographic partition coefficients. In most cases, the δ_H^2 term is only significant for highly structured solvents such as water, formamide, and 1,2-ethanediol.

In addition to the β parameter of solvent Lewis basicity, the co-ordinate covalency parameters ξ has been found useful in correlating certain types of so-called family-dependent solute basicity properties [226, 267]. Family-independent (FI) basicity properties are defined as those which have a linear relationship with β when *all* solute bases are considered together. Family-dependent (FD) basicity properties are those which exhibit a linear relationship with β only when different families of solutes having similar HBA sites are considered separately. Thus, FD properties can be correlated to FI properties if an empirical co-ordinate covalency parameter ξ is used in correlation equations such as $A = A_0 + b\beta + e\xi$. Values of ξ are equal to -0.2 for P=O bases (*e.g.* HMPT), 0.0 for C=O and S=O bases (arbitrary reference value), 0.1 for triple-bonded nitrogen bases (*e.g.* nitriles), 0.2 for single-bonded oxygen bases (*e.g.* ethers), 0.6 for sp^2-hybridized

* The subscript H is used to distinguish the Hildebrand solubility parameter δ from the polarizability correction term δ in Eq. (7-54). In earlier formulations of Eq. (7-54), δ_H instead of δ_H^2 was wrongly used [96, 97, 226, 255]. In linear Gibbs energy relationships all the terms of the regression equation should have the dimension of an energy, but δ_H has the dimension of the square root of the energy ($J^{1/2} \cdot cm^{-3/2}$); *cf.* Section 7.5 and references [256–258].

nitrogen bases (*e.g.* pyridines), and 1.0 for sp^3-hybridized nitrogen bases (*e.g.* amines). The ξ-values are interpreted as being approximate measures of the relative co-ordinate covalencies of the bonds which are formed between solute and solvent at the base centres of the solvent. Co-ordinate covalencies of solute/solvent adducts of a given solute acid decrease in strength (ξ decreases) as the electronegativity of the solvent base centre increases, because the positive charge created by co-ordinate covalency on an increasingly electronegative atom is unfavourable. The ξ-values have been found to be useful in correlating the basicity behaviour of neutral oxygen and nitrogen bases of widely differing properties [267]. It has been stressed that the correlation equation $A = A_0 + b\beta + e\xi$ formally and conceptually resembles the Drago E/C treatment, with b and β corresponding to E_A and E_B, and e and ξ corresponding to C_A and C_B; *cf.* Eq. (2-12) in Section 2.2.6.

The multiparameter equation (7-54) seems to be rather difficult to apply. However, in practice most of the linear solvation energy relationships that have been reported are simpler than indicated by Eq. (7-54) since one or more terms are inappropriate. For example, if the solute property A does not involve the creation of a cavity or a change in cavity volume between initial and activated or excited states (as is the case for solvent effects on spectral properties), the δ_H^2-term is dropped from Eq. (7-54). If the solvent-dependent process under study has been carried out in non-HBD solvents only, the α-term drops out. On the other hand, if the solutes are not hydrogen-bond donors or Lewis acids, the β-term drops out of Eq. (7-54). Thus, for many solvent-dependent processes, Eq. (7-54) can be reduced to a more manageable one-, two-or three-parameter correlation equation by a judicious choice of solutes and solvents [226].

Only one example will be mentioned*). Multiple regression of $\lg k_2$ of the reaction between benzoic acid and diazodiphenylmethane by virtue of the Kamlet-Taft equation (7-53), for the same set of 44 non-HBD solvents as used in the Koppel-Palm treatment of this reaction given in Eq. (7-52), leads to Eq. (7-55), with $n = 44$, $r = 0.980$, and $s = 0.171$ [15, 268].

$$\lg k_2 = 0.20 + 1.21\,\pi^* + 2.71\,\alpha - 3.70\,\beta \qquad (7\text{-}55)$$

A comparison of Eqs. (7-52) and (7-55) shows that both treatments give similar results, with opposite signs of the solvent Lewis acidity and basicity parameters. This is in agreement with the given reaction mechanism.

A few comments of the Kamlet-Taft treatment of solvent effects should be made. Some shortcomings and the sometimes insufficient precision in the determination of the solvatochromic parameters have been discussed by Laurence *et al.* [167, 237, 238], who also recommended improvements by virtue of a new *thermosolvatochromic comparison method*; *cf.* Section 7.4 for a further discussion of this method. – A study of linear Gibbs energy relationships (LGER) in a homologous series of *n*-alkane and *n*-alkylnitrile solvents, using Kamlet and Taft's solvatochromic indicator solutes, has shown that a single lumped parameter of solvent dipolarity/polarizability such as the π^*-values cannot

* Quite recently, a comprehensive analysis of the solvent-dependent solvolysis/dehydrohalogenation reaction of 2-chloro-2-methylpropane in terms of Eqs. (7-53) and (7-54) has been given by Abraham *et al.* [288].

be applied simultaneously to *n*-alkanes and *n*-alkylnitriles [269]. Therefore, it has been concluded by Carr *et al.* [269], that the hypothesis that solvent dipolarity and polarizability can be represented by a single parameter is certainly not generally valid. Kamlet and Taft have tried to consider this observation by incorporating the solvent polarizability correction term $d\delta$ in Eq. (7-53), which has different values for groups of aliphatic, chlorine-substituted aliphatic, and aromatic solvents*) – but the important distinction made is that the *solute* sensitivity to the two types of polarization can be quite different [269]. In this context, reference has to be made to Sjöström and Wold's results [236], who recommended, on statistical grounds, the use of a two-parameter equation of the form $A = A_0 + s_0\pi_0^* + s_d\pi_d^*$, where π_0^* and π_d^* represent, respectively, solvent orientational and distortional polarizability scales.

Furthermore, there has been considerable discussion on the question whether or not linear Gibbs energy relationships (LGER) such as linear solvation energy relationships (LSER) are really fundamental laws of chemistry, reflecting simple physicochemical relationships, – or rather local empirical rules of similarity, *i.e.* only locally valid linearizations of more complicated relationships [126, 127]; *cf.* Section 7.1 for a further discussion of this dialogue. Both parties, the chemometricians and the physical organic chemists, have tried to exemplify the merits of both types of treatment of solvent effects, *i.e.* principal component analysis (PCA) and linear solvation energy relationships (LSER), by means of the correlation and rationalization of solvent effects on ^{13}C-NMR chemical shifts of lithium indenide, a planar delocalized carbanion, measured in a set of 13 solvents; see references [270, 271] as well as [126, 127] for this lengthy discussion.

Kamlet, Taft, Abraham *et al.* have recently further modified Eq. (7-54) in order to correlate the solubility and distribution behaviour of nonelectrolyte solutes with solvent properties, according to Eq. (7-56) [271, 272],

$$A = A_0 + A'(\delta_H^2)_1(V_2/100) + B\pi_1^*\pi_2^* + C\alpha_1\beta_2 + D\beta_1\alpha_2 \qquad (7\text{-}56)$$

where subscript 1 refers to the solvent and subscript 2 to the solute, and A', B, C, and D are the regression coefficients for the endoergic cavity term, the exoergic dipolarity/polarizability term, and the exoergic hydrogen-bonding terms of adduct formation between HBD solvents and HBA solutes (measured by α_1 and β_2) as well as between HBA solvents and HBD solutes (measured by β_1 and α_2), respectively. V_2 is the molar volume of the solute, taken as its molecular mass divided by its liquid density at 20 °C. Whereas $(\delta_H^2)_1$ measures the solvent's contribution to the cavity term, $V_2/100$ represents the solute's contribution to the cavity term. Instead of V_2, $V_2/100$ is used so that the parameter measuring the cavity term will roughly cover the same numerical range as the other independent variables α, β, and π^* (*ca.* 0.0...1.0). This makes the evaluation of the relative contributions of the various terms of Eq. (7-56) to the property A easier.

When dealing with the effects of different solvents on properties of a single solute (*e.g.* solvent effects on reaction rates), the factors relating to the solute can be subsumed

* The description of different solvent polarizabilities by virtue of three discontinuous polarizability correction terms δ ($\delta = 0.0$, 0.5, or 1.0 for three solvent groups) as in Eqs. (7-53) and (7-54) is surely an oversimplification [167].

into the regression coefficients of Eq. (7-56), and the following equation results:

$$A = A_0 + h(\delta_{\mathrm{H}}^2)_1 + s\pi_1^* + a\alpha_1 + b\beta_1 \tag{7-57}$$

In Eq. (7-57), the dependence of the solute property A on each term is now solely given by solvent parameters. Furthermore, when the solute property A does not involve cavity formation or a cavity change, the cavity term drops out, and Eq. (7-57) takes the form of Eq. (7-53).

Conversely, when dealing with solubilities or other properties of a set of different solutes in a single solvent, or with distributions of different solutes between a pair of solvents, the resulting Eq. (7-58) relates property A specifically to the solute parameters V_2, π_2^*, α_2, and β_2.

$$A = A_0 + m(V_2/100) + s\pi_2^* + a\alpha_2 + b\beta_2 \tag{7-58}$$

For example, this equation has been successfully used to correlate the Hansch-Leo 1-octanol/water partition coefficient, $K_{o/w}$, of 102 aliphatic and aromatic solutes according to Eq. (7-59) with the indicated solute properties ($n = 102$; $r = 0.989$; $s = 0.175$) [149]; *cf.* Eq. (7-12) in Section 7.2 for the definition of $K_{o/w}$ [145–148].

$$\lg K_{o/w} = 2.74\ (V_2/100) - 0.92\ \pi_2^* - 3.49\ \beta_2 + 0.20 \tag{7-59}$$

Dealing with this type of multiple correlations, a series of nine articles entitled "Solubility Properties in Polymers and Biological Media" has been published up to now (May 1987): Part 1 [273] ... Part 9 [274]. The application of the linear solvation energy relationship (7-58) to the prediction of solubilities of organic nonelectrolytes in water, blood, and other body tissues has been recently reviewed [286].

A more simplified but likewise successful empirical two-parameter approach for the description of solvent effects has been proposed by Krygowski and Fawcett [113]. They assume that only specific solute/solvent interactions need to be considered. These authors postulated that the solvent effect on a solute property A can be represented as a linear function of only two independent but complementary parameters describing the Lewis acidity and Lewis basicity of a given solvent. Again, for reasons already mentioned, the $E_T(30)$-values were chosen as a measure of Lewis acidity. In addition, Gutmann's donor numbers *DN* [26, 27] were chosen as a measure of solvent basicity (*cf.* Table 2-3 and Eq. (7-10) in Sections 2.2.6 and 7.2, respectively). Thus, it is assumed that the solvent effect on A can be described in terms of Eq. (7-60)*⁾.

$$A = A_0 + \alpha \cdot E_T(30) + \beta \cdot DN \tag{7-60}$$

α and β are regression coefficients describing the sensitivity of the solute property A to electrophilic and nucleophilic solvent properties, respectively. Since α and β are not

* In the original paper Q and Q_0 were used instead A and A_0 [113]. For the sake of consistency with the general Eq. (7-49), A and A_0 are used in Eq. (7-60).

necessarily on the same scale, due to the fact that $E_T(30)$ and DN do not vary over the same range for a given set of data, the regression coefficients were normalized to give $\bar{\alpha}$ and $\bar{\beta}$. Application of Eq. (7-60) involves the supposition that non-specific solute/solvent interactions are negligible or nearly constant and can be included in the solvent Lewis acidity and basicity terms. Obviously, this is a serious simplification. But in spite of this simplification, the Krygowski-Fawcett treatment of solvent effects has been successfully applied in many cases [113]. Satisfactory correlations were obtained in 90% of the cases involving ion/solvent and ion/ion interactions, and in 75% of the dipole/dipole interactions [113]; see references [118–120, 232] for some more recent applications of Eq. (7-60) and its modifications.

A further interesting two-parameter treatment of solvent effects has been given recently by Swain *et al.* [265]. It is based on a computer calculation involving 1080 data sets for 61 solvents and 77 solvent-sensitive reactions and physicochemical properties, taken from the literature (*e.g.* rate constants, product ratios, equilibrium constants, UV/Vis, IR, ESR, and NMR spectra). According to these calculations, all solvent effects can be rationalized in terms of two complementary solvent property scales, *i.e.* A_j, measuring the solvent's anion-solvating tendency or *acity**), and B_j, measuring the solvent's cation-solvating tendency or *basity**), both combined in Eq. (7-61)**) [265].

$$A = A_0 + a_i \cdot A_j + b_i \cdot B_j \tag{7-61}$$

A_j and B_j characterize the solvent j. A and A_0 as well as the multiple regression coefficients a_i and b_i depend only on the solvent-sensitive solute property i under study. Constants a_i and b_i represent the sensitivity of solute property i to a solvent change. A nonlinear least squares procedure, using equal statistical weighting of the 1080 data, has been used to evaluate and to optimize all 353 $[=(2 \cdot 61)+(3 \cdot 77)]$ constants A_j, B_j, a_i, b_i, and A_0 in order to get the best possible fit consistent with Eq. (7-61). In order to obtain values of A_j and B_j which represent physically significant solvent influences that are cleanly separated, some scale-setting subsidiary conditions have to be fixed. As trivial, arbitrary conditions which set zeros and scale factors but do not affect rank orders, $A_j = B_j = 0.00$ for *n*-heptane and $A_j = B_j = 1.00$ for water have been chosen. As two non-trivial, critical conditions, $A_j = 0.00$ for hexamethylphosphoric acid triamide (HMPT) and $B_j = 0.00$ for trifluoroacetic acid (TFA), have been chosen. This choice is equivalent to the assumption that HMPT is almost as poor an anion solvator, and TFA is almost as poor a cation solvator, as *n*-heptane.

---

* The new names *acity* and *basity* were chosen because, although they are obviously kinds of acidity and basicity, they are neat (bulk) solvent properties involved in solute solvations. Such solvent properties cause specific local electrostatic solute/solvent interactions without major covalency changes and, therefore, are usually omitted from equations describing chemical reactions [265].

** In the original publication Eq. (7-61) takes the form

$$p_{ij} = c_i + a_i \cdot A_j + b_i \cdot B_j$$

with p_{ij} representing the solvent-dependent solute property i in solvent j, and c_i representing the predicted value for a reference solvent for which $A_j = B_j = 0$ [265]. For the sake of consistency with the general Eq. (7-49), A and A_0 instead of p_{ij} and c_i are used in Eq. (7-61), in spite of the unhappy cumulation of the letter A.

Table 7-5. Selection of values of solvent acity, A_j, and solvent basity, B_j, for 34 solvents, calculated according to Eq. (7-61) [265].

Solvents j	A_j	B_j	$(A_j + B_j)$ [a]
n-Heptane	*0.00*	*0.00*	0.00
Cyclohexane	0.02	0.06	0.09
Triethylamine	0.08	0.19	0.27
Tetrachloromethane	0.09	0.34	0.43
Diethyl ether	0.12	0.34	0.46
Carbon disulfide	0.10	0.38	0.48
Toluene	0.13	0.54	0.67
Benzene	0.15	0.59	0.73
Ethyl acetate	0.21	0.59	0.79
Tetrahydrofuran	0.17	0.67	0.84
Chlorobenzene	0.20	0.65	0.85
1,4-Dioxane	0.19	0.67	0.86
tert-Butanol	0.45	0.50	0.95
Methoxybenzene	0.21	0.74	0.96
2-Butanone	0.23	0.74	0.97
2-Propanol	0.59	0.44	1.03
Acetone	0.25	0.81	1.06
Acetic acid	0.93	0.13	1.06
Hexamethylphosphoric acid triamide	*0.00*	1.07	1.07
1-Propanol	0.63	0.44	1.08
Ethanol	0.66	0.45	1.11
1,2-Dichloroethane	0.30	0.82	1.12
Dichloromethane	0.33	0.80	1.13
Chloroform	0.42	0.73	1.15
Pyridine	0.24	0.96	1.20
Acetonitrile	0.37	0.86	1.22
N,N-Dimethylformamide	0.30	0.93	1.23
Methanol	0.75	0.50	1.25
Nitromethane	0.39	0.92	1.31
Dimethyl sulfoxide	0.34	1.08	1.41
1,2-Ethanediol	0.78	0.84	1.62
Formamide	0.66	0.99	1.65
Trifluoroacetic acid	1.72	*0.00*	1.72
Water	*1.00*	*1.00*	2.00

[a] The solvents are listed in order of their sum $(A_j + B_j)$, which is considered as reasonable measure of "solvent polarity" in terms of the overall solvation capability of a solvent [265].

Table 7-5 lists a selection of A_j and B_j values in order of their sum $(A_j + B_j)$, which may be considered as a measure of "solvent polarity" in terms of the overall solvation capability of a solvent; *cf.* Section 7.1 and references [1, 3] for this definition of solvent polarity. A plot of A_j against B_j shows that both parameters are highly variable but there is no correlation between A_j and B_j, *i.e.* an essential condition for the application of Eq. (7-61). Interestingly, the overall correlation coefficient between the 1080 input data and the predictions made by Eq. (7-61) with $r = 0.991$ is excellent. That means that the two unrelated solvent parameters A_j and B_j alone account for over 98% of the solvent effects in the set of 77 solvent-sensitive processes that have been examined by Swain *et al.* [265]. Since for the 1080 diverse input data the solvent properties can be adequately represented by only two parameters A_j and B_j, there must be a parallelism among each three neat solvent properties, *i.e.* anion-solvating tendency, hydrogen-bonding acidity (HBD acidity), and electrophilicity, and likewise cation-solvating tendency, hydrogen-bonding basicity (HBA basicity), and nucleophilicity, respectively.

The coefficients a_i and b_i, calculated for the 77 solvent-sensitive processes used to establish Eq. (7-61) as well as for 11 further solvent-dependent reactions (including another 75 data), have been discussed in detail in reference [265]. For example, $a_i = 1.87$ and $b_i = -0.05$ for the $n \rightarrow \pi^*$ absorption of benzophenone. With the high value of a_i and the negligible value of b_i, this solvent-dependent UV/Vis absorption comes close to measuring A_j in pure form. The $E_T(30)$-values (*cf.* Table 7-3) are characterized by $a_i = 30.36$ and $b_i = 4.45$ ($a_i/b_i = 6.8$), in agreement with other observations showing that the $E_T(30)$-values are mainly related to the solvent Lewis acidity, and not to the solvent Lewis basicity. – Application of Eq. (7-61) to several hundred increasingly diverse additional solvent-dependent reactions has not only lead to satisfactory correlations (with $r > 0.965$); four explanations for non-agreement have been found and discussed by Swain *et al.* [265].

In view of the success in correlating so many solvent effects by only two solvent parameters according to Eq. (7-61), Swain *et al.* concluded that all four solvent parameters of Eq. (7-53) are not necessary, additional parameters do not improve the fits already observed with Eq. (7-61), and, for example, the β-parameter of Eq. (7-53) is rendered superfluous [265]. These conclusions as well as Swain *et al.*'s general approach has been criticized by Taft, Abboud, and Kamlet [275], leading to a reply by Swain [276] dismissing this criticism. The reader is referred to references [275, 276] and to the June issue of the *Journal of Organic Chemistry* in 1984 for this interesting discussion (issue no. 11, p. 1989–2010).

Both, Kamlet and Taft *et al.*'s [224, 226] and Swain *et al.*'s [265] multiparameter solvent effect treatments have an inherent weakness in so far that the solvent parameters α, β, and π^* as well as A_j and B_j are averaged and statistically optimized parameters: the first are derived from various types of solvatochromic indicator dyes, the latter are calculated from a selection of 77 solvent-sensitive processes. Thus, they are no longer directly related to a distinct, carefully selected, well-understood single reference process [as, for example Y_{OTs}, Z, and $E_T(30)$]; *cf.* footnote on page 379. The significance of such averaged and statistically optimized solvent parameters depends above all on (i) the right choice of the various solvent-dependent processes used in the averaging procedure, which has to be done with a skilfull hand; and (ii) the right choice of the critical subsidiary conditions which have to be defined and justified in order to get solvent parameters with a clear physical meaning at the molecular-microscopic level*). For example, concerning point (i), according to Taft, Abboud, and Kamlet [275], most of the 77 solvent-dependent processes

* For a two-parameter treatment of solvent effects (with two independent solvent vectors) only two critical subsidiary conditions must be defined in order to force the two solvent parameters to represent physically significant solvent properties. Four other trivial arbitrary conditions have to be defined in order to fix zero reference points and scale-unit sizes. However, for a three-parameter treatment (with three independent solvent vectors) already six critical subsidiary conditions must be defined, in addition to the six trivial reference or scale-factor conditions. On the contrary, single-parameter treatments require no definition of critical subsidiary conditions, but only one reference (zero) condition and one standard (unit) condition, whose arbitrary assignment changes only the reference solvent and the scale-unit size [265, 276].

selected by Swain *et al.* [265] involve only non-HBD solutes (reactants or indicator dyes). Therefore, Swain *et al.*'s finding that Kamlet and Taft's β-parameter of solvent HBA basicity is superfluous can result from a somewhat unfortunate data selection. As far as point (ii) is concerned, Swain criticizes that in three-parameter equations such as (7-53) the six critical subsidiary conditions needed to assure that the derived solvent parameters are physically significant are often not properly defined. It has been stressed that optimization of the correlations through adjustment of the solvent parameters is not enough and has nothing to do with the physical significance of these calculated solvent parameters [276].

Another semiempirical multiparameter relationship for the description of solvent effects on the statics and kinetics of chemical reactions, according to Eqs. (7-62) and (7-63), has been introduced by Mayer [266].

$$\Delta G^S - \Delta G^R = a \cdot (DN^S - DN^R) + b \cdot (AN^S - AN^R) + c \cdot (\Delta G_{vp}^{0S} - \Delta G_{vp}^{0R}) \tag{7-62}$$

$$\Delta\Delta G = a \cdot \Delta DN + b \cdot \Delta AN + c \cdot \Delta\Delta G_{vp}^0 \tag{7-63}$$

ΔG represents the Gibbs energy of reaction or activation ($\Delta G^{\neq}$), DN the donor number [26, 27], AN the acceptor number [91], and ΔG_{vp}^0 the standard molar Gibbs energy of vaporization of a solvent S and a reference solvent R, respectively. Acetonitrile ($\Delta G_{vp}^0 = 5.31$ kJ/mol) has been used as a reference solvent [266]. The coefficients a and b are correlated to the donor and acceptor strengths of the reaction partners relative to those of the reference compounds $SbCl_5$ and $(C_2H_5)_3PO$, respectively. This approach is based on a model, developed for calculating the Gibbs energy change associated with the creation of cavities in the solvent the size of which corresponds to the volume occupied by the solute molecules. Experimental equilibrium and rate data, including solubility measurements, complex-formation equilibria, ion-pair equilibria, and an S_NAr reaction, have been successfully used to test Eqs. (7-62) and (7-63) by the method of multiple linear regression analysis [266]. A remarkable application of Eq. (7-63) (with $c = 0$) has been shown that for non-HBD solvents the logarithm of their dielectric constants can be represented by a linear combination of the acceptor number and the donor number [134, 207]; *cf.* also [139].

Eventually, a multiparameter correlation based on solvent ionization potentials (IP) and solvent electron affinities (EA) deserves attention [114]. Considering the interaction between the highest occupied orbitals (HOMO) of the solvent and the ions, and the corresponding lowest unoccupied orbitals (LUMO) of the ions and the solvent, and approximating the energies of these orbitals from ionization potentials and electron affinity data, Dougherty [114] developed Eq. (7-64); *cf.* also [139].

$$\delta E_{solv} = C_1 \cdot (IP_{solv} + EA_{solv}) + C_2 \cdot (IP_{solv}) + C_3 \cdot (EA_{solv})^2 + C_4 \tag{7-64}$$

The C's will depend on the IP's and EA's of the substrate. $(IP_{solv} + EA_{solv})$ can be thought of as the solvent ionizing power. The IP_{solv} term should reflect the solvent nucleophilicity, and the $(EA_{solv})^2$ term should represent the electrophilicity of the solvent. Obviously, there is some resemblance to Eqs. (7-60) and (7-63). Up until now, only a very limited data set could be used to test Eq. (7-64).

Finally, some special multiparameter correlations of solvent effects will be mentioned. A common multiple regression equation with seven fitting constants have been derived for the relation between the Gibbs energy of transfer of ions from water to polar solvents, ΔG_t°(X, W $\rightarrow$ S), and the properties of solvents *and* ions. Almost 200 data points can be described in terms of four solvent properties (*DN*, $E_T(30)$, ε_r, and δ_H^2) and three ionic solute properties (charge, size, and softness) [277]. – The substituent and solvent influence on the cationic polymerization of *p*-substituted styrenes can be described by four-parameter correlation equations involving terms for the substituent and solvent influence as well as the initiator activity [278, 279]. Such correlations are especially useful for controlling industrial polymerization processes as well as definite product formation [278, 279]. – Modified multiparameter equations involving solvent viscosity parameters have been used to correlate the solvent influence on quantum yields of radiative electronic desactivation processes [280]. – The dependence of the optical resolution of phenylglycine derivatives with L-(+)-tartaric acid on racemate structure and solvent polarity have been successfully described with multiparameter correlation equations [281].

Many different solvent parameters and multiparameter equations have been introduced in this Chapter 7. Certainly, only a few of them will survive the test of applicability and acceptance by organic chemists. Already now, the preference for certain time-tested solvent scales and multiparameter treatments is clearly discernible.

The multiparameter treatment of solvent effects can be criticized from at least three complementary points of view. First, the separation of solvent effects into various additive contributions is somewhat arbitrary, since different solute/solvent interaction mechanisms can cooperate in a non-independent way. Second, the choice of the best parameter for every type of solute/solvent interaction is critical because of the complexity of the corresponding empirical solvent parameters, and because of their susceptibility to more than one of the multiple facets of solvent polarity. Third, in order to establish a multiparameter regression equation in a statistically perfect way, so many experimental data points are usually necessary that there is often no room left for the prediction of solvent effects by extrapolation or interpolation. This helps to get a sound interpretation of the observed solvent effect for the process under study, but simultaneously it limits the value of such multiparameter equations for the chemist in its daily laboratory work.

In this context, one should be aware of the important remark made by the Austrian philosopher Karl Popper in his autobiography [284]: "It is always undesirable to make an effort to increase precision for its own sake ... since this usually leads to loss of clarity. ... One should never try to be more precise than the situation demands. ... Every increase in clarity is of intellectual value in itself; an increase in precision or exactness has only a pragmatic value as a means to some definite end."

Appendix

A. Properties, Purification, and Use of Organic Solvents

A.1 Physical Properties

The most useful organic solvents are listed according to decreasing polarity in Table A-1. Also given are their physical constants, *viz.* melting point, boiling point, dielectric constant, dipole moment, and index of refraction. The measure of polarity used is the empirical solvent parameter E_T^N, derived from the solvatochromism of a pyridinium-*N*-phenoxide betaine (*cf.* Section 7.4). Further physical data, including those of technically useful solvents, are found in references [1–8, 97–100]. In Table A-2 is given a selection of chiral organic solvents, together with some physical data. Such solvents have received much attention recently because of their use in determining optical purity [9], as media for stereoselective syntheses [10–12], and as NMR shift reagents [13].

Another important property of organic solvents is their miscibility with other organic liquids (*cf.* Fig. 2-2 in Chapter 2). The farther two solvents are located from each other in Hecker's "mixotropic" series of solvents [51], given in Table A-10, the less miscible they are.

According to Hildebrand's solubility parameter approach [101], two liquids are miscible if their solubility parameters δ differ by no more than 3.4 units [101, 102]; *cf.* Eqs. (2-1) and (5-77) for the definition of δ. That is, mutual miscibility decreases as the δ-values of two solvents become farther apart. Higher mutual solubility will follow if the δ-values of the solvents are closer. A comprehensive collection of δ-values is given by Barton [100].

An alternative, more empirical but more accurate method for predicting miscibility has been given by Godfrey [103], using so-called *miscibility numbers* (*M*-numbers). These are serial numbers of 31 classes of organic solvents, ordered empirically by means of their lipophilicity (*i.e.* their affinity for oil-like substances), using simple test tube miscibility experiments. All pairs of solvents whose *M*-numbers differ by 15 units or less are miscible in all proportions at 25 °C; a difference of ≥ 17 corresponds to immiscibility, and an *M*-number difference of 16 units indicates borderline behaviour (limited mutual miscibility) [103]. The central class of solvents with *M*-number equal to 16 (*e.g.* 2-*n*-butoxyethanol) comprises "universal" solvents which are miscible with less lipophilic as well as with more lipophilic solvents.

A.2 Purification of Organic Solvents

Normally it is necessary to purify a solvent before use. Naturally, the purity that can be achieved depends on the nature of the impurities [14, 15] and the desirable purity determined by the intended use [16]. The following is a practical definition of the purity of a solvent: "A material is sufficiently pure if it does not contain impurities of such nature and in such quantity as to interfere with the use for which it is intended" [1]. Detailed prescriptions for purification are available in standard texts [1, 17, 104, 105]. The most frequently found impurity in organic solvents is water. A water content of only 20 µg/g (20 ppm) is equivalent to the total amount of solute in a 10^{-3} molar solution! Since water

Table A-1. Compilation of hundred important organic solvents together with their physical constants[a], arranged by decreasing E_T^N-values as empirical parameter of solvent polarity[b].

Solvents	$t_{mp}/°C$[c]	$t_{bp}/°C$[d]	ε_r[e]	$\mu/(10^{-30}$ $\cdot$ C m)[f]	n_D^{20}[g]	E_T^N[h]
(1) Water	0.0	100.0	78.30	5.9	1.3330	1.000
(2) Formamide	2.55	210.5	111.0 (20°)	11.2	1.4475	0.799
(3) 1,2-Ethanediol	− 12.6	197.5	37.7	7.7	1.4318	0.790
(4) Methanol[i]	− 97.7	64.5	32.66	5.7	1.3284	0.762
(5) *N*-Methylformamide	− 3.8	180...185	182.4	12.9	1.4319	0.722
(6) Diethylene glycol	− 7.8	245.7	31.69 (20°)	7.7	1.4475	0.713
(7) Triethylene glycol	− 4.3	288.0	23.69 (20°)	10.0	1.4558	0.704
(8) 2-Methoxyethanol	− 85.1	124.6	16.93	6.8	1.4021	0.667
(9) Tetraethylene glycol	− 6.2	327.3	19.7	10.8	1.4577	0.664
(10) *N*-Methylacetamide[k]	30.6	206.7	191.3 (32°)	14.2	1.4253 (35°)	0.657
(11) Ethanol[l]	−114.5	78.3	24.55	5.8	1.3614	0.654
(12) 2-Aminoethanol	10.5	170.95	37.72	7.6	1.4545	0.651
(13) Acetic acid	16.7	117.9	6.17 (20°)	5.6	1.3719	0.648
(14) 1-Propanol[m]	−126.2	97.15	20.45	5.5	1.3856	0.617
(15) Benzyl alcohol	− 15.3	205.45	13.1 (20°)	5.5	1.5404	0.608
(16) 1-Butanol	− 88.6	117.7	17.51	5.8	1.3993	0.602
(17) 1-Pentanol	− 78.2	138.0	13.9	5.7	1.4100	0.568
(18) 3-Methyl-1-butanol, Isoamyl alcohol	−117.2	130.5	15.19	6.1	1.4072	0.565
(19) 2-Methyl-1-propanol, Isobutyl alcohol	−108	107.9	17.93	6.0	1.3959	0.552
(20) 2-Propanol[m]	− 88.0	82.2	19.92	5.5	1.3772	0.546
(21) 2-Butanol	−114.7	99.5	16.56	5.5	1.3971	0.506
(22) Cyclohexanol	25.15	161.1	15.0	6.2	1.4648 (25°)	0.500
(23) Propylene carbonate[k]	− 54.5	241.7	64.92	16.5	1.4215	0.491
(24) 2-Pentanol		119.0	13.71	5.5	1.4064	0.488
(25) Nitromethane[n]	− 28.55	101.2	35.94	11.9	1.3819	0.481
(26) 3-Pentanol	− 75	115.3	13.35	5.5	1.4104	0.463
(27) Acetonitrile[k]	− 43.8	81.6	35.94	11.8	1.3441	0.460
(28) Dimethyl sulfoxide[k]	18.5	189.0	46.45	13.5	1.4793	0.444
(29) Aniline	− 6.0	184.4	6.71 (30°)	5.0	1.5863	0.420
(30) Sulfolane[k]	28.45	287.3 (dec.)	43.3 (30°)	16.0	1.4816 (30°)	0.410
(31) Acetic anhydride	− 73.1	140.0	20.7 (19°)	9.4	1.3904	0.407
(32) *N,N*-Dimethyl-formamide[k]	− 60.4	153.0	36.71	10.8	1.4305	0.404
(33) *N,N*-Dimethyl-acetamide	− 20	166.1	37.78	12.4	1.4384	0.401
(34) Propanenitrile	− 92.8	97.35	28.86 (20°)	11.7	1.3658	0.401
(35) 2-Methyl-2-propanol, *t*-Butanol[m]	25.6	82.3	12.47	5.5	1.3877	0.389
(36) 1,3-Dimethyl-imidazolidin-2-one, DMEU[o]	8.2	225.5	37.60	13.6	1.4707 (25°)	0.364
(37) 1-Methylpyrrolidin-2-one[p]	− 24.4	202	32.2	13.6	1.4700	0.355
(38) Acetone[q]	− 94.7	56.1	20.56	9.0	1.3587	0.355
(39) 1,3-Dimethyl-2-oxo-hexahydropyrimidine, DMPU[o]	< − 20	230	36.12	14.1	1.4881 (25°)	0.352

Table A-1. (Continued)

Solvents	$t_{mp}/°C$[c)]	$t_{bp}/°C$[d)]	ε_r[e)]	$\mu/(10^{-30}\cdot C\,m)$[f)]	n_D^{20}[g)]	E_T^N[h)]
(40) 1,2-Diaminoethane[k)]	11.3	116.9	12.9	6.3	1.4568	0.349
(41) Cyanobenzene	− 12.75	191.1	25.20	13.4	1.5282	0.333
(42) 1,2-Dichloroethane[t)]	− 35.7	83.5	10.37	6.1	1.4448	0.327
(43) 2-Butanone	− 86.7	79.6	18.51 (20°)	9.2	1.3788	0.327
(44) Nitrobenzene	5.8	210.8	34.78	13.3	1.5562	0.324
(45) 2-Methyl-2-butanol *t*-Pentyl alcohol	− 8.8	102.0	5.78	5.7	1.4050	0.321
(46) 2-Pentanone	− 76.9	102.3	15.38 (20°)	9.0	1.3908	0.321
(47) Tetramethylurea[o)]	− 1.2	175.2	23.60	11.7	1.4493 (25°)	0.318
(48) Morpholine	− 4.8	128.9	7.42	5.2	1.4542	0.318
(49) Hexamethylphosphoric acid triamide, HMPT[k)]	7.2	233	29.6	18.5	1.4588	0.315
(50) 3-Methyl-2-butanone	− 92	94.2	15.87 (30°)	9.2	1.3880	0.315
(51) Dichloromethane[t)]	− 94.9	39.6	8.93	5.2	1.4242	0.309
(52) Acetophenone	19.6	202.0	17.39	9.8	1.5342	0.306
(53) Pyridine[k)]	− 41.55	115.25	12.91	7.9	1.5102	0.302
(54) Methyl acetate	− 98.05	56.9	6.68	5.7	1.3614	0.287
(55) Cyclohexanone	− 32.1	155.65	16.10 (20°)	10.3	1.4510	0.281
(56) 4-Methyl-2-pentanone	− 84.7	117.4	13.11 (20°)	2.7	1.3958	0.269
(57) 1,1-Dichloroethane[t)]	− 97.0	57.3	10.0 (18°)	6.1	1.4164	0.269
(58) Quinoline	− 14.85	237.1	8.95	7.3	1.6273	0.269
(59) 3-Pentanone	− 39.0	102.0	17.00 (20°)	9.4	1.3923	0.265
(60) Chloroform	− 63.5	61.2	4.81 (20°)	3.8	1.4459	0.259
(61) 3,3-Dimethyl-2-butanone	− 49.8	106.3	13.1 (14.5°)	9.3	1.3952	0.256
(62) Triethylene glycol dimethyl ether	− 45	216	7.5		1.4224	0.253
(63) 2,4-Dimethyl-3-pentanone	− 69.0	125.25	17.2 (20°)	9.1	1.3999	0.247
(64) Diethylene glycol dimethyl ether	− 64.0	159.8 (dec.)	5.8	6.6	1.4078	0.244
(65) 1,2-Dimethoxyethane[r)]	− 69	84.5	7.20	5.7	1.3796	0.231
(66) Ethyl acetate	− 83.55	77.1	6.02	6.1	1.3724	0.228
(67) 1,2-Dichlorobenzene	− 17.0	180.5	9.93	7.1	1.5515	0.225
(68) 2,6-Dimethyl-4-heptanone	− 46.0	168.2	9.91 (20°)	8.9	1.4122	0.225
(69) Diethylene glycol diethyl ether	− 44.3	188.9	5.70		1.4115	0.210
(70) Tetrahydrofuran[s)]	−108.4	66.0	7.58	5.8	1.4072	0.207
(71) Methoxybenzene	− 37.5	153.6	4.33	4.2	1.5170	0.198
(72) Diethyl carbonate	− 43.0	126.8	2.82 (20°)	3.0	1.3837	0.194
(73) Fluorobenzene	− 42.2	84.7	5.42	4.9	1.4684 (15°)	0.194
(74) 1,1-Dichloroethene	−122.6	31.6	4.82 (20°)	4.3	1.4247	0.194
(75) Chlorobenzene	− 45.6	131.7	5.62	5.4	1.5248	0.188
(76) Bromobenzene	− 30.8	155.9	5.40	5.2	1.5568	0.182
(77) Ethoxybenzene	− 29.5	169.8	4.22 (20°)	4.5	1.5074	0.182
(78) Iodobenzene	− 31.35	188.3	4.49 (20°)	4.7	1.6200	0.170
(79) 1,1,1-Trichloroethane	− 30.4	74.1	7.25 (20°)	5.7	1.4380	0.170
(80) 1,4-Dioxane[s)]	11.8	101.3	2.21	1.5	1.4224	0.164

Table A-1. (Continued)

Solvents	$t_{mp}/°C$ [c]	$t_{bp}/°C$ [d]	ε_r [e]	$\mu/(10^{-30}$ $\cdot$ C m) [f]	n_D^{20} [g]	E_T^N [h]
(81) Trichloroethene	− 86.4	87.2	3.42 (16°)	2.7	1.4773	0.160
(82) *t*-Butyl methyl ether	−108.6	55.2	4.5 (20°)	4.1	1.3690	0.148
(83) Piperidine	− 10.5	106.2	5.8 (20°)	4.0	1.4525	0.148
(84) Diethylamine	− 49.8	55.55	3.78	4.0	1.3846	0.145
(85) Diphenyl ether	26.9	258.1	3.69 (20°)	3.9	1.5763 (30°)	0.142
(86) Diethyl ether	−116.3	34.4	4.20	3.8	1.3524	0.117
(87) Benzene	5.5	80.1	2.27	0.0	1.5011	0.111
(88) Di-*n*-propyl ether	−123.2	90.1	3.39 (26°)	4.4	1.3805	0.102
(89) Toluene	− 95.0	110.6	2.38	1.0	1.4969	0.099
(90) 1,4-Dimethylbenzene	13.3	138.4	2.27 (20°)	0.0	1.4958	0.074
(91) Di-*n*-butyl ether	− 95.2	140.3	3.08 (20°)	3.9	1.3992	0.071
(92) Carbon disulfide	−111.6	46.2	2.64 (20°)	0.0	1.6275	0.065
(93) Tetrachloromethane	− 22.8	76.6	2.23	0.0	1,4602	0.052
(94) Triethylamine	−114.7	88.9	2.42 (20°)	2.9	1.4010	0.043
(95) Tri-*n*-butylamine	− 70.0	214.0		2.6	1.4291	0.043
(96) *cis*-Decahydro-naphthaline	− 43.0	195.8	2.20 (20°)	0.0	1.4810	0.015
(97) *n*-Heptane	− 90.6	98.4	1.92 (20°)	0.0	1.3876	0.012
(98) *n*-Hexane	− 95.3	68.7	1.88	0.0	1.3749	0.009
(99) *n*-Pentane	−129.7	36.1	1.84 (20°)	0.0	1.3575	0.009
(100) Cyclohexane	6.7	80.7	2.02 (20°)	0.0	1.4262	0.006

[a] The physical constants were taken from the following references: (1) R. C. Weast and M. J. Astle: *CRC Handbook of Data on Organic Compounds*. Vol. I and II, CRC Press, Boca Raton/Florida 1985; (2) R. C. Weast (ed.): *Handbook of Chemistry and Physics*, 66th edition, CRC Press, Boca Raton/Florida 1985/86; (3) J. A. Riddick, W. B. Bunger, and T. K. Sakano: *Organic Solvents. Physical Properties and Methods of Purification*. 4th edition. In A. Weissberger (ed.): *Techniques of Chemistry*, Vol. II, Wiley-Interscience, New York 1986; (4) A. L. McClellan: *Tables of Experimental Dipole Moments*. Freeman, San Francisco, London 1963; (5) A. A. Maryott and E. R. Smith: *Table of Dielectric Constants of Pure Liquids*. NBS Circular 514; Washington 1951; (6) *Beilstein's Handbook of Organic Chemistry*. Springer-Verlag, Berlin; (7) M. Windholz (ed.): *The Merck Index*. 10th edition, Rahway/New Jersey 1983.

[b] C. Reichardt and E. Harbusch-Görnert, Liebigs Ann. Chem. *1983*, 721; C. Laurence, P. Nicolet, M. Lucon, and C. Reichardt, Bull. Soc. Chim. Fr. *1987*, 125; ibid. *1987*, 1001; *cf.* also Table 7-3 in Chapter 7.

[c] Melting point.

[d] Boiling point at 1013 mbar.

[e] Relative permittivity (dielectric constant) for the pure liquid at 25 °C unless followed by another temperature in parentheses.

[f] Dipole moment in Coulombmeter (C m), measured in benzene, tetrachloromethane, 1,4-dioxane, or *n*-hexane at 20...30 °C. 1 Debye = $3.336 \cdot 10^{-30}$ C m.

[g] Refractive index at the average *D*-line of sodium (16969 cm^{-1}) at 20 °C unless followed by another temperature in parentheses.

[h] Normalised E_T^N-values, derived from the transition energy at 25 °C of the long-wavelength absorption of a standard pyridinium-*N*-phenoxide betaine dye, $E_T(30)$[b]; *cf.* Eqs. (7-27) and (7-29) in Section 7.4.

[i] Y. Marcus and S. Glikberg, Pure Appl. Chem. *57*, 855 (1985) (Methanol).

[k] J. F. Coetzee (ed.): *Recommended Methods for Purification of Solvents and Tests for Impurities* (Acetonitrile, Sulfolane, Propylene carbonate, Dimethyl sulfoxide, *N,N*-Dimethylformamide, Hexamethylphosphoric triamide, Pyridine, 1,2-Diaminoethane, *N*-Methylacetamide, and *N*-Methylpropionamide). Pergamon Press, Oxford 1982.

[l] Y. Marcus, Pure Appl. Chem. *57*, 860 (1985) (Ethanol).
[m] Y. Marcus, Pure Appl. Chem. *58*, 1411 (1986) (1-Propanol, 2-Propanol, *t*-Butanol).
[n] J. F. Coetzee and T.-H. Chang, Pure Appl. Chem. *58*, 1541 (1986) (Nitromethane).
[o] B. J. Barker, J. Rosenfarb, and J. A. Caruso, Angew. Chem. *91*, 560 (1979); Angew. Chem., Int. Ed. Engl. *18*, 503 (1979) (DMEU, DMPU, Tetramethylurea).
[p] M. Bréant, Bull. Soc. Chim. Fr. *1971*, 725 (1-Methylpyrrolidin-2-one).
[q] J. F. Coetzee and T.-H. Chang, Pure Appl. Chem. *58*, 1535 (1986) (Acetone).
[r] C. Agami, Bull. Soc. Chim. Fr. *1968*, 1205 (1,2-Dimethoxyethane).
[s] J. F. Coetzee and T.-H. Chang, Pure Appl. Chem. *57*, 633 (1985) (THF, 1,4-Dioxane).
[t] K. M. Kadish and J. E. Anderson, Pure Appl. Chem. *59*, 703 (1987) (Benzonitrile, Dichloromethane, 1,1-Dichloroethane, 1,2-Dichloroethane).

Table A-2. Selection of chiral solvents and cosolvents (in alphabetical order)[a)]

Solvents	t_{fp}/°C	t_{bp}/°C (mbar)	ε_r (25 °C)	$[\alpha]_D^{20}$ [b)]	Configuration
(1) 1,4-Bis(dimethylamino)-butane-2,3-diol (DBD)[c)]	43.0…43.6[c)]	68…70[c)] (0.7)		+34.3° (benzene)[c)]	(2*R*,2*R*)
				−34.8° (benzene)[c)]	(2*S*,3*S*)
(2) 1,4-Bis(dimethylamino)-2,3-dimethoxybutane (DDB)[d)]		62…64[d)] (4)		+14.7°[d)]	(2*S*,3*S*)
				−14.7°[d)]	(2*R*,3*R*)
(3) Butane-1,3-diol	< − 50	207.5 (1013)		+26.6° (23 °C)	(*S*)
				−26.6° (23 °C)	(*R*)
(4) Butane-2,3-diol	19	179.0 (1013)		−13.2° (25 °C)	(2*R*,3*R*)
				+13.2°	(2*S*,3*S*)
(5) 2-Chlorobutane	−140.5	68.25 (1013)	7.09 (30 °C)	−31.2°	(*R*)
				+31.2° (25 °C)	(*S*)
(6) Diethyl tartrate[e)]	18	280 (1013)		+ 8.2°[f)]	(2*R*,3*R*)
				− 7.6°	(2*S*,3*S*)
(7) 2,3-Dimethoxybutane[g,h)]	− 84.2… −83.5	109…110 (1000)		+ 3.7°[g)]	(2*R*,3*R*)
(8) 1-Dimethylamino-1-phenylethane[i)]		81 (21)		+61.8° (26 °C)	(*R*)
				−64.4°	(*S*)
(9) 2,3-Dimethylpentane		89.8 (1013)	1.94 (20 °C)	−11.4°	(*S*)
(10) 2-Heptanol		159.7 (1013)	9.21 (22 °C)	+10.3°	(*S*)

Table A-2. (Continued)

Solvents	t_{fp}/°C	t_{bp}/°C (mbar)	ε_r (25 °C)	$[\alpha]_D^{20}$ [b]	Configuration
(11) 3-Heptanol	− 70	156.7 (1013)	6.86 (22 °C)	+ 5.1° (25 °C)	(*S*)
(12) 2-Methyl-1-butanol	< − 70	128.7 (1013)	15.63	− 5.9°	(*S*)
(13) 3-Methyl-2-butanol		111.5 (1013)		+ 4.85°	(*S*)
(14) 3-Methylhexane	−119.4	91.85 (1013)	1.93 (20 °C)	+ 9.4°	(*S*)
(15) 4-Methyl-2-pentanol	− 90	131.7 (1013)		+21.2°	(*S*)
(16) 2-Methyltetrahydrofuran[j]	−137.2	79.9 (1013)	6.97	+27.0°[j,k] −27.5°	(*S*)[k] (*R*)[k]
(17) 1-(1-Naphthyl)-ethylamine[l]		153 (15)		+82.8° (17 °C) −80.8° (25 °C)	(*R*) (*S*)
(18) 2-Octanol[m]	− 32	179.8 (1013)	8.17 (20 °C)	+ 9.9° − 9.7°	(*S*) (*R*)
(19) 2-Pentanol		119.0 (1013)	13.71	+13.9	(*S*)
(20) 1-Phenyl-ethanol[n]	9…11	202…204 (1013)		+43.5°[o] −43.8°[o]	(*R*) (*S*)
(21) 1-Phenyl-ethylamine[p]	32.5	187 (1013)		+39.1°[q] −40.3°[q]	(*R*) (*S*)
(22) *N*-(1-Phenyl-ethyl)-formamide[r]	46…47	175…178[s] (20)		+180°[r] −172°[r]	(*R*) (*S*)
(23) Propane-1,2-diol	− 60	187.6 (1013)	32.0 (20 °C)	−15.4°	(*R*)
(24) 1,2,3,4-Tetra-methoxybutane[t]		70[t] (19)		− 5.9°[t]	(2*S*, 3*S*)
(25) 2,2,2-Trifluoro-1-(9-anthryl)-ethanol[u]	142…145[u]			+27.2° ($CHCl_3$)[u]	(*S*)

Table A-2. (Continued)

Solvents	t_{fp}/°C	t_{bp}/°C (mbar)	ε_r (25 °C)	$[\alpha]_D^{20}$ [b]	Con-figuration
(26) 2,2,2-Trifluoro-1-(1-naphthyl)-ethanol[v]	51.6...53.2[v]	83...85[v] (0.03)		−25.7° (EtOH)	(*R*)
(27) 2,2,2-Trifluoro-1-phenyl-ethanol[w]		82[x] (17)		−40.8° (25 °C)[w]	(*R*)

[a] The physical constants were taken from "Beilsteins Handbuch der organischen Chemie" (4th ed., 1.–4. Ergänzungswerk) and from J. A. Riddick, W. B. Bunger, and T. K. Sakano: *Organic Solvents. Physical Properties and Methods of Purification.* 4th ed., in A. Weissberger (ed.): *Techniques of Chemistry*, Vol. II, Wiley-Interscience, New York 1986, if no reference is given.

[b] Specific rotation of the neat solvent at 20 °C measured at the average *D*-line of sodium (16969 cm^{-1}; 589.3 nm) unless followed by another temperature or solvent in parentheses.

[c] D. Seebach and H. Daum, Chem. Ber. *107*, 1748 (1974).

[d] D. Seebach, H.-O. Kalinowski, W. Langer, G. Crass, and E.-M. Wilka, Org. Synth. *61*, 24, 42 (1983).

[e] H. Plieninger and H. P. Kraemer, Angew. Chem. *88*, 230 (1976); Angew. Chem., Int. Ed. Engl. *15*, 243 (1976).

[f] D. Seebach *et al.*, Helv. Chim. Acta *60*, 301 (1977).

[g] H. L. Cohen and G. F. Wright, J. Org. Chem. *18*, 432 (1953); N. Allentoff and G. F. Wright, ibid. *22*, 1 (1957).

[h] J. D. Morrison and R. W. Ridgeway, Tetrahedron Lett. *1969*, 569.

[i] W. H. Pirkle and M. S. Hoekstra, J. Magn. Reson. *18*, 396 (1975).

[j] D. C. Iffland and J. E. Davis, J. Org. Chem. *42*, 4150 (1977).

[k] D. Gagnaire and A. Butt, Bull. Soc. Chim. Fr. *1961*, 312; E. R. Novak and T. S. Tarbell, J. Am. Chem. Soc. *89*, 73 (1967).

[l] T. G. Burlingame and W. H. Pirkle, J. Am. Chem. Soc. *88*, 4294 (1966).

[m] E. Axelrod, G. Barth, and E. Bunnenberg, Tetrahedron Lett. *1969*, 5031.

[n] J. C. Jochims, G. Taigel, and A. Seeliger, Tetrahedron Lett. *1967*, 1901.

[o] A. J. H. Houssa and J. Kenyon, J. Chem. Soc. *1930*, 2260; E. Downer and J. Kenyon, ibid. *1939*, 1156.

[p] W. H. Pirkle, J. Am. Chem. Soc. *88*, 1837 (1966).

[q] W. Theilacker and H. G. Winkler, Chem. Ber. *87*, 690 (1954).

[r] P. Abley and F. J. McQuillin, J. Chem. Soc., Chem. Commun. *1969*, 477; J. Chem. Soc., Part C *1971*, 844.

[s] R. Huisgen and C. Rüchardt, Liebigs Ann. Chem. *601*, 21 (1956).

[t] D. Seebach *et al.*, Helv. Chim. Acta *60*, 301 (1977).

[u] W. H. Pirkle, D. L. Sikkenga, and M. S. Pavlin, J. Org. Chem. *42*, 384 (1977).

[v] W. H. Pirkle and M. S. Hoekstra, J. Org. Chem. *39*, 3904 (1974); J. Am. Chem. Soc. *98*, 1832 (1976).

[w] W. H. Pirkle, S. D. Beare, and T. G. Burlingame, J. Org. Chem. *34*, 470 (1969).

[x] R. Stewart and R. Van der Linden, Can. J. Chem. *38*, 399 (1960).

Table A-3. Some recommended simple physical and chemical drying methods for thirty-three common organic solvents: + + method gives super-dry solvents with less than 1 ppm water; + solvent sufficiently dry for most chemical applications; (+) often used but less efficient; − explosive hazard (!) or other chemical reaction; no entry means not recommended or no information in the literature. For extensive compilations of more sophisticated purification methods see references[a−f].

Drying methods		Static drying[k] with															Dynamic[l] drying with					Fractional distillation[m] [Other methods]
Solvents[Ref.]	S_{H_2O}[g] 25 °C	M.S. 0.3 nm[n]	M.S. 0.4 nm[n]	M.S. 0.5 nm[n]	P_4O_{10}	Natrium	Al/Hg[a,c,d]	$LiAlH_4$[a]	CaH_2	Al_2O_3 B—I	B_2O_3	KOH (powder)	$CaCl_2$	$CaSO_4 \cdot 0.5\ H_2O$[w]	Na_2SO_4	K_2CO_3	M.S. 0.3 nm[n]	M.S. 0.4 nm[n]	M.S. 0.5 nm[n]	Al_2O_3 B—I[l]	Al_2O_3 N—I[l]	
Acetic acid[a,b,d]	∞				+[d]	−	−	−	−	−		−	−			−				−	−	+(from Ac_2O or P_4O_{10})[b] [fractional freezing]
Acetone[a−d,p,q,β]	∞	(+)[q]	(+)[q]		−	−	−	−	−	−	+[q]	−	−	(+)	(+)	−				−	−	+(from $CaSO_4 \cdot 0.5\ H_2O$)
Acetonitrile[a−e,i,o,p,w]	∞	+[o,p]			+[o]	−		−	+[d]		+[o]	−	(+)[o]	(+)[w]	(+)[a]	(+)[o]	+[e,f]				+[e,i]	+(from P_4O_{10} and then from K_2CO_3)
Benzene[a−d,i,o,p]	(0.63)		++[o,p]	+[c]	+[a]	+[o]		++[o]	++[o]	++[q]			++[o]	+[d]	(+)			+[f]		+[i]		+(from Na)[a] [fractional freezing]
tert-Butanol[a,b,d,γ]	∞	+[u]	(+)[u]		−	(+)[u]	+	−	(+)[u]				−	(+)[d]		(+)[d]						+(from Mg/I_2)[d] [fractional freezing]
2-Butanone[a−d]	120.0				−	−	−	−	−	−		−	−	+[c]	+[d]	+[c]				−	−	+
Chloroform[a−d,i,x,y]	0.6		+[y]	+[x,y]	+[f]	−!	−			+[y]		−	+[y]	+[a]	+[d]	+[a]		+[f]		+[i]		+(from M.S. 0.4 nm[n])[d]
Cyclohexane[a,b,d,i]	0.1		+[d]		+[d]	+[d]				+			+[d]					+[f]		+[i]		+(from Na or $LiAlH_4$)[d,a]
1,2-Diaminoethane[b,d,e]	∞			+[e]	−	+[e]		−				+[e]	−									+(from Na or M.S. 0.5 nm under N_2)[d,e]
1,2-Dichloroethane[b,d]	1.5				+[d]	−!	−		+[d]	+[b]			+[d]	+[d]	+[b]	+[b]				+		+(from P_4O_{10})
Dichloromethane[a−d,i,p,r]	2.0		++[p]		+[d]	−!	−		+[d]	+		−	+[c]	+[d]				+[f]		+[i]		+(from P_4O_{10})
Diethyl ether[a−d,i,p,t,w,y,z]	13.0		+[p,t,z]	+[y]	−	+[c]		+		+			(+)[c,t]	(+)[w]		(+)[t]		+[f]		+[i]		+(from Na or $LiAlH_4$ under N_2)[a]
1,2-Dimethoxyethane[b−d]	∞				−	+[c]		+[c]	+[c]	+			+							+		+(under N_2)[c]
Di-*i*-propylamine[b,d,s]	(400)	+[s]	+[s]		−	+[s]		−	+[s]			(+)[s]	−									+(from NaOH)[d]
N,*N*-Dimethylformamide[a−e,p,q]	∞	+[q]	+[p,q]		+[q]	−		−	+[q]			−	−					+[f]				+(i. vac. under N_2 or as benzene azeotr.)
Dimethylsulfoxide[a−e,i,q]	∞	+[q]	+[q]	+[c]		−	−	−	+[d,e]				−	+[c,d]				+[d]				+(i. vac. from CaH_2)[c,q] [fractional freezing]
1,4-Dioxane[a−e,i,o,p,r,w,z]	∞	+[w]	+[p,z]	+[f]	−	+[o]			+[o]	+		+[o]	(+)[w]	(+)[w]			+[e]	+	+[f]	+[i]		+(from Na under N_2)[c,e]
1,2-Ethanediol[a,b,d,u]	∞	+[u]			−	−		−					−	(+)[d]	(+)[f]			+[d]				+(from Mg/I_2 or as benzene azeotr.)[u]
Ethanol[a−d,u,v]	∞	+[u]			−	−	+[a,d]	−	+[d,u]				−				+[f,v]					+(from Mg/I_2 or as benzene azeotr.)[u]
Ethyl acetate[a−d,i,p]	(29.4)		+[p]		+[d]	−	−	−	+[d]			−	−	+[d]	+[f]	+[c]		+[f]			+[i]	+(from P_4O_{10})[a]
HMPT[b−e,q]	∞	+[q]	+[q]						+[q]			+[q]					+[e]					+(i. vac. from CaH_2, under Ar)[e,q]
Methanol[a−d,u]	∞	+[u]	−	−	−	−	+[a]	−	+[u]				−	+[d]			+[f]					+(from Mg/I_2)[u]
N-Methylacetamide[b,d,e]	∞				+[e]	−		−				−	−									+(i. vac.)[e] [fractional freezing][e]
1-Methyl-2-pyrrolidinone[a−d]	∞					−		−				−	−									+(i. vac. or as benzene azeotrope)[c]
Nitromethane[a,b,d,i,δ]	19.0		+[f]		−[d]	−	−	−				−	+[d]	+[d]	+[d]							+(from M.S. 0.4 nm)
Propylenecarbonate[d,e]	(83)		+[e]	+[e]		−		−				−							+[c,d]			+(i. vac.)[c,d]
2-Propanol[a−d,γ]	∞	+[f]		+[c]	−	−	+[a]	−	+[d]				−	+			+[f]					+(from Mg/I_2)[a]
Pyridine[a−e,i,s]	∞	+[s]	+[s]					+[d]	+[s]	+		+[e,s]	−					+[f]		+[i]		+(from M.S. 0.5 nm/KOH)[e]
Sulfolane[a−e]	∞		+[c]							+		+		+[e]						+[c]		+(i. vac. from CaH_2)[e]
Tetrachloromethane[a−d,i]	0.1		+[f]	+[c]	+[i]	−!	−			+			+[i]		+	+[a]		+[f]		+[i]		+(from P_4O_{10})[d]
Tetrahydrofuran[a−e,i,p,z]	∞		+[p,z]	+[f]	−	+[d]		+[c,d]	+[d]	+		−![c]	+[c]	+[c]			+[e]	+	+[f]	+[i]		+(from $LiAlH_4$ or Na under N_2)[e,a]
Toluene[a−d,p,r]	0.5		++[p]	+[c]	+[d]	+[d]		+	+[d]	+			+[d]	+[d]				+[f]		+		+(from Na)[d]
Trichloroethene[a,b,d,i]	(0.25)					−!	−					−	+[d]		+[f]	+[d]				+[i]		+

[a] Houben-Weyl-Müller: *Methoden der organischen Chemie.* 4th edition, Thieme, Stuttgart 1959, Vol. I/2, p. 765...868 (W. Bunge: *Eigenschaften und Reinigung der wichtigsten organischen Lösungsmittel*) and p. 869...885 (H. Rickert and H. Schwarz: *Trockenmittel*).

[b] J. A. Riddick, W. B. Bunger, and T. K. Sakano: *Organic Solvents. Physical Properties and Methods of Purification.* 4th ed., in A. Weissberger (ed.): *Techniques of Chemistry*, Vol. II, Wiley-Interscience, New York 1986.

[c] B. S. Furniss, A. J. Hannaford, V. Rogers, P. W. G. Smith, and A. R Tatchell: *Vogel's Textbook of Practical Organic Chemistry.* 4th edition, Longman, London, New York 1978, p. 264...279.

[d] D. D. Perrin and W. L. F. Armarego: *Purification of Laboratory Chemicals.* 3rd edition, Pergamon Press, Oxford 1988.

[e] J. F. Coetzee (ed.): *Recommended Methods for Purification of Solvents and Tests for Impurities.* Pergamon Press, Oxford 1982; Pure Appl. Chem. *57*, 634 (1985).

[f] E. Merck AG: *Drying in the Laboratory.* D-6100 Darmstadt, Fed. Rep. Germany.

[g] Maximum solubility of water in solvent in g/l, except values in parentheses which are in g/kg; ∞ means completely miscible with water[b,h,i].

[h] Fluka AG: *Fluka Katalog 1988/89.* CH-9470 Buchs, Switzerland 1988, p. 1500.

[i] B. P. Engelbrecht: *Adsorptives Reinigen von Lösungsmitteln für die Chromatographie und Spektroskopie.* GIT Fachz. Lab. *23*, 681 (1979); Chem. Abstr. *91*, 133523y (1979); *cf.* also *Purification of Solvents on Adsorbents Woelm.* ICN Biomedicals GmbH (formerly Woelm Pharma GmbH), D-3440 Eschwege, Fed. Rep. Germany.

[k] Treatment with *ca.* 50...100 g/l desiccant at ambient temperature for about 24 h. Gentle agitation or stirring has an accelerating effect on drying. Sequential drying, accomplished by decanting monosiccated solvent onto a fresh charge of *ca.* 50 g/l desiccant, is more effective.

[l] Column drying by percolating the solvent through the desiccant contained in a glass column of 2...5 cm diameter and 40...150 cm long, and collecting the eluent in a storage container protected from atmospheric moisture by a drying tube filled with molecular sieve.

[m] Fractional distillation is often combined with static drying before or after the distillation.

[n] Zeolite molecular sieves (sodium and calcium aluminosilicates) of nominal pore size 0.3...0.5 nm, normally used as beads except in cases where the use of powdered molecular sieve is essential (marked with an asterisk).

[o] D. R. Burfield, K.-H. Lee, and R. H. Smithers, J. Org. Chem. *42*, 3060 (1977).

[p] D. R. Burfield, G.-H. Gan, and R. H. Smithers, J. Appl. Chem. Biotechnol. *28*, 23 (1978); Chem. Abstr. *89*, 12551f. (1978).

[q] D. R. Burfield and R. H. Smithers, J. Org. Chem. *43*, 3966 (1978).

[r] D. R. Burfield and R. H. Smithers, J. Chem. Technol. Biotechnol. *30*, 491 (1980); Chem. Abstr. *94*, 66822s (1981).

[s] D. R. Burfield, R. H. Smithers, and A. S. C. Tan, J. Org. Chem. *46*, 629 (1981).

[t] D. R. Burfield and R. H. Smithers, J. Chem. Educ. *59*, 703 (1982).

[u] D. R. Burfield and R. H. Smithers, J. Org. Chem. *48*, 2420 (1983); Y. Marcus and S. Glikberg, Pure Appl. Chem. *57*, 855, 860 (1985).

[v] D. R. Burfield, G. T. Hefter, and D. S. P. Koh, J. Chem. Technol. Biotechnol., Chem. Technol. Part A *34*, 187 (1984); Chem. Abstr. *101*, 133117u (1984).

[w] D. R. Burfield, J. Org. Chem. *49*, 3852 (1984).

[x] D. R. Burfield, J. Chem. Educ. *56*, 486 (1979); D. R. Burfield, E. H. Goh, E. H. Ong, and R. H. Smithers, Gazz. Chim. Ital. *113*, 841 (1983).

[y] D. R. Burfield and R. H. Smithers, Chem. Ind. (London) *1980*, 240.

[z] D. R. Burfield, J. Org. Chem. *47*, 3821 (1982); Deperoxidation of ethers with self-indicating molecular sieves 0.4 nm (Merck or Sigma).

[α] *Caution:* solutions of $LiAlH_4$ in oxygen-containing solvents may decompose at elevated temperatures ($\geqq$160 °C). Therefore, distillation should never be carried out to dryness, and solvents boiling above 100 °C should be distilled under reduced pressure.

[β] J. F. Coetzee and T.-H. Chang, Pure Appl. Chem. *58*, 1535 (1986).

[γ] Y. Marcus, Pure Appl. Chem. *58*, 1411 (1986).

[δ] J. F. Coetzee and T.-H. Chang, Pure Appl. Chem. *58*, 1541 (1986).

interferes undesirably with many reactions, its removal is one of the basic laboratory operations. Drying agents may bind water either physically or chemically [18, 19]. The best method depends in each case on the chemical nature of the solvent and the desired degree of dryness [20]. All organic solvents, possessing a dielectric constant of less than 15 can be freed, almost completely from water, alcohols, peroxides, and traces of acid, by simple adsorptive filtration through aluminium oxide (activity I) or silica gel (activity I), *e.g.* using a chromatography column of 2...5 cm diameter and 40...150 cm long [19, 21–24].

Recently, comparative studies of the drying efficiency of a number of common desiccants for different types of organic solvents have been carried out by Burfield and coworkers [106–117]. Using a new and very sensitive tritiated water tracer method for determining water content, Burfield *et al.* obtained rather unexpected results concerning the drying efficiency of commonly used desiccants. The results of this study, together with other recommended physical and chemical drying methods [1, 17, 18, 104, 105], are compiled in Table A-3. In particular, Burfield's results draw attention to the almost universal applicability of zeolithe molecular sieves as desiccants, which are capable of drying even the most difficult organic solvents [107].

A.3 Spectroscopic Solvents

Solvents used in ultraviolet, visible, infrared, microwave, and radiowave spectroscopy must meet the following requirements: transparency and stability toward the radiation used, solubility of and chemical stability to the substance to be examined, and a high and reproducible purity ("optical constancy"). Normally, intermolecular interaction with the solute should be minimal. On the other hand, important information about the solute can be obtained from the changes in the absorption spectrum arising from such interactions.

The number of solvents useful in UV/Vis spectroscopy decreases with decreasing wavelength (increasing wavenumber) since the absorption of all substances increases in this direction. The *cut-off point* depends on the chemical nature and to a large extent on the purity of the solvent. Hence, numerous procedures for the production of optically pure solvents have been developed [25–29]. Solvents for the measurement of fluorescence spectra must often be particularly pure [30]. The cut-off points of the solvents normally used in UV/Vis spectroscopy are collected in Table A-4. Saturated hydrocarbons are among the most useful because of the weak intermolecular interactions and the lack of excitable π-electrons. Perfluorinated hydrocarbons are recommended for the far-UV region (<200 nm) [31–33]. The UV-spectra of the more important organic solvents are reproduced in the "DMS-UV-Atlas of Organic Compounds" [34].

Solvents for infrared spectroscopy must meet the additional requirement that they do not attack the absorption cells themselves (normally made from alkali halides such as NaCl, KBr, and CsBr) [35]. The transparency regions of the IR-solvents within the mid-IR region (2...16 µm; 5000...625 cm^{-1}) are given in Table A-5. Complete IR-spectra of organic solvents are found in the "DMS-Working Atlas of Infrared Spectroscopy" [36] as well as in the "Sadtler IR Spectra Handbook of Common Organic Solvents" [118]. Transmission characteristics of organic solvents in the near-IR region (1...3 µm; 1000...3333 cm^{-1}) are given in references [37, 38], and for the far-IR region (15...35 µm; 667...286 cm^{-1}) in references [39, 40]. The IR-spectra of deuterated

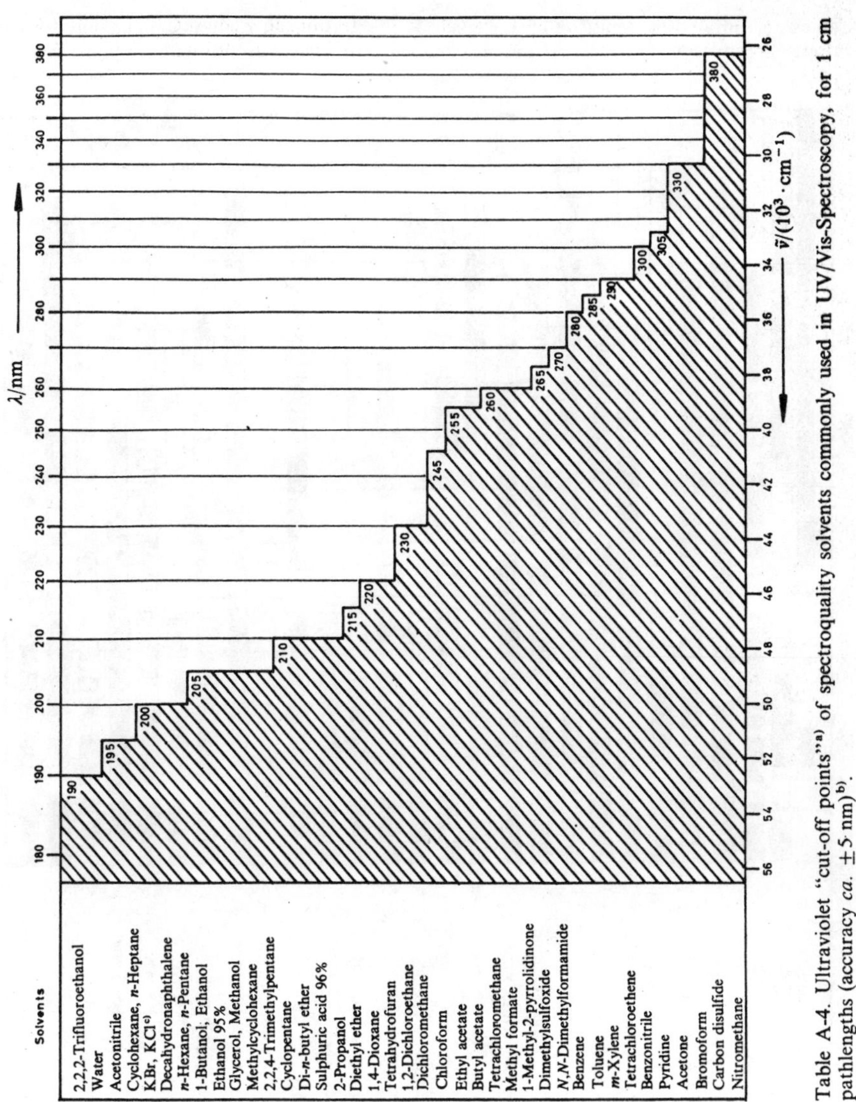

Table A-4. Ultraviolet "cut-off points"[a] of spectroquality solvents commonly used in UV/Vis-Spectroscopy, for 1 cm pathlengths (accuracy *ca.* ± 5 nm)[b].

[a] The "cut-off point" in the ultraviolet region is the wavelength at which the absorbance approaches 1.0 using a 1-cm cell path with water as the reference. Solvents should not be used for measurements below the cut-off point, even though a compensating reference cell is employed. The cut-off points are very dependent on the purity of the solvent used. Most of the solvents listed above are available in highly purified "spectrograde" quality.

[b] Compiled from the following references:
(1) Eastman Kodak Company: *Spectrophotometric Solvents*. Dataservice Catalog JJ-282, Rochester, New York 14650, USA, 1977;
(2) E. Merck: *UVASOLE® – Lösungsmittel und Substanzen für die Spektroskopie*. D-6100 Darmstadt, Fed. Rep. Germany;
(3) and from the reviews of Gordon and Ford [4] (p. 167), Pestemer [25], and Hampel [34].

[c] Values for solid, as used in a pellet for example.

Table A-5. Infrared transmission characteristics[a)] of spectroquality solvents commonly used in Infrared Spectroscopy in the 2...16 μm region (5000...625 cm^{-1}) for 0.1 mm solvent thickness[b)].

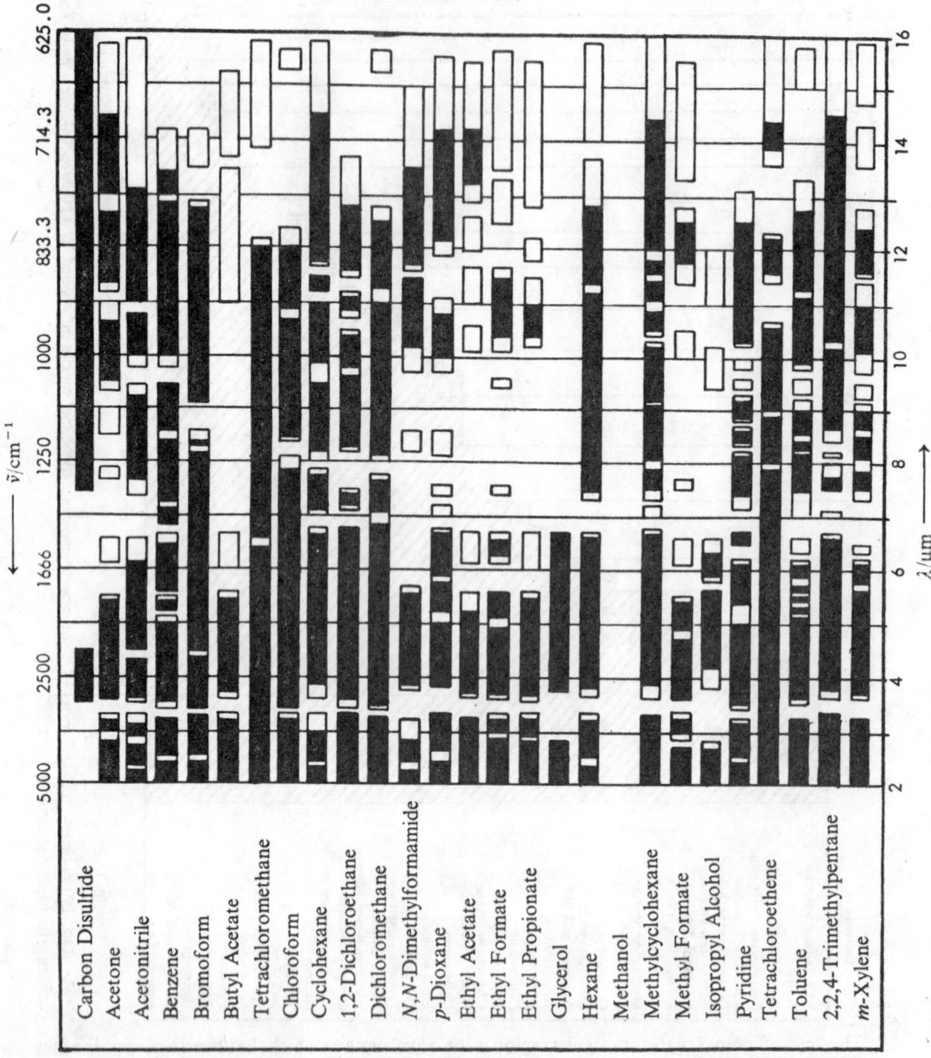

[a] ▬ 80...100% Transmission; ▭ 60...80% Transmission. The black lines represent the useful regions. In the areas where the solvent has absorption bands that are totally absorbing, no information can be obtained about the sample, even though a compensating reference cell is employed. There is no solvent that is completely transparent over the entire wavenumber range. The most useful solvents are tetrachloromethane and carbon disulfide.

[b] N. L. Alpert, W. E. Keiser, and H. A. Szymanski: *IR – Theory and Practice of Infrared Spectroscopy*. 2nd edition, first paperback printing, Plenum Publishing Corporation, New York 1973, p. 326.

organic solvents between 2.5 and 16.7 μm (4000...600 cm^{-1}) have also been measured [41]. The number of IR absorption bands, active in a covalent compound, decreases with the number of atoms and with increasing symmetry of the molecule. Therefore, small molecules of high symmetry are particularly useful IR-solvents, *e.g.* carbon disulfide (point group $D_{\infty h}$) and tetrachloromethane (point group T_d).

In ^{1}H-NMR spectroscopy, one uses solvents which either contain no hydrogens (*e.g.* CS_2, CCl_4, $Cl_2C{=}CCl_2$, hexachlorobutadiene) or deuterated solvents (*e.g.* C_6D_6, $CDCl_3$, D_2O, CD_3SOCD_3). Table A-6 gives the characteristic ^{1}H-NMR absorption bands of common organic solvents. More complete collections of ^{1}H-NMR chemical shifts of organic solvents are available in the literature [42, 43]. Table A-7 contains the ^{13}C-NMR absorption bands of compounds used as solvents as well as reference substances in ^{13}C-NMR spectroscopy [44]. In order to use ^{1}H- and ^{13}C-NMR chemical shifts diagnostically for certain functional groups, all solvent effects should be eliminated as far as possible. The ideal situation can be approached by extrapolation to infinite dilution from measurements at different concentrations. Solvent effects can generally be ignored in inert solvents such as CCl_4 or CS_2 at concentrations ≤ 50 mg/g.

Concerning solvents for matrix isolation see Section A.4.

A.4 Solvents as Reaction Media

The following criteria can be used in the proper choice of solvents for chemical reactions and kinetic investigations: a maximum working range between melting and boiling point, chemical and thermal stability in this temperature range, good solubility of reactants and products (or sometimes insolubility of the products), compatibility with the analytical methods employed, and usually high degree of purity. The most useful solvents for some frequent reaction types are indicated in Table A-8. When a solvent is used for technical purposes, other factors may play an additional role [45]: price, flammability (ignition point, flash point), explosiveness (upper and lower explosion limit), viscosity, miscibility (*e.g.* blending with extenders or diluents), toxicity, corrosive action, and feasibility of recycling.

Another interesting area concerning solvents as reaction media as well as for spectroscopy is the technique of *matrix isolation*, used when the molecule of interest is extremely unstable. Matrix isolation involves the trapping of the molecule in a rigid cage of a chemically inert solvent (the matrix) at low temperature. The main requirements are that the unstable molecule must have a stable precursor from which it can be liberated (usually by irradiation), and that the fragments so produced must not react with it or with the matrix. The reaction products can then be studied by IR or UV/Vis spectroscopy. Typical matrix materials are the noble gases (especially argon), the lighter hydrocarbons and halohydrocarbons, nitrogen, carbon dioxide, and other solvents and solvent mixtures which produce clear glasses when cooled to low temperatures. The properties of a number of matrix solvents can be found in references [93–96]. The temperature has to be sufficiently low to prevent diffusion of the solute into the matrix lattice. The useful temperature range is from 1.5 K to about 0.6 K of the matrix melting point. It should be remembered that the properties of a matrix isolated molecule may be different from those in the commonly used solvents because of the low temperature and unusual environment.

Table A-6. Characteristic ^{1}H-NMR-absorption bands[a)] of thirty spectroquality solvents, commonly used in ^{1}H-Nuclear Magnetic Resonance Spectroscopy[b)] (δ-scale: Tetramethylsilane $\delta_H = 0$ ppm).

Solvents	δ/ppm (multiplicity)[a)]
Acetic acid-d_4	11.53(1)[c)], 2.03(5)
Acetone-d_6	2.04(5)
Acetonitrile-d_3	1.93(5)
Benzene-d_6	7.15(br)
Bromoform-d_1	6.85(1)
Chloroform-d_1	7.24(1)
Cyclohexane-d_{12}	1.38(br)
Deuterium oxide	*ca.* 4.8[c)]
1,2-Dichloroethane-d_4	3.72(br)
Dichloromethane-d_2	5.32(3)
Diethyl ether-d_{10}	3.34(m), 1.07(m)
Diethylene glycol dimethyl ether-d_{14}	3.49(br), 3.40(br), 3.22(5)
1,2-Dimethoxyethane-d_{10}	3.40(m), 3.22(5)
N,N-Dimethylformamide-d_7	8.01(br), 2.91(5), 2.74(5)
Dimethylsulfoxide-d_6	2.49(5)
1,4-Dioxane-d_8	3.53(m)
Ethanol-d_6 (anhydrous)	5.19(1)[c)], 3.55(br), 1.11(m)
Hexafluoroacetone · 3/2 D_2O	*ca.* 9.0(br)[c)], 5.26(1)
Hexamethylphosphoric triamide-d_{18}	2.53(2 × 5)
Methanol-d_4	4.78(1)[c)], 3.30(5)
Nitrobenzene-d_5	8.11(br), 7.67(br), 7.50(br)
Nitromethane-d_3	4.33(5)
2-Propanol-d_8	5.12(1)[c)], 3.89(br), 1.10(br)
Pyridine-d_5	8.71(br), 7.55(br), 7.19(br)
Sulfuric acid-d_2	*ca.* 11.0[c)]
1,1,2,2-Tetrachloroethane-d_2	5.91(1)
Tetrahydrofuran-d_8	3.58(br), 1.73(br)
Toluene-d_8	7.09(m), 7.00(br), 6.98(m), 2.09(5)
Trifluoroacetic acid-d_1	11.50(1)[c)]
2,2,2-Trifluoroethanol-d_3	5.02(1)[c)], 3.88(4 × 3)

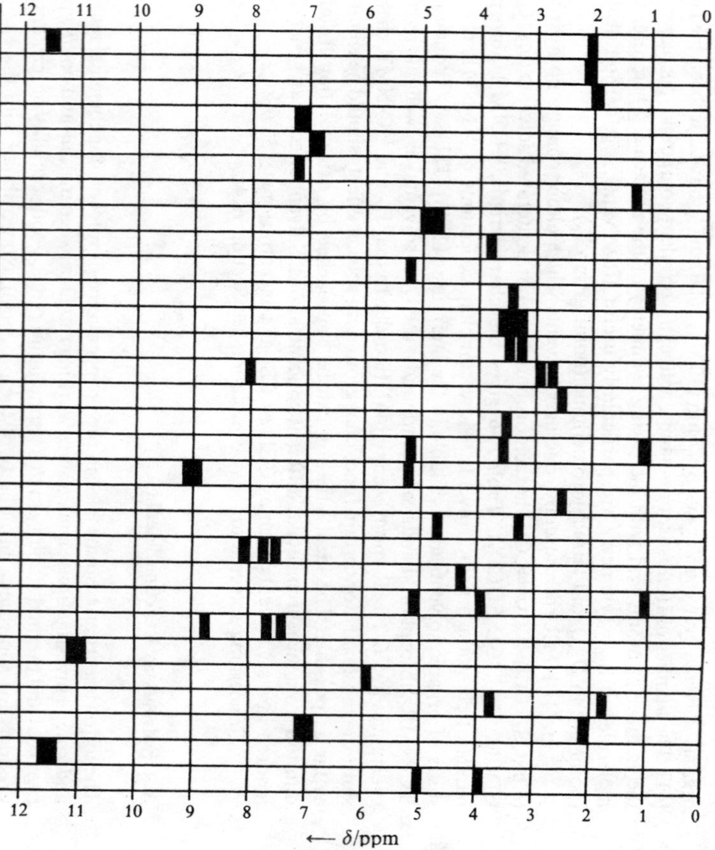

12 11 10 9 8 7 6 5 4 3 2 1 0

← δ/ppm

[a] The ^{1}H-chemical shifts of the residual hydrogen atoms are in ppm relative to tetramethylsilane, determined with solutions containing 5 ml/100 ml tetramethylsilane as the internal standard. The multiplicity of the peaks is given in parentheses; br indicates a broad peak without resolvable fine structure, while m denotes one with fine structure. Partial deuterium labeling results in broadening or splitting of the residual peaks, due either to deuterium-hydrogen coupling or non-equivalence of the residual hydrogens. If undeuterated solvents are used, the region obscured might be as wide as 2...3 ppm.

[b] Compiled from the following references:
(1) Merck Sharp and Dohme Canada Ltd., Isotope Division: *Deuterated Solvents – Handy Reference Data*. Pointe Claire-Dorval, Quebec, Canada;
(2) H.-O. Kalinowski, S. Berger, and S. Braun: *^{13}C-NMR Spektroskopie*. Thieme-Verlag, Stuttgart, New York 1984, p. 74.

[c] These peak positions may vary considerably depending on solute, concentration, and temperature.

Table A-7. Characteristic ^{13}C-NMR absorption bands[a)] of thirty spectroquality solvents, commonly used in ^{13}C-Nuclear Magnetic Resonance Spectroscopy[b)] (δ_C-scale: Tetramethylsilane $\delta_C = 0$ ppm).

Solvents	δ/ppm (multiplicity)[a)]
Acetic acid-d_4	178.4(br), 20.0(7)
Acetone-d_6	206.0(13), 29.8(7)
Acetonitrile-d_3	118.2(br), 1.3(7)
Benzene-d_6	128.0(3)
Bromoform-d_1	12.4(3)
Carbondisulfide	192.7(1)
Chloroform-d_1	77.0(3)
Cyclohexane-d_{12}	26.4(5)
Dichlorodifluoromethane	126.2(3)
1,2-Dichloroethane-d_4	43.6(5)
Dichloromethane-d_2	53.8(5)
Diethyl ether-d_{10}	65.3(5), 14.5(7)
Diethylene glycol dimethyl ether-d_{14}	70.7(5), 70.0(5), 57.7(7)
1,2-Dimethoxyethane-d_{10}	71.7(5), 57.8(7)
N,N-Dimethylformamide-d_7	162.7(3), 35.2(7), 30.1(7)
Dimethylsulfoxide-d_6	39.5(7)
1,4-Dioxane-d_8	66.5(5)
Ethanol-d_6 (anhydrous)	56.8(5), 17.2(7)
Hexafluoroacetone · 3/2 D_2O	122.5(4), 92.9(7)
Hexamethylphosphoric triamide-d_{18}	35.8(7)
Methanol-d_4	49.0(7)
Nitrobenzene-d_5	148.6(1), 134.8(3), 129.5(3), 123.5(3)
Nitromethane-d_3	62.8(7)
2-Propanol-d_8	62.9(3), 24.2(7)
Pyridine-d_5	149.9(3), 135.5(3), 123.5(3)
Tetrachloroethene	121.4(1)
Tetrahydrofuran-d_8	67.4(5), 25.3(br)
Toluene-d_8	137.5(1), 128.9(3), 128.0(3), 125.2(3), 20.4(7)
Trifluoroacetic acid-d_1	164.2(4), 116.6(4)
2,2,2-Trifluoroethanol-d_3	126.3(4), 61.5(4 × 5)

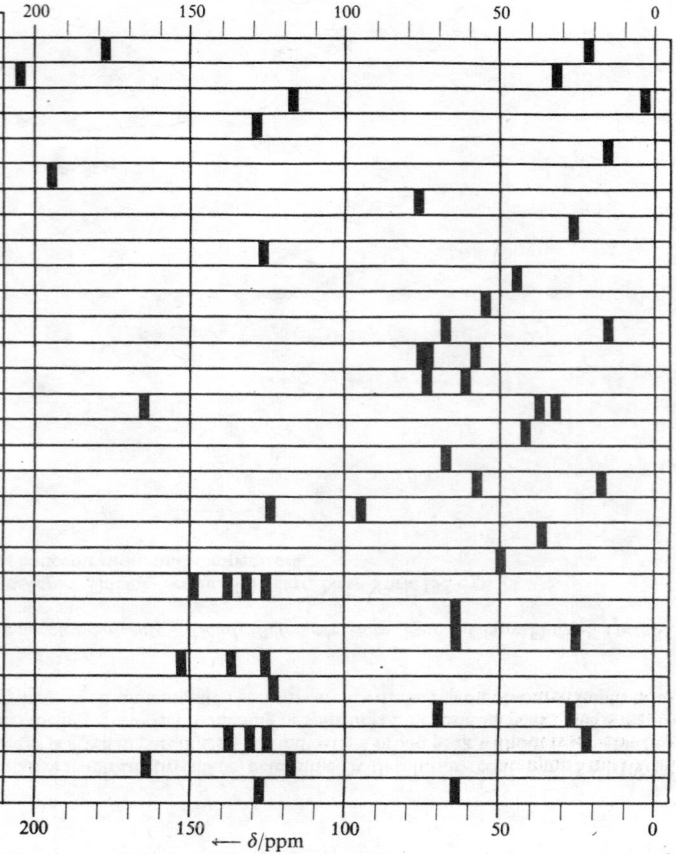

200 150 100 50 0
← δ/ppm

[a] The ^{13}C-chemical shifts are in ppm relative to tetramethylsilane, determined with solutions containing 5 ml/100 ml tetramethylsilane as the internal standard. The multiplicity of the peaks is given in parentheses; br indicates a broad peak without resolvable fine structure. The chemical shifts can be dependent on solute, concentration, and temperature.

[b] Compiled from the following references:
(1) Merck Sharp and Dohme Canada Ltd., Isotope Division: *Deuterated Solvents – Handy Reference Data*, Pointe Claire-Dorval, Quebec, Canada;
(2) H.-O. Kalinowski, S. Berger, and S. Braun: *^{13}C-NMR-Spektroskopie*. Thieme-Verlag, Stuttgart, New York 1984, p. 74.

A.5 Solvents for Recrystallization

The following requirements must be met by a solvent used for the recrystallization of an organic compound [46–48]: high solubility at high temperatures and low solubility at low temperatures of the substance to be recrystallised; impurities should be either highly soluble or totally insoluble; the boiling point should be as high as possible; the solvent should be chemically inert; the solvent must favor crystal growth, and be easily separated from the pure crystals by washing or drying. Hence, it should be either volatile or very soluble in a more volatile solvent and not form clathrates or solvates. In the first instance, the choice of solvent can be made according to the old rule "like dissolves like". The following list gives some guidelines [49]; a more complete collection of organic solvents used for recrystallization is given in Table A-9. When a compound is too soluble in one, and not soluble enough in another solvent, a binary mixture of the two may be a useful medium for recrystallization (*e.g.* acetic acid/water, ethanol/water, ethanol/benzene, acetone/petroleum ether, or chloroform/petroleum ether). For solvent selection *cf.* also references [119, 120].

Crystallization of a solid from solution cannot only be effected by changing the temperature (decreasing solubility with decreasing temperature) but also by changing the solvent polarity (decreasing solubility with decreasing solvent polarity) at constant temperature. A "cold" crystallization technique, using pairs of solvents of different polarity (*e.g.* methanol/water, acetone/water) has been considered in reference [121].

Compound to be recrystallised		Well soluble in
Hydrocarbons	*hydrophobic*	Hydrocarbons, Ethers, Halohydrocarbons
Halohydrocarbons	↓	
Ethers		
Amines		Carboxylic Esters
Esters		
Nitrohydrocarbons		
Nitriles		Alcohols, 1,4-Dioxane
Ketones		Acetic Acid
Aldehydes		
Phenols		Alcohols, Water
Amides		
Alcohols		
Carboxylic Acids		
Sulfonic Acids		
Salts	*hydrophilic*	Water

Table A-8. Compilation of solvents commonly used for some important chemical reactions[a, b].

Solvents \ Reactions	S_N1-Reactions[c]	S_N2-Reactions[c]	Oxidation reactions[d]	Ozonization reactions	Epoxidation reactions	Catalytic Hydrogenation[d, e]	Metal Hydride Reductions[f]	Aldol-Reactions[d, g]	Wittig-Reactions[h]	Diels-Alder Cycloadditions	Grignard-Reactions	Friedel-Crafts-Acylation and Alkylation Reactions	Halogenation	Nitration	Sulfonation	Diazotization	Diazo Coupling
Acetic acid	+		+	+	+	+		+					+	+	+	+	+
Acetone		+			+					+							
Acetonitrile		+							+	+							
Benzene			+		+		+	+	+	+		(+)				+	
t-Butanol	+		+						+								
Carbon disulfide												+					
Chloroform			+	+	+					+					+		
Cyclohexane						+											
Dichloromethane			+	+					+	+		+		+			
Diethyl ether					+		+	+	+	+	+						
Di-*n*-butyl ether											+						
1,2-Dichloroethane				+						+		+					
1,2-Dichlorobenzene										+		+	+	+			
1,2-Dimethoxyethane							+	+	+								
N,N-Dimethylformamide		+				+			+							+	
Dimethyl sulfoxide		+						+	+								
1,4-Dioxane			+		+	+	+			+					+		
Ethanol	+					+		+	+	+						+	+
Ethyl acetate			+	+		+											
HMPT		+							+								
Methanol	+			+		+		+	+	+							+
Nitrobenzene			+									+	+	+	+		
Nitromethane		+										+					
Petroleum ether						+						+			+		
Pyridine			+					+									+
Sulfuric acid			+											+	+	+	
Tetrachloroethene										+							
Tetrachloromethane				+	+								+	+	+		
Tetrahydrofuran						+	+	+	+	+	+						
Tetramethylene sulfone		+	+									+					
Toluene							+	+	+	+							
Trichloroethene										+		+					
Water	+		+			+		+		+						+	+

[a] *Cf.* A. K. Doolittle: *The Technology of Solvents and Plasticizers.* Wiley, New York 1954.
[b] Table A-8 contains a more arbitrary selection from a plurality of possibilities. If more than one solvent is given for a reaction, then binary mixtures of these solvents can also be used as reaction medium.
[c] C. K. Ingold: *Structure and Mechanism in Organic Chemistry.* 2nd edition, Cornell University Press, Ithaca/N.Y., and London 1969.
[d] H. O. House: *Modern Synthetic Reactions.* 2nd edition, Benjamin, Menlo Park/California 1972.
[e] P. N. Rylander: *Solvents in Catalytic Hydrogenation.* In W. H. Jones (ed.): *Catalysis in Organic Synthesis.* Academic Press, New York, London 1980.
[f] H. C. Brown: *Organic Syntheses via Boranes.* Wiley-Interscience, New York 1985.
[g] A. T. Nielsen and W. J. Houlikan: *The Aldol Condensation.* Org. React *16*, 1 (1968); *cf.* p. 76...77. – C. H. Heathcock: *Stereoselective Aldol Condensations.* In E. Buncel and T. Durst (eds.): *Comprehensive Carbanion Chemistry.* Part B, Chapter 5, p. 177ff., Elsevier, Amsterdam 1984; *cf.* p. 199.
[h] I. Gosney and A. G. Rowley in J. I. G. Cadogan (ed.): *Organophosphorous Reagents in Organic Synthesis.* Academic Press, London 1979, p. 24...25 and 41.

Table A-9. Compilation of solvents commonly used for crystallization of organic compounds[a)], listed in order of decreasing solvent polarity as measured by the E_T^N values[b)].

Solvent	E_T^N	t_{bp}/°C (1 bar)	Suitable solvent for	Second solvent for mixture[c)]
Water	1.000	100.0	Salts, amides, some carboxylic acids	Acetone, alcohols, 1,4-dioxane, acetonitrile
Methanol	0.762	64.7	General, esters, nitro and bromo compounds	Water, diethyl ether, benzene
2-Methoxyethanol	0.667	124.6	Sugars	Water, benzene, diethyl ether
Ethanol	0.654	78.3	General, esters, nitro and bromo compounds	Water, hydrocarbons, ethyl acetate
Acetic acid	0.648	117.9	Salts, amides, some carboxylic acids	Water
Acetonitrile	0.460	81.6	Dipolar compounds	Water, diethyl ether, benzene
Acetone	0.355	56.3	General, nitro and bromo compounds, osazones	Water, hydrocarbons, diethyl ether
Dichloromethane	0.309	39.8	General, low-melting compounds	Ethanol, hydrocarbons
Pyridine	0.302	115.3	High-melting, difficult soluble compounds	Water, methanol, hydrocarbons
Methyl acetate	0.287	56.3	General, esters	Water, diethyl ether
Chloroform	0.259	61.2	General, acid chlorides	Ethanol, hydrocarbons
Ethyl acetate	0.228	77.1	General, esters	Diethyl ether, benzene, hydrocarbons
1,4-Dioxane	0.164	101.3	Amides	Water, benzene, hydrocarbons

Table A-9. (Continued)

Solvent	E_T^N	t_{bp}/°C (1 bar)	Suitable solvent for	Second solvent for mixture[c]
Diethyl ether	0.117	34.6	General, low-melting compounds	Acetone, hydrocarbons
Benzene	0.111	80.1	Aromatics, hydrocarbons, molecular complexes	Diethyl ether, ethyl acetate, hydrocarbons
Toluene	0.099	110.6	Aromatics, hydrocarbons	Diethyl ether, ethyl acetate, hydrocarbons
Tetrachloromethane	0.052	76.8	Apolar compounds, acid chlorides, anhydrides	Diethyl ether, benzene, hydrocarbons
Ligroin	—	90–110	Hydrocarbons	Ethyl acetate, benzene, dichloromethane
Petroleum ether	—	40–60	Hydrocarbons	Any solvent on this list from ethanol down
n-Hexane	0.009	68.7	Hydrocarbons	Any solvent on this list from ethanol down
Cyclohexane	0.006	80.7	Hydrocarbons	Any solvent on this list from ethanol down

[a] *Cf.* A. J. Gordon and R. A. Ford: *The Chemist's Companion – A Handbook of Practical Data, Techniques, and References.* Wiley and Sons, New York, London, Sydney, Toronto 1972, p. 442.
[b] C. Reichardt and E. Harbusch-Görnert, Liebigs Ann. Chem. *1983*, 721.
[c] Trial and error are usually required in selecting a second solvent for a mixture. There are some generally successful mixtures, such as diethyl ether/methanol (or ethanol) for highly associated solids (especially amides and alcohols) and many natural products, and diethyl ether/petroleum ether (or benzene) for dipolar compounds (especially esters and alcohols) and hydrocarbons[a]. *Cf.* also J. B. Baumann: *Solvent Selection for Recrystallization.* J. Chem. Educ. *56*, 64 (1979).

A.6 Solvents for Extraction and Partitioning (Distribution)

The partitioning of a substance between two liquid phases (multistage partitioning, partition chromatography) and the extraction of solids, require similar properties of a solvent [50–55]. When a substance has to be partitioned, a solvent system with limited miscibility of the components is required in order that the substance dissolves to a different extent in the two phases. The greater the chemical differences between any two solvents, the more limited their miscibility. Other requirements that the solvent system must fulfill are, *inter alia*, a favorable partition coefficient (the average partition coefficient of the component mixture should be between *ca.* 0.2 and 5), as high a separation factor as possible (the ratio of the partition coefficients should not, in general, be smaller than 1.5), linearity of the partition isotherm (*i.e.* concentration independence of the partition isotherm), a sufficiently large capacity, high selectivity, no tendency to emulsion formation and rapid separation of the phases (this requires low viscosity, a large density difference, and a sufficient surface tension), absence of irreversible reactions between solvent and solute, and ease of recovery of the substance [51]. The optimization of these various requirements is difficult. Generally, a compromise between these, sometimes competing, factors must be made in the choice of phase partners. The solvents used most frequently in partitioning have been divided into five classes according to the intermolecular interactions between the solvent molecules. The difference lies predominantly in the number and strength of the intermolecular hydrogen bonds [51]. Typical representatives of these five classes of solvents are water, methanol, pyridine, chloroform, and *n*-heptane. Within each class the solvents can be ordered according to increasing solubility in water or decreasing solubility in *n*-heptane. Thus, a "mixotropic" series of solvents has been established [51], an extended version of which is presented in Table A-10. This series gives valuable information concerning both the miscibility of solvents and their use in partitioning methods (paper, column, and thin-layer chromatography). The farther that two solvents are from each other in Table A-10, the less miscible they are. – More than 400 liquid stationary phases for gas-liquid chromatography are commercially available. The selection of the proper stationary phase for a separation problem is often done by "trial and error", here again the rule "like dissolves like" is often useful. Collections of established liquid stationary phases are found in references [4, 56]. Methods for assessing the polarity and selectivity of stationary phases in gas-liquid chromatography have been reviewed [133].

For the optimization of solvent composition in high performance liquid chromatography (HPLC) see references [122–125].

A.7 Solvents for Adsorption Chromatography

The accumulation of an organic substance on the surface of an adsorbent is determined by its dipolarity as well as its polarizability and molecular size. This is also true for solvents which are adsorbed the more strongly, the more dipolar their molecules. The molecules of the adsorbed compound and the solvent compete for the active sites on the adsorbed surface. Thus, an organic compound will be adsorbed more strongly from an apolar than from a polar solvent. Conversely, a previously adsorbed material can only be replaced by a solvent when the latter has a higher affinity for the adsorbent [24, 52, 57–60, 126, 127].

The following requirements are desirable when choosing an eluant: high purity (in particular, absence of water and other strongly polar compounds), solubility for the crude mixture, low viscosity, ease of regeneration, and suitability for analytical method employed (*e.g.* when using UV/Vis-detection during elution, the solvent itself must be transparent at the wavelength used). Furthermore, the eluant must be chemically inert to the adsorbate; *e.g.* acetone and ethyl acetate on an alkaline adsorbent such as aluminium oxide are readily transformed into diacetone alcohol and acetic acid, respectively. The success of a separation problem often depends more on the proper choice of solvent than on the choice of stationary phase. The solvents used as eluants can be arranged according to their increasing eluting power, the so-called "eluotropic" series of solvents, by an empirical determination of the retention time for constant adsorbent and test-mixture [24, 52, 57–61]. The shorter the retention time of the test-sample, the higher the eluting power and hence the polarity of the solvent. Oxide adsorbents such as aluminium oxide and silica gel give almost identical eluotropic series. A generalised eluotropic series according to Pusey [61] is as follows: Saturated hydrocarbons < Aromatic hydrocarbons < Halogenated hydrocarbons < Ethers < Esters < Ketones < Alcohols < Carboxylic

Table A-10. Mixotropic Solvent Series[a–e].

(1) Water	(37) 1-Octanol
(2) Lactic acid	(38) Diethoxymethane
(3) Formamide	(39) Hexanoic acid
(4) Morpholine	(40) Butyl acetate
(5) Formic acid	(41) Di-*i*-propoxymethane
(6) Acetonitrile	(42) Nitromethane
(7) Methanol	(43) 1-Bromobutane
(8) Acetic acid	(44) Di-*i*-propyl ether
(9) Ethanol	(45) Butyl butyrate
(10) 2-Propanol	(46) 1-Bromopropane
(11) Acetone	(47) Di-*n*-butyl ether
(12) 1-Propanol	(48) Dichloromethane
(13) 1,4-Dioxane	(49) Chloroform
(14) Propanoic acid	(50) Di-*i*-amyl ether
(15) Tetrahydrofuran	(51) 1,2-Dichloroethane
(16) *t*-Butanol	(52) Bromobenzene
(17) 2-Methylpropanoic acid	(53) 1,1,2-Trichloroethane
(18) 2-Butanol	(54) 1,2-Dibromoethane
(19) 2-Butanone	(55) Bromoethane
(20) Cyclohexanone	(56) Benzene
(21) Phenol	(57) 1-Chloropropane
(22) *t*-Amyl alcohol	(58) Trichloroethene
(23) 1-Butanol	(59) Toluene
(24) 3-Methylphenol	(60) Xylenes
(25) Cyclohexanol	(61) Tetrachloromethane
(26) *i*-Amyl alcohol	(62) Carbon disulfide
(27) 1-Pentanol	(63) Decalin
(28) Benzyl alcohol	(64) Cyclopentane
(29) Ethyl acetate	(65) Cyclohexane
(30) 1-Hexanol	(66) *n*-Hexane
(31) 2,4,6-Trimethylpyridine	(67) *n*-Heptane
(32) Pentanoic acid	(68) Kerosine
(33) Ethyl formate	(69) Petroleum ether
(34) 3-Methylbutanoic acid	(70) Petroleum
(35) Furan	(71) Paraffin oil
(36) Diethyl ether	

[a] This series applies generally to partition separations on paper, column, or thin-layer partition chromatography. The solvents listed go from most to least hydrophilic behaviour.

[b] E. Hecker, Chimia *8*, 229 (1954); E. Hecker: *Verteilungsverfahren in Laboratorien*. Verlag Chemie, Weinheim 1955, p. 92 and 139.

[c] E. Heftmann (editor): *Chromatography*. 2nd edition, Reinhold Publishing Company, New York 1967.

[d] O. Mikeś: *Laboratory Handbook of Chromatographic Methods*. Van Nostrand, London 1970.

[e] *Cf.* also A. J. Gordon and R. A. Ford: *The Chemist's Companion – A Handbook of Practical Data, Techniques, and References*. Wiley-Interscience, New York 1972, p. 379.

acids. Table A-11 gives the eluotropic series for standard solvents in conjugation with hydrophilic adsorbents according to Snyder [59, 60]. For hydrophobic adsorbents such as charcoal and polyamides, the eluotropic series is almost reversed. Often mixtures of two or three solvents of different polarity cause a better separation than a pure solvent. Again, the multicomponent eluant can be ordered in an eluotropic series [59, 60]. For example, the

Table A-11. Eluotropic solvent series for hydrophilic adsorbents such as alumina or silica-gel listed in order of increasing eluting power of the solvent [a-d], quantitatively measured by Snyder's empirical eluant strength parameter ε° [e].

Solvents	ε° (Al_2O_3)[e]	Solvents	ε° (Al_2O_3)[e]
(1) Fluoroalkanes	−0.25	(25) 4-Methyl-2-pentanone	0.43
(2) *n*-Pentane	*0.00*	(26) 1,2-Dichloroethane	0.44
(3) 2,2,4-Trimethylpentane	0.01	(27) 2-Butanone	0.51
(4) Petroleum ether	0.01	(28) 1-Nitropropane	0.53
(5) *n*-Decane	0.04	(29) Triethylamine	0.54
(6) Cyclohexane	0.04	(30) Acetone	0.56
(7) Cyclopentane	0.05	(31) 1,4-Dioxane	0.56
(8) 2,4,4-Trimethyl-1-butene	0.06	(32) Tetrahydrofuran	0.57
(9) 1-Pentene	0.08	(33) Ethyl acetate	0.58
(10) Carbon disulfide	0.15	(34) Methyl acetate	0.60
(11) Tetrachloromethane	0.18	(35) 1-Pentanol	0.61
(12) 1-Chloro-*n*-pentane	0.26	(36) Aniline	0.62
(13) Xylene	0.26	(37) Diethylamine	0.63
(14) Di-*i*-propyl ether	0.28	(38) Nitromethane	0.64
(15) 2-Chloropropane	0.29	(39) Acetonitrile	0.65
(16) Toluene	0.29	(40) Pyridine	0.71
(17) 1-Chloropropane	0.30	(41) 2-*n*-Butoxyethanol	0.74
(18) Chlorobenzene	0.30	(42) Dimethyl sulfoxide	0.75
(19) Benzene	0.32	(43) 1- and 2-Propanol	0.82
(20) Bromoethane	0.37	(44) Ethanol	0.88
(21) Diethyl ether	0.38	(45) Methanol	0.95
(22) Diethylsulfane	0.38	(46) 1,2-Ethanediol	1.11
(23) Chloroform	0.40	(47) Acetic acid	large
(24) Dichloromethane	0.42		

[a] Because the relative elution power depends not only on the adsorbent, but in many cases on the compound types being separated, there exists no universal series of solvent strengths. This series is given by L. R. Snyder[d]. For other eluotropic solvent series see:
(1) W. Trappe, Biochem. Z. *305*, 150 (1940);
(2) H. H. Strain: *Chromatographic Adsorption Analysis*. Interscience, New York 1942; Ind. Eng. Chem., Anal. Ed. *14*, 245 (1942);
(3) J. Jacques and J. P. Mathieu, Bull. Soc. Chim. Fr. 94 (1946);
(4) E. M. Bickhoff, Anal. Chem. *20*, 51 (1948);
(5) H. S. Knight and S. Groennings, Anal. Chem. *26*, 1549 (1954);
(6) P. B. Moseley, A. L. LeRosen, and J. K. Carlton, Anal. Chem. *26*, 1563 (1954);
(7) J. J. Wren, J. Chromatogr. *4*, 173 (1960);
(8) S. Heřmánek, V. Schwarz, and Z. Čekan, Collect. Czech. Chem. Commun. *28*, 2031 (1963);
(9) J. C. Touchstone: *Practice of Thin-Layer Chromatography*. 2nd edition, Wiley, Chichester 1983.

[b] In binary solvent mixtures, additions of small amounts of one solvent to another less polar solvent rapidly increases the eluting power. The further away in the series the solvent pair is, the more drastic the change.

[c] For reverse-phase adsorbents such as charcoal or completely silanized silica-gel this eluotropic solvent series is valid in the reversed order. In this case the eluting power increases in the following sequence: water < methanol < ethanol < acetone < 1-propanol < 1-butanol ~ diethyl ether ~ ethyl acetate < benzene < *n*-hexane[d].

[d] L. R. Snyder: *Principles of Adsorption Chromatography*. M. Dekker, New York 1968; L. R. Snyder: *Solvent Selection for Separation Processes*. In E. S. Perry and A. Weissberger (eds.): *Techniques of Chemistry*. 3rd edition, Vol. XII, p. 25...75, Wiley-Interscience, New York 1978.

[e] ε° stands for the adsorption energy of solvent per unit area of adsorbent with unit activity, defined as $\varepsilon^{\circ} = 0$ for *n*-pentane on alumina[d].

eluting power increases steadily in the series: petroleum ether, petroleum ether/benzene (100, 200, and 500 ml/l), benzene, benzene/ethanol (20, 50, 100, and 200 ml/l). Since the eluting power of a solvent is the result of interactions between adsorbent, solvent, and sample, eluotropic series are generally valid only for the substance class for which they have been established. The mixotropic (Table A-10) as well as the eluotropic series (Table A-11) reflect, approximately, the series of increasing polarity given by the empirical polarity parameter E_T^N (*cf.* Table A-1). – The organization of solvents for analytical and preparative thin-layer chromatography appears to be related also to solvent viscosity [58, 126, 127].

A.8 Solvents for Acid-Base Titrations in Non-Aqueous Media

Many acids and bases are too weak to be titrated in aqueous solution (ionization constant $< 10^{-8}$). The use of non-aqueous amphiprotic differentiating titration solvents (*cf.* Section 3.3.1), however, often permits one to reach a sharp titration end-point [62–70]. Non-aqueous solvents for the titration of weak acids and bases should be obtainable water-free and in a high state of purity, be chemically inert to the titrant and the substance, readily dissolve the substance and its titration product, or if the latter precipitates favor the formation of compact, crystalline, non-voluminous material. Weak bases are frequently titrated with perchloric acid in acetic acid, and weak acids with tetraalkylammonium hydroxide in 1,2-diaminoethane, alcohols, or pyridine [62–70]. The autoprotolysis constant is a particularly important criterion for solvent selection, since this constant determines the acidity or basicity region available in the solvent used. The smaller the autoprotolysis constant, the greater is the range of acid or base strengths which can exist in a solvent and the greater is the likelihood that it will be a differentiating solvent. Hence, acid-base titrations are best carried out in solvents with small K_{auto}-values [64, 70]. The autoprotolysis constants, K_{auto}, of some amphiprotic solvents are given in Table A-12. These were often determined by conductivity measurements. It should be noted that the measured K_{auto}-values often represent only a minimum value since, for extremely weakly basic or acidic solvents, it is difficult to distinguish between electrical conductivity caused by autoprotolysis and by impurities. A solvent is amphiprotic when it shows well-defined self-ionization, such that an autoprotolysis constant can be reproducibly measured. It is recommended to use the term aprotic in place of amphiprotic for solvents with $K_{auto} < 10^{-20}$ ($pK_{auto} > 20$) [71]. In Table A-12 are also listed some solvents whose self-ionization cannot be due to protolysis, *e.g.* acetic anhydride. – Just as it is possible to define pK areas, one can also evaluate intervals of electrochemical potential for various solvents useful for titrations with glass and calomel electrodes [72].

A.9 Solvents for Electrochemistry

Many electrochemical reactions, especially of organic compounds, are better carried out in non-aqueous solvents and may even be excluded in water. The following requirements should be met by these solvents [73–77]: sufficient solubility of the compounds to be examined and, of necessity, the supporting electrolyte also (usually tetraalkylammonium

Table A-12. Autoprotolysis constants (auto-ionization ionic products) of some amphiprotic solvents at 25 °C[a] according to $2\,SH \rightleftharpoons SH_2^{\oplus} + S^{\ominus}$; $K_{auto} = [SH_2^{\oplus}] \cdot [S^{\ominus}]$[b].

Solvents	Lyonium-Ion	Lyate-Ion	$pK_{auto} = -\lg[K_{auto}/(mol^2 \cdot l^{-2})]$	References
Sulfuric acid	$H_3SO_4^{\oplus}$	$HSO_4^{\ominus}$	3.33	[c]
2-Aminoethanol	$HO{-}CH_2CH_2{-}NH_3^{\oplus}$	$H_2N{-}CH_2CH_2{-}O^{\ominus}$	5.7	[d]
Formic acid	$HC(OH)_2^{\oplus}$	$HCO_2^{\ominus}$	6.2	[e]
N-Methylformamide	$CH_3{-}\overset{\oplus}{N}H{=}CH{-}OH$	$CH_3{-}N{=}CH{-}O^{\ominus}$	10.74	[v]
Hydrogen fluoride	$H_2F^{\oplus}$	$F^{\ominus}$	12.5 (0 °C)	[w]
Water	$H_3O^{\oplus}$	$HO^{\ominus}$	14.00	[f]
Acetic acid	$CH_3C(OH)_2^{\oplus}$	$CH_3CO_2^{\ominus}$	14.45	[g]
Acetic anhydride	$CH_3CO^{\oplus}$	$CH_3CO_2^{\ominus}$	*ca.* 14.5 (20 °C)	[h]
Acetamide	$CH_3CO{-}NH_3^{\oplus}$	$CH_3CO^{\ominus}{=}NH$	14.6 (98 °C)	[i]
Deuterium oxide	$D_3O^{\oplus}$	$DO^{\ominus}$	14.96	[k]
1,2-Diaminoethane	$H_2N{-}CH_2CH_2{-}NH_3^{\oplus}$	$H_2N{-}CH_2CH_2{-}NH^{\ominus}$	15.2	[l]
1,2-Ethanediol	$HO{-}CH_2CH_2{-}OH_2^{\oplus}$	$HO{-}CH_2CH_2{-}O^{\ominus}$	15.84	[m]
Formamide	$HCO{-}NH_3^{\oplus}$	$HCO^{\ominus}{=}NH$	16.8 (20 °C)	[n]
Methanol	$H_3C{-}OH_2^{\oplus}$	$H_3C{-}O^{\ominus}$	17.20	[o]
Ethanol	$H_5C_2{-}OH_2^{\oplus}$	$H_5C_2{-}O^{\ominus}$	18.88	[p]
1-Propanol	$H_7C_3{-}OH_2^{\oplus}$	$H_7C_3{-}O^{\ominus}$	19.43	[p]
Hexamethylphosphoric triamide	$[(CH_3)_2N]_3\overset{\oplus}{P}{-}OH$		20.56	[o]
1-Pentanol	$H_{11}C_5{-}OH_2^{\oplus}$	$H_{11}C_5{-}O^{\ominus}$	20.65	[q]
2-Propanol	$(H_3C)_2C{-}OH_2^{\oplus}$	$(H_3C)_2C{-}O^{\ominus}$	20.80	[p]
1-Butanol	$H_9C_4{-}OH_2^{\oplus}$	$H_9C_4{-}O^{\ominus}$	21.56	[q]
Ethyl acetate	$H_3C{-}C(OH){=}\overset{\oplus}{O}C_2H_5$	$H_2C{=}C(O^{\ominus}){-}OC_2H_5$	22.83	[o]
N,*N*-Dimethylacetamide	$H_3C{-}C(OH){=}\overset{\oplus}{N}(CH_3)_2$	$H_2C{=}C(O^{\ominus}){-}N(CH_3)_2$	23.95	[o]

Table A-12. (Continued)

Solvents	Lyonium-Ion	Lyate-Ion	$pK_{auto} = -\lg[K_{auto}/(mol^2 \cdot l^{-2})]$	References
1-Methyl-2-pyrrolidinone			24.15	[o]
Tetrahydrothiophene-1,1-dioxide			25.45	[o]
2-Butanone	$H_3C-CH_2-C(=OH^{\oplus})-CH_3$	$H_3C-CH=C(O^{\ominus})-CH_3$	25.94	[o]
2-Methyl-2-propanol	$(H_3C)_3C-OH_2^{\oplus}$	$(H_3C)_3C-O^{\ominus}$	26.8	[r]
N,N-Dimethylformamide	$H-C(OH)=\overset{\oplus}{N}(CH_3)_2$		29.4	[r]
Ammonia	$H_4N^{\oplus}$	$H_2N^{\ominus}$	32.5 (−33 °C)	[s]
Acetone	$H_3C-C(=OH^{\oplus})-CH_3$	$H_2C=C(O^{\ominus})-CH_3$	32.5	[r]
Dimethyl sulfoxide	$H_3C-\overset{\oplus}{S}(OH)-CH_3$	$H_2C=S(O^{\ominus})-CH_3$	33.3	[t]
Acetonitrile	$H_3C-C\equiv NH^{\oplus}$	$H_2C=C=N^{\ominus}$	≥33.3	[u]

[a] Values at 25 °C unless stated otherwise.
[b] For more extensive compilations of autoprotolysis constants compare:
1. A. P. Kreshkov, Talanta *17*, 1029 (1970);
2. E. J. King: *Acid-Base Behaviour*, in A. K. Covington and T. Dickinson (eds.), *Physical Chemistry of Organic Solvent Systems*. Plenum Press, London, New York 1973, p. 342 and 391.
3. B. Trémillon: *Chemistry in Non-Aqueous Solvents*. Reidel Publishing Company, Dordrecht, Boston 1974, p. 70ff.
[c] P. A. H. Wyatt, Trans. Faraday Soc. *65*, 585 (1969).
[d] P. W. Brewster, F. C. Schmidt, and W. B. Schaap, J. Phys. Chem. *65*, 990 (1961).
[e] M. Bréant, C. Beguin, and C. Coulombeau, Anal. Chim. Acta *87*, 201 (1976).
[f] H. S. Harned, J. Am. Chem. Soc. *47*, 930 (1925); H. S. Harned and B. B. Owen: *The Physical Chemistry of Electrolytic Solutions*. 3rd edition, Reinhold, New York 1958.
[g] S. Bruckenstein and I. M. Kolthoff, J. Am. Chem. Soc. *78*, 2974 (1956).
[h] G. Jander and H. Surawski, Z. Elektrochem., Ber. Bunsenges. Physik. Chem. *65*, 527 (1961).

[i] S. Guiot and B. Trémillon, J. Electroanal. Chem. *18*, 261 (1968).
[k] A. K. Covington, R. A. Robinson, and R. G. Bates, J. Phys. Chem. *70*, 3820 (1966).
[l] W. B. Schaap, R. E. Bayer, J. R. Siefker, J. Y. Kim, P. W. Brewster, and F. C. Schmidt, Record Chem. Progr. *22*, 197 (1961); Chem. Abstr. *56*, 12664 (1962).
[m] K. K. Kundu, P. K. Chattopadhyay, D. Jana, and M. N. Das, J. Phys. Chem. *74*, 2633 (1970).
[n] F. H. Verhoek, J. Am. Chem. Soc. *58*, 2577 (1936).
[o] A. P. Kreshkov, N. Sh. Aldarova, and B. B. Tanganov, Zh. Fiz. Khim. *44*, 2089 (1970); Russian J. Phys. Chem. *44*, 1186 (1970); Chem. Abstr. *73*, 124160 d (1970); *cf.* also A. P. Kreshkov, N. T. Smolova, N. Sh. Aldarova, and N. A. Gabidulina, Zh. Fiz. Khim. *46*, 667 (1972); Chem. Abstr. *77*, 33798 j (1972).
[p] R. Schaal and A. Tézé, Bull. Soc. Chem. Fr. 1783 (1961).
[q] A. P. Kreshkov, N. Sh. Aldarova, and N. T. Smolova, Zh. Fiz. Khim. *43*, 2846 (1969); Chem. Abstr. *72*, 54413 s (1970).
[r] L. N. Bykova and S. I. Petrov, Zh. Anal. Khim. *27*, 1076 (1972); Chem. Abstr. *77*, 93626 m (1972); *cf.* also [o].
[s] L. V. Coulter, J. R. Sinclair, A. G. Cole, and G. C. Roper, J. Am. Chem. Soc. *81*, 2986 (1959).
[t] J. Courtot-Coupez and M. Le Démézet, Bull. Chim. Fr. 1033 (1969); *cf.* also [r].
[u] I. M. Kolthoff and M. K. Chantooni, J. Phys. Chem. *72*, 2270 (1968); *cf.* also [r].
[v] T. Oncescu, A.-M. Oancea, and L. de Maeyer, J. Phys. Chem. *84*, 3090 (1980).
[w] R. Gut, J. Fluorine Chem. *15*, 163 (1980).

For a recent review on autoprotolysis constants in nonaqueous solvents and aqueous organic solvent mixtures see S. Rondinini, P. Longhi, P. R. Mussini, and T. Mussini, Pure Appl. Chem. *59*, 1693 (1987).

salts), chemical inertness towards the electrolyte and the reactive intermediates formed (*e.g.* the frequently formed radical anions would immediately be protonated by protic solvents), and as high a dielectric constant as possible (usually $\varepsilon_r > 10$). The latter will increase the electrical conductivity by favoring the dissociation of the electrolyte and hence decreasing the electrical resistance of the solution. Furthermore, the solvent should be of low viscosity in order to guarantee rapid mass transport of ions to the electrodes. Absence of conducting impurities, especially water, is particularly important when measuring electrical conductivities [78, 79]. In addition, the solvent should have a large anodic and cathodic voltage limit, which defines the "window" of accessible electronic levels available for electron-transfer processes. Although organic solvents have intrinsic limits based on their chemical oxidation-reduction properties, the practical working limits also depend on the nature of the working electrode material and the composition of the supporting electrolyte. Therefore, the voltage limits are a system property and not only a solvent property [77]. Practical working limits in organic solvents are given in the literature [73, 74, 77, 80, 81].

Procedures for measuring and reporting electrode potentials in nonaqueous solvents are presented in reference [128]. Applications of ion-selective electrodes in nonaqueous solvents have been reviewed in reference [129].

A.10 Toxicity of Organic Solvents

Organic solvents show varied physiological and toxicological properties which all too often are neglected in the laboratory. All solvents influence the human organism to a greater or lesser extent [82–89, 130–132]. The extent of this influence depends on the time of exposure. High doses over short time intervals can lead to acute poisoning; small doses over prolonged periods can induce chronic damage.

Solvents can be ingested through the skin and the respiratory organs. Inhalation of solvent vapours affects not only the lungs but the whole circulatory system and hence the whole body. They accumulate principally in lipid and fat-rich cells in the nervous system, brain, bone marrow, liver, and body fat. Acute poisoning manifests itself by derangement of the central nervous system (euphoria, dizziness, unconsciousness). Chronic poisonings occur initially without any obvious symptoms and damage of the organs characteristic of the different solvents appears only later [89].

Many solvents are absorbed not only through the respiratory organs but also through the skin (*e.g.* tetrachloromethane, dimethyl sulfoxide, benzene). This leads to dehydration of the skin due to the removal of water and fat, thereby allowing the invasion of bacteria and dirt [89].

Hence, organic solvents should be handled with care. In the USA the *threshold limit values* (*TL*-values) are used as a measure of the inhalation toxicity for chronic interaction with solvent vapours [90]. In the Federal Republic of Germany the *maximum concentration values at the workplace* (*MAK*-values) are used [91, 92]. Threshold limit values refer to airborne concentrations of substances and represent conditions under which it is believed that workers may be repeatedly exposed daily without adverse effect. They refer to time-weighted concentrations for a normal 8-hour workday or 40-hour workweek [90]. – *MAK*-values represent the highest allowed airborne concentrations of gaseous,

Table A-13. *TL*-Values[a)], *MAK*-Values[b)], and Vapour Pressures[b, c)] for sixty-eight organic solvents.

Solvents[d)]	*TL*-Values (TWA)[e)]		*MAK*-Values[f)]		Vapour pressure at 20 °C in mbar
	ml/m³	mg/m³	ml/m³	mg/m³	
Acetic acid	10	25	10	25	15.3
Acetic anhydride	5	20	5	20	4.7
Acetone	1000	2400	1000	2400	233
Acetonitrile (skin)	40	70	40	70	97
1-Aminobutane (skin)	5	15	5	15	
2-Aminoethanol	3	6	3	8	
Ammonia	50	35	50	35	11600
Aniline (skin)[i)]	2	8	2	8	0.4
Benzene (skin)[g)]	–	–	–	–	101
n-Butane	1000	2350	1000	2350	
1-Butanol (skin)	100	300	100	300	4–40
tert-Butanol	100	300	100	300	
2-Butanone	200	590	200	590	105
n-Butyl acetate	200	950	200	950	12–21
Carbon disulfide (skin)	10	30	10	30	400
Chlorobenzene	50	230	50	230	12
Chloroform[i)]	10	50	10	50	210
Cyclohexane	300	1050	300	1050	104
Cyclohexanol	50	200	50	200	1.2
Cyclohexanone	50	200	50	200	5
1,2-Diaminoethane	10	25	10	25	
1,1-Dichloroethane	100	400	100	400	240
1,2-Dichloroethane[i)]	20	80	20	80	87
1,1-Dichloroethene[i)]	2	8	2	8	667
1,2-Dichloroethene	200	790	200	790	220
Dichloromethane[i)]	100	360	100	360	453
Diethylamine	10	30	10	30	253
Diethyl ether	400	1200	400	1200	587
Di-*i*-propyl ether	500	2100	500	2100	180
N,N-Dimethylacetamide (skin)	10	35	10	35	
N,N-Dimethylformamide (skin)	20	60	20	60	3.5
2,6-Dimethyl-4-heptanone, Di-*i*-butyl ketone	50	290	50	290	2.3
1,4-Dioxane (skin)[i)]	50	180	50	180	41
Diphenyl ether (vapour)	1	7	1	7	0.08
Ethanol	1000	1900	1000	1900	59
2-Ethoxyethanol (skin)	20	75	20	75	*ca.* 5
Ethyl acetate	400	1400	400	1400	97
Ethyl benzene (skin)	100	440	100	440	9.3
Ethyl formate	100	300	100	300	256
Formic acid	5	9	5	9	43
n-Heptane	500	2000	500	2000	48
Hexamethylphosphoric triamide[h)] (skin)	–	–	–	–	
n-Hexane	50	180	50	180	160
Methanol (skin)	200	260	200	260	128
2-Methoxyethanol (skin)	5	15	5	15	*ca.* 11
Methyl acetate	200	610	200	610	220
3-Methyl-1-butanol, *i*-Amyl alcohol	100	360	100	360	2.7

Table A-13. (Continued)

Solvents[d)]	*TL*-Values (TWA)[e)]		*MAK*-Values[f)]		Vapour pressure at 20 °C in mbar
	ml/m³	mg/m³	ml/m³	mg/m³	
Methylcyclohexane	500	2000	500	2000	48
Methyl formate	100	250	100	250	640
1-Methylpyrrolidin-2-one	100	400	100	400	1.3
Morpholine (skin)	20	70	20	70	10.7
Nitrobenzene (skin)	1	5	1	5	0.4
Nitromethane	100	250	100	250	37.1
n-Octane	500	2350	500	2350	15
n-Pentane	1000	2950	1000	2950	573
Phenol (skin)	5	19	5	19	0.2
2-Propanol	400	980	400	980	43
2-Propene-1-ol, Allyl alcohol (skin)	2	5	2	5	24
Pyridine	5	15	5	15	20
1,1,2,2-Tetrachloroethane (skin)[i)]	1	7	1	7	7
Tetrachloroethene	50	345	50	345	19
Tetrachloromethane (skin)[i)]	10	65	10	65	120
Tetrahydrofuran	200	590	200	590	200
Toluene	100	375	100	380	29
1,1,1-Trichloroethane	200	1080	200	1080	133
Trichloroethene[i)]	50	270	50	270	77
Triethylamine	10	40	10	40	61.6
Xylene (all three isomers)	100	440	100	440	7–9

[a] *Threshold Limit Values for Chemical Substances in the Work Environment*, adopted by the American Conference of Governmental Industrial Hygienists (ACGIH) for 1984–85, available by the Secretary Treasurer, ACGIH, 6500 Glenway Avenue, Cincinnati/Ohio 45211, U.S.A.

[b] *Maximale Arbeitsplatzkonzentrationen und Biologische Arbeitsstofftoleranzwerte 1987 (Maximum Concentrations at the Workplace and Biological Tolerance Values for Working Materials 1987)*, Mitteilung XXIII of the "Senatskommission zur Prüfung gesundheitsschädlicher Arbeitsstoffe" of the "Deutsche Forschungsgemeinschaft (DFG)", Kennedyallee 40, D-5300 Bonn 2, Fed. Rep. Germany.

[c] L. Roth: *Sicherheitsdaten – MAK-Werte – Krebserzeugende Stoffe.* 3rd edition Ecomed Verlagsgesellschaft, München 1984.

[d] Solvents followed by the designation "skin" refer to the potential contribution to the overall exposure by the cutaneous route including mucous membranes and eye, either by airborne, or more particularly, by direct contact with the solvent. This attention-calling designation is intended to suggest appropriate measures for the prevention of cutaneous absorption so that the threshold limit is not invalidated.

[e] Adopted *T*hreshold *L*imit Values – Time Weighted Average (TLV-TWA) – in parts of vapour or gas per million parts of contaminated air by volume (ml/m³; ppm) at 25 °C and 1013 mbar respectively in approximate milligrams of substance per cubic metre of air (mg/m³).

[f] *M*aximale *A*rbeitsplatz*k*onzentrationen (maximum concentrations at the workplace) in parts of vapour or gas per million parts of contaminated air by volume (ml/m³; ppm) at 20 °C and 1013 mbar respectively in approximate milligrams of substance per cubic metre of air (mg/m³).

[g] No declaration because the solvent is recognized as having carcinogenic potential, *i.e.* capable of inducing malignant tumors as shown through experience with humans.

[h] Solvents which have proven so far to be carcinogenic in animal experimentation only, however under conditions which are comparable to those for possible exposure of a human being at the workplace.

[i] Solvents suspect of carcinogenic potential for humans.

vapourous, or dusty chemical substances within a work area, which will neither, as far as is known, adversely affect the health of the workers nor disturb them, even by repeated long-term exposure. They refer to time-weighted concentrations for a 8-hour workday or 40-hour workweek [91, 92]. Table A-13 gives the *TL*- and *MAK*-values of frequently used organic solvents. These values are based on the best available information from industrial experience and from experimental human and animal studies. The basis on which the *MAK*-values are established is given in reference [92]. Since the amount and nature of the information available for establishing *TL*- and *MAK*-values varies, the precision of the estimated *TL*- and *MAK*-values is also subject to variation and the latest information should be consulted. The *TL*- and *MAK*-values of Table A-13 refer to those issued in 1984–1985 and 1987, respectively.

The vapour pressures of organic solvents are also listed in Table A-13, since these give an additional indication of their potential dangers caused by their volatility.

The designation "skin" is used to indicate that there is also a danger through skin ingestion either by airborne or by direct contact with the chemical substance [90–92]. When handling such compounds, meticulous care in avoiding contact with the skin, hair, and clothing is imperative for health protection.

The *TL*- and *MAK*-values should be used as guides in the control of health hazards. They are not constants which can be used to draw fine lines between safe and dangerous concentrations. It should be noted that these values are not intended for use as a relative index of hazard or toxicity. It is also not possible to calculate the *TL*- or *MAK*-values of solvent mixtures from the data in Table A-13, because antagonistic action or potentiation may occur with some combinations. Naturally, the safety intended through the establishing of *TL*- and *MAK*-values can only be guaranteed through systematic and quantitative measurements of the various substance concentrations in the work area by trained persons.

So-called *odor threshold values* are not included in Table A-13 since these often show great discrepancy in the literature. They depend strongly on the experimental technique and individual sensitivity. Nevertheless, an unpleasant smell together with any irritation should be taken as a warning signal.

References

Introduction

[1] Hermannus Boerhaave: *Elementa Chemiae*. Editio Altera, Leydensi multo correctior et accuratior, G. Cavelier, Parisii 1733, Tomus Primus, p. 558. [2] G. A. Lindeboom: *Herman Boerhaave– The Man and his Work*. Methuen, London 1968; *cf.* also Endeavour *28*, 2 (1969). [3] P. Walden: *Die Lösungstheorien in ihrer geschichtlichen Aufeinanderfolge*, in W. Herz (ed.): *Sammlung chemischer und chemisch-technischer Vorträge*. Enke, Stuttgart 1910, Vol. XV, Heft 8–12; Chem. Zentralbl. *1910 II*, 1352. [4] F. Szabadváry: *Geschichte der Analytischen Chemie*. Vieweg, Braunschweig, and Akadémiai Kiadó, Budapest 1966, p. 38. [5] J. H. van't Hoff: *Über die Theorie der Lösungen*, in F. W. Ahrens (ed.): *Sammlung chemischer und chemisch-technischer Vorträge*. Enke, Stuttgart 1900, Vol. V, Heft 1; Chem. Zentralbl. *1900 I*, 696. [6] J. R. Partington: *A History of Chemistry*. MacMillan, London, New York 1964, Vol. 4, Chapter XX, p. 637ff. [7] For a survey of older works on solvent effects on reaction rates see M. Magat, Z. Phys. Chem. *A 162*, 432 (1932); Chem. Zentralbl. *1933 I*, 1566. [8] J.H. van't Hoff: *Die chemische Dynamik*, in *Vorlesungen über theoretische und physikalische Chemie*. Vieweg, Braunschweig 1898, Heft 1; Chem. Zentralbl. *1899 II, 278*. [9] M. Berthelot and L. Péan de Saint-Gilles, Ann. Chim. et Phys., 3. Ser., *65*, 385 (1862); *66*, 5 (1862); *68*, 255 (1863). Translated into German and edited by M. and A. Ladenburg in: *Ostwald's Klassiker der exakten Naturwissenschaften*, Nr. 173, Engelmann, Leipzig 1910. [10] N. Menschutkin, Z. Phys. Chem. *1*, 611 (1887); ibid. *5*, 589 (1890); ibid. *6*, 41 (1890); ibid. *34*, 157 (1900).

[11] Th. Zincke and H. Bindewald, Ber. Dtsch. Chem. Ges. *17*, 3026 (1884); Th. Zincke and H. Thelen, ibid. *17*, 1809 (1884). [12] P. C. Laar, Ber. Dtsch. Chem. Ges. *18*, 648 (1885); ibid. *19*, 730 (1886). [13] C. K. Ingold: *Structure and Mechanism in Organic Chemistry*. 2nd edition, Cornell University Press, Ithaca, London 1969, p. 794ff. [14] L. Claisen, Liebigs Ann. Chem. *291*, 25 (1896); especially pp. 30, 43, and 86. [15] W. Wislicenus, Liebigs Ann. Chem. *291*, 147 (1896); especially p. 176 ff. [16] L. Knorr, Liebigs Ann. Chem. *293*, 70 (1896); especially p. 88. [17] A. Hantzsch and O. W. Schultze, Ber. Dtsch. Chem. Ges. *29*, 2251 (1896); especially p. 2256. [18] H. Stobbe, Liebigs Ann. Chem. *326*, 347 (1903); especially p. 357ff. [19] O. Dimroth, Liebigs Ann. Chem. *377*, 127 (1910); ibid. *399*, 91 (1913). [20] K. H. Meyer, Liebigs Ann. Chem. *380*, 212 (1911); Ber. Dtsch. Chem. Ges. *45*, 2843 (1912); ibid. *47*, 826 (1914).

[21] S. E. Sheppard, Rev. Modern Phys. *14* 303 (1942); Chem. Abstr. *37*, 1654 (1943). [22] A. Hantzsch, Ber. Dtsch. Chem. Ges. *55*, 953 (1922). [23] A. Kundt: *Über den Einfluß des Lösungsmittels auf die Absorptionsspectra gelöster absorbierender Medien*. Poggendorfs Ann. Phys. Chem. N.F. *4*, 34 (1878); Chem. Zentralbl. *1878*, 498. [24] W. R. Brode, J. Phys. Chem. *30*, 56 (1926); Chem. Zentralbl. *1926 I*, 2775. [25] G. Scheibe, E. Felger, and G. Rößler, Ber. Dtsch. Chem. Ges. *60*, 1406 (1927). [26] N. Menschutkin: *Sur les conditions de l'acte de la combinaison chimique; modifications déterminées la présence des dissolvants, soi-disant indifférents* (Extrait d'une lettre de M. Menschutkin, professeur de chimie à Saint-Petersburg, à M. Louis Henry, Louvain). Bulletin de l'Académie Royale des Sciences, des Lettres et des Beaux-Arts de Belgique [3] *19*, 513 ... 514 (1890); *Cf.* also A. Bruylants, Bulletin de l'Académie Royale de Belgique (Classe des Sciences) [5] *62*, 866 ... 882 (esp. p. 877) (1976); Chem. Abstr. *87*, 166770x (1977). – I thank Prof. Bruylants, Louvain-La-Neuve, for drawing my attention to this important letter of Menschutkin.

Chapter 2

[1] D. H. Whiffen (ed.): *Manual of Symbols and Terminology for Physicochemical Quantities and Units, Appendix I: Solutions*. Pure Appl. Chem. *51*, 1 (1979). [2] T. S. Ree, T. Ree, and H. Eyring, Angew. Chem. *77*, 993 (1965); Angew. Chem., Int. Ed. Engl. *4*, 923 (1965); H. Eyring and M. S. Jhon: *Significant Liquid Structures*. Wiley, New York, 1969. [3] J. S. Rowlinson and F. L. Swinton: *Liquids and Liquid Mixtures*. 3rd ed., Butterworths, London 1982; Essays in Chemistry *1*, 1 (1970). [4] J. H. Hildebrand, J. M. Prausnitz, and R. L. Scott: *Regular and Related Solutions*. Van Nostrand-Reinhold, Princeton 1970. [5] D. Henderson (ed.): *Liquid State*. Vol. 8 in "Physical Chemistry – An

Advanced Treatise" (ed. by H. Eyring, D. Henderson, and W. Jost), Academic Press, London, New York 1971. [6] F. Kohler: *The Liquid State*, Vol. 1 in *Monographs in Modern Chemistry*, Verlag Chemie, Weinheim 1972. [7] A. F. M. Barton: *The Dynamic Liquid State*. Longman, London 1974. [8] E. Wicke, Angew. Chem. *78*, 1 (1966); Angew. Chem., Int. Ed. Engl. *5*, 106 (1966). [9] R. A. Horne: *The Structure of Water and Aqueous Solutions*, in A. F. Scott (ed.): Survey of Progress in Chemistry *4*, 1 (1968). [10] D. Eisenberg and W. Kauzmann: *The Structure and Properties of Water*. Oxford University Press, New York 1969.

[11] P. Krindel and I. Eliezer; *Water Structure Models*, in Coord. Chem. Rev. *6*, 217 (1971). [12] R. A. Horne (ed.): *Water and Aqueous Solutions – Structure, Thermodynamics, and Transport Processes*. Wiley, New York 1972. [13] F. Franks (ed.): *Water – A Comprehensive Treatise*. Vol. 1–7, Plenum Press, London 1972–1982. [14] W. A. P. Luck (ed.): *Structure of Water and Aqueous Solutions*. Verlag Chemie, Weinheim 1974. [15] A. Ben-Naim: *Water and Aqueous Solutions – Introduction to a Molecular Theory*. Plenum Press, London 1975. [15a] M. Klose and J. I. Naberuchin: *Wasser – Struktur und Dynamik*. Akademie-Verlag, Berlin 1986. [16] H. S. Frank and W.-Y. Wen, Discussions Faraday Soc. *24*, 133 (1957); H. S. Frank, Proc. Roy. Soc. London Ser. A *247*, 481 (1958). [17] D. T. Hawkins: *A Bibliography on the Physical and Chemical Properties of Water (1969–1974)*, in J. Sol. Chem. *4*, 625 (1975). [18] K. Ziegler, in Houben-Weyl-Müller, "Methoden der Organischen Chemie". 4th ed., Thieme, Stuttgart 1955, Vol. IV/2, p. 738. [19] F. Franks and D. J. G. Ives, Quart. Rev., Chem. Soc. *20*, 1 (1966). [20] O. Fuchs, Fortschr. Chem. Forsch. *11*, 74 (1968); Deutsche Farben-Zeitschrift *22*, 548 (1968); *23*, 17, 57, 111 (1969); Chem. Abstr. *71*, 70949f (1969).

[21] K. C. James, Education in Chemistry *9*, 220 (1972). [22] G. C. Pimentel (ed.): *Chemistry – An Experimental Science*. Freeman, San Francisco 1963, p. 313 and 554. [23] G. Duve, O. Fuchs, and H. Overbeck: *Lösemittel Hoechst*. 6th ed., Hoechst AG, Frankfurt (Main) 1976. [24] A. F. M. Barton, *Handbook of Solubility Parameters and other Cohesion Parameters*. CRC Press, Boca Raton/Florida 1983. [25] M. R. J. Dack: *The Importance of Solvent Internal Pressure and Cohesion to Solution Phenomena*, Chem. Soc. Rev. (London) *4*, 211 (1975). [26] For recent reviews of intermolecular forces, see a) J. O. Hirschfelder: *Intermolecular Forces*. Interscience, New York 1967; b) H. A. Stuart: *Molekülstruktur*. 3rd edition, Springer, Berlin 1967, p. 40ff.; c) A. D. Buckingham, Pure Appl. Chem. *24*, 123 (1970); d) T. Kihara: *Intermolecular Forces*. Wiley, New York 1978. [27] Compilations of attempts to explain intermolecular interactions on a quantummechanical basis are given by a) G. Rudakoff, Z. Chem. *6*, 441 (1966); b) H. Winde, Z. Chem. *10*, 101 (1970); c) R. Daudel: *Quantum Theory of Chemical Reactivity*. Reidel, Dordrecht 1973, p. 41ff. [28] H. H. Jaffé, J. Chem. Educ. *40*, 649 (1963). [29] W. H. Keesom, Physik. Z. *22*, 129, 643 (1921); *23*, 225 (1922). [30] R.L. Amey, J. Phys. Chem. *72*, 3358 (1968).

[31] M. Rabinowitz and A. Pines, J. Am. Chem. Soc. *91*, 1585 (1969). [32] P. Debye, Physik. Z. *21*, 178 (1920); *22*, 302 (1921). [33] F. London, Z. Physik. Chem. (B) *11*, 222 (1931); Z. Physik *63*, 245 (1930); Trans. Faraday Soc. *33*, 8 (1937). [34] R. Ulbrich, Chemiker-Ztg. *99*, 320 (1975). [35] E. F. Meyer and R. E. Wagner, J. Phys. Chem. *70*, 3162 (1966). [35a] C. H. Yoder, J. Chem. Educ. *54*, 402 (1977). [36] A. J. Parker, Quart. Rev. (London) *16*, 163 (1962); Usp. Khim. *32*, 1270 (1963). [37] G. C. Pimentel and A. L. McClellan: *The Hydrogen Bond*. Freeman, San Francisco 1960. [38] H. Zimmermann, Angew. Chem. *76*, 1 (1964); Angew. Chem., Int. Ed. Engl. *3*, 157 (1964); Chimia *23*, 363 (1969); Chemie in unserer Zeit *4*, 69 (1970). [39] W. A. P. Luck, Naturwissenschaften *54*, 601 (1967). [40] S. H. Lin: *Hydrogen Bonding*, in H. Eyring, D. Henderson, and W. Jost, (eds.): *Physical Chemistry – An Advanced Treatise*. Academic Press, New York 1970, Vol. V, Chapter 8, p. 439ff.

[41] M. L. Huggins, Angew. Chem. *83*, 163 (1971); Angew. Chem., Int. Ed. Engl. *10*, 147 (1971). [42] S. N. Vinogradov and R. H. Linnell: *The Hydrogen Bond*. Van Nostrand-Reinhold, New York 1971. [43] P. A. Kollman and L. C. Allen, Chem. Rev. *72*, 283 (1972). [44] M. D. Joesten and L. J. Schaad: *Hydrogen Bonding*. Dekker, New York 1974. [45] G. Geiseler, Z. Chem. *15*, 417 (1975). [46] P. Schuster, G. Zundel, and C. Sandorfy (eds.): *The Hydrogen Bond – Recent Developments in Theory and Experiment*. North-Holland Publishing Company, Amsterdam 1976, Vol. I-III. [47] E. M. Arnett, L. Jores, E. Mitchell, T. S. S. R. Murty, T. M. Gorrie, and P. v. R. Schleyer, J. Am. Chem. Soc. *92*, 2365 (1970). [48] D. Hadži and N. Kobilarov, J. Chem. Soc. *A 1966*, 439. [49] G. Kortüm; *Lehrbuch der Elektrochemie*. 5th ed., Verlag Chemie, Weinheim 1972. [50] G. Briegleb, *Elektronen-Donator-Acceptor-Komplexe*. Springer, Berlin 1961.

[51] R. S. Mulliken and W. B. Person: *Molecular Complexes – A Lecture and Reprint Volume*. Wiley-Interscience, New York 1969. [52] O. Hassel, Angew. Chem. *82*, 821 (1970). [53] V. Gutmann: *Coordination Chemistry in Non-Aqueous Solvents*. Springer, Wien, New York 1968; V. Gutmann: *Chemische Funktionslehre*, Springer, Wien, New York 1971; V. Gutmann, *The Donor-Acceptor Approach to Molecular Interactions*. Plenum Publ. Corp., New York 1978. [54] H. A. Bent, Chem. Rev. *68*, 587 (1968). [55] E. N. Gur'yanova, Usp. Khim. *37*, 1981 (1968); Russian Chem. Rev. *37*, 863 (1968); E. N. Gur'yanova, I. P. Gol'dshtein, and I. P. Romm: *The Donor-Acceptor Bond*. Wiley, New York 1975. [56] J. Yarwood (ed.): *Spectroscopy and Structure of Molecular Complexes*. Plenum Press, New York 1973. [57] R. Foster: *Organic Charge-Transfer Complexes*. Academic Press, London, New York 1969; J. Phys. Chem. *84*, 2135 (1980). [58] R. Foster (ed.): *Molecular Complexes*. Vol. 1 and 2, Elek Science, London 1973/74. [59] R. Paetzold, Z. Chem. *15*, 377 (1975). [59a] W. B. Jensen: *The Lewis Acid-Base Definition – A Status Report*. Chem. Rev. *78*, 1 (1978). [59b] C. J. Bender: *Theoretical Models of Charge-Transfer Complexes*. Chem. Soc. Rev. *15*, 475 (1986). [60] R. S. Drago, Structure and Bonding *15*, 73 (1973); J. Chem. Educ. *51*, 300 (1974).

[61] D. W. Turner: *Ionization Potentials*, Adv. Phys. Org. Chem. *4*, 31 (1966). [62] G. Briegleb: *Elektronenaffinitäten organischer Moleküle*. Angew. Chem. *76*, 326 (1964); Angew. Chem., Int. Ed. Engl. *3*, 617 (1964). [63] V. Kampars and O. Neilands: *The Electron Affinities of Organic Electron Acceptors*. Usp. Khim. *46*, 945 (1977); Russian Chem. Rev. *46*, 503 (1977). [64] R. S. Mulliken and W. B. Person, J. Am. Chem. Soc. *91*, 3409 (1969). [65] D. W. Meek: *Lewis Acid-Base Interactions in Polar Non-Aqueous Solvents*, in J. J. Lagowski (ed.): *The Chemistry of Non-Aqueous Solvents*. Academic Press, London, New York 1966, Vol. I, p. 1ff. [66] R. S. Drago and K. F. Purcell: *Co-ordinating Solvents*, in T. C. Waddington (ed.): *Non-Aqueous Solvent Systems*. Academic Press, London, New York 1965, p. 211ff. [67] V. Gutmann, Coord. Chem. Rev. *2*, 239 (1967); ibid. *18*, 225 (1976); V. Gutmann and A. Scherhaufer, Monatsh. Chem. *99*, 335 (1968); V. Gutmann, Chimia *23*, 285 (1969); V. Gutmann, Electrochimica Acta *21*, 661 (1976). [67a] I. Lindqvist and M. Zackrisson, Acta Chem. Scand. *14*, 453 (1960). [68] U. Mayer, Pure Appl. Chem. *41*, 291 (1975). [69] V. Gutmann, Pure Appl. Chem. *27*, 73 (1971); Fortschr. Chem. Forsch. *27*, 59 (1972); Structure and Bonding *15*, 141 (1973). [70] U. Mayer, V. Gutmann, and W. Gerger, Monatsh. Chem. *106*, 1235 (1975); ibid. *108*, 489 (1977); U. Mayer, Pure Appl. Chem. *51*, 1697 (1979).

[71] R. S. Drago and B. B. Wayland, J. Am. Chem. Soc. *87*, 3571 (1965); R. S. Drago, Structure and Bonding *15*, 73 (1973); J. Chem. Educ. *51*, 300 (1974); R. S. Drago, L. B. Parr, and C. S. Chamberlain, J. Am. Chem. Soc. *99*, 3203 (1977). [72] T. Kagiya, Y. Sumida, and T. Inoue, Bull. Chem. Soc. Jpn. *41*, 767 (1968). [73] For a review see C. Agami, Bull. Soc. Chim. Fr. *1969*, 2183. [74] G. Nemethy and H. A. Scheraga, J. Chem. Phys. *36*, 3401 (1962). [75] G. Nemethy, Angew. Chem. *79*, 260 (1967); Angew. Chem., Int. Ed. Engl. *6*, 195 (1967). [76] T. S. Sarma and J. C. Ahluwalia, Chem. Soc. Rev. (London) *2*, 203 (1973). [77] E. Frieden, J. Chem. Educ. *52*, 754 (1975). [78] B. R. Baker, J. Chem. Educ. *44*, 610 (1967). [79] R. Cecil, Nature *214*, 369 (1967). [79a] R. D. Cramer, J. Am. Chem. Soc. *99*, 5408 (1977). [80] A. Ben-Naim, J. Chem. Phys. *54*, 1387 (1971); esp. footnote 5 on p. 1404.

[81] O. Wörz and G. Scheibe, Z. Naturforsch. *24b*, 381 (1969). [82] A. Ray, Nature *231*, 313 (1971). [83] S. N. Timasheff, Acc. Chem. Res. *3*, 62 (1970). [84] H. Zimmermann, Chemie in unserer Zeit *4*, 69 (1970). [85] O. Ya. Samoilov: *Structure of Aqueous Electrolyte Solutions and the Hydration of Ions*. Consultants Bureau, New York, Plenum Press, London 1965. [86] K. P. Mishchenko and G. M. Poltoratskii: *Problems of Thermodynamics and the Structure of Water and Non-Aqueous Electrolyte Solutions*. Khimiya, Leningrad 1968; *Thermodynamics and Structure of Electrolyte Solutions*. Plenum Press, London 1972. [87] G. Zundel: *Hydration and Intermolecular Interaction*. Academic Press, New York, London 1969; Angew. Chem. *81*, 507 (1969); Angew. Chem., Int. Ed. Engl. *8*, 499 (1969). [88] J. F. Coetzee and C. D. Ritchie (ed.): *Solute-Solvent Interactions*. Vol. 1 and 2, Dekker, New York, London 1969/1976. [89] E. S. Amis: *Solvent Composition and Chemical Phenomena*. Inorg. Chim. Acta Rev. *3*, 7 (1969). [90] L. P. Hammett: *Physical Organic Chemistry*. 2nd ed., McGraw-Hill, New York 1970; *Physikalische Organische Chemie*. Verlag Chemie, Weinheim 1973.

[91] H. G. Hertz: *Die Struktur der Solvathülle gelöster Teilchen*. Angew. Chem. *82*, 91 (1970); Angew. Chem., Int. Ed. Engl. *9*, 124 (1970). [92] M. J. Blandamer: *Structure and Properties of Aqueous Salt Solutions*. Quart. Rev. (London) *24*, 169 (1970). [93] A. K. Covington and T. Dickinson (eds.):

Physical Chemistry of Organic Solvent Systems. Plenum Press, London, New York 1973. [94] E. S. Amis and J. F. Hinton: *Solvent Effects on Chemical Phenomena*. Vol. 1, Academic Press, New York, London 1973. [95] H. L. Friedman: *Modern Aspects in Solvation Theory*. Chemistry in Britain *9*, 300 (1973). [96] J. E. Gordon: *The Organic Chemistry of Electrolyte Solutions*. Wiley, New York 1975. [97] E. S. Amis: *Solvation of Ions*, in M. R. J. Dack (ed.): *Solutions and Solubilities*, Vol. III, Part 1 of the series "Techniques of Chemistry". Wiley-Interscience, New York 1975. [98] P. Schuster. W. Jakubetz, and W. Marius: *Molecular Models for the Solvation of Small Ions and Polar Molecules*. Top. Curr. Chem. *60*, 1 (1975). [98a] A. M. Golub: *The Solvation of Inorganic Substances and Complex Formation in Non-Aqueous Solutions*. Usp. Khim. *45*, 961 (1976); Russian Chem. Rev. *45*, 479 (1976). [99] K. L. Wolf: *Theoretische Chemie*. 4th ed., J. A. Barth, Leipzig 1959, p. 592, Fig. 283. [100] R. Daudel: *Quantum Theory of Chemical Reactivity*. Reidel, Dordrecht, Boston 1973, p. 81ff.

[101] M. Davies, J. Chem. Educ. *48*, 591 (1971). [102] P. Sen and S. Basu, J. Chem. Phys. *48*, 4075 (1968). [103] S. F. Lincoln: *Solvent Coordination Numbers of Metal Ions in Solution*. Coord. Chem. Rev. *6*, 309 (1971). [104] J. F. Hinton and E. S. Amis: *Solvation Number of Ions*. Chem. Rev. *71*, 627 (1971). [105] J. A. Jackson, J. F. Lemons, and H. Taube, J. Chem. Phys. *32*, 553 (1960); M. Alei and J. A. Jackson, ibid. *41*, 3402 (1964). [106] J. F. Hinton and E. S. Amis: *NMR-Studies of Ions in Pure and Mixed Solvents*. Chem. Rev. *67*, 367 (1967). [107] A. Fratiello: *Nuclear Magnetic Resonance Cation Solvation Studies*. Progr. Inorg. Chem. *17*, 57 (1972). [108] J. Burgess and M. C. R. Symons: *The Study of Ion-Solvent and Ion-Ion Interactions by Magnetic Resonance Techniques*. Quart. Rev., Chem. Soc. *22*, 276 (1968); M. C. R. Symons: *Spectroscopic Studies of Solvation*. Annu. Rep. Progr. Chem., Sect. A *73*, 91 (1976). [109] C. Deverell: *NMR-Studies of Electrolyte Solutions*. Progr. Nucl. Magn. Resonance *4*, 235 (1969). [110] H. G. Hertz: *Magnetische Kernresonanzuntersuchungen zur Struktur von Elektrolytlösungen*, in H. Falkenhagen and W. Ebeling (ed.): *Theorie der Elektrolyte*. Hirzel, Stuttgart 1971.

[111] A. I. Popov: *The Use of Alkali Metal Nuclear Magnetic Resonance in the Study of Solvation and Complexation of Alkali Metal Ions*, in J. F. Coetzee and C. D. Ritchie (eds.): *Solute-Solvent Interactions*. Dekker, New York, London 1976, Vol. 2, p. 271ff.; A. I. Popov: *Alkali Metal, Magnesium-25, and Silver-109 NMR Studies of Complex Compounds in Nonaqueous Solvents*, in G. Mamantov (ed.): *Characterization of Solutes in Non-Aqueous Solvents*. Plenum Publ. Corp., New York 1978. [111a] P. Laszlo: *Kernresonanzspektroskopie mit Natrium-23*. Angew. Chem. *90*, 271 (1978); Angew. Chem., Int. Ed. Engl. *17*, 254 (1978). [112] N. A. Matwiyoff, P. E. Darley, and W. G. Movius, Inorg. Chem. *7*, 2173 (1968). [113] J. C. Boubel, J. J. Delpuech, M. R. Khaddar, and A. Peguy, J. Chem. Soc., Chem. Commun. *1971*, 1265. [114] J. F. Hinton and R. W. Briggs, J. Magn. Reson. *19*, 393 (1975). [115] Y. M. Cahen, P. R. Handy, E. T. Roach, and A. I. Popov, J. Phys. Chem. *79*, 80 (1975). [116] R. W. Gurney: *Ionic Processes in Solution*. Dover, New York 1962. [116a] G. W. Stewart and R. M. Morrow, Proc. Nat. Acad. Sci. Washington *13*, 222 (1927); Chem. Zbl. *1927 II*, 371; G. W. Stewart, Physical Rev. [2] *32*, 558 (1928); Chem. Zbl. *1929 I*, 17. [116b] J. R. Partington: *An Advanced Treatise on Physical Chemistry*. Longmans, Green, and Co., London 1951, Vol. 2, p. 2. [117] G. Engel and H. G. Hertz, Ber. Bunsenges. Phys. Chem. *72*, 808 (1968). [118] H. Schneider: *The Selective Solvation of Ions in Mixed Solvents*, in J. F. Coetzee and C. D. Ritchie (eds.): *Solute-Solvent Interactions*. Dekker, New York, London 1969/1976, Vol. 1, p. 301ff., Vol. 2, p. 155ff.; H. Schneider: *Ion Solvation in Mixed Solvents*. Top. Curr. Chem. *68*, 103 (1976). [119] H. Strehlow and H. Schneider: *Solvation of Ions in Pure and Mixed Solvents*. Pure Appl. Chem. *25*, 327 (1971); H. Strehlow, W. Knoche, and H. Schneider: *Ionensolvatation*. Ber. Bunsenges. Phys. Chem. *77*, 760 (1973). [120] J. I. Padova: *Solvation in Nonaqueous and Mixed Solvents*. Mod. Aspects Electrochem. *7*, 1 (1972); Chem. Abstr. *79*, 77566c (1973).

[121] H. Strehlow and H.-M. Koepp, Z. Elektrochem., Ber. Bunsenges. Phys. Chem. *62*, 373 (1958); H. Schneider and H. Strehlow, Z. Phys. Chem. (Frankfurt) *49*, 44 (1966). [122] H. Schneider and H. Strehlow, Z. Elektrochem., Ber. Bunsenges. Phys. Chem. *66*, 309 (1962). [123] H. Schneider and H. Strehlow, Ber. Bunsenges. Phys. Chem. *69*, 674 (1965). [123a] L. Rodehüser and H. Schneider, Z. Phys. Chem. (Frankfurt) *100*, 119 (1976). [124] K. Dimroth and C. Reichardt, Z. Anal. Chem. *215*, 344 (1966); Z. B. Maksimović, C. Reichardt, and A. Spirić, ibid. *270*, 100 (1974). [125] L. S. Frankel, T. R. Stengle, and C. H. Langford, Chem. Commun. *1965*, 393; Can. J. Chem. *46*, 3183 (1968); J. Phys. Chem. *74*, 1376 (1970). [126] R. H. Erlich, M. S. Greenberg, and A. I. Popov, Spectrochim.

Acta, Part A *29*, 543 (1973). [127] R. D. Gillard, E. D. McKenzie, and M. D. Ross, J. Inorg. Nucl. Chem. *28*, 1429 (1966). [128] M. E. L. McBain and E. Hutchison: *Solubilization and Related Phenomena*. Academic Press, New York 1955, Chapter 6. [129] E. M. Kosower: *An Introduction to Physical Organic Chemistry*. Wiley, New York 1968. [130] J. L. Kavanau: *Structure and Function in Biological Membranes*. Vol. 1, Holden Day, San Francisco 1965.

[131] K. Shinoda and S. Friberg: *Emulsions and Solubilization: Basis and Applications*. Wiley, Chichester, New York 1986. [132] J. H. Fendler and E. J. Fendler: *Catalysis in Micellar and Macromolecular Systems*. Academic Press, New York, 1975; J. H. Fendler, Acc. Chem. Res. *9*, 153 (1976). [133] J. H. Fendler, F. Nome, and H. C. Van Woert, J. Am. Chem. Soc. *96*, 6745 (1974). [134] R. Fuoss, J. Chem. Educ. *32*, 527 (1955); R. M. Fuoss and F. Accascina: *Electrolytic Conductance*. Wiley-Interscience, New York 1959. [135] E. Price: *Solvation of Electrolytes and Solution Equilibria*, in J. J. Lagowski (ed.): *The Chemistry of Non-Aqueous Solvents*. Academic Press, New York, London 1966, Vol. I, p. 67ff. [136] J. Barthel: *Leitfähigkeit von Elektrolytlösungen*. Angew. Chem. *80*, 253 (1968); Angew. Chem., Int. Ed. Engl. *7*, 260 (1968); J. Barthel: *Ionen in nichtwäßrigen Lösungen*. Steinkopff, Darmstadt 1976. [137] V. Gutmann, Chimia *23*, 285 (1969); Angew. Chem. *82*, 858 (1970); Angew. Chem., Int. Ed. Engl. *9*, 843 (1970); Chemie in unserer Zeit *4*, 90 (1970); Chemistry in Britain *7*, 102 (1971); Chimia *31*, 1 (1977). [138] M. Szwarc (ed.): *Ions and Ion Pairs in Organic Reactions*. Wiley-Interscience, New York 1972/1974, Vol. 1 and 2; M. Szwarc, Acc. Chem. Res. *2*, 87 (1969). [139] U. Mayer: *Solvent Effects on Ion-Pair Equilibria*. Coord. Chem. Rev. *21*, 159 (1976). [140] D. N. Glew and D. A. Hames, Can. J. Chem. *49*, 3114 (1971).

[141] J. Smid: *Die Struktur solvatisierter Ionenpaare*. Angew. Chem. *84*, 127 (1972); Angew. Chem., Int. Ed. Engl. *11*, 112 (1972). [141a] W. Nernst, Nachrichten der Akademie der Wissenschaften zu Göttingen, Mathematisch-naturwissenschaftliche Klasse *1893*, 491; Z. Phys. Chem. *13*, 531 (1894). [141b] J. J. Thomson, Philosophical Magazine (London) *36*, 313 (1893). [141c] E. Beckmann, Z. Phys. Chem. *6*, 437 (1890); esp. p. 470. [142] M. B. Reynolds and C. A. Kraus, J. Am. Chem. Soc. *70*, 1709 (1948). [143] C. C. Evans and S. Sugden, J. Chem. Soc. *1949*, 270. [144] S. Winstein, L. G. Savedoff, S. Smith, I. D. R. Stevens, and J. S. Gall, Tetrahedron Lett. *1960*, No. 9, p. 24. [145] H. Normant, Angew. Chem. *79*, 1029 (1967); Angew. Chem., Int. Ed. Engl. *6*, 1046 (1967); Bull. Soc. Chim. Fr. *1968*, 791; Usp. Khim. *39*, 990 (1970). [146] F. Madaule-Aubry, Bull. Soc. Chim. Fr. *1966*, 1456. [147] P. Walden, Ber. Dtsch. Chem. Ges. *35*, 2018 (1902). [148] N. N. Lichtin: *Ionization and Dissociation Equilibria in Solution in Liquid Sulfur Dioxide*. Progr. Phys. Org. Chem. *1*, 75 (1963). [149] W. Karcher and H. Hecht: *Chemie in flüssigem Schwefeldioxid*, in G. Jander, H. Spandau, and C. C. Addison (eds.): *Chemie in nichtwäßrigen ionisierenden Lösungsmitteln*. Vieweg, Braunschweig, Pergamon Press, Oxford 1967, Vol. III, Part 2, p. 79ff. [150] H. H. Freedman: *Arylcarbonium Ions*, in G. A. Olah and P. v. R. Schleyer (eds.): *Carbonium Ions*. Wiley-Interscience, New York 1973, Vol. IV, p. 1501ff.

[151] E. Price and N. N. Lichtin, Tetrahedron Lett. *1960*, No. 18, p. 10. [152] M. Baaz, V. Gutmann, and O. Kunze, Monatsh. Chem. *93*, 1142 (1962). [153] Private communication from R. Waack and M. A. Doran to H. H. Freedman, cited in [150]. [154] A. G. Evans, A. Price, and J. H. Thomas, Trans. Faraday Soc. *52*, 332 (1956). [155] Y. Pocker, J. Chem. Soc. *1958*, 240. [156] A. G. Evans, A. Price, and J. H. Thomas, Trans. Faraday Soc. *51*, 481 (1955). [157] A. G. Evans, I. H. McEwan, A. Price, and J. H. Thomas, J. Chem. Soc. *1955*, 3098. [158] U. Mayer and V. Gutmann, Monatsh. Chem. *102*, 148 (1971). [159] B. Chevrier, J. M. Le Carpentier, and R. Weiss, J. Am. Chem. Soc. *94*, 5718 (1972); B. Chevrier and R. Weiss, Angew. Chem. *86*, 12 (1974); Angew. Chem., Int. Ed. Engl. *13*, 1 (1974). [160] D. Cook, Can. J. Chem. *37*, 48 (1959).

[161] For recent reviews see I. P. Beletskaya: *Ions and Ion-Pairs in Nucleophilic Aliphatic Substitution*. Usp. Khim. *44*, 2205 (1975); Russian Chem. Rev. *44*, 1067 (1975); I. P. Beletskaya and A. A. Solov'yanov, Zh. Vses. Khim. O-va *22*, 286 (1977); Chem. Abstr. *87*, 83917z (1977). [162] S. Winstein, E. Clippinger, A. H. Fainberg, and G. C. Robinson, J. Am. Chem. Soc. *76*, 2597 (1954); S. Winstein, Experientia, Suppl. II, 137 (1955). [163] H. Sadek and R. M. Fuoss, J. Am. Chem. Soc. *76*, 5897, 5905 (1954). [164] T. E. Hogen-Esch and J. Smid, J. Am. Chem. Soc. *88*, 307, 318 (1966). [165] G. A. Olah and P. v. R. Schleyer (eds.): *Carbonium Ions*. Vol. I-IV, Wiley-Interscience, New York 1968–1973. [166] J. F. Garst: *Organoalkali Compounds in Ethers*, in J. F. Coetzee and C. D. Ritchie (eds.): *Solute-Solvent Interactions*. Dekker, New York, London 1969, Vol. 1, p. 539ff. [167] H. F.

Ebel: *Struktur und Reaktivität von Carbanionen und carbanionoiden Verbindungen.* Fortschr. Chem. Forsch. *12*, 387 (1969). [168] M. Schlosser: *Struktur und Reaktivität polarer Organometalle.* Springer, Berlin 1973, p. 26ff. [168a] T. E. Hogen-Esch: *Ion-Pairing Effects in Carbanion Reactions.* Adv. Phys. Org. Chem. *15*, 153 (1978). [169] N. M. Atherton and S. I. Weissman, J. Am. Chem. Soc. *83*, 1330 (1961). [170] P. Chang, R. V. Slates, and M. Szwarc, J. Phys. Chem. *70*, 3180 (1966).

[171] G. Jancsó and D. V. Fenby: *Thermodynamics of Dilute Solutions.* J. Chem. Educ. *60*, 382 (1983). [172] M. D. Zeidler: *Struktur einfacher molekularer Flüssigkeiten.* Angew. Chem. *92*, 700 (1980); Angew. Chem., Int. Ed. Engl. *19*, 697 (1980); *cf.* also Z. Phys. Chem., N. F. *133*, 1 (1982). [173] J. A. Barker and D. Henderson: *The Fluid Phases of Matter.* Scientific American *254*, 94 (1981); *Einheitliche Theorie der Flüssigkeiten und Gase.* Spektrum der Wissenschaft, Januar 1982, p. 80ff. [174] J. N. Murrell and E. A. Boucher: *Properties of Liquids and Solutions.* Wiley, New York 1982. [175] F. Franks: *Polywater.* MIT Press, Cambridge/Mass., London 1982; *Polywasser – Betrug oder Irrtum in der Wissenschaft.* Vieweg, Braunschweig, Wiesbaden 1984. [176] F. Franks: *Water.* The Royal Society of Chemistry Paperbacks, London 1983. [177] W. A. P. Luck: *Modellbetrachtung von Flüssigkeiten mit Wasserstoffbrücken.* Angew. Chem. *92*, 29 (1980); Angew. Chem., Int. Ed. Engl. *19*, 28 (1980). [178] M. C. R. Symons: *Structure in Solvents and Solutions.* Chem. Soc. Rev. *12*, 1 (1983). [179] P. Huyskens: *Molecular Structure of Liquid Alcohols.* J. Mol. Struct. *100*, 403 (1983). [180] J. H. Hildebrand, J. Phys. Colloid Chem. *53*, 944 (1949); Chem. Abstr. *43*, 7285c (1949).

[181] S. L. Kittsley and H. A. Goeden, J. Am. Chem. Soc. *72*, 4841 (1950). [182] G. C. Maitland, M. Rigby, E. B. Smith, and W. A. Wakeham: *Intermolecular Forces – Their Origin and Determination.* Oxford University Press, Oxford 1987. [183] H. Ratajczak and W. J. Orville-Thomas (eds.): *Molecular Interactions.* Vol. 1 . . . 3, Wiley, New York 1980 . . . 1982. [184] H. Gnamm and O. Fuchs: *Lösungsmittel und Weichmachungsmittel.* 8th edition, Wissenschaftliche Verlagsgesellschaft, Stuttgart 1980, Vol. I, Chapter 2, p. 9ff. [185] N. H. March and M. P. Tosi: *Coulomb Liquids.* Academic Press, New York, London 1984. [186] J. H. Mahanty and B. W. Ninham: *Dispersion Forces.* Academic Press, New York, London 1977. [187] G. Geiseler and H. Seidel: *Die Wasserstoffbrückenbindung.* Akademie-Verlag, Berlin, and Vieweg, Braunschweig 1977. [188] M. D. Joesten: *Hydrogen Bonding and Proton Transfer.* J. Chem. Educ. *59*, 362 (1982). [189] P. Schuster (ed.): *Hydrogen Bonds.* Top. Curr. Chem. *120*, 1 . . . 113 (1984) (*cf.* esp. p. 35ff.). [190] J. Emsley: *Very Strong Hydrogen Bonding.* Chem. Soc. Rev. *9*, 91 (1980).

[191] W. M. Latimer and W. H. Rodebush, J. Am Chem. Soc. *42*, 1419 (1920); *cf.* p. 1431. [192] N. Isenberg, J. Chem. Educ. *59*, 547 (1982). [193] W. Saenger, Nature (London) *279*, 343 (1979); W. Saenger and K. Lindner, Angew. Chem. *92*, 404 (1980); Angew. Chem., Int. Ed. Engl. *19*, 398 (1980); *cf.* also Nachr. Chem. Techn. Labor. *27*, 403 (1979). [194] R. Taylor and O. Kennard: *Hydrogen-Bond Geometry in Organic Crystals.* Acc. Chem. Res. *17*, 320 (1984). [195] M. Rospenk, J. Fritsch, and G. Zundel, J. Phys. Chem. *88*, 321 (1984). [196] M. J. Kamlet and R. W. Taft, J. Am. Chem. Soc. *98*, 377, 2886 (1976). [197] M. Mashima, R. T. McIver, R. W. Taft, F. G. Bordwell, and W. N. Olmstead, J. Am. Chem. Soc. *106*, 2717 (1984) (*cf.* footnote 12). [198] B. D. Gosh and R. Basu, J. Phys. Chem. *84*, 1887 (1980). [199] R. Schmid, J. Sol. Chem. *12*, 135 (1983). [200] Y. Marcus, J. Sol. Chem. *13*, 599 (1984).

[201] U. Mayer and V. Gutmann, Structure and Bonding *12*, 113 (1972); *cf.* p. 125. [202] A. I. Popov, Pure Appl. Chem. *41*, 275 (1975); and references cited therein. [203] T. Ogata, T. Fujisawa, N. Tanaka, and H. Yokoi, Bull. Chem. Soc. Jpn. *49*, 2759 (1976). [204] R. W. Soukop: *Farbreaktionen zur Klassifizierung von Lösungsmitteln.* Chemie in unserer Zeit *17*, 129, 163 (1983); R. W. Soukop and R. Schmid: *Metal Complexes as Color Indicators for Solvent Parameters.* J. Chem. Educ. *62*, 459 (1985). [205] R. Schmid, Rev. Inorg. Chem. *1*, 117 (1979); Chem. Abstr. *92*, 117098d (1980). [206] R. Schmid and V. N. Sapunov: *Non-Formal Kinetics in Search for Chemical Reaction Pathways.* Verlag Chemie, Weinheim 1982. [207] V. Gutmann and G. Resch: *The Unifying Impact of the Donor-Acceptor Approach.* In N. Tanaka, H. Ohtaki, and R. Tamamushi (eds.): *Ions and Molecules in Solution.* Elsevier, Amsterdam 1982, p. 203ff. [208] R. S. Drago, Coord. Chem. Rev. *33*, 251 (1980); Pure Appl. Chem. *52*, 2261 (1980). [209] R. W. Taft, N. J. Pienta, M. J. Kamlet, and E. M. Arnett, J. Org. Chem. *46*, 661 (1981). [210] G. Olofsson and I. Olofsson, J. Am. Chem. Soc. *95*, 7231 (1973).

[211] L. Elégant, G. Fratini, J.-F. Gal, and P.-C. Maria, Journ. Calorim. Anal. Therm. *11*, 3/21/1 . . . 3/21/9 (1980); Chem. Abstr. *95*, 121811k (1981). [212] P.-C. Maria and J.-F. Gal, J. Phys. Chem. *89*, 1296 (1985); P.-C. Maria, J.-F. Gal, J. de Franceschi, and E. Fargin, J. Am. Chem. Soc. *109*, 483 (1987). [213] U. Mayer: *NMR-Spectroscopic Studies on Solute-Solvent and Solute-Solute Interactions.* In N. Tanaka, H. Ohtaki, and R. Tamamushi (eds.): *Ions and Molecules in Solution.* Elsevier, Amsterdam 1983, p. 219ff. [214] J.-C. Bollinger, G. Yvernault, and T. Yvernault, Thermochim. Acta *60*, 137 (1983); Chem. Abstr. *98*, 118416s (1983). [215] R. S. Drago, G. C. Vogel, and T. E. Neddham, J. Am. Chem. Soc. *93*, 6014 (1971). [216] M. K. Kroeger and R. S. Drago, J. Am. Chem. Soc. *103*, 3250 (1981). [217] R. S. Drago: *The Coordination Model for Non-Aqueous Solvents.* Pure Appl. Chem. *52*, 2261 (1980). [218] R. S. Drago, M. K. Kroeger, and J. R. Stahlbush, Inorg. Chem. *20*, 306 (1981). [219] P. E. Doan and R. S. Drago, J. Am. Chem. Soc. *104*, 4524 (1982). [220] C. Tanford: *The Hydrophobic Effect.* 2nd ed., Wiley-Interscience, New York 1980.

[221] A. Ben-Naim: *Hydrophobic Interactions.* Plenum, New York 1980. [222] J.-Y. Huot and C. Jolicoeur: *Hydrophobic Effects in Ionic Hydration and Interactions.* In R. R. Dogonadze, E. Kálmán, A. A. Kornyshev, and J. Ulstrup (eds.): *The Chemical Physics of Solvation.* Part A, Elsevier, Amsterdam 1985. [223] H. van de Waterbeemd: *Hydrophobicity of Organic Compounds.* Compudrug International, Wien 1986.[224] L. R. Pratt: *Theory of Hydrophobic Effects.* Ann. Rev. Phys. Chem. *36*, 433 (1985). [225] J. B. F. N. Engberts, Pure Appl. Chem. *54*, 1797 (1982). [226] H. S. Frank and M. W. Evans, J. Phys. Chem. *13*, 507 (1945). [227] D. F. Evans, S.-H. Chen, G. W. Schriver, and E. M. Arnett, J. Am. Chem. Soc. *103*, 481 (1981); D. Mirejovsky and E. M. Arnett, ibid. *105*, 1112 (1983). [228] F. A. Greco, J. Phys. Chem. *88*, 3132 (1984). [229] J. H. Hildebrand: *Is there a "hydrophobic" effect?* Proc. Natl. Acad. Sci. U.S.A. *76*, 194 (1979); Chem. Abstr. *90*, 134237z (1979). [230] M. H. Abraham, J. Am. Chem. Soc. *102*, 5910 (1980).

[231] A. Hvidt, Acta Chem. Scand., Part A *37*, 99 (1983); and references cited therein. [232] W. Kauzmann, Adv. Protein Chem. *14*, 1 (1959). [233] J. Burgess: *Metal Ions in Solution.* Ellis Horwood, Chichester 1978. [234] B. G. Cox and W. E. Waghorne: *Thermodynamics of Ion-Solvent Interactions.* Chem. Soc. Rev. *9*, 381 (1980). [235] O. Popovych and R. P. T. Tomkins: *Nonaqueous Solution Chemistry.* Wiley-Interscience, New York 1981. [236] B. E. Conway: *Ionic Hydration in Chemistry and Biophysics.* Elsevier, Amsterdam 1981. [237] K. Burger: *Solvation, Ionic and Complex Formation Reactions in Non-Aqueous Solvents.* Elsevier, Amsterdam 1983. [238] R. R. Dogonadze, E. Kálmán, A. A. Kornyshev, and J. Ulstrup (eds.): *The Chemical Physics of Solvation.* Part A: *Theory of Solvation*; Part B: *Spectroscopy of Solvation*; Part C: *Solvation Phenomena in Specific Physical, Chemical, and Biological Systems.* Elsevier, Amsterdam 1985 . . . 1988. [239] J. Barthel, H.-J. Gores, G. Schmeer, and R. Wachter: *Non-Aqueous Electrolyte Solutions in Chemistry and Modern Technology.* Top. Curr. Chem. *111*, 33 (1983). [240] N. Tanaka, H. Ohtaki, and R. Tamamushi (eds.): *Ions and Molecules in Solution.* Elsevier, Amsterdam 1983.

[241] Y. Marcus: *Ion Solvation.* Wiley, Chichester 1985. [242] Y. Marcus, Pure Appl. Chem. *54*, 2327 (1982). [243] J. I. Kim, J. Phys. Chem. *82*, 191 (1978). [244] Y. Marcus, Rev. Anal. Chem. *5*, 53 (1980); Pure Appl. Chem. *55*, 977 (1983); ibid. *57*, 1103 (1985); ibid. *58*, 1721 (1986). [245] G. A. Vidulich and A. Fratiello, J. Chem. Educ. *55*, 672 (1978). [246] P. Laszlo and A. Stockis, J. Am. Chem. Soc. *102*, 7818 (1980) (^{59}Co-NMR). [247] P. S. Pregosin: *Platinium-195 NMR.* Coord. Chem. Rev. *44*, 247 (1982). [248] B. E. Conway: *Local Changes of Solubility induced by Electrolytes: Salting-out and Ionic Hydration.* Pure Appl. Chem. *57*, 263 (1985). [249] H. Langhals: *Polarität binärer Flüssigkeitsgemische.* Angew. Chem. *94*, 739 (1982); Angew. Chem., Int. Ed. Engl. *21*, 724 (1982). [250] S. Janardhanan and C. Kalidas: *Preferential Solvation of Ions in Binary Solvent Mixtures.* Rev. Inorg. Chem. *6*, 101 (1984).

[251] T. Ichikawa, H. Yoshida, A. S. W. Li, and L. Kevan, J. Am. Chem. Soc. *106*, 4324 (1984). [252] D. S. Gill, N. Kumari, and M. S. Chauhan, J. Chem. Soc., Faraday Trans. I *81*, 687 (1985). [253] A. J. Stace, J. Am. Chem. Soc. *106*, 2306 (1984). [254] H. Strehlow and G. Busse, Ber. Bunsenges. Phys. Chem. *88*, 467 (1984). [255] K. Remerle and J. B. F. N. Engberts, J. Phys. Chem. *87*, 5449 (1983). [256] G. González and N. Yutronic, Spectrochim. Acta Part A *39*, 269 (1983). [257] C. F. Wells, J. Chem. Soc., Faraday Trans. I *70*, 402, 694 (1974); and following papers in a series. [258] M. wa Muanda, J. B. Nagy, and O. B. Nagy, Tetrahedron Lett. *1974*, 3421; J. Chem. Soc., Faraday Trans. I *74*, 2210 (1978); J. Phys. Chem. *83*, 1961 (1979). [259] A. K. Covington and K. E. Newman, Pure Appl. Chem.

51, 2041 (1979). [260] A. Delville, C. Detellier, A. Gerstmans, and P. Laszlo, Helv. Chim. Acta *64*, 547, 556 (1981).

[261] Y. Marcus, Aust. J. Chem. *36*, 1719 (1983). [262] K. L. Mittal (ed.): *Micellization, Solubilization, and Microemulsions*. Vol. 1 and 2, Plenum Press, New York 1977. [263] E. J. R. Sudhölter, G. B. van de Langkruis, and J. B. F. N. Engberts: *Micelles. Structure and Catalysis*. Rec. Trav. Chim. Pays-Bas *99*, 73 (1980). [264] B. Lindman and H. Wennerström: *Micelles. Amphiphile Aggregation in Aqueous Solution*. Top. Curr. Chem. *87*, 1 (1980). [265] H.-F. Eicke: *Surfactants in Nonpolar Solvents. Aggregation and Micellization*. Top. Curr. Chem. *87*, 85 (1980). [266] J. H. Fendler: *Membrane Mimetic Chemistry*. Wiley-Interscience, New York 1982. [267] E. Wyn-Jones and J. Gormally (eds.): *Aggregation Processes in Solution*. Studies in Physical and Theoretical Chemistry, Vol. 26, Elsevier, Amsterdam 1983. [268] L. J. Cline Love, J. G. Habarts, and J. G. Dorsey, Anal. Chem. *56*, 1132 A (1984). [269] F. M. Menger: *On the Structure of Micelles*. Acc. Chem. Res. *12*, 111 (1979); F. M. Menger and D. W. Doll, J. Am. Chem. Soc. *106*, 1109 (1984). [270] C. W. von Nägeli and S. Schwendener: *Das Mikroskop. Theorie und Anwendung desselben*. 2nd edition, Verlag Engelmann, Leipzig 1877; C. W. von Nägeli: *Theorie der Gärung*. Verlag Oldenbourg, München 1879; C. W. von Nägeli: *Die Micellartheorie*. In A. Frey (ed.): *Ostwalds Klassiker der exakten Naturwissenschaften*. Vol. 227, Akademische Verlagsgesellschaft, Leipzig 1928.

[271] F. Krafft, Ber. Dtsch. Chem. Ges. *29*, 1334 (1896). [272] A. Reychler, Zeitschrift für Chemie und Industrie der Kolloide *12*, 277 (1913); Chem. Zbl. *1913*, II 1, 491. [273] P. Fromherz, Ber. Bunsenges. Phys. Chem. *85*, 891 (1981). [274] J. K. Thomas, Chem. Rev. *80*, 283 (1980). [275] D. G. Whitten, J. C. Russell, and R. H. Schmehl, Tetrahedron *38*, 2455 (1982). [276] J. H. Fendler, J. Chem. Educ. *60*, 872 (1983). [277] C. A. Bunton: *Reactions in Micelles and Similar Self-Organized Aggregates*. In M. I. Page (ed.): *The Chemistry of Enzyme Action*. Chapter 13, p. 461 ff., Elsevier, Amsterdam 1984. [278] J. Smid (ed.): *Ions and Ion Pairs and Their Role in Chemical Reactions*. Pergamon Press, Oxford 1979; *cf.* also Pure Appl. Chem. *51*, p. 49 ff. (1979). [279] M. Simonetta: *Structure of Ion Pairs in Solution*. Int. Rev. Phys. Chem. *1*, 31 (1981); Chem. Abstr. *95*, 176553h (1981); *cf.* also Chem. Soc. Rev. *13*, 1 (1984). [280] N. Bjerrum: *Untersuchungen über Ionenassoziation*. Kong. Danske Vidensk. Meddelelser. Math.-phys. Kl. 7, Nr. 9, p. 3... 48 (1926); Chem. Zentralbl. *1926* II, 1378; Svensk Kem. Tidskr. *38*, p. 2... 18 (1926); Chem. Zentralbl. *1926* I, 2174.

[281] V. Gutmann: *The Donor-Acceptor Approach to Molecular Interactions*. Plenum Press New York 1978. [282] M. Feigel and H. Kessler, Chem. Ber. *112*, 3715 (1979); Acc. Chem. Res. *15*, 2 (1980). [283] M. Feigel, H. Kessler, D. Leibfritz, and A. Walter, J. Am. Chem. Soc. *101*, 1943 (1979). [284] E. B. Troughton, K. E. Molter, and E. M. Arnett, J. Am. Chem. Soc. *106*, 6726 (1984); E. M. Arnett and K. E. Molter, Acc. Chem. Res. *18*, 339 (1985); J. Phys. Chem. *90*, 383 (1986). [285] M. Born, Z. Phys. *1*, 45 (1920). [286] V. Gold (ed.): *Glossary of Terms used in Physical Organic Chemistry*. Pure Appl. Chem. *55*, 1281 (1983); *cf.* p. 1324. [287] E. Buncel and B. Menon, J. Org. Chem. *44*, 317 (1979). [288] G. Boche and F. Heidenhain, J. Am. Chem. Soc. *101*, 738 (1979); G. Boche, F. Heidenhain, W. Thiel, and R. Eiben, Chem. Ber. *115*, 3167 (1982). [289] W. Bauer and D. Seebach, Helv. Chim. Acta *67*, 1972 (1984). [290] J. F. McGarrity and C. A. Ogle, J. Am. Chem. Soc. *107*, 1805 (1985).

[291] K. Takahashi, N. Hirata, and K. Takase, Tetrahedron Lett. *15*, 1285 (1970); K. Takahashi, K. Takase, and T. Sakae, Chem. Lett. *1980*, 1485. [292] K. Okamoto, T. Kitagawa, K. Takeuchi, K. Komatsu, and K. Takahashi, J. Chem. Soc., Chem. Commun. *1985*, 173. [293] S. Hahn, W. M. Miller, R. N. Lichtenthaler, and J. M. Prausnitz, J. Sol. Chem. *14*, 129 (1985). [294] J. F. Hinton, K. R. Metz, and R. W. Briggs: *Thallium NMR Spectroscopy*. In G. A. Webb (ed.): Annual Reports on NMR Spectroscopy. Vol. 13, p. 211 ff., Academic Press, London, New York 1982. [295] J. J. Dechter: *NMR of Metal Nuclides*. Progr. Inorg. Chem. *29*, 285 (1982); Fig. 7 on p. 320. [296] C. Hansch, Acc. Chem. Res. *2*, 232 (1969); A. Leo, J. Chem. Soc., Perkin Trans. II *1983*, 825. [297] F. M. Menger and U. V. Venkataram, J. Am. Chem. Soc. *108*, 2980 (1986). [298] M. J. Kamlet and R. W. Taft, J. Org. Chem. *47*, 1734 (1982). [299] G. Heublein and D. Bauernfeind, J. Prakt. Chem. *326*, 81 (1984). [300] C. Sandorfy, R. Buchet, L. S. Lussier, P. Ménassa, and L. Wilson, Pure Appl. Chem. *58*, 1115 (1986).

[301] I. Persson, Pure Appl. Chem. *58*, 1153 (1986). [302] W. L. Jorgensen, J. K. Buckner, S. E. Huston, and P. J. Rossky, J. Am. Chem. Soc. *109*, 1891 (1987). [303] I. Persson, M. Sandström, and P. L. Goggin, Inorg. Chim. Acta *129*, 183 (1987).

Chapter 3

[1] L. F. Audrieth and J. Kleinberg: *Non-Aqueous Solvents: Applications as Media for Chemical Reactions*. Wiley, New York, and Chapman and Hall, London 1953. [2] H. H. Sisler: *Chemistry in Non-Aqueous Solvents*. Reinhold, New York, and Chapman and Hall, London 1961. [3] G. Charlot and B. Trémillon: *Réactions Chimiques dans les Solvants et les Sels Fondus*. Gauthier-Villars, Paris 1963; *Chemical Reactions in Solvents and Melts*. Pergamon Press, Oxford 1969. [4] G. Jander, H. Spandau, and C. C. Addison (eds.): *Chemie in nichtwäßrigen Lösungsmitteln*. Vol. I-IV, Vieweg, Braunschweig, Wiley-Interscience, New York, and Pergamon Press, Oxford 1963–1967. [5] T. C. Waddington (ed.): *Non-Aqueous Solvent Systems*. Academic Press, London, New York 1965. [6] A. K. Holliday and A. G. Massey: *Non-Aqueous Solvents*. Pergamon Press, Oxford 1965. [7] J. J. Lagowski (ed.): *The Chemistry of Nonaqueous Solvents*. Vol. I-V, Academic Press, New York, London 1966–1978. J. J. Lagowski: *Non-Aqueous Ionizing Liquids*. Rev. Chim. Miner. *15*, 1 (1978). [8] J. Jander and C. Lafrenz: *Wasserähnliche Lösungsmittel*. Verlag Chemie, Weinheim 1968; *Ionizing Solvents*. Wiley, New York, London 1968. [9] R. A. Zingaro: *Nonaqueous Solvents*. Raytheon Education Company, Lexington/Mass. 1968. [10] T. C. Waddington: *Non-Aqueous Solvents*. 2nd ed., Nelson, London 1973; *Nichtwäßrige Lösungsmittel*. Hüthig, Heidelberg 1972.

[11] I. Mellan: *Industrial Solvents Handbook*. 2nd ed., Noyes Data Corporation, Park Ridge/New Jersey 1977. [12] J. A. Riddick, W. B. Bunger, and T. Sakano: *Organic Solvents*. 4rd ed., in A. Weissberger (ed.): *Techniques of Chemistry*, Vol. II, Wiley-Interscience, New York 1986. [13] T. H. Durrans (editor E. H. Davies): *Solvents*, 8th ed., Chapman and Hall, London 1971. [14] B. Trémillon: *La Chimie en Solvants Non-Aqueux* (Le Chimiste No. 3). Presses Universitaires de France, Paris 1971; *Chemistry in Non-Aqueous Solvents*. Reidel, Dordrecht, Boston 1974. [15] A. K. Covington and T. Dickinson (eds.): *Physical Chemistry of Organic Solvent Systems*. Plenum Press, London, New York 1973. [16] V. Gutmann: *Coordination Chemistry in Non-Aqueous Solutions*. Springer, Wien, New York 1968, p. 2 (and footnote 1). [17] A. Saupe: *Neuere Ergebnisse auf dem Gebiet der flüssigen Kristalle*. Angew. Chem. *80*, 99 (1968).; Angew. Chem., Int. Ed. Engl. *7*, 97 (1968). [18] G. H. Brown: *Liquid Crystals and some of their Applications in Chemistry*. Anal. Chem. *41*, 26 A (1969); Chemie in unserer Zeit *2*, 42 (1968). [19] R. Steinsträßer and L. Pohl: *Chemie und Verwendung flüssiger Kristalle*. Angew. Chem. *85*, 706 (1973); Angew. Chem., Int. Ed. Engl. *12*, 617 (1973). [20] E. Sackmann: *Scientific Applications of Liquid Crystals*, in G. Meier, E. Sackmann, and J. G. Grabmaier (eds.): *Applications of Liquid Crystals*. Springer, Berlin, Heidelberg, New York 1975, p. 21ff.

[21] D. Demus: *Eigenschaften, Theorie und Molekülbau flüssiger Kristalle*. Z. Chem. *15*, 1 (1975); D. Demus and H. Zaschke: *Flüssige Kristalle in Tabellen, I und II*. Deutscher Verlag für Grundstoffindustrie, Leipzig 1976 and 1987. [22] E. I. Kovshev, L. M. Blinov, and V. V. Titov: *Thermotropic Liquid Crystals and Their Application*. Usp. Khim. *46*, 753 (1977); Russ. Chem. Rev. *46*, 395 (1977). [22a] H. Kelker and R. Hatz, *Handbook of Liquid Crystals*. Verlag Chemie, Weinheim 1980. [23] H. Kelker and B. Scheurle, Angew. Chem. *81*, 903 (1969); Angew. Chem., Int. Ed. Engl. *8*, 884 (1969). [24] B. R. Sundheim (ed.): *Fused Salts*. McGraw-Hill, New York 1964. [25] W. Sundermeyer: *Salzschmelzen und ihre Verwendung als Reaktionsmedien*. Angew. Chem. *77*, 241 (1965); Angew. Chem., Int. Ed. Engl. *4*, 222 (1965); Chemie in unserer Zeit *1*, 150 (1967). [26] H. Bloom and J. W. Hastie: *Molten Salts as Solvents*, in T. C. Waddington (ed.): *Non-Aqueous Solvent Systems*. Academic Press, London, New York 1965, p. 353ff. [27] G. J. Janz: *Molten Salts Handbook*. Academic Press, London, New York 1967. [28] J. E. Gordon: *Applications of Fused Salts in Organic Chemistry*, in D. B. Denney (ed.): *Techniques and Methods of Organic and Organometallic Chemistry*. Dekker, New York, London 1969, p. 51ff. [29] J. Braunstein, G. Mamantov, and G. P. Smith (eds.): *Advances in Molten Salt Chemistry*. Vol. 1–4, Plenum Publ. Corp., London, New York 1971–1981. [30] D. H. Kerridge: *Molten Salts as Nonaqueous Solvents*, in J. J. Lagowski (ed.): *The Chemistry of Nonaqueous Solvents*. Vol. 5B, Academic Press, London, New York 1978.

[31] W. Sundermeyer, O. Glemser, and K. Kleine-Weischede, Chem. Ber. *95*, 1829 (1962). [31a] E. Channot, A. K. Sharma, and L. A. Paquette, Tetrahedron Lett. *1978*, 1963. [32] G. Duve, O. Fuchs, and H. Overbeck: *Lösemittel HOECHST*. 6th ed., Hoechst AG, Frankfurt (Main) 1976. [33] E. sz. Kováts: *Zu Fragen der Polarität – Die Methode der Linearkombination der Wechselwirkungskräfte*. Chimia *22*, 459 (1968). [34] C. Reichardt: *Optische Aktivität und Molekülsymmetrie*. Chemie in

unserer Zeit *4*, 188 (1970). [35] D. Seebach: *Neue links- und rechtshändige Werkzeuge für den Chemiker*, in "25 Jahre Fonds der Chemischen Industrie 1950–1975", Frankfurt (Main) 1975, p. 13ff. [36] H. Pracejus: *Asymmetrische Synthesen*. Fortschr. Chem. Forsch. *8*, 493 (1967). [37] T. D. Inch: *Asymmetric Syntheses*. Synthesis *1970*, 466. [38] J. D. Morrison and H. S. Mosher: *Asymmetric Organic Reactions*. Prentice Hall, Englewood Cliffs/New Jersey 1971, p. 411ff. [39] R. v. Ammon and R. D. Fischer, Angew. Chem. *84*, 737 (1972); Angew. Chem., Int. Ed. Engl. *11*, 675 (1972). [40] M. Raban and K. Mislow, Topics in Stereochemistry *2*, 199 (1967).

[41] R. Charles, U. Beither, B. Feibush, and E. Gil-Av, J. Chromatogr. *112*, 121 (1975). [42] H. Plieninger and H. P. Kraemer, Angew. Chem. *88*, 230 (1976); Angew. Chem., Int. Ed. Engl. *15*, 243 (1976); Tetrahedron *34*, 891 (1978). [43] W. F. Luder and S. Zuffanti: *The Electronic Theory of Acids and Bases*. 2nd ed., Dover Publications, New York 1961. [44] R. G. Bates: *Medium Effects and pH in Nonaqueous Solvents*, in J. F. Coetzee and C. D. Ritchie (eds.): *Solute-Solvent Interactions*. Dekker, New York, London 1969, Vol. 1, p. 46ff. [45] R. P. Bell: *Acids and Bases – Their Quantitative Behaviour*. Methuen, London 1969; *Säuren und Basen* (translated into German by F. Frickel). Verlag Chemie, Physik Verlag, Weinheim 1974. [46] R. P. Bell: *The Proton in Chemistry*. *2nd* ed., Chapman and Hall, London 1973. [47] M. M. Davies: *Brønsted Acid-Base Behaviour in "Inert" Organic Solvents*, in J. J. Lagowski (ed.): *The Chemistry of Nonaqueous Solvent Systems*. Academic Press, London, New York 1970, Vol. III, p. 1ff. [48] L. N. Bykova and S. I. Petrov: *Acid-Base Equilibria in Amphiprotic Solvents and Potentiometric Titration*. Usp. Khim. *41*, 2065 (1972); Russian Chem. Rev. *41*, 975 (1972). [49] E. J. King: *Acid-Base Behaviour*, in A. K. Covington and T. Dickinson (eds.): *Physical Chemistry of Organic Solvent Systems*. Plenum Press, London, New York 1973, p. 331ff. [50] R. G. Bates: *Acidity Functions in Aqueous and Nonaqueous Media*. Bull. Soc. Chim. Belg. *84*, 1139 (1975).

[51] J. N. Brønsted, Rec. Trav. Chim. Pays-Bas *42*, 718 (1923); Chem. Rev. *5*, 231 (1928); Angew. Chem. *43*, 229 (1930). [52] T. M. Lowry, Chem. Ind. (London) *42*, 43 (1923); Trans. Faraday Soc. *20*, 13 (1924). [53] For a compilation of important pK_a-values of acids and bases see A. J. Gordon and R. A. Ford: *The Chemist's Companion*. Wiley-Interscience, New York 1972, p. 54ff. [54] J. N. Brønsted, Ber. Dtsch. Chem. Ges. *61*, 2049 (1928). [55] I. M. Kolthoff, E. B. Sandell, E. J. Meehan, and S. Bruckenstein: *Quantitative Chemical Analysis*. 4th ed., MacMillan Company, London 1969 (especially Chap. 5 on Acid-Base Equilibria in Nonaqueous Solvents). [56] I. M. Kolthoff: *Acid-Base Equilibria in Dipolar Aprotic Solvents*. Anal. Chem. *46*, 1992 (1974). [57] V. I. Dulova, N. V. Lichkova, and L. P. Ivleva: *The Differentiating Action of Oxygen-containing Solvents on the Strengths of Acids*. Usp. Khim. *37*, 1893 (1968); Russian Chem. Rev. *37*, 818 (1968). [58] L. N. Bykova and S. I. Petrov: *Differentiating Effect of Amphiprotic Solvents in Relation to Phenol and Benzoic Acid Derivatives*. Usp. Khim. *39*, 1631 (1970); Russian Chem. Rev. *39*, 766 (1970). [59] I. M. Kolthoff, S. Bruckenstein, and M. K. Chantooni, J. Am. Chem. Soc. *83*, 3927 (1961). [60] R. H. Boyd: *Acidity Functions*, in J. F. Coetzee and C. D. Ritchie (eds.): *Solute-Solvent Interactions*. Vol. 1, Dekker, New York, London 1969, p. 98ff.

[61] C. Agami and M. Caillot, Bull. Soc. Chim. Fr. *1969*, 1990; C. Agami, ibid. *1969*, 2183. [62] M. L. Martin, Annales de Physique *7*, 35 (1962); G. J. Martin and M. L. Martin, J. Chim. Physique *61*, 1222 (1964). [63] G. N. Lewis: *Valence and the Structure of Atoms and Molecules*. Chemical Catalog Company, New York 1923, p. 141ff.; J. Franklin Inst. *226*, 293 (1938). [64] D. W. Meek: *Lewis Acid-Base Interactions in Polar Non-Aqueous Solvents*, in J. J. Lagowski (ed.): *The Chemistry of Non-Aqueous Solvents*. Academic Press, New York, London 1966, Vol. I, p. 1ff. [65] V. Gutmann: *Coordination Chemistry in Non-Aqueous Solvents*. Springer, Wien, New York 1968; V. Gutmann: *Chemische Funktionslehre*. Springer, Wien, New York 1971. [65a] W. B. Jensen: *The Lewis Acid-Base Concepts – An Overview*. Wiley, Chichester 1980. [66] R. G. Pearson, J. Am. Chem. Soc. *85*, 3533 (1963); Science *151*, 172 (1966); Chemistry in Britain *3*, 103 (1967); J. Chem. Educ. *45*, 581, 643 (1968); ibid. *64*, 561 (1987); R. G. Pearson and J. Songstad, J. Am. Chem. Soc. *89*, 1827 (1967); J. Org. Chem. *32*, 2899 (1967); R. G. Pearson, Survey of Progress in Chemistry *5*, 1 (1969). [67] R. G. Pearson (ed.): *Hard and Soft Acids and Bases*. Dowden, Hutchinson, and Ross, Stroudsburg/Pa. 1973. [68] H. Werner: *Harte und weiche Säuren und Basen – ein neues Klassifizierungsprinzip*. Chemie in unserer Zeit 1, 135 (1967). [69] J. Seyden-Penne: *Notions d'acides et de bases "durs et mous", applications en chimie organique*. Bull. Soc. Chim. Fr. *1968*, 3871. [70] T.-L. Ho: *Hard and Soft Acids and Bases Principle in*

Organic Chemistry. Academic Press, New York 1977; T.-L. Ho, Chem. Rev. *75*, 1 (1975); T.-L. Ho, Tetrahedron *41*, 3 (1985).

[71] R. Th. Myers, Inorg. Chem. *13*, 2040 (1974). [72] R. S. Drago, J. Chem. Educ. *51*, 300 (1974). [73] A. J. Parker, Quart. Rev. (London) *16*, 163 (1962); Usp. Khim. *32*, 1270 (1963); Adv. Org. Chem. *5*, 1 (1965); Adv. Phys. Org. Chem. *5*, 173 (1967); Usp. Khim. *40*, 2203 (1971); Chem. Rev. *69*, 1 (1969); Pure Appl. Chem. *25*, 345 (1971). [74] J. Miller and A. J. Parker, J. Am. Chem. Soc. *83*, 117 (1961); A. J. Parker, J. Chem. Soc. *1961*, 1328. [75] J. F. Coetzee: *Ionic Reactions in Acetonitrile*. Progr. Phys. Org. Chem. *4*, 45 (1967). [76] R. S. Kittila: *A Review of Catalytic and Synthetic Applications for DMF and DMAC*. Bulletin DuPont de Nemours and Co., Wilmington/USA 1960 (and Supplement 1962). [77] J. W. Vaughn: *Amides*, in J. J. Lagowski (ed.): *The Chemistry of Non-Aqueous Solvents*, Academic Press, New York, London 1967, Vol. II, p. 192ff. [78] S. S. Pizey: *Dimethylformamide*, in: *Synthetic Reagents*. Wiley, New York, London 1974, Vol. I. p. 1ff. [79] R. Sowada: *Darstellung, Eigenschaften und Verwendung von Dimethylsulfon*. Z. Chem. *8*, 361 (1968). [80] H. L. Schläfer and W. Schaffernicht: *Dimethylsulfoxid als Lösungsmittel für anorganische Verbindungen*. Angew. Chem. *72*, 618 (1960).

[81] C. Agami: *Le Diméthylsulfoxyde en Chimie Organique*. Bull. Soc. Chim. Fr. *1965*, 1021. [82] D. Martin, A. Weise, and H.-J. Niclas: *Das Lösungsmittel Dimethylsulfoxid*. Angew. Chem. *79*, 340 (1967); Angew. Chem., Int. Ed. Engl. *6*, 318 (1967). [83] D. Martin and H. G. Hauthal: *Dimethylsulfoxid*. Akademie-Verlag, Berlin 1971; *Dimethylsulphoxide*. Van Nostrand Reinhold, London 1975. [84] S. W. Jacob, E. E. Rosenbaum, and D. C. Wood (eds.): *Dimethyl Sulfoxide*. Dekker, New York 1971, Vol. 1. [85] H. Normant: *Hexamethylphosphorsäuretriamid*. Angew. Chem. *79*, 1029 (1967); Angew. Chem., Int. Ed. Engl. *6*, 1046 (1967); Bull. Soc. Chim. Fr. *1968*, 791; Usp. Khim. *39*, 990 (1970); Russian Chem. Rev. *39*, 457 (1970). [86] M. Bréant: *Propriétés chimiques et électrochimiques dans la N-méthylpyrrolidone*. Bull. Soc. Chim. Fr. *1971*, 725. [87] M. Bréant and G. Demange-Guerin: *Propriétés chimiques et électrochimiques dans la nitromethane*. Bull. Soc. Chim. Fr. *1975*, 163. [88] W. H. Lee: *Cyclic Carbonates*, in J. J. Lagowski (ed.): *The Chemistry of Non-aqueous Solvents*. Academic Press, New York, 1976, Vol. IV, p. 167ff. [89] E. M. Arnett and C. F. Douty: *Sulfolane – A Weakly Basic Aprotonic Solvent of High Dielectric Constant*. J. Am. Chem. Soc. *86*, 409 (1964). [90] T. Yamamoto: *Sulfolane*. Yuki Gosei Kagaku Kyokai Shi *28*, 853 (1970); Chem. Abstr. *73*, 109587c (1970). [90a] J. Martinmaa: *Sulfolane*, in J. J. Lagowski (ed.): *The Chemistry of Nonaqueous Solvents*. Academic Press, New York 1976, Vol. IV, p. 248ff.

[91] A. Lüttringhaus and H.-W. Dirksen: *Tetramethylharnstoff als Lösungsmittel und Reaktionspartner*. Angew. Chem. *75*, 1059 (1963). [91a] B. J. Barker, J. Rosenfarb, and J. A. Caruso: *Harnstoffe als Lösungsmittel in der chemischen Forschung*. Angew. Chem. *91*, 560 (1979); Angew. Chem., Int. Ed. Engl. *18*, 503 (1979). [91b] R. J. Lemire and P. G. Sears: *N-Methylacetamide as a Solvent*. Top. Curr. Chem. *74*, 45 (1978). [92] F. Madaule-Aubry: *Le rôle en chimie de certains solvants dipolaires aprotiques*. Bull. Soc. Chim. Fr. *1966*, 1456. [93] C. Agami: *Nouvelle classification des solvants aprotoniques*. Bull. Soc. Chim. Fr. *1967*, 4031. [94] C. D. Ritchie: *Interactions in Dipolar Aprotic Solvents*, in J. F. Coetzee and C. D. Ritchie (eds.): *Solute-Solvent Interactions*. Dekker, New York, London 1969, Vol. 1, p. 219ff. [95] H. Liebig: *Präparative Chemie in aprotonischen Lösungsmitteln*. Chemiker-Ztg. *95*, 301 (1971). [96] E. S. Amis and J. F. Hinton: *Solvent Effects on Chemical Phenomena*. Academic Press, New York, London 1973, Vol. 1, p. 271ff. [97] P. K. Kadaba: *Role of Protic and Dipolar Aprotic Solvents in Heterocyclic Syntheses via 1,3-Dipolar Cycloaddition Reactions*. Synthesis *1973*, 71. [98] J. H. Hildebrand and R. L. Scott: *The Solubility of Nonelectrolytes*, 3rd ed., Reinhold, New York 1950; Dover, New York 1964; J. H. Hildebrand and R. L. Scott: *Regular Solutions*. Prentice-Hall, Englewood Cliffs/New Jersey 1962; J. H. Hildebrand, J. M. Prausnitz, and R. L. Scott: *Regular and Related Solutions*. Van Nostrand-Reinhold, Princeton/New Jersey 1970. [99] A. F. M. Barton: *Handbook of Solubility Parameters and other Cohesion Parameters*. CRC Press, Boca Raton/Florida 1983. [100] M. R. J. Dack, Aust. J. Chem. *28*, 1643 (1975).

[101] M. R. J. Dack: *The Influence of Solvent on Chemical Reactivity*, in M. R. J. Dack (ed.): *Solutions and Solubilities*. Vol. VIII, Part II, p. 95ff., in A. Weissberger (ed.): *Techniques of Chemistry*. Wiley-Interscience, New York 1976. [102] M. Bohle, W. Kollecker, and D. Martin, Z. Chem. *17*, 161 (1977). [103] A. D. Buckingham, E. Lippert, and S. Bratos: *Organic Liquids. Structure, Dynamics, and*

Chemical Properties. Wiley-Interscience, Chichester 1978. [104] D. Stoye: *Lösungsmittel.* In: *Ullmann's Encyklopädie der technischen Chemie.* 4th ed., Vol. 16, p. 279...311, Verlag Chemie, Weinheim, New York 1979. [105] H. Gnamm and O. Fuchs: *Lösungsmittel und Weichmachungsmittel.* 8th ed., Vol. I and II, Wissenschaftliche Verlagsgesellschaft, Stuttgart 1980. [106] I. Bertini, L. Lunazzi, and A. Dei (eds.): *Advances in Solution Chemistry.* Plenum Press, New York, London 1981. [107] O. Popovych and R. P. T. Tomkins: *Nonaqueous Solution Chemistry.* Wiley-Interscience, New York 1981. [108] N. H. March and M. P. Tosi: *Coulomb Liquids.* Academic Press, New York, London 1984. [109] G. W. Gray and J. W. Goodby: *Liquid Crystals. Identification, Classification, and Structure.* Pergamon Press, Oxford 1983. [110] R. Eidenschink: *Flüssige Kristalle.* Chemie in unserer Zeit *18*, 168 (1984); B. Bahadur: *Liquid Crystal Displays.* Mol. Cryst. Liq. Cryst. *109*, 3 (1984).

[111] T. Nakano and H. Hirata, Bull. Chem. Soc. Jpn. *55*, 947 (1982). [112] D. Inman and D. G. Lovering (eds.): *Ionic Liquids.* Plenum Publ. Corp., New York 1981. [113] R. M. Pagni: *Organic and Organometallic Reactions in Molten Salts and Related Melts.* In G. Mamantov, C. B. Mamantov, and J. Braunstein (eds.): *Advances in Molten Salt Chemistry.* Vol 6, Elsevier, Amsterdam 1987. [114] C. L. Hussey: *Room-Temperature Molten Salt Systems.* In G. Mamantov (ed.): *Advances in Molten Salt Chemistry.* Vol. 5, p. 185ff., Elsevier, Amsterdam 1983. [115] L. G. Wade, K. J. Acker, R. A. Earl, and R. A. Osteryoung, J. Org. Chem. *44*, 3724 (1979). [116] L. Gattermann, H. Wieland, Th. Wieland, and W. Sucrow: *Die Praxis des organischen Chemikers.* 43th ed., de Gruyter, Berlin, New York 1982, p. 261. [117] L. K. Nash, J. Chem. Educ. *61*, 981 (1984). [118] L. R. Snyder, CHEMTECH *9*, 750 (1979); ibid. *10*, 188 (1980). [119] N. B. Godfrey, CHEMTECH *1972*, 359; *cf.* also I. Vavruch, Chemie für Labor und Betrieb *35*, 385 (1984). [120] American Society for Testing and Materials: *Annual Book of ASTM Standards 1982.* Part 29, p. 142...144 (ASTM D 1133–78), Philadelphia 1982.

[121] M. Claessens, L. Palombini, M.-L. Stien, and J. Reisse, Nouv. J. Chim. *6*, 595 (1982); M.-L. Stien, M. Claessens, A. Lopez, and J. Reisse, J. Am. Chem. Soc. *104*, 5902 (1982); J. Reisse: *Intermolecular Interaction and Molecular Multipolarity.* Plenary lecture, ESOC IV, Aix-en-Provence, September 4, 1985. [122] M.-E. Colnay, A. Vasseur, and M. Guérin, J. Chem. Res. (S) *1983*, 220; J. Chem. Res. (M) *1983*, 2050. [123] H. Rau: *Asymmetric Photochemistry in Solution.* Chem. Rev. *83*, 535 (1983); *cf.* p. 545. [124] W. H. Pirkle and D. J. Hoover: *NMR Chiral Solvating Agents.* Topics in Stereochemistry *13*, 263 (1982). [125] W. A. König, S. Sievers, and U. Schulze, Angew. Chem. *92*, 935 (1980); Angew. Chem., Int. Ed. Engl. *19*, 910 (1980). [126] M. Bucciarelli, A. Forni, I. Moretti, and G. Torre, J. Org. Chem. *48*, 2640 (1983). [127] L. Jalander and R. Strandberg, Acta Chem. Scand., Part B *37*, 15 (1983). [128] D. Seebach and A. Hidber, Org. Synth. *61*, 42 (1983). [129] P. A. Giguère, J. Chem. Educ. *56*, 571 (1979); *cf.* also Chemiker Ztg. *104*, 172 (1980). [130] L. P. Hammett and A. J. Deyrup, J. Am. Chem. Soc. *54*, 2721 (1932).

[131] G. A. Olah, G. K. Surya Prakash, and J. Sommer: *Superacids.* Science *206*, 13 (1979); *Superacids.* Wiley, Chichester 1985. [132] C. K. Jørgensen: *Super-Acids and Protonated Solvents.* Naturwissenschaften *67*, 188 (1980). [133] B. J. Barker, J. Rosenfarb, and J. A. Caruso, Angew. Chem. *91*, 560 (1979); Angew. Chem., Int. Ed. Engl. *18*, 503 (1979). [134] T. Mukhopadhyay and D. Seebach, Helv. Chim. Acta *65*, 385 (1982); *cf.* also Chimia *39*, 147 (1985), and Chem. Br. *21*, 632 (1985). [135] M. Mashima, R. T. McIver, R. W. Taft, F. G. Bordwell, and W. N. Olmstead, J. Am. Chem. Soc. *106*, 2717 (1984). [136] M. C. R. Symons and V. K. Thomas, J. Chem. Soc., Faraday Trans. I *1981*, 1891. [137] L. B. Kier: *Quantitation of Solvent Polarity Based on Molecular Structure.* J. Pharm. Sci. *70*, 930 (1981); Chem. Abstr. *96*, 6069 h (1982). [138] M. Chastrette, Tetrahedron *35*, 1441 (1979); M. Chastrette and J. Carretto, Tetrahedron *38*, 1615 (1982); M. Chastrette, M. Rajzmann, M. Chanon, and K. F. Purcell, J. Am. Chem. Soc. *107*, 1 (1985). [139] R. D. Cramer III, J. Am. Chem. Soc. *102*, 1837, 1849 (1980). [140] P. Svoboda, O. Pytela, and M. Večeřa, Coll. Czech. Chem. Comm. *48*, 3287 (1983).

[141] J. Elguero and A. Fruchier, Anales de Quimica, Ser. C. *79*, 72 (1983); Chem. Abstr. *100*, 5743 n (1984). [142] R. Carlson, T. Lundstedt, and C. Albano, Acta Chem. Scand., Part B *39*, 79 (1985). [143] P.-C. Maria, J.-F. Gal, J. de Franceschi, and E. Fargin, J. Am. Chem. Soc. *109*, 483 (1987). [144] J. Shorter: *Correlation Analysis of Organic Reactivity. With Particular Reference to Multiple Regression.* Wiley, Chichester 1982. [145] E. R. Malinowsky and D. G. Howery: *Factor Analysis in Chemistry.* Wiley, New York 1980. [146] M. A. Sharaf, D. L. Illman, and B. R. Kowalski:

Chemometrics. Wiley, Chichester, New York 1986. [147] B. G. M. Vandeginste: *Chemometrics – General Introduction and Historical Development*. Top. Curr. Chem. *141*, 1 (1987). [148] M. F. Delaney: *Chemometrics*. Anal. Chem. *56*, 261 R (1984), and references cited therein. [149] S. Wold and M. Sjöström: *LFE Relationships as Tools for Investigating Chemical Similarity. Theory and Practice*. In N. B. Chapman and J. Shorter (eds.): *Correlation Analysis in Chemistry. Recent Advances*. Plenum, New York 1978; Acta Chem. Scand., Part B *35*, 537 (1981). [150] C. Reichardt, Angew. Chem. *91*, 119 (1979); Angew. Chem., Int. Ed. Engl. *18*, 98 (1979).

[151] S. Wold and M. Sjöström in B. R. Kowalski (ed.): *Chemometrics. Theory and Practice*. ACS Symposium Series No. 52, American Chemical Society, Washington/D. C. 1977, Chapter 7. [152] A. Thielemans and D. Luc Massart, Chimia *39*, 236 (1985). [153] C. Hansch, Acc. Chem. Res. *2*, 232 (1969). [154] G. Allen, G. Gee, and G. J. Wilson, Polymer *1*, 456 (1960); *cf.* also reference [99], table 10 in chapter 7, p. 118. [155] W. L. Leigh, D. T. Frendo, and P. J. Klawunn, Can. J. Chem. *63*, 2131 (1985); ibid. *63*, 2736 (1985). [156] S. E. Fry and N. J. Pienta, J. Am. Chem. Soc. *107*, 6399 (1985). [157] D. Enders and R. W. Hoffmann: *Asymmetrische Synthese*. Chemie in unserer Zeit *19*, 177 (1985). [158] D. Seebach and W. Langer, Helv. Chim. Acta *62*, 1701 (1979). [159] C. H. Heathcock: *The Aldol Condensation in Stereoselective Organic Synthesis*. In H. Nozaki (ed.): *Current Trends in Organic Synthesis*. Pergamon Press, Oxford 1983, p. 27ff.; C. H. Heathcock: *Stereoselective Aldol Condensations*. In E. Buncel and T. Durst (eds.): *Comprehensive Carbanion Chemistry*. Part B. p. 177ff., Elsevier, Amsterdam 1984. [160] A. Streitwieser jr., J. I. Brauman, J. H. Hammons, and A. H. Pudjaatmaka, J. Am. Chem. Soc. *87*, 384 (1965); A. Streitwieser jr., E. Juaristi, and L. L. Nebenzahl: *Equilibrium Carbon Acidities in Solution*. In E. Buncel and T. Durst (eds.): *Comprehensive Carbanion Chemistry*. Part A, p. 323ff., Elsevier, Amsterdam 1980.

[161] F. G. Bordwell *et al.*, J. Am. Chem. Soc. *97*, 7006 (1975); F. G. Bordwell: *Equilibrium Acidities of Carbon Acids*. Pure Appl. Chem. *49*, 963 (1977); F. G. Bordwell, G. E. Drucker, N. H. Andersen, and A. D. Denniston, J. Am. Chem. Soc. *108*, 7310 (1986). [162] J. A. Boon, J. A. Levisky, J. L. Pflug, and J. S. Wilkes, J. Org. Chem. *51*, 480 (1986). [163] V. Ramesh and R. G. Weiss, J. Org. Chem. *51*, 2535 (1986). [164] H. U. Borgstedt: *Chemical Reactions in Alkali Metals*. Top. Curr. Chem. *134*, 125 (1986). [165] R. Carlson, T. Lundstedt, and R. Shabana, Acta Chem. Scand., Part B *40*, 694 (1986); ibid. *41*, 164 (1987). [166] J. Michl and E. W. Thulstrup: *Spectroscopy with Polarized Light – Solute Alignment by Photoselection, in Liquid Crystals, Polymers, and Membranes*. VCH Verlagsgesellschaft, Weinheim 1986. [167] D. Appleby, C. L. Hussey, K. R. Seddon, and J. E. Turp, Nature (London) *323*, 614 (1986). [168] L. Šafařik and Z. Stránský: *Titrimetric Analysis in Organic Solvents*. In G. Svehla (ed.): *Wilson and Wilson's Comprehensive Analytical Chemistry*. Vol. XXII, Chapters 1 . . . 3, Elsevier, Amsterdam, New York 1986. [169] S. D. Williams, J. P. Schoebrechts, J. C. Selkirk, and G. Mamantov, J. Am. Chem. Soc. *109*, 2218 (1987). [170] R. G. Pearson, J. Chem. Educ. *64*, 561 (1987). [171] Y. Marcus, J. Phys. Chem. *91*, 4422 (1987).

Chapter 4

[1] J. E. Leffler and E. Grunwald: *Rates and Equilibria of Organic Reactions*. Wiley, New York, London 1963, p. 15ff. [2] K. B. Wiberg: *Physical Organic Chemistry*. Wiley, New York 1964, p. 253ff. [3] E. A. Moelwyn-Hughes: *The Chemical Statics and Kinetics of Solutions*. Academic Press, London, New York 1971, chapter 3, p. 33ff. [4] J. Hine: *Structural Effects on Equilibria in Organic Chemistry*. Wiley-Interscience, New York 1975, p. 130ff. [5] H. Martin: *Über Beziehungen zwischen Lösungs- und Gasphasenreaktionen*. Angew. Chem. *78*, 73 (1966); Angew. Chem., Int. Ed. Engl. *5*, 78 (1966). [6] S. W. Benson and G. D. Mendenhall: *Comparison of Equilibrium Reactions in the Gaseous and Liquid Phases*. J. Am. Chem. Soc. *98*, 2046 (1976). [7] A. Ben-Naim: *Solvent and Solute Effects on Chemical Equilibria*. J. Chem. Phys. *63*, 2064 (1975). [8] I. M. Kolthoff, M. K. Chantoni, A. I. Popov, J. A. Caruso, and J. Steigman: *Acid-Base Equilibria in Non-Aqueous Solvents*. In I. M. Kolthoff and P. J. Elving (eds.): *Treatise on Analytical Chemistry*. 2nd ed., Part I, Vol. 2, p. 237ff., Wiley, New York 1979. [9] R. P. Bell: *The Proton in Chemistry*. Methuen, London 1959, p. 36ff. [10] E. J. King: *Acid-Base Equilibria*. Pergamon Press, Oxford 1965, Chapter 11, p. 280ff.

[11] G. Charlot and B. Trémillon: *Réactions Chimiques dans les Solvants et les Sels Fondus*. Gauthier-Villars, Paris 1963; *Chemical Reactions in Solvents and Melts*. Pergamon Press, Oxford 1969, Chapter 1 and 2, p. 10ff. and 45ff. [12] A. Coulombeau: *Effets de Solvant sur les Equilibres Acides-Bases*. Comissariat à l'Energie Atomique – France, Service Central de Documentation du C. E. A., Bibliographie CEA-BIB-147 (1969); Chem. Abstr. *72*, 6691j (1970). [13] See also references [44, 45, 49, 55, 56] in reference list to Chapter 3. [14] M. Born, Z. Physik *1*, 45 (1920); *cf.* also A. R. Rashin and B. Honig, J. Phys. Chem. *89*, 5588 (1985). [15] L. P. Hammett and A. J. Deyrup, J. Am. Chem. Soc. *54*, 2721 (1932); L. P. Hammett: *Physical Organic Chemistry*. McGraw-Hill, New York 1970; *Physikalische Organische Chemie*. Verlag Chemie, Weinheim 1973. [16] B. Gutbezahl and E. Grunwald, J. Am. Chem. Soc. *75*, 559, 565 (1953). [17] R. G. Bates: *Acidity Functions in Aqueous and Nonaqueous Media*. Bull. Soc. Chim. Belg. *84*, 1139 (1975). [18] E. M. Arnett: *Gas-Phase Proton Transfer – a Breakthrough for Solution Chemistry*. Acc. Chem. Res. *6*, 404 (1973). [19] C. Agami: *Acidités et Basicités en Phase Vapeur*. Bull. Soc. Chim. Fr. *1974*, 869. [20] O. A. Reutov, K. P. Butin, and I. P. Beletskaya: *Equilibrium Acidity of Carbon-hydrogen Bonds in Organic Compounds*. Usp. Khim. *43*, 35 (1974); Russian Chem. Rev. *43*, 17 (1974).

[21] R. W. Taft: *Gas Phase Proton Transfer Equilibria*, in E. F. Caldin and V. Gold (eds.): *Proton Transfer Reactions*. Chapman and Hall, London 1975. [22] M. S. B. Munson, J. Am. Chem. Soc. *87*, 2332 (1965). [23] P. Kebarle: *Ions and Ion-Solvent Molecule Interactions in the Gas Phase*, in M. Szwarc (ed.): *Ions and Ion Pairs in Organic Reactions*. Wiley-Interscience, New York 1972, Vol. I, p. 27ff. [24] H. Hartmann, K.-H. Lebert, and K.-P. Wanczek: *Ion Cyclotron Resonance Spectroscopy*. Fortschr. Chem. Forsch. *43*, 57 (1973). [25] R. C. Dunbar: *Ion Cyclotron Resonance and Fourier-Transform Mass Spectrometry*. In C. F. Bernasconi (ed.): *Investigation of Rates and Mechanisms of Reactions* (Volume VI of the Series *Techniques of Chemistry*). 4th ed., Wiley, New York 1986, Part I, p. 903ff. [26] T. A. Lehman and M. M. Bursey: *Ion Cyclotron Resonance Spectrometry*. Wiley-Interscience, New York 1976, p. 55ff. [27] H. C. Brown, H. Bartholomay, and M. D. Taylor, J. Am. Chem. Soc. *66*, 435 (1944). [28] J. I. Brauman and L. K. Blair, J. Am. Chem. Soc. *91*, 2126 (1969); J. I. Brauman, J. M. Riveros, and L. K. Blair, ibid. *93*, 3914 (1971). [29] E. M. Arnett, F. M. Jones, M. Taagepera, W. G. Henderson. J. L. Beauchamp, D. Holtz, and R. W. Taft, J. Am. Chem. Soc. *94*, 4724 (1972). [30] D. H. Aue, H. M. Webb, and M. T. Bowers, J. Am. Chem. Soc. *94*, 4726 (1972).

[31] A. F. Trotman-Dickenson, J. Chem. Soc. *1949*, 1293. [32] J. Long and B. Munson, J. Am. Chem. Soc. *95*, 2427 (1973). [33] M. Taagepera, W. G. Henderson, R. T. C. Brownlee, J. L. Beauchamp, D. Holtz, and R. W. Taft, J. Am. Chem. Soc. *94*, 1369 (1972). [34] J. I. Brauman and L. K. Blair, J. Am. Chem. Soc. *90*, 6561 (1968); ibid. *92*, 5986 (1970). [34a] R. Yamdagni and P. Kebarle, J. Am. Chem. Soc. *95*, 4050 (1973); K. Hiraoka, R. Yamdagni, and P. Kebarle, ibid. *95*, 6833 (1973); *cf.* also P. Haberfield and A. K. Rakshit, ibid. *98*, 4393 (1976). [35] J. Hine and M. Hine, J. Am. Chem. Soc. *74*, 5266 (1952). [36] *Cf.* references [11–19] in reference list to Chapter 1. [37] O. Dimroth, Liebigs Ann. Chem. *377*, 127 (1910); *399*, 91 (1913); *438*, 58 (1924). [38] H. Henecka: *Chemie der β-Dicarbonylverbindungen*. Springer, Berlin 1950, p. 7ff. [39] G. Briegleb and W. Strohmeier: *Prinzipielle kritische Bemerkungen zur Theorie der Keto-Enol-Umwandlung*. Angew. Chem. *64*, 409 (1952). [40] G. W. Wheland: *Advanced Organic Chemistry*. 3rd ed., Wiley, New York, London 1960, Chapter 14, p. 663ff.

[41] H. A. Staab: *Einführung in die theoretische organische Chemie*. 4th ed., Verlag Chemie, Weinheim 1964, p. 642ff. [42] S. Forsén and M. Nilsson: *Enolization*, in J. Zabicky (ed.): *The Chemistry of the Carbonyl Group*. Wiley-Interscience, New York 1970, Vol. 2, p. 157ff. [43] A. I. Kol'tsov and G. M. Kheifets: *Investigation of Keto-Enol-Tautomerism by NMR Spectroscopy*. Usp. Khim. *40*, 1646 (1971); Russian Chem. Rev. *40*, 773 (1971). [44] R. Matusch, Angew. Chem. *87*, 283 (1975); Angew. Chem., Int. Ed. Engl. *14*, 260 (1975). [45] R. S. Noy, V. A. Gindin, B. A. Ershov, A. I. Kol'tsov, and V. A. Zubkov, Org. Magn. Reson. *7*, 109 (1975). [46] B. Eistert and W. Riess, Chem. Ber. *87*, 92 (1954); B. Eistert and F. Geiss, Tetrahedron *7*, 1 (1959); B. Eistert and K. Schank, Tetrahedron Lett. 1964, 429. [47] M. T. Rogers and J. L. Burdett, J. Am. Chem. Soc. *86*, 2105 (1964); Can. J. Chem. *43*, 1516 (1965). [48] G. Allen and R. A. Dwek, J. Chem. Soc. *B 1966*, 161. [49] K. L. Lockwood, J. Chem. Educ. *42*, 481 (1965). [50] E. J. Drexler and K. W. Field, J. Chem. Educ. *53*, 392 (1976).

[51] G. Schwarzenbach and E. Felder, Helv. Chim. Acta *27*, 1044 (1944). [52] A. Yogev and Y. Mazur, J. Org. Chem. *32*, 2162 (1967). [53] P. B. Russell and J. Mentha, J. Am. Chem. Soc. *77*,4245 (1955). [54] A new approach deriving the van't Hoff-Dimroth relationship was given by S. Tanaka, Bull. Chem. Soc. Jpn. *39*, 84 (1966). [55] H. Mauser and B. Nickel, Chem. Ber. *97*, 1745, 1753 (1964). [56] J. G. Kirkwood, J. Chem. Phys. *2*, 351 (1934). [57] J. Powling and H. J. Bernstein, J. Am. Chem. Soc. *73*, 4353 (1951). [58] J. E. Dubois and J. Toullec, Tetrahedron *29*, 2859 (1973). [59] R. Hagen, E. Heilbronner, and P. A. Straub, Helv. Chim. Acta *50*, 2504 (1967). [59a] K. Hafner in R. N. Castle and M. Tišler (eds.): *Lectures in Heterocyclic Chemistry*. Vol. III, p. 33ff.; Supplementary Issue to J. Heterocycl. Chem. *13* (1976). [60] H. Baba and T. Takemura, Tetrahedron *24*, 4779 (1968).

[61] H. Sterk, Monatsh. Chem. *100*, 916 (1969). [62] A. I. Kol'tsov and G. M. Kheifets: *Study of Tautomerism by NMR Spectroscopy*. Usp. Khim. *41*, 877 (1972); Russian Chem. Rev. *41*, 452 (1972). [63] A. R. Katritzky and J. M. Lagowski: *Prototropic Tautomerism of Heteroaromatic Compounds I-IV*, in A. R. Katritzky (ed.): Advances in Heterocyclic Chemistry. Academic Press, New York, London 1963, Vol. 1, p. 311 and 339; Vol. 2, p. 1 and 27. [64] J. Elguero, C. Marzin, A. R. Katritzky, and P. Linda, *The Tautomerism of Heterocycles*, in A. R. Katritzky and A. J. Boulton (eds.): Advances in Heterocyclic Chemistry, Supplement 1. Academic Press, New York 1976. [65] P. Beak and F. S. Fry, J. Am. Chem. Soc. *95*, 1700 (1973); P. Beak, F. S. Fry, J. Lee, and F. Steele, ibid. *98*, 171 (1976); P. Beak, J. B. Covington, and S. G. Smith, ibid. *98*, 8284 (1976); P. Beak, Acc. Chem. Res. *10*, 186 (1977); P. Beak, J. B. Covington, and J. M. Zeigler, J. Org. Chem. *43*, 177 (1978). [65a] A. Maquestiau, Y. van Haverbeke, C. de Meyer, A. R. Katritzky, M. J. Cook, and A. D. Page, Can. J. Chem. *53*, 490 (1975); Bull. Soc. Chim. Belg. *84*, 465 (1975). [65b] M. J. Cook, S. El-Abbady, A. R. Katritzky, C. Guimon, and G. Pfister-Guillouzo, J. Chem. Soc., Perkin Trans. II *1977*, 1652. [66] G. Simchen, Chem. Ber. *103*, 398 (1970). [67] A. Gordon and A. R. Katritzky, Tetrahedron Lett. *1968*, 2767; J. Frank and A. R. Katritzky, J. Chem. Soc., Perkin Trans. II *1976*, 1428. [68] E. Daltrozzo, G. Hohlneicher, and G. Scheibe, Ber. Bunsenges. Phys. Chem. *69*, 190 (1965); G. Scheibe and E. Daltrozzo: *Diquinolylmethane and Its Analogs*, in A. R. Katritzky and A. J. Boulton (eds.): Advances in Heterocyclic Chemistry. Academic Press, New York, London 1966, Vol. 7, p. 153ff. [69] H. Ahlbrecht, Tetrahedron Lett. *1968*, 4421; H. Ahlbrecht and S. Fischer, Tetrahedron *26*, 2837 (1970). [69a] R. Scheffold, J. Löliger, H. Blaser, and P. Geisser, Helv. Chim. Acta *58*, 49 (1975). [69b] M. Dreyfus, G. Dodin, O. Bensaude, and J. E. Dubois, J. Am. Chem. Soc. *98*, 6338 (1976); ibid. *99*, 7027 (1977). [70] M. Mousseron-Canet, J. P. Boca, and V. Tabacik, Spectrochim. Acta, Part A *23*, 717 (1967).

[71] I. Ya. Bershtein and O. F. Ginsburg: *Tautomerism of Aromatic Azo-Compounds*. Usp. Khim. *41*, 177 (1972); Russian Chem. Rev. *41*, 97 (1972). [72] A. Burawoy and A. R. Thompson, J. Chem. Soc. *1953*, 1443. [73] B. L. Kaul, P. M. Nair, A. V. R. Rao, and K. Venkataraman, Tetrahedron Lett. *1966*, 3897. [74] R. Korewa and H. Urbanska, Rocz. Chem. *46*, 2007 (1972); Chem. Abstr. *78*, 147128 s (1973). [74a] M. Misuishi, R. Kamimura, K. Shinohara, and N. Ishii, Sen'i Gakkaishi *32*, T 382 (1976); Chem. Abstr. *85*, 194065y (1976). [74b] R. Hempel, H. Viola, J. Morgenstern, and R. Mayer, J. Prakt. Chem. *318*, 983 (1976). [75] R. K. Norris and S. Sternhell, Aust. J. Chem. *19*, 841 (1966). [76] H. Uffmann, Z. Naturforsch. *22b*, 491 (1967). [76a] Ya. I. Shpinel' and Yu. I. Tarnopol'skii, Zh. Org. Khim. *13*, 1030 (1977); J. Org. Chem. USSR *13*, 948 (1977). [77] R. Escale and J. Verducci: *Facteurs influencant la tautomérie cycle-chaîne*. Bull. Soc. Chim. Fr. *1974*, 1203. [78] H. Sterk, Monatsh. Chem. *99*, 1764 (1968). [79] J. Whiting and J. T. Edward, Can. J. Chem. *49*, 3799 (1971). [80] S. J. Angyal, Adv. Carbohydrate Chem. Biochem. *42*, 15 (1984). [80a] H. Möhrle, M. Lappenberg, and D. Wendisch, Monatsh. Chem. *108*, 273 (1977). [80b] H. B. Stegmann, R. Haller, and K. Scheffler, Chem. Ber. *110*, 3817 (1977).

[81] R. U. Lemieux, Pure Appl. Chem. *25*, 527 (1971); ibid. *27*, 527 (1971). [82] E. L. Eliel: *Konformationsanalyse an heterocyclischen Systemen – Neuere Ergebnisse und Anwendungen*. Angew. Chem. *84*, 779 (1972); Angew. Chem., Int. Ed. Engl. *11*, 739 (1972). [83] R. J. Abraham and E. Bretschneider: *Medium Effects on Rotational and Conformational Equilibria*, in W. J. Orville-Thomas (ed.): *Internal Rotation in Molecules*. Wiley-Interscience, New York, 1974, p. 481ff. [84] G. Maier: *Valenzisomerisierungen*. Verlag Chemie, Weinheim 1972. [85] U. Mayer: *Ionic Equilibria in Donor Solvents*. Pure Appl. Chem. *41*, 291 (1975). [86] S. D. Christian and E. H. Lane: *Solvent Effects on Molecular Complex Equilibria*, in M. R. J. Dack (ed.): *Solutions and Solubilities*. Wiley-Interscience,

New York 1975; Vol. VIII, Part 1 of A. Weissberger (ed.): *Techniques of Chemistry*, p. 327ff. [87] G. J. Karabatsos and D. J. Fenoglio, J. Am. Chem. Soc. *91*, 1124 (1969). [88] R. J. Abraham and M. A. Cooper, J. Chem. Soc. *B 1967*, 202. [89] E. L. Eliel, Pure Appl. Chem. *25*, 509 (1971); E. L. Eliel and O. Hofer, J. Am. Chem. Soc. *95*, 8041 (1973). [90] *Cf.* Table 13.23, p. 555, in reference [83].

[91] P. Klaboe, J. J. Lothe, and K. Lunds, Acta Chem. Scand. *11*, 1677 (1957). [92] K. Kozima and K. Sabashita, Bull. Chem. Soc. Jpn. *31*, 796 (1958). [93] R. U. Lemieux and J. W. Loan, Can. J. Chem. *42*, 893 (1964). [94] O. D. Ul'yanova, M. K. Ostrovskii, and Yu. A. Pentin, Zh. Fiz. Khim. *44*, 1013 (1970); Russian J. Phys. Chem. *44*, 562 (1970); and references cited therein. [94a] K. W. Baldry, M. H. Gordon, R. Hafter, and M. J. T. Robinson, Tetrahedron *32*, 2589 (1976). [94b] J. G. Vinter and H. M. R. Hoffmann, J. Am. Chem. Soc. *96*, 5466 (1974). [95] J. P. Snyder, Tetrahedron Lett. *1971*, 215. [96] C. J. Devlin and B. J. Walker, Tetrahedron Lett. *1971*, 4923. [97] R. E. Lutz and A. B. Turner, J. Org. Chem. *33*, 516 (1968). [98] R. Huisgen, W. Scheer, and H. Mäder, Angew. Chem. *81*, 619 (1969); Angew. Chem., Int. Ed. Engl. *8*, 602 (1969). [99] J. B. Flannery, J. Am. Chem. Soc. *90*, 5660 (1968); *cf.* also Y. Sueishi, M. Ohcho, and N. Nishimura, Bull. Chem. Soc. Jpn. *58*, 2608 (1985). [100] J. Elguero, R. Faure, J. P. Galy, and E. J. Vincent, Bull. Soc. Chim. Belg. *84*, 1189 (1975); L. A. Burke, J. Elguero, G. Leroy, and M. Sana, J. Am. Chem. Soc. *98*, 1985 (1976).

[101] E. Vogel, W. A. Böll, and H. Günther, Tetrahedron Lett. *1965*, 609; E. Vogel and H. Günther, Angew. Chem. *79*, 429 (1967); Angew. Chem., Int. Ed. Engl. *6*, 385 (1967). [102] E. Buncel and H. Wilson, Acc. Chem. Res. *12*, 42 (1979); J. Chem. Educ. *57*, 629 (1980). [103] M. H. Abraham, Progr. Phys. Org. Chem. *11*, 2 (1974). [104] R. Stewart: *The Proton: Applications to Organic Chemistry*. Academic Press, Orlando 1985. [105] K. M. Dyumaev and B. A. Korolev: *The Influence of Solvation on the Acid-Base Properties of Organic Compounds in Various Media*. Usp. Khim. *49*, 2065 (1980); Russian Chem. Rev. *49*, 1021 (1980). [106] M. I. Kabachnik: *Progress in the Theory of Acids and Bases*. Usp. Khim. *48*, 1523 (1979); Russian Chem. Rev. *49*, 814 (1979). [107] H. L. Finston and A. C. Rychtman: *A New View of Current Acid-Base Theories*. Wiley, Chichester 1982. [108] F. Strohbusch: *Neue Erkenntnisse der Säure-Basen-Theorie*. Chemie in unserer Zeit *16*, 103 (1982). [109] R. A. Cox and K. Yates: *Acidity Functions – An Update*. Can. J. Chem. *61*, 2225 (1983). [110] E. P. Serjeant and B. Dempsey: *Ionisation Constants of Organic Acids in Aqueous Solution*. Pergamon Press, Oxford 1979.

[111] A. Albert and E. P. Serjeant: *The Determination of Ionisation Constants – A Laboratory Manual*. 3rd ed., Chapman and Hall, London, New York 1984. [112] D. D. Perrin, B. Dempsey, and E. P. Serjeant: *pK_a-Prediction for Organic Acids and Bases*. Chapman and Hall, London 1981. [113] A. Streitwieser, E. Juaristi, and L. L. Nebenzahl: *Equilibrium Carbon Acidities in Solution*. In E. Buncel and T. Durst (eds.): *Comprehensive Carbanion Chemistry*. Part A, p. 323ff., Elsevier, Amsterdam 1980. [114] J. H. Futrell: *Gaseous Ion Chemistry and Mass Spectrometry*. Wiley-Interscience, New York 1986. [115] R. T. McIver: *Chemical Reactions without Solvation*. Scientific American *243*, 148 (1980); *Chemische Reaktionen ohne Lösungsmittel*. Spektrum der Wissenschaft, January 1981, p. 27ff. [116] R. W. Taft: *Protonic Acidities and Basicities in the Gas Phase and in Solution: Substituent and Solvent Effects*. Progr. Phys. Org. Chem. *14*, 247 (1983). [117] C. R. Moylan and J. I. Brauman: *Gas Phase Acid-Base Chemistry*. Annu. Rev. Phys. Chem. *34*, 187 (1983). [118] P. Kebarle: *Ion Thermochemistry and Solvation from Gas Phase Ion Equilibrium*. Annu. Rev. Phys. Chem. *28*, 445 (1977). [119] D. K. Bohme, E. Lee-Ruff, and L. B. Young, J. Am. Chem. Soc. *94*, 5153 (1972); D. K. Bohme in P. Ausloos (ed.): *Interactions between Ions and Molecules*. Plenum, New York 1974, p. 489ff. [120] M. J. Pellerite and J. I. Brauman: *Gas-Phase Acidities of Carbon Acids*. In E. Buncel and T. Durst (eds.): *Comprehensive Carbanion Chemistry*. Part A, p. 55ff., Elsevier, Amsterdam 1980.

[121] J. E. Bartmess and R. T. McIver: *The Gas-Phase Acidity Scale*. In M. T. Bowers (ed.): *Gas Phase Ion Chemistry*. Vol. 2, p. 87ff., Academic Press, New York 1979. [122] T. B. McMahon and P. Kebarle, J. Am. Chem. Soc. *107*, 2612 (1985); and references cited therein. [123] O. A. Reutov, I. P. Beletskaya, and K. P. Butin: *CH-Acids*. Pergamon Press, Oxford 1978. [124] F. G. Bordwell, Pure Appl. Chem. *49*, 963 (1977); F. G. Bordwell, G. E. Drucker, N. H. Andersen, and A. D. Denniston, J. Am. Chem. Soc. *108*, 7310 (1986). [125] J. E. Bartmess, J. A. Scott, and R. T. McIver, J. Am. Chem. Soc. *101*, 6046, 6056 (1979). [126] G. Boand, R. Houriet, and T. Gäumann, J. Am. Chem. Soc. *105*, 2203 (1983). [127] W. N. Olmstead, Z. Margolin, and F. G. Bordwell, J. Org. Chem. *45*, 3295 (1980).

[128] G. Caldwell, M. D. Rozeboom, J. P. Kiplinger, and J. E. Bartmess, J. Am. Chem. Soc. *106*, 4660 (1984). [129] R. T. McIver, J. A. Scott, and J. M. Riveros, J. Am. Chem. Soc. *95*, 2706 (1973). [130] D. K. Bohme, A. B. Rakshit, and G. I. Mackay, J. Am. Chem. Soc. *104*, 1100 (1982).

[131] M. Mashima, R. T. McIver, R. W. Taft, F. G. Bordwell, and W. N. Olmstead, J. Am. Chem. Soc. *106*, 2717 (1984). [132] G. Caldwell, T. B. McMahon, P. Kebarle, J. E. Bartmess, and J. P. Kiplinger, J. Am. Chem. Soc. *107*, 80 (1985). [133] M. J. Locke and R. T. McIver, J. Am. Chem. Soc. *105*, 4226 (1983). [134] S. G. Mills and P. Beak: *Solvent Effects on Keto-Enol-Equilibria: Tests of Quantitative Models.* J. Org. Chem. *50*, 1216 (1985); *cf.* also K. Almdal, H. Eggert, and O. Hammerich, Acta Chem. Scand., Part B *40*, 230 (1986), as well as M. Moriyasu, A. Kato, and Y. Hashimoto, J. Chem. Soc., Perkin Trans. II *1986*, 515. [135] J. N. Spencer, E. S. Holmboe, M. R. Kirshenbaum, D. W. Firth, and P. B. Pinto, Can. J. Chem. *60*, 1178 (1982). [136] K. D. Grande and S. M. Rosenfeld, J. Org. Chem. *45*, 1626 (1980). [137] H. Hart: *Simple Enols.* Chem. Rev. *79*, 515 (1979). [138] M. Melzig, S. Schneider, F. Dörr, and E. Daltrozzo, Ber. Bunsenges. Phys. Chem. *84*, 1108 (1980). [139] R. Roussel, M. Oteyza de Guerrero, P. Spegt, and J. C. Galin, J. Heterocycl. Chem. *19*, 785 (1982). [140] B. A. Shainyan and A. N. Mirskova: *The Carbon-Nitrogen Triad Prototropic Tautomerism.* Usp. Khim. *48*, 201 (1979); Russian Chem. Rev. *48*, 107 (1979).

[141] P. Beak, J. B. Covington, and J. M. White, J. Org. Chem. *45*, 1347 (1980). [142] P. Beak, J. B. Covington, S. G. Smith, J. M. White, and J. M. Zeigler, J. Org. Chem. *45*, 1354 (1980). [143] M. Chevrier, O. Bensaude, J. Guillerez, and J. E. Dubois, Tetrahedron Lett. *21*, 3359 (1980). [144] M. J. Scanlan, I. H. Hillier, and R. H. Davies, J. Chem. Soc., Chem. Commun. *1982*, 685. [145] G. Pfister-Guillouzo, C. Guimon, J. Frank, J. Ellison, and A. R. Katritzky, Liebigs Ann. Chem. *1981*, 366. [146] A. I. Artemenko, E. K. Anufriev, I. V. Tikunova, and O. Exner, Zh. Prikl. Spektrosk. *33*, 131 (1980); Chem. Abstr. *94*, 46316 b (1981). [147] H. Ahlbrecht and R.-D. Kalas, Liebigs Ann. Chem. *1979*, 102. [148] S. Ogawa and S. Shiraishi, J. Chem. Soc., Perkin Trans. I *1980*, 2527. [149] A. L. Weis, Tetrahedron Lett. *23*, 449 (1982). [150] P. Ball and C. H. Nicholls: *Azo-Hydrazone Tautomerism of Hydroxyazo Compounds – A Review.* Dyes Pigm. *3*, 5 (1982); Chem. Abstr. *96*, 124495s (1982).

[151] S. Kishimoto, S. Kitahara, O. Manabe, and H. Hiyama, J. Org. Chem. *43*, 3882 (1978). [152] E. Hofer, Chem. Ber. *112*, 2913 (1979). [153] G. V. Sheban, B. E. Zaitsev, and K. M. Dyumaev, Teor. Eksp. Khim. *16*, 249 (1980); Chem. Abstr. *93*, 113715 a (1980). [154] J. Kelemen, Dyes Pigm. *2*, 73 (1981); Chem. Abstr. *95*, 117036 u (1981). [155] O. I. Kolodyazhnyi, Tetrahedron Lett. *23*, 499 (1982). [156] A. Munoz, Bull. Soc. Chim. Fr. *1977*, 728. [157] F. G. Kamaev, N. I. Baram, A. I. Ismailov, V. B. Leont'ev, and A. S. Sadykov, Izv. Akad. Nauk SSSR, Ser. Khim. *1979*, 1003; Chem. Abstr. *91*, 55761 h (1979). [158] S. Chimichi, R. Nesi, M. Scotton, C. Mannucci, and G. Adembri, Gazz. Chim. Ital. *109*, 117 (1979); J. Chem. Soc., Perkin Trans. II *1980*, 1339. [159] F. Franks: *Physical Chemistry of Small Carbohydrates – Equilibrium Solution Properties.* Pure Appl. Chem. *59*, 1189 (1987); and references cited therein. [160] L. A. Fedorov, D. N. Kravtsov, and A. S. Peregudov: *Metallotropic Tautomeric Transformations of the σ,σ-Type in Organometallic and Complex Compounds.* Usp. Khim. *50*, 1304 (1981); Russian Chem. Rev. *50*, 682 (1981).

[161] I. F. Lutsenko, Yu. I. Baukov, and I. Yu. Belavin, J. Organomet. Chem. *24*, 359 (1970). [162] G. Boche, F. Heidenhain, and B. Staudigl, Angew. Chem. *91*, 228 (1979); Angew. Chem., Int. Ed. Engl. *18*, 218 (1979). [163] J. Malecki: *Solvent Effects on Molecular Complexes.* In H. Ratajczak and W. J. Orville-Thomas (eds.): *Molecular Interactions.* Vol. 3, Chapter 4, p. 183ff., Wiley-Interscience, New York 1982. [164] E. A. Yerger and G. M. Barrow, J. Am. Chem. Soc. *76*, 5211 (1954); ibid. *77*, 4474, 6206 (1955); ibid. *78*, 5802 (1956). [165] H. Baba, A. Matsuyama, and H. Kokubun, J. Chem. Phys. *41*, 895 (1964); Spectrochim. Acta, Part A *25*, 1709 (1969). [166] L. Elégant, J. Fidanza, J.-F. Gal, and S. N. Vinogradov, J. Chim. Phys. *75*, 914 (1978). [167] J. Fritsch and G. Zundel, J. Phys. Chem. *85*, 556 (1981); R. Krämer and G. Zundel, Z. Phys. Chem. (Frankfurt) N. F. *144*, 265 (1985). [168] Z. Dega-Szafran and E. Dulewicz, Org. Magn. Reson. *16*, 214 (1981). [169] A. Koll, M. Rospenk, and L. Sobczyk, J. Chem. Soc., Faraday Trans. I *1981*, 2309; A. Sucharda-Sobczyk and L. Sobczyk, J. Chem. Res. (S) *1985*, 208. [170] M. Rospenk, J. Fritsch, and G. Zundel, J. Phys. Chem. *88*, 321 (1984); M. Rospenk and T. Zeegers-Huyskens, Spectrochim. Acta, Part A *42*, 499 (1986).

[171] B.-L. Poh and H.-L. Siow, Aust. J. Chem. *33*, 491 (1980). [172] F. M. Menger and T. P. Singh, J. Org. Chem. *45*, 183 (1980). [173] H. Rosotti: *The Study of Ionic Equilibria. An Introduction.*

Longman, London 1978. [174] V. Gutmann: *The Donor-Acceptor Approach to Molecular Interactions.* Plenum Press, New York 1978. [175] I. Rosenthal, P. Peretz, and K. A. Muszkat, J. Phys. Chem. *83*, 351 (1979). [176] D. Fompeydie and P. Levillain, Bull. Soc. Chim. Fr. *1980*. I-459. [177] M. Feigel, H. Kessler, and A. Walter, Chem. Ber. *111*, 2947 (1978). [178] M. Feigel, H. Kessler, D. Leibfritz, and A. Walter, J. Am. Chem. Soc. *101*, 1943 (1979); *cf.* also H. Kessler and M. Feigel, Acc. Chem. Res. *15*, 2 (1982). [179] E. M. Arnett and E. B. Troughton, Tetrahedron Lett. *24*, 3299 (1983). [180] E. B. Troughton, K. E. Molter, and E. M. Arnett, J. Am. Chem. Soc. *106*, 6726 (1984); E. M. Arnett and K. E. Molter, Acc. Chem. Res. *18*, 339 (1985); J. Phys. Chem. *90*, 383 (1986).

[181] V. N. Sheinker, A. D. Garnovskii, and O. A. Osipov: *Advances in the Study of s-cis-trans-Isomerism – Stereochemistry of Carbonyl Derivatives of Five-Membered Heterocycles.* Usp. Khim. *50*, 632 (1981); Russian Chem. Rev. *50*, 336 (1981). [182] W. L. Jorgensen: *Theoretical Studies of Medium Effects on Conformational Equilibria.* J. Phys. Chem. *87*, 5304 (1983). [183] J. Fruwert, H. Böhlig, and G. Geiseler: *Gehemmte innere Rotation und Flexibilität kettenförmiger Moleküle.* Z. Chem. *25*, 41 (1985). [184] V. V. Samoshin and N. S. Zefirov: *Conformational Transformations of Organic Molecules in Solutions.* Zh. Vses. Khim. O-va. im. D. I. Mendeleeva (Moskva) *29*, 521 (1984); Chem. Abstr. *102*, 5205 e (1985). [185] R. J. Abraham and T. M. Siverns, Tetrahedron *28*, 3015 (1972). [186] G. Alberghina, F. A. Bottino, S. Fisichella, and C. Arnone, J. Chem. Res. (S) *1985*, 108; J. Chem. Res. (M) *1985*, 1201. [187] C. Lopez-Mardomingo, R. Perez-Ossorio, and J. Plumet, J. Chem. Res. (S) *1983*, 150. [188] Z. Friedl, P. Fiedler, and O. Exner, Coll. Czech. Chem. Commun. *45*, 1351 (1980); Z. Friedl, P. Fiedler, J. Biroš, V. Uchytilová, I. Tvaroška, S. Böhm, and O. Exner, ibid. *49*, 2050 (1984). [189] F. M. Menger and B. Boyer, J. Org. Chem. *49*, 1826 (1984). [190] L. Došen-Mićović and N. L. Allinger, Tetrahedron *34*, 3385 (1978).

[191] R. J. Abraham and L. Griffiths, Tetrahedron *37*, 575 (1981). [192] J. J. Moura-Ramos, L. Dumont, M. L. Stien, and J. Reisse, J. Am. Chem. Soc. *102*, 4150 (1980). [193] M. J. Cook, M. H. Abraham, L. E. Xodo, R. Cruz, Tetrahedron Lett. *22*, 2991 (1981). [194] M. H. Abraham, L. E. Xodo, R. J. Abraham, and M. J. Cook, Tetrahedron Lett. *22*, 5183 (1981). [195] M. Manoharan, E. L. Eliel, and F. I. Carroll, Tetrahedron Lett. *24*, 1855 (1983). [196] V. V. Samoshin and N. S. Zefirov, Zh. Org. Khim. *17*, 1319, 1771 (1981); J. Org. Chem. USSR *17*, 1170, 1585 (1981); Tetrahedron Lett. *22*, 2209 (1981). [197] L. Došen-Mićović and V. Žigman, J. Chem. Soc., Perkin Trans. II *1985*, 625. [198] F. Dietz, W. Förster, R. Thieme, and C. Weiss, Z. Chem. *22*, 144 (1982). [199] L. Onsager, J. Am. Chem. Soc. *58*, 1486 (1936). [200] L. Crombie: *Geometrical Isomerism about Carbon-Carbon Double Bonds.* Quart. Rev. (London) *6*, 101 (1952).

[201] R. E. Wood and D. P. Stevenson, J. Am. Chem. Soc. *63*, 1650 (1941). [202] M.-L. Stien, M. Claessens, A. Lopez, and J. Reisse, J. Am. Chem. Soc. *104*, 5902 (1982). [203] H. McNab, J. Chem. Soc., Perkin Trans. II *1981*, 1283. [204] M. V. Jovanovic, Heterocycles *20*, 1987 (1983). [205] Y. Inagaki, R. Okazaki, N. Inamoto, K. Yamada, and H. Kawazura, Bull. Chem. Soc. Jpn. *52*, 2008 (1979). [206] R. Thieme and C. Weiss, Stud. Biophys. *93*, 273 (1983); Chem. Abstr. *99*, 139087 w (1983). [207] A. Maercker and J. D. Roberts, J. Am. Chem. Soc. *88*, 1742 (1966). [208] G. Boche and F. Heidenhain, J. Am. Chem. Soc. *101*, 738 (1979); G. Boche, F. Heidenhain, W. Thiel, and R. Eiben, Chem. Ber. *115*, 3167 (1982). [209] J. N. Spencer, C. L. Campanella, E. M. Harris, and W. S. Wolbach, J. Phys. Chem. *89*, 1888 (1985). [210] R. E. Valters and W. Flitsch: *Ring-Chain Tautomerism.* Plenum Publishing Corp., New York 1985.

[211] O. Arjona, R. Pérez-Ossorio, A. Pérez-Rubalcaba, J. Plumet, and M. J. Santesmases, J. Org. Chem. *49*, 2624 (1984). [212] C. Reichardt, K.-Y. Yun, W. Massa, R. E. Schmidt, O. Exner, and E.-U. Würthwein, Liebigs Ann. Chem. *1985*, 1997. [213] E. M. Arnett: *Solvation Energies of Organic Ions.* J. Chem. Educ. *62*, 385 (1985). [214] G. Modena, C. Paradisi, and G. Scorrano: *Solvation Effects on Basicity and Nucleophilicity.* In F. Bernardi, I. G. Csizmadia, and A. Mangini (eds.): Organic Sulfur Chemistry. Elsevier, Amsterdam 1985, Chapter 10, p. 568ff. [215] M. Mohammad, A. Y. Khan, M. Iqbal, R. Iqbal, and M. Razzaq, J. Am. Chem. Soc. *100*, 7658 (1978); J. Phys. Chem. *85*, 2816 (1981). [216] E. M. Kosower: *Stable Pyridinyl Radicals.* Top. Curr. Chem. *112*, 117 (1983). [217] B. Fuchs, A. Ellencweig, E. Tartakovsky, and P. Aped, Angew. Chem. *98*, 289 (1986); Angew. Chem., Int. Ed. Engl. *98*, 287 (1986). [218] W. Schilf, L. Stefaniak, M. Witanowski, and G. A. Webb, Magn. Reson. Chem. *23*, 181 (1985), and references cited therein. [219] D. L. Hughes, J. J. Bergan, and E. J. J.

Grabowski, J. Org. Chem. *51*, 2579 (1986). [220] N. Stahl and W. P. Jencks, J. Am. Chem. Soc. *108*, 4196 (1986).

[221] D. A. Hinckley, P. G. Seybold, and D. P. Borris, Spectrochim. Acta, Part A *42*, 747 (1986); J. Chem. Educ. *64*, 362 (1987). [222] I. Lopez Arbeloa and K. K. Rohatgi-Mukherjee, Chem. Phys. Lett. *128*, 474 (1986). [223] H. Meier, W. Lauer, and V. Krause, Chem. Ber. *119*, 3382 (1986). [224] S. E. Biali and Z. Rappoport, J. Am. Chem. Soc. *106*, 5641 (1984). [225] J. F. Bunnett and F. P. Olsen, Can. J. Chem. *44*, 1899, 1917 (1966). [226] A. Bagno, G. Scorrano, and R. A. More O'Ferrall; *Stability and Solvation of Organic Cations*. Rev. Chem. Intermed. *7*, 313 (1987). [227] R. Wolfenden, Y.-L. Liang, M. Matthews, and R. Williams, J. Am. Chem. Soc. *109*, 463 (1987). [228] H. Mustroph: *Über die Azo-Hydrazon-Tautomerie*. Z. Chem. *27*, 281 (1987). [229] E. M. Arnett, K. E. Molter, E. C. Marchot, W. H. Donovan, and P. Smith, J. Am. Chem. Soc. *109*, 3788 (1987).

Chapter 5

[1] *Cf.* references [7–10] in reference list to Chapter 1. [2] S. Glasstone, K. J. Laidler, and H. Eyring: *The Theory of Rate Processes*. McGraw Hill, New York, London 1941, p. 400ff. [3] E. A. Moelwyn-Hughes; *Kinetics of Reactions in Solution*. 2nd ed., Oxford University Press, London 1947. [4] S. W. Benson: *The Foundations of Chemical Kinetics*. McGraw Hill, New York 1960, p. 493ff. [5] J. W. Moore and R. G. Pearson: *Kinetics and Mechanism – The Study of Homogeneous Chemical Reactions*. 3rd ed., Wiley, New York 1981; *Kinetik und Mechanismen homogener chemischer Reaktionen* (translated into German by F. Helfferich and U. Schindewolf). Verlag Chemie, Weinheim 1964, p. 114ff. [6] A. J. Parker, Quart. Rev. (London) *16*, 163 (1962); Usp. Khim. *32*, 1270 (1963); Adv. Org. Chem. *5*, 1 (1965); Adv. Phys. Org. Chem. *5*, 173 (1967); Usp. Khim. *40*, 2203 (1971); Chem. Rev. *69*, 1 (1969); Pure Appl. Chem. *53*, 1437 (1981). [7] J. E. Leffler and E. Grunwald: *Rates and Equilibria of Organic Reactions*. Wiley, New York 1963, p. 263ff. [8] K. B. Wiberg: *Physical Organic Chemistry*. Wiley, New York 1964, p. 374ff. [9] P. Baekelmans, M. Gielen, and J. Nasielski: *Quelques aspects des effects de solvants en chimie organique*. Ind. Chim. Belg. *29*, 1265 (1964); Chem. Abstr. *62*, 12495 (1965). [10] B. Tchoubar, *Quelques aspects du rôle des solvants en chimie organique*. Bull. Soc. Chim. Fr. *1964*, 2069; Usp. Khim. *34*, 1227 (1965).

[11] K. J. Laidler: *Chemical Kinetics*. 3rd ed. Harper and Row, Hilversum 1987. [12] E. S. Amis: *Solvent Effects on Reaction Rates and Mechanisms*. Academic Press, New York, London 1966. [13] V. A. Pal'm: *Osnovy kolichestvennoi teorii organicheskikh reaktsii*. Izdatel'stvo Khimiya, Leningrad 1967; *Grundlagen der quantitativen Theorie organischer Reaktionen* (translated into German by G. Heublein). Akademie-Verlag, Berlin 1971. [14] J. C. Jungers, L. Sajus, I. de Aguirre, and D. Decroocq: *L'Analyse cinétique de la transformation chimique*. Edition Technip, Paris 1967/68, Vol. II, Chapter V, p. 597ff. [15] E. M. Kosower: *An Introduction to Physical Organic Chemistry*. Wiley, New York 1968, p. 259ff. [16] C. K. Ingold: *Structure and Mechanism in Organic Chemistry*. 2nd ed., Cornell University Press, Ithaca/N.Y., and London 1969, p. 457ff. and 680ff. [17] L.P. Hammett: *Physical Organic Chemistry – Reaction Rates, Equilibria, and Mechanisms*. 2nd ed., McGraw Hill, New York 1970, Chapter 8; *Physikalische Organische Chemie – Reaktionsgeschwindigkeiten, Gleichgewichte, Mechanismen* (translated into German by P. Schmid). Verlag Chemie, Weinheim 1973, Chapter 8, p. 221ff. [18] D. Decroocq: *Analyse des effets du milieu sur les réactions chimiques*. 1. *Les réactions moléculaires crypto-ioniques*. Ind. Chim. Belg. *35*, 505 (1970); Chem. Abstr. *73*, 98075h (1970). [19] E. A. Moelwyn-Hughes: *The Chemical Statics and Kinetics of Solutions*. Academic Press, London, New York 1971. [20] K. Schwetlick: *Kinetische Methoden zur Untersuchung von Reaktionsmechanismen*. Deutscher Verlag der Wissenschaften, Berlin 1971, Chapter 4, p. 139ff.

[21] E. S. Amis and J. F. Hinton: *Solvent Effects on Chemical Phenomena*. Vol. I, Academic Press, New York, London 1973. [22] J. Burgess: *Solvent Effects*. Inorg. React. Mech. *3*, 312 (1974); ibid. *4*, 236 (1976); ibid. *5*, 260 (1977); ibid. *6*, 278 (1979); ibid. *7*, 287 (1981). [23] M. H. Abraham: *Solvent Effects on Transition States and Reaction Rates*, in A. Streitwieser and R. W. Taft (eds.): Progr. Phys. Org. Chem. *11*, 1 (1974); *Solvent Effects on Reaction Rates*. Pure Appl. Chem. *57*, 1055 (1985). [24] S. H. Lin, K. P. Li, and H. Eyring: *Theory of Reaction Rates in Condensed Phases*, in H. Eyring, D.

Henderson, and W. Jost (eds.): *Physical Chemistry – An Advanced Treatise*. Academic Press, New York 1975, Vol. VII, p. 1ff. [25] C. A. Eckert: *Molecular Thermodynamics of Reactions in Solution*, in M. R. J. Dack (ed.): *Solutions and Solubilities*. Vol. VIII, Part I, of A. Weissberger (ed.): *Techniques of Chemistry*. Wiley-Interscience, New York 1975, p. 1ff. [26] G. Illuminati: *Solvent Effects on Selected Organic and Organometallic Reactions. Guidelines to Synthetic Applications*, in M. R. J. Dack (ed.): *Solutions and Solubilities*. Vol. VIII, Part II, of A. Weissberger (ed.): *Techniques of Chemistry*. Wiley-Interscience, New York 1976, p. 159ff. [27] M. R. J. Dack: *The Influence of Solvent on Chemical Reactivity*, in M. R. J. Dack (ed.): *Solutions and Solubilities*. Vol. VIII, Part II, of A. Weisberger (ed.): *Techniques of Chemistry*. Wiley-Interscience, New York 1976, p. 95ff. [28] S. G. Entelis and R. P. Tiger: *Kinetika reaktsii v zhidkoi faze*. Izdatel'stvo Khimiya, Moscow 1973; *Reaction Kinetics in the Liquid Phase* (translated into English by R. Kondor). Wiley, New York, and Israel Program for Scientific Translations, Jerusalem 1976. [29] V. Gutmann: *Solvent Effects on the Reactivities of Organometallic Compounds*. Coord. Chem. Rev. *18*, 225 (1976). [30] N. Menschutkin, Z. Phys. Chem. *6*, 41 (1890); ibid. *34*, 157 (1900).

[31] D. J. Cram, B. Rickborn, C. A. Kingsbury, and P. Haberfield, J. Am. Chem. Soc. *83*, 3678 (1961). [32] H. Martin, *Über Beziehungen zwischen Lösungs- und Gasphasenreaktionen*. Angew. Chem. *78*, 73 (1966); Angew. Chem., Int. Ed. Engl. *5*, 78 (1966); Chimia *21*, 439 (1967). [33] J. B. Harkness, G. B. Kistiakowsky, and W. H. Mears, J. Chem. Phys. *5*, 682 (1937). [34] H. Kaufmann and A. Wassermann, J. Chem. Soc. *1939*, 870; A. Wassermann, Monatsh. Chem. *83*, 543 (1952). [35] G. Cöster and E. Pfeil, Chem. Ber. *101*, 4248 (1968); and references cited therein. [36] J. H. Raley, F. F. Rust, and W. E. Vaughan, J. Am. Chem. Soc. *70*, 88, 1336 (1948). [37] E. S. Huyser and R. M. VanScoy, J. Org. Chem. *33*, 3524 (1968). [38] A. Rembaum and M. Szwarc, J. Am. Chem. Soc. *76*, 5975 (1954); M. Levy, M. Steinberg, and M. Szwarc, ibid. *76*, 5978 (1954). [39] H. J. Shine, J. A. Waters, and D. M. Hofman, J. Am. Chem. Soc. *85*, 3613 (1963). [40] S. Winstein and A. H. Fainberg, J. Am. Chem. Soc. *78*, 2770 (1956); ibid. *79*, 5937 (1957).

[41] A. F. Moroni: *Die Abhängigkeit der thermodynamischen Größen des Übergangszustandes einer Reaktion vom Lösungsmittel*. Z. Phys. Chem. (Frankfurt) N. F. *59*, 1 (1968). [42] O. Dimroth, Liebigs Ann. Chem. *377*, 127 (1910); ibid. *399*, 91 (1913). [43] J. H. van't Hoff: *Vorlesungen über theoretische und physikalische Chemie*. Vieweg, Braunschweig 1898, 1. Heft, p. 217. [44] E. D. Hughes and C. K. Ingold, J. Chem. Soc. *1935*, 244; Trans. Faraday Soc. *37*, 603, 657 (1941); K. A. Cooper, M. L. Dhar, E. D. Hughes, C. K. Ingold, B. J. MacNulty, and L. I. Woolf, J. Chem. Soc. *1948*, 2043. [45] C. A. Bunton: *Nucleophilic Substitution at a Saturated Carbon Atom*. Elsevier, Amsterdam 1963, Chapter 4. [46] Y. Pocker: *Nucleophilic Substitutions at a Saturated Carbon Atom in Non-Hydroxylic Solvents*, in G. Porter and B. Stevens (eds.): Progress in Reaction Kinetics. Pergamon Press, Oxford 1961, Vol. 1, p. 215. [47] M. H. Abraham, J. Chem. Soc., Perkin Trans. II *1972*, 1343. [48] D. J. Raber, R. C. Bingham, J. M. Harris, J. L. Fry, and P. v. R. Schleyer, J. Am. Chem. Soc. *92*, 5977 (1970). [49] H. Meerwein and K. van Emster, Ber. Dtsch. Chem. Ges. *55*, 2500 (1922). [50] H. von Halban, Z. Phys. Chem. *67*, 129 (1909).

[51] A. von Hemptinne and A. Bekaert, Z. Phys. Chem. *28*, 225 (1899). [52] H. von Halban, Z. Phys. Chem. *84*, 129 (1913). [53] J. F. Norris and S. W. Prentiss, J. Am. Chem. Soc. *50*, 3042 (1928). [54] A. G. Grimm, H. Ruf, and H. Wolff, Z. Phys. Chem. *B 13*, 301 (1931). [55] N. J. T. Pickles and C. N. Hinshelwood, J. Chem. Soc. *1936*, 1353. [56] E. Tommila and P. Kauranen, Acta Chem. Scand. *8*, 1152 (1954); *13*, 622 (1959). [57] J. D. Reinheimer, J. D. Harley, and W. W. Meyers, J. Org. Chem. *28*, 1575 (1963). [58] H. Heydtmann, A. P. Schmidt, and H. Hartmann, Ber. Bunsenges. Phys. Chem. *70*, 444 (1966). [59] H. Hartmann and A. P. Schmidt, Z. Phys. Chem. (Frankfurt) N.F. *66*, 183 (1969). [60] C. Lassau and J.-C. Jungers, Bull. Soc. Chim. Fr. *1968*, 2678.

[61] Y. Drougard and D. Decroocq, Bull. Soc. Chem. Fr. *1969*, 2972; G. Berrebi and D. Decroocq, J. Chim. Phys., Phys. Chim. Biol. *71*, 673 (1974). [62] T. Matsui and N. Tokura, Bull. Chem. Soc. Jpn. *43*,1751 (1970); Y. Kondo, M. Ohnishi, and N. Tokura, ibid. *45*, 3579 (1972). [63] P. Haberfield, A. Nudelman, A. Bloom, R. Romm, and H. Ginsberg, J. Org. Chem. *36*, 1792 (1971). [64] M. H. Abraham, J. Chem. Soc. *B 1971*, 229; M. H. Abraham and R. J. Abraham, J. Chem. Soc., Perkin Trans. II *1975*, 1677; M. Abraham and P. C. Grellier, ibid. *1976*, 1735. [65] M. Auriel and E. de Hoffmann, J. Am. Chem. Soc. *97*, 7433 (1975); J. Chem. Soc., Perkin Trans. II *1979*, 325. [66] E. R.

Swart and L. J. LeRoux, J. Chem. Soc. *1956*, 2110; ibid. *1957*, 406. [67] J. J. Delpuech, Tetrahedron Lett. *1965*, 2111. [68] P. Müller and B. Siegfried, Helv. Chim. Acta *55*, 2400 (1972). [69] E. D. Hughes and D. J. Whittingham, J. Chem. Soc. *1960*, 806; A. V. Eltsov, N. V. Pavlish, and V. A. Ketlinski, Zh. Org. Khim. *14*, 1751 (1978); J. Org. Chem. USSR *14*, 1630 (1978). [70] J. L. Gleave, E. D. Hughes, and C. K. Ingold, J. Chem. Soc. *1935*, 236.

[71] S. Oae and Y. H. Khim, Bull. Chem. Soc. Jpn. *42*, 3528 (1969). [72] D. J. Raber and J. M. Harris, J. Chem. Educ. *49*, 60 (1971). [73] D. V. Banthorpe: *Elimination Reactions*. Elsevier, Amsterdam 1963, Chapter 2, p. 40. [74] W. H. Saunders and A. F. Cockerill: *Mechanisms of Elimination Reactions*. Wiley-Interscience, New York 1973. [75] A. Loupy and J. Seyden-Penne, Bull. Soc. Chim. Fr. *1971*, 2306. [76] M. Cocivera and S. Winstein, J. Am. Chem. Soc. *85*, 1702 (1963). [77] P. B. D. de la Mare and R. Bolton: *Electrophilic Additions to Unsaturated Systems*. 2nd ed. Elsevier, Amsterdam 1981; P. B. D. de la Mare: *Electrophilic Halogenation. Reaction Pathways Involving Attack by Electrophilic Halogens on Unsaturated Compounds*. Cambridge University Press, London 1976. [78] F. Freeman: *Possible Criteria for Distinguishing between Cyclic and Acyclic Activated Complexes in Addition Reactions*. Chem. Rev. *75*, 439 (1975). [79] G. Heublein: *Neuere Aspekte der elektrophilen Halogenaddition an Olefine*. Z. Chem. *9*, 281 (1969). [80] J. G. Hanna and S. Siggia, Anal. Chem. *37*, 690 (1965).

[81] J.-E. Dubois and F. Garnier, Chem. Commun. *1968*, 241; Bull. Soc. Chim. Fr. *1968*, 3797; M.-F. Ruasse and J.-E. Dubois, J. Am. Chem. Soc. *97*, 1977 (1975). [81a] G. Modena, F. Rivetti, and U. Tonellato, J. Org. Chem. *43*, 1521 (1978). [82] D. S. Campbell and D. R. Hogg, J. Chem. Soc. *B 1967*, 889. [83] L. Rasteikiene, D. Greiciute, M. G. Lin'kova, and I. L. Knunyants: *The Addition of Sulfenyl Chlorides to Unsaturated Compounds*. Usp. Khim. *46*, 1041 (1977); Russ. Chem. Rev. *46*, 548 (1977). [84] T. Beier, H. G. Hauthal, and W. Pritzkow, J. Prakt. Chem. *26*, 304 (1964); G. Collin, U. Jahnke, G. Just, G. Lorenz, W. Pritzkow, M. Röllig, L. Winguth, P. Dietrich, C. E. Doring, H. G. Hauthal, and A. Wiedenhoft, ibid. *311*, 238 (1969). [85] P. P. Kadzyauskas and N. S. Zefirov: *Nitrosochlorination of Alkenes*. Usp. Khim. *37*, 1243 (1968); Russian Chem. Rev. *37*, 543 (1968). [86] K. M. Ibne-Rasa and J. O. Edwards: *Role of Solvent in Some Oxygen Transfer Reactions of Peroxy Compounds*. Int. J. Chem. Kinet. *7*, 575 (1975); Chem. Abstr. *83*, 113436d (1975). [86a] R. Renolen and J. Ugelstad, J. Chim. Phys. *57*, 634 (1960). [87] H. Kropf and M. R. Yazdanbachsch, Tetrahedron *30*, 3455 (1974). [88] B. Giese and R. Huisgen, Tetrahedron Lett. *1967*, 1889. [89] M. Neuenschwander and P. Bigler, Helv. Chim. Acta *56*, 959 (1973). [90] R. B. Woodward and R. Hoffmann, Angew. Chem. *81*, 797 (1969); Angew. Chem., Int. Ed. Engl. *8*, 781 (1969).

[91] R. Huisgen: *Cycloadditionen – Begriff, Einteilung und Kennzeichnung*. Angew. Chem. *80*, 329 (1968); Angew. Chem., Int. Ed. Engl. *7*, 321 (1968). [92] R. Gompper: *Cycloadditionen mit polaren Zwischenstufen*. Angew. Chem. *81*, 348 (1969); Angew. Chem., Int. Ed. Engl. *8*, 312 (1968). [93] P. D. Bartlett: *Mechanisms of Cycloadditons*. Quart. Rev. (London) *24*, 473 (1970). [94] G. Steiner and R. Huisgen, Tetrahedron Lett. *1973*, 3763, 3769; J. Am. Chem. Soc. *95*, 5054, 5055, 5056 (1973); R. Huisgen, R. Schug, and G. Steiner, Bull. Soc. Chim. Fr. *1976*, 1813; R. Huisgen, Acc. Chem. Res. *10*, 117, 199 (1977). [95] J. v. Jouanne, H. Kelm, and R. Huisgen, J. Am. Chem. Soc. *101*, 151 (1979). [96] W. J. Le Noble and R. Mukhtar, J. Am. Chem. Soc. *97*, 5938 (1975). [97] R. Huisgen, R. Schug, and G. Steiner, Angew. Chem. *86*, 47, 48 (1974); Angew. Chem., Int. Ed. Engl. *13*, 80, 81 (1974). [98] I. Karle, J. Flippen, R. Huisgen, and R. Schug, J. Am. Chem. Soc. *97*, 5285 (1975); R. Huisgen, Chimia *31*, 13 (1976). [99] S. Proskow, H. E. Simmons, and T. L. Cairns, J. Am. Chem. Soc. *88*, 5254 (1966). [99a] M. Nakahara, Y. Tsuda, M. Sasaki, and J. Osugi, Chem. Lett. (Tokyo) *1976*, 731. [100] R. Huisgen, L. A. Feiler, and P. Otto, Tetrahedron Lett. *1968*, 4485; Chem. Ber. *102*, 3444 (1969); R. Huisgen, L. A. Feiler, and G. Binsch, Chem. Ber. *102*, 3460 (1969); G. Swieton, J. v. Jouanne, H. Kelm, and R. Huisgen, J. Chem. Soc., Perkin Trans. II *1983*, 37.

[101] R. Huisgen and P. Otto, J. Am. Chem. Soc. *90*, 5342 (1968). [102] R. Huisgen and P. Otto, J. Am. Chem. Soc. *91*, 5922 (1969). [103] R. Graf, Liebigs Ann. Chem. *661*, 111 (1963); Org. Synth. *46*, 51 (1966). [104] E. J. Moriconi and W. C. Meyer, J. Org. Chem. *36*, 2841 (1971). [105] J. K. Rasmussen and A. Hassner: *Recent Developments in the Synthetic Use of Chlorosulfonyl Isocyanates*. Chem. Rev. *76*, 389 (1976). [106] K. Clauß, Liebigs Ann. Chem. *722*, 110 (1969). [107] H. Bestian, Pure Appl. Chem. *27*, 611 (1971). [108] P. Bickart, F. W. Carson, J. Jacobus, E. G. Miller, and K.

Mislow, J. Am. Chem. Soc. *90*, 4869 (1968); R. Tang and K. Mislow, ibid. *92*, 2100 (1970). [109] C. A. Grob and P. W. Schiess: *Die heterolytische Fragmentierung als Reaktionstypus in der organischen Chemie*. Angew. Chem. *79*, 1 (1967); Angew. Chem., Int. Ed. Engl. *6*, 1 (1967); C. A. Grob: *Mechanismus und Stereochemie der heterolytischen Fragmentierung*. Angew. Chem. *81*, 543 (1969); Angew. Chem., Int. Ed. Engl. *8*, 535 (1969). [110] R. E. Pincock, J. Am. Chem. Soc. *86*, 1820 (1964); ibid. *87*, 1274 (1965).

[111] W. H. Richardson and R. S. Smith, J. Am. Chem. Soc. *91*, 3610 (1969). [112] H. Kwart and P. A. Silver, J. Org. Chem. *40*, 3019 (1975); H. Kwart and T. H. Lilley, ibid. *43*, 2374 (1978). [113] R. B. Woodward and R. Hoffmann: *Die Erhaltung der Orbitalsymmetrie*. Verlag Chemie, Weinheim 1970; *The Conservation of Orbital Symmetry*. Academic Press, New York, London 1970. [114] J. Sauer and R. Sustmann: *Mechanistische Aspekte der Diels-Alder-Reaktion. Ein kritischer Überblick*. Angew. Chem. *92*, 773 (1980); Angew. Chem., Int. Ed. Engl. *19*, 779 (1980). [115] S. Seltzer: *The Mechanism of the Diels-Alder Reaction*, in H. Hart and G. J. Karabatsos (eds.): *Advances in Alicyclic Chemistry*, Academic Press, New York, London 1968, Vol. 2, p. 1 ff. (especially p. 13 ff.). [116] M. J. S. Dewar, S. Olivella, and J. J. P. Stewart, J. Am. Chem. Soc. *108*, 5771 (1986). [117] R. Huisgen: *Kinetik und Mechanismus 1.3-dipolarer Cycloadditionen*. Angew. Chem. *75*, 604, 742 (1963); Angew. Chem., Int. Ed. Engl. *2*, 565, 633 (1963). [118] R. A. Firestone: *On the Mechanism of 1.3-Dipolar Cycloadditions*. J. Org. Chem. *33*, 2285 (1968); ibid. *37*, 2181 (1972); Tetrahedron *33*, 3009 (1977). [119] R. Huisgen: *The Concerted Nature of 1.3-Dipolar Cycloadditions and the Question of Diradical Intermediates*. J. Org. Chem. *33*, 2291 (1968); ibid. *41*, 403 (1976). [120] Y.-M. Chang, J. Sims, and K. N. Houk, Tetrahedron Lett. *1975*, 4445.

[121] R. A. Fairclough and C. N. Hinshelwood, J. Chem. Soc. *1938*, 236 (Cyclopentadiene + *p*-benzoquinone). [122] Y. Yukawa and A. Isohisa, Mem. Inst. Sci. Ind. Research Osaka Univ. *10*, 191 (1953); Chem. Abstr. *48*, 7598 (1954) (Cyclohexa-1,3-diene + maleic anhydride). [123] L. J. Andrews and R. M. Keefer, J. Am. Chem. Soc. *77*, 6284 (1955) (9,10-Dimethylanthracen + maleic anhydride). [124] J. A. Berson, Z. Hamlet, and W. A. Mueller, J. Am. Chem. Soc. *84*, 297 (1962) (Cyclopentadiene + methyl acrylate). [124a] H. Taniguchi, Y. Yoshida, and E. Imoto, Bull. Chem. Soc. Jpn. *50*, 3335 (1977); ibid. *51*, 2405 (1978). [125] P. Brown and R. C. Cookson, Tetrahedron *21*, 1977 (1965) (Anthracen + tetracyanoethylene). [126] K. F. Wong and C. A. Eckert, Ind. Eng. Chem., Process Design Develop. *8*, 568 (1969); Chem. Abstr. *71*, 116937 (1969); Trans. Faraday Soc. *66*, 2313 (1970); R. A. Grieger and C. A. Eckert, J. Am. Chem. Soc. *92*, 7149 (1970); see also ref. [25] (Isoprene + maleic anhydride). [127] M. J. S. Dewar and R. S. Pyron, J. Am. Chem. Soc. *92*, 3098 (1970) (Isoprene + maleic anhydride). [128] P. Haberfield and A. K. Ray, J. Org. Chem. *37*, 3093 (1972). [129] B. Blankenburg, H. Fiedler, M. Hampel, H. G. Hauthal, G. Just, K. Kahlert, J. Korn, K.-H. Müller, W. Pritzkow, Y. Reinhold, M. Röllig, E. Sauer, D. Schnurpfeil, and G. Zimmermann, J. Prakt. Chem. *316*, 804 (1974). [130] M. E. Burrage, R. C. Cookson, S. S. Gupte, and I. D. R. Stevens, J. Chem. Soc., Perkin Trans. II *1975*, 1325.

[131] R. Huisgen, H. Stangl, H. J. Sturm, and H. Wagenhofer, Angew. Chem. *73*, 170 (1961). [132] R. Huisgen, H. Seidl, and I. Brüning, Chem. Ber. *102*, 1102 (1969). [133] M. K. Meilahn, B. Cox, and M. E. Munk, J. Org. Chem. *40*, 819 (1975). [134] P. K. Kadaba, J. Heterocycl. Chem. *6*, 587 (1969); Tetrahedron *25*, 3053 (1969); J. Org. Chem. *41*, 1073 (1976). [134a] A. Eckell, M. V. George, R. Huisgen, and A. S. Kende, Chem. Ber. *110*, 578 (1977). [135] H. M. R. Hoffmann: *Die En-Reaktion*. Angew. Chem. *81*, 597 (1969); Angew. Chem., Int. Ed. Engl. *8*, 556 (1969). [136] R. Huisgen and H. Pohl, Chem. Ber. *93*, 527 (1960). [137] A. A. Frimer; *The Reaction of Singlet Oxygen with Olefins*. Chem. Rev. *79*, 359 (1979). [138] C. S. Foote, Pure Appl. Chem. *27*, 635 (1971); C. S. Foote and R. W. Denny, J. Am. Chem. Soc. *93*, 5168 (1971). [139] R. H. Young, K. Wehrly, and R. L. Martin, J. Am. Chem. Soc. *93*, 5774 (1971). [140] J. E. Baldwin and J. A. Kapecki, J. Am. Chem. Soc. *92*, 4868 (1970).

[141] E. Schaumann, Chem. Ber. *109*, 906 (1976). [142] G. Wittig: *From Diyls over Ylides to My Idyll*. Acc. Chem. Res. *7*, 6 (1974). [143] A. Maercker: *The Wittig Reaction*. Org. React. *14*, 270 (1965). [144] A. J. Speziale and D. E. Bissing, J. Am. Chem. Soc. *85*, 3878 (1963). [145] C. Rüchardt, P. Panse, and S. Eichler, Chem. Ber. *100*, 1144 (1967). [146] P. Frøyen, Acta Chem. Scand. *26*, 2163 (1972). [147] G. Aksnes and F. Y. Khalil, Phosphorus *2*, 105 (1972); ibid. *3*, 37, 79, 103 (1973). [148] E. Maccarone and G. Perrini, Gazz. Chim. Ital. *112*, 447 (1982). [149] F. Ramirez and S. Dershowitz, J. Org. Chem. *22*, 41 (1957). [150] D. I. Coomber and J. R. Partington, J. Chem. Soc. *1938*, 1444.

[151] D. C. Wigfield and S. Feiner, Can. J. Chem. *48*, 855 (1970). [152] D. C. Berndt, J. Chem. Eng. Data *14*, 112 (1969). [153] W. N. White and E. F. Wolfarth, J. Org. Chem. *35*, 2196, 3585 (1970). [154] S. J. Rhoads and N. R. Raulins: *The Claisen and Cope Rearrangements*. Org. React. *22*, 1 (1975). [155] R. Wehrli, D. Bellus, H.-J. Hansen, and H. Schmid, Chimia *30*, 416 (1976); Helv. Chim. Acta *60*, 1325 (1977). [156] S. G. Smith, J. Am. Chem. Soc. *83*, 4285 (1961). [157] R. Huisgen, A. Dahmen, and H. Huber, Tetrahedron Lett. *1969*, 1461, 1465. [158] H. Mayr and R. Huisgen, J. Chem. Soc., Chem. Commun. *1976*, 57. [158a] N. S. Isaacs and A. A. R. Laila, J. Chem. Soc., Perkin Trans. II *1976*, 1470. [158b] M. T. Reetz, Chem. Ber. *110*, 965 (1977); M. T. Reetz, N. Greif, and M. Kliment, ibid. *111*, 1095 (1978); M. T. Reetz, Angew. Chem. *91*, 185 (1979); Angew. Chem., Int. Ed. Engl. *18*, 173 (1979). [159] C. Walling: *Free Radicals in Solutions*. Wiley, New York, London 1957. [160] E. S. Huyser: *Solvent Effects in Radical Reactions*, in G. H. Williams (ed.): *Advances in Free-Radical Chemistry*. Logos Press and Academic Press, New York, London 1965, Vol. 1, p. 77ff.

[161] H. Sakurai and A. Hosomi: *Solvent Effects in Free-Radical Reactions*. Yuki Gosei Kagaku Kyokai Shi *25*, 1108 (1967); Chem. Abstr. *68*, 104 254 f (1968). [162] Yu. L. Spirin: *Reactivity of Radicals and Molecules in Radical Reactions*. Usp. Khim. *38*, 1201 (1969); Russian Chem. Rev. *38*, 529 (1969). [163] J. C. Martin: *Solvation and Association*, in J. K. Kochi (ed.): *Free Radicals*. Wiley, New York 1973, Vol. 1, Chapter 20, p. 493ff. [164] E. T. Denisov: *Konstanty Skorosti Gomoliticheskikh Zhidkofaznykh Reaktsii*. Nauka Press, Moscow 1971; *Liquid-Phase Reaction Rate Constants* (translated into English by R. K. Johnston). IFI/Plenum Data Company, New York 1974. [165] E. S. Huyser: *Kinetics of Free-Radical Reactions*, in H. Eyring, D. Henderson, and W. Jost (eds.): *Physical Chemistry – An Advanced Treatise*. Academic Press, New York 1975, Vol. VII, p. 299ff. (especially p. 340ff.). [166] See also reference [28], chapter IX, concerning homolytic reactions, and references cited therein. [167] K. Ziegler, P. Orth, and K. Weber, Liebigs Ann. Chem. *504*, 131 (1933); K. Ziegler, A. Seib, K. Knoevenagel, P. Herte, and F. Andreas, ibid. *551*, 150 (1942). [168] H. Lankamp, W. T. Nauta, and C. MacLean, Tetrahedron Lett. *1968*, 249. [169] H. A. Staab, H. Brettschneider, and H. Brunner, Chem. Ber. *103*, 1101 (1970). [170] P. Jacobson, Ber. Dtsch. Chem. Ges. *38*, 196 (1905).

[171] B. S. Tanaseichuk, L. G. Tikhonova, and A. P. Dydykina, Zh. Org. Khim. *9*, 1273 (1973); J. Org. Chem. USSR *9*, 1301 (1973). [172] E. S. Huyser and R. M. VanScoy, J. Org. Chem. *33*, 3524 (1968). [173] See reference [164], table 7 on page 37, and the accompanying references. [174] S. D. Ross and M. A. Fineman, J. Am. Chem. Soc. *73*, 2176 (1951). [175] M. Levy, M. Steinberg, and M. Szwarc, J. Am. Chem. Soc. *76*, 5978 (1954). [176] H. J. Shine, J. A. Waters, and D. M. Hoffman, J. Am. Chem. Soc. *85*, 3613 (1963). [177] K. Nozaki and P. D. Bartlett, J. Am. Chem. Soc. *68*, 1686 (1946). [178] C. G. Swain, W. H. Stockmayer, and J. T. Clarke, J. Am. Chem. Soc. *72*, 5426 (1950). [179] W. R. Foster and G. H. Williams, J. Chem. Soc. *1962*, 2862. [180] R. C. Lamb, J. G. Pacifici, and P. W. Ayers, J. Am. Chem. Soc. *87*, 3928 (1965).

[181] C. Walling, H. P. Waits, J. Milanovic, and C. G. Pappiaonnou, J. Am. Chem. Soc. *92*, 4927 (1970). [182] R. C. Petersen, J. H. Markgraf, and S. D. Ross, J. Am. Chem. Soc. *83*, 3819 (1961). [183] See reference [164], table 12 on page 57, and the accompanying references. [184] A. F. Moroni, Makromol. Chem. *105*, 43 (1967). [185] O. Yamamoto, J. Yamashita, and H. Hashimoto, Kogyo Kagaku Zasshi *71*, 223 (1968); Chem. Abstr. *69*, 66644 u (1968). [186] R. Huisgen and H. Nakaten, Liebigs Ann. Chem. *586*, 70 (1954). [187] M. G. Alder and J. E. Leffler, J. Am. Chem. Soc. *76*, 1425 (1954); J. E. Leffler and R. A. Hubbard, J. Org. Chem. *19*, 1089 (1954). [188] W. G. Bentrude and A. K. MacKnight, Tetrahedron Lett. *1966*, 3147. [189] R. Kerber, O. Nuyken, and R. Steinhausen, Makromol. Chem. *175*, 3225 (1975). [190] C. E. Dykstra and H. F. Schaefer, J. Am. Chem. Soc. *97*, 7210 (1975).

[191] W. Lobunez, J. R. Rittenhouse, and J. G. Miller, J. Am. Chem. Soc. *80*, 3505 (1958). [192] P. S. Engel, Chem. Rev. *80*, 99 (1980). [193] A. Schulz and C. Rüchardt, Tetrahedron Lett. *1976*, 3883; W. Duismann and C. Rüchardt, Chem. Ber. *111*, 596 (1978). [194] C. G. Overberger, J.-P. Anselme, and J. R. Hall, J. Am. Chem. Soc. *85*, 2752 (1963). [195] C. Rüchardt: *Nichtkatalysierte Perester-Zersetzungen*. Fortschr. Chem. Forsch. *6*, 251 (1966); Usp. Khim. *37*, 1402 (1968). [196] C. Rüchardt: *Zusammenhänge zwischen Struktur und Reaktivität in der Chemie freier Radikale*. Angew. Chem. *82*, 845 (1970); Angew. Chem., Int. Ed. Engl. *9*, 830 (1970). [197] P. D. Bartlett and R. R. Hiatt, J. Am.

Chem. Soc. *80*, 1398 (1958). [198] C. Rüchardt and H. Böck, Chem. Ber. *104*, 577 (1971). [199] M. Trachtman and J. G. Miller, J. Am. Chem. Soc. *84*, 4828 (1962). [200] J. P. Engstrom and J. C. DuBose, J. Org. Chem. *38*, 3817 (1973).

[201] P. D. Bartlett and C. Rüchardt, J. Am. Chem. Soc. *82*, 1756 (1960). [202] C. Rüchardt and H. Schwarzer, Chem. Ber. *99*, 1861 (1966). [203] A. A. Turovskii, R. V. Kucher, A. M. Ustinova, and A. E. Batog, Zh. Obshch. Khim. *45*, 860 (1975); J. Gen. Chem. USSR *45*, 844 (1975). [204] C. Rüchardt and H. Schwarzer, Chem. Ber. *99*, 1871 (1966). [205] T. W. Koenig and J. C. Martin, J. Org. Chem. *29*, 1520 (1964); J. Am. Chem. Soc. *86*, 1771 (1964). [206] D. L. Tuleen, W. G. Bentrude, and J. C. Martin, Tetrahedron Lett. *1962*, 229; J. Am. Chem. Soc. *85*, 1938 (1963); P. Livant and J. C. Martin, ibid. *98*, 7851 (1976). [207] C. Rüchardt and H.-J. Quadbeck-Seeger, Chem. Ber. *102*, 3525 (1969). [208] R. Criegee and R. Kaspar, Liebigs Ann. Chem. *560*, 127 (1948); R. Criegee, Fortschr. Chem. Forsch. *1*, 508 (1950), especially p. 551. [209] P. D. Bartlett and B. T. Storey, J. Am. Chem. Soc. *80*, 4954 (1958). [210] T. Koenig and W. Brewer, J. Am. Chem. Soc. *86*, 2728 (1964).

[211] E. Hedaya, R. L. Hinman, L. M. Kibler, and S. Theodoropulos, J. Am. Chem. Soc. *86*, 2727 (1964). [212] C. Rüchardt and H. Schwarzer, Chem. Ber. *99*, 1878 (1966). [213] E. M. Kosower and M. Mohammad, J. Am. Chem. Soc. *93*, 2709 (1971). [214] E. M. Kosower and I. Schwager, J. Am. Chem. Soc. *86*, 4493, 5528 (1964). [215] E. M. Kosower and M. Mohammad, J. Am. Chem. Soc. *90*, 3271 (1968); ibid. *93*, 2709, 2713 (1971). [216] R. Kerber, Z. Elektrochem., Ber. Bunsenges. Phys. Chem. *63*, 296 (1959). [217] G. E. Zaikov and Z. K. Maizus, Dokl. Akad. Nauk SSSR *150*, 116 (1963); Chem. Abstr. *59*, 5831 (1963); G. E. Zaikov, Z. K. Maizus, and N. M. Emanuel, Kinetika i Kataliz *7*, 401 (1966); Chem. Abstr. *65*, 10459 (1966); G. E. Zaikov, A. A. Vichutinskii, Z. K. Maizus, and N. M. Emanuel, Dokl. Akad. Nauk SSSR *168*, 1096 (1966); Chem. Abstr. *65*, 15184 (1966). [218] D. G. Hendry and G. A. Russell, J. Am. Chem. Soc. *86*, 2368 (1964). [219] J. A. Howard and K. U. Ingold, Can. J. Chem. *42*, 1044, 1250 (1964); ibid. *44*, 1119 (1966). [220] E. Niki, Y. Kamiya, and N. Ohta, Bull. Chem. Soc. Jpn. *42*, 3224 (1969).

[221] G. A. Russell, J. Am. Chem. Soc. *79*, 2977 (1957); ibid. *80*, 4987, 4997, 5002 (1958). [222] C. Walling and B. B. Jacknow, J. Am. Chem. Soc. *82*, 6108, 6113 (1960). [223] C. Walling: *Some Aspects of the Chemistry of Alkoxy Radicals*. Pure Appl. Chem. *15*, 69 (1967). [224] C. Walling and P. Wagner, J. Am. Chem. Soc. *85*, 2333 (1963); ibid. *86*, 3368 (1964). [225] C. D. Cook and B. E. Norcross, J. Am. Chem. Soc. *81*, 1176 (1959). [226] R. G. Pearson, J. Chem. Phys. *20*, 1478 (1952). [227] C. Walling and D. Bristol, J. Org. Chem. *36*, 733 (1971). [228] J. H. Hildebrand and R. L. Scott; *The Solubility of Nonelectrolytes*. 3rd ed., Dover, New York 1964. [229] J. H. Hildebrand, J. M. Prausnitz, and R. L. Scott: *Regular and Related Solutions*. Van Nostrand-Reinhold, Princeton 1970. [230] G. Scatchard, Chem. Rev. *8*, 321 (1931); *10*, 229 (1932).

[231] A. F. M. Barton: *Handbook of Solubility Parameters and other Cohesion Parameters*. CRC Press, Boca Raton/Florida 1983. [232] M. R. J. Dack: *The Importance of Solvent Internal Pressure and Cohesion to Solution Phenomena*. Chem. Soc. Rev. (London) *4*, 211 (1975); M. R. J. Dack, Aust. J. Chem. *28*, 1643 (1975). [233] M. Richardson and F. G. Soper, J. Chem. Soc. *1929*, 1873. [234] S. Glasstone, J. Chem. Soc. *1936*, 723. [235] R. J. Ouellette and S. H. Williams, J. Am. Chem. Soc. *93*, 466 (1971). [236] H. F. Herbrandson and F. R. Neufeld, J. Org. Chem. *31*, 1140 (1966). [237] R. C. Neuman, J. Org. Chem. *37*, 495 (1972). [238] H. Burrell: *Solubility Parameter Values*, in J. Brandrup and E. H. Immergut (eds.): *Polymer Handbook*. 2nd ed., Wiley-Interscience, New York 1975, p. IV-337ff. [239] M. R. J. Dack, J. Chem. Educ. *51*, 231 (1974). [240] J. G. Kirkwood: *Theory of Solutions of Molecules Containing Widely Separated Charges*. J. Chem. Phys. *2*, 351 (1934).

[241] J. G. Kirkwood and F. Westheimer, J. Chem. Phys. *6*, 506 (1938); C. Tanford and J. G. Kirkwood, J. Am. Chem. Soc. *79*, 5333, 5340, 5348 (1957). [242] K. J. Laidler and P. A. Landskroener, Trans. Faraday Soc. *52*, 200 (1956); K. J. Laidler, Suomen Kemistilehti *33*, A 44 (1960); Chem. Abstr. *56*, 2920 (1962). [243] K. Hiromi, Bull. Chem. Soc. Jpn. *33*, 1251, 1264 (1960). [244] E. S. Amis, J. Chem. Educ. *29*, 337 (1952); ibid. *30*, 351 (1953); Anal. Chem. *27*, 1672 (1955). [245] A. D. Stepukhovich, N. I. Lapshova, and T. D. Efimova, Zh. Fiz. Khim. *35*, 2532 (1961); Chem. Abstr. *59*, 3739 (1963). [246] M. Watanabe and R. M. Fuoss, J. Am. Chem. Soc. *78*, 527 (1965). [247] L. M. Litvinenko and V. A. Savelova, Reakts. Sposobnost Org. Soedin. *5*, 838 (1968); Chem. Abstr. *70*, 86878 x (1969). [248] E. T. Caldin and J. Peacock, Trans. Faraday Soc. *51*, 1217 (1955). [249] J. H.

Beard and P. H. Plesch, J. Chem. Soc. *1965*, 3682. [250] I. A. Koppel and V. A. Pal'm, Reakts. Sposobnost Org. Soedin. *4*, 862 (1967); Chem. Abstr. *69*, 99978 j (1968); ibid. *4*, 892 (1967); Chem. Abstr. *69*, 99980 d (1968).

[251] K. J. Laidler and H. Eyring, Ann. N. Y. Acad. Sci. *39*, 303 (1940); Chem. Abstr. *35*, 2777 (1941). [252] J. E. Quinlan and E. S. Amis, J. Am. Chem. Soc. *77*, 4187 (1955). [253] G. Scatchard, Chem. Rev. *10*, 229 (1932). [254] E. S. Amis and V. K. LaMer, J. Am. Chem. Soc. *61*, 905 (1939). [255] E. S. Amis and J. E. Price, J. Phys. Chem. *47*, 338 (1943). [256] J. N. Brønsted, Z. Phys. Chem. *102*, 169 (1922); *115*, 337 (1925). [257] N. Bjerrum, Z. Phys. Chem. *108*, 82 (1924); *118*, 251 (1925). [258] J. A. Christiansen, Z. Phys. Chem. *113*, 35 (1924). [259] P. Debye and E. Hückel, Phys. Z. *24*, 185, 305 (1923); E. Hückel, ibid. *26*, 93 (1925). [260] C. W. Davies; *Salt Effects in Solution Kinetics*, in G. Porter (ed.): *Progress in Reaction Kinetics*, Vol. 1, p. 161 ff., Pergamon Press, Oxford 1961.

[261] A. Streitwieser jr.: *Solvolytic Displacement Reactions*. McGraw Hill, New York 1962; Chem. Reviews *56*, 571 (1956). [262] E. S. Gould: *Mechanism and Structure in Organic Chemistry*. Holt, New York 1959; *Mechanismus und Struktur in der organischen Chemie* (translated by G. Koch). 2nd ed., Verlag Chemie, Weinheim 1969. [263] H. G. O. Becker: *Einführung in die Elektronentheorie organisch-chemischer Reaktionen*. 3rd ed., Deutscher Verlag der Wissenschaften, Berlin 1974. [264] C. G. Swain and R. W. Eddy, J. Am. Chem. Soc. *70*, 2989 (1948). [265] S. G. Smith, A. H. Fainberg, and S. Winstein, J. Am. Chem. Soc. *83*, 618 (1961). [266] J. A. Leary and M. Kahn, J. Am. Chem. Soc. *81*, 4173 (1959). [267] E. A. S. Cavell, J. Chem. Soc. *1958*, 4217; E. A. S. Cavell and J. A. Speed, ibid. *1960*, 1453. [268] S. Wideqvist, Arkiv Kemi *9*, 475 (1956); Chem. Abstr. *50*, 15187 (1956). [269] J. Murto, Suomen Kemistilehti *34*, 92 (1961); Chem. Abstr. *58*, 7803 (1963). [270] S. R. Palit, J. Org. Chem. *12*, 752 (1947).

[271] J. F. Bunnett: *Nucleophilic Reactivity*, in H. Eyring, C. J. Christensen, and H. S. Johnston (eds.): Ann. Rev. Phys. Chem. *14*, 271 (1963); Chem. Abstr. *59*, 14588 (1963). [272] C. G. Swain and C. B. Scott, J. Am. Chem. Soc. *75*, 141 (1953). [273] J. O. Edwards, J. Am. Chem. Soc. *76*, 1540 (1954); J. O. Edwards and R. G. Pearson, ibid. *84*, 16 (1962). [274] C. D. Ritchie and P. O. I. Virtanen, J. Am. Chem. Soc. *94*, 4966 (1972); ibid. *95*, 1882 (1973); ibid. *97*, 1170 (1975); ibid. *105*, 7313 (1983); Acc. Chem. Res. *5*, 348 (1972). [275] R. G. Pearson: *The Influence of the Reagent on Organic Reactivity*, in N. B. Chapman and J. Shorter (eds.): *Advances in Linear Free Energy Relationships*. Plenum Press, London, New York 1972, Chapter 6, p. 281 ff. [276] R. G. Pearson, H. Sobel, and J. Songstad, J. Am. Chem. Soc. *90*, 319 (1968). [277] R. G. Pearson and J. Songstad, J. Org. Chem. *32*, 2899 (1967). [278] W. M. Weaver and J. D. Hutchison, J. Am. Chem. Soc. *86*, 261 (1964). [279] S. Winstein, L. G. Savedoff, S. Smith, I. D. R. Stevens, and J. S. Gall, Tetrahedron Lett. Nr. 9, 24 (1960). [280] R. Fuchs and K. Mahendran, J. Org. Chem. *36*, 730 (1971).

[281] R. F. Rodewald, K. Mahendran, J. L. Bear, and R. Fuchs, J. Am. Chem. Soc. *90*, 6698 (1968); R. Fuchs, J. L. Bear, and R. F. Rodewald, ibid. *91*, 5797 (1969). [282] C. L. Liotta, E. E. Grisdale, and H. P. Hopkins, Tetrahedron Lett. *1975*, 4205. [283] W. T. Ford, R. J. Hauri, and S. G. Smith, J. Am. Chem. Soc. *96*, 4316 (1974). [284] J. E. Gordon and P. Varughese, J. Chem. Soc., Chem. Commun. *1971*, 1160. [285] J. I. Brauman, W. N. Olmstead, and C. A. Lieder, J. Am. Chem. Soc. *96*, 4030 (1974); W. N. Olmstead and J. I. Brauman, ibid. *99*, 4219 (1977). [286] M. S. Puar: *Nucleophilic Reactivities of the Halide Anions*. J. Chem. Educ. *47*, 473 (1970). [287] C. Minot and N. T. Anh: *Reversal of Nucleophilic Orders by Solvent Effects*. Tetrahedron Lett. *1975*, 3905. [288] G. Choux and R. L. Benoit, J. Am. Chem. Soc. *91*, 6221 (1969). [289] J. E. Gordon: *The Organic Chemistry of Electrolyte Solutions*. Wiley-Interscience, New York 1975, p. 299 ff. and 471 ff. [290] D. K. Bohme, G. I. Mackay, and J. D. Payzant, J. Am. Chem. Soc. *96*, 4027 (1974); K. Tanaka, G. I. Mackay, J. D. Payzant, and D. K. Bohme, Can. J. Chem. *54*, 1643 (1976).

[291] R. Alexander, E. C. F. Ko, A. J. Parker, and T. J. Broxton, J. Am. Chem. Soc. *90*, 5049 (1968); see also B. G. Cox, G. R. Hedwig, A. J. Parker, and D. W. Watts, Aust. J. Chem. *27*, 477 (1974). [292] B. G. Cox and A. J. Parker, J. Am. Chem. Soc. *95*, 408 (1973). [292a] A. J. Parker, U. Mayer, R. Schmid, and V. Gutmann, J. Org. Chem. *43*, 1843 (1978). [293] H. Suhr, Ber. Bunsenges. Phys. Chem. *67*, 893 (1963); Chem. Ber. *97*, 3277 (1964); Liebigs Ann. Chem. *687*, 175 (1965). [294] F. Madaule-Aubry: *Le Rôle en Chimie de Certaines Solvants Dipolaires Aprotiques*. Bull. Soc. Chim. Fr. *1966*, 1456. [295] H. Liebig: *Präparative Chemie in aprotischen Lösungsmitteln*, Chemiker Ztg. *95*, 301

(1971). [296] J. F. Coetzee: *Ionic Reactions in Acetonitrile*. Progr. Phys. Org. Chem. *4*, 45 (1967). [297] R. S. Kittila: *A Review of Catalytic and Synthetic Applications for DMF and DMAC*. Bulletin DuPont de Nemours and Co., Wilmington/USA 1960 (and Supplement 1962). [298] S. S. Pizey: *Dimethylformamide*, in S. S. Pizey, *Synthetic Reagents*. Wiley, New York, London 1974, Vol. I, p. 1ff. [299] D. Martin and H. G. Hauthal: *Dimethylsulfoxid*. Akademie-Verlag, Berlin 1971; *Dimethylsulphoxide*. Van Nostrand-Reinhold, London 1975. [299a] E. Buncel and H. Wilson: *Physical Organic Chemistry of Reactions in Dimethylsulphoxide*. Adv. Phys. Org. Chem. *14*, 133 (1977). [300] H. Normant: *Hexamethylphosphorsäuretriamid*. Angew. Chem. *79*, 1029 (1967); Angew. Chem., Int. Ed. Engl. *6*, 1046 (1967); Bull. Soc. Chim. Fr. *1968*, 791; Usp. Khim. *39*, 990 (1970); Russian Chem. Rev. *39*, 457 (1970)

[301] P. E. Pfeffer, T. A. Foglia, P. A. Barr, I. Schmeltz, and L. S. Silbert, Tetrahedron Lett. *1972*, 4063. [302] J. E. Shaw, D. C. Kunerth, and J. J. Sherry, Tetrahedron Lett. *1973*, 689. [303] R. G. Smith, A. Vanterpool, and H. J. Kulak, Can. J. Chem. *47*, 2015 (1969). [304] D. J. Cram, B. Rickborn, and G. R. Knox, J. Am. Chem. Soc. *82*, 6412 (1960). [305] A. J. Parker, J. Chem. Soc. *1961*, 1328. [306] T. J. Wallace, J. E. Hoffmann, and A. Schriesheim, J. Am. Chem. Soc. *85*, 2739 (1963). [307] J. T. Maynard, J. Org. Chem. *28*, 112 (1963). [308] J. F. Normant and H. Deshayes, Bull. Soc. Chim. Fr. *1967*, 2455. [309] S. Bank, J. Org. Chem. *37*, 114 (1972). [310] A. I. Shatenshtein, E. A. Gvozdeva, and Yu. I. Ranneva, Zh. Obshch. Khim. *41*, 1818 (1971); J. Gen. Chem. USSR *41*, 1827 (1971).

[311] P. K. Kadaba, Synthesis *1973*, 71; J. Org. Chem. *41*, 1073 (1976). [312] E. Tommila and M.-L. Murto, Acta Chem. Scand. *17*, 1947 (1963). [313] R. Goitein and T. C. Bruice, J. Phys. Chem. *76*, 432 (1972). [314] H.-L. Pan and T. L. Fletcher, Chem. Ind. (London) *1969*, 240. [315] G. Bähr and G. Schleitzer, Chem. Ber. *88*, 1771 (1955). [316] O. Popovych: *Estimation of Medium Effects for Single Ions in Nonaqueous Solvents*. Critical Reviews in Analytical Chemistry *1*, 73 (1970); Chem. Abstr. *75*, 133680 s (1971). [317] B. G. Cox: *Electrolyte Solutions in Dipolar Aprotic Solvents*. Annual Reports on the Progress of Chemistry *A 70*, 249 (1973), and references cited therein. [318] B. G. Cox and A. J. Parker, J. Am. Chem. Soc. *95*, 402 (1973). [319] D. W. Watts: *Reaction Kinetics and Mechanism*, in A. K. Covington and T. Dickinson (eds.): *Physical Chemistry of Organic Solvent Systems*. Plenum Press, London, New York 1973, Chapter 6, p. 681 ff. [320] R. Alexander and A. J. Parker, J. Am. Chem. Soc. *90*, 3313 (1968); R. Alexander, A. J. Parker, J. H. Sharp, and W. E. Waghorne, ibid. *94*, 1148 (1972).

[321] D. J. Cram; *Fundamentals of Carbanion Chemistry*. Academic Press, New York, London 1965, p. 32ff. [322] E. Buncel: *Carbanions: Mechanistic and Isotopic Aspects*. Elsevier. Amsterdam 1975. [323] A. J. Hubert and H. Reimlinger: *The Isomerization of Olefins. Part I. Base-Catalysed Isomerization of Olefins*. Synthesis *1969*, 97. [324] H. Pines: *Base-Catalysed Carbon-Carbon Addition of Hydrocarbons and of Related Compounds*. Acc. Chem. Res. *7*, 155 (1974). [325] C. C. Price and W. H. Snyder, J. Am. Chem. Soc. *83*, 1773 (1961); Tetrahedron Lett. No. 2, 69 (1962). [326] J. Sauer and H. Prahl, Chem. Ber. *102*, 1917 (1969). [327] R. Greenwald, M. Chaykovsky, and E. J. Corey, J. Org. Chem. *28*, 1128 (1963). [328] D. J. Cram, M. R. V. Sahyun, and G. R. Knox, J. Am. Chem. Soc. *84*, 1734 (1962). [329] H. H. Szmant: *Der Mechanismus der Wolff-Kishner-Reaktionen: Reduktion, Eliminierung und Isomerisierung*. Angew. Chem. *80*, 141 (1968); Angew. Chem., Int. Ed. Engl. *7*, 120 (1968). [330] H. H. Szmant and M. N. Roman, J. Am. Chem. Soc. *88*, 4034 (1966).

[331] M. R. V. Sahyun and D. J. Cram, J. Am. Chem. Soc. *85*, 1263 (1963). [332] S. F. Acree, Am. Chem. J. *48*, 352 (1912). [333] M. Swarcz (ed.): *Ions and Ion Pairs in Organic Reactions*. Wiley-Interscience, New York 1972/74, Vol. 1 and 2. [334] I. P. Beletskaya: *Ions and Ion-Pairs in Nucleophilic Aliphatic Substitutions*. Usp. Khim. *44*, 2205 (1975); Russian Chem. Rev. *44*, 1067 (1975). [335] F. Guibe and G. Bram: *Réactivité S_N2 des formes dissociée et associée aux cations alcalins des nucléophiles anioniques*. Bull. Soc. Chim. Fr. *1975*, 933. [336] E. A. Kovrizhnykh and A. I. Shatenshtein: *Effect of Electron-Donor Solvents on the Reactivity of Lithium Alkyls*. Usp. Khim. *38*, 1836 (1969); Russian Chem. Rev. *38*, 840 (1969). [337] V. S. Petrosyan and O. A. Reutov: *Effect of Solvent upon the Rates and Mechanisms of Organometallic Reactions. General Aspects*. J. Organomet. Chem. *52*, 307 (1973). [338] V. Gutmann: *Solvent Effects on the Reactivities of Organometallic Compounds*. Coord. Chem. Rev. *18*, 225 (1976). [339] *Cf.* also references [166–168] to Chapter 2. [340] J. Ugelstad, T. Ellingsen, and A. Berge, Acta Chem. Scand, *20*, 1593 (1966).

[341] P. Müller and B. Siegfried, Helv. Chim. Acta *55*, 2965 (1972). [342] U. Mayer, V. Gutmann, and A. Lodzinski, Monatsh. Chem. *104*, 1045 (1973). [343] C. J. Pedersen, J. Am. Chem. Soc. *89*, 2495, 7017 (1967); C. J. Pedersen and H. K. Frensdorff: *Makrocyclische Polyäther und ihre Komplexe*. Angew. Chem. *84*, 16 (1972); Angew. Chem., Int. Ed. Engl. *11*, 16 (1972). [344] B. Dietrich, J.-M. Lehn, and J.-P. Sauvage, Tetrahedron Lett. *1969*, 2885, 2889; Chemie in unserer Zeit *7*, 120 (1973); J.-M. Lehn: *Supramolekulare Chemie*. Angew. Chem. *100*, 92 (1988); Angew. Chem., Int. Ed. Engl. *27* (1988); J.-M. Lehn: *Cryptates: The Chemistry of Macropolycyclic Inclusion Complexes*. Acc. Chem. Res. *11*, 49 (1978). [345] F. Vögtle and E. Weber (eds.): *Host Guest Complex Chemistry – Macrocycles – Synthesis, Structures, Applications*. Springer, Berlin 1985. [346] R. Izatt and J. J. Christensen (eds.): *Synthetic Multidentate Macrocyclic Compounds: Synthesis, Properties, and Uses*. Academic Press, New York, London 1978. [347] G. W. Gokel and H. D. Durst: *Principles and Synthetic Applications in Crown Ether Chemistry*. Synthesis *1976*, 168. [348] F. Vögtle and E. Weber: *Progress in Crown Ether Chemistry*. Part IV A...IV E. Kontakte (Merck, Darmstadt) *1980* (2), 36ff.; *1981* (1), 24ff.; *1982* (1), 24ff.; *1983* (1), 38ff.; *1984* (1), 26ff. [349] E. Graf and J.-M. Lehn, J. Am. Chem. Soc. *97*, 5022 (1975); B. Metz, J. M. Rosalky, and R. Weiss, J. Chem. Soc., Chem. Commun. *1976*, 533. [350] H. E. Zaugg, B. W. Horrom, and S. Borgwardt, J. Am. Chem. Soc. *82*, 2895 (1960); H. E. Zaugg, ibid. *83*, 837 (1961).

[351] H. E. Zaugg, J. F. Ratajczyk, J. E. Leonard, and A. D. Schaefer, J. Org. Chem. *37*, 2249 (1972). [352] H. D. Zook, T. J. Russo, E. F. Ferrand, and D. S. Stotz, J. Org. Chem. *33*, 2222 (1968). [353] H. D. Zook and T. J. Russo, J. Am. Chem. Soc. *82*, 1258 (1960); H. D. Zook and W. L. Gumby, ibid. *82*, 1386 (1960). [354] L. M. Thomasson, T. Ellingsen, and J. Ugelstad, Acta Chem. Scand. *25*, 3024 (1971). [355] C. L. Liotta and H. P. Harris, J. Am. Chem. Soc. *96*, 2250 (1974). [356] F. L. Cook, C. W. Bowers, and C. L. Liotta, J. Org. Chem. *39*, 3416 (1974). [357] D. J. Sam and H. E. Simmons, J. Am. Chem. Soc. *96*, 2252 (1974). [358] M. Cinquini, F. Montanori, and P. Tundo, J. Chem. Soc., Chem. Commun. *1975*, 393. [359] C. L. Liotta, H. P. Harris, M. McDermott, T. Gonzalez, and K. Smith, Tetrahedron Lett. *1974*, 2417. [360] H. D. Durst, Tetrahedron Lett. *1974*, 2421; H. D. Durst, M. Milano, E. J. Kikta, S. A. Connelly, and E. Grushka, Anal. Chem. *47*, 1797 (1975).

[361] J. W. Zubrick, B. I. Dunbar, and H. D. Durst, Tetrahedron Lett. *1975*, 71. [362] R. A. Bartsch: *Ionic Association in Base-Promoted β-Elimination Reactions*. Acc. Chem. Res. *8*, 239 (1975). [363] N. Kornblum, R. A. Smiley, R. K. Blackwood, and D. C. Iffland, J. Am. Chem. Soc. *77*, 6269 (1955). [364] R. Gompper: *Beziehungen zwischen Struktur und Reaktivität ambifunktioneller nucleophiler Verbindungen*. Angew. Chem. *76*, 412 (1964); Angew. Chem., Int. Ed. Engl. *3*, 560 (1964); Usp. Khim. *36*, 803 (1967). [365] W. J. Le Noble: *Conditions for the Alkylation of Ambident Anions*. Synthesis *1970*, 1. [366] S. A. Shevelev: *Dual Reactivity of Ambident Anions*. Usp. Khim. *39*, 1773 (1970); Russian Chem. Rev. *39*, 844 (1970). [367] R. Gompper and H.-U. Wagner: *Das Allopolarisierungs-Prinzip. Substituenteneinflüsse auf Reaktionen ambifunktioneller Anionen*. Angew. Chem. *88*, 389 (1976); Angew. Chem., Int. Ed. Engl. *15*, 321 (1976). [367a] L. M. Jackman and B. C. Lange: *Structure and Reactivity of Alkali Metal Enolates*. Tetrahedron *33*, 2737 (1977). [367b] O. A. Reutov, I. P. Beletskaya, and A. L. Kurts: *Ambident Anions* (translated from Russian by J. P. Michael), Consultants Bureau, Plenum Publishing Corp., New York 1983. [368] S. Hünig: *Die Reaktionsweise ambidenter Kationen*. Angew. Chem. *76*, 400 (1964); Angew. Chem., Int. Ed. Engl. *3*, 548 (1964); Usp. Khim. *36*, 693 (1967). [369] L. Claisen, Z. Angew. Chem. *36*, 478 (1923); L. Claisen, F. Kremers, F. Roth, and E. Tietze, Liebigs Ann. Chem. *442*, 210 (1925). [370] N. Kornblum, P. J. Berrigan, and W. J. Le Noble, J. Am. Chem. Soc. *82*, 1257 (1960); ibid. *85*, 1141 (1963).

[371] N. Kornblum, R. Seltzer, and P. Haberfield, J. Am. Chem. Soc. *85*, 1148 (1963). [372] A. L. Kurts, A. Macias, N. K. Genkina, I. P. Beletskaya, and O. A. Reutov, Dokl. Akad. Nauk SSSR *187*, 807 (1969); Chem. Abstr. *71*, 112364u (1969). [373] A. L. Kurts, P. I. Dem'yanov, A. Macias, I. P. Beletskaya, and O. A. Reutov, Tetrahedron Lett. *1968*, 3679; Tetrahedron *27*, 4759, 4769, 4777 (1971). [374] G. Brieger and W. M. Pelletier, Tetrahedron Lett. *1965*, 3555. [375] S. J. Rhoads and R. W. Holder, Tetrahedron *25*, 5443 (1969). [376] S. G. Smith and M. P. Hanson, J. Org. Chem. *36*, 1931 (1971). [377] A. L. Kurts, P. I. Dem'yanov, I. P. Beletskaya, and O. A. Reutov, Zh. Org. Khim. *9*, 1313 (1973); J. Org. Chem. USSR *9*, 1341 (1973). [378] W. J. Le Noble and S. K. Palit, Tetrahedron Lett. *1972*, 493. [379] G. Klopman, J. Am. Chem. Soc. *90*, 223 (1968); *Chemical Reactivity and Reaction Phaths*. Wiley, New York 1974. [380] A. Bertho, Liebigs Ann. Chem. *714*, 155 (1968).

[381] P. G. Duggan and W. S. Murphy, J. Chem. Soc., Perkin Trans. II *1975*, 1291. [382] O. A. Reutov: *Mechanism of Reactions of Electrophilic Substitution at a Saturated Carbon Atom*. Usp. Khim. *36*, 414 (1967); Russian Chem. Rev. *36*, 163 (1967); Fortschr. Chem. Forsch. *8*, 61 (1967); J. Organomet. Chem. *100*, 219 (1975). [383] M. H. Abraham: *Constitutional Effects, Salt Effects, and Solvent Effects in Electrophilic Substitution at Saturated Carbon*. Compr. Chem. Kinet. *12*, 211 (1973); Chem. Abstr. *83*, 8562u (1975). [384] O. A. Reutov, V. I. Sokolov, I. P. Beletskaya, Yu. S. Ryabokobylko, and B. Prajsnar, Izvest. Akad. Nauk SSSR *1963*, 966, 970; Chem. Abstr. *59*, 5837 (1963). [385] E. D. Hughes, C. K. Ingold, and R. M. G. Roberts, J. Chem. Soc. *1964*, 3900. [386] G. Heublein, J. Prakt. Chem. *31*, 84 (1966). [387] J. H. Rolston and K. Yates, J. Am. Chem. Soc. *91*, 1477 (1969). [388] S. P. McManus and P. E. Peterson, Tetrahedron Lett. *1975*, 2753; S. P. McManus and S. D. Worley, ibid. *1977*, 555; and references cited therein. [389] K. Nakagawa, Y. Ishii, and M. Ogawa, Chemistry Lett. (Tokyo) *1976*, 511; Tetrahedron *32*, 1427 (1976). [390] G. Mark, F. Mark, and O. E. Polansky, Liebigs Ann. Chem. *719*, 151 (1968).

[391] R. Steinmetz, W. Hartmann, and G. O. Schenck, Chem. Ber. *98*, 3854 (1965); I.-M. Hartmann, W. Hartmann, and G. O. Schenck, ibid. *100*, 3146 (1967). [392] F. D. Lewis and R. J. DeVoe, Tetrahedron *38*, 1069 (1982). [393] H. van Zwet and E. C. Kooyman, Rec. Trav. Chim. Pays-Bas *86*, 1143 (1967). [394] J. Brokken-Zijp and H. v. d. Bogaert, Tetrahedron *29*, 4169 (1973). [395] J. Sicher: *Der syn- und anti-koplanare Verlauf bimolekularer olefin-bildender Eliminierungen*. Angew. Chem. *84*, 177 (1972); Angew. Chem., Int. Ed. Engl. *11*, 200 (1972). [396] J. Avraamides and A. J. Parker, Tetrahedron Lett. *1971*, 4043. [397] H. Kessler: *Nachweis gehinderter Rotationen und Inversionen durch NMR-Spektroskopie*. Angew. Chem. *82*, 237 (1970); Angew. Chem., Int. Ed. Engl. *9*, 219 (1970). [398] U. Berns, G. Heinrich, and H. Güsten, Z. Naturforsch. *31b*, 953 (1976); and references cited therein. [399] *Cf.* also references [74–77, 80, 128, 131, 132, 220–224] in reference list to Chapter 2. [400] C. A. Bunton and G. Savelli: *Organic Reactivity in Aqueous Micelles and Similar Assemblies*. Adv. Phys. Org. Chem. *22*, 213 (1986).

[401] E. H. Cordes (ed.): *Reaction Kinetics in Micelles*. Plenum Press, New York, London 1973. [402] E. J. Fendler and J. H. Fendler: *Micellar Catalysis in Organic Reactions: Kinetic and Mechanistic Implications*. Adv. Phys. Org. Chem. *8*, 271 (1970). [403] J. H. Fendler and E. J. Fendler: *Catalysis in Micellar and Macromolecular Systems*. Academic Press, New York 1975; J. H. Fendler: *Membrane Mimetic Chemistry*. Wiley, New York 1982. [404] H. Chaimovich, A. Blanco, L. Chayet, L. M. Costa, P. M. Monteiro, C. A. Bunton, and C. Paik, Tetrahedron *31*, 1139 (1975); C. A. Bunton, Pure Appl. Chem. *49*, 969 (1977). [405] C. A. Blyth and J. R. Knowles, J. Am. Chem. Soc. *93*, 3017 (1971). [406] D. Oakenfull, J. Chem. Soc., Perkin Trans. II *1973*, 1006. [407] J. H. Fendler, E. J. Fendler, and L. W. Smith, J. Chem. Soc., Perkin Trans. II *1972*, 2097. [408] R. M. Noyes: *Effects of Diffusion Rates on Chemical Kinetics*, in G. Porter und B. Stevens (eds.): *Progress in Chemical Kinetics*. Pergamon Press, Oxford 1961, Vol. 1, p. 129ff. [409] A. M. North: *Diffusion-Controlled Reactions*. Quart. Rev. *20*, 421 (1966). [409a] S. A. Rice (ed.): *Diffusion-limited Reactions*. In C. H. Bamford, C. F. H. Tipper, and R. G. Compton (eds.): *Comprehensive Chemical Kinetics*. Vol. 25, Elsevier, Amsterdam 1985. [410] K. U. Ingold: *Rate Constants for Free Radical Reactions in Solution*, in J. K. Kochi (ed.): *Free Radicals*. Wiley-Interscience, New York 1973, Vol. 1, Chapter 2, p. 37ff.

[411] M. Eigen: *Protonenübertragung, Säure-Base-Katalyse und enzymatische Hydrolyse*. Angew. Chem. *75*, 489 (1963). [412] J. Frank and E. Rabinowitsch, Trans. Faraday Soc. *30*, 120 (1934); E. Rabinowitsch and W. C. Wood, ibid. *32*, 1381 (1936). [413] T. Koenig and H. Fischer: *Cage Effects*, in J. K. Kochi (ed.): *Free Radicals*. Wiley-Interscience, New York 1973, Vol. 1, Chapter 4, p. 157ff. [414] M. J. Gibian and R. C. Corley: *Organic Radical-Radical Reactions. Disproportionation vs. Combination*. Chem. Rev. *73*, 441 (1973). [415] L. Herk, M. Feld, and M. Szwarc, J. Am. Chem. Soc. *83*, 2998 (1961); O. Dobis, J. M. Pearson, and M. Szwarc, ibid. *90*, 278 (1968). [416] R. K. Lyon and D. H. Levy, J. Am. Chem. Soc. *83*, 4290 (1961); R. K. Lyon, ibid. *86*, 1907 (1964). [417] R. E. Rebbert and P. Ausloos, J. Phys. Chem. *66*, 2253 (1962). [418] G. S. Hammond, J. N. Sen, and C. E. Boozer, J. Am. Chem. Soc. *77*, 3244 (1955). [419] G. Ayrey, Chem. Rev. *63*, 645 (1963). [420] J. W. Taylor and J. C. Martin, J. Am. Chem. Soc. *88*, 3650 (1966); ibid. *89*, 6904 (1967).

[421] C. Rüchardt and M. Grundmaier, Chem. Ber. *108*, 2448 (1975). [422] J. K. Kochi, J. Am. Chem. Soc. *92*, 4395 (1970). [423] N. Nodelman and J. C. Martin, J. Am. Chem. Soc. *98*, 6597 (1976). [424] R. Criegee: *Mechanismus der Ozonolyse*. Angew. Chem. *87*, 765 (1975); Angew. Chem., Int. Ed. Engl.

14, 745 (1975). [425] L. D. Loan, R. W. Murray, and P. R. Story, J. Am. Chem. Soc. *87*, 737 (1965). [426] O. Lorenz and C. R. Parks, J. Org. Chem. *30*, 1976 (1965). [427] R. Criegee: *The Course of Ozonization of Unsaturated Compounds*. Record Chem. Progr. (Kresge-Hooker Sci. Lib.) *18*, 111 (1957); Chem. Abstr. *51*, 11982 (1957). [428] R. W. Murray: *The Mechanism of Ozonolysis*. Acc. Chem. Res. *1*, 313 (1968). [428a] V. Ramachandran and R. W. Murray, J. Am. Chem. Soc. *100*, 2197 (1978). [429] E. Whalley: *Chemical Reactions in Solutions under High Pressure*. Ber. Bunsenges. Phys. Chem. *70*, 958 (1966). [430] W. J. Le Noble: *Kinetics of Reactions in Solutions under Pressure*. Progr. Phys. Org. Chem. *5*, 207 (1967); J. Chem. Educ. *44*, 729 (1967); T. Asano and W. J. Le Noble: *Activation and Reaction Volumes in Solution*. Chem. Rev. *78*, 407 (1978); W. J. Le Noble and H. Kelm: *Chemie in komprimierten Lösungen*. Angew. Chem. *92*, 887 (1980); Angew. Chem., Int. Ed. Engl. *19*, 841 (1980); W. J. Le Noble, Chemie in unserer Zeit *17*, 152 (1983).

[431] H. Heydtmann: *Reaktionen in Lösungen unter erhöhten statischen Drucken*, in H. Hartmann (ed.): *Chemische Elementarprozesse*. Springer-Verlag, Berlin 1968, p. 331 ff. [432] C. A. Eckert: *High-Pressure Kinetics in Solution*. Annu. Rev. Phys. Chem. *23*, 239 (1972). [433] G. Jenner: *Hochdruck-kinetische Untersuchungen in der Organischen und Makromolekularen Chemie*. Angew. Chem. *87*, 186 (1975); Angew. Chem., Int. Ed. Engl. *14*, 137 (1975). [434] M. G. Evans and M. Polanyi, Trans. Faraday Soc. *31*, 875 (1935); ibid. *32*, 1333 (1936). [435] R. C. Neuman: *Pressure Effects as Mechanistic Probes of Organic Radical Reactions*. Acc. Chem. Res. *5*, 381 (1972). [436] J. Buchanan and S. D. Hamann, Trans. Faraday Soc. *49*, 1425 (1953). [437] R. A. Grieger and C. A. Eckert, Trans. Faraday Soc. *66*, 2579 (1970). [438] R. J. Ouellete and S. H. Williams, J. Am. Chem. Soc. *93*, 466 (1971). [439] K. R. Brower, J. Am. Chem. Soc. *85*, 1401 (1963). [440] H. Hartmann, H. Kelm, and G. Rinck, Z. Phys. Chem. (Frankfurt) N. F. *44*, 335 (1965).

[441] H. Hartmann, H.-D. Brauer, H. Kelm, and G. Rinck, Z. Phys. Chem. (Frankfurt) N.F. *61*, 47, 53 (1968). [442] H.-D. Brauer and H. Kelm, Z. Phys. Chem. (Frankfurt) N.F. *76*, 98 (1971); ibid. *79*, 96 (1972). [443] Y. Kondo, M. Ohnishi, M. Uchida, and N. Tokura, Bull. Chem. Soc. Jpn. *41*, 992 (1968); ibid. *45*, 3579 (1972). [444] J. R. McCabe, R. A. Grieger, and C. A. Eckert, Ind. Eng. Chem., Fundam. *9*, 156 (1970); Chem. Abstr. *72*, 71368 y (1970). [445] K. Tamura, Y. Ogo, and T. Imoto, Bull. Chem. Soc. Jpn. *46*, 2988 (1973); ibid. *48*, 369 (1975). [446] E. M. Arnett and D. R. McKelvey: *Solvent Isotope Effects on Thermodynamics of Nonreacting Solutes*, in J. F. Coetzee and C. D. Ritchie (eds.): *Solute-Solvent Interactions*. Dekker, New York 1969, Vol. 1, p. 344ff. [447] P. M. Laughton and R. E. Robertson: *Solvent Isotope Effects for Equilibria and Reactions*, in J. F. Coetzee and C. D. Ritchie (eds.): *Solute-Solvent Interactions*. Dekker, New York 1969, Vol. 1, p. 399ff. [448] C. J. Collins and N. S. Bowman (eds.): *Isotope Effects in Chemical Reactions*. Van Nostrand Reinhold, New York 1970, p. 246 and 266ff. [449] R. L. Schowen: *Mechanistic Deductions from Solvent Isotope Effects*. Progr. Phys. Org. Chem. *9*, 275 (1972). [450] C. G. Swain and R. F. W. Bader, Tetrahedron *10*, 182 (1960).

[451] R. C. Weast (ed.): *Handbook of Chemistry and Physics*, 67th edition, CRC Press, Boca Raton/Florida 1986/87. [452] E. F. Caldin: *Reaction Kinetics and Solvation in Non-Aqueous Solvents*. Pure Appl. Chem. *51*, 2067 (1979). [453] M. J. Blandamer and J. Burgess: *Initial State and Transition State Solvation in Inorganic Reactions*. Coord. Chem. Rev. *31*, 93 (1980); M. J. Blandamer, J. Burgess, and J. B. F. N. Engberts: *Activation Parameters for Chemical Reactions in Solution*. Chem. Soc. Rev. *14*, 237 (1985). [454] O. Popovych and R. P. T. Tompkins: *Nonaqueous Solution Chemistry*. Wiley, New York 1981. [455] I. Bertini, L. Lunazzi, and A. Dei (eds.): *Advances in Solution Chemistry*. Plenum Press, New York, London 1981. [456] C. Reichardt: *Der Lösungsmitteleinfluß auf chemische Reaktionen*. Chemie in unserer Zeit *15*, 139 (1981). [457] R. Schmid and V. N. Sapunov: *Non-Formal Kinetics*. Verlag Chemie, Weinheim 1982. [458] K. Burger: *Solvation, Ionic and Complex Formation Reactions in Non-Aqueous Solvents*. Elsevier, Amsterdam 1983. [459] R. A. Y. Jones: *Physical and Mechanistic Organic Chemistry*. 2nd edition, Cambridge University Press, Cambridge 1984, Chapter 5, p. 94ff. [460] N. M. Emanuel, G. E. Zaikov, and Z. K. Maizus: *Oxidation of Organic Compounds. Medium Effects in Radical Reactions*. Pergamon Press, Oxford 1984.

[461] J. Hayami: *Solvent Effects in the Organic Reactions*. Yuki Gosei Kagaku Kyokaishi *42*, 1107 (1984); Chem. Abstr. *102*, 148339 (1985). [462] A. L. Kurts: *Role of the Solvent in Organic Reactions*. Zh. Vses. Khim. Obshchestva im. D. I. Mendeleeva *29*, 530 (1984); Chem. Abstr. *102*, 5206 f (1985).

[463] J. T. Hynes: *Chemical Reaction Dynamics in Solution.* Annu. Rev. Phys. Chem. *36*, 573 (1985); G. van der Zwan and J. T. Hynes, J. Phys. Chem. *89*, 4181 (1985). [464] K. J. Laidler and M. C. King: *The Development of Transition-State Theory.* J. Phys. Chem. *87*, 2657 (1983). [465] M. M. Kreevoy and D. G. Truhlar: *Transition State Theory.* In C. F. Bernasconi (ed.): *Investigation of Rates and Mechanisms of Reactions* (Volume VI of the Series *Techniques of Chemistry*). 4th ed., Wiley, New York 1986, Part I, p. 13 ff. [466] J. Bertrán and F. S. Burgos: *The Question of Equilibrium in Transition State Solvation.* J. Chem. Educ. *61*, 416 (1984). [467] E. Buncel and H. Wilson: *Initial-State and Transition-State Solvent Effects on Reaction Rates and the Use of Thermodynamic Transfer Functions.* Acc. Chem. Res. *12*, 42 (1979); *Solvent Effects on Rates and Equilibria.* J. Chem. Educ. *57*, 629 (1980). [468] C. Reichardt: *Solvent Effects on Chemical Reactivity.* Pure Appl. Chem. *54*, 1867 (1982). [469] R. T. McIver: *Chemical Reactions without Solvation.* Scientific American *243*, 148 (1980); Spektrum der Wissenschaft, Januar 1981, p. 27ff. [470] P. Ausloos (ed.): *Kinetics of Ion-Molecule Reactions.* (NATO Advanced Study Institute Series B, Vol. 40). Plenum Publishing Company, New York 1979.

[471] M. T. Bowers (ed.): *Gas Phase Ion Chemistry.* Vol. 1 and 2, Academic Press, New York 1979. [472] J. H. Bowie: *Bimolecular Reactions of Nucleophiles in the Gas Phase.* Acc. Chem. Res. *13*, 76 (1980). [473] T. H. Morton: *Gas Phase Analogues of Solvolysis Reactions.* Tetrahedron *38*, 3195 (1982). [474] W. N. Olmstead and J. I. Brauman: *Gas-Phase Nucleophilic Displacement Reactions.* J. Am. Chem. Soc. *99*, 4219 (1977); M. J. Pellerite and J. I. Brauman: *Intrinsic Barriers in Nucleophilic Displacements.* J. Am. Chem. Soc. *102*, 5993 (1980); J. I. Brauman *et al.*, Pure Appl. Chem. *56*, 1809 (1984); J. Phys. Chem. *90*, 471 (1986). [475] D. K. Bohme: *Gas-Phase Studies of the Influence of Solvation on Ion Reactivity.* In M. A. Almoster Ferreira (ed.): *Ionic Processes in the Gas Phase.* Reidel Publishing Company, Dordrecht 1984, p. 111ff. [476] P. Kebarle, G. Caldwell, T. Magnera, and J. Sunner: *Ions-Gas Phase and Solution-Dipolar Aprotic Solvents.* Pure Appl. Chem. *57*, 339 (1985). [477] J. M. Riveros, S. M. José, and K. Takashima; *Gas-Phase Nucleophilic Displacement Reactions.* Adv. Phys. Org. Chem. *21*, 197 (1985). [478] E. M. Arnett: *Solvation Energies of Organic Ions.* J. Chem. Educ. *62*, 385 (1985). [479] H. A. Kramers: *Brownian Motion in a Field of Force and the Diffusion Model of Chemical Reactions.* Physica 7, 284 (1940); Chem. Abstr. *34*, 4968 (1940). [480] K. Tanaka, G. I. Mackay, J. D. Payzant, and D. K. Bohme, Can. J. Chem. *54*, 1643 (1976).

[481] J. Chandrasekhar, S. F. Smith, and W. L. Jorgensen, J. Am. Chem. Soc. *106*, 3049 (1984); ibid. *107*, 154 (1985). [482] F. Carrion and M. J. S. Dewar, J. Am. Chem. Soc. *106*, 3531 (1984); M. J. S. Dewar and D. M. Storch, J. Chem. Soc., Chem. Commun. *1985*, 94. [483] S. S. Shaik, J. Am. Chem. Soc. *106*, 1227 (1984); Progr. Phys. Org. Chem. *15*, 197 (1985); Israel J. Chem. *26*, 367 (1985). [484] D. K. Bohme and L. B. Young, J. Am. Chem. Soc. *92*, 7354 (1970). [485] D. K. Bohme and G. I. Mackay, J. Am. Chem. Soc. *103*, 978 (1981). [486] D. K. Bohme and A. B. Raksit, J. Am. Chem. Soc. *106*, 3447 (1984); Can. J. Chem. *63*, 3007 (1985). [487] M. Henchman, J. F. Paulson, and P. M. Hierl, J. Am. Chem. Soc. *105*, 5509 (1983); P. M. Hierl, A. F. Ahrens, M. Henchman, A. A. Viggiano, J. F. Paulson, and D. C. Clary, ibid. *108*, 3142 (1986). [488] G. Caldwell, M. D. Rozeboom, J. P. Kiplinger, and J. E. Bartmess, J. Am. Chem. Soc. *106*, 809 (1984). [489] K. Morokuma, J. Am. Chem. Soc. *104*, 3732 (1982). [490] D. N. Kevill and G. M. L. Liu, J. Am. Chem. Soc. *101*, 3916 (1979).

[491] F. Quemeneur and B. Bariou, J. Chem. Res. (S) *1979*, 187, 188; J. Chem. Res. (M) *1979*, 2344, 2357. [492] E. M. Arnett and R. Reich, J. Am. Chem. Soc. *102*, 5892 (1980). [493] D. N. Kevill, J. Chem. Soc., Chem. Commun. *1981*, 421. [494] M. H. Abraham and A. Nasehzadeh, J. Chem. Soc., Chem. Commun. *1981*, 905. [495] Y. Kondo, M. Ogasa, and S. Kusabayashi, J. Chem. Soc., Perkin Trans. II *1984*, 2093. [496] V. Bekárek, T. Nevěčná, and J. Vymětalová, Collect. Czech. Chem. Commun. *50*, 1928 (1985); ibid. *51*, 2071 (1986). [497] Y. Kondo, M. Ittoh, and S. Kusabayashi, J. Chem. Soc., Faraday Trans. I *78*, 2793 (1982). [498] K. C. Westaway, Can. J. Chem. *56*, 2691 (1978). [499] H. Maskill; *The Physical Basis of Organic Chemistry.* Oxford University Press, Oxford, New York 1985, p. 409ff. [500] Y. Kondo, A. Zanka, and S. Kusabayashi, J. Chem. Soc., Perkin Trans. II 1985, 827.

[501] C. F. Bernasconi in H. Zollinger (ed.): *MTP Rev. Sci., Org. Chem. Ser. One*, Vol. 3, p. 35ff., Butterworths, London 1973; Acc. Chem. Res. *11*, 147 (1978). [502] L. Forlani, J. Chem. Res. (S) *1984*, 260; J. Chem. Res. (M) *1984*, 2379. [503] P. M. E. Mancini, R. D. Martinez, L. R. Vottero, and N. S. Nudelman, J. Chem. Soc., Perkin Trans. II *1984*, 1133, ibid. *1986*, 1427; ibid. *1987*, 951. [504] I. L. Bagal, S. A. Skvortsov, and A. V. El'tsov, Zh. Org. Khim. *14*, 361 (1978); J. Org. Chem. USSR *14*, 328

(1978). [505] Y. Hashida, F. Tanabe, and K. Matsui, Nippon Kagaku Kaishi *1980*, 865. [506] I. Szele and H. Zollinger: *Azo Coupling Reactions. Structures and Mechanisms.* Top. Curr. Chem. *112*, 1 (1983). [507] M. Gielen, Acc. Chem. Res. *6*, 198 (1973). [508] J. M. Fukuto and F. R. Jensen, Acc. Chem. Res. *16*, 177 (1983). [509] S. Fukuzumi and J. K. Kochi, J. Phys. Chem. *84*, 2254 (1980); J. Am. Chem. Soc. *102*, 2141 (1980). [510] V. S. Petrosyan, J. Organomet. Chem. *250*, 157 (1983); Advances in the USSR Organometallic Chemistry. Chapter 3, p. 73ff., MIR Publishers, Moscow 1985.

[511] K. Ziegler *et al.*, Liebigs Ann. Chem. *473*, 1 (1929); ibid. *479*, 135 (1930). [512] G. Belluci, R. Bianchini, and R. Ambrosetti, J. Am. Chem. Soc. *107*, 2464 (1985). [513] M.-F. Ruasse and B.-L. Zhang, J. Org. Chem. *49*, 3207 (1984). [514] G. A. Jones, C. J. M. Stirling, and N. G. Bromby, J. Chem. Soc., Perkin Trans. II *1983*, 385. [515] W. A. Smit, N. S. Zefirov, I. V. Bodrikov, and M. Z. Krimer, Acc. Chem. Res. *12*, 282 (1979). [516] H. Mimoun, Angew. Chem. *94*, 750 (1982); Angew. Chem., Int. Ed. Engl. *21*, 734 (1982). [517] V. M. Vishnyakov, A. A. Bezdenezhnykh, Z. S. Zamchuk, and O. M. Kuznetsova, Zh. Org. Khim. *14*, 1238 (1978); J. Org. Chem. USSR *14*, 1146 (1978). [518] H. Graf and R. Huisgen, J. Org. Chem. *44*, 2594, 2595 (1979). [519] R. Huisgen and J. P. Ortega, Tetrahedron Lett. *1978*, 3975. [520] J. Drexler, R. Lindermeyer, M. A. Hassan, and J. Sauer, Tetrahedron Lett. *26*, 2559 (1985).

[521] H. K. Hale, L. C. Dunn, and A. B. Padias, J. Org. Chem. *45*, 835 (1980). [522] J. Mulzer and M. Zippel, Tetrahedron Lett. *21*, 751 (1980). [523] N. S. Isaacs and A. H. Laila, Tetrahedron Lett. *24*, 2897 (1983). [524] T. Minato and S. Yamabe, J. Org. Chem. *48*, 1479 (1983). [525] C. H. Heathcock; *Stereoselective Aldol Condensations.* In E. Buncel and T. Durst (eds.): *Comprehensive Carbanion Chemistry.* Part B, p. 177ff. (particularly p. 198), Elsevier, Amsterdam 1984. [526] C. H. Heathcock and J. Lampe, J. Org. Chem. *48*, 4330 (1983). [527] P. Wildes, J. G. Pacifici, G. Irick, and D. G. Whitten, J. Am. Chem. Soc. *93*, 2004 (1971); K. S. Schanze, T. F. Mattox, and D. G. Whitten, ibid. *104*, 1733 (1982); J. Org. Chem. *48*, 2808 (1983). [528] T. Asano and T. Okada, J. Org. Chem. *49*, 4387 (1984); ibid. *51*, 4454 (1986); Chem. Lett. (Tokyo) *1987*, 695. [529] N. Nishimura, T. Tanaka, M. Asano, and Y. Sueishi, J. Chem. Soc., Perkin Trans. II *1986*, 1839. [530] Y. Sueishi, K. Ohtani, and N. Nishimura, Bull. Chem. Soc. Jpn. *58*, 810 (1985).

[531] G. Swieton and H. Kelm, J. Chem. Soc., Perkin Trans. II *1979*, 519. [532] A. I. Konovalov, I. P. Breus, I. A. Sharagin, and V. D. Kiselev, Zh. Org. Chem. *15*, 361 (1979); J. Org. Chem. USSR *15*, 315 (1979). [533] C. Rücker, D. Lang, J. Sauer, H. Friege, and R. Sustmann, Chem. Ber. *113*, 1663 (1980). [534] F. P. Ballistreri, E. Maccarone, G. Perrini, G. A. Tomaselli, and M. Torre, J. Chem. Soc., Perkin Trans. II *1982*, 273. [535] R. A. Firestone and S. G. Saffar, J. Org. Chem. *48*, 4783 (1983). [536] L. Pardo, V. Branchadell, A. Oliva, and J. Bertran, J. Mol. Struct. *93*, 255 (1983). [537] G. Jenner, M. Papadopoulos, and J. Rimmelin, J. Org. Chem. *48*, 748 (1983). [538] M. Lotfi and R. M. G. Roberts, Tetrahedron *35*, 2137 (1979). [539] V. D. Kiselev, G. V. Mavrin, and A. I. Konovalov, Zh. Org. Khim. *16*, 1435 (1980); J. Org. Chem. USSR *16*, 1233 (1980). [540] O. B. Nagy, Can. J. Chem. *63*, 1382 (1985).

[541] R. Huisgen; *Cycloaddition Mechanism and the Solvent Dependence on Rate.* Pure Appl. Chem. *52*, 2283 (1980). [542] R. Huisgen: *1,3-Dipolar Cycloadditions – Introduction, Survey, Mechanism.* In A. Padwa (ed.): *1,3-Dipolar Cycloaddition Reactions.* Wiley-Interscience, New York 1984, Vol. 1, p. 1ff. (particularly p. 76ff.). [543] M. S. Haque: *Concertedness of 1,3-Dipolar Cycloadditions.* J. Chem. Educ. *61*, 490 (1984). [544] R. Huisgen, H.-U. Reissig, and H. Huber, J. Am. Chem. Soc. *101*, 3647 (1979). [545] R. Huisgen, H.-U. Reissig, H. Huber, and S. Voss, Tetrahedron Lett. *1979*, 2987. [546] G. Bianchi, C. de Micheli, and R. Gandolfi: *1,3-Dipolar Cycloreversions.* Angew. Chem. *91*, 781 (1979); Angew. Chem., Int. Ed. Engl. *18*, 721 (1979). [547] R. Huisgen, R. Grashey, M. Seidel, G. Wallbillich, H. Knupfer, and R. Schmidt, Liebigs Ann. Chem. *653*, 105 (1962). [548] J.-M. Vandensavel, G. Smets, and G. L'abbé, J. Org. Chem. *38*, 675 (1973); G. L'abbé, Bull. Soc. Chim. Fr. *1975*, 1127. [549] J. P. Snyder and D. N. Horpp, J. Am. Chem. Soc. *98*, 7821 (1976); H. Olsen and J. P. Snyder, J. Am. Chem. Soc. *100*, 285 (1978). [550] L. M. Stephenson, M. J. Grdina, and M. Orfanopoulos, Acc. Chem. Res. *13*, 419 (1980).

[551] C. W. Jefford and S. Kohmoto, Helv. Chim. Acta *65*, 133 (1982). [552] A. Dondoni, A. Battaglia, F. Bernardi, and P. Giorgianni, J. Org. Chem. *45*, 3773 (1980). [553] M. Papadopoulos and G. Jenner, Nouv. J. Chim. *7*, 463 (1983). [554] H.-J. Bestmann, Pure Appl. Chem. *52*, 771 (1980);

H.-J. Bestmann and O. Vostrowsky: *The Mechanism of the Wittig Reaction*. Top. Curr. Chem. *109*, 85 (1983). [555] N. S. Isaacs and O. H. Abed, Tetrahedron Lett. *27*, 995 (1986). [556] M. Schlosser and B. Schaub, J. Am. Chem. Soc. *104*, 5821 (1982); A. Piskala, A. H. Rehan, and M. Schlosser, Collect. Czech. Chem. Commun. *48*, 3539 (1983). [557] G. S. Bates and S. Ramaswamy, Can. J. Chem. *59*, 3120 (1981). [558] P. Schiess and H. Stalder, Tetrahedron Lett. *21*, 1413 (1980). [559] K. Harano and T. Taguchi, Chem. Pharm. Bull. (Tokyo) *23*, 467 (1975); Chem. Abstr. *83*, 96061 h (1975). [560] R. Braun, F. Schuster, and J. Sauer, Tetrahedron Lett. *27*, 1285 (1986).

[561] B. Marcandalli, L. Pellicciari-Di Liddo, C. Di Fede, and I. R. Bellobono, J. Chem. Soc., Perkin Trans. II *1984*, 589. [562] M. G. Kulkarni, R. A. Mashelkar, and L. K. Doraiswamy, Chem. Eng. Sci. *35*, 823 (1980); Chem. Abstr. *93*, 94665 r (1980). [563] C. Walling, R. R. W. Humphreys, J. P. Sloan, and T. Miller, J. Org. Chem. *46*, 5261 (1981). [564] R. Hiatt and P. M. Rahimi, Int. J. Chem. Kinet. *10*, 185 (1978); Chem. Abstr. *89*, 128858 r (1978). [565] M. Schmittel, A. Schulz, C. Rüchardt, and E. Hädicke, Chem. Ber. *114*, 3533 (1981). [566] R. van Eldik, H. Kelm, M. Schmittel, and C. Rüchardt, J. Org. Chem. *50*, 2998 (1985). [567] R. C. Neuman and G. A. Binegar, J. Am. Chem. Soc. *105*, 134 (1983). [568] W. D. Hinsberg, P. G. Schultz, and P. B. Dervan, J. Am. Chem. Soc. *104*, 766 (1982). [569] W. Adam: *Oxygen Diradicals Derived from Cyclic Peroxides*. Acc. Chem. Res. *12*, 390 (1979). [570] E. M. Kosower: *Stable Pyridinyl Radicals*. Top. Curr. Chem. *112*, 117 (1983).

[571] J. P. Soumillion: *La Substitution Radicalaire des Composés Aliphatiques par le Chlore*. Ind. Chim. Belge *35*, 851, 1065 (1970). [572] N. J. Bunce, K. U. Ingold, J. P. Landers, J. Lusztyk, and J. C. Scaiano, J. Am. Chem. Soc. *107*, 5464 (1985). [573] A. Potter, J. M. Tedder, and J. C. Walton, J. Chem. Soc., Perkin Trans. II *1982*, 143. [574] A. Potter and J. M. Tedder, J. Chem. Soc., Perkin Trans. II *1982*, 1689. [575] S. Y. Atto, J. M. Tedder, and J. C. Walton, J. Chem. Soc., Perkin Trans. II *1983*, 629. [576] O. Ito and M. Matsuda, J. Phys. Chem. *88*, 1002 (1984); J. Am. Chem. Soc. *104*, 568 (1982). [577] W. Offermann and F. Vögtle, Angew. Chem. *92*, 471 (1980); Angew. Chem., Int. Ed. Engl. *19*, 464 (1980). [578] O. Tapia: *Quantum Theories of Solvent-Effect Representation: An Overview of Methods and Results*. In H. Ratajczak and W. J. Orville-Thomas (eds.): *Molecular Interactions*. Vol. 3, Chapter 2, p. 47ff., Wiley, New York 1982. [579] B. Ya. Simkin and I. I. Sheikhet; *Theoretical Methods of Calculating Solvation Effects*. J. Mol. Liquids *27*, 79 (1983). [580] B. M. Ladanyi and J. T. Hynes, J. Am. Chem. Soc. *108*, 585 (1986).

[581] L. M. Epshtein: *Hydrogen Bonds and the Reactivity of Organic Compounds in Proton Transfer and Nucleophilic Substitution Reactions*. Usp. Khim. *48*, 1600 (1979); Russian Chem. Rev. *48*, 854 (1979). [582] K. Okamoto: *Solvent Molecules and Carbocation Intermediates in Solvolysis*. Pure Appl. Chem. *56*, 1797 (1984). [583] D. Landini, A. Maia, F. Montanari, and F. Rolla, J. Org. Chem. *48*, 3774 (1983). [584] M. Auriel and E. de Hoffmann, J. Chem. Soc., Perkin Trans. II *1979*, 325. [585] M. C. R. Symons, J. Chem. Soc., Chem. Commun. *1978*, 418; M. J. Blandamer, J. Burgess, P. P. Duce, M. C. R. Symons, R. E. Robertson, and J. W. M. Scott, J. Chem. Res. (S) *1982*, 130. [586] J. O. Edwards and R. G. Pearson, J. Am. Chem. Soc. *84*, 16 (1962). [587] C. H. Depuy, E. W. Della, J. Filley, J. J. Grabowski, and V. M. Bierbaum, J. Am. Chem. Soc. *105*, 2481 (1983). [588] S. Hoz and E. Buncel, Tetrahedron Lett. *25*, 3411 (1984); Israel J. Chem. *26*, 313 (1985). [589] D. S. Kemp and K. G. Paul, J. Am. Chem. Soc. *97*, 7305, 7312 (1975); D. S. Kemp, J. Reczek, and F. Vellaccio, Tetrahedron Lett. *1978*, 741. [590] M. Shirai and J. Smid, J. Am. Chem. Soc. *102*, 2863 (1980).

[591] J.-P. Kintzinger, J.-M. Lehn, E. Kauffmann, J. L. Dye, and A. I. Popov, J. Am. Chem. Soc. *105*, 7549 (1983). [592] J.-L. Pierre and P. Baret: *Complexes Moleculaires d'Anions*. Bull. Soc. Chim. Fr. *1983*, II-367. [593] T. Mitsuhashi, J. Am. Chem. Soc. *108*, 2400 (1986). [594] M. Mashima, R. R. McIver, R. W. Taft, F. G. Bordwell, and W. N. Olmstead, J. Am. Chem. Soc. *106*, 2717 (1984). [595] F. G. Bordwell and D. L. Hughes, J. Org. Chem. *45*, 3314 (1980). [596] T. F. Magnera, G. Caldwell, J. Sunner, S. Ikuta, and P. Kebarle, J. Am. Chem. Soc. *106*, 6140 (1984). [597] C. D. Ritchie, Can. J. Chem. *64*, 2239 (1986). [598] J. M. Harris and S. P. McManus (eds.): *Nucleophilicity*. Adv. Chem. Ser. 215 (1987); Am. Chem. Soc., Washington/D.C. [599] C. M. Sharts and W. A. Sheppard, Org. Reactions *21*, 125 (1974). [600] J. H. Clark, Chem. Rev. *80*, 429 (1980).

[601] R. O. Hutchins and I. M. Taffer, J. Org. Chem. *48*, 1360 (1983). [602] A. Lüttringhaus and H.-W. Dirksen; *Tetramethylharnstoff als Lösungsmittel und Reaktionspartner*. Angew. Chem. *75*, 1059 (1963). [603] B. J. Barker, J. Rosenfarb, and J. A. Caruso: *Harnstoffe als Lösungsmittel in der*

chemischen Forschung. Angew. Chem. *91*, 560 (1979); Angew. Chem., Int. Ed. Engl. *18*, 503 (1979). [604] M. Bréant and G. Demange-Guerin: *Propriétés chimiques et électrochimiques dans le nitromethane*. Bull. Soc. Chim Fr. *1975*, 163. [605] V. Gutmann and A. Scherhaufer: *Äthylensulfit als Lösungsmittel*. Monatsh. Chem. *99*, 1686 (1968). [606] R. S. Kittila: *A Review of Catalytic and Synthetic Applications for DMF and DMAC*. Bulletin DuPont de Nemours and Co., Wilmington/USA 1960 (and Supplement 1962). [607] S. S. Pizey: *Dimethylformamide*, in S. S. Pizey (ed.): *Synthetic Reagents*. Wiley, New York, London 1974, Vol. I, p. 1ff. [608] J. F. Coetzee (ed.): *Recommended Methods for Purification of Solvents and Tests for Impurities*. Pergamon Press, Oxford 1982. [609] M. Bréant: *Propriétés chimiques et électrochimiques dans le N-méthylpyrrolidone*. Bull. Soc. Chim. Fr. *1971*, 725. [610] H. L. Huffman and P. G. Sears, J. Sol. Chem. *1*, 187 (1972).

[611] N. Saleh and J. A. Caruso, J. Sol. Chem. *8*, 197 (1979). [612] J. F. Coetzee: *Ionic Reactions in Acetonitrile*. Progr. Phys. Org. Chem. *4*, 45 (1967). [613] C. Agami: *Le Dimethylsulfoxyde en Chimie Organique*. Bull. Soc. Chim. Fr. *1965*, 1021. [614] D. Martin, A. Weise, and H.-J. Niclas: *Das Lösungsmittel Dimethylsulfoxid*. Angew. Chem. *79*, 340 (1967); Angew. Chem., Int. Ed. Engl. *6*, 318 (1967). [615] S. W. Jacob, E. E. Rosenbaum, and D. C. Wood (eds.): *Dimethyl Sulfoxide*. M. Dekker, New York 1971, Vol. 1. [616] D. Martin and H. G. Hauthal: *Dimethylsulfoxid*. Akademie-Verlag, Berlin 1971; *Dimethylsulphoxide*. Van Nostrand-Reinhold, London 1975. [617] T. Mukhopadhyay and D. Seebach: *Substitution of HMPT by the Cyclic Urea DMPU als a Cosolvent for Highly Reactive Nucleophiles and Bases*. Helv. Chim. Acta *65*, 385 (1982). [618] D. Seebach, Chimia *39*, 147 (1985); Chemistry in Britain *21*, 632 (1985); *cf.* also Nachr. Chem. Techn. Labor *33*, 396 (1985). [619] R. Sowada: *Darstellung, Eigenschaften und Verwendung von Dimethylsulfon*. Z. Chem. *8*, 361 (1968). [620] N. Furukawa, F. Takahashi, T. Yoshimura, H. Morita, and S. Oae: *Enhanced Reactivities in Substitution and Elimination Reactions in Dimethyl Sulphoximide*. J. Chem. Soc., Perkin Trans. II *1981*, 432.

[621] H. G. Richey, R. D. Smith, B. A. King, T. C. Kester, and E. P. Squiller, J. Org. Chem. *46*, 2823 (1981); H. G. Richey and J. Farkas: *Sulfamides and Sulfonamides as Polar Aprotic Solvents*. J. Org. Chem. *52*, 479 (1987). [622] E. M. Arnett and C. F. Douty; *Sulfolane – A Weakly Basic Aprotonic Solvent of High Dielectric Constant*. J. Am. Chem. Soc. *86*, 409 (1964). [623] T. Yamamoto: *Sulfolane*. Yuki Gosei Kagaku Kyokai Shi *28*, 853 (1970); Chem. Abstr. *73*, 109587 c (1970). [624] J. Martinma: *Sulfolane*, in J. J. Lagowski (ed.): *The Chemistry of Nonaqueous Solvents*. Academic Press, New York 1976, Vol. IV, p. 248ff. [625] W. H. Lee: *Cyclic Carbonates*, in J. J. Lagowski (ed.): *The Chemistry of Nonaqueous Solvents*. Academic Press, New York 1976, Vol. IV, p. 167ff. [626] R. H. Wood and Q. D. Craft, J. Sol. Chem. *7*, 799 (1978). [627] Yu. I. Ranneva, I. S. Temnova, E. S. Petrov, A. I. Shatenshtein, E. N. Tsvetkov, and M. I. Kabachnik, Izv. Akad. Nauk SSSR, Ser. Khim. *1967*, 2129. [628] K. Weissermel, H.-J. Kleiner, M. Finke, and U.-H. Felcht, Angew. Chem. *93*, 256 (1981); Angew. Chem., Int. Ed. Engl. *20*, 223 (1981). [629] Jefferson Chemical Company, Inc.: *Propylene Carbonate*. Technical Bulletin, Houston/Texas 1960. [630] J. Barthel and F. Feuerlein, J. Sol. Chem. *13*, 393 (1984).

[631] H. Normant; *Hexamethylphosphorsäuretriamid*. Angew. Chem. *79*, 1029 (1967); Angew. Chem., Int. Ed. Engl. *6*, 1046 (1967); Bull. Soc. Chim. Fr. *1968*, 791; Usp. Khim. *39*, 990 (1970); Russian Chem. Rev. *39*, 457 (1970). [632] D. T. Sawyer, E. J. Nanni, and J. L. Roberts: *The Reaction Chemistry of Superoxide Ion in Aprotic Media*. In K. M. Kadish (ed.): Advances in Chemistry Series, No. 201, Electrochemical and Spectrochemical Studies of Biological Redox Components. American Chemical Society, Washington/D.C., *1982*, p. 585ff. [633] A. F. Sowinski and G. M. Whitesides, J. Org. Chem. *44*, 2369 (1979). [634] Y. Marcus, Rev. Anal. Chem. *5*, 53 (1980); Pure Appl. Chem. *55*, 977 (1983); ibid. *57*, 1103 (1985). [635] Y. Marcus: *Ion Solvation*. Wiley, Chichester, New York 1985. [636] E. Buncel and E. A. Symons: *Initial State and Transition State Solvent Effects: Reactions in Protic and Dipolar Aprotic Media*. In I. Bertini, L. Lunazzi, and A. Dei (eds.): *Advances in Solution Chemistry*. Plenum Press, New York 1981, p. 355ff. [637] M. H. Abraham: *Solvent Effects on Some Nucleophilic Substitutions*. In I. Bertini, L. Lunazzi, and A. Dei (eds.): *Advances in Solution Chemistry*. Plenum Press, New York 1981, p. 341ff. [638] P. Haberfield and J. Pessin, J. Am. Chem. Soc. *105*, 526 (1983); P. Haberfield and D. Fortier, J. Org. Chem. *48*, 4554 (1983). [639] M. J. Blandamer *et al.*, J. Chem. Soc., Faraday Trans. I *1986*, 1471. [640] I. N. Rozhkov and I. L. Knunyants, Dokl. Akad. Nauk SSSR *199*, 614 (1971); Chem. Abstr. *76*, 7291 (1972).

[641] R. A. Bartsch, J. Org. Chem. *35*, 1023 (1970). [642] G. P. Schiemenz, J. Becker, and J. Stöckigt, Chem. Ber. *103*, 2077 (1970). [643] P. Viout: *Effects of Macrocyclic Cation Ligands and Quaternary Onium Salts on the Anionic Reactivity*. J. Mol. Catalysis *10*, 231 (1981). [644] D. J. Cram: *Cavitands: Organic Hosts with Enforced Cavities*. Science *219*, 1177 (1983); D. J. Cram: *Präorganisation – von Solventien zu Sphäranden*. Angew. Chem. *98*, 1041 (1986); Angew. Chem., Int. Ed. Engl. *25*, 1039 (1986). [645] F. De Jong and D. N. Reinhoudt: *Stability and Reactivity of Crown-Ether-Complexes*. Adv. Phys. Org. Chem. *17*, 279 (1980). [646] E. Weber and F. Vögtle: *Crown-Type Compounds – An Introductory Overview*. Top. Curr. Chem. *98*, 1 (1981). [647] M. Hiraoka: *Crown Compounds. Their Characteristics and Applications*. Elsevier, Amsterdam 1982. [648] J.-M. Lehn: *Supramolekulare Chemie*. Angew. Chem. *100*, 92 (1988); Angew. Chem., Int. Ed. Engl. *27* (1988). [649] D. J. Cram, T. Kaneda, R. C. Helgeson, and G. M. Lein, J. Am. Chem. Soc. *101*, 6752 (1979). [650] B. G. Cox, J. Garcia-Rosas, and H. Schneider, J. Am. Chem. Soc. *103*, 1054 (1981); *cf.* also S. F. Lincoln, I. M. Brereton, and T. M. Spotswood, ibid. *108*, 8134 (1986).

[651] P. A. Mosier-Boss and A. I. Popov, J. Am. Chem. Soc. *107*, 6168 (1985); M. Shamsipur and A. I. Popov, J. Phys. Chem. *90*, 5997 (1986). [652] B. Dietrich and J. M. Lehn, Tetrahedron Lett. *1973*, 1225. [653] D. Landini, A. Maia, and F. Montanari, J. Am. Chem. Soc. *100*, 2796 (1978); J. Chem. Soc., Perkin Trans. II *1980*, 46. [654] H. Handel and J. L. Pierre, Tetrahedron Lett. *1976*, 741. [655] J. L. Pierre, H. Handel, and R. Perraud, Tetrahedron Lett. *1977*, 2013. [656] G. W. Gokel and W. P. Weber, J. Chem. Educ. *55*, 350, 429 (1978). [657] C. M. Starks and C. Liotta; *Phase Transfer Catalysis: Principles and Techniques*. Academic Press, New York 1978. [658] E. V. Dehmlow and S. Dehmlow: *Phase Transfer Catalysis*. 2nd edition, Verlag Chemie, Weinheim 1983. [659] P. Sarthou, G. Bram, and F. Guibe, Can. J. Chem. *58*, 786 (1980). [660] S. Akabori and H. Tuji, Bull. Chem. Soc. Jpn. *51*, 1197 (1978).

[661] E. M. Arnett, S. G. Maroldo, G. W. Schriver, S. L. Schilling, and E. B. Troughton, J. Am. Chem. Soc. *107*, 2091 (1985). [662] C. Wesdemiotis and F. W. McLafferty, Tetrahedron *37*, 3111 (1981). [663] S. Hünig and G. Wehner, Chem. Ber. *113*, 302 (1980). [664] T. W. Bentley and P. v. R. Schleyer, J. Am. Chem. Soc. *98*, 7658 (1976); F. L. Schadt, T. W. Bentley, and P. v. R. Schleyer, ibid. *98*, 7667 (1976). [665] T. W. Bentley, C. T. Bowen, D. H. Morten, and P. v. R. Schleyer, J. Am. Chem. Soc. *103*, 5466 (1981); T. W. Bentley and G. E. Carter, ibid. *104*, 5741 (1982); T. W. Bentley, G. E. Carter, and H. C. Harris, J. Chem. Soc., Chem. Commun. *1984*, 387. [666] D. Fărcasiu, J. Jähme, and C. Rüchardt, J. Am. Chem. Soc. *107*, 5717 (1985). [667] A. D. Allen, V. M. Kanagasabapathy, and T. T. Tidwell, J. Am. Chem. Soc. *107*, 4513 (1985). [668] J. P. Richard and W. P. Jencks, J. Am. Chem. Soc. *106*, 1373, 1383, 1396 (1984). [669] G. A. Olah and G. K. S. Prakash, Chemistry in Britain *19*, 916 (1983). [670] A. P. Johnson and V. Vajs, J. Chem. Soc., Chem. Commun. *1979*, 817.

[671] E. K. Fukuda and R. T. McIver, J. Am. Chem. Soc. *101*, 2498 (1979). [672] K. Takashima and J. M. Riveros, J. Am. Chem. Soc. *100*, 6128 (1978). [673] V. S. Petrosyan: *Solvent Effects on the Kinetics and Mechanism of Organomercury Reactions*. Advances in the USSR Organometallic Chemistry. MIR Publishers, Moscow 1985, p. 73ff. [674] V. S. Petrosyan, J. Organomet. Chem. *250*, 157 (1983). [675] P. Kolsaker, H.-J. Storesund, T. Gulbrandsen, and G. Wøien, Acta Chem. Scand., Part B *37*, 187 (1983). [676] K. Müller and J. Sauer, Tetrahedron Lett. *25*, 2541 (1984). [677] W. Eberbach and J. C. Carré, Tetrahedron Lett. *21*, 1145 (1980). [678] N. Abe, T. Nishiwaki, and N. Komoto, Chem. Lett. (Tokyo) *1980*, 223. [679] R. Herges and I. Ugi, Angew. Chem. *97*, 596 (1985); Angew. Chem., Int. Ed. Engl. *24*, 594 (1985); Chem. Ber. *119*, 829 (1986). [680] S. Takamuku, T. Kuroda, and H. Sakurai, Chem. Lett. (Tokyo) *1982*, 377.

[681] A. A. Frimer: *The Reaction of Singlet Oxygen with Olefins: The Question of Mechanism*. Chem. Rev. *79*, 359 (1979). [682] K. Gollnick and G. O. Schenck: *Oxygen as Dienophile*. In J. Hamer (ed.): *1,4-Cycloaddition Reactions*. Academic Press, New York, London 1967, Chapter 10, p. 255ff. [683] C. W. Jefford and S. Kohmoto, Helv. Chim. Acta *65*, 133 (1982). [684] C. W. Jefford, H. G. Grant, D. Jaggi, J. Boukouvalas, and S. Kohmoto, Helv. Chim. Acta *67*, 2210 (1984). [685] A. A. Frimer, P. D. Bartlett, A. F. Boschung, and J. G. Jewett, J. Am. Chem. Soc. *99*, 7977 (1977). [686] L. B. Harding and W. A. Goddard, J. Am. Chem. Soc. *102*, 439 (1980). [687] W. Adam and H. Rebollo, Israel J. Chem. *23*, 399 (1983). [688] M. L. Graziano, M. R. Iesce, S. Chiosi, and R. Scarpati, J. Chem. Soc., Perkin Trans. I *1983*, 2071. [689] E. L. Clennan and R. P. L'Esperance, J. Am. Chem. Soc. *107*, 5178 (1985); J. Org. Chem. *50*, 5424 (1985). [690] I. Szele and H. Zollinger, Helv. Chim. Acta *61*, 1721 (1978).

[691] H. Sawada, H. Hagii, K. Aoshima, M. Yoshida, and M. Kobayashi, Bull. Chem. Soc. Jpn. *58*, 3448 (1985). [692] Yu. I. Ovchinnikova, V. A. Fomin, A. I. D'yachkov, and V. S. Etlis, Zh. Obshch. Khim. *51*, 2355 (1981); J. Gen. Chem. USSR *51*, 2031 (1981). [693] R. A. Bartsch and J. Závada, Chem. Rev. *5*‿, 453 (1980). [694] M. Schlosser and T. D. An, Helv. Chim. Acta *62*, 1194 (1979). [695] S. Krishnamurthy, J. Org. Chem. *45*, 2550 (1980). [696] M. P. Hartshorn, R. S. Thompson, and J. Vaughan, Aust. J. Chem. *30*, 865 (1977); J. W. Blunt, M. P. Hartshorn, M. H. G. Munro, L. T. Soong, R. S. Thompson, and J. Vaughan, J. Chem. Soc., Chem. Commun. *1980*, 820.[697] S. Kiyooka, F. Goto, and K. Suzuki, Chem. Lett. (Tokyo) *1981*, 1429. [698] H. O. House and T. V. Lee, J. Org. Chem. *43*, 4369 (1978). [699] J. B. Ousset, C. Mioskowski, and G. Solladié, Synth. Commun. *13*, 1193 (1984). [700] D. Enders and R. W. Hoffmann: *Asymmetrische Synthese*. Chemie in unserer Zeit *19*, 177 (1985).

[701] J. D. Morrison (ed.): *Asymmetric Synthesis*. Vol. 1...5, Academic Press, Orlando 1983...1985. [702] R. Pérez-Ossorio, A. Pérez-Rubalcaba, M. L. Quiroga, and M. Lasperas, Tetrahedron Lett. *21*, 1565 (1980); O. Arjona, R. Pérez-Ossorio, A. Pérez-Rubalcaba, and M. L. Quiroga, J. Chem. Soc., Perkin Trans. II *1981*, 597.[703] J. Berlan, Y. Besace, D. Prat, and G. Pourcelot, J. Organomet. Chem. *264*, 399 (1984). [704] K. Soai, K. Komiya, Y. Shigematsu, H. Hasegawa, and A. Ookawa, J.Chem. Soc., Chem. Commun. *1982*, 1282. [705] R. Schmierer, G. Grotemeier, G. Helmchen, and A. Selim, Angew. Chem. *93*, 209 (1981); Angew. Chem., Int. Ed. Engl. *20*, 207 (1981); G. Helmchen, A. Selim, D. Dorsch, and I. Taufer, Tetrahedron Lett. *24*, 3213 (1983). [706] R. E. Ireland, R. H. Mueller, and A. K. Willard, J. Am. Chem. Soc. *98*, 2868 (1976). [707] J. Sauer and J. Kredel, Tetrahedron Lett. *1966*, 6359. [708] T. Poll, G. Helmchen, and B. Bauer, Tetrahedron Lett. *25*, 2191 (1984). [709] K. Soai and A. Ookawa, J. Org. Chem. *51*, 4000 (1986). [710] A. R. Katritzky and B. Brycki, J. Am. Chem. Soc. *108*, 7295 (1986); and references cited therein.

[711] J. B. F. N. Engberts: *Organic Reactions in Highly Aqueous Binaries*. Pure Appl. Chem. *54*, 1797 (1982). [712] N. J. Turro, G. S. Cox, and M. A. Paczkowski: *Photochemistry in Micelles*. Top. Curr. Chem. *129*, 57 (1985). [713] V. Ramamurthy: *Organic Photochemistry in Organized Media*. Tetrahedron *42*, 5753 (1986). [714] D. C. Rideout and R. Breslow, J. Am. Chem. Soc. *102*, 7816 (1980); R. Breslow, U. Maitra, and D. Rideout, Tetrahedron Lett. *24*, 1901 (1983). [715] P. A. Grieco, P. Garner, and Z. He, Tetrahedron Lett. *24*, 1897 (1983); P. A. Grieco, P. Garner, K. Yoshida, and J. C. Huffman, ibid. *24*, 3807 (1983). [716] P. A. Grieco, K. Yoshida, and P. Garner, J. Org. Chem. *48*, 3137 (1983). [717] S. D. Larsen and P. A. Grieco, J. Am. Chem. Soc. *107*, 1768 (1985). [718] H.-U. Reißig: *C-C-Verknüpfungen in Wasser*. Nachr. Chem. Tech. Lab. *34*, 1169 (1986). [719] P. Laszlo and J. Lucchetti, Tetrahedron Lett. *25*, 2147 (1984); P. Laszlo, Acc. Chem. Res. *19*, 121 (1986). [720] Y. Nakamura, Y. Imakura, T. Kato, and Y. Morita, J. Chem. Soc., Chem. Commun. *1977*, 887.

[721] H. Mayer, F. Schuster, and J. Sauer, Tetrahedron Lett. *27*, 1289 (1986). [722] K. A. Zachariasse, N. V. Phuc, and B. Kozankiewicz, J. Phys. Chem. *85*, 2676 (1981). [723] P. Plieninger and H. Baumgärtel, Ber. Bunsenges. Phys. Chem. *86*, 161 (1982). [724] C. J. Drummond, F. Grieser, and T. W. Healy, Faraday Discuss. Chem. Soc. *81*, 95 (1986); Chem. Phys. Lett. *140*, 493 (1987). [725] See references [17–22, 22a, 109, 110] to Chapter 3. [726] T. Svedberg: *Chemische Reaktionen in anisotropen Flüssigkeiten*. Kolloid Z. *18*, 54, 101 (1916); Chem. Zentralbl. *1916* I, 958; ibid. *1916* II, 211. [727] S. Ganapathy, R. G. Zimmermann, and R. G. Weiss, J. Org. Chem. *51*, 2529 (1986). [728] W. J. Leigh, D. T. Frendo, and P. J. Klawunn, Can. J. Chem. *63*, 2131 (1985). [729] J. P. Otruba and R. G. Weiss, J. Org. Chem. *48*, 3448 (1983). [730] D. A. Hrovat, J. H. Liu, N. J. Turro, and R. G. Weiss, J. Am. Chem. Soc. *106*, 7033 (1984); R. L. Treanor and R. G. Weiss, Tetrahedron *43*, 1371 (1987).

[731] V. Ramesh and R. G. Weiss, J. Org. Chem. *51*, 2535 (1986). [732] J. M. Nerbonne and R. G. Weiss, J. Am. Chem. Soc. *100*, 2571 (1978); ibid. *101*, 402 (1979). [733] T. Nakano and H. Hirata, Bull. Chem. Soc. Jpn. *55*, 947 (1982). [734] W. J. Leigh, Can. J. Chem. *63*, 2736 (1985). [735] M. Nakazaki, K. Yamamoto, and K. Fujiwara, Chem. Lett. (Tokyo) *1978*, 863. [736] A. Dondoni, A. Medici, S. Colonna, G. Gottarelli, and B. Samori, Mol. Cryst. Liq. Cryst. *55*, 47 (1979). [737] C. Eskenazi, J. F. Nicoud, and H. B. Kagan, J. Org. Chem. *44*, 995 (1979). [738] P. P. S. Saluja and E. Whalley, J. Chem. Soc., Chem. Commun. *1983*, 552. [739] P. S. Bailey: *Ozonation in Organic Chemistry*. Vol. 1 and 2, Academic Press, New York 1978 and 1982; *cf.* p. 89ff. in Vol. 1. [740] R. L.

Kuczkowski: *Ozone and Carbonyl Oxides*. In A. Padwa (ed.): *1,3-Dipolar Cycloaddition Chemistry*. Wiley-Interscience, New York 1984, Vol. 2, p. 197ff.; R. L. Kuczkowski: *Formation and Structure of Ozonides*. Acc. Chem. Res. *16*, 42 (1983).

[741] G. D. Fong and R. L. Kuczkowski, J. Am. Chem. Soc. *102*, 4763 (1980). [742] M. Miura, T. Fujisaka, M. Nojima, S. Kusabayashi, and K. J. McCullough, J. Org. Chem. *50*, 1504 (1985). [743] A. W. Castleman and R. G. Keesee: *Clusters – Bridging the Gas and Condensed Phases*. Acc. Chem. Res. *19*, 413 (1986). [744] A. Lubineau and Y. Queneau, Tetrahedron Lett. *26*, 2653 (1985). [745] H.-J. Schneider and N. K. Sangwan, J. Chem. Soc., Chem. Commun. *1986*, 1787; Angew. Chem. *99*, 924 (1987); Angew. Chem., Int. Ed. Engl. *26*, 896 (1987). [746] A. Lubineau, J. Org. Chem. *51*, 2144 (1986). [747] N. S. Isaacs: *Liquid Phase High Pressure Chemistry*. Wiley, New York 1981; N. S. Isaacs and A. V. George, Chem. Br. *23*, 47 (1987). [748] M. J. Blandamer, J. Burgess, R. E. Robertson, and J. M. W. Scott: *Dependence of Equilibrium and Rate Constants on Temperature and Pressure*. Chem. Rev. *82*, 259 (1982). [749] K. Matsumoto, A. Sera, and T. Uchida: *Organic Synthesis under High Pressure, I and II*. Synthesis *1985*, 1, 999. [750] R. van Eldik: *High Pressure Kinetics – Fundamental and Experimental Aspects*. In R. van Eldik (ed.): *Inorganic High Pressure Chemistry – Kinetics and Mechanisms*. Studies in Inorganic Chemistry, Vol. 7, p. 1ff., Elsevier, Amsterdam 1986.

[751] B. Raistrick, R. H. Sapiro, and D. M. Newitt, J. Chem. Soc. *1939*, 1761; *cf.* also M. G. Gonikberg and L. F. Vereshchagin, Zh. Fiz. Khim. *23*, 1447 (1949); Chem. Abstr. *44*, 2832 f (1950). [752] G. Swieton, J. v. Jouanne, H. Kelm, and R. Huisgen, J. Org. Chem. *48*, 1035 (1983). [753] Y. Yoshimura, J. Osugi, and M. Nakahara, Bull. Chem. Soc. Jpn. *56*, 680 (1983). [754] E. M. Schulman, A. E. Merbach, M. Turin, R. Wedinger, and W. J. Le Noble, J. Am. Chem. Soc. *105*, 3988 (1983). [755] Y. Kondo, A. Zanka, and S. Kusabayashi, J. Chem. Soc., Perkin Trans. II *1985*, 827. [756] J. B. Hyne, H. S. Golinkin, and W. G. Laidlaw, J. Am. Chem. Soc. *88*, 2104 (1966). [757] G. Wilke, Angew. Chem. *90*, 747 (1978); Angew. Chem., Int. Ed. Engl. *17*, 701 (1978); and the following articles. [758] M. E. Paulaitis and G. C. Alexander, Pure Appl. Chem. *59*, 61 (1987). [759] J. A. Hyatt, J. Org. Chem. *49*, 5097 (1984); see also M. E. Sigman, S. M. Lindley, and J. E. Leffler, J. Am. Chem. Soc. *107*, 1471 (1985); as well as C. R. Yonker, S. L. Frye, D. R. Kalkwarf, and R. D. Smith, J. Phys. Chem. *90*, 3022 (1986). [760] E. Buncel and C. C. Lee (eds.): *Secondary and Solvent Isotope Effects* (Vol. 7 of the series *Isotopes in Organic Chemistry*), Elsevier, Amsterdam 1987.

[761] A. V. Willi: *Isotopieeffekte bei chemischen Reaktionen*. Thieme, Stuttgart 1983. [762] R. A. More O'Ferrall, G. W. Koeppl, and A. J. Kresge, J. Am. Chem. Soc. *93*, 9 (1971). [763] C. A. Bunton and V. J. Shiner, J. Am. Chem. Soc. *83*, 42, 3207, 3214 (1961). [764] C. G. Mitton, M. Gressner, and R. L. Schowen, J. Am. Chem. Soc. *91*, 2045 (1963). [765] L. Verbit, T. R. Halbert, and R. B. Patterson, J. Org. Chem. *40*, 1649 (1975). [766] F. D. Saeva, P. E. Sharpe, and G. R. Olin, J. Am. Chem. Soc. *97*, 204 (1975). [767] R. M. Coates, B. D. Rogers, S. J. Hobbs, D. R. Peck, and D. P. Curran, J. Am. Chem. Soc. *109*, 1160 (1987). [768] H. Nöth, R. Rurländer, and P. Wolfgarth, Z. Naturforsch., Part B *35*, 31 (1981). [769] C. F. Bernasconi, R. D. Bunnell, D. A. Kliner, A. Mullin, P. Paschalis, and F. Terrier: *The Effect of Solvation on Intrinsic Rates of Proton-Transfer*. In M. Kobayashi (ed.): *Physical Organic Chemistry 1986*. Elsevier, Amsterdam 1987, p. 583ff.; C. F. Bernasconi, Acc. Chem. Res. *20*, 301 (1987).

Chapter 6

[1] A. E. Lutskii, V. V. Prezhdo, L. I. Degtereva, and V. G. Gordienko: *Spectroscopy of Intermolecular Field Interactions in Solutions*. Usp. Khim. *51*, 1398 (1982); Russian Chem. Rev. *51*, 802 (1982) [2] T. E. Gough, D. E. Irish, and I. R. Lantzke: *Spectroscopic Measurements (electron absorption, infrared and Raman, ESR and NMR spectroscopy)*, in A. K. Covington and T. Dickinson (eds.): *Physical Chemistry of Organic Solvent Systems*. Plenum Press, London, New York 1973, Chapter 4, p. 405ff. [3] M. Jauquet and P. Laszlo: *Influence of Solvents on Spectroscopy*, in M. R. J. Dack (ed.): *Solutions and Solubilities*. Vol. VIII, Part I, of A. Weissberger (ed.): *Techniques of Chemistry*. Wiley-Interscience, New York 1975, p. 195ff. [4] C. N. R. Rao, S. Singh, and V. P. Senthilnathan: *Spectroscopic Studies of Solute-Solvent Interactions*. Chem. Soc. Rev. *5*, 297 (1976).

[5] K. Dimroth: *Über den Einfluß von Lösungsmitteln auf die Farbe organischer Verbindungen*, Sitzungsberichte der Gesellschaft zur Beförderung der gesamten Naturwissenschaften zu Marburg *76*, Heft 3, p. 3ff. (1953); Chem. Zentralbl. *1954*, 9481; Chimia *15*, 80 (1961). [6] A. I. Kiprianov: *Effect of the Solvent on the Color of Dyes (Solvatochromism)*. Usp. Khim. *29*, 1336 (1960); Russian Chem. Rev. *29*, 618 (1960). [7] W. Foerst (ed.): *Optische Anregung organischer Systeme*. 2. International Symposium on Color Chemistry, 1964 in Schloß Elmau. Verlag Chemie, Weinheim 1966: a) W. Liptay: *Die Lösungsmittelabhängigkeit der Wellenzahl von Elektronenbanden und die chemisch-physikalischen Grundlagen*, p. 263ff.; b) E. Lippert: *Die Medienabhängigkeit der Fluoreszenzfarbe*, p. 342ff.; c) G. Briegleb: *Lösungsmittelabhängigkeit der Lichtabsorption und -emission und Elektronen-Donator-Acceptor-Wechselwirkung*, p. 391ff.; d) A. Schweig, K. Dimroth, and H. Kuhn: *Zur Ursache der Solvatochromie des Pyridinium-cyclopentadien-ylids*, p. 101 and 765. [8] H. A. Staab: *Einführung in die theoretische organische Chemie*. 4th edition, Verlag Chemie, Weinheim 1964, p. 336ff. and p. 385ff. [9] S. Hünig, G. Bernhard, W. Liptay, and W. Brenninger: *Zum Problem der Solvatochromie bei Merocyaninen*. Liebigs Ann. Chem. *690*, 9 (1965); and references cited therein. [10] C. Reichardt: *Empirische Parameter der Lösungsmittelpolarität*. Angew. Chem. *77*, 30 (1965); Angew. Chem., Int. Ed. Engl. *4*, 29 (1965); C. Reichardt: *Empirische Parameter der Lösungsmittelpolarität als Lineare Freie-Enthalpie-Beziehungen*. Angew. Chem. *91*,119 (1979); Angew. Chem., Int. Ed. Engl. *18*, 98 (1979).

[11] L. G. S. Brooker: *Sensitizing and Desensitizing Dyes*, in C. E. K. Mees and T. H. James (eds.): *The Theory of the Photographic Process*. 3rd ed., MacMillan, New York, London 1966, Chapter II, p. 198ff. (especially p. 219ff.); Chimia *15*, 87 (1961); ibid. *20*, 327 (1966). [12] A. van Dormael: *Solvatochromism, Tautochromism, and Metachromism*. Ind. Chim. Belges *31*, 1 (1966); Chem. Abstr. *64*, 19833c (1966). [13] W. Liptay: *Elektrochromie – Solvatochromie*. Angew. Chem. *81*, 195 (1969); Angew. Chem., Int. Ed. Engl. *8*, 177 (1969); Z. Naturforsch. *20a*, 1441 (1965); ibid. *21a*, 1605 (1966). [14] N. Mataga and T. Kubota: *Molecular Interactions and Electronic Spectra*. Dekker, New York1970.[15] A. T. Amos and B. L. Burrows: *Solvent Shift Effects on Electron Spectra and Excited State Dipole Moments and Polarizabilities*. Adv. Quantum Chem. *7*, 289 (1973); Chem. Abstr. *78*, 141890n (1973). [16] M. F. Nicol: *Solvent Effects on Electronic Spectra*. Applied Spectroscopy Reviews *8*, 183 (1974); Chem. Abstr. *82*, 9412m (1975). [17] K. M. C. Davis: *Solvent Effects on Charge-Transfer Complexes*, in R. Foster (ed.): *Molecular Association*. Academic Press, London, New York 1975, Vol. 1, p. 151ff. [18] S. Dähne: *Systematik und Begriffserweiterung der Polymethinfarbstoffe*. Z. Chem. *5*,441 (1965); S. Dähne and D. Leupold: *Der Polymethinzustand*. Ber. Bunsenges. Phys. Chem. *70*, 618 (1966); S. Dähne and F. Moldenhauer, Progr. Phys. Org. Chem. *15*, 1 (1985). [19] J. Fabian and H. Hartmann: *π-Elektron Structure of Polymethines*. J. Mol. Structure *27*, 67 (1975). [20] S. Dähne, D. Leupold, H.-E. Nikolajewski, and R. Radeglia, Z. Naturforsch. *20b*, 1006 (1965); R. Radeglia and S. Dähne, J. Mol. Structure *5*, 399 (1970).

[21] K. Lauer and R. Oda, Ber. Dtsch. Chem. Ges. *69*, 851 (1936). [22] N. S. Bayliss and L. Hulme, Aust. J. Chem. *6*, 257 (1953). [23] A. L. Le Rosen and C. E. Reid, J. Chem. Phys. *20*, 233 (1952). [24] F. Feichtmayr, E. Heilbronner, A. Nürrenbach, H. Pommer, and J. Schlag, Tetrahedron *25*, 5383 (1969). [25] H. Christen and P. A. Straub, Helv. Chim. Acta *56*, 739, 1752 (1973). [26] W. West and A. L. Geddes, J. Phys. Chem. *68*, 837 (1964). [27] N. A. Derevyanko, G. G. Dyadyusha, A. A. Ishchenko, and A. I. Tolmachev, Teor. Eksp. Khim. *19*, 169 (1983); Chem. Abstr. *99*, 39795g (1983). [28] M. Wähnert, S. Dähne, R. Radeglia, A. M. Alperovich, and I. I. Levkoev, Z. Chem. *16*, 76 (1976); Adv. Mol. Relaxation Interact. Processes *11*, 263 (1977); Chem. Abstr. *88*, 75291w (1978); M. Wähnert and S. Dähne, J. Signal AM *4*, 403 (1976). [29] K. Dimroth, C. Reichardt, T. Siepmann, and F. Bohlmann, Liebigs Ann. Chem. *661*, 1 (1963); K. Dimroth and C. Reichardt, ibid. *727*, 93 (1969); C. Reichardt, ibid. *752*, 64 (1971); C. Reichardt and E. Harbusch-Görnert, ibid. *1983*, 721. [30] K. Dimroth and C. Reichardt, Z. Anal. Chem. *215*, 344 (1966); Z. B. Maksimović, C. Reichardt, and A. Spirić, ibid. *270*, 100 (1974).

[31] S. Kumoi, K. Oyama, T. Yano, H. Kobayashi, and K. Ueno, Talanta *17*, 319 (1970). [32] N. G. Bakshiev, M. I. Knyazhanskii, V. I. Minkin, O. A. Osipov, and G. V. Saidov: *Experimental Determination of the Dipole Moments of Organic Molecules in Excited Electronic States*. Usp. Khim. *38*, 1644 (1969); Russian Chem. Rev. *38*, 740 (1969). [33] W. Liptay: *Dipole Moments and Polarizabilities of Molecules in Excited Electronic States*, in E. C. Linn (ed.): *Excited States*. Academic Press,

New York, London 1974, Vol. I, p. 129ff. [34] L. G. S. Brooker, A. C. Craig, D. W. Heseltine, P. W. Jenkins, and L. L. Lincoln, J. Am. Chem. Soc. *87*, 2443 (1965). [35] M. J. Kamlet, E. G. Kayser, J. W. Eastes, and W. H. Gilligan,J. Am. Chem. Soc. *95*, 5210 (1973); M. J. Kamlet and R. W. Taft, ibid. *98*, 377, 2886, 3233 (1976). [36] P. Scheibe, S. Schneider, F. Dörr, and E. Daltrozzo, Ber. Bunsenges. Phys. Chem. *80*, 630 (1976); S. Schneider, ibid *80*, 218 (1976). [37] S. Dähne, F. Schob, and K.-D. Nolte, Z. Chem. *13*, 471 (1973); S. Dähne, F. Schob, K.-D. Nolte, and R. Radeglia, Ukr. Khim. Zh. (Russ. Ed.) *41*, 1170 (1975); Chem. Abstr. *84*, 43086j (1976). [38] D. J. Cowley, J. Chem. Soc., Perkin Trans. II *1975*, 287. [39] L. G. S. Brooker and R. H. Sprague, J. Am. Chem. Soc. *63*, 3214 (1941). [40] J. Figueras, J. Am. Chem. Soc. *93*, 3255 (1971).

[41] O. W. Kolling and J. L. Goodnight, Anal. Chem. *46*, 482 (1974); ibid. *53*, 54 (1981). [42] E. M. Kosower, J. Am. Chem. Soc. *80*, 3261 (1958). [43] S. Dähne and M. Seebacher, Z. Chem. *12*, 141 (1972). [44] M. M. Davies and H. B. Hetzer, Anal. Chem. *38*, 451 (1966). [45] O. W. Kolling, Anal. Chem. *50*, 212 (1978). [46] M. Klessinger and W. Lüttke, Chem. Ber. *99*, 2136 (1966); M. Klessinger, Tetrahedron *22*, 3355 (1966). [47] E. Lippert, Z. Elektrochem., Ber. Bunsenges. Phys. Chem. *61*, 962 (1957); Angew. Chem. *73*, 695 (1961). [47a] W. Baumann, H. Deckers, K.-D. Loosen, and F. Petzke, Ber. Bunsenges. Phys. Chem. *81*, 799 (1977). [48] L. G. S. Brooker, G. H. Keyes, and D. W. Heseltine, J. Am. Chem. Soc. *73*, 5350 (1951). [49] S. Hünig and O. Rosenthal, Liebigs Ann. Chem. *592*, 161 (1955). [50] H. G. Benson and J. N. Murrell, J. Chem. Soc., Faraday Trans. II *1972*, 137. [50a] M. J. Minch and S. S. Shah, J. Chem. Educ. *54*, 709 (1977).

[51] H. W. Gibson and F. C. Bailey, Tetrahedron *30*, 2043 (1974); Can. J. Chem. *53*, 2162 (1975); J. Chem. Soc., Perkin Trans. II *1976*, 1575. [52] L. G. S. Brooker, G. H. Keyes, R. H. Sprague, R. H. VanDyke, E. VanLare, G. VanZandt, F. L. White, H. W. J. Cressman, and S. G. Dent, J. Am. Chem. Soc. *73*, 5332 (1951). [53] N. S. Bayliss and E. G. McRae, J. Am. Chem. Soc. *74*, 5803 (1952). [54] T. Eichler and V. Schäfer, Tetrahedron *30*, 4025 (1974). [55] C. Reichardt and W. Mormann, Chem. Ber. *105*, 1815 (1972). [56] P. H. Vandewyer, J. Hoefnagels, and G. Smets, Tetrahedron *25*, 3251 (1969). [57] C. Reichardt, Tetrahedron Lett. *1965*, 429; Liebigs Ann. Chem. *715*, 74 (1968). [58] E. M. Kosower and B. G. Ramsay, J. Am. Chem. Soc. *81*, 856 (1959). [59] A. Schweig, Z. Naturforsch. *22a*, 724 (1967). [60] H. Prinzbach and E. Woischnik, Helv. Chim. Acta *52*, 2472 (1969).

[61] A. Schweig and C. Reichardt, Z. Naturforsch. *21a*, 1373 (1966). [62] W. Liptay, H.-J. Schlosser, B. Dumbacher, and S. Hünig, Z. Naturforsch. *23a*, 1613 (1968). [63] R. Foster: *Organic Charge-Transfer Complexes*. Academic Press, London, New York 1969, p. 62ff. [64] S. Dupire, J. M. Mulindabyuma, J. B. Nagy, and O. B. Nagy, Tetrahedron *31*, 135 (1975). [65] E. M. Kosower, J. Am. Chem. Soc. *80*, 3253 (1958); J. Chim. Phys. *61*, 230 (1964); E. M. Kosower and M. Mohammad, J. Am. Chem. Soc. *90*, 3271 (1968); J. Phys. Chem. *74*, 1153 (1970). [66] E. M. Kosower: *Molecular Biochemistry*. McGraw Hill, New York 1962, p. 180ff. [67] E. M. Kosower: *An Introduction to Physical Organic Chemistry*. Wiley, New York 1968, p. 293ff. [68] M. J. Blandamer and M. F. Fox: *Theory and Applications of Charge-Transfer-to-Solvent Spectra*. Chem. Rev. *70*, 59 (1970). [69] N. S. Bayliss and E. G. McRae, J. Phys. Chem. *58*, 1002, 1006 (1954). [70] E. G. McRae, J. Phys. Chem. *61*, 562 (1957).

[71] O. E. Weigang and D. D. Wild, J. Chem. Phys. *37*, 1180 (1962). [72] Th. Förster, Z. Elektrochem., Angew. Phys. Chem. *45*, 548 (1939); especially p. 570. [73] G. Scheibe, W. Seiffert, G. Hohlneicher, C. Jutz, and H. J. Springer, Tetrahedron Lett. *1966*, 5053. [74] R. Radeglia, D. Leupold, and S. Dähne, Ber. Bunsenges. Phys. Chem. *70*, 745, 1167 (1966). [75] R. Radeglia, Z. Phys. Chem. (Leipzig) *235*, 335 (1967); Z. Chem. *15*, 355 (1975). [76] S. Dähne and R. Radeglia, Tetrahedron *27*, 3673 (1971). [77] R. Radeglia, G. Engelhardt, E. Lippmaa, T. Pehk, K.-D. Nolte, and S. Dähne, Org. Magn. Reson. *4*, 571 (1972). [78] M. Wähnert and S. Dähne, J. Prakt. Chem. *318*, 321 (1976). [78a] M. Wähnert and S. Dähne, J. Signal AM *4*, 403 (1976). [79] S. Dähne and K.-D. Nolte, J. Chem. Soc., Chem. Commun. *1972*, 1056; J. Prakt. Chem. *318*, 643 (1976); Acta Chim. Acad. Sci. Hung. *97*, 147 (1978); Chem. Abstr. *90*, 38406g (1979). [80] L. Onsager, J. Am. Chem. Soc. *58*, 1486 (1936).

[81] W. Liptay, Ber. Bunsenges. Phys. Chem. *80*, 207 (1976). [82] H. Labhart, Ber. Bunsenges. Phys. Chem. *80*, 240 (1976). [83] A. Schweig, Mol. Phys. *15*, 1 (1968). [84] G. Scheibe, E. Daltrozzo, O. Wörz, and J. Heiss, Z. Phys. Chem. (Frankfurt) N. F. *64*, 97 (1969). [85] W. Liptay, Z. Naturforsch. *21a*, 1605 (1966). [86] Y. Oshika, J. Phys. Soc. Jpn. *9*, 594 (1954); Chem. Abstr. *49*, 15483c (1955). [87]

N. G. Bakshiev, Optika i Spektroskopiya *10*, 717 (1961); Optics and Spectroscopy (Engl. Ed.) *10*, 379 (1961); Chem. Abstr. *58*, 4027h (1963). [88] L. Bilot and A. Kawski, Z. Naturforsch. *17a*, 621 (1962). [89] T. Abe, Bull. Chem. Soc. Jpn. *54*, 327 (1981). [90] W. Liptay in O. Sinanoğlu (ed.): *Modern Quantum Chemistry*. Academic Press, New York, London 1965, Part II, p. 173ff.; Z. Naturforsch. *20a*, 1441 (1965).

[91] H. Kuhn and A. Schweig, Chem. Phys. Lett. *1*, 255 (1967). [92] M. F. Nicol, J. Swain, Y.-Y. Shum, R. Merin, and R. H. H. Chen, J. Chem. Phys. *48*, 3587 (1968). [93] P. Suppan, J. Chem. Soc. *A 1968*, 3125. [94] W. Liptay and G. Walz, Z. Naturforsch. *26a*, 2007 (1971). [95] F. J. Kampas, Chem. Phys. Lett. *26*, 334 (1974). [95a] H. A. Germer, Theor. Chim. Acta *34*, 145 (1974); ibid. *35*, 273 (1974). [95b] K. D. Nolte and S. Dähne, Advances in Molecular Relaxation and Interaction Processes *10*, 299 (1977); Chem. Abstr. *87*, 151529b (1977). [96] R. M. Hochstrasser and L. J. Noe, J. Mol. Spectrosc. *38*, 175 (1971). [97] H. McConnell, J. Chem. Phys. *20*, 700 (1952). [98] G. J. Brealey and M. Kasha, J. Am. Chem. Soc. *77*, 4462 (1955). [99] G. C. Pimentel, J. Am. Chem. Soc. *79*, 3323 (1957). [100] R. S. Becker, J. Mol. Spectrosc. *3*, 1 (1959).

[101] E. Lippert: *Der Einfluß von Wasserstoffbrücken auf Elektronenspektren*, in D. Hadži and H. W. Thompson (eds.): *Hydrogen Bonding*. Pergamon Press, London 1959, p. 217ff. [102] M. Ito, K. Inuzuka, and S. Imanishi, J. Am. Chem. Soc. *82*, 1317 (1960); Bull. Chem. Soc. Jpn. *34*, 467 (1961). [103] J. E. Dubois, E. Goetz, and A. Bienvenüe, Spectrochim. Acta *20*,1815 (1964). [103a] W. P. Hayes and C. J. Simmons, Spectrochim. Acta *21*, 529 (1965). [104] W. L. Dilling, J. Org. Chem. *31*, 1045 (1966). [105] J. E. Del Bene, J. Am. Chem. Soc. *95*, 6517 (1973); ibid. *96*, 5643 (1974). [106] P. Haberfield *et al.*, J. Am. Chem. Soc. *96*, 6526 (1974); ibid. *99*, 6828 (1977); ibid. *101*, 645, 3196 (1979). [107] O. Exner: *Dipole Moments in Organic Chemistry*. Thieme, Stuttgart 1975, p. 122 [108] N. J. Turro: *Modern Molecular Photochemistry*. Benjamin, Menlo Park 1978, p. 108. [109] G. Porter and P. Suppan, Trans. Faraday Soc. *61*, 1664 (1965); ibid. *62*, 3375 (1966). [110] J. H. Fendler, E. J. Fendler, G. A. Infante, P.-S. Shih, and L. K. Patterson, J. Am. Chem. Soc. *97*, 89 (1975).

[111] D. P. Stevenson, G. M. Coppinger, and J. W. Forbes, J. Am. Chem. Soc. *83*, 4350 (1961). [112] B. L. van Duuren: *Effects of the Environment on the Fluorescence of Aromatic Compounds in Solution*. Chem. Rev. *63*, 325 (1963). [113] R. S. Becker: *Theory and Interpretation of Fluorescence and Phosphorescence*. Wiley-Interscience, New York, London 1969, Chapter 10, p. 111ff. [114] S. G. Schulman: *Fluorescence and Phosphorescence of Heterocyclic Molecules*, in A. R. Katrizky (ed.): *Physical Methods in Heterocyclic Chemistry*. Academic Press, New York, London 1974, Vol. VI, p. 147ff. (especially p. 171ff.). [115] A. Kawski: *Die Solvathülle und ihr Einfluß auf die Fluoreszenz*. Chimia *28*, 715 (1974). [116] A. Dienes, C.V. Shank, and A. M. Trozzolo: *Dye Lasers*, in W. R. Ware (ed.): *Creation and Detection of the Excited State*. Dekker, New York 1974, Vol. 2, p. 149ff. [117] K. M. C. Davis, Nature *223*, 728 (1969). [118] J. W. Eastman, Spectrochim. Acta, Part A *26*, 1545 (1970). [119] E. M. Kosower, H. Dodiuk, K. Tanizawa, M. Ottolenghi, and N. Orbach, J. Am. Chem. Soc. *97*, 2167 (1975). [120] E. M. Kosower, Acc. Chem. Res. *15*, 259 (1982).

[121] F. Ciardelli and P. Salvadori (eds.): *Fundamental Aspects and Recent Developments in ORD and CD*. Heyden and Son, London 1973. [122] G. Snatzke: *Anwendung von Polarimetrie, CD und ORD in der organischen Stereochemie*. In F. Korte (ed.): *Methodicum Chimicum*. Vol. I, Part 1, p. 426ff., Thieme, Stuttgart, and Academic Press, New York 1973; G. Snatzke, Angew. Chem. *91*, 380 (1979); Angew. Chem., Int. Ed. Engl. *18*, 363 (1979). [123] E. Charney: *The Molecular Bases of Optical Activity: ORD and CD*. Wiley, New York 1979. [124] N. Tokura: *Solvent Effects and Circular Dichroism*. Kagaku No Ryoiki *28*, 682 (1974); Chem. Abstr. *82*, 147175 (1975). [125] C. Coulombeau and A. Rassat, Bull. Soc. Chim. Fr. *1966*, 3752. [126] D. N. Kirk, W. Klyne, and S. R. Wallis, J. Chem. Soc. *C 1970*, 350. [127] R. Corriu and J. Masse, Tetrahedron *26*, 5123 (1970). [128] W. Klyne, D. N. Kirk, J. Tilley, and H. Suginome, Tetrahedron *36*, 543 (1980). [129] D. W. Miles, M. J. Robins, R. K. Robins, M. W. Winkley, and H. Eyring, J. Am. Chem. Soc. *91*, 824 (1969). [130] M. Cinquini, S. Colonna, I. Moretti, and G. Torre, Tetrahedron Lett. *1970*, 2773.

[131] B. Bosnich, J. Am. Chem. Soc. *89*, 6143 (1967). [132] N. Tokura, T. Nagai, S. Takenaka, and T. Oshima, J. Chem. Soc., Perkin Trans. II *1974*, 337. [133] Pham Van Huong and J. Lascombe, C. R. H. Acad. Sci., Ser *C 254*, 2543 (1962); Chem. Abstr. *57*, 297 (1962). [134] H. E. Hallam: *Hydrogen Bonding and Solvent Effects*, in M. Davis (ed.): *Infra-Red Spectroscopy and Molecular Structure*.

Elsevier, Amsterdam 1963, Chapter XII, p. 405ff. [135] M. D. Joesten and L. J. Schaad: *Hydrogen Bonding*. Dekker, New York 1974. [136] L. J. Bellamy: *Solvent Effects on Infra-Red Group Frequencies*. In E. Thornton and H. W. Thompson (eds.): *Proceedings of the Conference on Molecular Spectroscopy*, held in London 1958. Pergamon Press, London 1959, p. 216ff.; Spectrochim. Acta *14*, 192 (1959). [137] R. L. Williams: *Infrared Spectra and Molecular Interactions*. Ann. Rept. Progr. Chem. (Chem. Soc. London) *58*, 34 (1961); Chem. Abstr. *57*, 9219e (1962). [138] M.-L. Josien: *Infrared-Spectroscopy Study of Molecular Interactions in Liquids*. Pure Appl. Chem. *4*, 33 (1962); J. Chim. Phys. *61*, 245 (1964). [139] C. N. R. Rao: *Chemical Applications of Infrared Spectroscopy*. Academic Press, New York, London 1963, p. 577ff. [140] M. Horák and J. Plíva: *Studies of Solute-Solvent Interactions. General Considerations*. Spectrochim. Acta *21*, 911 (1965).

[141] W. Brügel: *Einführung in die Ultrarotspektroskopie*. 4th ed., Steinkopff, Darmstadt 1969, p. 260 and 375. [142] N. Oi: *Infrared Solvents Shifts and Molecular Interactions*. Bunko Kenkyu *18*, 194 (1969); Chem. Abstr. *72*, 49158 (1970). [143] C. Agami: *Méthodes spectrographiques d'étude des solvants*. Bull. Soc. Chim. Fr. *1969*, 2183. [144] L. J. Bellamy: *The Infrared Spectra of Complex Molecules*. 3rd ed., Methuen, London 1975; *Ultrarot-Spektrum und chemische Konstitution* (translated into German by W. Brügel). 2nd ed., Steinkopff, Darmstadt 1966. [145] H. W. Thompson and D. J. Jewell, Spectrochim. Acta *13*, 254 (1958). [146] L. J. Bellamy and R. L. Williams, Trans. Faraday Soc. *55*, 14 (1959); Proc. Roy. Soc., London, Ser. A *255*, 22 (1960). [147] A. D. E. Pullin, Spectrochim. Acta *16*, 12 (1960). [148] C. Agami, Bull. Soc. Chim. Fr. *1968*, 2033. [149] A. G. Burden, G. Collier, and J. Shorter, J. Chem. Soc., Perkin Trans. II *1976*, 1627. [150] N. Oi and J. F. Coetzee, J. Am. Chem. Soc. *91*, 2473 (1969).

[151] L. J. Bellamy and P. E. Rogasch, Spectrochim. Acta *16*, 30 (1960). [152] Y. Ikegami, Bull. Chem. Soc. Jpn. *35*, 972 (1962). [153] T. Gramstad, Spectrochim. Acta *19*, 1363 (1963). [154] A. Allerhand and P. v. R. Schleyer, J. Am. Chem. Soc. *85*, 371 (1963); ibid. *86*, 5709 (1964). [155] T. Kagiya, Y. Sumida, and T. Inoue, Bull. Chem. Soc. Jpn. *41*, 767 (1968). [156] M. Horák and J. Moravec, Collect. Czech. Chem. Commun. *36*, 2757 (1971). [157] J. S. Byrne, P. F. Jackson, and K. J. Morgan, J. Chem. Soc., Perkin Trans. II *1972*, 1291; ibid. *1976*, 1800. [158] M.-L. Josien and N. Fuson, J. Chem. Phys. *22*, 1169, 1264 (1954). [159] J. Devaure, Pham. Van Huong, and J. Lascombe, J. Chim. Phys. *65*, 1686 (1968). [160] W. Lesch and R. Ulbrich, Z. Naturforsch. *23a*, 1639 (1968).

[161] M. F. El Bermani, A. J. Woodward, and N. Jonathan, J. Am. Chem. Soc. *92*, 6750 (1970). [162] L. J. Bellamy, H. E. Hallam, and R. L. Williams, Trans. Faraday Soc. *54*, 1120 (1958). [163] H. E. Hallam, J. Mol. Structure *3*, 43 (1969). [164] L. J. Bellamy and P. E. Rogasch, J. Chem. Soc. *1960*, 2218. [165] H. Götz, E. Heilbronner, A. R. Katrizky, and R. A. Jones, Helv. Chim. Acta *44*, 387 (1961). [166] J. G. Kirkwood, in W. West and R. T. Edwards, J. Chem. Phys. *5*, 14 (1937). [167] E. Bauer and M. Magat, J. Physique Radium *9*, 319 (1938); Chem. Abstr. *32*, 8856 (1938). [168] N. S. Bayliss, A. R. H. Cole, and L. H. Little, Aust. J. Chem. *8*, 26 (1955); Spectrochim. Acta *15*, 12 (1959). [169] A. D. E. Pullin, Spectrochim. Acta *13*, 125 (1958); ibid. *16*, 12 (1960). [170] A. D. Buckingham, Proc. Roy. Soc. London, Ser. A *248*, 169 (1957); ibid. A *255*, 32 (1960); Trans. Faraday Soc. *56*, 753 (1960).

[171] R. Ulbrich, Z. Naturforsch. *23a*, 1323 (1968). [172] J. Barriol, P. Bonnet, and J. Devaure, J. Chim. Phys. *71*, 107 (1974). [173] J. Fruwert and G. Geiseler: *Infrarotintensitäten*. Z. Chem. *20*, 157 (1980). [174] S. R. Polo and M. K. Wilson, J. Chem. Phys. *23*, 2376 (1955). [175] P. Mirone, Spectrochim. Acta *22*, 1897 (1966). [176] T. L. Brown, Spectrochim. Acta *10*, 149 (1957). [177] G. L. Caldow, D. C. Jones, and H. W. Thompson, Proc. Roy. Soc. London, Ser. A *254*, 17 (1960). [178] R. Heess and H. Kriegsmann, Spectrochim. Acta, Part A *24*, 2121 (1968). [179] J. Fruwert, G. Geiseler, and D. Luppa, Z. Chem. *8*, 238 (1968). [180] H. P. Figeys and V. Mahieu, Spectrochim. Acta, Part A *24*, 1553 (1968).

[181] M. Kakimoto and T. Fujiyama, Bull. Chem. Soc. Jpn. *48*, 2258 (1975). [182] C. M. Huggins and G. C. Pimentel, J. Chem. Phys. *23*, 896 (1955). [183] J. Gendell, J. H. Freed. and G. K. Fraenkel: *Solvent Effects in Electron Spin Resonance Spectra*. J. Chem. Phys. *37*, 2832 (1962). [184] A. Carrington and A. D. McLachlan: *Introduction to Magnetic Resonance*. Harper and Row, New York, and John Weatherhill, Tokyo 1967, p. 96ff. [185] M. C. R. Symons: *Application of ESR Spectroscopy to the Study of Solvation*. Pure Appl. Chem. *49*, 13 (1977). [186] K. Scheffler and H. B.

Stegmann: *Elektronenspinresonanz*. Springer, Berlin 1970. [187] G. R. Luckhurst and L. E. Orgel, Mol. Phys. *8*, 117 (1964). [188] N. Hirota: *Metal Ketyls and Related Radical Ions*, in E. T. Kaiser and L. Kevan (eds.): *Radical Ions*. Wiley-Interscience, New York 1968, Chapter 2, p. 35ff. [189] E. W. Stone and A. H. Maki, J. Chem. Phys. *36*, 1944 (1962); J. Am. Chem. Soc. *87*, 454 (1969). [190] E. A. C. Lucken, J. Chem. Soc. *1963*, 5123.

[191] P. J. Zandstra, J. Chem. Phys. *41*, 3655 (1964). [192] T. A. Claxton, J. Oakes, and M. C. R. Symons, Trans. Faraday Soc. *63*, 2125 (1967). [193] J. Oakes and M. C. R. Symons, Trans. Faraday Soc. *64*, 2579 (1968). [194] T. A. Claxton and D. McWilliams, Trans. Faraday Soc. *64*, 2593 (1968). [195] J. A. Pedersen and J. Spanget-Larsen, Chem. Phys. Lett. *35*, 41 (1975). [196] G. A. Russell: *Semidione Radical Anions*, in E. T. Kaiser and L. Kevan (eds.): *Radical Ions*. Wiley-Interscience, New York 1968, Chapter 3, p. 87ff. [197] P. J. Zandstra and E. M. Evleth, J. Am. Chem. Soc. *86*, 2664 (1964). [198] K. Scheffler and H. B. Stegmann, Z. Phys. Chem. (Frankfurt) N. F. *44*, 353 (1965). [199] K. Mukai, H. Nishiguchi, K. Ishizu, Y. Deguchi, and H. Takaki, Bull. Chem. Soc. Jpn. *40*, 2731 (1967). [200] S. Aono and M. Suhara, Bull. Chem. Soc. Jpn. *41*, 2553 (1968).

[201] J. Q. Chambers, P. Ludwig, T. Layloff, and R. N. Adams, J. Phys. Chem. *68*, 661 (1964); J. Am. Chem. Soc. *86*, 4568 (1964). [202] J. Pannel, Mol. Phys. *7*, 317 (1964). [203] M. T. Jones: *Recent Advances in the Chemistry of Aromatic Anion Radicals*, in E. T. Kaiser and L. Kevan (eds.): *Radical Ions*. Wiley-Interscience, New York 1968, p. 245ff. (especially p. 261ff.). [204] J. H. Sharp and M. C. R. Symons: *Electron Spin Resonance Studies of Ion Pairs*, in M. Szwarc (ed.): *Ions and Ion Pairs in Organic Reactions*. Wiley-Interscience, New York 1972, Vol. 1, Chapter 5, p. 177ff. (especially fig. 4). [205] G. R. Stevenson and H. Hidalgo, J. Phys. Chem. *77*, 1027 (1973). [206] H. G. Aurich and W. Weiss: *Formation and Reactions of Aminyloxides (Nitroxides)*. Fortschr. Chem. Forsch. *59*, 65 (1975); and references cited therein. [207] Y. Deguchi, Bull. Chem. Soc. Jpn. *35*, 260 (1962); K. Mukai, H. Nishiguchi, K. Ishizu, Y. Deguchi, and H. Takaki, ibid. *40*, 2731 (1967). [208] A. V. Il'yasov, Zh. Strukt. Khim. *3*, 95 (1962); Chem. Abstr. *58*, 7523d (1963). [209] A. L. Buchachenko, Dokl. Akad. Nauk USSR *158*, 932 (1964); Chem. Abstr. *62*, 2390g (1965). [210] R. Briére, G. Chapelet-Letourneux, H. Lemaire, and A. Rassat, Tetrahedron Lett. *1964*, 1775; Bull. Soc. Chim. Fr. *1965*, 444, 3273.

[211] P. B. Ayscough and F. P. Sargent, J. Chem. Soc., Part B *1966*, 907. [212] T. Kawamura, S. Matsunami, and T. Yonezawa, Bull. Chem. Soc. Jpn. *40*, 1111 (1967). [213] Th. A. J. W. Wajer, A. Mackor, and Th. J. de Boer, Tetrahedron *25*, 175 (1969). [214] G. H. Dodd, M. D. Barratt, and L. Rayner, FEBS Letters *8*, 286 (1970). [215] E. F. Ullmann, L. Call, and J. H. Osiecki, J. Org. Chem. *35*, 3623 (1970); E. F. Ullmann and L. Call, J. Am. Chem. Soc. *92*, 7210 (1970). [216] H. Hayat and B. L. Silver, J. Phys. Chem. *77*, 72 (1973). [217] G. Stout and J. B. F. N. Engberts, J. Org. Chem. *39*, 3800 (1974). [218] B. R. Knauer and J. J. Napier, J. Am. Chem. Soc. *98*, 4395 (1976). [219] H. G. Aurich and J. Trösken, Liebigs Ann. Chem. *745*, 159 (1971). [220] H. G. Aurich, W. Dersch, and H. Forster, Chem. Ber. *106*, 2854 (1973).

[221] F. W. Heineken, M. Bruin, and F. Bruin J. Chem. Phys. *37*, 452 (1962). [222] L. Grossi, F. Minisci, and G. F. Pedulli, J. Chem. Soc., Perkin Trans. II *1977*, 943; M. Guerra, F. Bernardi, and G. F. Pedulli, Chem. Phys. Lett. *48*, 311 (1977). [223] M. C. R. Symons, J. Phys. Chem. *71*, 172 (1967). [224] F. Gerson: *Hochauflösende ESR-Spektroskopie – dargestellt anhand aromatischer Radikal-Ionen*. Verlag Chemie, Weinheim 1967, p. 30ff., 128, and 173ff. [225] J. E. Gordon: *The Organic Chemistry of Electrolyte Solutions*. Wiley-Interscience, New York 1975, p. 384ff. [226] A. A. Bothner-By and R. E. Glick, J. Chem. Phys. *26*, 1651 (1957). [227] L. W. Reeves and W. G. Schneider, Can. J. Chem. *35*, 251 (1957). [228] E. O. Bishop: *Chemical Shifts – Intermolecular Effects: Solvent Dependent Studies*. Annu. Rep. Progr. Chem. for 1961 (Chem. Soc. London) *58*, 71 (1962). [229] G. Mavel: *NMR and Molecular Interactions in the Liquid Phase*. J. Chim. Phys. *61*, 182 (1964). [230] I. Yamaguchi: *Solvent Effects on the High Resolution NMR Spectrum*. Kogyo Kagaku Zasshi *68*, 1328 (1965); Chem. Abstr. *64*, 2884g (1966); Kagaku (Kyoto) *20*, 328 (1965); Chem. Abstr. *64*, 10618g (1966).

[231] J. W. Emsley, J. Feeney, and L. H. Sutcliffe: *High Resolution Nuclear Magnetic Resonance Spectroscopy*. Pergamon Press, Oxford 1965/66, Vol. II, p. 841ff. [232] P. Laszlo: *Solvent Effects and Nuclear Magnetic Resonance*, in J. W. Emsley, F. Feeney, and L. H. Sutcliffe (eds.): *Progress in*

Nuclear Magnetic Resonance Spectroscopy. Pergamon Press, Oxford 1967, Vol. 3, Chapter 6, p. 231 ff. [233] J. Ronayne and D. H. Williams: *Solvent Effects in Proton Magnetic Resonance Spectroscopy*, in E. F. Mooney (ed.): *Annual Review of NMR Spectroscopy*. Academic Press, London, New York 1969, Vol. 2, p. 83 ff. [234] M. I. Foreman: *Medium Effects in Nuclear Magnetic Resonance* (Specialist Periodical Report, Chem. Soc. London) *2*, 355 (1973); ibid. *5*, 292 (1976); Nucl. Magn. Reson. *6*, 233 (1977). [235] J. Homer: *Solvent Effects on Nuclear Magnetic Resonance Chemical Shifts*. Applied Spectroscopy Reviews *9*, 1 (1975). [236] S. L. Smith: *Solvent Effects and NMR Coupling Constants*. Fortschr. Chem. Forsch. *27*, 117 (1972). [237] M. Barfield and M. D. Johnston: *Solvent Dependence of Nuclear Spin-Spin Coupling Constants*. Chem. Rev. *73*, 53 (1973). [238] J. J. Dechter and J. I. Zink, J. Am. Chem. Soc. *97*, 2937 (1975). [239] J. F. Hinton ans R. W. Briggs, J. Magn. Reson. *19*, 393 (1975); J. Sol. Chem. *6*, 827 (1977); J. F. Hinton, K. R. Metz, and R. W. Briggs, Progr. NMR Spectrosc. *13*, 211 (1982). [240] U. Mayer, V. Gutmann, and W. Gerger, Monatsh. Chem. *106*, 1235 (1975); ibid. *108*, 489 (1977).

[241] R. W. Taft, E. Price, I. R. Fox, I. C. Lewis, K. K. Andersen, and G. T. Davis, J. Am. Chem. Soc. *85*, 709, 3146 (1963). [242] M. R. Bacon and G. E. Maciel, J. Am. Chem. Soc. *95*, 2413 (1973). [243] D. Ziessow and M. Carroll, Ber. Bunsenges. Phys. Chem. *76*, 61 (1972). [244] Y. M. Cahen, P. R. Handy, E. T. Roach, and A. I. Popov, J. Phys. Chem. *79*, 80 (1975). [245] M. S. Greenberg, R. L. Bodner, and A. I. Popov, J. Phys. Chem. *77*, 2449 (1973). [246] W. J. DeWitte, R. C. Schoening, and A. I. Popov, Inorg. Nucl. Chem. Lett. *12*, 251 (1976); W. J. DeWitte, L. Liu, E. Mei, J. L. Dye, and A. I. Popov, J. Sol. Chem. *6*, 337 (1977). [247] See references [106–111] to Chapter 2. [248] W. F. Reynolds and U. R. Priller, Can. J. Chem. *46*, 2787 (1968). [249] R. G. Anderson and M. C. R. Symons, Trans. Faraday Soc. *65*, 2537 (1969). [250] T. G. Beaumont and K. M. C. Davis, J. Chem. Soc., Part B *1970*, 592.

[251] G. E. Maciel and R. V. James, J. Am. Chem. Soc. *86*, 3893 (1964); G. E. Maciel and J. J. Natterstad, J. Chem. Phys. *42*, 2752 (1965). [252] G. L. Nelson, G. C. Levy, and J. D. Cargioli, J. Am. Chem. Soc. *94*, 3089 (1972). [253] S. Ueji and M. Nakamura, Tetrahedron Lett. *1976*, 2549. [254] H. A. Christ and P. Diehl, Helv. Phys. Acta *36*, 170 (1963); Chem. Abstr. *59*, 4702f (1963). [255] T. Schaefer and W. G. Schneider, J. Chem. Phys. *32*, 1218, 1224 (1960). [256] G. V. Rao, M. Balakrishnan, and N. Venkatasubramanian, Indian J. Chem. *13*, 1090 (1975). [257] R. L. Lichter and J. D. Roberts, J. Phys. Chem. *74*, 912 (1970). [258] F. A. L. Anet and I. Yavari, J. Org. Chem. *41*, 3589 (1976). [259] C. S. Giam and J. L. Lyle, J. Am. Chem. Soc. *95*, 3235 (1973). [260] S. K. Dayal and R. W. Taft, J. Am. Chem. Soc. *95*, 5595 (1973).

[261] G. E. Maciel and R. V. James, Inorg. Chem. *3*, 1650 (1964). [262] A. D. Buckingham, T. Schaefer, and W. G. Schneider, J. Chem. Phys. *32*, 1227 (1960); ibid. *34*, 1064 (1961). [263] F. H. A. Rummens: *Van der Waals Forces in NMR – Intermolecular Shielding Effects*, in P. Diehl, E. Fluck, and R. Kosfield (eds.): *NMR – Basic Principles and Progress*. Springer, Berlin 1975, Vol. 10, p. 1 ff. [264] A. D. Buckingham, Can. J. Chem. *38*, 300 (1960). [265] I. D. Kuntz and M. D. Johnston, J. Am. Chem. Soc. *89*, 6008 (1967). [266] F. H. A. Rummens, J. Am. Chem. Soc. *92*, 3214 (1970). [267] H. J. Bernstein: *Solvent Effects on NMR Spectra of Gases and Liquids*. Pure Appl. Chem. *32*, 79 (1972). [268] J. T. Arnold and M. E. Packard, J. Chem. Phys. *19*, 1608 (1951). [269] U. Liddel and N. F. Ramsay, J. Chem. Phys. *19*, 1608 (1951). [270] E. Lippert: *Wasserstoffbrückenbindung und magnetische Protonenresonanz*. Ber. Bunsenges. Phys. Chem. *67*, 267 (1963).

[271] A. S. N. Murthy and C. N. R. Rao: *Spectroscopic Studies of the Hydrogen Bond*. In E. G. Brame (ed.): *Applied Spectroscopy Reviews*. Dekker, New York, London 1969, Vol. 2, p. 69 ff. (especially p. 106). [272] J. C. Davis and K. K. Deb: *Analysis of Hydrogen Bonding and Related Association Equilibria by Nuclear Magnetic Resonance*, in J. S. Waugh (ed.): *Advances in Magnetic Resonance*. Academic Press, New York, London 1970, Vol. 4, p. 201 ff. [273] J. V. Hatton and R. E. Richards, Trans. Faraday Soc. *57*, 28 (1961). [274] M. L. Martin, Annales de Physique *7*, 35 (1962); G. J. Martin and M. L. Martin, J. Chim. Phys. *61*, 1222 (1964). [275] C. Agami, M. Andrac-Taussig, and C. Prévost, Bull. Soc. Chim. Fr. *1968*, 952, 4467. [276] J. Reuben, J. Am. Chem. Soc. *91*, 5725 (1969). [277] See also references [37–46] to Chapter 2. [278] P. Laszlo, Bull. Soc. Chim. Fr. *1964*, 2658. [279] E. M. Engler and P. Laszlo, J. Am. Chem. Soc. *93*, 1317 (1971). [280] See also references [17–22] to Chapter 3.

[281] G. R. Luckhurst: *Liquid Crystals as Solvents in NMR.* Quart. Rev. (London) *22*, 179 (1968). [282] J. Bulthuis: *NMR in Liquid Crystalline Solvents.* Amsterdam 1974; Chem. Abstr. *82*, 148359n (1975). [283] J. W. Emsley and J. C. Lindon: *NMR-Spectroscopy Using Liquid Crystal Solvents.* Pergamon Press, Oxford 1975. [284] W. H. Pirkle and S. D. Beare, J. Am. Chem. Soc. *89*, 5485 (1967); W. H. Pirkle, S. D. Beare, and T. G. Burlingame, J. Org. Chem. *34*, 470 (1969). [285] D. F. Evans, J. Chem. Soc. *1963*, 5575. [286] P. Laszlo, Bull. Soc. Chim. Fr. *1966*, 558. [287] V. S. Watts and J. H. Goldstein, J. Phys. Chem. *70*, 3887 (1966). [288] S. L. Smith and R. H. Cox, J. Chem. Phys. *45*, 2848 (1966). [289] K. A. McLauchlan, L. W. Reeves, and T. Schaefer, Can. J. Chem. *44*, 1473 (1966). [290] R. L. Schmidt, R. S. Butler, and J. H. Goldstein, J. Phys. Chem. *73*, 1117 (1969).

[291] A. M. Ihrig and S. L. Smith, J. Am. Chem. Soc. *94*, 34 (1972). [292] Y. Tang, W. Cao, X. Song, W. Liu, and F. Zhang, Ganguang Kexue Yu Kuang Huaxue *1985*, 32; Chem. Abstr. *104*, 7212v (1986). [293] D. J. Lougnot, P. Brunero, J. P. Fouassier, and J. Faure, J. Chim. Phys. Phys.-Chim. Biol. *79*, 343 (1982). [294] C. Reichardt: *Empirical Parameters of Solvent Polarity and Chemical Reactivity.* In H. Ratajczak and W. J. Orville-Thomas (eds.): *Molecular Interactions.* Wiley, Chichester 1982, Vol. 3, Chapter 5, p. 241 ff. [295] H. Langhals, Z. Anal. Chem. *305*, 26 (1981); ibid. *308*, 441 (1981). [296] H. Langhals: *Polarität binärer Flüssigkeitsgemische.* Angew. Chem. *94*, 739 (1982); Angew. Chem., Int. Ed. Engl. *21*, 724 (1982). [297] B. P. Johnson, B. Gabrielsen, M. Matulenko, J. G. Dorsey, and C. Reichardt: *Solvatochromic Solvent Polarity Measurements: Synthesis and Applications of ET-30.* Analytical Letters *19*, 939 (1986). [298] P. Plieninger and H. Baumgärtel, Ber. Bunsenges. Phys. Chem. *86*, 161 (1982); Liebigs Ann. Chem. *1983*, 860. [299] K. A. Zachariasse, N. Van Phuc, and B. Kozankiewicz, J. Phys. Chem. *85*, 2676 (1981). [300] C. J. Drummond, F. Grieser, and T. W. Healy, Faraday Discuss. Chem. Soc. *81*, 95 (1986); Chem. Abstr. *106*, 126332k (1987).

[301] A. J. M. Van Beijnen, R. J. M. Nolte, and W. Drenth, Rec. Trav. Chim. Pays-Bas *105*, 255 (1986). [302] B. P. Johnson, M. G. Khaledi, and J. G. Dorsey, J. Chromatogr. *384*, 221 (1987); Anal. Chem. *58*, 2354 (1986); Chimia oggi (Milano) *1986* (11), 23. [303] For new methods for the determinatioon of excited-state dipole moments by solvatochromic effects see: (a) J. J. Moura Ramos, M.-L. Stien, and J. Reisse, Rec. Trav. Chim. Pays-Bas *98*, 338 (1979); (b) P. Suppan, Chem. Phys. Lett. *94*, 272 (1983); (c) T. Abe and I. Iweibo, Bull. Chem. Soc. Jpn. *58*, 3415 (1985); (d) N. H. Ayachit, D. K. Deshpande, M. A. Shashidhar, and K. S. Rao, Spectrochim. Acta, Part A *42*, 585 (1986). [304] C. A. G. O. Varma and E. J. G. Groenen, Rec. Trav. Chim. Pays-Bas *91*, 296 (1972). [305] K. F. Donchi, G. P. Robert, B. Ternai, and P. J. Derrick, Aust. J. Chem. *33*, 2199 (1980). [306] I. Gruda and F. Bolduc, J. Org. Chem. *49*, 3300 (1984). [307] M. M. Habashy, F. El-Zawawi, M. S. Antonious, A. K. Sherif, and M. S. A. Abdel-Mottaleb, Indian J. Chem., Part A *24*, 908 (1985). [308] A. Botrel, A. Le Beuze, P. Jacques, and H. Straub, J. Chem. Soc., Faraday Trans. II *80*, 1235 (1984). [309] P. Jacques, J. Phys. Chem. *90*, 5535 (1986). [310] A. Le Beuze, A. Botrel, A. Samat, P. Appriou, and R. Guglielmetti, J. Chim. Phys. *75*, 255, 267 (1978).

[311] *Heterocyclic betaine dyes (mesoionic zwitterions):* (a) Z. Pawelka and L. Sobczyk, J. Chem. Soc., Faraday Trans. I *76*, 43 (1980); (b) J.-H. Finkentey and H. W. Zimmermann, Liebigs Ann. Chem. *1981*, 1; (c) J. Bendig, D. Kreysig, E. Sauer, and M. Just, J. Prakt. Chem. *323*, 529 (1981); Z. Phys. Chem. (Leipzig) *263*, 442 (1982); (d) M. S. A. Abdel-Mottaleb, Z. Naturforsch., Part A *37*, 1353 (1982); Z. Phys. Chem. (Leipzig) *265*, 154 (1984); (e) S. Arai, M. Yamazaki, K. Nagakura, M. Ishikawa, and M. Hida, J. Chem. Soc., Chem. Commun. *1983*, 1037; Sen-I Gakkaishi *42*, T-74 (1986); (f) V. Kampars, V. Kokars, Z. Bruvers, and O. Neilands, Zh. Obshch. Khim. *53*, 2299 (1983); Chem. Abstr. *100*, 50949v (1984); (g) P. de Mayo, A. Safarzadeh-Amiri, and S. K. Wong, Can. J. Chem. *62*, 1001 (1984); (h) H. Gotthardt and M. Oppermann, Chem. Ber. *119*, 2094 (1986). [312] *Substituted azobenzenes:* (a) H. Mustroph and J. Epperlein, J. Prakt. Chem. *322*, 305 (1980); (b) B. Marcandalli, E. Dubini-Paglia, and L. Pellicciari-Di Liddo, Gazz. Chim. Ital. *111*, 187 (1981); (c) K. S. Schanze, T. F. Mattox, and D. G. Whitten, J. Org. Chem. *48*, 2808 (1983); (d) Y. Sueishi, M. Asano, S. Yamamoto, and N. Nishimura, Bull. Chem. Soc. Jpn. *58*, 2729 (1985). [313] *Quinone methides:* L. Pavličková, B. Koutek, V. Jehlička, and M. Souček, Collect. Czech. Chem. Commun. *48*, 2376 (1983). [314] *Thiolate ions and thiyl radicals:* (a) G. H. Morine and R. R. Kuntz, Chem. Phys. Lett. *67*, 552 (1979); (b) O. Ito and M. Matsuda, J. Phys. Chem. *88*, 1002 (1984); (c) T. J. Novak, S. G. Pleva, and J. Epstein, Anal. Chem. *52*, 1851 (1980). [315] J. W. Larsen, A. G. Edwards, and P. Dobi, J.

Am. Chem. Soc. *102*, 6780 (1980). [316] T. R. Griffiths and R. H. Wijayanayake, Trans. Faraday Soc. *66*, 1563 (1970); J. Chem. Soc., Faraday Trans. I *69*, 1899 (1973). [317] J. L. Dye, Angew. Chem. *91*, 613 (1979); Angew. Chem., Int. Ed. Engl. *18*, 587 (1979). [318] J. E. Brady and P. W. Carr: *An Analysis of Dielectric Models of Solvatochromism*. J. Phys. Chem. *89*, 5759 (1985); and references cited therein, particularly S. Ehrenson, J. Am. Chem. Soc. *103*, 6036 (1981). [319] A. A. Ishchenko, N. A. Derevyanko, and A. I. Tolmachev, Dokl. Akad. Nauk SSSR *274*, 106 (1984); Chem. Abstr. *100*, 193516n (1984). [320] J. S. Ham, J. Chem. Phys. *21*, 756 (1953); G. Durocher and C. Sandorfy, J. Mol. Spectrosc. *20*, 410 (1966).

[321] A. B. Myers and R. R. Birge, J. Chem. Phys. *73*, 5314 (1980); ibid. *77*, 1075 (1982). [322] T. Abe, J. Chem. Phys. *77*, 1074 (1982); ibid. *83*, 1546 (1985); T. Abe and I. Iweibo, Bull. Chem. Soc. Jpn. *59*, 2381 (1986). [323] T. Shibuya, J. Chem. Phys. *78*, 5175 (1983); Bull. Chem. Soc. Jpn. *57*, 2991 (1984). [324] V. Bekárek *et al.*, Collect. Czech. Chem. Commun. *45*, 2063 (1980); ibid. *47*, 1060 (1982); ibid. *51*, 746 (1986). [325] Yu. T. Mazurenko, Opt. i Spektrosk. *55*, 471 (1983); Chem. Abstr. *100*, 42159x (1984). [326] W. Förster, F. Dietz, B. Mikutta, and C. Weiss, Z. Chem. *19*, 234 (1979). [327] A. Mehlhorn, J. Signal AM *6*, 211 (1978). [328] A. F. Beecham, A. C. Hurley, and C. H. Johnson, Aust. J. Chem. *32*, 1643 (1979); ibid. *33*, 699 (1980). [329] P. R. Taylor, J. Am. Chem. Soc. *104*, 5248 (1982). [330] J. J. Moura Ramos, M.-L. Stien, and J. Reisse, Chem. Phys. Lett. *42*, 373 (1976).

[331] C. Haessner and H. Mustroph, Z. Chem. *26*, 170 (1986). [332] Y. Jinnouchi, M. Kohno, and A. Kuboyama, Bull. Chem. Soc. Jpn. *57*, 1147 (1984). [333] C. Benamou and L. Bellon, Bull. Soc. Chim. Fr. *1984*, I-321. [334] P. Mukerjee, C. Ramachandran, and R. A. Pyter, J. Phys. Chem. *86*, 3189 (1982). [335] M. C. R. Symons and A. S. Pena-Nuñez, J. Chem. Soc., Faraday Trans. I *81*, 2421 (1985). [336] A. Janowski, I. Turowska-Tyrk, and P. K. Wrona, J. Chem. Soc., Perkin Trans. II *1985*, 821. [337] W. D. Hinsberg, P. G. Schultz, and P. B. Dervan, J. Am. Chem. Soc. *104*, 766 (1982). [338] C. Ramachandran, R. A. Pyter, and P. Mukerjee, J. Phys. Chem. *86*, 3198 (1982). [339] G. van der Zwan and J. T. Hynes, J. Phys. Chem. *89*, 4181 (1985). [340] E. M. Kosower, J. Am. Chem. Soc. *107*, 1114 (1985); E. M. Kosower *et al.*, J. Phys. Chem. *87*, 2479 (1983).

[341] G. Jones, W. R. Jackson, C. Choi, and W. R. Bergmark, J. Phys. Chem. *89*, 294 (1985). [342] E. Lippert, W. Lüder, and H. Boos, Adv. Mol. Spectrosc., Proc. Int. Meet., 4th, *1959* (1962), 443; E. Lippert *et al.*, J. Photochem. *17*, 237 (1981). [343] K. Rotkiewicz, K. H. Grellmann, and Z. R. Grabowski, Chem. Phys. Lett. *19*, 315 (1973); ibid. *21*, 212 (1973); Z. R. Grabowski *et al.*, Nouv. J. Chim. *3*, 443 (1979). [344] W. Rettig, Angew. Chem. *98*, 969 (1986); Angew. Chem., Int. Ed. Engl. *25*, 971 (1986). [345] G. F. Mes, B. de Jong, H. J. van Ramesdonk, J. W. Verhoeven, J. M. Warman, M. P. de Haas, and L. E. W. Horsman-van den Dool, J. Am. Chem. Soc. *106*, 6524 (1984). [346] E. M. Kosower and H. Kanety, J. Am. Chem. Soc. *105*, 6236 (1983). [347] E. M. Kosower, R. Giniger, A. Radkowsky, D. Hebel, and A. Shusterman, J. Phys. Chem. *90*, 5552 (1986). [348] J. Bendig, B. Henkel, and D. Kreysig, Ber. Bunsenges. Phys. Chem. *85*, 38 (1981). [349] J. Bendig, D. Kreysig, E. Sauer, and M. Just, J. Prakt. Chem. *323*, 538 (1981). [350] T. C. Werner and D. B. Lyon, J. Phys. Chem. *86*, 933 (1982).

[351] A. C. Capomacchia, F. L. White and T. L. Sobol, Spectrochim. Acta, Part A *38*, 513 (1982). [352] M. Swaminathan and S. K. Dogra, J. Am. Chem. Soc. *105*, 6223 (1983). [353] G. E. Johnson and W. W. Limburg, J. Phys. Chem. *88*, 2211 (1984). [354] Y. Wang, J. Phys. Chem. *89*, 3799 (1985). [355] M. Nosowitz and A. M. Halpern, J. Phys. Chem. *90*, 906 (1986). [356] A. Nakajima, Bull. Chem. Soc. Jpn. *44*, 3272 (1971); J. Mol. Spectrosc. *61*, 467 (1976). [357] D. C. Dong and M. A. Winnik, Photochem. Photobiol. *35*, 17 (1982); Chem. Abstr. *96*, 161966s (1982); Can. J. Chem. *62*, 2560 (1984). [358] K. W. Street and W. E. Acree, Analyst *111*, 1197 (1986). [359] I. Kristjánsson and J. Ulstrup, Chemica Scripta *25*, 49 (1985). [360] K. Kalyanasundaram: *Photophysics of Molecules in Micelle-forming Surfactant Solutions*. Chem. Soc. Rev. *7*, 453 (1978).

[361] D. N. Kirk: *The Chiroptical Properties of Carbonyl Compounds*. Tetrahedron *42*, 777 (1986). [362] F. M. Menger and B. Boyer, J. Org. Chem. *49*, 1826 (1984). [363] S. Ohta *et al.*, Chem. Lett. (Tokyo) *1985*, 1331; Bull. Chem. Soc. Jpn. *59*, 1181 (1986). [364] M. Legrand: *Use of Solvent and Temperature Effects*. In F. Ciardelli and P. Salvadori (eds.): *Fundamental Aspects and Recent Developments in ORD and CD*. Heyden and Son, London 1973, Chapter 4.2, p. 285ff.; and references cited therein. [365] B. Bosnich: *Induced Optical Activity*. In F. Ciardelli and P. Salvadori (eds.):

Fundamental Aspects and Recent Developments in ORD and CD. Heyden and Son, London 1973, Chapter 3.8, p. 254ff. [366] T. Hensel, J. Fruwert, M. Szombately, and H. Kriegsmann, Z. Phys. Chem. (Leipzig) *267*, 641, 650 (1986); T. Hensel, J. Fruwert, and K. Dathe, Coll. Czech. Chem. Commun. *52*, 22 (1987). [367] M. C. R. Symons, Chem. Soc. Rev. *12*, 1 (1983); Pure Appl. Chem. *58*, 1121 (1986). [368] M. C. R. Symons and G. Eaton, J. Chem. Soc., Faraday Trans. I *81*, 1963 (1985). [369] J. Manzur and G. Gonzalez, Z. Naturforsch., Part B *36*, 763 (1981). [370] K. B. Patel, G. Eaton, and M. C. R. Symons, J. Chem. Soc., Faraday Trans. I *81*, 2775 (1985).

[371] T. Yoshimura, Arch. Biochem. Biophys. *220*, 167 (1983). [372] M. C. R. Symons and G. Eaton, J. Chem. Soc., Faraday Trans. I *78*, 3033 (1982). [373] V. Bekárek and T. Nevěčná, Coll. Czech. Chem. Commun. *50*, 1928 (1985). [374] R. G. Makitra, Ya. N. Pirig, Ya. Tsikanchuk, and V. Ya. Zhukovskii, Ukr. Khim. Zh. *46*, 729 (1980); Chem. Abstr. *93*, 203859e (1980). [375] D. G. Cameron, S. C. Hsi, J. Umemura, and H. H. Mantsch, Can. J. Chem. *59*, 1357 (1981). [376] G. González and E. Clavijo, J. Chem. Soc., Perkin Trans. II *1985*, 1751. [377] R. G. Makitra, V. Ya. Zhukovskii, Ya. N. Pirig, and B. I. Chernyak, Zh. Prikl. Spektrosk. *36*, 962 (1982); Chem. Abstr. *97*, 82005v (1982). [378] C. Laurence, M. Berthelot, M.Lucon, M. Helbert, D. G. Morris, and J.-F. Gal, J. Chem. Soc., Perkin Trans. II *1984*, 705. [379] V. Bekárek and A. Mikulecká, Coll. Czech. Chem. Commun. *43*, 2879 (1978). [380] V. Bekárek, Acta Univ. Palacki. Olomuc., Fac. Rerum Nat. *61/65*, 87 (1980); Chem. Abstr. *94*, 147932e (1981).

[381] S. Kubota and Y. Ikegami, J. Phys. Chem. *82*, 2739 (1978). [382] D. M. Holton and D. Murphy, J. Chem. Soc., Faraday Trans. I *78*, 1223 (1982). [383] J. Spanget-Larsen, Theor. Chim. Acta *51*, 65 (1979). [384] H. G. Aurich, K. Hahn, K. Stork, and W. Weiss, Tetrahedron *33*, 969 (1977). [385] A. H. Reddoch and S. Konishi, J. Chem. Phys. *70*, 2121 (1979). [386] E. G. Janzen, G. A. Coulter, U. M. Oehler, and J. P. Bergsma, Can. J. Chem. *60*, 2725 (1982). [387] Y. Liu and Y. Wang, Huaxue Xuebao *43*, 232 (1985); Chem. Abstr. *103*, 177687a (1985). [388] Z. Kecki, B. Łyczkowski, and W. Kołodziejski, J. Sol. Chem. *15*, 413 (1986). [389] O. W. Kolling, Anal. Chem. *55*, 143 (1983). [390] B. J. Tabner: *Organic Radicals in Solution*. In M. C. R. Symons (ed.): *Electron Spin Resonance*. Specialist Periodical Reports, Vol. 10A, Chapter 1, Royal Society of Chemistry, London 1986; and previous volumes.

[391] K. Nakamura, Chem. Lett. (Tokyo) *1980*, 301; J. Am. Chem. Soc. *103*, 6973 (1981). [392] T. Abe, S. Tero-Kubota, and Y. Ikegami, J. Phys. Chem. *86*, 1358 (1982). [393] H. Block and S. M. Walker, Chem. Phys. Lett. *19*, 363 (1973). [394] P. Suppan, J. Chem. Soc., Faraday Trans. I *83*, 495 (1987). [395] P. Laszlo and A. Stockis, J. Am. Chem. Soc. *102*, 7818 (1980); B. Allard and E. Casadevall, Noveau J. Chim. *9*, 565 (1985). [396] P. Laszlo: *^{23}Na-NMR-Spektroskopie*. Nachr. Chem. Tech. Lab. *27*, 710 (1979). [397] B. Tiffon and J.-E. Dubois, Org. Magn. Reson. *11*, 295 (1978); B. Tiffon and B. Ancian, ibid. *16*, 247 (1981). [398] M. Jallali-Heravi and G. A. Webb, Org. Magn. Reson. *13*, 116 (1980). [399] W. Freitag and H.-J. Schneider, J. Chem. Soc., Perkin Trans. II *1979*, 1337. [400] V. Bekárek and V. Simánek, Coll. Czech. Chem. Commun. *46*, 1549 (1981).

[401] C. W. Fong and H. G. Grant, Z. Naturforsch., Part B *36*, 585 (1981). [402] R. Radeglia, R. Wolff, B. Bornowski, and S. Dähne, Z. Phys. Chem. (Leipzig) *261*, 502 (1980). [403] P. Plieninger and H. Baumgärtel, Liebigs Ann. Chem. *1983*, 860. [404] J. G. Dawber and R. A. Williams, J. Chem. Soc., Faraday Trans. I *82*, 3097 (1986). [405] J.-L. M. Abboud, A. Auhmani, H. Bitar, M. El Mouhtadi, J. Martin, and M. Rico, J. Am. Chem. Soc. *109*, 1332 (1987). [406] I. Ando and G. A. Webb, Org. Magn. Reson. *15*, 111 (1981). [407] M. Sugiura, N. Takao, and S. Ueji, Org. Magn. Reson. *18*, 128 (1982). [408] N. Kalyanam: *Application of Aromatic Solvent Induced Shifts in Organic Chemistry*. J. Chem. Educ. *60*, 635 (1983). [409] S. Ueji, M. Sugiura, and N. Takao, Tetrahedron Lett. *21*, 475 (1980). [410] J. D. Connolly and R. McCrindle, Chem. Ind. (London) *1965*, 379.

[411] D. H. Williams and N. S. Bhacca, Tetrahedron *21*, 2021 (1965). [412] K. Tori, I. Horibe, H. Sigemoto, and K. Umemoto, Tetrahedron Lett. *1975*, 2199. [413] K. Nikki and N. J. Nakagawa, Bull. Chem. Soc. Jpn. *51*, 3267 (1978); Org. Magn. Reson. *23*, 432 (1985). [414] M. Jutila, U. Edlund, D. Johnels, and E. Johansson, Acta Chem. Scand., Part B *38*, 131 (1984). [415] T. C. Morill (ed.): *Lanthanide Shifts Reagents in Stereochemical Analysis*. In A. P. Marchand (ed.): *Methods in Stereochemical Analysis*. Vol. 5, VCH Publishers, New York 1986. [416] I. M. Dean and R. S. Varma, Tetrahedron *38*, 2069 (1982). [417] J. F. Bertrán and M. Rodríguez, Org. Magn. Reson. *21*, 1, 6

(1983). [418] J. C. Jochims: *Die magnetische Kernresonanz als Werkzeug des Stereochemikers*. In G. Hess (ed.): *Konstanzer Universitätsreden*. No. 56, Universitätsverlag, Konstanz 1973, p. 26ff.; Chem. Abstr. *81*, 43982m (1974). [419] P. E. Hansen: *Carbon-Hydrogen Spin-Spin Coupling Constants*. Progr. NMR Spectrosc. *14*, 175 (1982). [420] I. Ando and S. Watanabe, Bull. Chem. Soc. Jpn. *53*, 1257 (1980); ibid. *54*, 1241 (1981).

[421] T. Schaefer, H. M. Hutton, and S. R. Salman, Can. J. Chem. *57*, 1877 (1979). [422] S. Watanabe and I. Ando, J. Mol. Struct. *84*, 77 (1982). [423] J. Burgess, J. G. Chambers, and R. I. Haines, Transition Met. Chem. *6*, 145 (1981). [424] W. Kaim, S. Ernst, and S. Kohlmann: *Farbige Komplexe: das Charge-Transfer-Phänomen*. Chemie in unserer Zeit *21*, 50 (1987). [425] D. M. Manuta and A. J. Lees, Inorg. Chem. *22*, 3825 (1983); ibid. *25*, 3212 (1986). [426] S. Ernst, Y. Kurth, and W. Kaim, J. Organomet. Chem. *302*, 211 (1986); W. Kaim and S. Kohlmann, Inorg. Chem. *25*, 3306 (1986). [427] R. E. Shepherd, M. F. Hoq, N. Hoblack, and C. R. Johnson, Inorg. Chem. *23*, 3249 (1984). [428] W. Kaim, S. Kohlmann, S. Ernst, B. Olbrich-Deussner, C. Bessenbacher, and A. Schulz, J. Organomet. Chem. *321*, 215 (1987). [429] H. A. Staab, C. Krieger, P. Wahl, and K.-Y. Kay, Chem. Ber. *120*, 551 (1987). [430] G. L. Gaines, Angew. Chem. *99*, 346 (1987); Angew. Chem., Int. Ed. Engl. *26*, 341 (1987).

[431] A. M. Kjaer and J. Ulstrup, J. Am. Chem. Soc. *109*, 1934 (1987). [432] V. Bekárek and V. Bekárek, jr., Coll. Czech. Chem. Commun. *52*, 287 (1987). [433] H. Langhals, Z. Phys. Chem. (Frankfurt) N. F. *127*, 45 (1981); Tetrahedron *43*, 1771 (1987). [434] M. V. Garcia and M. I. Redondo, Spectrochim. Acta, Part A *43*, 879 (1987).

Chapter 7

[1] C. Reichardt: *Empirische Parameter der Lösungsmittelpolarität*, Angew. Chem. *77*, 30 (1965); Angew. Chem., Int. Ed. Engl. *4*, 29 (1965); C. Reichardt: *Empirical Parameters of Solvent Polarity as Linear Free-Energy Relationships*. Angew. Chem. *91*, 119 (1979); Angew. Chem., Int. Ed. Engl. *18*, 98 (1979). [2] C. Reichardt and K. Dimroth: *Lösungsmittel und empirische Parameter zur Charakterisierung ihrer Polarität*. Fortschr. Chem. Forsch. *11*, 1 (1968). [3] C. Reichardt: *Lösungsmittel-Effekte in der organischen Chemie*. 2nd ed., Verlag Chemie, Weinheim 1973; *Effets de solvant en chimie organique* (translated into French by I. Tkatchenko), Flammarion, Paris 1971; *Rastvoriteli v organicheskoi khimii* (translated into Russian by E. R. Zakhsa), Izdatel'stvo Khimiya, Leningrad 1973. [4] V. A. Palm: *Osnovy kolichestvennoi teorii organicheskikh reaktsii*. 2nd ed. Izdatel'stvo Khimiya, Leningrad 1977; *Grundlagen der quantitativen Theorie organischer Reaktionen* (translated into German by G. Heublein). Akademie-Verlag, Berlin 1971. [5] E. M. Kosower: *An Introduction to Physical Organic Chemistry*. Wiley, New York 1968, p. 293ff. [6] I. A. Koppel and V. A. Palm: *The Influence of the Solvent on Organic Reactivity*, in N. B. Chapman and J. Shorter (eds.): *Advances in Linear Free Energy Relationships*. Plenum Press, London, New York 1972, Chapter 5, p. 203ff. [7] J. A. Hirsch: *Concepts in Theoretical Organic Chemistry*. Allyn and Bacon, Boston 1974, Chapter 9, p. 207ff. [8] M. R. J. Dack: *The Influence of Solvent on Chemical Reactivity*, in M. R. J. Dack (ed.): *Solutions and Solubilities*. Vol. VIII, Part II, of A. Weissberger (ed.): *Techniques of Chemistry*. Wiley-Interscience, New York 1976, p. 95ff. [9] V. Gutmann: *The Donor-Acceptor Approach to Molecular Interactions*. Plenum Publ. Corp., New York 1978. [10] L. P. Hammett, J. Am. Chem. Soc. *59*, 96 (1937); Trans. Faraday Soc. *34*, 156 (1938); *cf.* also J. Shorter: *Die Hammett-Gleichung – und was daraus in fünfzig Jahren wurde*. Chemie in unserer Zeit *19*, 197 (1985).

[11] P. R. Wells: *Linear Free Energy Relationships*. Academic Press, London, New York 1968. [12] N. B. Chapman and J. Shorter (eds.): *Advances in Linear Free Energy Relationships*. Plenum Press, London, New York 1972. [13] N. B. Chapman and J. Shorter (eds.): *Correlation Analysis in Chemistry – Recent Advances*. Plenum Publ. Corp., New York 1978. [14] J. Shorter: *Correlation Analysis in Organic Chemistry – An Introduction to Linear Free Energy Relationships*. Clarendon Press, Oxford 1973. [15] J. Shorter: *Correlation Analysis of Organic Reactivity – With Particular Reference to Multiple Regression*. Research Studies Press, Chichester 1982. [16] C. Reichardt and R.

Müller, Liebigs Ann. Chem. *1976*, 1953. [17] J. N. Brønsted and K. J. Pedersen. Z. Phys. Chem. *108*, 185 (1924); Chem. Rev. *5*, 231 (1928). [18] J. E. Leffler and E. Grunwald: *Rates and Equilibria of Organic Reactions*. Wiley, New York, London 1963. [19] A. R. Katritzky and R. D. Topsom: *Linear Free Energy Relationships and Optical Spectroscopy*, in N. B. Chapman and J. Shorter (eds.): *Advances in Linear Free Energy Relationships*. Plenum Press, London, New York 1972, Chapter 3, p. 119ff. [20] M. T. Tribble and J. G. Traynham: *Linear Correlations of Substituent Effects in ^1H, ^{19}F, and ^{13}C Nuclear Magnetic Resonance Spectroscopy*, in N. B. Chapman and J. Shorter (eds.): *Advances in Linear Free Energy Relationships*. Plenum Press, London, New York 1972, Chapter 4, p. 143ff.

[21] E. M. Kosower, D. Hofmann, and K. Wallenfels, J. Am. Chem. Soc. *84*, 2755 (1962). [22] C. Reichardt and W. Grahn, Chem. Ber. *103*, 1072 (1970). [23] C. Reichardt and R. Müller, Liebigs Ann. Chem. *1976*, 1937. [24] K. H. Meyer, Ber. Dtsch. Chem. Ges. *47*, 826 (1914); ibid. *53*, 1410 (1920); ibid. *54*, 579 (1921). [25] E. L. Eliel, Pure Appl. Chem. *25*, 509 (1971); E. L. Eliel, Supplement to Pure Appl. Chem. 1971, p. 219ff.; E. L. Eliel and O. Hofer, J. Am. Chem. Soc. *95*, 8041 (1973). [26] V. Gutmann and E. Wychera, Inorg. Nucl. Chem. Lett. *2*, 257 (1966); V. Gutmann, Coordination Chem. Rev. *2*, 239 (1967); ibid. *18*, 225 (1976); V. Gutmann and A. Scherhaufer, Monatsh. Chem. *99*, 335 (1968); V. Gutmann, Chimia *23*, 285 (1969); V. Gutmann, Electrochimica Acta *21*, 661 (1976). [27] V. Gutmann: *Coordination Chemistry in Non-Aqueous Solvents*. Springer, Wien, New York 1968; V. Gutmann: *Chemische Funktionslehre*. Springer, Wien, New York 1971; V. Gutmann: *The Donor-Acceptor Approach to Molecular Interactions*. Plenum, New York 1978. [28] G. Olofson and I. Olofson, J. Am. Chem. Soc. *95*, 7231 (1973). [29] M. S. Greenberg, R. L. Bodner, and A. I. Popov, J. Phys. Chem. *77*, 2449 (1973). [30] V. Gutmann, Pure Appl. Chem. *27*, 73 (1971); Fortschr. Chem. Forsch. *27*, 59 (1972); Structure and Bonding *15*, 141 (1973); Chimia *31*, 1 (1977); CHEMTECH *7*, 255 (1977); Pure Appl. Chem. *51*, 2197 (1979); A. J. Parker, U. Mayer, R. Schmid, and V. Gutmann, J. Org. Chem. *73*, 1843 (1978).

[31] J. M. Kelly, D. V. Bent, H. Hermann, D. Schulte-Frohlinde, E. Koerner von Gustorf, J. Organomet. Chem. *69*, 259 (1974). [32] R. S. Drago and B. B. Wayland, J. Am. Chem. Soc. *87*, 3571 (1965); R. S. Drago, J. Chem. Educ. *51*, 300 (1974). [32a] J. Llor and M. Cortijo, J. Chem. Soc., Perkin Trans. II *1977*, 1111. [33] E. S. Gould: *Mechanism and Structure in Organic Chemistry*. Henry Holt and Co., New York 1959, p. 299ff.; *Mechanismus und Struktur in der organischen Chemie* (translated into German by G. Koch). 2nd ed., Verlag Chemie, Weinheim 1962, p. 353ff. [34] K. Schwetlick: *Kinetische Methoden zur Untersuchung von Reaktionsmechanismen*. Deutscher Verlag der Wissenschaften, Berlin 1971, p. 162ff. [35] E. Grunwald and S. Winstein, J. Am. Chem. Soc. *70*, 846 (1948). [36] S. Winstein, E. Grunwald, and H. W. Jones, J. Am. Chem. Soc. *73*, 2700 (1951); A. H. Fainberg and S. Winstein, ibid. *78*, 2770 (1956); ibid. *79*, 1597, 1602, 1608 (1957); S. Winstein, A. H. Fainberg, and E. Grunwald, ibid. *79*, 4146, 5937 (1957). [37] S. G. Smith, A. H. Fainberg, and S. Winstein, J. Am. Chem. Soc. *83*, 618 (1961). [38] I. A. Koppel and V. A. Palm, Reakts. Sposobnost Org. Soedin. *6*, 504 (1969); Chem. Abstr. *72*, 6732 y (1970). [39] D. J. Raber, R. C. Bingham, J. M. Harris, J. L. Fry, and P. v. R. Schleyer, J. Am. Chem. Soc. *92*, 5977 (1970). [40] D. N. Kevill, K. C. Kolwyck, and F. L. Weitl, J. Am. Chem. Soc. *92*, 7300 (1970).

[41] T. W. Bentley and P. v. R. Schleyer, J. Am. Chem. Soc. *98*, 7658 (1976); Adv. Phys. Org. Chem. *14*, 2 (1977). [42] S. Winstein, E. Grunwald, and H. W. Jones, J. Am. Chem. Soc. *73*, 2700 (1951); S. Winstein, A. H. Fainberg, and E. Grunwald, ibid. *79*, 4146 (1957). [43] P. E. Peterson and F. J. Waller, J. Am. Chem. Soc. *94*, 991 (1972); P. E. Peterson, D. W. Vidrine, F. J. Waller, P. M. Henrichs, S. Magaha, and B. Stevens, ibid. *99*, 7968 (1977). [44] T. W. Bentley, F. L. Schadt, and P. v. R. Schleyer, J. Am. Chem. Soc. *94*, 992 (1972); ibid. *98*, 7667 (1976). [45] C. G. Swain, R. B. Mosely, and D. E. Bown, J. Am. Chem. Soc. *77*, 3731 (1955). [46] A. Streitwieser jr.: *Solvolytic Displacement Reactions*. McGraw-Hill, New York 1962. [47] P. D. Bartlett, J. Am. Chem. Soc. *94*, 2161 (1972). [48] Y. Drougard and D. Decroocq, Bull. Soc. Chim. Fr. *1969*, 2972. [49] C. Lassau and J. C. Jungers, Bull. Soc. Chim. Fr. *1938*, 2678. [50] J. C. Jungers, L. Sajus, I. de Aguirre, and D. Decroocq: *L'Analyse cinétique de la transformation chimique*. Editions Technip, Paris 1967/68, Vol. II, p. 627, 771, and 1220.

[51] M. Gielen and J. Nasielski, J. Organomet. Chem. *1*, 173 (1963); ibid. *7*, 273 (1967). [52] J. A. Berson, Z. Hamlet, and W. A. Mueller, J. Am. Chem. Soc. *84*, 297 (1962). [53] B. Blankenburg, H.

Fiedler, M. Hampel, H. G. Hauthal, G. Just, K. Kahlert, J. Korn, K.-H. Müller, W. Pritzkow, Y. Reinhold, M. Röllig, E. Sauer, D. Schnurpfeil, and G. Zimmermann, J. Prakt. Chem. *316*, 804 (1974). [54] L. G. S. Brooker, G. H. Keyes, and D. W. Heseltine, J. Am. Chem. Soc. *73*, 5350 (1951). [55] E. M. Kosower, J. Am. Chem. Soc. *80*, 3253 (1958); J. Chim. Phys. *61*, 230 (1964); E. M. Kosower and M. Mohammad, J. Am. Chem. Soc. *90*, 3271 (1968); ibid. *93*, 2713 (1971); J. Phys. Chem. *74*, 1153 (1970). [56] J. S. Gowland and G. H. Schmid, Can. J. Chem. *47*, 2953 (1969): *Z*-values for water/dimethylsulfoxide and water/ethanol mixtures. [57] M. Mohammad and E. M. Kosower, J. Phys. Chem. *74*, 1153 (1970). [58] J. H. Fendler and L.-J. Liu, J. Am. Chem. Soc. *97*, 999 (1975). [59] A. Ray, J. Am. Chem. Soc. *93*, 7146 (1971). [60] K. Tamura and T. Imoto, Chem. Lett. (Tokyo) *1973*, 1251.

[61] E. M. Kosower, M. Ito, and P.-K. Huang, Trans. Faraday Soc. *57*, 1662 (1961). [62] E. M. Kosower, J. Am. Chem. Soc. *80*, 3261, 3267 (1958). [63] E. M. Kosower, J. Org. Chem. *29*, 956 (1964). [64] N. S. Isaacs: *Experiments in Physical Organic Chemistry*. MacMillan Company, London 1969, p. 161ff. [65] S. Brownstein, Can. J. Chem. *38*, 1590 (1960). [66] K. Dimroth, C. Reichardt, T. Siepmann, and F. Bohlmann, Liebigs Ann. Chem. *661*, 1 (1963); C. Reichardt, ibid. *752*, 64 (1971). [67] K. Dimroth and C. Reichardt, Liebigs Ann. Chem. *727*, 93 (1969). [68] K. Dimroth and C. Reichardt, Z. Anal. Chem. *215*, 344 (1966); Z. B. Maksimović, C. Reichardt, and A. Spirić, ibid. *270*, 100 (1974). [69] T. G. Beaumont and K. M. C. Davis, J. Chem. Soc., Part B *1968*, 1010 (dichloromethane-acetonitrile mixtures). [70] W. Koehler, P. Froelich, and R. Radeglia, Z. Phys. Chem. (Leipzig) *242*, 220 (1969) (acetone-, 1,4-dioxane- and tetrahydrofuran-water mixtures).

[71] K. Tamura, Y. Ogo, and T. Imoto, Bull. Chem. Soc. Jpn. *46*, 2988 (1973) (nitrobenzene-benzene mixtures). [72] E. M. Kosower, H. Dodiuk, K. Tanizawa, M. Ottolenghi, and N. Orbach, J. Am. Chem. Soc. *97*, 2167 (1975) (1,4-dioxane-water mixtures). [73] K. Dimroth, C. Reichardt, and A. Schweig, Liebigs Ann. Chem. *669*, 95 (1963). [74] K. Tamura, Y. Ogo, and T. Imoto, Chem. Lett. (Tokyo) *1973*, 625; K. Tamura and T. Imoto, Bull. Chem. Soc. Jpn. *48*, 369 (1975). [75] R. Allmann, Z. Kristallogr. *128*, 115 (1969); Chem. Abstr. *70*, 100691q (1969). [76] H. C. Andersen: *Complete Fairy-Tales and Stories* (translated by E. Haugard). Gollantz, London 1974. [77] L. G. S. Brooker, A. C. Craig, D. W. Heseltine, P. W. Jenkins, and L. L. Lincoln, J. Am. Chem. Soc. *87*, 2443 (1965). [78] S. Dähne, F. Schob, K.-D. Nolte, and R. Radeglia, Ukr. Khim. Zh. (Russ. Ed.) *41*, 1170 (1975); Chem. Abstr. *84*, 43086j (1976). [79] J.-E. Dubois, E. Goetz, and A. Bienvenüe, Spectrochim. Acta *20*, 1815 (1964); J.-E. Dubois and A. Bienvenüe, J. Chim. Phys. *65*, 1259 (1968). [80] I. A. Zhmyreva, V. V. Zelinskii, V. P. Kolobkov, and N. D. Krasnitskaya, Dokl. Akad. Nauk SSSR, Ser. Khim. *129*, 1089 (1959); Chem. Abstr. *55*, 26658 e (1961).

[81] D. Walther, J. Prakt. Chem. *316*, 604 (1974). [82] D. M. Manuta and A. J. Lees, Inorg. Chem. *22*, 3825 (1983); ibid. *25*, 3212 (1986). [83] W. Kaim, S. Kohlmann, S. Ernst, B. Olbrich-Deussner, C. Bessenbacher, and A. Schulz, J. Organomet. Chem. *321*, 215 (1987). [84] M. J. Kamlet and R. W. Taft, J. Am. Chem. Soc. *98*, 377, 2886 (1976). [84a] M. J. Kamlet, J. L. Abboud, and R. W. Taft, J. Am. Chem. Soc. *99*, 6027 (1977); ibid. *99*, 8325 (1977). [85] A. Allerhand and P. v. R. Schleyer, J. Am. Chem. Soc. *85*, 371 (1963); ibid. *86*, 5709 (1964). [86] T. Kagiya, Y. Sumida, and T. Inoue, Bull. Chem. Soc. Jpn. *41*, 767 (1968). [87] A. G. Burden, G. Collier, and J. Shorter, J. Chem. Soc., Perkin Trans. II *1976*, 1627. [88] I. A. Koppel and A. I. Paju, Reakts. Sposobnost. Org. Soedin. *11*, 121 (1974); Chem. Abstr. *82*, 42805 q (1975). [89] B. R. Knauer and J. J. Napier, J. Am. Chem. Soc. *98*, 4395 (1976). [90] R. W. Taft, G. B. Klingensmith, E. Price, and I. R. Fox, Preprints of papers presented at the Symposium on LFE Correlations, Durham/N.C., October 1964, p. 265; R. T. C. Brownlee, S. K. Dayal, J. L. Lyle, and R. W. Taft, J. Am. Chem. Soc. *94*, 7208 (1972); and references cited therein.

[91] U. Mayer, V. Gutmann, and W. Gerger, Monatsh. Chem. *106*, 1235 (1975); ibid. *108*, 489, 757 (1977); V. Gutmann, Electrochimica Acta *21*, 661 (1976); V. Gutmann, CHEMTECH *7*, 255 (1977); Chem. Abstr. *86*, 178193 d (1977); A. J. Parker, U. Mayer, R. Schmid, and V. Gutmann, J. Org. Chem. *43*, 1843 (1978). [92] C. D. Ritchie: *Interactions in Dipolar Aprotic Solvents*, in J. F. Coetzee and C. D. Ritchie (eds.): *Solute-Solvent Interactions*. Dekker, New York, London 1969, Vol. 1, Table 4–13 on p. 281. [93] J. H. Hildebrand, J. M. Prausnitz, and R. L. Scott: *Regular and Related Solutions*. Van Nostrand-Reinhold, Princeton 1970. [94] A. F. M. Barton: *Handbook of Solubility Parameters*

and other Cohesion Parameters. CRC Press, Boca Raton/Florida 1983. [95] H. Burrell: *Solubility Parameters,* in J. Brandrup and E. H. Immergut (eds.): *Polymer Handbook.* 2nd ed., Wiley-Interscience, New York 1975, p. IV-337ff. [96] H. F. Herbrandson and F. R. Neufeld, J. Org. Chem. *31,* 1140 (1966). [97] J. E. Gordon, J. Phys. Chem. *70,* 2413 (1966). [98] L. Rohrschneider: *The Polarity of Stationary Liquid Phases in Gas-Chromatography,* in J. C. Giddings and R. A. Keller (eds.): *Advances in Chromatography.* Dekker, New York 1967, Vol. 4, p. 333ff.; L. Rohrschneider: *Der Lösungsmitteleinfluß auf die gas-chromatographische Retention gelöster Stoffe.* Fortschr. Chem. Forsch. *11,* 146 (1968). [99] R. V. Golovnaya and Yu. N. Arsen'ev: *System of Retention Indices and Its Physicochemical Applications.* Usp. Khim. *42,* 2221 (1973); Russian Chem. Rev. *42,* 1034 (1973). [100] E. Kováts, Z. Anal. Chem. *181,* 351 (1961); Adv. Chromatogr. *1,* 229 (1965).

[101] E. Kováts and P. B. Weiß, Ber. Bunsenges. Phys. Chem. *69,* 812 (1965). [102] L. Rohrschneider, J. Chromatogr. *22,* 6 (1966); ibid. *38,* 383 (1969); Z. Anal. Chem. *236,* 149 (1968). [103] W. O. McReynolds, J. Chromatogr. Sci. *8,* 685 (1970). [104] L. Rohrschneider, Anal. Chem. *45,* 1241 (1973). [105] L. R. Snyder: *The Role of the Mobile Phase in Liquid Chromatography,* in J. J. Kirkland (ed.): *Modern Practice of Liquid Chromatography.* Wiley-Interscience, New York 1971, Chapter 4, p. 125. [106] L. R. Snyder and J. J. Kirkland; *Introduction to Modern Liquid Chromatography.* Wiley-Interscience, New York 1974, p. 215ff. and p. 444ff. [107] L. R. Snyder, J. Chromatogr. Sci. *16,* 223 (1978). [108] P. H. Vandewyer, J. Hoefnagels, and G. Smets, Tetrahedron *25,* 3251 (1969). [109] P. Bickart, F. W. Carson, J. Jacobus, E. G. Miller, and K. Mislow, J. Am. Chem. Soc. *90,* 4869 (1968); R. Tang and K. Mislow, ibid. *92,* 2100 (1970). [110] M. H. Abraham and P. L. Grellier, J. Chem. Soc., Perkin Trans. II *1976,* 1735.

[111] F. W. Fowler, A. R. Katritzky, and R. J. D. Rutherford, J. Chem. Soc., Part B *1971,* 460. [112] I. A. Koppel and V. A. Palm, Reakts. Sposobnost Org. Soedin. *8,* 291 (1971); Organic Reactivity (Tartu) *8,* 296 (1971); Chem. Abstr. *76,* 28418 k (1972). [113] T. M. Krygowski and W. R. Fawcett, J. Am. Chem. Soc. *97,* 2143 (1975); Aust. J. Chem. *28,* 2115 (1975); Can. J. Chem. *54,* 3283 (1976). [114] R. C. Dougherty, Tetrahedron Lett. *1975,* 385. [115] I. A. Koppel and A. Paju, Reakts. Sposobnost Org. Soedin. *11,* 139 (1974); Organic Reactivity (Tartu) *11,* 137 (1974); Chem. Abstr. *82,* 42806 r (1975). [116] A. G. Burden, N. B. Chapman, H. F. Duggua, and J. Shorter, J. Chem. Soc., Perkin Trans II *1978,* 296; M. H. Aslam, G. Collier, and J. Shorter, ibid. *1981,* 1572. [117] N. S. Isaacs and E. Rannala, J. Chem. Soc., Perkin Trans. II *1974,* 902. [118] W. G. Jackson, G. A. Lawrence, P. A. Lay, and A. M. Sargeson, Aust. J. Chem. *35,* 1561 (1982). [119] S. Glikberg and Y. Marcus, J. Solution Chem. *12,* 255 (1983). [120] Z. Zhong-yuan, S. Cheng-E, and H. Te-kang, J. Chem. Soc., Perkin Trans. II *1985,* 929.

[121] M. Claessens, L. Palombini, M.-L. Stien, and J. Reisse, Nouv. J. Chim. *6,* 595 (1982). [122] T. R. Griffiths and D. C. Pugh, Coord. Chem. Rev. *29,* 129 (1979). [123] J.-L. M. Abboud, M. J. Kamlet, and R. W. Taft, Progr. Phys. Org. Chem. *13,* 485 (1981). [124] C. Reichardt: *Empirical Parameters of Solvent Polarity and Chemical Reactivity.* In H. Ratajczak and W. J. Orville-Thomas (eds.): *Molecular Interactions.* Vol. 3, p. 241ff., Wiley, Chichester 1982. [125] E. S. Lewis: *Linear Free Energy Relationships.* In C. F. Bernasconi (ed.): *Investigation of Rates and Mechanisms of Reactions* (Vol. IV of the Series *Techniques in Chemistry*) 4th ed., Part I, p. 871ff. , Wiley, New York 1986. [126] M. Sjöström and S. Wold: *Linear Free Energy Relationships. Local Empirical Rules or Fundamental Laws of Chemistry?* Acta Chem. Scand., Part B *35,* 537 (1981); ibid. *40,* 270 (1986). [127] M. J. Kamlet and R. W. Taft: *Linear Solvation Energy Relationships. Local Empirical Rules – or Fundamental Laws of Chemistry? A Reply to the Chemometricians.* Acta Chem. Scand., Part B *39,* 611 (1985); ibid. *40,* 619 (1986); M. J. Kamlet, R. M. Doherty, G. R. Famini, and R. W. Taft, ibid. *41,* 589 (1987). [128] J.-C. Bollinger, G. Yvernault, and T. Yvernault, Thermochim. Acta *60,* 137 (1983). [129] H. D. Schädler, M. Riemer, and W. Schroth, Z. Chem. *24,* 407 (1984). [130] M. C. Day, J. H. Medley, and N. Ahmad, Can. J. Chem. *61,* 1719 (1983).

[131] S. Hahn, W. M. Miller, R. N. Lichtenthaler, and J. M. Prausnitz, J. Sol. Chem. *14,* 129 (1985). [132] R. W. Soukop, Chemie in unserer Zeit *17,* 129, 163 (1983); R. W. Soukop and R. Schmid, J. Chem. Educ. *62,* 459 (1985). [133] Y. Marcus, J. Solution Chem. *13,* 599 (1984); *cf.* also Y. Marcus: *Ion Solvation.* Chapter 6, Wiley, Chichester 1985. [134] R. Schmid, J. Solution Chem. *12,* 135 (1983). [135] U. Mayer and V. Gutmann, Structure and Bonding *12,* 113 (1972); *cf.* p. 125. [136] A. I. Popov,

Pure Appl. Chem. *41*, 275 (1975); and references cited therein. [137] T. Ogata, T. Fujisawa, N. Tanaka, and H. Yokoi, Bull. Chem. Soc. Jpn. *49*, 2759 (1976). [138] P.-C. Maria and J.-F. Gal, J. Phys. Chem. *89*, 1296 (1985); P.-C. Maria, J.-F. Gal, J. de Franceschi, and E. Fargin, J. Am. Chem. Soc. *109*, 483 (1987). [139] A. Sabatino, G. La Manna, and L. Paoloni, J. Phys. Chem. *84*, 2641 (1980). [140] K. Hiraoka, Bull. Chem. Soc. Jpn. *59*, 2571 (1986).

[141] R. S. Drago, Coord. Chem. Rev. *33*, 251 (1980); Pure Appl. Chem. *52*, 2261 (1980). [142] L. Elégant, G. Fratini, J.-F. Gal, and P.-C. Maria, Journ. Calorim. Anal. Therm. *11*, 3/21/1...3/21/9 (1980); Chem. Abstr. *95*, 121811 k (1981). [143] M. Berthelot, J.-F. Gal, M. Helbert, C. Laurence, and P.-C. Maria, J. Chim. Phys. *82*, 427 (1985). [144] I. Persson, Pure Appl. Chem. *58*, 1153 (1986). [145] A. Leo, C. Hansch, and D. Elkins: *Partition Coefficients and Their Uses*. Chem. Rev. *71*, 525 (1971). [146] C. Hansch: *Recent Advances in Biochemical QSAR*. In N. B. Chapman and J. Shorter (eds.): *Correlation Analysis in Organic Chemistry – Recent Advances*. Plenum Press, New York, London 1978, Chapter 9, p. 397ff. [147] C. Hansch and A. Leo: *Substituent Constants for Correlation Analyses in Chemistry and Biology*. Wiley-Interscience, New York 1979. [148I A. Leo, J. Chem. Soc., Perkin Trans. II *1983*, 825. [149] R. W. Taft, M. H. Abraham, G. F. Famini, R. M. Doherty, J.-L. M. Abboud, and M. J. Kamlet, J. Pharm. Sci. *74*, 807 (1985). [150] F. M. Menger and U. V. Venkataram, J. Am. Chem. Soc. *108*, 2980 (1986).

[151] D. Fărcasiu, J. Jähme, and C. Rüchardt, J. Am. Chem. Soc. *107*, 5717 (1985). [152] M. H. Abraham, R. W. Taft, and M. J. Kamlet, J. Org. Chem. *46*, 3053 (1981). [153] T. W. Bentley and G. E. Carter, J. Am. Chem. Soc. *104*, 5741 (1982); and references cited therein. [154] T. W. Bentley, C. T. Bowen, D. H. Morten, and P. v. R. Schleyer, J. Am. Chem. Soc. *103*, 5466 (1981). [155] D. N. Kevill, W. A. Kamil, and S. W. Anderson, Tetrahedron Lett. *1982*, 4635. [156] C. A. Bunton, M. M. Mhala, and J. R. Moffatt, J. Org. Chem. *49*, 3639 (1984). [157] S. P. McManus and S. E. Zutaut, Tetrahedron Lett. *25*, 2859 (1984). [158] T. W. Bentley and G. E. Carter, J. Org. Chem. *48*, 579 (1983). [159] T. W. Bentley, S. Jurczyk, K. Roberts, and D. J. Williams, J. Chem. Soc., Perkin Trans II *1987*, 293.[160] D. N. Kevill and G. M. L. Lin, J. Am. Chem. Soc. *101*, 3916 (1979); D. N. Kevill *et al.*: *Studies of Displacement Reactions Using* $R—X^{\oplus}$ *Substrates*. In M. Kobayashi (ed.): *Physical Organic Chemistry 1986*. Elsevier, Amsterdam 1987, p. 311ff.

[161] T. W. Bentley, G. E. Carter, and K. Roberts, J. Org. Chem. *49*, 5183 (1984). [162] D. N. Kevill, M. S. Bahari, and S. W. Anderson, J. Am. Chem. Soc. *106*, 2895 (1984). [163] D. N. Kevill and S. W. Anderson, J. Am. Chem. Soc. *108*, 1579 (1986). [164] D. N. Kevill and S. W. Anderson, J. Org. Chem. *50*, 3330 (1985). [165] X. Creary and S. R. McDonald, J. Org. Chem. *50*, 474 (1985). [166] T. W. Bentley and K. Roberts, J. Org. Chem. *50*, 4821 (1985). [167] T. W. Bentley and K. Roberts, J. Org. Chem. *50*, 5852 (1985). [168] F. L. Schadt, C. J. Lancelot, and P. v. R. Schleyer, J. Am. Chem. Soc. *100*, 228 (1978). [169] T. Oshima, S. Arikata, and T. Nagai, J. Chem. Res. *1981* (S) 204; (M) 2518. [170] T. Oshima and T. Nagai, Bull. Chem. Soc. Jpn. *55*, 555 (1982); Tetrahedron Lett. *26*, 4785 (1985).

[171] J. W. Larsen, A. G. Edwards, and P. Dobi, J. Am. Chem. Soc. *102*, 6780 (1980). [172] T. R. Griffiths and D. C. Pugh, J. Sol. Chem. *8*, 247 (1979); Coord. Chem. Rev. *29*, 129 (1979). [173] P. Štrop, F. Mikeš, and J. Kálal, J. Phys. Chem. *80*, 694 (1976); F. Mikeš, J. Labský, P. Štrop, and D. Vyprachticky: *On the Microenvironment of Soluble and Crosslinked Polymers*. Sb. Vys. Sk. Chem.-Technol. Praze [Oddil] S *1984*, S11, 83–144; Chem. Abstr. *102*, 167386 s (1985). [174] C. Reichardt and E. Harbusch-Görnert, Liebigs Ann. Chem. *1983*, 721. [175] C. Laurence, P. Nicolet, and C. Reichardt, Bull. Soc. Chim. Fr. *1987*, 125. [176] C. Laurence, P. Nicolet, M. Lucon, and C. Reichardt, Bull. Soc. Chim. Fr. *1987*, 1001. [177] Y. Marcus: *Ion Solvation*. Chapter 7, Table 7.3 on page 188, Wiley, New York 1985. [178] T. M. Krygowski, P. K. Wrona, U. Zielkowska, and C. Reichardt, Tetrahedron *41*, 4519 (1985). [179] M. De Vijlder, Bull. Soc. Chim. Belge *91*, 947 (1982). [180] J. R. Haak and J. B. F. N. Engberts, Rec. Trav. Chim. Pays-Bas *105*, 307 (1986).

[181] B. P. Johnson, M. G. Khaledi, and J. G. Dorsey, J. Chromatogr. *384*, 221 (1987). [182] J. v. Jouanne, D. A. Palmer, and H. Kelm, Bull. Chem. Soc. Jpn. *51*, 463 (1978). [183] S. Balakrishnan and A. J. Easteal, Aust. J. Chem. *34*, 933, 943 (1981). [184] H. Langhals, Nouv. J. Chim. *5*, 97, 511 (1981); ibid. *6*, 265 (1982). [185] I. A. Koppel and J. B. Koppel, Organic Reactivity (Tartu) *20*, 547 (1983); Chem. Abstr. *101*, 110274 d (1984). [186] H. Elias, G. Gumbel, S. Neitzel, and H. Volz, Z. Anal.

Chem. *306*, 240 (1981). [187] I. A. Koppel and J. B. Koppel, Organic Reactivity (Tartu) *20*, 523 (1983); Chem. Abstr. *101*, 110180 v (1984). [188] V. Bekárek and T. Nevěčná, Coll. Czech. Chem. Commun. *50*, 1928 (1985); ibid. *51*, 1942, 2071 (1986). [189] J. Hicks, M. Vandersall, Z. Babrogic, and K. B. Eisenthal, Chem. Phys. Lett. *116*, 18 (1985). [190] P. Nagy and R. Herzfeld, Acta Universitatis Szegediensis. Acta Physica et Chimica *31*, 736 (1985); Chem. Abstr. *107*, 22905d (1987).

[191] H. Langhals, Tetrahedron Lett. *27*, 339 (1986); Z. Phys. Chem. (Leipzig) *286*, 91 (1987). [192] H. Langhals: *Polarität binärer Flüssigkeitsgemische*. Angew. Chem. *94*, 739 (1982); Angew. Chem., Int. Ed. Engl. *21*, 724 (1982). [193] M. Chastrette and J. Carretto, Tetrahedron *38*, 1615 (1982); Can. J. Chem. *63*, 3492 (1985). [194] V. Bekárek and J. Juřina, Coll. Czech. Chem. Commun. *47*, 1060 (1982). [195] H. Langhals, Angew. Chem. *94*, 452 (1982); Angew. Chem., Int. Ed. Engl. *21*, 432 (1982); Angew. Chem. Suppl. *1982*, 1138. [196] H. Langhals, Chem. Ber. *114*, 2907 (1981). [197] M. C. Rezende, C. Zucco, and D. Zanette, Tetrahedron Lett. *25*, 3423 (1984); Tetrahedron *41*, 87 (1985). [198] S. Kumoi, K. Oyama, T. Yano, H. Kobayashi, and K. Ueno, Talanta *17*, 319 (1970). [199] H. Langhals, Z. Anal. Chem. *305*, 26 (1981); ibid. *308*, 441 (1981). [200] C. Reichardt: *Solvent Scales and Chemical Reactivity*. In A. D. Buckingham, E. Lippert, and S. Bratos (eds.): *Organic Liquids: Structure, Dynamics, and Chemical Properties*. Chapter 16, p. 269ff., Wiley, Chichester 1978.

[201] C. Reichardt: *Solvent Effects on Chemical Reactivity*. Pure Appl. Chem. *54*, 1867 (1982). [202] C. Han: E_T *Empirical Parameters of Solvent Polarity and Their Applications*. Huaxue Tongbao [Chemistry, Beijing] *1985*, 40; Chem. Abstr. *104*, 147904 x (1986). [203] B. P. Johnson, B. Gabrielsen, M. Matulenko, J. G. Dorsey, and C. Reichardt: *Solvatochromic Solvent Polarity Measurements in Analytical Chemistry: Synthesis and Applications of ET-30*. Analytical Letters *19*, 939 (1986). [204] M. S. Tunuli, M. A. Rauf, and Farhataziz, J. Photochem. *24*, 411 (1984). [205] F. Ibáñez, J. Photochem. *30*, 245 (1985). [206] Y. Marcus, E. Pross, and J. Hormadaly, J. Phys. Chem. *84*, 2708 (1980). [207] R. Schmid and V. N. Sapunov: *Non-Formal Kinetics in Search for Chemical Reaction Pathways*. Verlag Chemie, Weinheim 1982; R. Schmid, J. Solution Chem. *12*, 135 (1983). [208] C. Reichardt, E. Harbusch, and R. Müller: *Pyridinium-N-phenoxide Betaine Dyes as Solvent Polarity Indicators. Some New Findings*. In I. Bertini, L. Lunazzi, and A. Dei (eds.): *Advances in Solution Chemistry*. Plenum Press, New York 1981, p. 275ff. [209] C. Reichardt, unpublished results. [210] S. Dähne and R. Radeglia, Tetrahedron *27*, 3673 (1971).

[211] G. Hollmann and F. Vögtle, Chem. Ber. *117*, 1355 (1984). [212] R. Braun and J. Sauer, Chem. Ber. *119*, 1269 (1986). [213] H. Langhals, Tetrahedron *43*, 1771 (1987). [214] A. v. Baeyer and V. Villiger, Ber. Dtsch. Chem. Ges. *35*, 1189 (1902). [215] P. Plieninger and H. Baumgärtel, Liebigs Ann. Chem. *1983*, 860. [216] J. J. Dawber and R. A. Williams, J. Chem. Soc., Faraday Trans. I *82*, 3097 (1986). [217] P. de Mayo, A. Safarzadeh-Amiri, and S. K. Wong, Can. J. Chem. *62*, 1001 (1984). [218] P. Mukerjee, C. Ramachandran, and R. A. Pyter, J. Phys. Chem. *86*, 3189 (1982). [219] A. Janowski, I. Turowska-Tyrk, and P. K. Wrona, J. Chem. Soc., Perkin Trans. II *1985*, 821. [220] W. Walter and O. H. Bauer, Liebigs Ann. Chem. *1977*, 421.

[221] R. W. Soukup, Chemie in unserer Zeit *17*, 129, 163 (1983); R. W. Soukup and R. Schmid, J. Chem. Educ. *62*, 459 (1985). [222] D. C. Dong and M. A. Winnik, Photochem. Photobiol. *35*, 17 (1982); Can. J. Chem. *62*, 2560 (1984). [223] K. W. Street and W. E. Acree, Analyst. *111*, 1197 (1986). [224] J.-L. M. Abboud, M. J. Kamlet, and R. W. Taft, Progr. Phys. Org. Chem. *13*, 485 (1981); R. W. Taft, J.-L. M. Abboud, M. J. Kamlet, and M. H. Abraham, J. Solution Chem. *14*, 153 (1985). [225] H. C. Brown, G. K. Barbaras, H. L. Berneis, W. H. Bonner, R. B. Johannesen, M. Grayson, and K. L. Nelson, J. Am. Chem. Soc. *75*, 1 (1953). [226] M. J. Kamlet, J.-L. M. Abboud, M. H. Abraham, and R. W. Taft, J. Org. Chem. *48*, 2877 (1983). [227] J. E. Brady and P. W. Carr, Anal. Chem. *54*, 1751 (1982); J. Phys. Chem. *86*, 3053 (1982). [228] M. Essfar, G. Guihéneuf, and J.-L. M. Abboud, J. Am. Chem. Soc. *104*, 6786 (1982). [229] M. J. Kamlet and R. W. Taft, J. Chem. Soc., Perkin Trans. II *1979*, 337. [230] M. J. Kamlet, R. M. Doherty, M. H. Abraham, P. W. Carr, R. F. Doherty, and R. W. Taft, J. Phys. Chem. *91*, 1996 (1987).

[231] T. M. Krygowski, E. Milczarek, and P. K. Wrona, J. Chem. Soc., Perkin Trans. II *1980*, 1563. [232] T. M. Krygowski *et al.*, Tetrahedron *37*, 119 (1981); J. Chem. Res. (S) *1983*, 116. [233] O. W. Kolling, Anal. Chem. *53*, 54 (1981); ibid. *56*, 430 (1984). [234] V. Bekárek, J. Phys. Chem. *85*, 722 (1981); J. Chem. Soc., Perkin Trans. II *1983*, 1293; ibid. *1986*, 1425; V. Bekárek and M. Stolařová,

Coll. Czech. Chem. Commun. *48*, 1237 (1983); ibid. *49*, 2332 (1984). [235] J.-L. M. Abboud, R. W. Taft, and M. J. Kamlet, J. Chem. Soc., Perkin Trans. II *1985*, 815. [236] M. Sjöström and S. Wold, J. Chem. Soc., Perkin Trans. II *1981*, 104. [237] P. Nicolet and C. Laurence, J. Chem. Soc., Perkin Trans. II *1986*, 1071. [238] C. Laurence, P. Nicolet, and M. Helbert, J. Chem. Soc., Perkin Trans. II *1986*, 1081. [239] P. Nicolet, C. Laurence, and M. Lucon, J. Chem. Soc., Perkin Trans. II *1987*, 483. [240] C. Somolinos, I. Rodriguez, M. I. Redondo, and M. V. Garcia, J. Mol. Struct. *143*, 301 (1986).

[241] M. I. Redondo, C. Somolinos, and M. V. Garcia, Spectrochim. Acta, Part A *42*, 677 (1986). [242] R. G. Makitra, Ya. M. Vasyutin, and V. Ya. Zhukowskii, Org. React. (Tartu) *21*, 190 (1984); Chem. Abstr. *103*, 122883 f (1985). [243] R. G. Makitra and Ya. N. Pirig, Org. React. (Tartu) *16*, 103 (1979); Chem. Abstr. *92*, 41142 r (1980); Org. React. (Tartu) *17*, 180 (1980); Chem. Abstr. *95*, 79388 a (1981). [244] O. W. Kolling, Anal. Chem. *49*, 591 (1977); ibid. *55*, 143 (1983). [245] A. H. Reddoch and S. Konishi, J. Chem. Phys. *70*, 2121 (1979). [246] M. C. R. Symons and A. S. Pena-Nuñez, J. Chem. Soc., Faraday Trans. I *81*, 2421 (1985). [247] Y. Liu and Y. Wang, Huaxue Xuebo *43*, 232 (1985); Chem. Abstr. *103*, 177687 a (1985). [248] Z. Kecki, B. Łyczkowski, and W. Kołodziejski, J. Solution Chem. *15*, 413 (1986). [249] B. Chawla, S. K. Pollack, C. B. Lebrilla, M. J. Kamlet, and R. W. Taft, J. Am. Chem. Soc. *103*, 6924 (1981). [250] H. Elias, M. Dreher, S. Neitzel, and H. Volz, Z. Naturforsch., Part B *37*, 684 (1982).

[251] U. Mayer: *NMR-Spectroscopic Studies on Solute-Solvent and Solute-Solute Interactions*. In N. Tanaka, H. Ohtaki, and R. Tamamushi (eds.): *Ions and Molecules in Solution*. Elsevier, Amsterdam 1983, p. 219ff. [252] M. C. R. Symons and G. Eaton, J. Chem. Soc., Faraday Trans. I *78*, 3033 (1982). [253] P. Bacelon, J. Corset, and C. de Loze, J. Solution Chem. *12*, 23 (1983). [254] B. Allard and E. Casadevall, Nouv. J. Chim. *9*, 565 (1985). [255] M. J. Kamlet, P. W. Carr, R. W. Taft, and M. H. Abraham, J. Am. Chem. Soc. *103*, 6062 (1981). [256] R. W. Taft, M. H. Abraham, R. M. Doherty, and M. J. Kamlet, J. Am. Chem. Soc. *107*, 3105 (1985). [257] M. H. Abraham, R. M. Doherty, M. J. Kamlet, J. M. Harris, and R. W. Taft, J. Chem. Soc., Perkin Trans. II *1987*, 913, 1097. [258] R. G. Makitra and Ya. N. Pirig, Zh. Obshch. Khim. *56*, 657 (1986); J. Gen. Chem. USSR *56*, 584 (1986). [259] F. Vernon: *The Characterization of Solute-Solvent Interactions in Gas-Liquid Chromatography*. In C. E. H. Knapman (ed.): *Developments in Chromatography*. Applied Science Publ., London 1978, Vol. 1, Chapter 1, p. 1ff.; Chem. Abstr. *92*, 65125 p (1980). [260] R. V. Golovnaya and T. A. Misharina: *Thermodynamic Treatment of the Polarity and Selectivity of Solvents in Gas Chromatography*. Usp. Khim. *49*, 171 (1980); Russian Chem. Rev. *49*, 88 (1980).

[261] G. Tarján, Á. Kiss, G. Kocsis, S. Mészáros, and J. M. Takács, J. Chromatogr. *119*, 327 (1976). [262] S. Wold and K. Andersson, J. Chromatogr. *80*, 43 (1973). [263] J.-L. M. Abboud, R. W. Taft, and M. J. Kamlet, J. Chem. Res. (S) *1984*, 98. [264] J. E. Brady and P. W. Carr, J. Phys. Chem. *88*, 5796 (1984). [265] C. G. Swain, M. S. Swain, A. L. Powell, and S. Alunni, J. Am. Chem. Soc. *105*, 502 (1983). [266] U. Mayer, Monatsh. Chem. *109*, 421, 775 (1978). [267] M. J. Kamlet, J.-F. Gal, P.-C. Maria, and R. W. Taft, J. Chem. Soc., Perkin Trans. II *1985*, 1583. [268] G. Collier, I. A. Pickering, and J. Shorter, unpublished work; *cf.* I. A. Pickering, Thesis, University of Hull 1985, and reference [15], p. 164, equation (5.22). [269] J. E. Brady and P. W. Carr, J. Phys. Chem. *89*, 1813 (1985). [270] B. Eliasson, D. Johnels, S. Wold, U. Edlund, and M. Sjöström, Acta Chem. Scand., Part B *36*, 155 (1982); ibid. *41*, 291 (1987).

[271] M. J. Kamlet and R. W. Taft, Acta Chem. Scand., Part B *40*, 619 (1986). [272] M. J. Kamlet, M. H. Abraham, R. M. Doherty, and R. W. Taft, J. Am. Chem. Soc. *106*, 464 (1984); Nature (London) *313*, 384 (1985). [273] M. H. Abraham, M. J. Kamlet, R. W. Taft, and P. K. Weathersby, J. Am. Chem. Soc. *105*, 6797 (1983). [274] M. J. Kamlet, R. M. Doherty, V. Fiserova-Bergerova, P. W. Carr, M. H. Abraham, and R. W. Taft, J. Pharm. Sci. *76*, 14 (1987). [275] R. W. Taft, J.-L. M. Abboud, and M. J. Kamlet, J. Org. Chem. *49*, 2001 (1984). [276] C. G. Swain, J. Org. Chem. *49*, 2005 (1984). [277] S. Glikberg and Y. Marcus, J. Solution Chem. *12*, 255 (1983). [278] G. Heublein, P. Hallpap, R. Wondraczek, and P. Adler, Z. Chem. *20*, 11 (1980). [279] G. Heublein, J. Macromol. Sci., Part A *22*, 1277 (1985). [280] M. Kupfer, A. Henrion, and W. Abraham, Z. Phys. Chem. (Leipzig) *267*, 705 (1986).

[281] E. Fogassy, A. Lopata, F. Faigl., F. Darvas, M. Ács, and L. Töke, Tetrahedron Lett. *21*, 647 (1980); A. Lopata, F. Faigl, E. Fogassy, and F. Darvas, J. Chem. Res. (S) *1984*, 322, 324. [282] M. H.

Abraham, P. L. Grellier, and R. A. McGill, J. Chem. Soc., Perkin Trans. II *1988*, 339. [283] H.-J. Schneider and N. K. Sangwan, J. Chem. Soc., Chem. Commun. *1986*, 1787; Angew. Chem. *99*, 924 (1987); Angew. Chem., Int. Ed. Engl. *26*, 896 (1987). [284] K. Popper: *Unended Quest. An Intellectual Autobiography*. Revised edition, Open Court, La Salle/Illinois 1982, p. 28. [285] Y. Marcus, J. Phys. Chem. *91*, 4422 (1987). [286] M. J. Kamlet, R. M. Doherty, J.-L. M. Abboud, M. H. Abraham, and R. W. Taft: *Solubility – A New Look*. CHEMTECH *16*, 566 (1986). [287] I. Persson, M. Sandström, and P. L. Goggin, Inorg. Chim. Acta *129*, 183 (1987). [288] M. H. Abraham, R. M. Doherty, M. J. Kamlet, J. M. Harris, and R. W. Taft, J. Chem. Soc., Perkin Trans. II *1987*, 913, 1097.

Appendix

[1] J. A. Riddick, W. B. Bunger, and T. K. Sakano: *Organic Solvents. Physical Properties and Methods of Purification*. 4th ed., in A. Weissberger (ed.): *Techniques of Chemistry*, Vol. II, Wiley-Interscience, New York 1986. [2] E. W. Flick (ed.): *Industrial Solvents Handbook*. 3rd ed. Noyes Publications, Park Ridge 1985. [3] T. H. Durrans (editor E. H. Davies): *Solvents*. 8th ed., Chapman and Hall, London 1971. [4] A. J. Gordon and R. A. Ford: *The Chemist's Companion – A Handbook of Practical Data, Techniques, and References*. Wiley-Interscience, New York 1972. [5] G. P. Nilles and R. D. Schuetz, J. Chem. Educ. *50*, 267 (1973). [6] R. L. Schneider: *Physical Properties of Some Organic Solvents*. Eastman Organic Chemical Bulletin *47*, No. 1 (1975). [7] R. C. Weast: *Handbook of Chemistry and Physics*. 64th ed., CRC Press, Boca Raton/Florida 1983/84. [8] G. Duve, O. Fuchs, and H. Overbeck: *Lösemittel Hoechst*. 6th ed., Hoechst AG, Frankfurt (Main) 1976, Federal Republic of Germany. [9] M. Raban and K. Mislow, Topics in Stereochemistry *2*, 199 (1967). [10] H. Pracejus, Fortschr. Chem. Forsch. *8*, 493 (1967).

[11] T. D. Inch, Synthesis 466 (1970). [12] J. D. Morrison and H. S. Mosher; *Asymmetric Organic Reactions*. Prentice Hall, Englewood Cliffs/New Jersey 1971, pp. 411. [13] W. H. Pirkle and D. J. Hoover: *NMR Chiral Solvating Agents*. Topics in Stereochemistry *13*, 263 (1982). [14] V. Sedivec and J. Flek: *Handbook of Analysis of Organic Solvents*. Ellis Horwood, Chichester, and Wiley, New York 1976. [15] M. Spiteller and G. Spiteller: *Massenspektrensammlung von Lösungsmitteln, Verunreinigungen, Säulenbelegmaterialien und einfachen aliphatischen Verbindungen*. Springer-Verlag, Wien, New York 1973. [16] B. Hampel and K. Maas, Chemiker-Ztg. *95*, 316 (1971). [17] W. Bunge in Houben-Weyl-Müller, *Methoden der organischen Chemie*, 4th ed., Thieme, Stuttgart 1959, Vol. I/2, pp. 765...868. [18] H. Rickert and H. Schwarz in Houben-Weyl-Müller, *Methoden der organischen Chemie*, 4th ed., Thieme, Stuttgart 1959, Vol. I/2, pp. 869...885. [19] G. Wohlleben, Angew. Chem. *67*, 741 (1955); ibid. *68*, 752 (1956). [20] B. D. Pearson and J. E. Ollerenshaw, Chem. Ind. (London) *1966*, 370.

[21] G. Hesse and H. Schildknecht, Angew. Chem. *67*, 737 (1955); G. Hesse, B. P. Engelbrecht, H. Engelhardt, and S. Nitsch, Z. Anal. Chem. *241*, 91 (1968); G. Hesse: *Chromatographisches Praktikum*. Akademische Verlagsgesellschaft, Frankfurt (Main) 1968. [22] *Purification of Solvents on Adsorbents Woelm*. ICN Biochemicals GmbH, D-3440 Eschwege, Federal Republic of Germany. [23] H. Engelhardt: *Purification of Organic Solvents by Frontal-Analysis Chromatography*, in M. Zief and R. M. Speights (eds.): *Ultrapurity-Methods and Techniques*. Dekker, New York 1972, p. 71f. [24] B. P. Engelbrecht, GIT Fachz. Lab. *23*, 681 (1979); Chem. Abstr. *91*, 133523 y (1979). [25] M. Pestemer, Angew. Chem. *63*, 118 (1951); ibid. *67*, 740 (1955); M. Pestemer: *Anleitung zum Messen von Absorptionsspektren im Ultravioletten und Sichtbaren*. Thieme, Stuttgart 1964. [26] J. Schurz and H. Stübchen, Z. Elektrochem., Ber. Bunsenges. Physik. Chem. *61*, 754 (1957). [27] D. D. Tunnicliff, Talanta *2*, 341 (1959). [28] W. Esselborn and B. Hampel, Photoelectric Spectrometry Group Bulletin Nr. 17, 498 (1967); Chem. Abstr. *69*, 35055 y (1968). [29] J. Ratcliffe: *Practical Hints on Absorption Spectrometry*. Plenum Publishing Corporation, New York 1967. [30] N. Fisher and R. M. Cooper, Chem. Ind. (London) *1968*, 619.

[31] G. Belanger, P. Sauvageau, and C. Sandorfy, Chem. Phys. Lett. *3*, 649 (1969). [32] H. R. Dickinson and W. C. Johnson, Appl. Opt. *10*, 681 (1971); Chem. Abstr. *74*, 117764 w (1971). [33] M. F. Fox and E. Hayon, J. Phys. Chem. *76*, 2703 (1972). [34] B. Hampel in: *DMS-UV-Atlas organischer Verbindungen*. Butterworths, London, and Verlag Chemie, Weinheim 1966, Hauptgruppe M. [35] N.

L. Alpert, W. E. Keiser, and H. A. Szymanski; *IR-Theory and Practice of Infrared Spectroscopy.* 2nd ed. Plenum Press, New York 1973. [36] *DMS-Working Atlas of Infrared Spectroscopy.* Butterworths, London, and Verlag Chemie, Weinheim 1972. [37] R. F. Goddu and D. A. Delker, Anal. Chem. *32*, 140 (1960). [38] A. Visapää, Kemian Teollisuus *22*, 487 (1965), Chem. Abstr. *63*, 15732 (1965); Tek. Tutkimuslaitos, Tiedotus, Sarja IV *76* (1965), Chem. Abstr. *64*, 18701 (1966). [39] F. F. Bentley, E. F. Wolfarth, N. E. Srp, and W. R. Powell, Spectrochim. Acta *13*, 1 (1958). [40] H. R. Wyss, R. D. Werder, and H. H. Günthard, Spectrochim. Acta *20*, 573 (1964).

[41] N. L. McNiven and R. Court: *IR-Spectra of Deuterated Solvents*, in L. May (ed.): *Spectroscopic Tricks.* Vol. 3, Plenum Press, New York, London 1974, p. 148. [42] D. W. Henty and S. Vary, Chem. Ind. (London) *1967*, 1782. [43] R. A. Fletton and J. E. Page, The Analyst *96*, 370 (1971). [44] H.-O. Kalinowski, S. Berger, and S. Braun: *^{13}C-NMR-Spektroskopie.* Thieme-Verlag, Stuttgart, New York 1984, p. 74. [45] O. Fuchs, W. Schmieder, F. Oschatz, and H. Oettel in W. Foerst (ed.): *Ullmann's Encyklopädie der technischen Chemie.* 3rd ed., Urban and Schwarzenberg, München, Berlin 1960, Vol. 12, pp. 1...49. [46] A. Lüttringhaus in Houben-Weyl-Müller, *Methoden der organischen Chemie.* 4th ed., G. Thieme, Stuttgart 1958, Vol. I/1, p. 341 (especially p. 369). [47] R. S. Tipson in A. Weissberger (ed.): *Technique of Organic Chemistry.* 2nd ed., Interscience, New York, London 1956, Vol. III, Part I, p. 395 (especially p. 549). [48] B. Cornils and J. Falbe, in F. Korte (ed.): *Methodicum Chimicum.* G. Thieme, Stuttgart, and Academic Press, New York, London 1973, Vol. I/1, pp. 39...45. [49] *Organikum – Organisch-chemisches Grundpraktikum.* 16th ed., Deutscher Verlag der Wissenschaften, Berlin 1986, p. 32. [50] F. A. von Metzsch, Angew. Chem. *65*, 586 (1953).

[51] E. Hecker, Chimia *8*, 229 (1954); E. Hecker: *Verteilungsverfahren in Laboratorien.* Verlag Chemie, Weinheim 1955, p. 92 and 139. [52] L. R. Snyder: *Solvent Selection for Separation Processes*, in E. S. Perry and A. Weissberger (eds.): *Techniques of Chemistry.* 3rd ed., Vol. XII, p. 25ff., Wiley-Interscience, New York 1978. [53] T.C. Lo, M. H. J. Baird, and C. Hanson (eds.): *Handbook of Solvent Extraction.* Wiley, Chichester 1983. [54] M. Hampe: *Flüssig/Flüssig-Extraktion: Einsatzgebiete und Lösungsmittel-Auswahl.* Chem.-Ing.-Tech. *50*, 647 (1978); Chem. Abstr. *90*, 8134a (1979). [55] F. J. Wolf: *Separation Methods in Organic and Biochemistry.* Academic Press, New York, London 1969. [56] H. Pauschmann and E. Bayer, in F. Korte (ed.): *Methodicum Chimicum.* G. Thieme, Stuttgart, and Academic Press, New York, London 1973, Vol. I/1, p. 148 (espicially p. 156) and references cited in this review. [57] G. Wohlleben, Fortschr. Chem. Forsch. *6*, 640 (1966). [58] E. Stahl: *Dünnschicht-Chromatographie.* 2nd ed., Springer-Verlag, Berlin, Heidelberg, New York 1967; E. Stahl, Z. Anal. Chem. *236*, 294 (1968). [59] L. R. Snyder: *Principles of Adsorption Chromatography.* Dekker, New York 1968. [60] L. R. Snyder and J. J. Kirkland: *Introduction to Modern Liquid Chromatography.* 2nd ed., Wiley-Interscience, New York 1979.

[61] D. F. G. Pusey, Chem. Brit. *5*, 408 (1969). [62] W. Huber: *Titrationen in nichtwäßrigen Lösungsmitteln.* Akademische Verlagsgesellschaft, Frankfurt (Main) 1964; *Titrations in Nonaqueous Solvents.* Academic Press, New York, London 1967. [63] J. Kucharský and L. Safařik: *Titrations in Non-Aqueous Solvents.* Elsevier, Amsterdam 1965. [64] L. N. Bykova, A. P. Kreshkov, N. A. Kazaryan, and I. D. Pevzner, Usp. Khim. *37*, 677 (1968); Russian Chem. Rev. *37*, 278 (1968). [65] L. Šafařik and Z. Stránský: *Titrimetric Analysis in Organic Solvents.* In G. Svehla (ed.): *Wilson and Wilson's Comprehensive Analytical Chemistry.* Vol. XXII, Elsevier, Amsterdam, New York 1986. [66] G. Charlot and B. Trémillon: *Réactions Chimiques dans les Solvants et les Sels Fondus.* Gauthier-Villars, Paris 1963; *Chemical Reactions in Solvents and Melts.* Pergamon Press, Oxford 1969. [67] C. A. Streuli: *Acid-Base Titration in Nonaqueous Solvents*, in I. M. Kolthoff, P. J. Elving, and E. B. Sandell (eds.): *Treatise on Analytical Chemistry.* Wiley-Interscience, New York 1975, Part I, Vol. 11, p. 7035...7116. [68] I. Gyenes: *Titrationen in nichtwäßrigen Medien.* 3rd ed., Akadémiai Kiadó, Budapest, and F. Enke, Stuttgart 1970; *Titrations in Non-Aqueous Media.* Van Nostrand, New York 1967. [69] B. Kratochvil, Anal. Chem. *54*, 105 R (1982). [70] L. N. Bykova and S. I. Petrov, Usp. Khim. *41*, 2065 (1972); Russian Chem. Rev. *41*, 975 (1972).

[71] R. G. Bates: *Medium Effects and pH in Nonaqueous Solvents*, in J. F. Coetzee and C. D. Ritchie (eds.): *Solute-Solvent Interactions.* Dekker, New York, London 1969, Vol. 1, p. 50ff. [72] H. B. van der Heijde, Anal. Chim. Acta *17*, 512 (1957). [73] C. K. Mann: *Nonaqueous Solvents for Electrochemical Use*, in A. J. Bard (ed.): *Electroanalytical Chemistry*, Dekker, New York 1969, Vol. 3,

p. 57ff. [74] G. J. Janz and R. P. T. Tomkins (eds.): *Nonaqueous Electrolytes Handbook*. Academic Press, London, New York 1972/73, Vol. 1 and 2. [75] G. Kortüm: *Lehrbuch der Elektrochemie*. 5th ed., Verlag Chemie, Weinheim 1972. [76] M. Salomon: *Electrochemical Measurements*, in A. K. Covington and T. Dickinson (eds.): *Physical Chemistry of Organic Solvent Systems*. Plenum Press, London, New York 1973, Chapter 2, Part 2, p. 137ff. [77] D. T. Sawyer and J. L. Roberts: *Experimental Electrochemistry for Chemists*. Wiley, New York 1974, p. 167ff. [78] B. Kratochvil and H. L. Yeager: *Conductance of Electrolytes in Organic Solvents*. Fortschr. Chem. Forsch. *27*, 1 (1972). [79] R. Fernández-Prini: *Conductance*, in A. K. Covington and T. Dickinson (eds.): *Physical Chemistry of Organic Solvent Systems*. Plenum Press, London, New York 1973, Chapter 5, Part 1, p. 525ff. [80] H. Strehlow: *Electrode Potentials in Non-Aqueous Solvents*, in J. J. Lagowski (ed.): *The Chemistry of Non-Aqueous Solvents*. Academic Press, New York, London 1966, Vol. I, p. 129ff.

[81] J. Badoz-Lambling and G. Cauquis; *Voltammetry in Non-Aqueous Solvents and Melts*, in H. W. Nürnberg (ed.): *Electroanalytical Chemistry*. Vol. 10 in the series: *Advances in Analytical Chemistry and Instrumentation* (ed. C. N. Reilley and R. W. Murray). Wiley-Interscience, London 1974, p. 335...419 (especially p. 397ff.). [82] K. Blumrich, H. Schwarz, and A. Wingler, in Houben-Weyl-Müller. *Methoden der organischen Chemie*. 4th ed., Thieme, Stuttgart 1959, Vol. I/2, p. 887...942. [83] D. Stoye: *Lösemittel*, in *Ullmann's Encyklopädie der technischen Chemie*. 4th ed., Verlag Chemie, Weinheim 1978, Vol. 16, p. 279ff. (esp. p. 291ff.: physiological properties of solvents). [84] E. Browning: *Toxicity and Metabolism of Industrial Solvents*. 2nd ed., Elsevier, Amsterdam 1965. [85] N. I. Sax: *Dangerous Properties of Industrial Materials*. 5th ed., Van Nostrand Reinhold, New York 1979. [86] Berufsgenossenschaft der chemischen Industrie: *Lösemittel*. Merkblatt M 017 (1/85), Jedermann-Verlag Dr. Pfeffer, Heidelberg, Fed. Rep. Germany. [87] N.-P. Lüpke *et al.*: *Toxikologie*, in E. Weise (ed.): *Ullmann's Encyklopädie der technischen Chemie*. 4th ed., Verlag Chemie, Weinheim 1981, Vol. 6, p. 65ff. [88] L. Bretherick (ed.): *Hazards in the Chemical Laboratory*. 4th ed., The Royal Society of Chemistry, London 1986. [89] G. Duve, O. Fuchs, and H. Overbeck: *Lösemittel Hoechst*. 6th ed., Hoechst AG, Frankfurt (Main) 1976, p. 107ff. [90] American Conference of Governmental Industrial Hygienists (ACGIH): *Threshold Limit Values for Chemical Substances in Workroom Air*, adopted by ACGIH for 1984/85. Cincinnati/Ohio 45211, USA.

[91] Deutsche Forschungsgemeinschaft, Senatskommission zur Prüfung gesundheitsschädlicher Arbeitsstoffe: *Maximale Arbeitsplatzkonzentrationen und Biologische Arbeitsstofftoleranzwerte 1987*. Mitteilung XXIII, VCH Verlagsgesellschaft, Weinheim 1987. [92] D. Henschler (ed.): *Gesundheitsschädliche Arbeitsstoffe – Toxikologisch-arbeitsmedizinische Begründung von MAK-Werten*. Verlag Chemie, Weinheim 1972...1984. [93] B. Meyer: *Low Temperature Spectroscopy*. Elsevier, New York 1971, p. 191ff. [94] H. E. Hallam (ed.): *Vibrational Spectroscopy of Trapped Species*. Wiley-Interscience, London 1973, p. 45. [95] S. Cradock and A. J. Hinchcliffe: *Matrix Isolation – A Technique for the Study of Reactive Inorganic Species*. Cambridge University Press, Cambridge 1975. [96] M. Moskovits and G. A. Ozin (eds.): *Cryochemistry*. Wiley-Interscience, New York 1976. [97] D. Stoye: *Lösemittel*, in *Ullmann's Encyklopädie der technischen Chemie*. 4th ed., Verlag Chemie, Weinheim, New York 1978, Vol. 16, p. 279...311. [98] H. Gnamm and O. Fuchs: *Lösungsmittel und Weichmachungsmittel*. 8th ed., Vol. I and II, Wissenschaftliche Verlagsgesellschaft, Stuttgart 1980. [99] G. Kakabadse (ed.): *Solvent Problems in Industry*. Elsevier Applied Science Publishers, London, New York 1984. [100] A. F. M. Barton: *Handbook of Solubility Parameters and Other Cohesion Parameters*, CRC Press, Boca Raton/Florida 1983.

[101] J. H. Hildebrand and R. L. Scott: *The Solubility of Nonelectrolytes*. 3rd ed., Dover, New York 1964. [102] L. R. Snyder: *Solutions to Solution Problems*. CHEMTECH *9*, 750 (1979); ibid. *10*, 188 (1980). [103] N. B. Godfrey: *Solvent Selection via Miscibility Numbers*. CHEMTECH *1972*, 359; *cf.* also I. Vavruch, Chemie für Labor und Betrieb *35*, 385 (1984). [104] D. D. Perrin, W. L. F. Armarego, and D. R. Perrin: *Purification of Laboratory Chemicals*. 2nd ed., Pergamon Press, Oxford 1980. [105] J. F. Coetzee (ed.): *Recommended Methods for Purification of Solvents and Tests for Impurities*. Pergamon Press, Oxford 1982; *cf.* also Pure Appl. Chem. *57*, 634, 855, 860 (1985). [106] D. R. Burfield, K.-H. Lee, and R. H. Smithers, J. Org. Chem. *42*, 3060 (1977) (acetonitrile, benzene, 1,4-dioxane). [107] D. R. Burfield, G.-H. Gan, and R. H. Smithers, J. Appl. Chem. Biotechnol. *28*, 23 (1978); Chem. Abstr. *89*, 12551 f (1978) (molecular sieves as desiccants). [108] D. R. Burfield and R. H. Smithers, J. Org. Chem. *43*, 3966 (1978) (dipolar aprotic solvents). [109] D. R. Burfield and R. H.

Smithers, J. Chem. Technol. Biotechnol. *30*, 491 (1980); Chem. Abstr. *94*, 66822 s (1981) (cation exchange resins as desiccants). [110] D. R. Burfield, R. H. Smithers, and A. S. C. Tan, J. Org. Chem. *46*, 629 (1981) (amines).

[111] D. R. Burfield and R. H. Smithers, J. Chem. Educ. *59*, 703 (1982) (diethyl ether). [112] D. R. Burfield and R. H. Smithers, J. Org. Chem. *48*, 2420 (1983) (alcohols). [113] D. R. Burfield, G. T. Hefter, and D. S. P. Koh, J. Chem. Technol. Biotechnol.,Chem. Technol. Part A *34*, 187 (1984); Chem. Abstr. *101*, 133117 u (1984) (ethanol). [114] D. R. Burfield, J. Org. Chem. *49*, 3852 (1984) (calcium sulfate as desiccant). [115] D. R. Burfield, J. Chem. Educ. *56*, 486 (1979); D. R. Burfield, E. H. Goh, E. H. Ong, and R. H. Smithers, Gazz. Chim. Ital. *113*, 841 (1983) (chloroform). [116] D. R. Burfield and R. H. Smithers, Chem. Ind. (London) *1980*, 240 (chloroform, diethyl ether). [117] D. R. Burfield, J. Org. Chem. *47*, 3821 (1982) (deperoxidation of ethers). [118] Sadtler Research Laboratories: *Infrared Spectra Handbook of Common Organic Solvents*. Philadelphia/USA 1983. [119] J. B. Baumann: *Solvent Selection for Recrystallization*. J. Chem. Educ. *56*, 64 (1979). [120] R. J. Davey: *Solvent Effects in Crystallization Processes*. Curr. Top. Mater. Sci. *8*, 429 (1982); Chem. Abstr. *97*, 154023 z (1982).

[121] D. Bierne, S. Smith, and B. E. Hoogenboom, J. Chem. Educ. *51*, 602 (1974). [122] J. T. Baker Chemical Company: *HPLC Solvent Reference Manual*. Phillipsburg/New Jersey, USA 1985. [123] R. E. Majors: *Optimization of Solvent Composition in High Performance Liquid Chromatography*, VIA, Varian Instrum. Appl. *13* (1), 10 (1979); Chem. Abstr. *91*, 9881z (1979). [124] H. Engelhardt: *Hochdruck-Flüssigkeits-Chromatographie*. 2nd ed., Springer, Berlin 1977; *High Performance Liquid Chromatography* (translated from German by G. Gutnikov), Springer, Berlin 1979. [125] V. Meyer: *Praxis der Hochleistungs-Flüssigchromatographie*. 3rd ed., Diesterweg, Salle, Sauerländer, Frankfurt (Main) and Aarau 1984 (esp. p. 35ff.). [126] J. G. Kirchner: *Thin-Layer Chromatography*, 2nd ed., In E. S. Perry and A. Weissberger (eds.): *Techniques of Chemistry*, Vol. XIV, Wiley-Interscience, New York 1978 (esp. Chapter V on p. 105ff.). [127] J. C. Touchstone: *Practice of Thin-Layer Chromatography*. 2nd ed., Wiley, New York 1983; J. C. Touchstone and J. Sherma (eds.): *Techniques and Applications of Thin-Layer Chromatography*. Wiley, New York 1985. [128] G. Gritzner and J. Kuta: *Recommendations on Reporting Electrode Potentials in Nonaqueous Solvents*. Pure Appl. Chem. *56*, 461 (1984). [129] E. Pungor, K. Tóth, P. G. Klatsmányi, and K. Izutsu (eds.): *Applications of Ion-Selective Electrodes in Nonaqueous and Mixed Solvents*. Pure Appl. Chem. *55*, 2029 (1983). [130] A. J. Collings and S. G. Luxon (eds.): *Safe Use of Solvents*. Academic Press, London, New York 1982.

[131] A. Sato and T. Nakajima: *Metabolism and Toxicity of Organic Solvents*. Tokishikoroji Foramu *6*, 105 (1983); Chem. Abstr. *98*, 192777b (1983). [132] D. J. De Renzo (ed.): *Solvents Safety Handbook*. Noyes, Park Ridge/USA 1986. [133] R. V. Golovnaya and T. A. Misharina: *Thermodynamic Treatment of the Polarity and the Selectivity of Sorbents in Gas Chromatography*. Usp. Khim. *49*, 171 (1980); Russian Chem. Rev. *49*, 88 (1980).

Figure and Table Credits

Fig. 2-1 is reproduced from R. A. Horne, *Survey of Progress in Chemistry 4*, 1 (1968), by permission of the copyright owner, Academic Press, London, New York.

Fig. 2-2 is reproduced from G. Duve, O. Fuchs, and H. Overbeck, *Lösemittel Hoechst*. 6th edition, Hoechst Aktiengesellschaft, Frankfurt (Main) 1976. Used with permission of Hoechst AG, Federal Republic of Germany.

Fig. 2-7 is reproduced from K. L. Wolf, *Theoretische Chemie*. 4th edition, Johann Ambrosius Barth Verlag, Leipzig 1959, by permission of the copyright owner.

Fig. 3-2 is reproduced from H. H. Sisler, *Chemistry in Non-aqueous Solvents*. Reinhold Publishing Corporation, New York, and Chapman and Hall, London 1961, with permission of the copyright owners, Litton Educational Publishing Corporation, New York, and the author.

Fig. 4-3 is reproduced from E. J. Drexler and K. W. Field, Journal of Chemical Education *53*, 392 (1976), by permission of the Division of Chemical Education of the American Chemical Society, Tucson/Arizona, and the authors.

Fig. 5-16 is reproduced from J. A. Leary and M. Kahn, Journal of the American Chemical Society *81*, 4173 (1959); by permission of the American Chemical Society, Washington, D. C., and the authors.

Fig. 6-3 is reproduced from E. M. Kosower, *Molecular Biochemistry*. McGraw-Hill Book Co., New York 1962, by permission of the copyright owner and the author.

Fig. 6-5 is reproduced from G. J. Brealey and M. Kasha, Journal of the American Chemical Society *77*, 4462 (1955), by permission of the American Chemical Society, Washington D. C., and the authors.

Fig. 6-6 is reproduced from N. J. Turro, *Molecular Photochemistry*. Benjamin, New York 1965, by permission of the publishers, Addison-Wesley and W. A. Benjamin Inc., Reading/Massachusetts, and the author.

Table 5-25 is reproduced from M. R. J. Dack, Journal of Chemical Education *51*, 231 (1974), by permission of the Division of Chemical Education of the American Chemical Society, Tucson/Arizona, and the author.

Table A-5 is reprinted from N. L. Alpert, W. E. Keiser, and H. A. Szymanski, *IR – Theory and Practice of Infrared Spectroscopy*. 2nd edition 1973, by courtesy of Plenum Press Publishing Corporation, New York.

The *MAK-values* of Table A-13 are reproduced with permission of the 'Senatskommission zur Prüfung gesundheitsschädlicher Arbeitsstoffe' of the 'Deutsche Forschungsgemeinschaft', Bonn-Bad Godesberg, Federal Republic of Germany.

The *TL-values* of Table A-13 are reprinted with permission of the 'American Conference of Governmental Industrial Hygienists (ACGIH)', Cincinnati/Ohio, U.S.A.

Figure and Table Credits

Subject Index

Author Index